Vibrio cholerae and Cholera

MOLECULAR TO GLOBAL PERSPECTIVES

Vibrio cholerae and Cholera

MOLECULAR TO GLOBAL PERSPECTIVES

Edited by

I. KAYE WACHSMUTH

Division of Bacterial and Mycotic Diseases
Centers for Disease Control and Prevention
Atlanta, Georgia

PAUL A. BLAKE

Foodborne and Diarrheal Diseases Branch
Division of Bacterial and Mycotic Diseases
Centers for Disease Control and Prevention
Atlanta, Georgia

ØRJAN OLSVIK

Foodborne and Diarrheal Diseases Branch
Division of Bacterial and Mycotic Diseases
Centers for Disease Control and Prevention
Atlanta, Georgia

ASM PRESS
Washington, D.C.

Library of Congress Cataloging-in-Publication Data

Vibrio cholerae and cholera: molecular to global perspectives/edited by I. Kaye Wachsmuth, Paul A. Blake,
   Ørjan Olsvik.
      p.   cm.
   Includes index.
   ISBN 1-55581-067-5
   1. Cholera.   I. Wachsmuth, Kaye.   II. Blake, Paul A.   III. Olsvik, Ørjan.
   [DNLM: 1. Cholera.   2. Vibrio cholerae.   3. Cholera Vaccine.   WC262 V626 1994]
RC126.V53   1994
616.9′32—dc20
DNLM/DLC
for Library of Congress   93-39024
                                                                                                                          CIP

Cover photograph © Tom Van Sant/Photo Researchers.

Contents

Contributors

Timothy J. Barrett
Foodborne and Diarrheal Diseases Branch, Division of Bacterial and Mycotic Diseases, National Center for Infectious Diseases, Centers for Disease Control and Prevention, Atlanta, GA 30333

Michael L. Bennish
Division of Geographic Medicine and Infectious Diseases, Departments of Pediatrics and Medicine, New England Medical Center, Tufts University School of Medicine, Boston, MA 02111

Paul A. Blake
Foodborne and Diarrheal Diseases Branch, Division of Bacterial and Mycotic Diseases, National Center for Infectious Diseases, Centers for Disease Control and Prevention, Atlanta, GA 30333

Cheryl A. Bopp
Foodborne and Diarrheal Diseases Branch, Division of Bacterial and Mycotic Diseases, National Center for Infectious Diseases, Centers for Disease Control and Prevention, Atlanta, GA 30333

Daniel N. Cameron
Foodborne and Diarrheal Diseases Branch, Division of Bacterial and Mycotic Diseases, National Center for Infectious Diseases, Centers for Disease Control and Prevention, Atlanta, GA 30333

John Clemens
Division of Epidemiology, Statistics, and Prevention Research, National Institute of Child Health and Human Development, Bethesda, MD 20892

Mitchell L. Cohen
Division of Bacterial and Mycotic Diseases, National Center for Infectious Diseases, Centers for Disease Control and Prevention, Atlanta, GA 30333

Rita R. Colwell
Center of Marine Biotechnology, Maryland Biotechnology Institute, Baltimore, MD 21202, and Department of Microbiology, University of Maryland, College Park, MD 20742

Gracia M. Evins
Division of Bacterial and Mycotic Diseases, National Center for Infectious Diseases, Centers for Disease Control and Prevention, Atlanta, GA 30333

Alessio Fasano
Center for Vaccine Development, University of Maryland School of Medicine, Baltimore, MD 21201

John C. Feeley
Foodborne and Diarrheal Diseases Branch, Division of Bacterial and Mycotic Diseases, National Center for Infectious Diseases, Centers for Disease Control and Prevention, Atlanta, GA 30333

Patricia Fields
Foodborne and Diarrheal Diseases Branch, Division of Bacterial and Mycotic Diseases, National Center for Infectious Diseases, Centers for Disease Control and Prevention, Atlanta, GA 30333

Kathy Greene
Foodborne and Diarrheal Diseases Branch, Division of Bacterial and Mycotic Diseases, National Center for Infectious Diseases, Centers for Disease Control and Prevention, Atlanta, GA 30333

Rajesh Gupta
Massachusetts Public Health Biologic Laboratories, Jamaica Plain, MA 02130

Walter E. Hill
Food and Drug Administration, Bothell, WA 98041-3012

Jan Holmgren
Department of Medical Microbiology and Immunology, University of Goteborg, S-41346 Goteborg, Sweden

Anwarul Huq
Department of Microbiology, University of Maryland, College Park, MD 20742

Margaritha Isaäcson
Department of Tropical Diseases, School of Pathology, South African Institute for Medical Research, University of the Witwatersrand, Johannesburg 2000, South Africa

Gunhild Jonson
Department of Medical Microbiology and Immunology, University of Goteborg, S-41346 Goteborg, Sweden

James B. Kaper
Center for Vaccine Development, University of Maryland School of Medicine, Baltimore, MD 21201

Melissa R. Kaufman
Hopkins Marine Station, Stanford University, Pacific Grove, California 93950

Bradford A. Kay
Foodborne and Diarrheal Diseases Branch, Division of Bacterial and Mycotic Diseases, National Center for Infectious Diseases, Centers for Disease Control and Prevention, Atlanta, GA 30333

Charles A. Kaysner
Food and Drug Administration, Bothell, WA 98041-3012

Myron M. Levine
Center for Vaccine Development, University of Maryland School of Medicine, Baltimore, MD 21201

Marlo Libel
Pan American Health Organization, Washington, DC 20037

Paul A. Manning
Department of Microbiology and Immunology, University of Adelaide, Adelaide S.A. 5005, Australia

John J. Mekalanos
Department of Microbiology and Molecular Genetics, Harvard Medical School, Boston, MA 02115

Eric D. Mintz
Foodborne and Diarrheal Diseases Branch, Division of Bacterial and Mycotic Diseases, National Center for Infectious Diseases, Centers for Disease Control and Prevention, Atlanta, GA 30333

Renato Morona
Department of Microbiology and Immunology, University of Adelaide, Adelaide S.A. 5005, Australia

J. Glenn Morris
Divisions of Geographic Medicine and Infectious Diseases, Department of Medicine,
University of Maryland School of Medicine, and Veterans Affairs Medical Center,
Baltimore, MD 21201

G. Balakrish Nair
National Institute of Cholera and Enteric Diseases, Beliaghata, Calcutta 700010, India

Ørjan Olsvik
Diarrheal and Foodborne Diseases Branch, Division of Bacterial and Mycotic Diseases,
National Center for Infectious Diseases, Centers for Disease Control and Prevention,
Atlanta, GA 30333

J. Osek
Department of Medical Microbiology and Immunology, University of Goteborg, S-14346
Goteborg, Sweden

Karen M. Otteman
Department of Microbiology and Molecular Genetics, Harvard Medical School, Boston, MA
02115

Tanja Popovic
Diarrheal and Foodborne Diseases Branch, Division of Bacterial and Mycotic Diseases,
National Center for Infectious Diseases, Centers for Disease Control and Prevention,
Atlanta, GA 30333

Stephen H. Richardson
Department of Microbiology and Immunology, Bowman Gray School of Medicine, Wake
Forest University, Winston-Salem, NC 27157

John B. Robbins
Laboratory of Developmental and Molecular Immunity, National Institute of Child Health
and Human Development, Bethesda, MD 20892

Daniel C. Rodrigue
Department of Medicine, Division of Infectious Diseases, University of Southern California
School of Medicine, Los Angeles, CA 90033

David Sack
Department of International Health, Johns Hopkins School of Public Health, Baltimore, MD
21205

Luis Seminario
General Office of Epidemiology, Ministry of Health, Lima, Peru

Dale Spriggs
Virus Research Institute, 61 Moulton Street, Cambridge, MA 02138

Uwe H. Stroeher
Department of Microbiology and Immunology, University of Adelaide, Adelaide S.A. 5005,
Australia

Ann-Mari Svennerholm
Department of Medical Microbiology and Immunology, University of Goteborg, S-41346
Goteborg, Sweden

David L. Swerdlow
Foodborne and Diarrheal Diseases Branch, Division of Bacterial and Mycotic Diseases,
National Center for Infectious Diseases, Centers for Disease Control and Prevention,
Atlanta, GA 30333

Shousun C. Szu
Laboratory of Developmental and Molecular Immunity, National Institute of Child Health
and Human Development, Bethesda, MD 20892

Carol O. Tacket

Center for Vaccine Development, University of Maryland School of Medicine, Baltimore, MD 21201

Yoshifumi Takeda

Department of Microbiology, Faculty of Medicine, Kyoto University, Sakyo-ku, Kyoto 606-01, Japan

Roberto Tapia

Direccion General de Epidemiologia, Col. Lomas de Plateros, Mexico

Robert A. Tauxe

Foodborne and Diarrheal Diseases Branch, Division of Bacterial and Mycotic Diseases, National Center for Infectious Diseases, Centers for Disease Control and Prevention, Atlanta, GA 30333

Ronald K. Taylor

Department of Microbiology, Dartmouth Medical School, Vail Building, Hanover, NH 03755-3842

Michele Trucksis

Center for Vaccine Development, University of Maryland School of Medicine, Baltimore, MD 21201

Duc J. Vugia

Foodborne and Diarrheal Diseases Branch, Division of Bacterial and Mycotic Diseases, National Center for Infectious Diseases, Centers for Disease Control and Prevention, Atlanta, GA 30333

I. Kaye Wachsmuth

Division of Bacterial and Mycotic Diseases, National Center for Infectious Diseases, Centers for Disease Control and Prevention, Atlanta, GA 30333

Joy G. Wells

Foodborne and Diarrheal Diseases Branch, Division of Bacterial and Mycotic Diseases, National Center for Infectious Diseases, Centers for Disease Control and Prevention, Atlanta, GA 30333

Introduction

From the time the cholera proclamation was issued, the local garrison shot a cannon from the fortress every quarter hour, day and night, in accordance with the local superstition that gun powder purified the atmosphere.
—Gabriel Garcia Marquez

This quotation from the novel *Love in the Time of Cholera* nicely reflects the prevailing misperceptions concerning the cause, transmission, and prevention of cholera at the end of the 19th century. The cholera of Garcia Marquez's novel was an often irreversible illness that more frequently affected the poor but was feared by all because treatment was inadequate and illness more often than not led to death. Although science was beginning to understand the cause of cholera and how the organism was transmitted, fear and superstition were understandable and common in the public's and government's response to the disease.

In the century since this novel's setting, we have learned much about the bacterium and the disease. In this book, this information is examined in comprehensive and integrated discussions by some of the field's preeminent microbiologic, clinical, and epidemiologic experts. The amount of knowledge that has accumulated in the last several decades is most impressive. Microbiologists have developed isolation techniques, refined the taxonomy and subtyping of the vibrios, and determined the importance and biochemistry of cholera toxin as well as the potential of a series of traditional and recombinant cholera vaccines. Clinicians have better understood the pathogenic process and established the importance of rehydration therapy as the cornerstone of treatment. Epidemiologists have defined the who, what, where, and when of disease, as well as, in the tradition of John Snow, the why, by examining the mechanisms and vehicles of transmission and the risk factors of affected persons. These data have provided the public health community with the scientific basis for developing control and prevention strategies. Nevertheless, the advances in microbiology and clinical treatment and the implementation of appropriate control measures have not led to the effective control of cholera worldwide.

In fact, cholera has remained a persistent problem for the developing and, occasionally, the developed world. The seventh pandemic had seen the spread of cholera since the 1970s throughout southeast Asia and into Africa. Although conditions in South and Central America were conducive to the transmission of cholera, except for a few cases associated with the U.S. Gulf Coast, the Western Hemisphere had remained unaffected. In January 1991, all this suddenly changed with the emergence of cholera in Peru and its subsequent spread to many countries in the hemisphere. With hundreds of thousands of cases, cholera had emerged in a few months as one of the hemisphere's most important public health problems. As is often the case when cholera affects a new area, microbiologists, clinicians, and public health officials were frequently unprepared. Delayed or inadequate diagnosis coupled with often inappropriate treatment led to unnecessary morbidity and mortality. This occurred not only in developing countries but also among travelers who became ill after returning to developed countries that had not seen cases of cholera for almost a century. Thus, the information included in this book has become important for clinicians, researchers, clinical microbiologists, and public health officials worldwide.

Cholera has had additional surprises. In many parts of the world, strains of *Vibrio cholerae* have become resistant to multiple antimicrobial agents, affecting both treatment and chemoprophylaxis. Perhaps the greatest surprise, however, has been the emergence in the last year of a previously unrecognized strain of epidemic *V. cholerae*, serotype O139, in India and Bangladesh. Although at the time this book was published a number of studies were under way to explain the emergence of this new strain, preliminary data suggested an evolution from epidemic *V. cholerae* O1.

The more we learn about the organism and the disease, the more we are confronted by new scientific and public health dilemmas. This greatly compounds the difficulties that already impede the general prevention and control of this disease. Barring tremendous expenditures to improve sanitation and hygiene in large parts of the world or the rapid development of an effective vaccine, it is likely that cholera will hold additional surprises and remain an important public health problem. The scientific information provided by this book is critical to confronting this challenge and defining responses that are more effective than the firing of cannons.

Mitchell L. Cohen
Division of Bacterial and Mycotic Diseases
National Center for Infectious Diseases
Centers for Disease Control and Prevention
Atlanta, Georgia

I. THE BACTERIUM *VIBRIO CHOLERAE*

Vibrio cholerae and Cholera: Molecular to Global Perspectives
Edited by I. Kaye Wachsmuth, Paul A. Blake, and Ørjan Olsvik
© 1994 American Society for Microbiology, Washington, DC 20005

Chapter 1

Isolation and Identification of *Vibrio cholerae* O1 from Fecal Specimens

Bradford A. Kay, Cheryl A. Bopp, and Joy G. Wells

The introduction of cholera into Latin America during 1991 and 1992 has underscored the need for public health and clinical microbiologists worldwide to be familiar with methods for the isolation and characterization of *Vibrio cholerae*. The majority of these techniques have been established and used for years by microbiologists, particularly those who work in areas where cholera is endemic. While treatment of cholera does not require laboratory confirmation of etiology, epidemiologic studies and cholera control programs rely heavily on laboratory results.

This chapter summarizes basic laboratory techniques for cholera and is intended to provide a reference for the microbiologist who has not specialized in these procedures. The reader is encouraged to consult other sources for a more detailed description of methods for isolation and characterization of *V. cholerae* O1.

CHARACTERISTICS OF *V. CHOLERAE*

V. cholerae is the type species of the genus *Vibrio,* which is the type genus of the family

Bradford A. Kay, Cheryl A. Bopp, and Joy G. Wells
• Foodborne and Diarrheal Diseases Branch, Division of Bacterial and Mycotic Diseases, National Center for Infectious Diseases, Centers for Disease Control and Prevention, 1600 Clifton Road, Atlanta, Georgia 30333.

Vibrionaceae (20, 21). Members of this genus are facultatively anaerobic, asporogenous, motile, curved or straight gram-negative rods 1.4 to 2.6 μm in length. *Vibrio* spp. either require salt (NaCl) or have their growth stimulated by its addition. *V. cholerae* requires 5 to 15 mM Na^+ for optimum growth but will usually grow in complex medium without added salt (nutrient broth, for example), although growth is reduced to 50 to 80% of optimal growth (107). *V. cholerae* will grow in alkaline conditions up to a maximum of pH 10 but is inhibited when the pH drops to 6.0 or below (101). To date, all current members of the genus *Vibrio* except *V. metschnikovii* and *V. gazogenes* are oxidase positive.

Currently, more than 130 serogroups (serovars) of *V. cholerae* have been identified; however, before 1992, only the O1 serogroup was associated with epidemic and pandemic cholera (Table 1). Within the O1 serogroup, the ability to produce cholera toxin (CT) is an essential determinant of virulence. *V. cholerae* O1 strains that do not produce CT may be isolated from patients with sporadic diarrheal, extraintestinal sites, environmental specimens, or food but have not yet demonstrated the potential to cause epidemic cholera (69, 81, 87).

Major outbreaks of choleralike illness occurred in India and Bangladesh in 1992 and 1993 (4, 25, 104). The agent responsible for

Table 1. Characteristics of *V. cholerae*

Classification	Characteristic			
	Serogroups	Biotypes	Serotypes	CT production
Epidemic associated	O1, O139	Classical and El Tor (not applicable to O139 strains)	Inaba, Ogawa, and Hikojima	Yes[a]
Not epidemic associated	137 exist	Not applicable to non-O1 strains	The three O1 serotypes are not applicable to non-O1 strains	Usually no; other toxins sometimes produced

[a]Nontoxigenic O1 strains exist but are not epidemic associated.

these outbreaks was identified as a strain of *V. cholerae* that did not agglutinate in *V. cholerae* O1 antiserum or in any of the 137 non-O1 serogroup antisera. This strain was therefore determined to be a new serogroup, O139 (118). *V. cholerae* O139 produced CT but not the heat-stable enterotoxin sometimes associated with *V. cholerae* non-O1 strains (8).

Isolates of *V. cholerae* serogroups other than O1 or O139 are occasionally associated with sporadic diarrheal illness, limited outbreaks of diarrhea, and extraintestinal infections (9). *V. cholerae* non-O1, non-O139 strains can possess a variety of possible virulence factors, including production of CT (33). However, evidence has not yet demonstrated that these organisms are a significant public health threat. See chapter 7 for a further discussion of these organisms.

Isolates of *V. cholerae* O1 can be divided into two biotypes, El Tor and classical, on the basis of several phenotypic characteristics. Isolates associated with the fifth and sixth pandemics have been identified as the classical biotype of *V. cholerae* O1. However, with the onset of the seventh pandemic, the El Tor biotype became predominant and currently is responsible for a majority of the cholera cases throughout the world. Classical biotype isolates are now rarely encountered outside of Bangladesh (97). The classical and El Tor biotypes are not applicable to *V. cholerae* O139, and they are not valid for all other non-O1 serogroups.

ISOLATION TECHNIQUES

Collection of Specimens

Fecal specimens (stool or rectal swabs) should be collected as early as possible in the illness, as the number of *V. cholerae* O1 in the stool begins to decrease soon after the onset of symptoms (13, 125). Since initiation of antibiotic therapy also results in a rapid decline in the numbers of *V. cholerae* in the stool, stool specimens should not be collected after treatment has started, and other diagnostic methods such as serologic testing for vibriocidal or anti-CT antibodies should be used (73) (see chapter 10 for a discussion of these methods). If there will be a delay of more than 2 h before the specimen reaches the laboratory, the stool or a swab of the stool may be placed in transport medium. Rectal swabs should always be inoculated into transport medium to reduce the risk of the specimen drying out. Fecal specimens (whole stool or swabs in transport media) may be transported at room temperature unless there is a chance of encountering elevated temperatures ($>40°C$) (36, 138).

Transport of Specimens

Specialized transport media

A number of transport media have been described and used in cholera studies. Two liquid transport media widely used in Asia for transport of *V. cholerae* are Venkatraman-

Ramakrisnan sea salt medium and Monsur's transport medium (83, 133). Both media provide excellent results with *Vibrio* spp., but they are not suitable for transport of most other enteric organisms, and neither medium is commercially available.

Cary-Blair transport medium

Cary-Blair transport medium is useful for preserving a spectrum of pathogenic bacteria and is one of the most widely used transport media for fecal specimens (26, 48, 90). Cary-Blair is well suited for transport of *V. cholerae* because of its high pH. In addition, Cary-Blair's semisolid consistency facilitates transport. Once prepared, the medium can be stored and used for more than a year as long as contamination or loss of volume does not occur (85). Cary-Blair is commercially available in both dehydrated and prepared forms.

APW

Alkaline peptone water (APW) is a simple and inexpensive medium that has been used for transport as well as enrichment of specimens for *V. cholerae* (36). However, APW is not recommended for transport if subculture will be delayed more than 6 to 8 h from the time of inoculation. Beyond this time, other organisms may overgrow the vibrios and hinder their isolation. Several similar formulas for APW, varying slightly in NaCl concentration and pH, have been described in the literature. One commonly used formulation contains 1% peptone and 1% sodium chloride with the pH adjusted to 8.5. APW is not commercially available.

Other nonspecialized transport media

Stuart (pH 7.3) or Amies (pH 7.4) transport medium, both of which are commercially available, may be used if Cary-Blair cannot be obtained. However, these media should be used to transport suspect cholera specimens for only short periods (1 to 2 days) because of their lower pH (54, 127). Buffered glycerol saline medium is not suitable for transport of *V. cholerae* (36).

Unpreserved specimens

If the use of transport medium is not possible when fecal specimens are collected, a piece of filter paper, gauze, or cotton may be soaked in liquid stool and placed into a leak-proof plastic bag (64). The bag should be tightly sealed so that the specimen will remain moist and not leak. Several drops of sterile saline may be added to the bag to help prevent drying of the specimen. Refrigeration during transport is not necessary unless exposure to extremes of temperature is possible.

Enrichment

Most investigators agree that enrichment enhances the isolation of *V. cholerae* from convalescent patients and is helpful particularly in the later stages of acute illness, when numbers of *V. cholerae* O1 in the stool have declined (17, 63, 108). However, enrichment probably offers no real advantage with specimens from persons in the very early stages of acute cholera, since the numbers of *V. cholerae* O1 present in these specimens are sufficiently high to offer excellent isolation from direct plating alone (39, 51, 63).

APW is by far the most widely used enrichment broth for *V. cholerae*. Since *Vibrio* spp. grow rapidly in APW, at 6 to 8 h they will usually be present in greater numbers than non-*Vibrio* organisms. Other enrichment media, such as Monsur's Trypticase tellurite taurocholate-peptone (TTP) (83), have been formulated. TTP has been reported to increase the isolation of *V. cholerae* compared to APW, owing to the inhibition of non-vibrios (99). A modification of APW (PWT), to which potassium tellurite at concentrations of 1:100,000 to 1:200,000 was added, was almost as effective as TTP in reducing the growth of nonvibrios, particularly *Pseudomonas* spp., after overnight incubation

(99). However, the use of a selective enrichment medium like TTP or PWT may not offer any advantage over APW with a short incubation time (6 to 8 h).

Selective Plating Media

Although *V. cholerae* O1 will grow on a variety of commonly used agar media, isolation from fecal specimens is more easily accomplished by use of a selective plating medium. Except when cultures of patient specimens are done in the very early stages of illness, a selective plating medium should always be used, particularly when specimens from convalescent-patients or patients with suspected asymptomatic infections are cultured or whenever high numbers of competing organisms are likely to be present in the specimen.

Several specialized selective plating media have been developed for *V. cholerae*, since most routine enteric screening media are unsuitable for these organisms. These specialized media include thiosulfate-citrate-bile salts-sucrose agar (TCBS), tellurite taurocholate gelatin agar (TTGA), *Vibrio* agar, sucrose tellurite teepol medium, and polymyxin mannose tellurite agar (27, 74, 82, 121, 131). Each of these media has advantages and disadvantages, and each has a constituency of users.

TCBS and TTGA are two of the most commonly used and most widely studied selective plating media for cholera, and each will be discussed later in this chapter. In three studies, the relative efficacies of TCBS and TTGA for the isolation of *V. cholerae* O1 were reported to be essentially equal (51, 63, 86). The two media were equally efficient for the isolation of the classical and El Tor biotypes (65).

TCBS

TCBS is perhaps the most widely used selective plating medium for vibrios (Fig. 1). It is available commercially, does not require autoclaving, is highly selective, and contains sucrose, which allows the differentiation of *V. cholerae* (sucrose positive) and *V. parahaemolyticus* (sucrose negative) (74). However, TCBS is subject to brand-to-brand and lot-to-lot variations in selectivity, and growth on TCBS is not suitable for direct testing with *V. cholerae* O1 antisera (51, 86, 92). TCBS may also be prohibitively expensive for laboratories that process large numbers of stool cultures. A formulation of TCBS with salt added to increase the selectivity has been described (84).

Overnight growth (18 to 24 h) of *V. cholerae* on TCBS will produce large (2 to 4 mm in diameter), slightly flattened, yellow colonies with opaque centers and translucent peripheries. The yellow color is due to the fermentation of sucrose in the medium. Nonsucrose-fermenting organisms such as *V. parahaemolyticus* will produce large green to blue-green colonies. Typical colonies should be subcultured to a noninhibitory medium for further testing.

TTGA

TTGA, also referred to as Monsur's agar, is a selective and differential agar specifically designed for the isolation of *V. cholerae* (82) (Fig. 2). TTGA is inexpensive to prepare and allows the performance of slide agglutination and oxidase tests directly from the plate. However, TTGA is not commercially available and requires autoclaving during its preparation.

TTGA contains Trypticase, 10.0 g; sodium chloride, 10.0 g; sodium taurocholate, 5.0 g; sodium carbonate, 1.0 g; gelatin, 30.0 g; agar, 15.0 g; distilled water, 1,000.0 ml; and a varying amount of 1% solution of potassium tellurite. The amount of the 1% potassium tellurite solution added to TTGA may vary with each lot of the chemical. Before regular use, each lot of potassium tellurite should be examined to determine the proper working concentration to use. Most

Figure 1. Overnight colonies of *V. cholerae* on TCBS agar are yellow because of sucrose fermentation. The colonies are typically large (2 to 4 mm in diameter), circular, smooth, glistening, and slightly flattened.

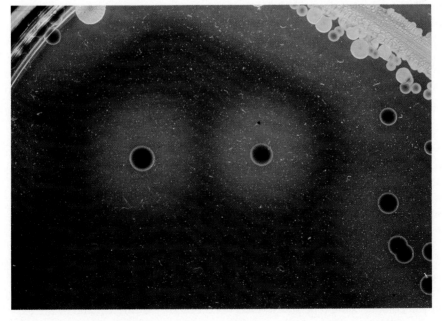

Figure 2. On TTGA, colonies of *V. cholerae* are gray, flattened, and surrounded by a cloudy zone caused by the production of gelatinase.

lots of potassium tellurite are satisfactory at dilutions of 1:100,000 to 1:200,000 (13).

Overnight growth of *V. cholerae* on TTGA appears as small, opaque colonies with slightly darkened centers. At 24 h the centers of the colonies become darker, and eventually the entire colony will become gunmetal gray because of the reduction of tellurite. The development of an opaque zone resembling a halo around colonies is characteristic of vibrios and is due to digestion of gelatin by the enzyme gelatinase. Gelatinase activity is usually visible after incubation for 24 h or longer. The halo effect may be intensified by brief (15- to 30-min) refrigeration of the plate. After incubation of the TTGA plate for 48 h, the colonies become larger and develop black centers and more distinct gelatinase halos. Since many members of the genus *Vibrio* have similar characteristics on TTGA, isolates suspected to be *V. cholerae* must be examined by using biochemical tests or polyvalent *V. cholerae* O1 antisera. Colonies may be tested directly from TTGA for agglutination in O1 antisera and oxidase reaction (86).

A modification of TTGA containing 4-methylumbelliferyl-β-D-galactoside has been described elsewhere (93). The ability of *V. cholerae* to utilize β-D-galactoside differentiates it from *V. parahaemolyticus,* which is unable to utilize this substrate.

Enteric plating media

MacConkey agar is used widely for the isolation of members of the family *Enterobacteriaceae*. This medium has been used for isolation of *V. cholerae* even though it is not as efficient as specialized media and is inhibitory to some strains of *V. cholerae* (39, 101). Colonies of *V. cholerae* that grow on MacConkey agar after overnight incubation are usually 1 to 3 mm in diameter and colorless to slightly pink, often resembling colonies of "late" or "slow" lactose-fermenting organisms. Typical colonies should be subcultured to screening media or to a noninhibitory medium for further testing. SS agar, another commonly used selective medium for enteric pathogens, is not suitable for isolation of *V. cholerae* (39).

Nonselective Plating Media

Gelatin agar (GA) is a widely used nonselective growth medium for *V. cholerae* (44, 126). GA contains peptone, 4.0 g; yeast extract, 1.0 g; gelatin, 15.0 g; sodium chloride, 10.0 g; agar, 15.0 g; and distilled water, 1,000.0 ml. Colonies of *V. cholerae* on GA are smooth, opaque, white, and 2 to 4 mm in diameter after overnight incubation at 35 to 37°C. Gelatinase production, a general characteristic of vibrios, is manifested as the development of a hazy white halo around colonies. The halo effect can be intensified by brief (15- to 30-min) refrigeration of the plate. Colonies may be tested directly from this medium for agglutination with antisera as well as oxidase and string reactions.

Meat extract agar (MEA), also referred to as alkaline nutrient agar, is a nonselective isolation medium that has been successfully used in cholera studies (44). MEA contains peptone, 10.0 g; sodium chloride, 10.0 g; meat extract, 3.0 g; agar, 20.0 g; and distilled water, 1,000.0 ml. Overnight colonies of *V. cholerae* on MEA are 2 to 4 mm in diameter and are smooth, opaque, and cream colored. When viewed by low-power stereoscopic microscopy with oblique light, colonies appear finely granular and iridescent with a greenish bronze sheen. The oxidase and string tests may be performed directly from growth on the plate, as can screening with antisera.

Isolation Procedures

Inoculation of enrichment media

Enrichment media may be inoculated with liquid stool, a fecal suspension, or a rectal swab (Fig. 3). The stool inoculum should not exceed 10% of the volume of the enrichment medium. If APW is used, it should be incubated at 35 to 37°C for 6 to 8 h and then

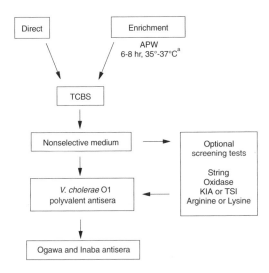

Figure 3. Procedure for recovery of *V. cholerae* from fecal specimens. [a], If the APW cannot be streaked after 6 to 8 h of incubation, subculture at 18 h or sooner to a fresh tube of APW, incubate for 6 to 8 h, and then streak to TCBS.

streaked to a selective plating medium. Subcultures should be from the surface and topmost portion of APW, as vibrios preferentially grow in this area. If the APW cannot be streaked to plating media after 6 to 8 h of incubation, it should be subcultured to a fresh tube of APW no later than 18 to 24 h after inoculation. This second APW tube should be inoculated to selective plating media within 6 to 8 h of incubation.

Inoculation of plating media

Highly selective media (TCBS, TTGA) are inoculated with a large inoculum equivalent to 1 large loopful or 2 smaller loopfuls of liquid stool or enrichment medium. Media of low selectivity (GA, MEA) are inoculated with smaller amounts (equivalent to 1 small loopful). All plates are streaked for isolated colonies. After inoculation, plates are incubated for 18 to 24 h at 35 to 37°C. If *Vibrio*-like colonies are not visible after overnight incubation on TTGA, plates may be incubated for an additional 18 to 24 h and reexamined.

Isolation of *V. cholerae* O139

The procedures described above may be used for isolation of *V. cholerae* O139, since its morphologic and physiologic characteristics appear to be similar to those of the O1 serogroup as well as to those of all other serogroups of *V. cholerae*.

IDENTIFICATION OF ISOLATES

Serological Identification

Serogroups (serovars)

There are currently 139 specific O antigens of *V. cholerae* (118, 120). O antigens are thermostable polysaccharides that are part of the cell wall lipopolysaccharide (61, 106). *V. cholerae* belonging to different O serogroups cannot be biochemically differentiated. The O1 serogroup of *V. cholerae* has at least one unique somatic (O) antigen and is, by definition, agglutinated only by O1 antiserum, although cross-reactions with other *Vibrio* spp. have been observed (53, 113, 122).

All O groups of *V. cholerae* share a common heat-labile flagellar (H) antigen (24). A variety of other H antigens exist and have been identified in other *Vibrio* species (132). A common R antigen has been identified and is identical among all serogroups of *V. cholerae* (119).

Serotypes

Isolates belonging to the O1 serogroup of *V. cholerae* have historically been divided into three serotypes or subtypes: Ogawa, Inaba, and Hikojima. These serotypes have been designated by the antigenic formulas AB, AC, and ABC, respectively, with the A antigen common to all three serotypes (113) (Table 2). No serotypes have been described for the O139 serogroup. Antigenic shifts of Ogawa serotypes converting to Inaba have been described elsewhere (23, 52, 113). The opposite shift, Inaba to Ogawa, has been ob-

Table 2. Serotypes of *V. cholerae* serogroup O1

Serotype	Major O-antigen determinants	Agglutination in absorbed antiserum	
		Ogawa	Inaba
Ogawa	A, B	+	−
Inaba	A, C	−	+
Hikojima	A, B, C	+	+

served in vivo (112, 117). Antigenic shifts have also been known to occur during outbreaks. In the early stages of the Latin American outbreak, only the Inaba serotype was isolated, but serotype Ogawa strains, which are genotypically indistinguishable from the original Inaba strain, appeared as the epidemic spread (134). Although the identification of the serotype may have limitations as an epidemiologic tool because of antigenic shift, it is unlikely that serotyping isolates of *V. cholerae* O1 will soon be abandoned. The detection of a second serotype or a shift in the predominant serotype may be a valid reflection of important epidemiologic events. Also, the identification of the serotype with monospecific (absorbed) antisera remains the definitive serologic confirmatory test for *V. cholerae* isolates that agglutinate in polyvalent O1 antisera. See chapter 6 for a discussion of antigenic shift.

Presumptive identification with polyvalent antisera

Agglutination in polyvalent *V. cholerae* O1 antiserum is sufficient to provide a presumptive identification of *V. cholerae* O1. Specific O139 antiserum must be used to presumptively identify this serogroup (118). Isolates that agglutinate in polyvalent O1 antiserum may be confirmed by agglutination with monovalent Ogawa and Inaba antisera.

Confirmatory identification with monovalent antisera

Identification of Inaba and Ogawa antigens has significance only with serogroup O1 iso-

lates, and serotype-specific antisera should not be used with strains that are negative with polyvalent O1 antisera. Agglutination reactions with both Inaba and Ogawa antisera should be examined simultaneously. The strongest and most rapid reaction should be used to identify the serotype, since *V. cholerae* O1 strains frequently cross-react slowly or weakly in monospecific antiserum for the other serotype. Strains that agglutinate equally in both Ogawa and Inaba antisera may be serotype Hikojima. *V. cholerae* O1 serotype Hikojima is rarely encountered and if suspected should be referred to a reference laboratory for further examination. However, isolates that weakly or slowly agglutinate with serogroup O1 antisera but do not agglutinate with either Inaba or Ogawa antisera are not considered serogroup O1.

Slide agglutination

Agglutination tests for *V. cholerae* somatic O antigens may be carried out using a saline suspension of fresh growth. Growth taken directly from TTGA or a nonselective medium such as GA or MEA may be used for slide agglutination (86). From TCBS, several typical colonies should be selected and inoculated to a noninhibitory medium (nutrient agar should not be used, as it has insufficient NaCl to allow optimal growth of *V. cholerae*) (Fig. 3). After 5 to 6 h of incubation at 35 to 37°C, there is usually sufficient growth on the slant to perform serological testing with *V. cholerae* O1 polyvalent antisera. There is no need to heat the suspension before slide agglutination unless the O antiserum contains antibodies against H antigens (37). If the culture is rough, it can be serotyped only if a smooth colony can be obtained. Positive slide agglutination reactions will appear within 30 s to 1 min. A positive control strain should always be tested to verify the quality of the antiserum.

Preparation and quality control of antisera

Procedures for the preparation and quality control of polyclonal *V. cholerae* O1 antisera have been given elsewhere (3, 13, 37, 47). In addition, methods for the production of monoclonal antibodies for cholera studies have been described by several investigators (2, 57, 66, 67, 128).

Polyclonal antisera, both polyvalent and monovalent, for identification of *V. cholerae* O1 are available commercially. In addition, a slide agglutination kit (*Vibrio cholerae* O1 AD Seiken) consisting of three reagents containing latex particles sensitized with monoclonal antibodies to the antigenic factors A, B, and C is being marketed by Denka Seiken Co., Ltd., Tokyo, Japan. Specific O139 antiserum is being produced and is likely to be available commercially.

Antisera, regardless of source and type, should be examined for specificity and titer by testing with standard strains of quality control organisms (37).

Biochemical Identification of *V. cholerae*

The biochemical identification of *V. cholerae* may be accomplished using the tests listed in Table 3. If the test results are the same as those shown in Table 3, the isolate can confidently be identified as *V. cholerae*. However, if the isolate does not give results as shown, additional tests may be necessary to determine its identity (72). All biochemical tests should be incubated at 35 to 37°C. Caps on all tubes must be loosened.

Biochemical screening tests

Biochemical tests may be useful in screening isolates before testing the isolates with antisera or performing a full set of biochemical tests for identification. Kligler's iron agar (KIA), triple sugar iron agar (TSI), arginine, lysine, oxidase, or string tests may be helpful in the initial screening of isolates resembling *V. cholerae* from selective media.

KIA and TSI

KIA and TSI are carbohydrate-containing media commonly used for screening enteric pathogens. While similar in use, KIA and TSI vary in the carbohydrates they contain. The reactions of *V. cholerae* on KIA, which contains the carbohydrates glucose and lactose, are similar to those of non-lactose-

Table 3. Differential characteristics of selected members of the *Vibrionaceae* and *Enterobacteriaceae*

Test	Reaction[a]						
	Vibrio cholerae	*Vibrio mimicus*	Halophilic vibrios	*Aeromonas hydrophila*	*Aeromonas veronii*	*Plesiomonas shigelloides*	*Enterobacteriaceae*
KIA	K/A	K/A	V	V	K/AG	K/A	V
TSI	A/A	K/A	V	V	A/AG	K/A	V
String	+	+	+[b]	−	−	−	−
Oxidase	+	+	+	+	+	+	−
Gas from glucose	−	−	−[c]	+	+	−	V
Sucrose	+	−	V	V	+	−	V
Lysine	+	+	V	V	+	+	V
Arginine	−	−	V	+	−	+	V
Ornithine	+	+	V	−	+	+	V
VP	V	−	V	V	+	−	V
Growth in 0% NaCl[d]	+	+	−	+	+	+	+
Growth in 1% NaCl[d]	+	+	+	+	+	+	+

[a] +, positive; −, negative; V, variable reaction; K, alkaline; A, acid; G, gas produced.
[b] *V. parahaemolyticus*, *V. cincinnatiensis*, and *V. damsela* show variable reactions.
[c] *V. furnissii* and *V. damsela* are variable.
[d] Nutrient broth base.

fermenting enteropathogens (K/A [K, alkaline; A, acid], no gas, no H_2S). In addition to glucose and lactose, TSI contains sucrose, which is fermented by *V. cholerae* to give reactions of A/A, no gas, and no H_2S.

String test

The string test is performed on a glass microscope slide or in a plastic petri dish by suspending a large loop of growth from a noninhibitory agar medium in a drop of 0.5% aqueous solution of sodium deoxycholate (Fig. 4) (124). In a positive test, bacterial cells are lysed by the sodium deoxycholate, the suspension loses turbidity, and DNA is released, causing the mixture to become viscous. A mucoid "string" is formed when a bacteriological loop is drawn slowly away from the suspension. *V. cholerae* is string test positive, as are most other vibrios. *Aero-*

Figure 4. The string test is a simple method for differentiating *Vibrio* species, which are almost always positive, from *Aeromonas* species, which are negative.

monas and *Plesiomonas* spp. and *Enterobacteriaceae* are usually negative.

Oxidase test

The oxidase test is conducted with fresh growth from a non-carbohydrate-containing agar medium. Several different oxidase reagents are available. A 1% aqueous solution of *n,n,n',n'*-tetramethyl-*p*-phenylenediamine is used for the Kovac oxidase test, which is more sensitive than the indophenol oxidase method (60). Positive and negative controls should be tested at the same time. Organisms of the genera *Vibrio, Neisseria, Campylobacter, Aeromonas, Plesiomonas, Pseudomonas,* and *Alcaligenes* are all oxidase positive; all *Enterobacteriaceae* are oxidase negative.

Fermentation of carbohydrates

Typically, strains of *V. cholerae* ferment glucose and sucrose without producing gas. Fermentation tests should be read at 24 h but if negative may be incubated for up to 7 days.

Decarboxylase/dihydrolase tests

Arginine, lysine, ornithine, and control broths may be supplemented with 1% NaCl to enhance growth. *V. cholerae* is usually positive within 2 days for lysine and ornithine decarboxylase but if negative should be incubated for up to 7 days. *V. cholerae* is typically negative for arginine dihydrolase.

VP test

A modification of the Voges-Proskauer (VP) test is recommended for *V. cholerae*. In this modified test, the test medium (MR-VP broth) incorporates 1% NaCl, test reagent A consists of 5% α-naphthol in absolute ethanol, and test reagent B incorporates 0.3% creatine in 40% KOH (potassium hydroxide) (72). VP broth with 1% NaCl for vibrios (VP–1% NaCl) contains peptone, 7.0 g;

K_2HPO_4, 5.0 g; glucose, 5.0 g; sodium chloride, 10.0 g; and distilled water, 1,000 ml. Inoculate the VP–1% NaCl broth with growth from a nonselective medium, and incubate the tube for 48 h at 35 to 37°C. Add approximately 0.6 ml of MR-VP reagent A and 0.2 ml of MR-VP reagent B. The production of a dark pink to cherry red color within 5 min is a positive test. No color development or yellow to orange color development within this time indicates a negative test.

Salt requirement tests

Many *Vibrio* species are halophilic and require added NaCl for growth (20). *V. cholerae* and *V. mimicus* are the most notable exceptions to this requirement. Accordingly, the determination of NaCl requirements is an important part of identification of *V. cholerae* (Table 3). Salt requirement tests are performed in a tube of nutrient broth with no

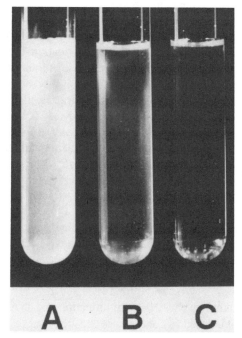

Figure 5. *V. cholerae* will grow in nutrient broth without added NaCl (0% NaCl; tube B), but growth of this species is stimulated by the addition of 1% NaCl (tube A). *V. parahaemolyticus* and other halophilic vibrios will not grow in nutrient broth with 0% NaCl (tube C).

added NaCl and in tubes of nutrient broth to which various amounts of NaCl are added (Fig. 5) (72). Salt broths should be inoculated very lightly with fresh growth from a non-selective medium. Care must be taken to ensure that the inoculum is light enough that no turbidity is visible before incubation. Salt broths should be incubated at 35 to 37°C and examined at 24 h. In the absence of overnight growth, the salt broths may be examined for up to 7 days.

Susceptibility to vibriostatic compound O/129

Susceptibility to 2,4-diamino-6,7-diisopropyl-pteridine phosphate (referred to as O/129, or vibriostatic compound) was considered a hallmark characteristic of the genus *Vibrio*. This test has been used as a primary means for differentiating between *Vibrio* and *Aeromonas* spp.

Although almost all *V. cholerae* are sensitive to O/129, occasional reports over the years have described O/129-resistant strains (71, 72, 79, 129). Several recent reports have described clinical and environmental isolates of *V. cholerae* resistant to O/129 (1, 62, 105). In these studies a majority of *V. cholerae* isolates belonging to O1 and non-O1 serogroups were resistant to 10 and 150 μg of O/129, thereby resembling *Aeromonas* spp. The new O139 serogroup is also reported to be resistant to O/129 (4). Accordingly, caution should be exercised when relying on this test for separating *Vibrio* and *Aeromonas* spp.

BIOTYPE DETERMINATION

The differentiation of *V. cholerae* O1 into classical and El Tor biotypes is not necessary for control or treatment of patients but may be of public health or epidemiologic importance in helping identify the source of the infection, particularly when cholera is first isolated in a country or geographic area. Biotyping is not appropriate for *V. cholerae* non-O1, including

Table 4. Differentiation of classical and El Tor biotypes of *V. cholerae* serogroup O1

Biotype	Reaction				
	VP test (modified, with 1% NaCl)	Zone around polymyxin B (50 U)	Agglutination of chicken, goat, or sheep erythrocytes	Lysis by:	
				Classical IV phage	El Tor phage
Classical	−	+	−	+	−
El Tor	+	−	+	−	+

the O139 serogroup, and the tests can give atypical results for nontoxigenic *V. cholerae* O1. At least two or more of the tests listed in Table 4 should be used to determine biotype, since results can vary for individual isolates.

VP Test

The VP test, previously described in this chapter, has been used to differentiate between the El Tor and classical biotypes of *V. cholerae*. Classical biotypes usually give negative results; El Tor isolates are generally positive (42, 101).

Polymyxin B Sensitivity

Sensitivity to the antibiotic polymyxin B as determined by the agar disk diffusion procedure can be used to differentiate between the biotypes of *V. cholerae* O1 (12, 59). Classical strains are usually sensitive to polymyxin B and will give a zone of inhibition. The size of the zone is not important; any zone is interpreted as a positive result. The El Tor biotype is usually resistant to this concentration of polymyxin B and will not give a zone of inhibition; however, polymyxin B-sensitive variants have been reported (56, 98, 111).

Errors in determining sensitivity to polymyxin B may occur if a selective or differential medium such as TCBS or TTGA is used. Mueller-Hinton agar has been reported to give the most consistent results (50). Nutrient agar should not be used, because the poor growth of *V. cholerae* on this medium leads to difficulty in determining the presence of a zone of inhibition.

Hemagglutination

Strains of the El Tor biotype of *V. cholerae* O1 agglutinate erythrocytes from several animal species, while classical strains do not (15, 46). Chicken erythrocytes are usually used for this assay, although erythrocytes from sheep and goats also give satisfactory results (139). Guinea pig erythrocytes should not be used (16).

Blood for hemagglutination is collected in Alsever's solution or saline, packed by centrifugation, and washed three times, and then the packed erythrocytes are suspended in normal saline to give a 2.5% (vol/vol) suspension. The test should be performed only with growth from cultures that have been grown on solid medium, as variable hemagglutination reactions have been encountered with broth cultures (46). The hemagglutination test is performed on a glass slide with a large loopful of the 2.5% suspension of erythrocytes. A loopful of overnight growth of *V. cholerae* from a noninhibitory, non-carbohydrate-containing medium is added to the erythrocyte suspension and mixed well. In a positive test, agglutination will occur within 30 to 60 s.

While strains belonging to the classical biotype are usually hemagglutination negative, hemagglutinating strains have been reported (40, 110). Moreover, isolates of classical *V. cholerae* O1 that have aged in the laboratory or undergone repeated passage in broth may cause hemagglutination and should not be used as controls (34, 139).

Bacteriophage Susceptibility

Strains of the classical biotype of *V. cholerae* O1 are usually sensitive to the bacteriophage classical IV, while El Tor isolates are usually resistant (88). Similarly, El Tor isolates are susceptible to bacteriophage El Tor 5, also referred to as El Tor V (19). Some phage-resistant classical strains have been reported (109). It has been reported that all of the serogroup O139 strains are resistant to the bacteriophages that constitute Mukerjee's El Tor and classical biotyping schemes (4).

The isolate to be tested is grown overnight on a noninhibitory, non-carbohydrate-containing medium. From the overnight growth, brain heart infusion broth (or Trypticase soy broth) is inoculated and incubated for 6 h at 35 to 37°C. A lawn of bacteria in log-phase growth (optical density = 0.1) is then seeded onto the surface of a brain heart infusion agar plate by dipping a cotton swab into the 6-h broth and lightly inoculating (swabbing) the entire surface of the plate. The inoculum size is important, because false negatives can occur if the lawn is too heavy. A drop of the bacteriophage diluted to routine test dilution (a measure of concentration of active bacteriophage particles) is applied to the bacterial lawn. Positive and negative control strains should be included. The plate is incubated overnight and read the next day. If the bacteria are susceptible to the bacteriophage, they will be lysed, and there will be a zone of clearing in the bacterial lawn.

Although phage susceptibility is reliable for biotyping, the propagation, titration, storage, and use of bacteriophage are technically demanding, and the procedure is usually performed only in specialized reference laboratories.

Hemolysis Testing

Historically, the El Tor biotype was identified by its ability to hemolyze sheep erythrocytes, while classical strains were nonhemolytic. However, by 1972, nearly all El Tor isolates worldwide had become nonhemolytic (10, 41, 130) (Table 5). Exceptions are found in isolates from the U.S. Gulf Coast and from Australia (10, 35). Since El

Table 5. Hemolytic activity of *V. cholerae* O1 classical and El Tor biotypes

Biotype and location	Hemolytic activity
Classical	Negative
El Tor	
Australia	Strongly positive
U.S. Gulf Coast	Strongly positive
Latin America	Negative
Asia, Africa, Europe, Pacific[a]	Negative

[a]Strains isolated between 1963 and 1992.

Tor isolates from these two areas are strongly hemolytic when assayed by either plate or tube hemolysis methods, hemolysis of erythrocytes continues to be a useful phenotypic characteristic for differentiating the Gulf Coast and Australian clones of *V. cholerae* O1 from El Tor strains from the rest of the world, including Latin America.

A genetic probe for the El Tor biotype based on the *hlyA* structural gene for the *V. cholerae* hemolysin has been developed (5). While this probe is specific for El Tor hemolysin genes, it is not specific for *V. cholerae* O1, since other serogroups may possess El Tor hemolysin genes.

Plate hemolysis

Blood agar plates containing 5 to 10% sheep blood may be used to test *V. cholerae* O1 for hemolysis (10). Test strains should be streaked to obtain isolated colonies. Plates may be incubated aerobically, but anaerobic growth conditions enhance hemolytic activity (114). On aerobic sheep blood agar plates, nonhemolytic *V. cholerae* frequently produce a greenish clearing around areas of heavy growth but not around well-isolated colonies. This phenomenon, often described as hemodigestion, is produced by metabolic by-products that are inhibited by anaerobic incubation of the blood agar plate (78). For this reason, when aerobic growth conditions are used, hemolysis should be determined around isolated colonies, not in areas of confluent growth. Also, aerobic blood agar plates should be incubated for no more than 18 to 24 h, since the hemodigestion effect is accentuated during longer incubations.

Aerobic incubation of the plate for no longer than 24 h, although not optimum for detection of hemolysis, will permit differentiation of strongly hemolytic strains such as the U.S. Gulf Coast and Australian clones from the non-hemolytic Latin American strains. Hemolytic colonies should have clear zones around them where erythrocytes have been totally lysed, and a suspected hemolytic strain should be compared with a strongly hemolytic control strain. Strains that give incomplete hemolysis (incomplete clearing of the erythrocytes) should not be reported as hemolytic. If the results of the aerobic plate hemolysis assay are not conclusive, test the strain by the anaerobic plate hemolysis method or by tube hemolysis, which is less susceptible to misinterpretation than the plate method.

Tube hemolysis

If rigorously followed, the tube hemolysis method should minimize performance variables that affect results (10, 14, 41). Positive and negative control strains should be used with each test run.

A 1% (vol/vol) suspension of washed and packed sheep erythrocytes in phosphate-buffered saline (PBS) is prepared. Goat erythrocytes may also be used (101). Test and control organisms are grown overnight in heart infusion broth with 1% glycerol. Supernatants are removed from cultures and divided into two aliquots, one of which is heated to 56°C for 30 min. Serial twofold dilutions (to 1:1,024) of heated and unheated supernatants are made in PBS. To each 0.5-ml dilution of supernatant is added 0.5 ml of the 1% suspension of sheep erythrocytes in PBS. The tubes are incubated in a water bath at 37°C for 2 h, after which they are removed and held overnight at 4°C. The next day, all tubes are examined for evidence of hemolysis. Nonhemolyzed erythrocytes will settle to the bottom of the test tube and form a "button." No button will form if the cells were lysed by hemolysin. Hemolysin titers should be recorded as the highest dilution at which com-plete hemolysis has occurred. Heated supernatants should show no hemolysis, as the hemolysin of *V. cholerae* is heat labile and, if present, will be inactivated by incubation at 56°C. Titers of 1:2 to 1:8 are considered weakly positive, and titers of 1:16 or above are strongly positive.

CT PRODUCTION

Although nontoxigenic O1 strains have been isolated from diarrheal stools of patients with sporadic disease, from extraintestinal sites, and from the environment, only CT-producing strains of *V. cholerae* O1 are associated with epidemic cholera. Therefore, from a public health viewpoint (control, intervention, or prevention), it is important to establish the toxin-producing capability of isolates. During an outbreak of cholera, the isolation of *V. cholerae* possessing the O1 antigen from symptomatic patients correlates well with toxin production and virulence, and there is no need to routinely test every isolate for CT. This is also true in most endemic-cholera situations with a reasonably high frequency of disease. However, in areas where the incidence of cholera is low or during the early stages of an outbreak, most *V. cholerae* O1 strains isolated from diarrheal stools should be tested for CT production.

Since *V. cholerae* serogroup O139 strains produce CT, suspect isolates should be tested for CT production after agglutination in O139 antiserum (4, 25, 104). Other non-O1 serogroups of *V. cholerae* may produce CT or other toxins such as heat-stable enterotoxin or Shiga-like toxin, but toxin-producing strains are very rare. Therefore, testing sporadic isolates of non-O1 *V. cholerae* for CT or other possible toxins is not usually necessary.

There are several approaches to assaying for CT, including tests for toxin activity, toxin antigens, and toxin-coding genes. Please see chapters 3 and 4 for a description of CT assays. However, at least one kit for CT testing is available commercially (7). The

VET-RPLA kit (Oxoid Limited, Hampshire, England) is designed for the detection of CT or *Escherichia coli* heat-labile toxin (LT) in culture supernatant fluids. This test procedure is known as reversed passive latex agglutination. Polystyrene latex particles are sensitized (coated) with purified antiserum produced in rabbits immunized with purified *V. cholerae* enterotoxin. These latex particles will agglutinate in the presence of CT or LT. A control reagent, which consists of latex particles coated or "sensitized" with nonimmune rabbit globulins, is provided. The test is performed in V- or U-bottom microtiter plates (V-bottom plates are preferred; flat-bottom plates are not suitable for this procedure). Dilutions of culture supernatant to be tested are made in two rows of wells. Supernatants should not be prepared by the method described in the kit's instructions, since *V. cholerae* O1 does not usually produce CT in sufficient quantities in APW (137). Using modified Kusama and Craig medium for preparation of supernatants has produced satisfactory results (7, 137). A suspension of antibody-coated latex particles is added to the first column, and unsensitized control latex is added to the second column. After an overnight incubation, agglutination will occur if sufficient amounts of either CT or LT are present in the supernatant.

SUBTYPING METHODS

Prevention and control strategies for cholera depend on understanding the origin, vehicles of transmission, and other characteristics associated with the epidemic spread of *V. cholerae*. This need has stimulated the search for effective means of subtyping isolates. Several bacteriophage and molecular subtyping methods for cholera have been described and used with various degrees of success.

Bacteriophage Typing

Bacteriophages have been used since the 1950s to differentiate among isolates of *V.* *cholerae*. Current phage-typing systems are based on host range as demonstrated by lysis of the organism. The first phage-typing system for cholera was developed in Calcutta and divided classical isolates into five phage types by using four different phages (89). With the advent of the seventh pandemic and the near-total worldwide replacement of classical strains with El Tor strains, the utility of the Calcutta scheme was diminished, and several other phage-typing schemes were developed (18, 38, 49, 77, 91). Nearly all phage-typing schemes now include phages specific for typing both classical and El Tor vibrios.

To date, all phage-typing schemes for *V. cholerae* O1 have the drawback of limited discrimination, since two or three phage types usually account for the majority of strains examined. However, a new phage-typing scheme for *V. cholerae* biotype El Tor strains has been proposed (28). This scheme is a modification of Basu and Mukerjee's El Tor scheme (18), with five new phages added to the original five phages used. This method was able to type 99.6% of the isolates, and the most common type constituted 11.9% of all isolates tested.

Phage-typing methods can test large numbers of strains rapidly, and the methods are generally less laborious than molecular techniques. However, most laboratories find the propagation and maintenance of phages to be too rigorous and demanding to employ phage typing as a routine test. At present, phage typing of *V. cholerae* O1 is performed by only a few specialized laboratories worldwide.

Molecular Subtyping Methods

The development and proliferation of molecular methods have revolutionized the means by which bacterial isolates may be characterized. Several methodologies are available for molecular characterization of *V. cholerae*. These include plasmid profiles; restriction fragment length polymorphisms of rRNA, *ctx,* and bacteriophage genes; multi-locus enzyme electrophoresis; pulsed-field gel electrophoresis; and DNA sequencing (6, 11, 29, 32, 70, 75, 96, 102, 134, 136). Clinical and public health microbiologists who have reagents and equipment available for molecular methods may find these techniques useful for determining the epidemiologic significance of *V. cholerae* O1 isolates and differentiating these isolates from other strains. These methods are described in greater detail in chapter 23.

DETERMINATION OF ANTIMICROBIAL SUSCEPTIBILITY

Historically, antimicrobial resistance of *V. cholerae* has not been widely reported, and routine susceptibility testing was not recommended (47, 94). However, resistance to a broad spectrum of antimicrobial agents has been identified with increasing frequency in isolates from around the world (43, 55, 80, 105). Most of the *V. cholerae* O139 isolates from recent outbreaks in Bangladesh and India were reported to be resistant to trimethoprim-sulfamethoxazole, streptomycin, furazolidone, and the vibriostatic agent O/129 (1, 3). It now seems prudent to recommend the periodic monitoring of suscep-

tibility of isolates of toxigenic *V. cholerae* O1 and O139.

Although the disk diffusion technique is the most commonly used method for antimicrobial susceptibility testing, interpretive criteria have not been established for *V. cholerae* O1 and O139, and the method's reliability for these organisms is unknown. To ensure the accuracy of susceptibility results for *V. cholerae*, agar or broth dilution methods should be used. If a laboratory cannot routinely perform one of the dilution techniques, the disk diffusion method may be used to screen for antimicrobial resistance. Table 6 lists the interpretive criteria for the antimicrobial agents that are presently recommended by the World Health Organization for treatment of cholera (tetracycline, doxycycline, furazolidone, erythromycin, chloramphenicol, and trimethoprim-sulfamethoxazole) as well as some other commonly used antimicrobial agents (135). These criteria, which have been standardized for the *Enterobacteriaceae,* may be used as tentative zone size standards for screening for antimicrobial resistances in *V. cholerae* O1 until interpretive criteria have been validated (12). When screening with the disk diffusion method, any isolates that fall within the intermediate or resistant ranges should be tested by a dilution method to obtain the MIC of the drug in question.

Table 6. Zone size interpretative standards for the *Enterobacteriaceae* for selected antimicrobial disks (not validated for *V. cholerae* O1)

Antimicrobial agent	Disk potency (μg)	Zone diam (mm)[a]			Limits for *E. coli* ATCC 25922
		Resistant	Intermediate	Sensitive	
Chloramphenicol	30	≤ 12	13–17	≥ 18	21–27
Doxycycline	30	≤ 12	13–15	≥ 16	18–24
Erythromycin	15	≤ 13	14–22	≥ 23	8–14[b]
Furazolidone	100	≤ 13	14–17	≥ 18	22–26[c]
Trimethoprim-sulfamethoxazole	1.25/23.75	≤ 10	11–15	≥ 16	24–32
Tetracycline	30	≤ 14	15–18	≥ 19	18–25
Ciprofloxacin	5	≤ 15	16–20	≥ 21	30–40
Nalidixic acid	30	≤ 13	14–18	≥ 19	22–28

[a]Source: National Committee for Clinical Laboratory Standards (NCCLS), 1992. Zone sizes for *V. cholerae* have not been standardized by NCCLS.
[b]Source: World Health Organization.
[c]Source: manufacturer.

STORAGE OF ISOLATES

V. cholerae will usually remain viable for several weeks on solid medium held at room temperature (22 to 25°C) unless the medium is allowed to dry out. However, if cultures are to be maintained for longer periods, they should be appropriately prepared for storage. Most nonselective agars, except nutrient agar (which has no added salt and does not allow optimal growth) and carbohydrate-containing media, are good storage media for *V. cholerae*. Storage in agar deeps or stabs is preferable to storage in slants. Tubes must be sealed or the growth may be covered with sterile mineral oil to prevent evaporation. Stock cultures should be stored at room temperature (about 22°C) in the dark (47).

V. cholerae isolates may be stored indefinitely if maintained in an ultra-low-temperature freezer (−70°C) or in liquid nitrogen (−196°C). *V. cholerae* may be frozen in a variety of suspending media, such as skim milk, blood, or Trypticase soy broth with 15 to 20% glycerol. Lyophilization (freeze-drying) of a suspension of fresh growth in a suitable medium such as skim milk or polyvinylpyrrolidone is also a useful method for long-term storage of *V. cholerae* O1. Freeze-dried cultures are best maintained at 4°C, but successful room temperature storage of lyophilized isolates of *V. cholerae* has been reported (47).

RAPID DIAGNOSTIC METHODS

Prompt laboratory diagnosis of cholera is often advantageous for monitoring the spread of the disease and allowing rapid institution of control measures. Several rapid-detection methods have been developed and used for the detection of *V. cholerae* O1 directly from stools of acutely ill patients or from enrichment broth. In certain situations, these rapid methods may be practical; however, they provide only a preliminary diagnosis. Despite their advantages of speed and (sometimes) simpler requirements for reagents and equipment, rapid diagnostic methods for the most part cannot replace traditional culture methods. Traditional techniques must be relied on when an isolate is required for further tests, such as assays for CT production, antimicrobial sensitivity, hemolysis, biotype, or molecular subtype. Rapid methods may be most useful in remote field situations where bedside diagnosis is required to confirm the existence of an outbreak of cholera.

Dark-Field and Phase-Contrast Methods

Dark-field and phase-contrast microscopy have been used for screening fecal specimens for the presence of *V. cholerae* (17, 22, 58). In these techniques, liquid stools are microscopically examined for the presence of organisms with typical darting ("shooting star") motility. In this method, stools or enrichment broths are examined with and without the addition of specific *V. cholerae* O1 antisera. If the addition of antiserum results in the cessation of motility, as observed by either dark-field or phase-contrast microscopy, the test is considered positive. The presence in feces of organisms with typical darting motility correctly predicted a positive culture in about 80% of cases of cholera (22).

Diluents for the antisera and stool must be selected carefully to avoid nonspecific inhibition of motility (for example, certain preservatives such as sodium azide or merthiolate should not be used). Distilled water is not a suitable diluent for stool specimens, since it is known to extinguish the motility of *V. cholerae* (30). Other disadvantages include the requirements for a dark-field or phase-contrast microscope and a technician experienced in this technique. Despite these disadvantages, this technique has been widely used.

Immunofluorescence

Immunofluorescence techniques using antisera conjugated to fluorescein isothiocyanate have been used to visualize *V. cholerae*

O1 cells in a variety of specimen types (45, 58). The use of highly specific monoclonal antibodies directed towards the O1 antigen of *V. cholerae* has been reported elsewhere (2). Immunofluorescence methods are becoming popular as primary diagnostic tools, although the requirements for expensive equipment, high-quality immunologic reagents, and trained technicians are often obstacles for laboratories.

Latex Agglutination

A commercially manufactured slide agglutination kit (Denka Seiken) has been developed for serotyping of *V. cholerae* O1 isolates but has been used for examination of specimens. The kit uses latex particles coated with monoclonal antibodies directed against the A, B, and C antigens of *V. cholerae* O1. During an investigation of an epidemic of cholera, the kit was evaluated for its ability to confirm the diagnosis of cholera at bedside by using rectal swabs (116). The latex agglutination test detected 63% of culture-positive patients from rectal swabs but gave false positives for 12% of culture-negative patients. The sensitivity and specificity of this test with liquid stool specimens have not been determined.

Coagglutination

Coagglutination methods for rapid diagnosis of cholera directly from stool or enrichment broths have been reported by a number of authors (68, 103). In the coagglutination method, antibodies against *V. cholerae* are bound to the surface of *Staphylococcus aureus* (Cowan 1) while retaining their binding capacity and specificity.

Recently, a commercially prepared monoclonal antibody-based coagglutination test (CholeraScreen; New Horizons Diagnostics, Columbia, Md.) has been marketed. Reports of evaluations of the product with culture collections in the United States and with clinical specimens in Guatemala and Bangladesh have been encouraging (31).

Polymerase Chain Reaction

The polymerase chain reaction, in which multiple copies of a specific portion of the bacterial genomic DNA are synthesized, has been applied to rapid diagnosis of *V. cholerae* O1. Investigators used a set of oligonucleotide primers specific for the *ctx* operon to detect *V. cholerae* cells containing the CT gene directly in stool samples (123). While not extensively evaluated, this method shows promise and may allow for diagnosis of cholera in stool specimens that do not contain viable or culturable *V. cholerae* O1 cells.

Other Rapid Isolation and Identification Techniques

Other techniques for the rapid diagnosis of *V. cholerae* O1 have been described elsewhere and include methods based on the multiplication of bacteriophage, addition of antiserum to growth medium to precipitate *V. cholerae* cells in broth, and use of antibody-coated magnetic beads to physically aggregate *V. cholerae* bacterial cells (76, 95, 100, 115).

REFERENCES

1. **Abbott, S., W. K. Cheung, B. Portoni, and J. M. Janda.** 1992. Isolation of vibriostatic agent O/129-resistant *Vibrio cholerae* non-O1 from a patient with gastroenteritis. *J. Clin. Microbiol.* **30:**1598–1599.
2. **Adams, L. B., M. C. Henk, and R. J. Siebeling.** 1988. Detection of *Vibrio cholerae* with monoclonal antibodies specific for serovar O1 lipopolysaccharide. *J. Clin. Microbiol.* **26:**1801–1809.
3. **Adams, L. B., and R. J. Siebeling.** 1984. Production of *Vibrio cholerae* O1 and non-O1 typing sera in rabbits immunized with polysaccharide-protein carrier conjugates. *J. Clin. Microbiol.* **19:**181–186.
4. **Albert, M. J., A. K. Siddique, M. S. Islam, A. S. G. Faruque, M. Ansaruzzaman, S. M. Faruque, and R. B. Sack.** 1993. Large outbreak of clinical cholera due to *Vibrio cholerae* non-O1 in Bangladesh. *Lancet* **341:**704.
5. **Alm, R. A., and P. A. Manning.** 1990. Biotype-specific probe for *Vibrio cholerae* serogroup O1. *J. Clin. Microbiol.* **28:**823–824.

112. Sack, R. B., and C. E. Miller. 1969. Progressive changes in *Vibrio* serotypes in germ-free mice infected with *Vibrio cholerae*. *J. Bacteriol.* **99:**688–695.

113. Sakazaki, R., and K. Tamura. 1971. Somatic antigen variation in *Vibrio cholerae*. *Jpn. J. Med. Sci. Biol.* **24:**93–100.

114. Sakazaki, R., K. Tamura, and M. Murase. 1971. Determination of the haemolytic activity of *V. cholerae*. *Jpn. J. Med. Sci. Biol.* **24:**83–91.

115. Sechter, I., C. H. B. Gerichter, and D. Cahan. 1975. Method for detecting small numbers of *Vibrio cholerae* in very polluted substrates. *Appl. Microbiol.* **29:**814–818.

116. Shaffer, N., E. S. do Santos, P.-A. Andreason, and J. J. Farmer III. 1989. Rapid laboratory diagnosis of cholera in the field. *Trans. R. Soc. Trop. Med. Hyg.* **83:**119–120.

117. Sheehy, T. W., H. Sprinz, W. S. Augerson, and S. B. Formal. 1966. Laboratory *Vibrio cholerae* infection in the United States. *JAMA* **197:**99–104.

118. Shimada, T., G. B. Nair, B. C. Deb, M. J. Albert, R. B. Sack, and Y. Takeda. 1993. Outbreak of *Vibrio cholerae* non-O1 in India and Bangladesh. *Lancet* **341:**1346–1347.

119. Shimada, T., and R. Sakazaki. 1973. R antigen of *Vibrio cholerae*. *Jpn. J. Med. Sci. Biol.* **26:**155–160.

120. Shimada, T., and R. Sakazaki. 1977. Additional serovars and inter-O antigenic relationships of *Vibrio cholerae*. *Jpn. J. Med. Sci. Biol.* **30:**275–277.

121. Shimada, T., R. Sakazaki, S. Fujimura, K. Niwano, M. Mishina, and K. Takizawa. 1990. A new selective differential agar medium for isolation of *Vibrio cholerae* O1: PMT (polymyxin mannose-tellurite agar). *Jpn. J. Med. Sci. Biol.* **43:**37–41.

122. Shimada, R., R. Sakazaki, and M. Oue. 1987. A bioserogroup of marine vibrios possessing somatic antigen factors in common with *Vibrio cholerae* O1. *J. Appl. Bacteriol.* **62:**453–456.

123. Shirai, H., M. Nishibuchi, T. Ramamurthy, S. K. Bhattacharya, S. C. Pal, and Y. Takeda. 1991. Polymerase chain reaction for detection of the cholera enterotoxin operon of *Vibrio cholerae*. *J. Clin. Microbiol.* **29:**2517–2521.

124. Smith, H. L., Jr. 1970. A presumptive test for vibrios: the "string" test. *Bull. W.H.O.* **42:**817–818.

125. Smith, H. L., Jr., R. Freter, and F. J. Sweeney, Jr. 1961. Enumeration of cholera vibrios in fecal samples. *J. Infect. Dis.* **109:**31–34.

126. Smith, H. L., Jr., and K. Goodner. 1958. Detection of bacterial gelatinases by gelatin-agar plate methods. *J. Bacteriol.* **76:**662–665.

127. Stuart, R. D., S. R. Toshach, and T. M. Patsula. 1954. The problem of transport of specimens for culture of gonococci. *Can. J. Public Health* **45:**73–83.

128. Sugiyama, J., F. Gondaira, J. Matsuda, M. Soga, and Y. Terada. 1987. New method for serological typing of *Vibrio cholerae* 1:O using a monoclonal antibody-sensitized latex agglutination test. *Microbiol. Immunol.* **31:**387–391.

129. Sundaram, S., and K. Murthy. 1983. Occurrence of 2,4-diamino-6,7-diisopropylpteridine resistance of human isolates of *Vibrio cholerae*. *FEMS Microbiol. Lett.* **19:**115–117.

130. Swanson, R. W., and J. D. Gillmore. 1964. Biochemical characteristics of recent cholera isolates in the far east. *Bull. W.H.O.* **31:**422–425.

131. Tamura, K., S. Shimada, and L. M. Prescott. 1971. *Vibrio* agar: a new plating medium for isolation of *V. cholerae*. *Jpn. J. Med. Sci. Biol.* **24:**125–127.

132. Tassin, M. G., R. J. Siebeling, N. C. Roberts, and A. D. Larson. 1983. Presumptive identification of *Vibrio* species with H antiserum. *J. Clin. Microbiol.* **18:**400–407.

133. Venkatraman, K. V., and C. S. Ramakrishnan. 1941. A preserving medium for the transmission of specimens for the isolation of *Vibrio cholerae*. *Indian J. Med. Res.* **29:**681–684.

134. Wachsmuth, I. K., G. M. Evins, P. I. Fields, Ø. Olsvik, T. Popovic, C. A. Bopp, J. G. Wells, C. Carrillo, and P. A. Blake. 1993. The molecular epidemiology of cholera in Latin America. *J. Infect. Dis.* **167:**621–626.

135. World Health Organization. 1993. *Guidelines for Cholera Control*. World Health Organization, Geneva.

136. Yam, W. C., M. L. Lung, and M. H. Ng. 1991. Restriction fragment length polymorphism analysis of *Vibrio cholerae* strains associated with a cholera outbreak in Hong Kong. *J. Clin. Microbiol.* **29:**1058–1059.

137. Yam, W. C., M. L. Lung, and M. H. Ng. 1992. Evaluation and optimization of a latex agglutination assay for detection of cholera toxin and *Escherichia coli* heat-labile toxin. *J. Clin. Microbiol.* **30:**2518–2520.

138. Zafari, Y., A. Zarifi, and F. Zomorodi. 1968. A comparative study of sea-salt water and Cary-Blair media for transportation of stool specimens. *J. Trop. Med. Hyg.* **71:**178–179.

139. Zinnaka, Y., S. Shimodori, and K. Takeya. 1964. Haemagglutinating activity of *Vibrio comma*. *Jpn. J. Microbiol.* **8:**97–103.

Vibrio cholerae and Cholera: Molecular to Global Perspectives
Edited by I. Kaye Wachsmuth, Paul A. Blake, and Ørjan Olsvik
© 1994 American Society for Microbiology, Washington, DC 20005

Chapter 2

Toxigenic *Vibrio cholerae* O1 in Food and Water

Charles A. Kaysner and Walter E. Hill

When *Vibrio cholerae* infects the human intestinal tract, it can cause the disease cholera. Large numbers of *V. cholerae* are discharged in the feces of infected individuals. Inadequate sewage treatment allows water systems to become contaminated. Generally, cholera epidemics occur in areas of the world that have poor sanitation practices. When water becomes contaminated with toxigenic strains of *V. cholerae* O1 and is then used by other individuals for drinking or in food preparation, the result is often widespread disease. Studies of the organism in the environment have found that it is able to survive for extended periods in aquatic and estuarine areas.

In developed countries, the threat of large-scale cholera epidemics has essentially been eliminated thanks to collection, treatment, and adequate discharge of sewage. In addition, drinking water is treated, usually by chlorination, which is effective in eliminating this and many other human pathogenic microbes. Also, personal hygiene is taught in order to alleviate problems such as secondary spread of disease. Although sporadic outbreaks have occurred in the developed countries during the past seven pandemics, the number of individuals involved is dramatically lower than in developing countries.

Strictly speaking, cholera is the name of the disease caused exclusively by serogroup O1 of *V. cholerae*. Although non-O1 serogroups of the organism can cause a similar illness, other terms such as "choleralike" have been used to differentiate non-O1 infections from the true Asiatic or epidemic cholera. Only recently have there been reports of illness from non-O1 *V. cholerae* that could be classified as epidemic in proportion (1, 70).

The disease is considered an ancient affliction of humans (22, 68). The distinct symptomatology and the explosive onslaught of cholera epidemics are believed to be depicted in Arabian, Hindu, Greek, and Roman writings (59). The word "cholera" has been used for over 2,500 years to describe any diarrhea and vomiting, not just that caused by bacteria. Whatever the origin of the word, each year cholera affects a large number of people. Over 40,000 cases were recorded in 1989, mostly in Africa (13), and an epidemic began in South America in 1991. The risk of outbreaks of cholera is greater where the disease is endemic if large groups of people are crowded together without adequate sanitary facilities and adequate food-handling practices. Cholera was one of the first epidemic diseases to be controlled by public health measures in the 19th century (76). Even though modern sanitation practices in developed countries have virtually eliminated epidemic cholera from them, those countries still

Charles A. Kaysner and Walter E. Hill • Food and Drug Administration, 22201 23rd Drive S.E., P.O. Box 3012, Bothell, Washington 98041-3012.

contain areas of endemic toxigenic strains, for example, the United States Gulf Coast. Modern transportation, such as ships and airplanes, can also be a mechanism for transporting *V. cholerae* O1 from areas of the world in which the disease is endemic to other countries and populations. International travel has long been the cause of sporadic cases of cholera in the United States. In fact, food service aboard a flight originating in Argentina was suspected of causing 75 cases and 1 death after the plane made a stop in Peru before landing in the United States (16, 17).

WATER CONTAMINATION AND OUTBREAKS

As mentioned above, cholera is generally spread by poor sanitation practices resulting in contamination of water supplies used for drinking and food preparation. Epidemic cholera has been well documented in many countries as being spread by contaminated water (6, 7, 31, 76). This has been demonstrated quite clearly in the spread of the disease in South American communities in 1991. Sporadic cases of cholera have been associated with the aquatic environment as well (8).

In the large epidemics that have occurred during the last century, contaminated water has undoubtedly been the major vehicle of transmission, with food playing a lesser but significant role in the spread of the disease. As the Pan American Health Organization recently reported (66), over 750,000 cases of cholera (including almost 6,500 deaths) have been reported since January 1991. These figures represent reports from 21 countries in South, Central, and North America, with over 70% of the cases and approximately 60% of the deaths occurring in Peru. The origin of the El Tor strain responsible for the outbreak remains unknown; however, this strain can be clearly distinguished from the strains causing the present seventh pandemic worldwide and from the strains associated

with the environmental reservoirs in the U.S. Gulf Coast area and Australia. The genetic differences of these strains will be discussed in detail in other chapters. The spread of the "Latin American" strain has for the most part been caused by contamination of drinking water, wells, rivers, and other sources that are used by many of the people of South America. Even the marine environment has been contaminated, thus making the fish and other seafood gathered from the estuaries potential vehicles of transmission (82). Until recently, the survival of *V. cholerae* in the environment was thought to be short (less than 1 week in seawater [18]); thus contaminated water required continual recontamination with feces from carriers of the organism. However, since the late 1970s, *V. cholerae* O1 has been isolated from occasional sites in cholera-free areas. It is fortunate that the majority of the O1 environmental strains are nontoxigenic, as are most of the non-O1 strains that reside in the environment. However, studies in Australia, along the Gulf Coast of the United States, and in England have shown that even toxigenic *V. cholerae* O1 is a resident of some aquatic areas and that its presence is not associated with fecal contamination from cholera patients (77, 85).

During large epidemics it is often difficult to determine whether water or a food was the source of the organism infecting a particular person or a family. Direct person-to-person contact is not a major factor in the spread of the disease in these epidemics; rather, contact with some common factor such as a water cistern contaminated by dirty hands followed by spread to family members seems to be the principal means of contamination.

Bottled mineral water was the vehicle of transmission of cholera in Portugal in 1974 (7). Spring water used at a bottling plant had become contaminated with *V. cholerae* suspected to have come from a nearby river carrying sewage. Untreated water that was bottled and distributed in that country caused some of over 2,000 cases reported in Portugal that year. Interestingly, the plant also bottled carbonated water, which was not documented to have caused the disease. The acidic conditions produced by the carbonation may effectively eliminate the organism owing to its intolerance to acid. With the current trend for "pure" bottled water to complement the "health food" diets that are now popular, the potential for spread of disease has increased, so control measures must be strictly adhered to.

An interesting epidemiologic investigation was reported by Johnston et al. (41). In this case, the potable water system on a floating oil rig off the coast of Texas was contaminated by a toxigenic strain of *V. cholerae* O1 El Tor Inaba. Contact with one of the workers, who was ill prior to reporting to work on the oil rig, led to an outbreak among 15 coworkers, the largest cholera outbreak in the United States in over a century. A cross-connection between the drinking water and sewage disposal systems was discovered. The potable water system then became contaminated with feces from the infected worker. Cooked rice was rinsed with the contaminated water and held unrefrigerated for most of the day, permitting *V. cholerae* to grow. Ironically, the cross-connection in the piping system was discovered that same day and repaired; however, the damage had already been done. As with many foodborne-disease outbreaks, a combination of factors was involved, but this case demonstrates that both water and food can be directly involved in the transmission of the disease.

The isolation of the Latin American strain of El Tor Inaba *V. cholerae* O1 from oysters taken from waters along the coast of Alabama was the first occurrence of this particular strain in the environment outside of South and Central Americas (25). This sparked an interest in the spread of the organism to other countries by means other than the exportation of food products from affected countries. One theory involved transportation of the strain via ships from South America to the Northern Hemisphere. The release of sewage from these ships and the bilgewater used for ballast might be a likely source. Indeed, upon analysis of several samples of water from ships, an identical strain was isolated (58), indicating that the source of the strain in the Alabama marine environment might have been bilgewater released in the coastal areas of the United States.

FOOD CONTAMINATION AND ILLNESSES

The transmission of cholera by ingestion of contaminated food has been documented in numerous instances discussed below (83). Investigators in the United States make most of these reports, perhaps because in the United States there are resources to aggressively investigate outbreaks, better recognition of the disease and reporting of it to health authorities, more public awareness of diseases of this nature and public demand for prevention, and, certainly, laws enacted to prevent the contamination and distribution of food that could cause illness.

From the reports of illness involving foods, it is evident that vehicles of transmission of *V. cholerae* O1 are the foods that are nearly neutral in pH (72). This is not surprising when one considers the characteristics of the organism. *V. cholerae* O1 does not tolerate acidic conditions. In fact, its tolerance to alkaline pH is used in procedures to isolate the organism. Thus, one can consider the food history of a patient, eliminate the foods of acidic nature, and concentrate on neutral or alkaline foods. It has also been shown that *V. cholerae* O1 survives refrigeration and freezing, and contaminated food products, including foods being shipped internationally, can thus be distributed far from the point of origin (15). Thus, the source of an epidemic may be a product far removed from the location of illnesses. Cooked foods are a more favorable milieu than raw foods for growth of *V. cholerae* O1, since cooking destroys many of the competing bacteria.

As mentioned above, it may be difficult in some outbreaks of cholera to determine whether food or water was the vehicle of transmission. Several reports indicate that both were involved or that water was the primary vehicle, with food then becoming a secondary vehicle within a household or even in food preparation by a street vendor. These findings do not preclude the source of the contamination being the food preparer, who may be an asymptomatic carrier. An epidemiologic study by St. Louis et al. (81) determined that a funeral-associated outbreak of cholera in Africa may have resulted from contamination of a rice and meat sauce dish prepared for those attending the funeral of a cholera victim by the three women who prepared the body for burial. A part of such preparation for burial includes evacuating the victim's bowels.

One of the food types frequently mentioned in reports as a vehicle is molluscan shellfish. Such transmission is nicely summarized by DePaola (24). Investigations have shown that these shellfish are contaminated with toxigenic *V. cholerae* O1 in the marine environment in which they grow, where the toxigenic strain of the organism has become endemic or where the environment has recently been contaminated with sewage. Since several species of molluscan shellfish, i.e., oysters, clams, and mussels, are consumed raw or with minimal heat treatment, sporadic illnesses have resulted. Molluscan shellfish are filter feeders and thus can accumulate *V. cholerae* O1 when it is present in the water column, which usually occurs during the summer months. Therefore, it is not surprising that most reports of cholera from shellfish consumption are recorded during the warmer months. Since individual mollusca filter feed at different rates, it is not uncommon to find one animal with a higher accumulation or concentration of *V. cholerae* O1 than another animal close to it. This may be one reason that the outbreaks of cholera caused by shellfish are sporadic and have not involved a large number of cases per outbreak. Additionally, the number of shellfish consumed at one sitting varies considerably among individuals.

One of the first reports of shellfish as a vehicle of transmission resulted from an outbreak in Malaysia in 1969, when mussels were implicated in a 135-case outbreak (28). Mussels were again implicated in an outbreak involving 278 cases in Italy in 1973 (3). The following year, raw cockles were implicated in Portugal, where over 2,400 individuals were affected (6). The suspect shellfish were imported from a North African country into Portugal. Consumption of contaminated raw oysters in Texas was thought to be the cause of one case of cholera in 1973 (84). This was the first cholera case in the United States since the early 1900s. Raw oysters were again implicated in a single case in the state of Alabama in 1977 (11). The El Tor biotype of *V. cholerae* O1 was identified in all of the above outbreaks; the Inaba serogroup was identified in all but one of them (3), with the Ogawa serogroup also isolated from patients (28).

Information regarding U.S. shellfish-borne illnesses dating back to the early 1900s has been compiled by Rippey (71). His report records single episodes and outbreaks (two or more cases) for many etiologic agents, both identified and unknown. From his data, oysters, not mussels and clams, are implicated in over 90% of the cases caused by *Vibrio* species and nearly one-third of those caused by O1 and non-O1 *V. cholerae*. Table 1 lists only some recently documented cholera outbreaks in the United States, several of which were associated with the consumption of seafood.

Cholera outbreaks implicating shellfish stimulated interest in exploring the estuarine environment to determine the incidence of *V. cholerae* O1. Several environmental studies found areas of the United States harboring *V. cholerae* O1 to some degree; however, it occurred much less frequently than non-O1 serogroups. Kaper et al. (43) reported results for Chesapeake Bay, from which only non-O1 strains of *V. cholerae* were isolated. However, these findings prompted further investigations, not only along the Gulf Coast but in other coastal areas as well (10, 20, 21, 27,

Table 1. Food products recently associated with cholera in the United States

Year	Food type (no. of cases)	Location of illness	Source	Reference(s)
1984	Crabs (1)	Maryland	Texas	52
1986	Oysters (1)	Georgia	Texas	67
1986	Oysters (1)	Florida	Louisiana	46
1988	Oysters (1)	Colorado	Louisiana	12
1991	Crab meat (8)	New Jersey	Ecuador	14
1991	Coconut milk (3)	Maryland	Thailand	15
1992	Crab salad (75)	Airplane	Peru	16, 17

34, 38, 39, 45, 51, 55, 73). These studies found that essentially all investigated estuarine areas contained *V. cholerae* non-O1 as natural inhabitants, with occasional pockets of the O1 serogroup; most areas yielded only nontoxigenic strains, but in a few areas, toxigenic O1 Inaba was considered endemic (5, 10, 20, 77).

Many foods imported from South and Central America into the United States were sampled by the U.S. Food and Drug Administration after the Peruvian epidemic of cholera was reported. This was done to determine the presence of *V. cholerae* O1 and to reduce the probability of this organism being spread to the United States. Although sampling of imported products, particularly seafood, continues, results of the intensified effort found few samples that contained *V. cholerae*. Of about 900 samples from 14 countries in the Americas, almost 600 were seafood. Non-seafood samples included fruits, vegetables, water, and ice. Thirty-one samples were reported to contain nontoxigenic non-O1 *V. cholerae* with 27 of those samples being seafood and the majority of these being frozen shrimp. Although none of the samples were found to contain toxigenic *V. cholerae* O1, the presence of non-O1 strains in imported products indicates that the O1 strain could be introduced into the United States as well as other countries.

Other seafoods (crustaceans and finfish) have been implicated in the spread of cholera. In 1961, 330 cholera cases stemming from the consumption of raw shrimp were recorded in the Philippines; however, the serogroup of the El Tor strain was not noted, and the source of the shrimp was not known (42). In 1974, raw fish contaminated by sewage were responsible for six cases of cholera on the island of Guam (60). Home-boiled crabs contaminated with O1 Inaba *V. cholerae* caused two separate outbreaks involving 11 individuals in Louisiana in 1978 (5). This outbreak sparked interest among public health officials in the United States because it was the first outbreak of cholera in the United States since the early 1900s that involved more than one individual. That was the beginning of increased research efforts to find a focus of endemic toxigenic *V. cholerae* O1 Inaba.

There are reports of the spread of the disease in South America due to the consumption of ceviche, a dish made with raw, marinated fish that may have been contaminated in the estuary from which they were harvested or during preparation. The rapid spread of the disease in the coastal area of Peru during January and February 1991 is consistent with multiple vectors. A number of foci of infection may have resulted from seafood being harvested during a *V. cholerae* marine bloom (30, 35). The disease rapidly spread throughout Peru and into bordering countries via contaminated water and food and infected individuals. Vegetables irrigated with raw sewage were suspected vehicles in Chile (35). *V. cholerae* O1, often at high levels, was found in many freshwater and marine sites, including the major source of drinking water for the city of Lima, Peru, and also on the skin and in the intestines of fish from coastal areas (82).

Since foods were suspected in the transmission of cholera, studies were undertaken to determine the growth and/or survival of *V. cholerae* O1 in various foods. *V. cholerae* O1 grew to high levels in several cooked foods (rice, lentils, chicken, mussels) incubated at 22, 30, or 37°C (49) and in infant milk formula stored at 25°C (63). Survival of *V. cholerae* O1 in various foods has been studied and can range from as little as 1 h up to several weeks, depending on the type of food and the storage conditions. A short survival period was common in acidic food, such as fruits, whereas survival could be several weeks in cooked and raw vegetables and shellfish. The survival data have been nicely summarized and reported by several investigators (32, 69, 72). Proper cooking of foods also effectively eliminated this pathogen (9, 56). Factors affecting survival of *V. cholerae* in foods are summarized in Table 2. It was concluded after a review of the data on survival of *V. cholerae* in food and after inspection of processing plants in South America that an excessive restriction on foods imported from countries in which cholera is endemic is unwarranted, as the risk of transmission by imported food is small (65).

One of the more recent axioms that can be applied to the prevention of the spread of cholera (and other bacterial diseases, for that matter) by food and water is the "boil it, cook it, peel it, or forget it" philosophy (50), particularly regarding supplies of untreated water and the consumption of shellfish.

ISOLATION OF *V. CHOLERAE* O1 FROM WATER AND FOOD

As with many procedures for the microbiologic examination of food, some of the techniques for isolation of *V. cholerae* O1 have been borrowed from the clinical laboratory. However, in contrast to the analysis of a clinical specimen, the isolation of a sought-after organism from a food or water sample can be difficult. The cells may be under more stress because of temperature (heating, freezing), pH, water activity, nutrient deprivation, and competition from other bacteria present. Methods of enhancing the isolation of *V. cholerae* O1 have been developed and optimized, primarily by taking advantage of the organism's rapid generation time under certain conditions and its tolerance to alkali.

LABORATORY OBSERVATIONS

Alkaline peptone water, an enrichment medium used by many laboratories, was developed to resuscitate the organism from samples. Peptone water (pH 6.8) was first used in water analysis in 1914 (19), and a modified version consisting of 1% NaCl and 1% peptone is relatively inexpensive and easy to prepare. The pH is adjusted to 8.5 for selective purposes, and samples are incubated at 35 to 37°C (29). *V. cholerae* strains have short gen-

Table 2. Factors influencing survival of *V. cholerae* in food and the environment[a]

Factor	Comment
Temp	Refrigeration extends survival, freezing more so; growth occurs above 10°C
pH	Optimum, 7.0–8.5; range, 6–10
Water activity (a_w)	> 0.93
Salt content	Best survival, 0.25–3%; optimum, 2%; requires Na+
Degree of contamination	Longer survival with higher dose
Exposure to sunlight	Reduces survival
Presence of organic matter	Extends survival
Osmotic pressure	Unfavorable at high pressure
Carbohydrate content	Reduced in high sugar content
Other microflora	Suppressed by competing flora
Humidity, moisture	Survives longer in high humidity

[a]Adapted from reference 72.

eration times and can usually outgrow the normal microflora of the sample. In fact, incubating the enrichment phase for 6 to 8 h at 35 to 37°C is recommended if the sample is suspected of having been recently contaminated. Samples of frozen product may have to be incubated overnight (18 to 24 h) to fully resuscitate the organism.

V. cholerae O1 is relatively easy to isolate from water samples if the contamination is recent. Filtration techniques for examining known quantities of water and estimating the amount of contamination have been developed. Also, the most-probable-number procedure (19) can be used to estimate the number of organisms. Moore swabs have been used successfully to isolate *V. cholerae* O1, because bacterial cells will adhere to the cotton gauze, thereby concentrating the organism from large volumes of water (62). After extended periods in the environment, *V. cholerae* O1 enters a viable but nonculturable state (86) that will be discussed in greater detail in another chapter. When the pathogen is in this quiescent physiologic state, other techniques must be used to determine its presence. An immunofluorescence microscopy technique can be used (40), but this method requires special equipment to which laboratories may not have access.

A successful modification of the method used in some environmental work is the incubation of a separate portion of the sample in alkaline peptone water at 42°C (26). This elevated enrichment temperature enhances the isolation of *V. cholerae* O1 and non-O1 strains from water, sediment, and shellfish samples (26; unpublished observations). The tolerance of *V. cholerae* O1 to this elevated temperature increases the recovery rate, because many of the environmental bacteria present in these samples cannot grow at all or grow slowly. In several instances, apparently pure cultures of *V. cholerae* were observed on the selective agar. The elevated temperature is now routinely used in testing oysters for this organism (29).

After the enrichment is completed, a loopful of the contents is streaked to a selective and differential agar medium. Thiosulfate-citrate-bile salts-sucrose agar (TCBS) developed for *V. cholerae* (47) is probably the agar most widely used for *Vibrio* isolation. The commercially available TCBS used for clinical samples is also used for food and water analyses. The fermentation of sucrose is the differential aspect of this medium. Acid production will change the color indicator from green to yellow. Thus, *V. cholerae* O1 appears as a flat, yellow colony with an opaque center and a clear periphery. Non-sucrose-fermenting colonies are green.

An additional medium, modified cellobiose-polymyxin B-colistin agar (29, 57), is selective for El Tor strains of *V. cholerae* O1 because of the addition of polymyxin B; classical strains are sensitive to this antibiotic. The 40°C incubation temperature prevents many environmental strains from growing. Typical *V. cholerae* colonies will be medium sized, opaque, and green to purple because cellobiose is not fermented. Unfortunately, this agar is not available commercially.

Monsur's agar (taurocholate-tellurite-gelatin agar [61]) is a selective and differential medium specifically designed for the isolation of *V. cholerae*. Typical colonies after overnight incubation of this medium are small and opaque, with darkened centers. After additional incubation, the colonies become darker in the center and gunmetal gray in color. The coloration is due to the reduction of tellurite. The production of gelatinase by *Vibrio* species will result in a cloudy white halo around a colony that can be intensified by placing the plate in the refrigerator for 15 to 30 min. This medium is not commercially available but offers the advantage that colonies from the plate can be directly used for oxidase and agglutination tests. This medium is used by the Centers for Disease Control and is described in their published handbooks for the isolation of *V. cholerae*.

Several screening media are quite useful for the identification of this pathogen. Arginine glucose slants (29) are used to differentiate *V. cholerae* from arginine dihydrolase-positive organisms such as *V. fluvialis, V. furnissii,* and *Aeromonas* species. *V. cholerae* is arginine negative and produces a yellow butt in the tube owing to fermentation of glucose and resultant acid production. Arginine glucose slants are a modification of the presumptive *V. parahaemolyticus* medium developed by Kaper et al. (44). Mannitol was replaced by glucose in the formulation, and sucrose and lactose were eliminated so that the sole test would be for arginine degradation.

V. cholerae requires sodium ions for growth (79), and most commercially available media will contain this critical component. In contrast, other species of *Vibrio* require both Na^+ and Cl^- ions for growth. Therefore, most media require the addition of NaCl to the formula, usually a 2 to 3% final concentration, for optimum growth of marine species. Thus, a medium such as tryptophan broth without added NaCl (1% tryptophan, 0% NaCl [T_1N_0]) can be used to screen for *V. cholerae.* Halophilic *Vibrio* species will not grow in this medium, but they do grow in T_1N_1, in which 1% of each ingredient is included. The growth of an isolate in T_1N_0 and the lack of arginine dihydrolase may be the most useful screening characteristics on which to focus. When these two results occur, the isolate can be tested for the oxidase reaction (positive) and agglutination of the cells in poly-O1 antisera, which is commercially available. Another simple test is the "string" test (80), which is based on the development of a mucoid colony in the presence of sodium deoxycholate. The colony forms a "string" when a loopful of the colony is withdrawn from the suspension. Although this reaction occurs with many *Vibrio* species, the test could be used for presumptive identification of the genus. The test is inexpensive and easy to perform. A commercial diagnostic identification test strip, the API-20E (Analytab Products, Plainview, N.Y.), has been used successfully for the identification of *V. cholerae.* These test strips may be a cost-effective means of identification for some laboratories that are involved in "sentinel" sample analysis, because it cuts down on the number of tubes, amount of medium, and analyst time that are involved.

If the isolate is identified as *V. cholerae* O1, it must then be tested for the ability to produce cholera toxin (CT) to determine its disease-producing potential. Several laboratory tests can be used to detect the presence of CT. Tissue cell lines of Y-1 mouse adrenal cells and Chinese hamster ovary cells have been used to demonstrate the production of CT by *V. cholerae* O1 isolates. These cell lines react in a distinct manner in the presence of enterotoxins such as CT. The suckling mouse assay has also been used to demonstrate the presence of CT by injection of cell extracts into the stomach of 3- to 5-day-old mice (23). After 3 to 4 h, the mice are sacrificed; fluid accumulation in the intestine, noted by a swollen condition, denotes a positive test. To quantitate the result, the weight of the intestinal tract is compared with the weight of the carcass.

More promising, less time-consuming, and requiring less-specialized facilities and techniques and less expense are the immunologic tests now available for the detection of CT. The VET-RPLA TD-20 (Oxoid USA, Columbia, Md.) is a commercially available reversed passive latex agglutination test kit. Also, a membrane enzyme-linked immunosorbent assay procedure for CT detection is described in the *Bacteriological Analytical Manual* (29). These two procedures give equivalent results (2) and can be performed more rapidly than the tissue cell and animal bioassays.

In addition to the two screening reactions and the serologic identification of suspect *V. cholerae* isolates, 12 other traditional biochemical tests are recommended for confirmation of the strain (29). These additional tests include Gram stain (gram-negative rod), appearance on triple sugar iron agar (acid slant and butt, no H_2S or gas), oxidative-fermentative reaction in glucose (fermentative), production of cytochrome oxidase and lysine decarboxylase, Voges-Proskauer reaction, growth at 42°C, salt tolerance test (0, 3, 6, 8, and 10% NaCl), sucrose fermentation (positive), *o*-nitrophenyl-β-D-galactopyranoside test (positive), arabinose fermentation reaction (negative), and sensitivity to vibriostatic agent O/129 (2,4-diamino-6,7-diisopropyl pteridine) (78).

If *V. cholerae* O1 is identified, the serologic subgroup, Inaba or Ogawa, should be determined for epidemiologic purposes by agglutination in specific antisera that are available commercially. In addition, the biotype (classical or El Tor) should be determined. This differential analysis can be accomplished by using the results of at least two of the tests listed in Table 3 in procedures described by Elliot et al. (29).

RAPID METHODS

Traditional bacteriologic isolation and identification procedures usually take from several days to a couple of weeks to complete. When human illness is involved, a quick diagnosis and identification of the causative agent can be critical for a number of obvious reasons. The advent of genetics-based identification methods for pathogenic bacteria has great advantages in speed and specificity. Recombinant DNA techniques have allowed the cloning, isolation, purification, and nucleotide sequencing of specific fragments of DNA, such as the CT gene (53). Thus, one may focus on specific strains of *V. cholerae* O1 to determine their disease-producing potential. The use of genetic probes has allowed for the rapid detection of the CT gene by colony hybridization (36). This procedure can also be used to enumerate CT-gene-containing strains within a sample by a direct-plating procedure. At present, colony hybridization is not widely used owing to the radioisotope labeling of the gene probe. This procedure has a practical lower limit of detection of about 20 to 100 cells of the target organism, partly because of dilution and blending of the sample and the small amount of sample that can be plated. However, the analysis can be completed in 3 to 4 days compared to the traditional analysis of a week or more or even the probing of an isolate obtained by the traditional method. Recent advances in the use of nonisotopic labels for DNA should lead to a broader application of the colony hybridization procedure.

Amplification of specific DNA fragments by a technique termed the polymerase chain reaction (PCR) (74, 75) has presented laboratories with a means of rapidly determining the presence of a virulence gene of a bacterial strain from a food or water sample. Though it requires specialized equipment, PCR can detect toxigenic *V. cholerae* within 24 h (48). In fact, PCR has been used to detect the CT gene in clinical strains (64) and in South American epidemic strains (33). Research on procedures for preparing bacterial DNA from food samples for PCR analysis is continuing. Food homogenates appear to have inhibiting substances that interfere with DNA ampli-

Table 3. Biochemical differentiation of biotypes of *V. cholerae* O1[a]

Test	Reaction	
	Classical	El Tor
Sensitivity to El Tor phage V	−	+
Sensitivity to classical phage IV	+	−
Sensitivity to polymyxin B (50 U)	+	−
Hemolysis (sheep erythrocytes)	−	V[b]
Hemagglutination (chicken erythrocytes)	−	+
Voges-Proskauer	−	+

[a]From references 4 and 54.
[b]V, variable.

fication (37). A preliminary, short enrichment period prior to PCR may be necessary to increase the number of target copies. This enrichment step may then show that the pathogen was viable in the sample when tested. Even when enrichment is necessary, however, the introduction of PCR to food analysis brings a promising future for the rapid detection of pathogens within a sample.

SUMMARY

While the disease cholera may never be completely eliminated from the world as long as there are countries that have poor sanitation and crowed conditions, several measures that reduce the number of individuals involved in outbreaks can be taken. Education is critical. Since food and water are intimately involved in transmission, people must be taught hygienic practices that limit fecal-oral transmission. Sewage treatment would also greatly reduce the risk and spread of cholera, as would sanitary preparation of food, but providing safe drinking water and adequate sewage disposal is cost-prohibitive in many countries. Disinfection of water supplies would also prevent a great many cases, and disinfection of the home water supply may be economically feasible. Teaching people to at least boil drinking water and cook all foods that may be contaminated may be the cheapest means at present of lowering the incidence of disease. While *V. cholerae* can be rather easily isolated from food and water samples and identified, in a few years rapid DNA-based procedures will provide an analysis of a sample within a day of collection. The rapid identification of outbreaks and sources of *V. cholerae* O1 will aid in overall reduction of the spread of the disease.

REFERENCES

1. **Albert, M. J., A. K. Siddique, M. S. Islam, A. S. G. Faruque, M. Ansaruzzaman, S. M. Faruque, and R. B. Sack.** 1993. Large outbreak of clinical cholera due to *Vibrio cholerae* non-O1 in Bangladesh. *Lancet* **341**:704.

2. **Almeida, R. J., F. W. Hickman-Brenner, E. G. Sowers, N. D. Puhr, J. J. Farmer III, and I. K. Wachsmuth.** 1990. Comparison of a latex agglutination assay and an enzyme-linked immunosorbent assay for detecting cholera toxin. *J. Clin. Microbiol.* **28**:128–130.

3. **Baine, W. B., M. Mazzotti, D. Greco, E. Izzo, A. Zampieri, G. Angioni, M. DiGrola, E. J. Gangarosa and F. Pocchiari.** 1974. Epidemiology of cholera in Italy in 1973. *Lancet* **ii**:1370–1374.

4. **Baumann, P., and R. H. W. Schubert.** 1984. Section 5. Facultatively anaerobic gram-negative rods, Family II. Vibrionaceae, p. 516-550. *In* J. G. Holt and N. R. Krieg (ed.), *Bergey's Manual of Systematic Bacteriology,* vol. 1. Williams & Wilkins, Baltimore.

5. **Blake, P. A., D. T. Allegra, J. D. Snyder, T. J. Barrett, L. McFarland, C. T. Caraway, J. C. Feeley, J. P. Craig, J. V. Lee, N. D. Puhr, and R. A. Feldman.** 1980. Cholera—a possible endemic focus in the United States. *N. Engl. J. Med.* **6**:305–309.

6. **Blake, P. A., M. L. Rosenberg, J. B. Costa, P. S. Ferreira, C. L. Guimaraes, and E. J. Gangarosa.** 1977. Cholera in Portugal. I. Modes of Transmission. *Am. J. Epidemiol.* **105**:337–343.

7. **Blake, P. A., M. L. Rosenberg, J. Florencia, J. B. Costa, L. D. P. Quintino, and E. J. Gangarosa.** 1977. Cholera in Portugal, 1974. II. Transmission by bottled mineral water. *Am. J. Epidemiol.* **105**:344–348.

8. **Bourke, A. T. C., Y. N. Cossins, B. R. W. Gray, T. J. Lunney, N. A. Rostron, R. V. Holmes, E. R. Griggs, D. J. Larsen, and V. R. Kelk.** 1986. Investigation of cholera acquired from the riverine environment in Queensland. *Med. J. Aust.* **144**:229–234.

9. **Boutin, B. K., J. G. Bradshaw, and W. H. Stroup.** 1982. Heat processing of oysters naturally contaminated with *Vibrio cholerae* serotype O1. *J. Food Prot.* **45**:169–171.

10. **Bradford, H. B., Jr.** 1984. An epidemiological study of *V. cholerae* in Louisiana, p. 59–72. *In* R. R. Colwell (ed.), *Vibrios in the Environment.* John Wiley & Sons, New York.

11. **Cameron, J. M., K. Hester, W. L. Smith, E. Caviness, T. Hosty, and F. S. Wolf.** 1977. *Vibrio cholerae*—Alabama. *Morbid. Mortal. Weekly Rep.* **26**:159–160.

12. **Centers for Disease Control.** 1989. Toxigenic *Vibrio cholerae* O1 infection acquired in Colorado. *Morbid. Mortal. Weekly Rep.* **38**:19–20.

13. **Centers for Disease Control.** 1990. Cholera—worldwide, 1989. *Morbid. Mortal. Weekly Rep.* **39**:365–367.

14. **Centers for Disease Control.** 1991. New York, 1991. *Morbid. Mortal. Weekly Rep.* **40**:516–518.

15. **Centers for Disease Control.** 1991. Cholera associated with imported frozen coconut milk—Mary-

land, 1991. *Morbid. Mortal. Weekly Rep.* **40:**844–845.

16. **Centers for Disease Control.** 1992. Cholera associated with an international airline flight, 1992. *Morbid. Mortal. Weekly Rep.* **41:**134–135.

17. **Centers for Disease Control.** 1992. Cholera associated with international travel, 1992. *Morbid. Mortal. Weekly Rep.* **41:**664–667.

18. **Chevalier, M., and M. Vandekerkove.** 1980. Experimental studies on the survival of *Vibrio cholerae* in untreated seawater. *Bull. Soc. Pathol. Exot.* **73:**364–371. (In French.)

19. **Clesceri, L. S., A. F. Greenberg, and R. R. Trussel (ed.).** 1989. *Standard Methods for the Examination of Water and Wastewater, Part 9000. Microbiological Examination of Water,* 17th ed. American Public Health Association, Washington D.C.

20. **Colwell, R. R., R. J. Siedler, J. Kaper, S. W. Joseph, S. Garges, H. Lockman, D. Maneval, H. Bradford, N. Roberts, E. Remmers, I. Huq, and A. Huq.** 1981. Occurrence of *Vibrio cholerae* serotype O1 in Maryland and Louisiana estuaries. *Appl. Environ. Microbiol.* **41:**555–558.

21. **Davey, G. R., J. K. Prendergast, and M. J. Eyles.** 1982. Detection of *Vibrio cholerae* in oysters, water and sediment from the Georges River. *Food Technol. Aust.* **34:**334–336.

22. **De, S. N.** 1961. *Cholera. Its Pathology and Pathogenesis.* Oliver and Boyd, Edinburgh.

23. **Dean, A. G., Y. Ching, R. G. Williams, and L. B. Harden.** 1972. Test for *Escherichia coli* enterotoxin using infant mice: application in a study of diarrhea in children in Honolulu. *J. Infect. Dis.* **125:**407–411.

24. **DePaola, A.** 1981. *Vibrio cholerae* in marine foods and environmental waters: a literature review. *J. Food Sci.* **46:**66–70.

25. **DePaola, A., G. M. Capers, M. L. Motes, Ø. Olsvik, P. I. Fields, J. Wells, I. K. Wachsmuth, T. A. Cebula, W. H. Koch, F. Khambaty, M. H. Kothary, W. L. Payne, and B. A. Wentz.** 1992. Isolation of Latin American epidemic strain of *Vibrio cholerae* O1 from U.S. Gulf Coast. *Lancet.* **339:**624.

26. **DePaola, A., C. A. Kaysner, and R. M. McPhearson.** 1987. Elevated temperature method for recovery of *Vibrio cholerae* from oysters (*Crassostrea gigas*). *Appl. Environ. Microbiol.* **53:**1181–1182.

27. **Desmarchelier, P. M., and J. L. Reichelt.** 1981. Phenotypic characterization of clinical and environmental isolates of *Vibrio cholerae* from Australia. *Curr. Microbiol.* **5:**123–127.

28. **Dutt, A. K., S. Alwi, and T. Velauthan.** 1971. A shellfish-borne cholera outbreak in Malaysia. *Trans. R. Soc. Trop. Med. Hyg.* **65:**815–816.

29. **Elliot, E. L., C. A. Kaysner, and M. L. Tamplin.** 1992. Vibrio cholerae, V. parahaemolyticus,

V. vulnificus and other Vibrio spp., p. 111–140. *In Food and Drug Administration, Bacteriological Analytical Manual,* 7th ed. Association of Official Analytical Chemists, Arlington, Va.

30. **Epstein, P. R.** 1992. Cholera and the environment. *Lancet* **339:**1167–1168.

31. **Feldman, R. A.** 1992. Transmission of cholera: modes of spread and vehicles. *Microbiol. Dig.* **9:**37–41.

32. **Felsenfeld, O.** 1965. Notes on food, beverages and fomites contaminated with *Vibrio cholerae*. *Bull. W.H.O.* **33:**725–734.

33. **Fields, P. I., T. Popovic, K. Wachsmuth, and Ø. Olsvik.** 1992. Use of polymerase chain reaction for detection of toxigenic *Vibrio cholerae* O1 strains from the Latin American cholera epidemic. *J. Clin. Microbiol.* **30:**2118–2121.

34. **Garay, E., A. Arnau, and C. Amaro.** 1985. Incidence of *Vibrio cholerae* and related vibrios in a coastal lagoon and seawater influenced by lake discharges along an annual cycle. *Appl. Environ. Microbiol.* **50:**426–430.

35. **Glass, R. I., M. Libel, and A. D. Brandling-Bennett.** 1992. Epidemic cholera in the Americas. *Science* **256:**1524–1525.

36. **Hill, W. E., A. R. Datta, P. Feng, K. A. Lampel, and W. L. Payne.** 1992. Identification of foodborne bacterial pathogens by gene probes, p. 383–418. *In Food and Drug Administration, Bacteriological Analytical Manual.* Association of Official Analytical Chemists, Arlington, Va.

37. **Hill, W. E., S. P. Keasler, M. W. Trucksess, P. Feng, C. A. Kaysner, and K. A. Lampel.** 1991. Polymerase chain reaction identification of *Vibrio vulnificus* in artificially contaminated oysters. *Appl. Environ. Microbiol.* **57:**707–711.

38. **Hood, M. A., G. E. Ness, and G. E. Rodrick.** 1981. Isolation of *Vibrio cholerae* serotype O1 from the Eastern oyster, *Crassostrea virginica*. *Appl. Environ. Microbiol.* **41:**559–560.

39. **Hood, M. A., G. E. Ness, G. E. Rodrick, and N. J. Blake.** 1983. Distribution of *Vibrio cholerae* in two Florida estuaries. *Microb. Ecol.* **9:**65–75.

40. **Huq, A., R. R. Colwell, R. Rahman, A. Ali, M. A. R. Chowdhury, S. Parveen, D. A. Sack, and E. Russek-Cohen.** 1990. Detection of *Vibrio cholerae* O1 in the aquatic environment by fluorescent-monoclonal antibody and culture methods. *Appl. Environ. Microbiol.* **56:**2370–2373.

41. **Johnston, J. M., D. L. Martin, J. Perdue, L. M. McFarland, C. T. Caraway, E. C. Lippey, and P. A. Blake.** 1983. Cholera on a Gulf Coast oil rig. *N. Engl. J. Med.* **309:**523–526.

42. **Joseph, P. R., J. F. Tamayo, W. H. Mosley, M. G. Alvero, J. J. Dizon, and D. A. Henderson.** 1965. Studies of cholera El Tor in the Philippines. 2. A retrospective investigation of an explosive outbreak in Bacolod City and Talisay, November 1961. *Bull. W.H.O.* **33:**637–640.

43. **Kaper, J., H. Lockman, R. R. Colwell, and S. W. Joseph.** 1979. Ecology, serology, and enterotoxin production of *Vibrio cholerae* in Chesapeake Bay. *Appl. Environ. Microbiol.* **37**:91–103.

44. **Kaper, J., E. F. Remmers, and R. R. Colwell.** 1980. A presumptive medium for identification of *Vibrio parahaemolyticus*. *J. Food Prot.* **43**:956–958.

45. **Kaysner, C. A., C. Abeyta, Jr., M. M. Wekell, A. DePaola, Jr., R. F. Stott, and J. M. Leitch.** 1987. Incidence of *Vibrio cholerae* from estuaries of the United States west coast. *Appl. Environ. Microbiol.* **53**:1344–1348.

46. **Klontz, K. C., R. V. Tauxe, W. L. Cook, W. H. Riley, and I. K. Wachsmuth.** 1987. Cholera after consumption of raw oysters. *Ann. Intern. Med.* **107**:846–848.

47. **Kobayashi, T., S. Enomoto, and R. Sakazaki.** 1963. A new selective isolation medium for the vibrio group on a modified Nakanishi's medium (TCBS agar medium). *Jpn. J. Bacteriol.* **18**:387–392.

48. **Koch, W. H., W. L. Payne, B. A. Wentz, and T. A. Cebula.** 1993. Rapid polymerase chain reaction method for detection of *Vibrio cholerae* in foods. *Appl. Environ. Microbiol.* **59**:556–560.

49. **Kolvin, J. L., and D. Roberts.** 1982. Studies on the growth of *Vibrio cholerae* biotype eltor and biotype classical in foods. *J. Hyg.* (Cambridge) **89**:243–252.

50. **Kozicki, M., R. Steffen, and M. Schar.** 1985. "Boil it, cook it, peel it, or forget it": does this rule prevent travelers diarrhoea? *Int. J. Epidemiol.* **14**:169–172.

51. **Lee, J. V., D. J. Bashford, T. J. Donovan, A. L. Furniss, and P. A. West.** 1982. The incidence of *Vibrio cholerae* in water, animals and birds in Kent, England. *J. Appl. Bacteriol.* **52**:281–291.

52. **Lin, F. C., J. G. Morris, Jr., J. B. Kaper, T. Gross, J. Michalski, C. Morrison, J. P. Libonati, and E. Israel.** 1986. Persistence of cholera in the United States: Isolation of *Vibrio cholerae* O1 from a patient with diarrhea in Maryland. *J. Clin. Microbiol.* **23**:624–626.

53. **Lockman, H., and J. B. Kaper.** 1983. Nucleotide sequence analysis if the A2 and B subunits of *Vibrio cholerae* enterotoxin. *J. Biol. Chem.* **258**:13722–13726.

54. **Madden, J. M., B. A. McCardell, and J. G. Morris, Jr.** 1989. Vibrio cholerae, p. 525–542. *In* M. P. Doyle (ed.), *Foodborne Bacterial Pathogens*. Marcel Dekker, New York.

55. **Madden, J. M., B. A. McCardell, and R. B. Read, Jr.** 1982. *Vibrio cholerae* in shellfish from U.S. coastal waters. *Food Technol.* **36**:93–96.

56. **Makukutu, C. A., and R. F. Guthrie.** 1986. Behavior of *Vibrio cholerae* in hot foods. *Appl. Environ. Microbiol.* **52**:824–831.

57. **Massad, G., and J. D. Oliver.** 1987. New selective and differential medium for *Vibrio cholerae* and *Vibrio vulnificus*. *Appl. Environ. Microbiol.* **53**:2262–2264.

58. **McCarthy, S. A., R. M. McPhearson, A. M. Guarino, and J. L. Gaines.** 1992. Toxigenic *Vibrio cholerae* O1 and cargo ships entering Gulf of Mexico. *Lancet* **339**:624–625.

59. **McNicol, L. A., and R. M. Doetsch.** 1983. A hypothesis accounting for the origin of pandemic cholera: a retrograde analysis. *Perspect. Biol. Med.* **26**:547–552.

60. **Merson, M. H., W. T. Martin, J. P. Craig, G. K. Morris, P. A. Blake, G. F. Craun, J. C. Feeley, J. C. Camacho, and E. J. Gangarosa.** 1977. Cholera on Guam, 1974. *Am. J. Epidemiol.* **105**:349–351.

61. **Monsur, K. A.** 1961. A highly selective gelatin-taurocholate-tellurite medium for the isolation of *Vibrio cholerae*. *Trans. R. Soc. Med. Hyg.* **55**:440–442.

62. **Moore, B.** 1948. The detection of paratyphoid carriers in towns by means of sewage examination. *Mon. Bull. Minist. Health. Public Health Lab. Serv.* **7**:241–248.

63. **Odugbo, M. O., S. I. Onuorah, and A. A. Adesiyun.** 1990. Survival and multiplication of *Vibrio cholerae* serotype Ogawa in reconstituted infant milk. *J. Food Prot.* **53**:1071–1072.

64. **Olsvik, Ø., T. Popovic, and P. I. Fields.** 1992. Polymerase chain reaction for detection of toxin genes in strains of *Vibrio cholerae* O1, p. 573–583. *In* D. H. Persing, F. C. Tenover, T. F. Smith, and T. J. White (ed.) *Diagnostic Molecular Microbiology*. American Society for Microbiology, Washington, D.C.

65. **Pan American Health Organization.** 1991. *Risks of Transmission of Cholera by Food*, Pan American Health Organization, Washington, D.C.

66. **Pan American Health Organization.** 1993. *Cholera in the Americas*. Pan American Health Organization, Washington, D.C.

67. **Pavia, A. T., J. F. Campbell, P. A. Blake, J. D. Smith, T. W. McKinley, and D. I. Martin.** 1987. Cholera from raw oysters shipped interstate. *JAMA* **258**:2374.

68. **Pollitzer, R.** 1959. *Cholera.* Monograph no. 43. World Health Organization, Geneva. Switzerland.

69. **Prescott, L. M., and N. K. Bhattacharjee.** 1969. Viability of El Tor Vibrios in common foodstuffs found in an endemic cholera area. *Bull. W.H.O.* **40**:980–982.

70. **Ramamurthy, T., S. Garg, R. Sharma, S. K. Bhattacharya, G. B. Nair, T. Shimada, T. Takeda, T. Karasawa, H. Kurazano, A. Pal, and Y. Takeda.** 1993. Emergence of a novel strain of *Vibrio cholerae* with epidemic potential in southern and eastern India. *Lancet* **341**:703–704.

71. **Rippey, S. R.** 1991. *Shellfish Borne Disease Outbreaks.* Food and Drug Administration, North Kingstown, R.I.

72. **Roberts, D.** 1992. Growth and survival of *Vibrio cholerae* in foods. *Microbiol. Dig.* **9:**24–31.

73. **Roberts, N. C., R. J. Seibeling, J. B. Kaper, and H. B. Bradford, Jr.** 1982. Vibrios in the Louisiana Gulf Coast environment. *Microb. Ecol.* **8:**299–312.

74. **Saiki, R. K., D. H. Gelfand, S. Stoffel, S. J. Scharf, R. Higuchi, G. T. Horn, K. B. Mullis, and H. A. Erlich.** 1988. Primer-directed enzymatic amplification of DNA with a thermostable DNA polymerase. *Science* **239:**487–491.

75. **Saiki, R. K., S. Scharf, F. Faloona, K. B. Mullis, G. T. Horn, and H. A. Erlich.** 1985. Enzymatic amplification of beta-globin genomic sequence and restriction analysis for diagnosis of sickle cell anemia. *Science* **230:**1350–1354.

76. **Schoenberg, B. E.** 1974. Snow on the water of London. *Mayo Clin. Proc.* **49:**680–684.

77. **Shandera, W. X., B. Hafkin, D. L. Martin, J. P. Taylor, D. L. Maserang, J. G. Wells, M. Kelly, K. Ghandi, J. B. Kaper, J. V. Lee, and P. A. Blake.** 1983. Persistence of cholera in the United States. *Am. J. Trop. Med. Hyg.* **32:**812–817.

78. **Shewan, J. M., W. Hodgkiss, and J. Liston.** 1954. A method for the rapid identification of certain non-pathogenic, asporogenous bacilli. *Nature* (London) **173:**208–209.

79. **Singleton, F. L., R. Attwell, S. Jangi, and R. R. Colwell.** 1982. Effects of temperature and salinity on *Vibrio cholerae* growth. *Appl. Environ. Microbiol.* **44:**1047–1058.

80. **Smith, H. L.** 1970. A presumptive test for Vibrios: the "string" test. *Bull. W.H.O.* **42:**817–818.

81. **St. Louis, M. E., J. D. Porter, A. Helal, K. Drame, N. Hargrett-Bean, J. G. Wells, and R. V. Tauxe.** 1990. Epidemic cholera in West Africa: the role of food handling and high-risk foods. *Am. J. Epidemiol.* **131:**719–728.

82. **Tamplin, M. L., and C. C. Parodi.** 1991. Environmental spread of *Vibrio cholerae* in Peru. *Lancet* **338:**1216–1217.

83. **Tauxe, R. V., and P. A. Blake.** 1992. Epidemic cholera in Latin America. *JAMA* **267:**1388–1390.

84. **Weissman, J. B., W. E. DeWitt, J. Thompson, C. N. Muchnick, B. L. Portnoy, J. C. Feeley, and E. J. Gangarosa.** 1974. A case of cholera in Texas, 1973. *Am. J. Epidemiol.* **100:**487–490.

85. **West, P. A.** 1992. The ecology and survival of *Vibrio cholerae* in natural aquatic environments. *Microbiol. Dig.* **9:**20–23.

86. **Xu, H-S., N. Roberts, F. L. Singleton, R. W. Attwell, D. J. Grimes, and R. R. Colwell.** 1982. Survival and viability of nonculturable *Escherichia coli* and *Vibrio cholerae* in the estuarine and marine environment. *Microb. Ecol.* **8:**313–323.

Vibrio cholerae and Cholera: Molecular to Global Perspectives
Edited by I. Kaye Wachsmuth, Paul A. Blake, and Ørjan Olsvik
© 1994 American Society for Microbiology, Washington, DC 20005

Chapter 3

Detection of Cholera Toxin Genes

Tanja Popovic, Patricia I. Fields, and Ørjan Olsvik

Laboratory diagnosis of cholera has traditionally been based on the phenotypic characteristics of *Vibrio cholerae* O1, expressed as its morphologic, physiologic, and biochemical properties, including its antigenic composition. Specific assays have been developed for detection of such antigens, enzymes, toxins, and other gene products. Methods for detection of cholera enterotoxin (CT) were especially valuable because of the profound role that the CT plays in the pathogenesis of the disease. Despite significant methodologic progress, assays based on the organism's phenotypic characterization, including CT detection, are often limited in their specificity and/or sensitivity. In addition, they are subject to changes in gene expression and may be influenced by environmental and growth-related conditions. In the process of bacterial isolation and enrichment from clinical samples, the bacterium goes from a hard life in the human body to a lazy life on an agar plate, and specific genes may be turned on and off, rearranged, or even lost. Consequently, more direct, specific, sensitive, and rapid methods have been sought to meet the growing demands of a precise and timely laboratory diagnosis.

Nucleic acid-based methods appeared to provide some of the answers, as they are characterized by specificity, sensitivity, and speed. The genotypic approach to detection of specific genes is unique because it focuses on the information found in the nucleic acid of the organism rather than the gene products; it allows detection of specific genes involved in the virulence of *V. cholerae*. These techniques were initially used in research studies of pathogenicity and vaccine development and have recently been applied to the laboratory diagnosis of cholera. Advances in molecular biology techniques have enabled even routine clinical laboratories to diagnose disease on the basis of the organism's nucleic acids. In addition, by reducing the required time and the amount of labor necessary for obtaining precise and reliable information, the clinical and public health responses to cholera outbreaks have been greatly facilitated.

Although *V. cholerae* produces a substantial number of other potential virulence factors, CT is acknowledged as the factor most responsible for the massive outpouring of fluid that characterizes severe forms of cholera. The identification of CT is important in the diagnosis of cholera, since only toxin-producing strains have been associated with epidemics and pandemics of severe, watery

Tanja Popovic, Patricia I. Fields, and Ørjan Olsvik
• Foodborne and Diarrheal Diseases Branch, Division of Bacterial and Mycotic Diseases, National Center for Infectious Diseases, Centers for Disease Control and Prevention, Atlanta, Georgia 30333.

diarrhea. Also, nontoxigenic strains of *V. cholerae* are isolated often from environmental samples and occasionally from patients, but generally they do not cause severe watery diarrhea.

CT belongs to a family of related enterotoxins that consists of two functional polypeptides (39; see also chapter 11). The B subunit is composed of five identical elements responsible for binding to the epithelial cell surface receptor, GM_1. The A subunit is transported through the cell membrane and cleaved into two fragments, A1 and A2. The A1 fragment is then responsible for adenylate cyclase activation of small-intestine epithelial cells, inducing the active secretion of water and salts that causes the tremendous fluid loss associated with cholera. The chromosomal genes encoding the A and B subunits are designated *ctxA* and *ctxB*, respectively (28, 36, 40) (Fig. 1). They are expressed as a single transcriptional unit. *ctxA* contains 777 bp: 636 bp code for the A1 subunit, and 141 code for the A2 subunit. The last four nucleotides of *ctxA* (A TGA), including the stop codon (TGA), are also the first nucleotides of *ctxB* (ATG A) (12). The B subunit of *ctx* contains 375 bp: 63 bp code for 21 amino acids of the leader peptide, and the remaining 312 bp code for 103 amino acids of the B-unit monomer and the stop codon (28).

Demonstration of CT production using animal models, cell cultures, or immunologic procedures is reviewed in another chapter (see chapter 4). Assays based on DNA hybridization, polymerase chain reaction (PCR), and DNA sequencing have been developed for the detection and characterization of the *ctx* gene. We will here describe and discuss the advantages and disadvantages of using these techniques in the diagnosis of cholera.

DNA PROBES FOR DETECTION OF *V. CHOLERAE ctx*

Nucleic acid hybridization technology offers distinct advantages over phenotypic iden-

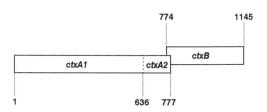

Figure 1. CT genes. Numbers designate nucleotide positions, starting with the first nucleotide of the initiation codon ATG of *ctxA* (nucleotide 1) and ending with the third nucleotide of the last codon AAT of *ctxB* (nucleotide 1145).

tification systems because of its specificity and sensitivity. In addition, the potential problems of gene expression, e.g., translational errors or exportation obstacles of the gene product, do not occur in genotypic assays. However, as with other nucleic acid-based assays, the significance of detection of toxin genes in the absence of the organism or the gene product must be carefully evaluated. Nucleic acid probes can provide information not only on the presence or absence of a particular gene sequence but also on the location and copy number of such genes by using endonuclease digestion of chromosomal DNA and the Southern blot procedure.

DNA probes for detection of *ctxAB* have played a significant role in epidemiologic studies of *V. cholerae* O1. Cloned DNA fragments, synthetic oligonucleotide probes, and PCR-generated amplicons from *ctx* have been used as probes in colony blot and Southern blot hybridization assays (Table 1).

The format of the hybridization test is important for the practical application of DNA probes for diagnostic purposes. The colony blot system has a great potential for screening a large number of colonies but gives little additional molecular information. Probes have traditionally been labeled with radioactive isotopes, making the procedure too hazardous for routine clinical laboratories. Several nonradioactive molecules, such as digoxigenin, or enzymes, such as alkaline phosphatase, have been recently used to directly label DNA and to serve as the basis

Table 1. DNA probes for detection of CT genes

Source of probe	Target gene[a] (nucleotides)	Size (bp) of probe	Labeling system	Reference(s)
Cloned HincII DNA fragment of E. coli LT-A from plasmid pEWD299	ctxA	1,275	32P	11, 14, 16
Cloned EcoRI-HindIII DNA fragment of E. coli LT-B from plasmid pEWD299	ctxB	590	32P	11
Cloned HindIII DNA fragment of E. coli LT-B from plasmid pEWD299	ctxB	850	32P	5, 16
Cloned XbaI-ClaI DNA fragment of V. cholerae O1 from plasmid pCVD27	ctxA	554	32P	15, 17, 29, 42
Cloned PstI-BglII DNA fragment of V. cholerae O1 from plasmid pCVD002	ctxAB	4,000	32P	9, 25, 48
Cloned AccI DNA fragment of V. cholerae O1 from plasmid pCVD002	ctxA	800	32P	2, 6
Cloned XbaI-HincII DNA fragment of V. cholerae O1 from plasmid pJM17	ctxAB	950	32P	27
PCR-generated CT amplicon	ctxA (73–636)	564	Digoxigenin	46
Oligonucleotide CT probe	ctxA (187–206)	20	32P or enhanced chemiluminescence	1, 19, 21
	ctxA (557–579)	23	Alkaline phosphatase	48
	ctxA (478–507)	30	Alkaline phosphatase	50

[a]Numbers in parentheses refer to the nucleotide positions, starting with the first nucleotide of the initiation codon ATG of ctxA (nucleotide 1) and ending with the third nucleotide of the last codon AAT of ctxB (nucleotide 1145).

for immune or enzymatic detection systems (43).

Cloned DNA Fragments Encoding the Closely Related Heat-Labile Enterotoxin of Enterotoxigenic *E. coli* (LT Probes)

The first probes used to detect *ctx* were a 1,275-bp *Hinc*II fragment encoding the LT-A subunit and a 590-bp *Eco*RI-*Hin*dII fragment encoding the LT-B subunit from *Escherichia coli* LT-producing strains (31). There is a 78% DNA sequence homology between *E. coli* LT genes *(eltAB)* and *ctxAB* (see chapter 11). Moseley and Falkow (31) reported that these probes hybridized with *V. cholerae* under conditions of low stringency. Also, their work gave the first indication that *V. cholerae* O1 strains might carry multiple copies of *ctx*, as they observed that at least four *Eco*RI restriction fragments of the chromosomal DNA of the classical biotype strains, each greater than 20 kb, hybridized with the LT-A and LT-B probes. Additionally, two *Hin*dIII restriction fragments of the chromosomal DNA of *V. cholerae* O1 classical biotype strains hybridized with LT-A probe (16). Unlike the majority of the El Tor isolates, *V. cholerae* strains associated with the U.S. Gulf Coast possessed a unique Southern blot pattern in which two small *Hin*dIII fragments, 6 and 7 kbp, hybridized with the LT-A probe (14, 16). Subsequently, it was demonstrated that these strains have two or more copies of *ctxAB* (27).

In separate studies, an 850-bp *Hin*dIII fragment of *E. coli*, containing a DNA sequence encoding the entire B subunit and a portion of the A subunit, was used to characterize environmental and nontoxigenic *V. cholerae* strains (5, 16) (Table 1). The use of these probes indicated that nontoxigenic strains lack structural genes for both the A and the B subunits of CT. One drawback of the use of LT probes to detect *ctx* is that LT probes are useful for *ctx* only in the Southern blot format. Because of the low-stringency hybridization condition that must be used for detec-tion of *ctx,* background hybridization is frequently too high in the colony blot format.

Cloned DNA Fragments Encoding *ctx* (CT Probes)

For most diagnostic purposes, LT probes have been replaced by specific CT probes. Initial studies used an *Eco*RI-*Bgl*II fragment containing both *ctxA* and *ctxB* (25) that was successfully used in a study of the classical *V. cholerae* O1 strains (6). Restriction of genomic DNA with *Hin*dIII indicated that two *Hin*dIII fragments hybridized with the CT probe. These two *ctx* copies are widely separated in the genome but are located at the same chromosomal site in all isolates, with variations of 2 or 3 kbp in restriction fragment sizes (6). The same probe was also used in a molecular study of *V. cholerae* O1 strains associated with environmental reservoirs of *V. cholerae* O1 in northeastern Australia (9). The probe detected *ctx* in 169 (81%) of 208 *V. cholerae* O1 strains and in 4 (1%) of 454 *V. cholerae* non-O1 isolates of both human and environmental origin. All 169 toxigenic *V. cholerae* O1 Australian isolates had a common 23-kbp *Hin*dIII fragment that hybridized with the probe and were of the Inaba serotype. At the same time, no local Australian Ogawa serotype isolates hybridized with the probe. The correlation between the probe test and the enzyme-linked immunosorbent assay (ELISA) for toxicity in this Australian strain collection was rather low. Thirty-eight of 220 strains tested were only probe positive, while 3 strains were only ELISA positive. The paper offers no explanation, as the employed probe and the techniques have been well evaluated by others (25, 48). However, the Australian strains have been shown to possess toxin and toxin genes that differ from those in strains from other continents (34), and additional, presently unknown heterogeneity might be due to these unexpected findings.

To assess the public health significance of the presence of *V. cholerae* O1, especially those strains that produce CT, in the environment, 235 strains of *V. cholerae* O1 isolated from natural and environmental waters were assayed by a probe consisting of a 554-bp *Xba*I-*Cla*I fragment from the plasmid pCVD27, which encodes most of the *ctxA* gene (29). The *ctxA* gene was not detected in any of 225 isolates from natural waters. However, all 10 isolates from environmental waters presumed to be contaminated with human feces and 162 of 172 strains isolated from stool samples of cholera patients implicated in domestic and imported cholera cases hybridized with the CT probe in this colony blot assay.

PCR-Generated CT Probe

Wachsmuth et al. developed a PCR-generated probe that is a digoxigenin-11-dUTP-labeled 564-bp fragment of *ctxA* (46). The primers used were 5'-CGGGCAGATTCTAGACCTCCTG-3' and 5'-CGATGATCTTGGAGCATTCCCAC-3'. The probe is very simple to prepare, can be stored at −20°C for extended periods, and has the advantage of a nonradioactive label. This probe was used to assay over 200 *V. cholerae* O1 strains as a part of the molecular epidemiologic study of the cholera epidemic in Latin America. By using this probe, it has recently been demonstrated that with few exceptions, the majority of strains associated with the Latin American epidemic have a single copy of *ctxAB* (45).

Oligonucleotide CT Probes

Cloned DNA fragments and PCR-generated probes are fairly large pieces of DNA. The development of the technology to synthesize DNA in the early 1980s has allowed the development of oligonucleotide probes 15 to 40 nucleotides long. In general, these probes have several advantages over longer probes. They are very stable and relatively easy to prepare. Also, they tend to hybridize quickly to the target DNA (<30 min), unlike the long probes, which may require hours before stable hybridization is achieved (47). However, the oligonucleotide probes are more difficult to label, and the number of reporter molecules that can be attached to the probe is limited (43). Use of probes for diagnostic purposes and eventually in commercially available kits will favor the oligonucleotide probes because of their easy preparation, greater stability, and rapid hybridization. To date, three different oligonucleotide probes have been used to detect *ctx* (Table 1).

An oligonucleotide probe developed by Hill (12a) was used to detect *ctx* in 70 of 90 *V. cholerae* strains (1). This probe detected only one small *Hin*dIII restriction fragment of chromosomal DNA of a *V. cholerae* O1 isolate, recovered from a patient in Florida who ate raw oysters (19). This strain is the only U.S. Gulf Coast strain reported to have only one *Hin*dIII restriction fragment that hybridized with either CT or LT probes.

A colony blot hybridization assay employing an alkaline phosphatase-labeled 23-bp oligonucleotide probe (CTAP) derived from a specific sequence of *ctxA* was used by Wright et al. on a strain collection representing 11 *Vibrio* species and toxigenic and nontoxigenic strains of *E. coli* (48). The CTAP sequence was selected on the basis of the dissimilarity between *ctxA* and *eltA* in that specific region. The whole-gene CT probe described earlier (25) was used for comparison. Both probes hybridized with complementary sequences in 66 of 69 *V. cholerae* O1 strains and 3 of 202 non-O1 strains. However, under the same stringency conditions, the whole-gene CT probe hybridized with complementary sequences in 3 of 7 *V. mimicus* strains and 22 of 52 LT-positive *E. coli* strains. The CTAP was 100% specific for CT-producing *V. cholerae* strains of both O1 and non-O1 serogroups, and it hybridized with complementary sequences in none of 46 strains representing 10 other *Vibrio* species and in none of 52 strains of LT-producing *E. coli*. *ctxAB* could be detected in a 10-μl sample that contained 10^7 organisms. An initial inoculation of three bacteria in 10 μl incubated for 16 h was detectable by this assay. This probe was also evaluated by directly screening stool samples from human volunteers who were orally administered *V. cholerae* 569B (classical, Inaba); 20 of 22 samples were positive. Two stool samples in which *ctx* was not detected by CTAP had concentrations of $<3 \times 10^2$/g of stool as determined by growth on thiosulfate-citrate-bile salts sucrose agar (TCBS).

The nucleotide sequence of the third probe (CT-ENLOP), developed by Yoh et al., corresponded to nucleotides 478 to 507 of *ctxA* (50). The oligonucleotide was covalently linked to alkaline phosphatase. The probe exhibited a sensitivity of 100% and a specificity of 94% when tested on 88 clinical and 20 environmental strains of *V. cholerae* O1. The only discrepancies between CT-ENLOP and the latex agglutination kit reverse passive latex agglutination (RPLA) results were observed in two *V. cholerae* O1 strains of human origin that were positive with CT-ENLOP but negative in RPLA. These two strains were additionally tested by ELISA, in which one was weakly positive while the other remained negative. The probe showed cross-reactivity with complementary sequences in 1 of 53 *V. mimicus* isolates, but it hybridized with complementary sequences in none of 48 LT-producing *E. coli* strains. The minimum number of detectable organisms in a 2-μl sample was 2,000. This oligonucleotide probe was successfully used for detection of *ctx* in a patient at the Osaka Airport quarantine who had just returned from Indonesia (30). Results of the hybridization were available within 4 h from the time the colonies were harvested on TCBS, allowing for prompt and adequate clinical and epidemiologic responses.

PCR

The use of PCR is becoming increasingly important in the diagnosis of many infectious diseases, as this method has the potential to give quicker, more specific, and possibly more sensitive results than currently available diagnostic tests. PCR has increased the application of nucleic acid-based tests in the diagnosis of infectious diseases, primarily because of its ability to specifically amplify a gene or a segment of a gene directly from clinical materials by using standardized reagents and specific equipment.

Unlike traditional cultivation methods, which are limited to detecting viable organisms, PCR requires only that DNA, not viable cells, remain intact (35). In PCR, a specifically defined DNA segment is amplified through repeated synthesis. The ends of the amplified DNA are determined by the location of two short, specific DNA oligonucleotides that are used as primers for DNA synthesis.

Three sets of primers were initially used in a PCR assay developed by Shirai et al (38) (Table 2). All three sets amplified regions of *ctxA*. PCR reactions were carried out with purified chromosomal DNA from *V. cholerae* isolates, with broth cultures, and directly with stool samples. The best results were obtained with the first set of primers amplifying a 302-bp region. The authors report that the assay could detect as little as 1 pg of DNA or three viable cells from broth cultures after 40 cycles of amplification. The PCR results for broth cultures were compared with those of a colony hybridization test using specific CT and LT oligonucleotide probes on clinical and environmental strains of *V. cholerae* O1 and related organisms isolated worldwide. Only strains that were probe positive were positive in the PCR assay. Other organisms belonging to five other *Vibrio* species and LT-producing *E. coli* strains were negative in both tests.

Fecal samples from 25 patients with secretory diarrhea were also examined by Shirai et al. (38) using this PCR assay with the first set of primers; 7 patients were positive for *V. cholerae* O1 by direct and enrichment culture. Five of those seven samples yielded positive results in the PCR assay after 30 cy-

Table 2. PCR for detection of CT genes

Target gene	5'–3' DNA sequence of primer[a]		No. of bases	Positions[b]	Amplicon size (bp)	Reference
ctxA	F:	T CAA ACT ATA TTG TCT GGT C	20	273–292	380	20
	R:	C GCA AGT ATT ACT CAT CGA	19	634–652		
ctxA	F:	CT CAG ACG GGA TTT GTT AGG CAC G	24	197–220	302	38
	R:	TCT ATC TCT GTA GCC CCT ATT ACG	24	475–498		
ctxA	F:	T GAC CGA GGT ACT CAA ATG AAT AT	24	147–170	270	38
	R:	TA TGG AAT CCC ACC TAA AGC AGA A	24	393–416		
ctxA	F:	T GTT AGG CAC GAT GAT GGA TAT GT	24	209–232	175	38
	R:	TC ATC TGG ATG AGG ACT GTA TGC C	24	360–383		
ctxA	F:	CGG GCA GAT TCT AGA CCT CCT G	22	73–94	564	11
	R:	CGA TGA TCT TGG AGC ATT CCC AC	23	614–636		
ctxA	F:	AG ACG GGA TTT GTT AGG CAC GAT	23	200–222	431	44
	R:	TCT TGG AGC ATT CCC ACA ACC CG	23	608–630		
ctxAB	F:	T AGA GCT TGG AGG GAA GAG CCG T	23	567–589	567	44
	R:	AT TGC GGC AAT CGC ATG AGG CGT	23	1110–1132		
ctxAB	F:	T GAA ATA AAG CAG TCA GGT G	20	96–115	779	21
	R:	G TGA TTC TGC ACA CAA ATC AG	21	853–873		

[a]F, forward primer; R, reverse primer.

[b]Nucleotide positions start with the first nucleotide of the initiation codon ATG of ctxA (nucleotide 1) and end with the third nucleotide of the last codon AAT of ctxB (nucleotide 1145).

cles of amplification both of undiluted samples and of samples diluted up to 1:10,000. The remaining two culture-positive samples were positive in PCR after 40 cycles of amplification of the undiluted samples only. Stool samples have been reported to contain substances that can be inhibitory to the PCR assay (13). However, this did not appear to be the case with the rice-watery stool of the cholera patients, since undiluted stool samples were successfully used for this PCR amplification.

A PCR assay developed by Fields et al. was applied as a part of the routine diagnostic procedure to detect *ctxA* in over 300 *V. cholerae* strains from the present Latin American epidemic (11, 33) (Fig. 2). A collection of 30 toxigenic and 18 nontoxigenic *V. cholerae* O1 isolates of both biotypes and serotypes isolated worldwide over the past 60 years was used to validate the assay. Also, several different bacterial species have been reported to possess toxins or proteins that are genetically cross-reactive with CT (37). Therefore, 35

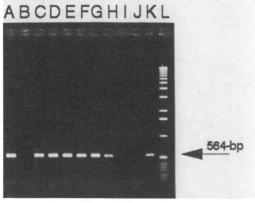

ABCDEFGHIJKL

564-bp

Figure 2. Analysis of PCR products from the *ctxA* genes. Lanes: A, *V. cholerae* O1, C9505, human, Mexico, 1992; B, *V. cholerae* O1, C9380, human, Mexico, 1992; C, *V. cholerae* O1, A1261, human, Bangladesh, 1980; D, *V. cholerae* O1, A1263, human, Bangladesh, 1980; E, *V. cholerae* O1, C9606, human, Paraguay, 1993; F, *V. cholerae* O1, C9607, human, Paraguay, 1993; G, *V. cholerae* O1, C9608, human, United States, 1993 H, *V. cholerae* O1, C8129, human, Peru, 1992; I, *V. cholerae* O1, C8155, human, Peru, 1992; J, *C. jejuni,* D116, human, United States, 1982; K, *V. cholerae* O1, ATCC 14035, human, India, 1949; L, molecular size ladder.

LT-producing *E. coli* strains of different serotypes and origins, 8 *Aeromonas hydrophila* strains, 12 *Campylobacter jejuni* strains, and 10 *Campylobacter coli* strains, all having been reported to produce choleralike toxin, were negative in this PCR. PCR was tested on stool specimens taken directly from the transport media, and these results corresponded 100% with amplification after cultivation from a single colony and with the ELISA results. A nonspecific reaction (i.e., DNA amplicon of the incorrect size) was observed for some of the enterotoxigenic *E. coli* strains at low annealing temperatures. These data are consistent with the concept of a family of LT genes with various degrees of DNA relatedness (7, 49). However, the nonspecific PCR product was eliminated by increasing the annealing temperature to 60°C.

This PCR assay was also used to detect *ctxA* in crabs incriminated as the source of cholera in a small outbreak of imported cholera in New York in April 1991 (4, 32). A 57-year-old man was diagnosed with cholera several days after he had returned from Ecuador. Three of his recent contacts who ate crabs bought at a pier in Guayaquil had laboratory evidence of infection. *V. cholerae* O1 was isolated from the patients but not from the crab meat incriminated as the source of infection. However, when assayed with the *ctxA* PCR, one of four crab samples was positive (32).

Forty-three *V. cholerae* O1 isolates of human and environmental origin from Argentina were tested in a PCR assay developed by Varela et al. (44). Two distinct regions of *ctx* were amplified by two different sets of primers. The first set amplified a 431-bp region of *ctxA*. Weak and nonspecific amplicons observed in some reactions could be eliminated by increasing the annealing temperature. The authors report 100% correlation between the PCR assay and the ELISA for all of the *V. cholerae* O1 strains tested.

Direct detection of specific genes from specific pathogens in food samples is one of the major goals of PCR technology. Conventional methods for detection of *V. cholerae* in foods are reliable, but they all require considerable amounts of time, and this may result in significant loss of perishable foods. Koch et al. reported a PCR assay for detection of *V. cholerae* in foods (21). This PCR employs primers that amplify a 779-bp region covering both *ctxA* and *ctxB*. PCR assays were performed on various seeded food samples. Amplification of the specific fragment was observed with seeded fruit and vegetables. However, seeded shellfish homogenates (10% blends of oyster or shrimp) often inhibited the PCR reaction. When the proportion of oyster or shrimp homogenate was reduced to 1%, the specific *ctx* fragment was readily amplified. The sensitivity of this PCR assay was tested by seeding crab or oysters with *V. cholerae* O1 before homogenization. Approximately 10 viable cells in the inoculum were detected after 35 cycles of amplification. Also, oysters seeded with as few as 10 toxigenic *V. cholerae* organisms per g were positive in the PCR after an 8-h enrichment. The identities of all the amplified fragments were confirmed by hybridization with an internal *ctxA* oligonucleotide probe (1) labeled with enhanced chemiluminescence.

An important factor in evaluating any DNA-based test is the specificity of the DNA sequences chosen for the gene and strains of interest. The size of the amplicon is often used as evidence that the PCR is positive. However, the identity of the amplified fragment should be confirmed by hybridization to a DNA probe, digestion with specific restriction enzymes, or DNA sequencing.

Rapid testing procedure is an important advantage of PCR over ELISA. During the investigation of the present Latin American cholera epidemic, the presence of *ctxA* could be determined within 4 h, while the toxin ELISA required at least 4 days (11). Important public health decisions were made on the basis of the PCR data, and control strategies were implemented in some situations within 24 h.

Like any new technology, PCR assays must be thoroughly controlled and evaluated

before they are applied to clinical specimens. Applications directly to crude fecal, food, or environmental samples often require special attention. PCR is a simple and fast alternative to immunologic assays that are currently employed in microbiology laboratories for identifying toxigenic *V. cholerae* strains. The PCR seems to be a suitable test method for detection of *ctx*. However, heterogeneity in the DNA sequences of the genes might be responsible for the false-negative results. Therefore, information on the DNA sequence of the gene is of extreme value in designing the primers for *ctx* PCR.

DNA SEQUENCING

Nucleic acid-sequencing methods have developed so rapidly over the past 2 decades that comparative sequencing of homologous genes has become almost a standard method in systematic classification of bacteria. Even though DNA sequencing is not commonly used as a direct diagnostic tool at present, it greatly facilitates PCR assays by generating the data for new specific primers and confirms the correctness of the obtained amplicon. DNA sequencing has recently been automated, and such sequencing of the PCR amplicons has made it possible to obtain sequences from a number of strains within 1 day. An advanced laboratory equipped with robots and an automatic DNA analyzer might be able to sequence up to 400 bases from each of 10 to 20 strains within 12 h. Sequence data are powerful tools in both identification and characterization of strains and in epidemiologic investigations.

The primary structure of the B subunit was initially revealed in the late 1970s (10, 22–24). On the basis of DNA and protein sequencing, heterogeneity within the B subunit was first reported in the early 1980s (3, 14). It was initially reported that classical strains differed from El Tor strains at five nucleotides, which translated into five amino acid differences; two additional base pair differ-

Table 3. Sequence differences in B subunits of CT genes

Genotype and strain description	Nucleotide at position[a]:			Amino acid at position[b]:		
	888	911	976	18	25	47
Genotype 1						
Classical, worldwide	C	T	C	His	Phe	Thr
El Tor, U.S. Gulf Coast						
Genotype 2						
El Tor, Australia	C	G	C	His	Leu	Thr
Genotype 3						
El Tor, seventh pandemic	T	T	T	Tyr	Phe	Ile
El Tor, Latin American epidemic						

[a]Starting with the first nucleotide of the initiation codon ATG of *ctxA* (nucleotide 1).
[b]In the final B-subunit polypeptide sequence.

ences in the *ctxB* sequence of strain 2125, initially reported by Mekalanos et al., are most likely misprints or sequence errors (28). It was subsequently shown that there were three amino acid substitutions. Classical strains had His replacing Tyr at position 39 and Thr replacing Ile at position 68 (numbers refer to the amino acid positions in the final B-subunit monomer). The third amino acid difference was reported for El Tor strain 2125 at position 75 (Ser instead of Gly) (18, 28, 41). Recently, DNA sequencing of the same strain revealed Gly at this position (8). In addition, in over 40 analyzed strains of the El Tor biotype, Lockman et al. (26) and Olsvik et al. (34) also reported Gly at that particular position, bringing the number of amino acid differences between the classical and El Tor strains down to two. Heterogeneity, especially in the B subunit, can have significant implications for development of vaccines, use of diagnostic tests for CT antibodies in patients, and design of primers for *ctx* PCR.

Recently, a 300-bp region of *ctxB* was sequenced in 45 *V. cholerae* O1 strains of both biotypes and serotypes isolated worldwide over a 60-year period (34). The DNA sequence data showed specific mutations in the 300-base segment at three positions (Table 3). Two of these mutations were in agreement with previously published sequences (27), suggesting that random muta-

tions in this gene are not common. These three base changes resulted in a change in the deduced amino acid sequence. Three genotypes of *ctxB* were established. All classical and El Tor strains from the U.S. Gulf Coast had *ctxB* of genotype 1, El Tor strains associated with the Australian environmental reservoir had genotype 2, and El Tor strains from the seventh pandemic and the recent Latin American epidemic had *ctxB* genotype 3. An interesting and unexpected observation was identical *ctxB* sequences for the classical and U.S. Gulf Coast El Tor strains. In addition, these strains had *ctxB* present in two copies, although strains from these two groups have different phenotypic characteristics (28).

Information on the DNA sequence of the A subunit is scarce. Three strains, 2125, 62746, and 3038, all of the El Tor biotype, had identical DNA sequences in their A1 and A2 subunits (3, 25, 28). A single base substitution in Texas Star-SR, a *ctxA* mutant derivative of El Tor strain 3038, resulted in the substitution of Thr for Ala at position 191 of the 258-amino-acid *ctxA* precursor (3). El Tor strains had six amino acid residues that were different from those determined from partial peptide sequences of classical strain 569B (8, 25, 28).

The progress in nucleic acid-based methods in recent years has demonstrated that the molecular approach to diagnosis of cholera could very well be the diagnostic choice of the future. Automation and standardization of molecular methods have also increased reproducibility and improved the quality of the laboratory test results. However, as for all the new diagnostic approaches, these results have to be carefully evaluated and compared with the clinical data in order to further ensure that laboratory results are of benefit for diagnosis, treatment, and prevention of diseases like cholera.

REFERENCES

1. **Almeida, R. J., D. N. Cameron, W. L. Cook, and I. K. Wachsmuth.** 1992. Vibriophage VcA-3 as an epidemic strain marker for the U.S. Gulf Coast *Vibrio cholerae* O1 clone. *J. Clin. Microbiol.* **30:**300–304.

2. **Blake, P. A., K. Wachsmuth, B. R. Davis, C. A. Bopp, B. P. Chaiken, and J. V. Lee.** 1983. Toxigenic *V. cholerae* O1 strains from Mexico identical to United States isolates. *Lancet* **ii:**912.

3. **Brickman, T. J., M. Boesman-Finkelstein, R. A. Finkelstein, and M. A. McIntosh.** 1990. Molecular cloning and nucleotide sequence analysis of cholera toxin genes of the CtxA⁻ *Vibrio cholerae* strains Texas Star-SR. *Infect. Immun.* **58:**4142–4144.

4. **Centers for Disease Control.** 1991. Cholera—New York, 1991. *Morbid. Mortal. Weekly Rep.* **40:**516–518.

5. **Colwell, R. R., R. J. Seidler, J. B. Kaper, S. W. Joseph, S. Garges, H. Lockman, D. Maneval, H. Bradford, N. Roberts, E. Remmers, I. Huq, and A. Huq.** 1981. Occurrence of *Vibrio cholerae* serotype O1 in Maryland and Louisiana estuaries. *Appl. Environ. Microbiol.* **41:**555–558.

6. **Cook, W. L., K. Wachsmuth, S. R. Johnson, K. A. Birkness, and A. R. Samadi.** 1984. Persistence of plasmids, cholera toxin genes and prophage DNA in classical *Vibrio cholerae* O1. *Infect. Immun.* **45:**222–226.

7. **Dallas, W. S., and S. Falkow.** 1980. Amino acid sequence homology between cholera toxin and *Escherichia coli* heat-labile toxin. *Nature* (London) **288:**499–501.

8. **Dams, E., M. De Wolf, and W. Dierick.** 1991. Nucleotide sequence analysis of the CT operon of the *Vibrio cholerae* classical strain 569B. *Biochim. Biophys. Acta* **1090:**139–141.

9. **Desmarchelier, P. M., and C. R. Senn.** 1989. A molecular epidemiological study of *V. cholerae* in Australia. *Med. J. Austr.* **150:**631–634.

10. **Duffy, J. K., J. W. Peterson, and A. Kurosky.** 1981. Isolation and characterization of a precursor form of the A subunit of cholera toxin. *FEBS Lett.* **126:**187–190.

11. **Fields, P. I., T. Popovic, K. Wachsmuth, and Ø. Olsvik.** 1992. Use of polymerase chain reaction for detection of toxigenic *Vibrio cholerae* O1 strains from the Latin American cholera epidemic. *J. Clin. Microbiol.* **30:**2118–2121.

12. **Gennaro, M. L., and P. J. Greenaway.** 1983. Nucleotide sequence within the cholera toxin operon. *Nucleic Acids Res.* **12:**3855–3861.

12a. **Hill, W. (Food and Drug Administration, Seattle, Wash.).** Unpublished data.

13. **Hornes, E., Y. Wasteson, and Ø. Olsvik.** 1991. Detection of *Escherichia coli* heat-stable enterotoxin genes in pig stool specimens by an immobilized, colorimetric nested polymerase chain reaction. *J. Clin. Microbiol.* **29:**2375–2379.

14. **Kaper, J. B., H. B. Bradford, N. C. Roberts, and S. Falkow.** 1982. Molecular epidemiology of *Vibrio cholerae* in the U.S. Gulf Coast. *J. Clin. Microbiol.* **16:**129–134.

15. **Kaper, J. B., J. G. Morris, and M. Nishibuchi.** 1989. DNA probes for pathogenic *Vibrio* species, p. 65–77. *In* F. C. Tenover (ed.), *DNA Probes for Infectious Diseases.* CRC Press, Inc., Boca Raton, Fl.

16. **Kaper, J. B., S. L. Moseley, and S. Falkow.** 1981. Molecular characterization of environmental and nontoxigenic strains of *Vibrio cholerae. Infect. Immun.* **32:**661–667.

17. **Karasawa, T., T. Mihara, H. Kurazono, G. B. Nair, S. Garg, T. Ramamurthy, and Y. Takeda.** 1993. Distribution of the *zot* (zonula occludens toxin) gene among strains of *Vibrio cholerae* O1 and non-O1. *FEMS Microbiol. Lett.* **106:**143–146.

18. **Kazemi, M., and R. A. Finkelstein.** 1990. Study of epitopes of cholera enterotoxin-related enterotoxins by checkerboard immunoblotting. *Infect. Immun.* **58:**2352–2360.

19. **Klontz, K. C., R. V. Tauxe, W. L. Cook, W. H. Riley, and I. K. Wachsmuth.** 1987. Cholera after consumption of raw oysters. *Ann. Intern. Med.* **107:**846–848.

20. **Kobayashi, K., K. Seto, and M. Makino.** 1990. Detection of toxigenic *Vibrio cholerae* O1 using polymerase chain reaction for amplifying the cholera enterotoxin gene. *J. Jpn. Assoc. Infect. Dis.* **64:**1323–1329. (In Japanese.)

21. **Koch, W. H., W. L. Payne, B. A. Wentz, and T. A. Cebula.** 1993. Rapid polymerase chain reaction method for detection of *Vibrio cholerae* in foods. *Appl. Environ. Microbiol.* **59:**556–560.

22. **Kurosky, A., D. E. Markel, and W. M. Finch.** 1977. Primary structure of cholera toxin beta chain. A glycoprotein analog? *Science* **195:**299–301.

23. **Kurosky, A., D. E. Markel, and J. W. Peterson.** 1977. Covalent structure of the beta chain of cholera enterotoxin. *J. Biol. Chem.* **252:**7257–7264.

24. **Lai, C.-Y.** 1977. Determination of the primary structure of cholera toxin B subunit. *J. Biol. Chem.* **252:**7249–7256.

25. **Lockman, H., and J. B. Kaper.** 1983. Nucleotide sequence analysis of the A2 and B subunits of *Vibrio cholerae* enterotoxin. *J. Biol. Chem.* **258:**13722–13726.

26. **Lockman, H. A., J. E. Galen, and J. B. Kaper.** 1984. *Vibrio cholerae* enterotoxin genes: nucleotide sequence analysis of DNA encoding ADP-ribosyltransferase. *J. Bacteriol.* **159:**1086–1089.

27. **Mekalanos, J. J.** 1983. Duplication and amplification of toxin genes in *Vibrio cholerae. Cell* **35:**253–263.

28. **Mekalanos, J. J., D. J. Swartz, G. D. N. Person, N. Harford, F. Groyne, and M. deWilde.** 1983. Cholera toxin genes: nucleotide sequence, deletion analysis and vaccine development. *Nature* (London) **306:** 551–557.

29. **Minami, A., S. Hashimoto, H. Abe, M. Arita, T. Taniguchi, T. Honda, T. Miwatani, and M. Nishibuchi.** 1991. Cholera enterotoxin production in *Vibrio cholerae* O1 strains isolated from the environment and from humans in Japan. *J. Environ. Microbiol.* **57:**2152–2157.

30. **Miyagi, K., Y. Matsumoto, K. Hayashi, Y. Takarada, S. Shibata, M. Yoh, K. Yamamoto, and T. Honda.** 1992. Cholera diagnosed in clinical laboratory by DNA hybridization. *Lancet* **339:**988–989.

31. **Moseley, S. L., and S. Falkow.** 1980. Nucleotide sequence homology between the heat-labile enterotoxin of *Escherichia coli* and *Vibrio cholerae* deoxyribonucleic acid. *J. Bacteriol.* **144:**444–446.

32. **Olsvik, Ø.** Unpublished data.

33. **Olsvik, Ø., T. Popovic, and P. I. Fields.** 1993. Polymerase chain reaction for detection of toxin genes in strains of *Vibrio cholerae* O1, p. 573–583. *In* D. H. Persing, F. C. Tenover, T. F. Smith, and T. J. White (ed.), *Diagnostic Molecular Microbiology,* American Society for Microbiology, Washington, D.C.

34. **Olsvik, Ø., J. Wahlberg, B. Petterson, M. Uhlen, T. Popovic, I. K. Wachsmuth, and P. I. Fields.** 1993. Use of automated sequencing of PCR-generated amplicons to identify three types of cholera toxin subunit B in *Vibrio cholerae* O1 strains. *J. Clin. Microbiol.* **31:**22–25.

35. **Peter, J. B.** 1990. The polymerase chain reaction: amplifying our options. *Rev. Infect. Dis.* **13:**166–171.

36. **Saunders, D. W., K. J. Schanbacher, and M. G. Bramucci.** 1982. Mapping of a gene in *Vibrio cholerae* that determines the antigenic structure of cholera toxin. *Infect. Immun.* **38:**1109–1116.

37. **Schultz, A. J., and B. A. McCardell.** 1988. DNA homology and immunological cross-reactivity between *Aeromonas hydrophila* cytotoxic toxin and cholera toxin. *J. Clin. Microbiol.* **26:**57–61.

38. **Shirai, H., M. Nishibuchi, T. Ramamurthy, S. K. Bhattacharya, S. C. Pal, and Y. Takada.** 1991. Polymerase chain reaction for detection of cholera enterotoxin operon of *Vibrio cholerae. J. Clin. Microbiol.* **29:**2517–2521.

39. **Spangler, B. D.** 1992. Structure and function of cholera toxin and related *Escherichia coli* heat-labile enterotoxin. *Microbiol. Rev.* **56:**622–647.

40. **Sporecke, I., D. Castro, and J. J. Mekalanos.** 1984. Genetic mapping of *Vibrio cholerae* enterotoxin structural genes. *J. Bacteriol.* **157:**253–261.

41. **Takao, Y., H. Watanabe, and Y. Shimonishi.** 1985. Facile identification of protein sequences by mass spectrometry. *Eur. J. Biochem.* **146:**503–508.

42. **Takeda, T., Y. Peina, A. Ogawa, S. Dohi, H. Abe, G. B. Nair, and S. C. Pal.** 1991. Detection of heat-stable enterotoxin in a cholera toxin gene-positive strain of *V. cholerae* O1. *FEMS Microbiol. Lett.* **80:**23–28.

43. **Tenover, F. C.** 1988. Diagnostic deoxyribonucleic acid probes for infectious diseases. *Clin. Microbiol. Rev.* **1:**82–101.

44. **Varela, P., M. Rivas, N. Binsztein, M. L. Cremona, P. Herrmann, O. Burrone, R. A. Ugalde, and A. C. C. Frasch.** 1993. Identification of toxigenic *Vibrio cholerae* from the Argentine outbreak by PCR for *ctxA1* and *ctxA2-B. FEBS Lett.* **315:**74–76.

45. **Wachsmuth, I. K., C. A. Bopp, and P. I. Fields.** 1991. Difference between toxigenic *Vibrio cholerae* O1 from South America and the US Gulf Coast. *Lancet* **i:**1097–1098.

46. **Wachsmuth, I. K., G. M. Evins, P. I. Fields, Ø. Olsvik, T. Popovic, C. A. Bopp, J. G. Wells, C. Carrillo, and P. A. Blake.** 1993. The molecular epidemiology of cholera in Latin America. *J. Infect. Dis.* **167:**621–626.

47. **Wolcott, M. J.** 1992. Advances in nucleic acid-based detection methods. *Clin. Microbiol. Rev.* **5:**370–386.

48. **Wright, A. C., Y. Guo, J. Johnson, J. P. Nataro, and J. G. Morris.** 1992. Development and testing of a nonradioactive DNA oligonucleotide probe that is specific for *Vibrio cholerae* cholera toxin. *J. Clin. Microbiol.* **30:**2302–2306.

49. **Yamamoto, T., T. Nakazawa, T. Miyata, and T. Yokota.** 1984. Evolution and structure of two ADP-ribosylation enterotoxins, *Escherichia coli* heat-labile toxin and cholera toxin. *FEBS Lett.* **169:**241–246.

50. **Yoh, M., K. Miyagi, Y. Matsumoto, K. Hayashi, Y. Takarada, K. Yamamoto, and T. Honda.** 1993. Development of an enzyme-labeled oligonucleotide probe for the cholera toxin gene. *J. Clin. Microbiol.* **31:**1312–1314.

Vibrio cholerae and Cholera: Molecular to Global Perspectives
Edited by I. Kaye Wachsmuth, Paul A. Blake, and Ørjan Olsvik
© 1994 American Society for Microbiology, Washington, DC 20005

Chapter 4

Detection of Toxins of *Vibrio cholerae* O1 and Non-O1

G. Balakrish Nair and Yoshifumi Takeda

Although cellular components are involved in the pathogenesis of cholera in ways not yet clearly discerned, the pathogenicity of *Vibrio cholerae* serovar O1 is largely due to the synthesis of an exotoxin that triggers the secretion of water and electrolytes into the small intestine. Since Robert Koch hypothesized, around the turn of the century, that cholera is caused by an enteropathogenic exotoxin (62), a variety of laboratory animals, such as guinea pigs, suckling rabbits, and dogs, as well as ligated intestinal rabbit loops have been used to study the toxin(s) produced by *V. cholerae*. For reasons that remain unclear, none of the models proved successful in the hands of those early investigators, although all studies included some animals that developed a choleralike reaction or illness (80).

Only after the epochal discovery by De and Chatterjee (18), who injected living *V. cholerae* or cell-free filtrates into the lumen of a ligated rabbit ileal loop and thereby caused a large amount of fluid to accumulate, did it become clear that cholera is caused by a potent cell-free heat-labile enterotoxin now known as cholera toxin. The ligated rabbit ileal loop model therefore became the first

succcessful means of detecting cholera toxin. Recognition that cholera is a toxin-mediated infection heralded the development of a variety of laboratory tests to establish the presence of the toxin among strains of *V. cholerae*. The first generation of these tests used by clinical laboratories was almost exclusively dependent on animals, particularly rabbits. About 17 years after De's discovery, these tests were followed by tissue culture assays. The purification of cholera toxin and the availability of an antitoxin ushered in the rapid development of a variety of immunologic assays in an impressive array of formats ranging from simple agglutination tests (8) to highly complex immunoassays directed toward precise quantitation of the toxin (9). The period between 1970 and 1980 witnessed significant advances, including the discovery of enterotoxigenic *Escherichia coli* (ETEC) (39, 90) and the recognition that cholera toxin is the prototype of a family of heat-labile enterotoxin (LT) produced by ETEC (11). Unlike cholera, the need to identify LTs for delineating ETEC-mediated infection accounted for a shift in emphasis to the LT of ETEC. All this occurred within 40 years, during which cholera toxin became one of the most extensively studied bacterial protein toxins.

Until recently, it was thought that the clinical manifestations of cholera result primarily from the interaction between cholera toxin

G. Balakrish Nair • National Institute of Cholera and Enteric Diseases, Beliaghata, Calcutta 700010, India. *Yoshifumi Takeda* • Department of Microbiology, Faculty of Medicine, Kyoto University, Sakyo-ku, Kyoto 606-01, Japan.

though whether or not the characteristic edematous induration reaction was a consequence of cholera toxin was not explicit when PF was discovered, the PF activity of cholera toxin later formed the basis of a useful assay that was widely used for the detection and titration of cholera toxin.

Infant mouse model. Introduced by Ujiiye and Kobari (114), who demonstrated that infant mice responded to oral inoculation with viable *V. cholerae* with diarrhea and death, the infant mouse model has been used by several cholera workers who measured the 50% lethal dose as the parameter for disease production (10, 41). Later, the diarrheal response of orally inoculated infant mice to viable *V. cholerae* and purified cholera toxin was quantified by means of the fluid accumulation ratio, which was both time and dose dependent (4). These investigators observed that the onset of fluid accumulation with viable cells of strains of *V. cholerae* O1 occurred 8 h postinoculation and reached near maximum at 16 h, whereas onset of fluid accumulation with cholera toxin occurred 6 to 8 h postinoculation and reached a maximum at 10 h. A minimum dose of 0.5 μg was required to evoke a positive response 10 to 12 h postinoculation, and the positive response of the toxin was neutralized by preincubation with a specific antitoxin (4).

Other animal models. Several other animal models that use rhesus monkeys (32), dogs (43, 88, 89), germfree mice (95, 96), suckling hamsters (109), and chinchillas (7) have been developed for studying the pathogenesis of *V. cholerae* and the pathophysiology of cholera. These animal models, including the reversible intestinal tie adult rabbit diarrhea model (100), were used for the study of the kinetics and magnitude of immunologic events in patients with cholera.

Tissue culture assays

The effect of cholera toxin on the cell morphology of Chinese hamster ovary (CHO) cells (42) and of Y1 mouse adrenal tumor cells (22) formed the basis for developing sensitive tissue culture methods for detecting cholera toxin. S49 mouse lymphosarcoma cells are also used to detect cholera toxin (85). The production of Δ^4,3-ketosteroids induced by cholera toxin in Y1 mouse adrenal tumor cells (23, 24, 67) is accompanied by changes in cellular shape from flat to round that are mediated by adenyl cyclase stimulation (56, 79). Likewise, cholera toxin induces characteristic changes in the shapes of CHO cells, from oval to spindle forms; the cells are sensitive to 10 pg of the toxin (42). In S49 mouse lymphosarcoma cells, intracellular cyclic AMP levels increased by cholera toxin inhibit the growth of the cells (85). These characteristic changes in cells have been used to develop a rapid, clinically useful screening method (23). The adaptation of adrenal cells to miniculture plates (87) has allowed large numbers of strains to be screened.

Immunologic assays

Passive hemagglutination test. Among the first immunologic tests developed after cholera toxin was highly purified was the passive hemagglutination test, which involves sensitized tanned chicken erythrocytes. This test was directed chiefly toward the detection of antibody in cholera patients but was also modified and used in an inhibition format to detect and assay cholera toxin (37). The titration of purified cholera toxin indicated that levels as low as 3 ng were sufficient to cause detectable inhibition of 4 U of antibody in the passive hemagglutination inhibition test.

ELISA. Development of a variety of enzyme-linked immunosorbent assays (ELISAs) was initially directed toward the study of cholera serology. Using the ELISA principle of Engvall and Perlmann (29), Holmgren and Svennerholm (47) developed ELISAs that permitted immunoglobulin class-specific determination of antibody titers and average antibody avidity to cholera toxin as well as sensitive in vitro quantitation of cholera toxin. The lowest level of cholera toxin detectable by this method was 0.09 μg/ml (47).

Later, the sandwich ELISA, which used anti-cholera toxin adsorbed to wells of microtiter plates as the capture antibody, became more popular. Test samples are added to the coated wells, and any cholera toxin in the sample binds to the capture antibody, which in turn is reacted with a second rabbit anti-cholera toxin antibody. Alkaline phosphatase-labeled goat anti-rabbit immunoglobulin is added; it acts as the detector. When the enzyme substrate *p*-nitrophenyl phosphate is added, the bound alkaline phosphatase degrades the substrate, forming a yellow product that indicates the presence of cholera toxin. Color changes can be observed visually or measured in an ELISA reader. A culture is considered positive or toxigenic when the test well has an optical density at least twice that of the negative control wells. Adequate positive and negative controls should be used in every plate.

GM$_1$ ganglioside ELISA. The findings that GM$_1$ ganglioside is the membrane receptor for cholera toxin (16) and that GM$_1$ ganglioside neutralized cholera toxin in about equimolar proportions (45) led investigators to utilize these properties to develop a diagnostic assay. These efforts were further assisted by the finding that GM$_1$ gangliosides can readily be attached to a plastic surface, which is well suited for use in conjunction with solid-phase immunologic detection (46). The microtiter plate-based GM$_1$ ganglioside ELISA was developed to detect LT and cholera toxin (86, 102) and was based on the specific binding of these toxins to polystyrene-adsorbed GM$_1$ ganglioside and subsequently to enzyme to demonstrate the presence of the bound toxin by immunologic means. The GM$_1$ ganglioside ELISA compared well with the antiserum ELISA (118), CHO cell assay (42), and Y1 adrenal cell assay (87) and was an acceptable means of screening bacterial cultures

for toxin and of quantifying toxin production. The assay could also be used to quantify antibodies that block attachment of the toxin to the GM$_1$ ganglioside (86).

Bead ELISA. The ELISA in its current format, using microtiter plates, precludes the use of this technique in less-sophisticated laboratories in areas where quality microtiter plates are expensive and not readily available. To augment simplicity, Oku et al. (77) developed an alternative highly sensitive ELISA that used 6-mm-diameter polystyrene beads as the solid phase (bead ELISA). This assay could detect cholera toxin and other bacterial protein toxins within several hours in standard test tubes. The sensitivity and specificity of the bead ELISA were compared with those of the commonly used and commercially available reversed passive latex agglutination (RPLA) test for detection of cholera toxin in strains of *V. cholerae* O1 (including both biotypes and serotypes) examined with a DNA probe specific for the A1 subunit of cholera toxin. This revealed that the bead ELISA was much more sensitive than the RPLA test (113). An absorption assay using the bead-ELISA can be performed to determine whether cholera toxin can be absorbed with anti-cholera toxin immunoglobulin G (113).

Monoclonal antibody-based ELISA. Monoclonal antibodies against cholera toxin have been raised (66, 74, 83, 103, 105) in order to develop immunologic methods that would permit differentiation between cholera toxin and other related heat-labile enterotoxins produced by different enteropathogens as well as to evaluate the relationship among the toxins. Using a monoclonal antibody with full cross-reactivity with cholera toxin and *E. coli* LTs of human and porcine origins and other monoclonal antibodies with full or partial cross-reactivity with LTs, Svennerholm et al. (105) developed simple but sensitive immunodiagnostic assays based on the GM$_1$ ELISA format. These assays fulfilled the criteria of excellent sensitivity and specific-

ity for the detection of any heat-labile enterotoxin by using the cross-reactive monoclonal antibody or for the detection of species-specific toxin by using the unique
epitope. Having species-specific antibodies
in the immunodetection step would not only
detect heat-labile enterotoxin in the test sample but would also provide information on
the origin of the toxin produced (105). Likewise, a sensitive sandwich ELISA for specific detection of cholera toxin has recently
been developed by using a high-affinity
monoclonal antibody (6). This ELISA detected cholera toxin from all culture supernatants of *V. cholerae* serovar O1 but failed
to detect LT of *E. coli.*

GM$_1$ erythroassay. A GM$_1$ erythroassay
has been develpoed to detect LT and cholera
toxin (38). This assay uses GM$_1$ ganglioside-
coated polystyrene plates and is based on the
competition between the toxin to be assayed
and the cholera toxin covalently bound to
sheep erythrocytes. The assay, developed for
use in poorly equipped laboratories in developing countries, can detect 0.9 or 0.5 ng of
cholera toxin per ml depending on the method
of sensitization of erythrocytes. This test,
however, has not been suitably evaluated
since its development.

Passive immune hemolysis. Initially developed for the detection of the LT of ETEC
(30, 112), passive immune hemolysis was
later used to detect the cholera toxin-like enterotoxin of *V. cholerae* non-O1 (115). In this
assay, the test material and a 2% sheep erythrocyte suspension are incubated, anti-cholera
toxin is added and incubated, and then guinea
pig complement is added. The amount of
cholera toxin in the sample is estimated by
reading the A$_{420}$ of the supernatant fluid of
the mixture.

Solid-phase   radioimmunoassay.   To
determine very low amounts of cholera toxin,
Ceska et al. (9) developed a direct solid-
phase radioimmunoassay that used anti-cholera toxin attached to polystyrene tubes as a
solid binder for cholera toxin. The binding of
radioiodinated cholera toxin to its immo-

bilized antibody was inhibited by unlabeled
cholera toxin and anti-cholera toxin antibody.
The sensitivity of this method reached to concentration ranges of a few nanograms per milliliter. Similar standard radioimmunoassay
curves were obtained when anti-cholera toxin
antibodies raised in horses or sheep were
used. This solid-phase radioimmunoassay
was also used to measure cholera toxin antibodies. At about the same time, another
group of workers developed radioim-
munoassays capable of defining the antigenic
structure of cholera toxin and detecting LT
(40, 44). The practical value of the assay was
in detecting differences between antibodies
elicited by toxoid immunization and antibodies to cholera toxin after natural infection
(44).

The combined use of antibody-enzyme
conjugates and radioactive substrates remarkably enhances the sensitivity, since the
detection of radioisotope-labeled enzyme reaction products is generally more sensitive
than the measurement of nonlabeled enzyme
reaction products by fluorescence or lumines-
cence (50). Although the radioimmunoassay
is very sensitive and specific, the inherent
disadvantage of working with radioactive isotopes has relegated this technique to research
laboratories only.

RPLA. In terms of simplicity, the RPLA
test perhaps surpasses all other immunologic
assays thus far developed for the detection of
cholera toxin. Commercially available, the
test is rapid, simple to perform, easy to read,
and inexpensive. However, simplicity is
achieved at the expense of sensitivity, and the
RPLA assay is rather insensitive compared
with tissue culture assays, radioimmunoassay, and ELISA. Automatic and manual
RPLA methods developed by Ito et al. (51)
measure cholera toxin and its subunits A and
B. In this method, polystyrene latex particles
(diameter, 0.22 μm) and polystyrene-chloro-
styrene latex particles (diameter, 1 μm) are
sensitized by rabbit anti-cholera toxin immunoglobulin and used as the reagents for automated and manual agglutination tests, respec-

tively. Levels as low as 1,000 and 31 pg of cholera toxin per ml were estimated by the automated and manual RPLA assays, respectively (51). The manual RPLA test is less expensive and more sensitive than the automatic method, although an overnight reaction is required.

Recently, an RPLA kit that can detect 1 to 2 ng of cholera toxin per ml has become commercially available and offers the potential of a rapid, simple diagnostic assay for cholera toxin and *E. coli* LT. When the commercial kit was evaluated with the routine ELISA for detection of cholera toxin from strains of *V. cholerae* O1 and non-O1, the RPLA test was found to be 98% accurate for serogroup O1 and 100% accurate for non-O1 (2). On the basis of the findings of the study, Almeida et al. (2) have recommended the use of the Oxoid RPLA kit for routine detection of cholera toxin from culture supernatants of strains of *V. cholerae* in clinical as well as reference laboratories.

Enzymatic assays

The enzymatic method takes advantage of the catalytic property of a cholera toxin subunit, which transfers an ADP-ribose group from NAD^+ to a guanidine group of a substituted benzylidine-aminoguanidine, resulting in ADP-ribosylated benzylidine-aminoguanidine. The product of the ADP-ribosyltransferase has distinct ionic and spectral properties that allow it to be separated from unreacted substrate and quantitated (75). Spectrophotometric and high-performance liquid chromatographic (HPLC) assays that detect cholera toxin have been developed (98, 99). A recent evaluation of the enzymatic assay has shown that as little as 25 ng (spectrofluorometry) or 125 ng (spectrometry or HPLC) of cholera toxin can be detected in an assay volume of 250 μl (75). Although the enzymatic method adds a new dimension to the assay of cholera toxin and other toxins that ribosylate arginine residues of proteins, this method is more academic

than pragmatic and therefore has not been widely applied.

Detection of cholera toxin directly from stool specimens

The direct detection of enterotoxins in stool specimens of patients with acute diarrhea represents a greatly simplified approach for the rapid diagnosis of toxin-mediated diarrhea. Detection of cholera toxin directly from stool specimens of patients suffering from acute diarrhea is especially important in epidemiologic surveys and drug evaluation studies and for following and evaluating new vaccination programs for cholera. The enterotoxicity of cholera stools using the ligated rabbit ileal loop assay was demonstrated as early as 1965 (26). Since then, attempts to detect cholera toxin and *E. coli* LT in human diarrheic stools by using the Y1 adrenal cell culture assay (28, 68, 70), ELISA (104), and counterimmunoelectrophoresis (70) have had various degrees of success. A preliminary effort was made using the monoclonal antibody-GM_1 ELISA to directly detect cholera toxin from stool specimens collected from cholera patients admitted to the cholera hospital at Dhaka, Bangladesh (104). Despite the long delay of 3 to 5 days before analyses, as many as 13 of 17 stools without prior culture from cholera patients were positive in the GM_1 ELISA when a cholera toxin-specific monoclonal antibody was used in the immunodetection step. Tissue culture assays are less sensitive than ELISA in detecting fecal cholera toxin and LT (68), and the need for sterile stool filtrates further limits the universal applicability of this method. Besides, the process of passing stools through membrane filters (0.22-μm pore size) is likely to destroy or inactivate the toxin (70).

A recent evaluation of the ability of the highly sensitive and simple bead ELISA to detect cholera toxin showed that the assay directly detected cholera toxin in 84.7% of culture-positive cholera stool samples, thereby allowing a potential diagnosis of cholera to be made within several hours of receipt of the sample at the laboratory (81). The extent of cross-reactivity of the cholera toxin bead ELISA with purified LT of *E. coli* was also assessed, and it was found that about 10 times more LT was needed to give optical density values similar to those obtained with cholera toxin. This was reflected in the evaluation study, because the cholera toxin bead ELISA did not yield a positive result in two samples in which ETEC (one was LT positive, and the other was both LT and heat-stable enterotoxin positive) was isolated as the sole enteropathogen. The concentration of cholera toxin present in the bead ELISA-positive stool samples ranged between 26 pg/ml and >100 ng/ml. The concentrations of cholera toxin in 42.4% of the cholera toxin-positive stool samples and 57.6% of the *V. cholerae* O1-positive stool samples were below 1 ng/ml. The enterotoxin-binding effect of free gangliosides present in mucin found in stools is probably an important reason for the low or undetectable levels of enterotoxins in stools (78, 101).

The commercially available RPLA assay for detection of cholera toxin (2) is a simpler alternative to ELISA, but the lower minimum detection ability (1 to 2 ng/ml) precludes its use for assaying free fecal cholera toxin in stool specimens. For instance, the RPLA assay would not be able to detect cholera toxin in 42.2% of the bead ELISA-positive stool samples examined in the above-mentioned evaluation study, in which the concentration of cholera toxin was below 1 ng/ml. Using the bead ELISA to demonstrate the presence of anti-cholera toxin immunoglobulin G absorbable fecal cholera toxin in the stools (82), an outbreak of cholera was declared recently even though *V. cholerae* O1 could not be isolated from stool samples taken during the outbreak. The applications of an assay like the bead ELISA, which can directly detect fecal cholera toxin, are manifold, and the bead ELISA is thus far the only rigidly evaluated method that could be used for the direct detection of cholera toxin in nonsterile stool samples.

New Cholera Toxin

The new cholera toxin reported by Sanyal et al. (93) is an enterotoxin present in cholera toxin gene-negative strains of *V. cholerae* O1 that cause diarrhea in infant rabbits, fluid accumulation in ligated rabbit ileal loops, and increased permeability in rabbit skin (94). Cholera toxin gene-positive strains of *V. cholerae* O1 belonging to serotypes Ogawa and Inaba also produce the new cholera toxin (91). The enterotoxic activity of the new cholera toxin is completely neutralized by an antitoxin raised against the enterotoxin that is prepared from the cholera toxin-positive *V. cholerae* O1 strain 569B (92). Animal models such as the rabbit ileal loop, infant rabbit, and permeability in rabbit skin models were used to establish the presence of the new cholera toxin. The exact structural identity of this new toxin is still unknown, and therefore it is unclear which method would be most suitable for detecting it, although it is reportedly active in the conventional animal models developed for cholera toxin.

Zonula Occludens Toxin

The most recent addition to the toxin family of *V. cholerae* is zonula occludens toxin, which increases the permeability of the small-intestine mucosa by affecting the structure of the intercellular tight junction, or zonula occludens (31). The zonula occludens toxin can be detected by using rabbit intestinal tissue mounted in Ussing chambers and measuring the increases in intestinal-tissue conductance (33). Detection of the zonula occludens toxin among strains of *V. cholerae* by an appropriate DNA probe (55, 60) is also feasible, since the sequence of the gene encoding the zonula occludens toxin has been determined (5).

DETECTION OF *V. CHOLERAE* NON-O1 TOXINS

Heat-Labile Enterotoxin

V. cholerae non-O1 strains produce heat-labile enterotoxins that are identical to cholera toxin (15, 76, 116, 120) or are similar but not identical to cholera toxin (57, 117). Although the toxin that is similar but not identical to cholera toxin of *V. cholerae* non-O1 has been purified to homogeneity (117), no specific immunologic methods have been developed to detect it. The production of cholera toxin-like enterotoxin among *V. cholerae* non-O1 strains is low, but incorporation of 300 μg of lincomycin per ml of CAYE-glucose medium substantially stimulates production of cholera toxin-like enterotoxin (115).

Heat-Stable Enterotoxin

The best studied enterotoxin of *V. cholerae* non-O1 is the heat-stable enterotoxin (NAG-ST) (3, 119), a 17-amino-acid peptide that exhibits remarkable similarity, especially in the carboxyl-terminal toxic domain, to the methanol-soluble suckling mouse-active heat-stable enterotoxins STh (1) and STp (106) produced by ETEC. A recent human volunteer study has clearly shown that NAG-ST plays a role in the pathogenesis of non-O1 *V. cholerae* gastroenteritis (71). However, the epidemiologic importance of NAG-ST-producing strains of *V. cholerae* non-O1 and the magnitude of occurrence of these strains in the environment, in foods, and among clinical cases remain uncertain. This hiatus in information stems in part from the lack of a simple, accurate, highly sensitive assay system for detection of NAG-ST.

Suckling mouse assay

Until recently, the only means of detecting NAG-ST-producing strains of *V. cholerae* non-O1 was the suckling mouse assay. First developed to detect the methanol-soluble infant mouse-active heat-stable enterotoxins (STas) of ETEC (20), the suckling mouse has been used successfully with relative ease and without controversy for the detection of methanol-soluable heat-stable enterotoxins of ETEC (110). In contrast, the detection of NAG-ST among *V. cholerae* non-O1 strains is complicated by the presence of additional suckling mouse-active factors, which confuse the interpretation of the results obtained by the suckling mouse assay. For instance, the El Tor hemolysin liberated by most strains of *V. cholerae* non-O1 is active in the suckling mouse (49, 73). The El Tor hemolysin is heat labile; its activity can therefore be destroyed by heating, which theoretically should not affect NAG-ST. However, crude NAG-ST is coprecipitated with and masked by heat-denaturing products (3).

The suckling mouse assay is a delicate, cumbersome, time-consuming test and is unsuitable for simultaneously screening several strains. Many of the variations in the detection rates of NAG-ST-producing *V. cholerae* non-O1 strains have stemmed from inaccuracies related to the suckling mouse assay for detection of NAG-ST. A recent study defined the optimal conditions under which to obtain reliable results when the suckling mouse assay is used to detect NAG-ST from *V. cholerae* non-O1 strains. To obtain unequivocal and reliable results, the culture supernatant of *V. cholerae* non-O1 strains grown in brain heart infusion broth containing 0.5% NaCl at 37°C needs to be concentrated at least 20-fold and then heated at 100°C for 5 min before being intragastrically inoculated into infant mice (72).

Monoclonal antibody-based competitive ELISA

Immunodetection of heat-stable enterotoxins has been problematic owing to the nonimmunogenic nature of the toxin in its natural state and to the consequent difficulties in producing homogeneous, high-titer antibodies. Monoclonal antibodies against *E.*

coli STh and STp are valuable and sensitive reagents used in the development of immunoassays for *E. coli* heat-stable enterotoxins (21, 104, 111). However, these assays cannot be used to detect NAG-ST because of the high specificity of the monoclonal antibodies. Recently, a high-affinity monoclonal antibody against NAG-ST was raised. The contact residue defining the activity of the monoclonal antibody was determined to be aspartic acid, located at position 2 from the N terminus of NAG-ST (107). A competitive ELISA using this monoclonal antibody has been developed to detect NAG-ST from culture supernatants of strains of *V. cholerae* non-O1 (72).

Evaluation of the monoclonal antibody-based ELISA for the detection of NAG-ST, however, has revealed the need to concentrate secondary broth cultures of *V. cholerae* non-O1 strains to obtain reliable and reproducible results because of the inherent low production of NAG-ST by *V. cholerae* non-O1 strains (72).

APPLICATION VERSATILITY OF DIFFERENT METHODS FOR DETECTING *V. CHOLERAE* TOXINS

A variety of methods for detection of cholera toxin and other toxins elaborated by *V. cholerae* O1 and non-O1 strains have evolved over the last decades. However, the applicability of several of the methods is limited to research laboratories because of the need for materials or reagents not readily available in Third World countries or epidemic settings. In this context, the development of the bead ELISA for detection of cholera toxin and its direct application to stool samples have been major advances. The fact that the bead ELISA can be performed in several hours in a setting with minimal infrastructure further enhances the acceptability of the test. The commercially available RPLA test for detection of cholera toxin is one step ahead of the bead ELISA in terms of simplicity and is ideal for field or peripheral laboratories where infrastructure is at a bare minimum. The dependence on laboratory animals, especially rabbits, for detection of cholera toxin is on the decline with the development of more precise and easier assays. However, laboratory animals still constitute the principal means for the demonstration of biological activity of cholera toxin and other toxins of *V. cholerae*.

CONCLUSION

There is no denying that since the early days of De's discovery, there have been spectacular advances in our knowledge of the variety of toxins produced by *V. cholerae* O1 and non-O1 strains. These advances have been paralleled by the development of an assortment of assays for detecting the toxins. From the perspective of cholera diagnosis, the current emphasis is to transcend the dependence on bacterial culture for diagnostic purposes and to effect a diagnosis directly from stool samples, using cholera toxin as the signature molecule. If this approach is successful, then it will eventually become possible to clinically detect cholera by determining the relevant toxin directly in the stool at the bedside. This capability would represent a quantum leap in the diagnosis of cholera that would have far-reaching consequences in the care and management of cholera patients.

REFERENCES

1. **Aimoto, S., T. Takao, Y. Shimonishi, S. Hara, T. Takeda, Y. Takeda, and T. Miwatani.** 1982. Amino-acid sequence of a heat-stable enterotoxin produced by human enterotoxigenic *Escherichia coli. Eur. J. Biochem.* **129:**257–263.
2. **Almeida, R. J., F. W. Hickman-Brenner, E. G. Sowers, N. D. Puhr, J. J. Farmer III, and I. K. Wachsmuth.** 1990. Comparison of a latex agglutination assay and an enzyme-linked immunosorbent assay for detecting cholera toxin. *J. Clin. Microbiol.* **28:**128–130.
3. **Arita, M., T. Takeda, T. Honda, and T. Miwatani.** 1986. Purification and characterization of *Vibrio cholerae* non-O1 heat-stable enterotoxin. *Infect. Immun.* **52:**45–49.

4. Baselski, V., R. Briggs, and C. Parker. 1977. Intestinal fluid accumulation induced by oral challenge with *Vibrio cholerae* or cholera toxin in infant mice. *Infect. Immun.* **15:**704–712.

5. Baudry, B., A. Fasano, J. Ketley, and J. B. Kaper. 1992. Cloning of a gene (*zot*) encoding a new toxin produced by *Vibrio cholerae*. *Infect. Immun.* **60:**428–434.

6. Bhadra, R. K., T. Biswas, S. C. Pal, T. Takeda, and G. B. Nair. 1991. A polyclonal-monoclonal antibody based sensitive sandwich enzyme-linked immunosorbent assay for specific detection of cholera toxin. *Zentralbl. Bakteriol.* **275:**467–473.

7. Blachman, U., S. J. Goss, and M. J. Pickett. 1974. Experimental cholera in the chinchilla. *J. Infect. Dis.* **129:**376–384.

8. Brill, B. M., B. L. Wasilauskas, and S. H. Richardson. 1979. Adaptation of the staphylococcal coagglutination technique for detection of heat-labile enterotoxin of *Escherichia coli*. *J. Clin. Microbiol.* **9:**49–55.

9. Ceska, M., F. Effenberger, and F. Grossmüller. 1978. Highly sensitive solid-phase radioimmunoassay suitable for determination of low amounts of cholera toxin and cholera toxin antibodies. *J. Clin. Microbiol.* **7:**209–213.

10. Chaicumpa, W., and D. Rowley. 1972. Experimental cholera in infant mice: protective effects of antibody. *J. Infect. Dis.* **125:**480–485.

11. Clements, J. D., and R. A. Finkelstein. 1979. Isolation and characterization of homogeneous heat-labile enterotoxins with high specific activity from *Escherichia coli* cultures. *Infect. Immun.* **24:**760–769.

12. Craig, J. P. 1965. A permeability factor (toxin) found in cholera stools and culture filtrates and its neutralization by convalescent cholera sera. *Nature* (London) **207:**614–616.

13. Craig, J. P. 1966. Preparation of the vascular permeability factor of *Vibrio cholerae*. *J. Bacteriol.* **92:**793–795.

14. Craig, J. P. 1985. The vibrio diseases in 1982: an overview, p. 11–23. *In* Y. Takeda and T. Miwatani (ed.), *Bacterial Diarrheal Diseases.* KTK Scientific Publishers, Tokyo.

15. Craig, J. P., K. Yamamoto, Y. Takeda, and T. Miwatani. 1981. Production of cholera-like enterotoxin by a *Vibrio cholerae* non-O1 strain isolated from the environment. *Infect. Immun.* **34:**90–97.

16. Cuatrecasas, P. 1973. Gangliosides and membrane receptors for cholera toxin. *Biochemistry* **12:**3558–3566.

17. De, S. N. 1959. Enterotoxicity of bacteria-free culture-filtrate of *Vibrio cholerae*. *Nature* (London) **183:**1533–1534.

18. De, S. N., and D. N. Chatterje. 1953. An experimental study of the mechanism of action of *Vibrio cholerae* on the intestinal mucous membrane. *J. Pathol. Bacteriol.* **66:**559–562.

19. De, S. N., M. L. Ghose, and A. Sen. 1960. Activities of bacteria-free preparations from *Vibrio cholerae*. *J. Pathol. Bacteriol.* **79:**373–380.

20. Dean, A. G., Y.-C. Ching, R. G. Williams, and L. B. Harden. 1972. Test for *Escherichia coli* enterotoxin using infant mice: application in a study of diarrhea in children in Honolulu. *J. Infect. Dis.* **125:**407–411.

21. DeMol, P., W. Hemelhof, P. Retoré, T. Takeda, T. Miwatani, Y. Takeda, and J. P. Butzler. 1985. A competitive immunosorbent assay for the detection of heat-stable enterotoxin of *Escherichia coli*. *J. Med. Microbiol.* **20:**69–74.

22. Donta, S. T., M. King, and K. Sloper. 1973. Induction of steroidogenesis in tissue culture by cholera enterotoxin. *Nature* (London) *New Biol.* **243:**246–247.

23. Donta, S. T., H. W. Moon, and S. C. Whipp. 1974. Detection of heat-labile *Escherichia coli* enterotoxin with the use of adrenal cells in tissue culture. *Science* **183:**334–336.

24. Donta, S. T., and D. M. Smith. 1974. Stimulation of steroidogenesis in tissue culture by enterotoxigenic *Escherichia coli* and its neutralization by specific antiserum. *Infect. Immun.* **9:**500–505.

25. Dubey, R. S., M. Lindblad, and J. Holmgren. 1990. Purification of El Tore cholera enterotoxins and comparisons with classical toxin. *J. Gen. Microbiol.* **136:**1839–1847.

26. Dutt, A. R. 1965. Enterotoxic activity in cholera stool. *Indian J. Med. Res.* **53:**605–609.

27. Dutta, N. K., and M. K. Habbu. 1955. Experimental cholera in infant rabbits: a method for chemotherapeutic investigation. *Br. J. Pharmacol. Chemother.* **10:**153–159.

28. Echeverria, P., L. Verheart, C. V. Ulyanco, and L. T. Santiago. 1978. Detection of heat-labile enterotoxin-like activity in stools of patients with cholera and *Escherichia coli* diarrhea. *Infect. Immun.* **19:**343–344.

29. Engvall, E., and P. Perlmann. 1972. Enzyme-linked immunosorbent assay, ELISA. III. Quantitation of specific antibodies by enzyme-labeled anti-immunoglobulin in antigen-coated tubes. *J. Immunol.* **109:**129–135.

30. Evans, D. J., Jr., and D. G. Evans. 1977. Direct serological assay for the heat-labile enterotoxin of *Escherichia coli*, using passive immune hemolysis. *Infect. Immun.* **16:**604–609.

31. Fasano, A., B. Baudry, D. W. Pumplin, S. S. Wasserman, B. D. Tall, J. M. Ketley, and J. B. Kaper. 1991. *Vibrio cholerae* produces a second enterotoxin, which affects intestinal tight junctions. *Proc. Natl. Acad. Sci. USA* **88:**5242–5246.

32. Felsenfeld, O., W. E. Greer, and A. D. Felsenfeld. 1967. Cholera toxin neutralization and some

cellular sites of immune globulin formation in *Cercopithecus aethiops*. *Nature* (London) **213**:1249–1251.

33. **Field, M., D. Fromm, and I. McColl.** 1971. Ion transport in rabbit ileal mucosa. I. Na and Cl fluxes and short circuit current. *Am. J. Physiol.* **220**:1388–1396.

34. **Finkelstein, R. A., P. Atthasampunna, M. Chulasamaya, and P. Charunmethee.** 1966. Pathogenesis of experimental cholera: biologic activities of purified procholeragen A. *J. Immunol.* **96**:440–449.

35. **Finkelstein, R. A., and J. J. LoSpalluto.** 1969. Pathogenesis of experimental cholera: preparation and isolation of choleragen and choleragenoid. *J. Exp. Med.* **130**:185–202.

36. **Finkelstein, R. A., H. T. Norris, and N. K. Dutta.** 1964. Pathogenesis of experimental cholera in infant rabbits. I. Observations on the intraintestinal infection and experimental cholera produced with cell-free products. *J. Infect. Dis.* **114**:203–216.

37. **Finkelstein, R. A., and J. W. Peterson.** 1970. In vitro detection of antibody to cholera enterotoxin in cholera patients and laboratory animals. *Infect. Immun.* **1**:21–29.

38. **Germani, Y., E. Bégaud, J. L. Guesdon, J. P. Moreau.** 1987. Assay of *Vibrio cholerae/Escherichia coli* heat-labile enterotoxin by inhibition of erythroadsorption on the GM_1 ganglioside. *Ann. Inst. Pasteur Microbiol.* **138**:223–234.

39. **Gorbach, S. L., and C. M. Khurana.** 1972. Toxigenic *Escherichia coli:* a cause of infantile diarrhea in Chicago. *N. Engl. J. Med.* **287**:791–795.

40. **Greenberg, H. B., D. A. Sack, W. Rodriguez, R. B. Sack, R. G. Wyatt, A. R. Kalica, R. L. Horswood, R. M. Chanock, and A. Z. Kapikian.** 1977. Microtiter solid-phase radioimmunoassay for detection of *Escherichia coli* heat-labile enterotoxin. *Infect. Immun.* **17**:541–545.

41. **Guentzel, M. N., and L. J. Berry.** 1975. Motility as a virulence factor for *Vibrio cholerae*. *Infect. Immun.* **11**:890–897.

42. **Guerrant, R. L., L. L. Brunton, T. C. Schnaitman, L. I. Rebhun, and A. G. Gilman.** 1974. Cyclic adenosine monophosphate and alteration of Chinese hamster ovary cell morphology: a rapid, sensitive in vitro assay for the enterotoxins of *Vibrio cholerae* and *Escherichia coli*. *Infect. Immun.* **10**:320–327.

43. **Guerrant, R. L., L. C. Chen, and G. W. G. Sharp.** 1972. Intestinal adenyl-cyclase activity in canine cholera: correlation with fluid accumulation. *J. Infect. Dis.* **125**:377–381.

44. **Hejtmancik, K. E., J. W. Peterson, D. E. Markel, and A. Kurosky.** 1977. Radioimmunoassay for the antigenic determinants of cholera toxin and its components. *Infect. Immun.* **17**:621–628.

45. **Holmgren, J.** 1973. Comparison of the tissue receptors for *Vibrio cholerae* and *Escherichia coli* enterotoxins by means of gangliosides and natural cholera toxoid. *Infect. Immun.* **8**:851–859.

46. **Holmgren, J., I. Lönnroth, and L. Svennerholm.** 1973. Fixation and inactivation of cholera toxin by GM_1 ganglioside. *Scand. J. Infect. Dis.* **5**:77–78.

47. **Holmgren, J., and A.-M. Svennerholm.** 1973. Enzyme-linked immunosorbent assays for cholera serology. *Infect. Immun.* **7**:759–763.

48. **Honda, T., and R. A. Finkelstein.** 1979. Selection and characteristics of a *Vibrio cholerae* mutant lacking the A (ADP-ribosylating) portion of the cholera enterotoxin. *Proc. Natl. Acad. Sci. USA* **76**:2052–2056.

49. **Ichinose, Y., K. Yamamoto, N. Nakasone, M. J. Tanabe, T. Takeda, T. Miwatani, and M. Iwanaga.** 1987. Enterotoxicity of El Tor-like hemolysin of non-O1 *Vibrio cholerae*. *Infect. Immun.* **55**:1090–1093.

50. **Ishikawa, E., M. Imagawa, and S. Hashida.** 1983. Ultrasensitive enzyme immunoassay using fluorogenic, luminogenic, radioactive and related substrates and factors to limit the sensitivity, p. 219–232. *In* S. Avrameas, P. Duret, R. Masseyeff, and G. Geldmann (ed.), *Immunoenzymatic Techniques*. Elsevier Science Publishers B. V., Amsterdam.

51. **Ito, T., S. Kuwahara, and T. Yokota.** 1983. Automatic and manual latex agglutination tests for measurement of cholera toxin and heat-labile enterotoxin of *Escherichia coli*. *J. Clin. Microbiol.* **17**:7–12.

52. **Iwanaga, M., and T. Kuyyakanond.** 1987. Large production of cholera toxin by *Vibrio cholerae* O1 in yeast extract peptone water. *J. Clin. Microbiol.* **25**:2314–2316.

53. **Iwanaga, M., and K. Yamamoto.** 1985. New medium for the production of cholera toxin by *Vibrio cholerae* O1 biotype El Tor. *J. Clin. Microbiol.* **22**:405–408.

54. **Iwanaga, M., K. Yamamoto, N. Higa, Y. Ichinose, N. Nakasone, and M. Tanabe.** 1986. Culture conditions for stimulating cholera toxin production by *Vibrio Cholerae* O1 El Tor. *Microbiol. Immunol.* **30**:1075–1083.

55. **Johnson, J. A., J. G. Morris, Jr., and J. B. Kaper.** 1993. Gene encoding zonula occludens toxin (*zot*) does not occur independently from cholera enterotoxin genes (*ctx*) in *Vibrio cholerae*. *J. Clin. Microbiol.* **31**:732–733.

56. **Kantor, H. S., P. Tao, and S. L. Gorbach.** 1974. Stimulation of intestinal adenyl cyclase by *Escherichia coli* enterotoxin: comparison of strains from an infant and an adult with diarrhea. *J. Infect. Dis.* **129**:1–9.

57. **Kaper, J. B., H. B. Bradford, N. C. Roberts, and S. Falkow.** 1982. Molecular epidemiology and *Vibrio cholerae* in the U.S. gulf coast. *J. Clin. Microbiol.* **16:**129–134.

58. **Kaper, J. B., H. Lockman, M. M. Baldini, and M. M. Levine.** 1984. Recombinant live oral cholera vaccine. *Bio/Technology* **2:**345–349.

59. **Kaper, J. B., H. Lockman, M. M. Baldini, and M. M. Levine.** 1984. Recombinant nontoxinogenic *Vibrio cholerae* strains as attenuated cholera vaccines candidates. *Nature* (London) **308:**655–658.

60. **Karasawa, T., T. Mihara, H. Kurazono, G. B. Nair, S. Garg, T. Ramamurthy, and Y. Takeda.** 1993. Distribution of the *zot* (zonula occludens toxin) gene among strains of *Vibrio cholerae* O1 and non-O1. *FEMS Microbiol. Lett.* **106:**143–146.

61. **Kasai, G. J., and W. Burrows.** 1966. The titration of cholera toxin and antitoxin in the rabbit ileal loop. *J. Infect. Dis.* **116:**606–614.

62. **Koch, R.** 1919. Erste Konferenz zur Eröterung der Cholerafrage am 26 Juli 1884 in Berlin, p. 20–60. *In Gesammelte Werke von Robert Koch,* 2er Band, 1ste Teil. Verlag von Georg Thieme, Leipzig, Gremany.

63. **Kusama, H., and J. P. Craig.** 1970. Production of biologically active substances by two strains of *Vibrio cholerae. Infect. Immun.* **1:**80–87.

64. **Levine, M. M., J. B. Kaper, D. Herrington, G. Losonsky, J. G. Morris, M. L. Clemets, R. E. Black, B. Tall, and R. Hall.** 1988. Volunteer studies of deletion mutants of *Vibrio cholerae* O1 prepared by recombinant techniques. *Infect. Immun.* **56:**161–167.

65. **Levner, M., F. P. Wiener, and B. A. Rubin.** 1977. Induction of *Escherichia coli* and *Vibrio cholerae* enterotoxins by an inhibitor of protein synthesis. *Infect. Immun.* **15:**132–137.

66. **Lindholm, L., J. Holmgren, M. Wikström, U. Karlsson, K. Andersson, and N. Lycke.** 1983. Monoclonal antibodies to cholera toxin with special references to cross-reactions with *Escherichia coli* heat-labile enterotoxin. *Infect. Immun.* **40:**570–576.

67. **Maneval, D. R., R. R. Colwell, S. W. Joseph, R. Grays, and S. T. Donta.** 1980. A tissue culture method for the detection of bacterial enterotoxins. *J. Tissue Culture Methods* **6:**85–90.

68. **Merson, M. H., R. H. Yolken, R. B. Sack, J. L. Froehlich, H. B. Greenberg, I. Huq, and R. W. Black.** 1980. Detection of *Escherichia coli* enterotoxins in stools. *Infect. Immun.* **29:**108–113.

69. **Minami, A., S. Hashimoto, H. Abe, M. Arita, T. Taniguchi, T. Honda, T. Miwatani, and M. Nishibuchi.** 1991. Cholera enterotoxin production in *Vibrio cholerae* O1 strains isolated from the environment and from humans in Japan. *Appl. Environ. Microbiol.* **57:**2152–2157.

70. **Morgan, D. R., H. L. DuPont, L. V. Wood, and C. D. Ericsson.** 1983. Comparison of methods to detect *Escherichia coli* heat-labile enterotoxin in stool and cell-free culture supernatants. *J. Clin. Microbiol.* **18:**798–802.

71. **Morris, J. G., Jr., T. Takeda, B. D. Tall, G. A. Losonsky, S. K. Bhattacharya, B. D. Forrest, B. A. Kay, and M. Nishibuchi.** 1990. Experimental non-O group 1 *Vibrio cholerae* gastroenteritis in humans. *J. Clin. Invest.* **85:**697–705.

72. **Nair, G. B., S. K. Bhattacharya, and T. Takeda.** 1993. Identification of heat-stable enterotoxin-producing strains of *Yersinia enterocolitica* and *Vibrio cholerae* non-O1 by a monoclonal antibody-based enzyme-linked immunosorbent assay. *Microbiol. Immunol.* **37:**181–186.

73. **Nair, G. B., Y. Oku, Y. Takeda, A. Ghosh, R. K. Ghosh, S. Chattopadhyay, S. C. Pal, J. B. Kaper, and T. Takeda.** 1988. Toxin profiles of *Vibrio cholerae* non-O1 from environmental sources in Calcutta, India. *Appl. Environ. Microbiol.* **54:**3180–3182.

74. **Nair, G. B., A. Pal, R. K. Bhadra, S. Misra, P. Das, P. P. Chaudhuri, and S. C. Pal.** 1989. Production, subunit specificity and neutralizing activity of murine monoclonal antibodies to cholera toxin, p. 141–145. *In Proceedings of the Indo-U.K. Workshop on Diarrhoeal Diseases, Calcutta, India.* National Institute of Cholera and Enteric Diseases, Indian Medical Council.

75. **Narayanan, J., P. A. Hartman, and D. J. Graves.** 1989. Assay of heat-labile enterotoxins by their ADP-ribosyltransferase activities. *J. Clin. Microbiol.* **27:**2414–2419.

76. **Ohashi, M., T. Shimada, and H. Fukumi.** 1972. In vitro production of enterotoxin and hemorrhagic principle by *Vibrio cholerae,* NAG. *Jpn. J. Med. Sci. Biol.* **25:**179–194.

77. **Oku, Y., Y. Uesaka, T. Hirayama, and Y. Takeda.** 1988. Development of a highly sensitive bead-ELISA to detect bacterial protein toxins. *Microbiol. Immunol.* **32:**807–816.

78. **Pierce, N. F.** 1973. Differential inhibitory effects of cholera toxoids and ganglioside on the enterotoxins of *Vibrio cholerae* and *Escherichia coli. J. Exp. Med.* **137:**1009–1023.

79. **Pierce, N. F., W. B. Greenough III, and C. C. J. Carpenter, Jr.** 1971. *Vibrio cholerae* enterotoxin and its mode of action. *Bacteriol. Rev.* **35:**1–13.

80. **Pollitzer, R.** 1959. *Cholera.* Monograph no. 43. World Health Organization, Geneva.

81. **Ramamurthy, T., S. K. Bhattacharya, Y. Uesaka, K. Horigome, M. Paul, D. Sen, S. C. Pal, T. Takeda, Y. Takeda, and G. B. Nair.** 1992. Evaluation of the bead enzyme-linked immunosorbent assay for detection of cholera toxin directly from stool specimens. *J. Clin. Microbiol.* **30:**1783–1786.

82. Ramamurthy, T., A. Pal, G. B. Nair, S. C. Pal, T. Takeda, and Y. Takeda. 1990. Experience with toxin bead ELISA in cholera outbreak. *Lancet* **ii:**375–376.

83. Remmers, E. F., R. R. Colwell, and R. A. Goldsby. 1982. Production and characterization of monoclonal antibodies to cholera toxin. *Infect. Immun.* **37:**70–76.

84. Richardson, S. H. 1969. Factors influencing in vitro skin permeability factor production by *Vibrio cholerae*. *J. Bacteriol.* **100:**27–34.

85. Ruch, F. E., Jr., J. R. Murphy, L. H. Graf, and M. Field. 1978. Isolation of nontoxinogenic mutants of *Vibrio cholerae* in a colorimetric assay for cholera toxin using the S49 mouse lymphosarcoma cell line. *J. Infect. Dis.* **137:**747–755.

86. Sack, D. A., S. Huda, P. K. B. Neogi, R. R. Daniel, and W. M. Spira. 1980. Microtiter ganglioside enzyme-linked immunosorbent assay for *Vibrio* and *Escherichia coli* heat-labile enterotoxins and antitoxin. *J. Clin. Microbiol.* **11:**35–40.

87. Sack, D. A., and R. B. Sack. 1975. Test for enterotoxigenic *Escherichia coli* using Y1 adrenal cells in miniculture. *Infect. Immun.* **11:**334–336.

88. Sack, R. B., and C. C. J. Carpenter. 1969. Experimental canine cholera. I. Development of the model. *J. Infect. Dis.* **119:**138–149.

89. Sack, R. B., and C. C. J. Carpenter. 1969. Experimental canine cholera. II. Production by cell-free culture filtrates of *Vibrio cholerae*. *J. Infect. Dis.* **119:**150–157.

90. Sack, R. B., S. L. Gorbach, J. G. Banwell, B. Jacobs, B. D. Chatterjee, and R. C. Mitra. 1971. Enterotoxigenic *Escherichia coli* isloated from patients with severe cholera-like disease. *J. Infect. Dis.* **123:**378–385.

91. Saha, S., and S. C. Sanyal. 1988. Cholera toxin gene-positive *Vibrio cholerae* O1 Ogawa and Inaba strains produce the new cholera toxin. *FEMS Microbiol. Lett.* **50:**113–116.

92. Saha, S., and S. C. Sanyal. 1989. Neutralization of the new cholera toxin by antiserum against crude enterotoxin of cholera toxin gene-positive *Vibrio cholerae* O1 in rabbit ileal loop model. *Indian J. Med. Res.* **89:**117–120.

93. Sanyal, S. C., K. Alam, P. K. B. Neogi, M. I. Huq, and K. A. Al-Mahmud. 1983. A new cholera toxin. *Lancet* **i:**1337.

94. Sanyal, S. C., P. K. B. Neogi, K. Alam, M. I. Huq, and K. A. Al-Mahamud. 1984. A new enterotoxin produced by *Vibrio cholerae* O1. *J. Diarrhoeal Dis. Res.* **2:**3–12.

95. Shimamura, T. 1972. Immune response in germfree mice orally immunized with *Vibrio cholerae*. *Keio J. Med.* **21:**113–126.

96. Shimamura, T., S. Tazume, K. Hashimoto, and S. Sasaki. 1981. Experimental cholera in germfree suckling mice. *Infect. Immun.* **34:**296–298.

97. Shimamura, T., S. Watanabe, and S. Sasaki. 1985. Enhancement of enterotoxin production by carbon dioxide in *Vibrio cholerae*. *Infect. Immun.* **49:**455–456.

98. Soman, G., J. Narayanan, B. L. Martin, and D. J. Graves. 1986. Use of substituted (benzylidineamino) guanidines in the study of guanidino group specific ADP-ribosyltransferase. *Biochemistry* **25:**4113–4119.

99. Soman, G., K. B. Tomer, and D. J. Graves. 1983. Assay of mono ADP-ribosyltransferase activity by using guanylhydrazones. *Anal. Biochem.* **134:**101–110.

100. Spira, W. M., R. B. Sack, and J. L. Froehlich. 1981. Simple adult rabbit model for *Vibrio cholerae* and enterotoxigenic *Escherichia coli* diarrhea. *Infect. Immun.* **32:**739–747.

101. Strombeck, D. R., and D. Harrold. 1974. Binding of cholera toxin to mucins and inhibition by gastric mucin. *Infect. Immun.* **10:**1266–1272.

102. Svennerholm, A.-M., and J. Holmgren. 1978. Identification of *Escherichia coli* heat-labile enterotoxin by means of a ganglioside immunosorbent assay (GM$_1$-ELISA) procedure. *Curr. Microbiol.* **1:**19–23.

103. Svennerholm, A.-M., M. Wikström, M. Lindblad, and J. Holmgren. 1986. Monoclonal antibodies to *Escherichia coli* heat-labile enterotoxins: neutralizing activity and differentiation of human and porcine LTs and cholera toxin. *Med. Biol.* **64:**23–30.

104. Svennerholm, A.-M., M. Wikström, M. Lindblad, and J. Holmgren. 1986. Monoclonal antibodies against *Escherichia coli* heat-stable toxin (STa) and their use in a diagnostic ST ganglioside GM$_1$-enzyme-linked immunosorbent assay. *J. Clin. Microbiol.* **24:**585–590.

105. Svennerholm, A.-M., M. Wikström, L. Lindholm, and J. Holmgren. 1986. Monoclonal antibodies and immunodetection methods for *Vibrio cholerae* and *Escherichia coli* enterotoxins, p. 77–97. *In* A. J. Macario and E. Conway de Macario (ed.), *Monoclonal Antibodies against Bacteria*, vol III. Academic Press, New York.

106. Takao, T., T. Hitouji, S. Aimoto, Y. Shimonishi, S. Hara, T. Takeda, Y. Takeda, and T. Miwatani. 1983. Amino acid sequence of a heat-stable enterotoxin isolated from enterotoxigenic *Escherichia coli* strain 18D. *FEBS Lett.* **152:**1–5.

107. Takeda, T., G. B. Nair, K. Suzuki, and Y. Shimonishi. 1990. Production of a monoclonal antibody to *Vibrio cholerae* non-O1 heat-stable enterotoxin (ST) which is cross-reactive with *Yersinia enterocolitica* ST. *Infect. Immun.* **58:**2755–2759.

108. Takeda, T., Y. Peina, A. Ogawa, S. Dohi, H. Abe, G. B. Nair, and S. C. Pal. 1991. Detection

of heat-stable enterotoxin in a cholera toxin gene positive strains of *Vibrio cholerae* O1. *FEMS Microbiol. Lett.* **80:**23–28.

109. Takeda, T., Y. Takeda, T. Miwatani, and N. Ohtomo. 1978. Detection of cholera enterotoxin activity in suckling hamsters. *Infect. Immun.* **19:**752–754.

110. Takeda, Y., T. Takeda, T. Yano, K. Yamamoto, and T. Miwatani. 1979. Purification and partial characterization of heat-stable enterotoxin of enterotoxigenic *Escherichia coli*. *Infect. Immun.* **25:**978–985.

111. Thompson, M. R., H. Brandwein, M. LaBine-Racke, and R. A. Giannella. 1984. Simple and reliable enzyme-linked immunosorbent assay with monoclonal antibodies for detection of *Escherichia coli* heat-stable enterotoxins. *J. Clin. Microbiol.* **20:**59–64.

112. Tsukamoto, T., Y. Kinoshita, S. Taga, Y. Takeda, and T. Miwatani. 1980. Value of passive immune hemolysis for detection of heat-labile enterotoxin produced by enterotoxigenic *Escherichia coli*. *J. Clin. Microbiol.* **12:**768–771.

113. Uesaka, Y., Y. Otsuka, M. Kashida, Y. Oku, K. Horigome, G. B. Nair, S. C. Pal, S. Yamasaki, and Y. Takeda. 1992. Detection of cholera toxin by a highly sensitive bead-enzyme linked immunosorbent assay. *Microbiol. Immunol.* **36:**43–53.

114. Ujiiye, A., and K. Kobari. 1970. Protective effect on infections with *Vibrio cholerae* in suckling mice caused by the passive immunization with milk of immune mothers. *J. Infect. Dis.* **121**(Suppl.):S50–S55.

115. Yamamoto, K., Y. Takeda, T. Miwatani, and J. P. Craig. 1981. Stimulation by lincomycin of production of cholera-like enterotoxin in *Vibrio cholerae* non-O1. *FEMS Microbiol. Lett.* **12:**245–248.

116. Yamamoto, K., Y. Takeda, T. Miwatani, and J. P. Craig. 1982. Purification and some properties of a non-O1 *Vibrio cholerae* enterotoxin that is identical to cholera enterotoxin. *Infect. Immun.* **39:**1128–1135.

117. Yamamoto, K., Y. Takeda, T. Miwatani, and J. P. Craig. 1983. Evidence that a non-O1 *Vibrio cholerae* produces enterotoxin that is similar but not identical to cholera enterotoxin. *Infect. Immun.* **41:**896–901.

118. Yolken, R. H., H. B. Greenberg, M. H. Merson, R. B. Sack, and A. Z. Kapikian. 1977. Enzyme-linked immunosorbent assay for detection of *Escherichia coli* heat-labile enterotoxin. *J. Clin. Microbiol.* **6:**439–444.

119. Yoshimura, S., T. Takao, Y. Shimonishi, M. Arita, T. Takeda, H. Imaishi, T. Honda, and T. Miwatani. 1986. A heat-stable enterotoxin of *Vibrio cholerae* non-O1: chemical synthesis, and biological and physicochemical properties. *Biopolymers* **25:**S69–S83.

120. Zinnaka, Y., and C. C. J. Carpenter, Jr. 1972. An enterotoxin produced by non-cholera vibrios. *Johns Hopkins Med. J.* **131:**403–411.

Vibrio cholerae and Cholera: Molecular to Global Perspectives
Edited by I. Kaye Wachsmuth, Paul A. Blake, and Ørjan Olsvik
© 1994 American Society for Microbiology, Washington, DC 20005

Chapter 5

Nontoxigenic *Vibrio cholerae* O1 Infections in the United States

Daniel C. Rodrigue, Tanja Popovic, and I. Kaye Wachsmuth

During the past two decades, strains of nontoxigenic (NT) *Vibrio cholerae* O1 have been isolated in several countries, including Bangladesh, Guam, Brazil, Peru, Japan, England, and the United States (13–15, 20, 22). The strains were isolated from humans, sewage, oysters, and clean and fecally contaminated surface water. Their pathogenic and epidemic potentials as well as their clinical and public health significance remain unclear and controversial (13–15, 20, 22). In this chapter, we summarize the available epidemiologic, microbiologic, and clinical information on NT *V. cholerae* O1 isolates from humans in the United States received by the Centers for Disease Control and Prevention (CDC) between 1977 and 1991.

PHENOTYPIC CHARACTERISTICS

NT *V. cholerae* of O group 1 is biochemically and serologically indistinguishable from toxigenic *V. cholerae* O1, the cause of epidemic cholera, but does not produce cholera toxin (CT). Strains are considered NT when CT assays for activity, antigenicity, and the presence of toxin structural genes are negative (3, 10, 30, 31). Phenotypic characteristics of 14 isolates are summarized in Table 1.

NT *V. cholerae* O1 strains have been isolated from stool specimens by direct inoculation of thiosulfate-citrate-bile salts-sucrose agar. Isolation of NT *V. cholerae* O1 from extraintestinal sites did not require use of selective media. Detailed descriptions of methods for isolating vibrios have been published elsewhere (18).

PATHOGENIC CHARACTERISTICS

As mentioned above, the pathogenic mechanism by which NT *V. cholerae* O1 strains cause illness in humans is unclear. The pathogenicity of toxigenic *V. cholerae* depends on a combination of properties including presence of a colonization factor (the TcpA pilus), production of CT, and presence of associated outer membrane proteins (12). After colonization by toxigenic *V. cholerae* O1, production of CT activates the membrane-bound adenylate cyclase of target enterocytes. The elevated levels of cyclicAMP in the mucosal cells of the small bowel cause a cascade that ultimately results in hypersecretion of chloride and bicarbonate into the intestinal lumen and hence a net osmotic efflux

Daniel C. Rodrigue • Department of Medicine, Division of Infectious Diseases, University of Southern California School of Medicine, Los Angeles, California 90033. *Tanja Popovic and I. Kaye Wachsmuth* • Division of Bacterial and Mycotic Diseases, National Center for Infectious Diseases, Centers for Disease Control and Prevention, Atlanta, Georgia 30333.

Table 1. Laboratory characteristics of 14 NT
V. cholerae O1 isolates, United States, 1977 to 1991

Characteristic	No. (%) with characteristic
Gram negative, fermenter	14 (100)
Growth on TCBS[a]	14 (100)
String test positive	14 (100)
Oxidase positive	14 (100)
Nitrate-to-nitrate reduction	14 (100)
L-Lysine decarboxylase	14 (100)
L-Ornithine decarboxylase	14 (100)
Voges-Proskauer positive	14 (100)
Gas from D-glucose	0 (0)
Acid production in sucrose	14 (100)
L-Arginine dihydrolase	0 (0)
Growth in 0% Nacl	14 (100)
Resistance to polymyxin B (50 U) ..	14 (100)
Hemolysis of sheep erythrocytes (tube test)	12 (86)
Susceptibility to classical phage IV	4 (29)
Production of VcA-3 phage	4 (29)

[a]TCBS, thiosulfate-citrate-bile salts-sucrose agar.

of water. The genes encoding the CT are located on the chromosome. Strains can possess one or more copies of the CT genes, which appear to be associated with a large transposable genetic element (9). The coordinate expression of CT, TcpA pilus, and outer membrane proteins is mediated by various genes such as *toxR* (8, 27).

Pathogenicity studies have given conflicting data regarding the ability of culture supernatant of NT strains to cause fluid accumulation in various animal models (25, 34–36). In one study, when volunteers were infected with NT *V. cholerae* O1 strains, they failed to develop either disease or protective immunity to challenge with a toxigenic strain (20). In another study, volunteers developed mild illness when infected with derivatives of toxigenic *V. cholerae* O1 in which structural genes for both the A and B subunits of CT had been deleted (16, 21). Yet another volunteer study demonstrated that NT derivatives (deletion mutants) that also lacked the *toxR* gene or the TcpA pilus did not cause diarrhea or colonize human volunteers (12). As shown in Table 2, some of the NT *V. cholerae* O1 strains received by CDC

were probe negative for the TcpA pilus and the *toxR* gene yet were isolated from persons with a diarrheal illness. However, the isolation of toxin-negative *V. cholerae* O1 from an ill person does not necessarily mean that this organism caused the illness. It appears that there are factors other than the O1 antigen and the CT gene complex that contribute to the pathogenicity of NT strains for humans. Some have postulated that the El Tor hemolysin/cytotoxin or a Shiga-like toxin may be part of the pathogenic mechanism of diarrhea due to NT *V. cholerae* O1 (26, 33, 34, 39). It is also possible that some NT strains have an entirely different pathogenic mechanism responsible for extraintestinal illnesses. Table 2 summarizes pathogenic properties of the U.S. isolates in this study.

EPIDEMIOLOGY

The Diarrheal Diseases Laboratory at CDC serves as a World Health Organization reference laboratory for identification, serotyping, and toxin determination of *V. cholerae* isolates. No NT *V. cholerae* O1 isolates from humans were sent to CDC for confirmation until 1977. Between January 1977 and January 1991, NT *V. cholerae* O1 isolates from 14 patients were confirmed at CDC. All of the strains were tested with a commercial monoclonal antibody kit for detection of *V. cholerae* O1 (1). All were biotype El Tor; 11 (79%) of 14 were serotype Inaba, and 3 were serotype Ogawa. NT *V. cholerae* O1 was isolated from feces in 10 cases, the biliary tract in 1 case, and other tissues or body fluids in 3 cases. Two of these cases have been reported previously (15, 23). Other enteric organisms besides NT *V. cholerae* O1 were isolated from 3 (30%) of 10 stool cultures (two intances of *Campylobacter fetus* and one of *Plesiomonas shigelloides*).

As shown in Table 3, seven patients were male, and seven were female. The median age was 46 years, with a range of 17 to 97 years. The median age of 10 patients with

Table 2. Genotypic characteristics of NT *V. cholerae* O1 strains

| Isolate | *toxR* gene | TcpA pili | *ctxA* PCR[a] | Sam[b] | Cytotoxicity[c] | | | VcA3 phage[d] | | ET[e] | RT[f] |
					CHO cells	Vero cells	HeLa cells	Production	Probe		
1165–77	−	−	−	+	−	+	+	−	−	8	E
1077–79	−	−	−	+	+	+	+	−	−	8	C
2741–80	−	−	−	−	+	+	+	+	ii	1	A (14[g])
884–82	−	−	−	−	+	+	+	−	vi	8	C
123–83	−	+	ND[h]	−	+	+	−	−	−	ND	A
468–83	−	−	−	−	+	−	−	−	−	9 (5[i])	D
917–84	−	+	−	−	+	+	+	+	ii	1	A (2[g])
1029–84	−	−	−	+	+	+	+	−	−	7	C
2483–85	ND	ND	−	+	+	+	+	−	−	8	C
2496–85	ND	ND	−	−	+	+	+	+	ii	1	A
2527–86	ND	ND	ND	ND	ND	ND	ND	−	ix	ND	D
2540–87	ND	ND	ND	−	+	+	−	+	vi	ND	A
2452–88	ND	ND	−	+	+	+	+	−	−	7	C
2569–88	ND	ND	−	ND	ND	ND	ND	−	−	7	C

[a]PCR, polymerase chain reaction for *ctxA* (10).
[b]SAM, immunocompetent sealed adult mouse model (25, 29).
[c]See reference 25.
[d]See reference 2.
[e]ET, electrophoretic type (6).
[f]RT, ribotype (1).
[g]Ribotype (28).
[h]ND, not determined.
[i]Electrophoretic type (37).

fecal isolates of NT *V. cholerae* O1 was 38 years compared to 73 years for the four patients with extraintestinal isolates. Thirteen (93%) of the 14 patients lived in coastal states. The majority of intestinal strains were isolated during the summer and fall months, with 7 (70%) of 10 intestinal strains coming from persons with onset of illness during the 5-month period July through November (Fig. 1).

Seven of 10 patients with stool isolates of NT *V. cholerae* O1 reported exposure to raw oysters (six ate oysters, and one gathered oysters as a fisherman), and 2 others traveled to Mexico during the week before onset of illness (Table 3). Both of the Ogawa strains from persons with known exposures were from persons who had traveled to Mexico. No information about other cases of infection with NT *V. cholerae* O1 occurring in Mexico at the time of exposure was available.

MOLECULAR EPIDEMIOLOGY

In a study of the genetic relationships of toxigenic and NT *V. cholerae* O1 strains iso-lated in the western hemisphere, Chen et al. used multilocus enzyme electrophoresis to show impressive diversity among NT isolates (6). However, they also showed that some NT isolates were indistinguishable from toxigenic ones, suggesting that these strains may be precursors or derivatives of the Gulf Coast clone of toxigenic *V. cholerae* O1. Supporting evidence for this concept was reported by Klontz et al., who showed that a toxigenic Gulf Coast isolate had only one copy of the CT gene (19); the Gulf Coast clone had previously been shown to have two copies of the gene (11). Further support came from a study of the vibriophage VcA3, which has a specific insertion site and a restriction enzyme digest pattern for all toxigenic Gulf Coast isolates (11). A study by Almeida et al. again demonstrated diversity among the NT *V. cholerae* O1 isolates and showed that some were indistinguishable from the toxigenic Gulf Coast clone (2).

As shown in Table 2, the U.S. isolates of *V. cholerae* O1 are diverse by at least three different molecular subtyping methods that

Table 3. Features of 14 patients from whom NT *V. cholerae* O1 was isolated, United States, 1977 to 1991

Isolate	Patient age (yr)	Patient	State	Diagnosis	Source	Probable exposure
1165–77[a,b]	61	Male	Alabama	Cholecystitis	Gallbladder	Raw oysters
1077–79[a,b]	46	Male	Louisiana	Leg ulcer	Wound	Raw oysters
2741–80[a]	46	Female	Florida	Gastroenteritis	Stool	Raw oysters
884–82[a,b]	22	Male	Florida	Gastroenteritis	Stool	Raw oysters
123–83[a]	17	Male	Louisiana	Gastroenteritis	Stool	Unknown
468–83[c]	23	Female	Pennsylvania	Gastroenteritis	Stool	Travel to Mexico
917–84[a]	48	Female	Georgia	Gastroenteritis	Stool	Raw oysters
1029–84[a]	31	Female	Florida	Gastroenteritis	Stool	Raw oysters
2483–85[a,b]	44	Male	California	Gastroenteritis	Stool	Raw oysters
2496–85[a]	68	Female	Florida	Gastroenteritis	Stool	Raw oysters
2527–86[c]	66	Female	Michigan	Gastroenteritis	Stool	Travel to Mexico
2540–87[a]	20	Male	Louisiana	Gastroenteritis	Stool	Oysters[d]
2452–88[a]	85	Male	Mississippi	Sepsis	Blood	Raw oysters
2569–88[a]	97	Female	Mississippi	Postsurgical wound drainage	Wound	Unknown

[a]El Tor, Inaba.
[b]El Tor, Ogawa.
[c]Sensitive to classical IV phage.
[d]Patient was a fisherman who harvested oysters.

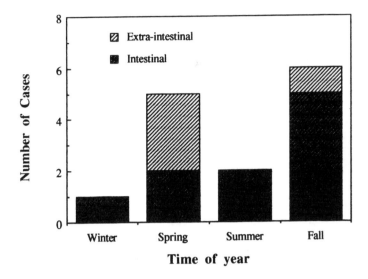

Figure 1. Seasonal distribution and isolation sites of 14 NT *V. cholerae* O1 infections, United States, 1977 to 1991.

have been used to study the molecular epidemiology of cholera (37). Previously published multilocus enzyme electrophoresis data (5) placed all toxigenic Gulf Coast *V. cholerae* O1 isolates in one electrophoretic type (ET 1); three of the NT strains (2741-80, 917-84, and 2496-85) fall into this electrophoretic type. The same three isolates are also indistinguishable from the toxigenic Gulf Coast clone by VcA3 pattern (ii) and by ribotype (RT A). The epidemiologic data presented in Table 3 show that the three isolates described above are all from stools of females with gastroenteritis associated with raw-oyster exposure. There were no data, however, to determine whether the ET 1 isolates were associated with a more severe clinical illness than the illness associated with the other electrophoretic types.

CLINICAL FEATURES

NT *V. cholerae* O1 isolates were associated with a broad spectrum of illnesses. Symptoms recorded in medical records and epidemiologic reporting forms showed that 6 (43%) of 14 patients were hospitalized but

that none died. All 10 patients whose stool specimens yielded NT *V. cholerae* O1 reported watery diarrhea (Table 4); abdominal cramps, nausea, vomiting, and fever were common associated symptoms. Seven (70%) of 10 patients were treated with antibiotics, but only 1 (10%) of 10 patients received intravenous fluids, and 2 (20%) required hospitalization. None of the illnesses were considered life threatening. The duration of illness was known for nine patients; the median duration was 7 days (range, 2 to 7 days). No histories of gastric surgery or underlying chronic illness were reported.

Table 4. Symptoms of 10 patients with NT *V. cholerae* O1 isolated from stool

Symptom	No. with data[a]	No. (%) with symptom
Diarrhea	10	10 (100)
Abdominal cramps	10	9 (90)
Nausea	10	8 (80)
Fever[b]	9	6 (67)
Vomiting	10	5 (50)
Mucus in stools	10	1 (10)
Blood in stools	10	0 (0)

[a]Data on some symptoms are unavailable for some patients.
[b]Documented > 101 °F (> 38.3 °C) in four patients.

All four patients with extraintestinal disease were hospitalized but not always as a consequence of infection with NT *V. cholerae* O1. The patient with the first reported extraintestinal case presented with right upper quadrant abdominal pain and had the organism isolated from his gallbladder on surgery for acute cholecystitis (5). The second patient reported on presented with fever and had an erythematous leg ulcer with serosanguinous drainage from which the organism was isolated (15). Of the remaining two patients, one was asymptomatic and had the organism isolated from serosanguinous drainage from an operative site after surgery for a broken hip, and the other presented with clinical sepsis and blood cultures that yielded NT *V. cholerae* O1. Three (75%) of four patients were treated with antibiotics. The durations of illness for two patients (cholecystitis and sepsis) were 12 and 7 days, respectively. No one reported recent foreign travel, chronic underlying illness, or use of antacids; the patient with sepsis had a Bilroth I surgery in 1966 and a cholecystectomy in 1980. Three (75%) of four patients reported eating raw oysters in the week before onset of illness; the exception was the 97-year-old woman with a postsurgical wound who had not eaten raw oysters for years.

NT *V. cholerae* O1 isolates were susceptible to a wide range of antimicrobial agents, including tetracycline, chloramphenicol, trimethoprim-sulfamethoxazole, nalidixic acid, gentamicin, ampicillin, and cephalothin. Thirteen (93%) of 14 isolates were resistant to colistin; 3 (21%) of 14 were resistant to sulfadiazine. Optimal antimicrobial treatment of NT *V. cholerae* O1 infections has not been established, since the small number of reported cases and the variety of antibiotics administered preclude reaching any definitive conclusions about their clinical efficacy. Tetracycline has been shown to decrease the duration of disease caused by toxigenic *V. cholerae* O1 (24, 38). In one small study of toxigenic *V. cholerae* O1, norfloxacin was reported to significantly decrease stool output and duration of diarrhea compared to trimethoprim-sulfamethoxazole or placebo (4).

SUMMARY

There have been a few clinical reports of illness associated with NT *V. cholerae* O1 organisms. Although the small number of cases makes it difficult to describe the clinical spectrum, 71% of patients from whom NT *V. cholerae* O1 was isolated reported gastrointestinal illness characterized by watery diarrhea, abdominal cramps, and nausea. Unlike toxigenic *V. cholerae* O1 (32), 29% of NT *V. cholerae* O1 strains isolated from patients were associated with extraintestinal infections. While it is possible that these organisms were commensal, the extraintestinal infections, such as the case of sepsis, strongly suggest that NT *V. cholerae* O1 can be a pathogen for humans. Definitive proof that these organisms can cause gastroenteritis may require further study.

Public health practitioners should regard NT *V. cholerae* O1 as a potential human pathogen but not as a pathogen with the same lethality and epidemic potential as toxigenic *V. cholerae* O1. The clinical spectrum and epidemiology of NT *V. cholerae* O1 infections are more analogous to those of non-O1 *V. cholerae*, which do not produce CT. It is likely, judging from molecular subtyping data, that some of these organisms represent either derivatives or precursors of toxigenic *V. cholerae* O1 strains, while some may be *V. cholerae* strains that cross-react in current serologic assays for the O1 antigen. Similar to diseases caused by many other *Vibrio* species, disease caused by NT *V. cholerae* O1 is associated with eating raw oysters. NT *V. cholerae* O1 have been isolated from estuarine waters, sediment, and shellfish along the Gulf Coast (7). People in coastal states should be aware of the risk of illness associated with eating raw shellfish (19); physicians should include these and other vibrios in the differential diagnosis of watery diarrhea.

REFERENCES

1. **Almeida, R. J.** 1990. Identification and differentiation of pathogenic and epidemic *Vibrio cholerae* O1 strains. Ph.D. dissertation. University of North Carolina, Chapel Hill.

2. **Almeida, R. J., D. N. Cameron, W. L. Cook, and I. K. Wachsmuth.** 1992. Vibriophage VcA-3 as an epidemic strain marker for the U.S. Gulf Coast *V. cholerae* O1 clone. *J. Clin. Microbiol.* **30:**300–304.

3. **Almeida, R. J., F. W. Hickman-Brenner, E. G. Sowers, N. D. Puhr, J. J. Farmer III, and I. K. Wachsmuth.** 1990. Comparison of a latex agglutination assay and an enzyme-linked immunosorbent assay for detecting cholera toxin. *J. Clin. Microbiol.* **28:**128–130.

4. **Bhattacharya, S. K., M. K. Bhattacharya, P. Dutta, D. Dutta, S. P. De, S. N. Sikdar, A. Maitra, A. Dutta, and S. C. Pal.** 1990. Double-blind, randomized, controlled clinical trial of norfloxacin for cholera. *Antimicrob. Agents Chemother.* **34:**939–940.

5. **Centers for Disease Control.** 1977. *Vibrio cholerae*—Alabama. *Morbid. Mortal. Weekly Rep.* **26:**159–160.

6. **Chen, F., G. M. Evins, W. L. Cook, R. Almeida, N. Hargrett-Bean, and I. K. Wachsmuth.** 1991. Genetic diversity among toxigenic and nontoxigenic *Vibro cholerae* O1 isolated from the western hemisphere. *Epidemiol. Infect.* **107:**225–233.

7. **Colwell, R. R., R. J. Seidler, J. B. Kaper, S. W. Joseph, S. Garges, H. Lockman, D. Maneval, H. Bradford, N. Roberts, E. Remmers, I. Huq, and A. Huq.** 1981. Occurrence of *Vibro cholerae* serotype O1 in Maryland and Louisiana estuaries. *Appl. Environ. Microbiol.* **41:**555–557.

8. **DiRita, V., C. Parsot, G. Jander, and J. J. Mekalanos.** 1991. Regulatory cascade controls virulence in *Vibrio cholerae. Proc. Natl. Acad. Sci USA* **88:**5403–5407.

9. **Falkow, S., and J. J. Mekalanos.** 1990. The enteric bacilli and vibrios, p. 561–587. *In* B. D. Davis (ed.), *Microbiology*, 4th ed. J. B. Lippincott Co., Philadelphia.

10. **Fields, P. I., T. Popovic, K. Wachsmuth, and Ø. Olsvik.** 1992. Use of polymerase chain reaction for detection of toxigenic *Vibrio cholerae* O1 strains from the Latin American cholera epidemic. *J. Clin. Microbiol.* **30:**2118–2121.

11. **Goldberg, S., and J. R. Murphy.** 1983. Molecular epidemiological studies of the United States Gulf Coast *Vibrio cholerae* strains: integration site of mutator vibriophage VcA-3. *Infect. Immun.* **42:**224–230.

12. **Herrington, D. A., R. H. Hall, G. Losonsky, J. J. Mekalanos, R. K. Taylor, and M. M. Levine.** 1988. Toxin, toxin coregulated pili, and the *toxR* regulon are essential for *Vibrio cholerae* pathogenesis in humans. *J. Exp. Med.* **168:**1487–1492.

13. **Honda, S. I., K. Shimoirisa, and A. Adachi.** 1988. Clinical isolates of *Vibrio cholerae* O1 not producing cholera toxin. *Lancet* **ii:**1486.

14. **Hood, M. A., G. E. Ness, and G. E. Rodrick.** Isolation of *Vibrio cholerae* serotype O1 from the eastern oyster, *Crassostrea virginica. Appl. Environ. Microbiol.* **41:**559–560.

15. **Johnston, J. M., L. M. McFarland, H. B. Bradford, and C. T. Caraway.** 1983. Isolation of nontoxigenic *Vibrio cholerae* O1 from a human wound infection. *J. Clin. Microbiol.* **17:**918–920.

16. **Kaper, J. B.** 1989. *Vibrio cholerae* vaccines. *Rev. Infect. Dis. Suppl* **3:**S568–S573.

17. **Kaper, J. B.** 1989. *Vibrio cholerae* vaccines. *Rev. Infect. Dis. Suppl.* **3:**S568–S573.

18. **Kelly, M. T., F. W. Hickman-Brenner, and J. J. Farmer III.** 1991. *Vibrio*, p. 384–395. *In* A. Balows, W. J. Hausler, Jr., K. L. Herrmann, H. D. Isenberg, and H. J. Shadomy (ed.), *Manual of Clinical Microbiology*, 5th ed. American Society for Microbiology, Washington, D.C.

19. **Klontz, K. C., R. V. Tauxe, W. L. Cook, W. H. Riley, and I. K. Wachsmuth.** 1987. Cholera after the consumption of raw oysters. *Ann. Intern. Med.* **107:**846–848.

20. **Levine, M. M., R. E. Black, M. L. Clements, L. Cisneros, A. Saah, D. R. Nalin, D. M. Gill, J. P. Craig, C. R. Young, and P. Ristaino.** 1982. The pathogenicity of nonenterotoxigenic *Vibrio cholerae* serogroup O1 biotype E. Tor isolated from sewage water in Brazil. *J. Infect. Dis.* **145:**296–299.

21. **Levine, M. M., J. B. Kaper, D. Herrington, J. Ketley, G. Losonsky, C. O. Tacket, B. Tall, and S. Cryz.** 1988. Safety, immunogenicity, and the efficacy of recombinant live oral cholera vaccines, CVD 103 and CVD 103 HgR. *Lancet* **I:**467–470.

22. **Morris, J. G., and R. E. Black.** 1985. Cholera and other vibrioses in the United States. *N. Engl. J. Med.* **312:**343–350.

23. **Morris, J. G., J. L. Picardi, S. Lieb, J. V. Lee, A. Roberts, M. Hood, R. A. Gunn, and P. A. Blake.** 1984. Isolation of nontoxigenic *Vibrio cholerae* O group 1 from a patient with severe gastrointestinal disease. *J. Clin. Microbiol.* **19:**296–297.

24. **Morris, J. G., Jr., J. H. Tenney, and G. L. Drusano.** 1985. In vitro susceptibility of pathogenic *Vibrio* species to norfloxacin and six other antimicrobial agents. *Antimicrob. Agents Chemother.* **28:**442–445.

25. **Moyenuddin, M., K. Wachsmuth, S. H. Richardson, and W. L. Cook.** 1992. Enteropathogenicity of non-toxigenic *Vibrio cholerae* O1 for adult mice. *Microb. Pathog.* **12:**451–458.

26. **O'Brien, A. D., M. E. Chen, R. K. Holmes, J. B. Kaper, and M. M. Levine.** 1984. Environmen-

tal and human isolates *Vibrio cholerae* and *Vibrio parahemolyticus* produce a *Shigella dysenteriae* 1 (Shiga)-like cytotoxin. *Lancet* **i**:77–78.

27. Peterson, K. M., R. K. Taylor, V. I. Miller, V. J. DiRita, and J. J. Mekalanos. 1988. Coordinate regulation of virulence determinants in *Vibrio cholerae*, p. 165–172. *In* M. Schwartz (ed.), *Molecular biology and infectious diseases.* Elsevier, Paris.

28. Popovic, T., C. A. Bopp, Ø. Olsvik, and I. K. Wachsmuth. 1993. A standardized ribotype scheme for *Vibrio cholerae* O1: epidemiologic application. *J. Clin. Microbiol.* **31**:2474–2482.

29. Richardson S. H., J. C. Giles, and K. S. Kruger. 1984. Sealed adult mice: new model for enterotoxin evaluation. *Infect. Immun.* **43**:482–486.

30. Sack, D. A., S. Huda, P. K. B. Neogi, R. R. Danial, and W. M. Spira. 1980. Microtiter ganglioside enzyme-linked immunosorbent assay for *Vibrio* and *Escherichia* heat-labile enterotoxins and antioxin. *J. Clin. Microbiol.* **11**:35–40.

31. Sack, D. A., and R. B. Sack. 1975. Test for enterotoxigenic *Escherichia coli* using Y1 adrenal cells in miniculture. *Infect. Immun.* **11**:334–336.

32. Safrin, S., J. G. Morris, M. Adams, V. Pons, R. Jacobs, and J. E. Conte, Jr. 1988. Non-O1 *Vibrio cholerae* bacteremia: case report and review. *Rev. Infect. Dis.* **10**:1012–1017.

33. Sanyal, S. C., K. Alam, P. K. B. Neogi, M. I. Huq, and K. A. Al-Mahmud. 1983. A new cholera toxin. *Lancet* **i**:1337.

34. Sanyal, S. C., P. K. B. Neogi, K. Alam, H. I. Huq, and K. A. Al-Mahmud. 1984. A new enterotoxin produced by *Vibrio cholerae* O1, *J. Diarrhoeal Dis. Res.* **2**:3–12.

35. Sigel, S. P., S. Lanier, V. S. Baselski, and C. D. Parker. 1980. In vivo evaluation of pathogenicity of clinical and environmental isolates of *Vibrio cholerae. Infect. Immun.* **28**:681–687.

36. Spira, W. M., D. A. Sack, P. J. Fedorka-Cray, S. Sanyal, J. Madden, and B. McCardell. 1986. Description of a possible new extracellular virulence factor in nontoxigenic *Vibrio cholerae* O1, p. 263–270. *In* S. Kuwahara and N. F. Pierce (ed.), *Advances in research on cholera and related diarrheas*, 3rd ed. KTK Scientific Publishers, Tokyo.

37. Wachsmuth, I. K., G. M. Evins, P. I. fields, Ø. Olsvik, T. Popovic, J. G. Wells, C. Carrillo, and P. A. Blake. 1993. The molecular epidemiology of cholera in Latin America. *J. Infect. Dis.* **167**:621–626.

38. Wallace, C. K., P. N. Anderson, T. C. Brown, S. R. Khanra, G. W. Lewis, N. F. Pierce, S. N. Sanyal, G. V. Segre, and R. H. Waldman. 1968. Optimal antibiotic therapy in cholera. *Bull W.H.O.* **39**:239–245.

39. Yamomoto, K., Y. Ichinose, N. Nakasone, M. Tanabe, M. Nagahama, J. Sakurai, and M. Iwanaga. 1986. Identity of hemolysins produced by *Vibrio cholerae* non-O1 and *Vibrio cholerae* O1, biotype El Tor. *Infect. Immun.* **51**:927–937.

Vibrio cholerae and Cholera: Molecular to Global Perspectives
Edited by I. Kaye Wachsmuth, Paul A. Blake, and Ørjan Olsvik
© 1994 American Society for Microbiology, Washington, DC 20005

Chapter 6

Molecular Basis for O-Antigen Biosynthesis in *Vibrio cholerae* O1: Ogawa-Inaba Switching

Paul A. Manning, Uwe H. Stroeher, and Renato Morona

Vibrio cholerae is divided into more than 130 O serogroups; however, only organisms of the O1 serogroup have so far been associated with cholera in humans. The other serogroups are usually referred to as non-O1 or noncholera vibrios. The O1 serotype exists as two biotypes, classical and El Tor, which can be distinguished by a number of characteristics. The two biotypes were originally differentiated by the ability of El Tor strains to produce a soluble hemolysin for sheep erythrocytes, but marked variability is seen in the hemolytic phenotype. Strains of the El Tor biotype can agglutinate chicken erythrocytes, whereas classical strains cannot. There are, however, a number of exceptions (25, 26, 78, 86, 87). The most accurate and reliable test is the susceptibility of classical *V. cholerae* to Mukerjee's group IV phages, to which El Tor strains are resistant (65, 69). Furthermore, the two biotypes can be distinguished by the fact that El Tor strains are resistant to polymyxin B.

The classical biotype is thought to have been responsible for the first six of the seven great cholera pandemics that have occurred since the early 19th century. Prior to the seventh pandemic, the El Tor biotype was thought not to cause cholera, since it was associated with only mild diarrhea. However, in 1962, with the outbreak of the seventh pandemic, originating in Southeast Asia, the disease caused by El Tor strains was defined as true cholera. Thus, it was clear that both *V. cholerae* biotypes are pathogenic for humans.

V. cholerae O1 strains of both biotypes have been further subdivided into three serotypes, designated Inaba, Ogawa, and Hikojima, grouped according to the structure of the O antigens on the lipopolysaccharide (LPS). The three serotypes have in common an antigenic determinant referred to as the A antigen. In addition, there are two specific antigens, B and C, that are expressed to various degrees on the different serotypes: Inaba strains express only C, while Ogawa strains express both B and C, although C is present in a much reduced amount compared to the amount in Inaba (13, 14, 84, 85, 89). Redmond et al. (85) showed that after absorption of an Ogawa antiserum with Inaba cells, 5 to 10% of activity against Ogawa cells remained, indicating an Ogawa-specific antigen termed B antigen. The third serotype, termed Hikojima, is extremely rare and unstable and has been described as expressing all three antigens in high amounts (71). The exact nature of Hikojima strains is unknown. It has been proposed that such strains may be segregating diploids or strains undergoing high-frequency serotype conversion (6–8).

Paul A. Manning, Uwe H. Stroeher, and Renato Morona • Department of Microbiology and Immunology, The University of Adelaide, Adelaide, South Australia 5005, Australia.

V. CHOLERAE LPS

The LPS of gram-negative bacteria is the most abundant molecule on the cell surface, where it provides a protective barrier to hydrophobic agents and detergents (76). The best and most extensively characterized LPS is that of *Salmonella typhimurium*, and studies with this organism together with studies with *Escherichia coli* have provided a model for LPS biosynthesis (79).

The chemical structure of LPS has been determined. It consists of three distinct regions: lipid A, the core oligosaccharide, and the O antigen (Fig. 1) (56). Lipid A forms part of the lipid bilayer of the outer membrane, and the core is subsequently attached to lipid A via an acid-labile linkage with keto-3-deoxy-D-mannose-octulosonic acid (KDO) (79). The outermost region is the O antigen, which is linked to the core oligosaccharide. It carries the heat-stable O-serotype specificity of gram-negative bacteria and is widely used as the basis of serology (43, 57). The O antigen usually consists of a polysaccharide or monosaccharide polymer of variable length.

The LPS of *V. cholerae* seems to fit this model structurally; however, details of the mechanisms of biosynthesis and assembly of LPS and its component parts are still very scant. The structure of lipid A has been determined, but many questions about the structures of the core and the O antigen are still unanswered (12, 80).

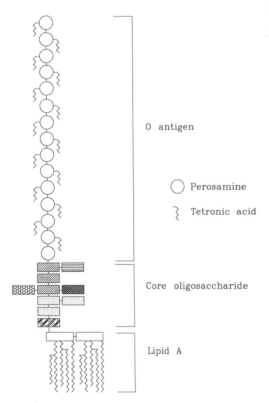

O antigen

◯ Perosamine

⌇ Tetronic acid

Core oligosaccharide

Lipid A

Figure 1. Schematic representation of *V. cholerae* O1 LPS, which comprises three distinct regions: lipid A, the core oligosaccharide, and the O antigen. The O antigen is a polymer of perosamine substituted with 3-deoxy-*L*-tetronic acid. The A, B, and C antigenic determinants are associated with the O antigen, but the nature of the structures they recognize is unknown.

Lipid A

The lipid A component of *V. cholerae* LPS consists of a β-1,6-linked D-glucosamine oligosaccharide that is substituted with a phosphate group ester bound to the nonreducing glucosamine residue and a pyrophosphorylethanolamine (PP-Etn) linked to C-1 of the reducing glucosamine residue (12, 80). Thus, the backbone structure of lipid A is P-GlcN(β 1-6)GluN-1-PP-Etn.

The glucosamine residues are further substituted at their hydroxyl and amino groups by a number of fatty acids (1). Three of these fatty acids, tetradecanoic, hexadecanoic, and 3-hydroxydodecanoic acid, are involved in ester linkages, whereas 3-hydroxytetradecanoic acid (3-hydroxymyristic acid) is involved in an amide linkage. The fatty acids found in lipid A occur in both Ogawa and Inaba serotypes as well as in non-O1 *Vibrio* strains (12). The exact positions of the ester-linked fatty acids are not known, but *V. cholerae* lipid A is similar to most lipid A species studied, although it differs from that of *Salmonella* spp. and other species in containing D-3-hydroxydodecanoic acid, which is 3-O-acylated by

D-3-hydroxydodecanoic acid. As in *Salmonella* spp., the lipid A is linked to KDO, although the existence of KDO has been a point of controversy (11).

Core Oligosaccharide

The core oligosaccharide of *V. cholerae* contains glucose, heptose, fructose, ethanolamine phosphate, and KDO (11). Initial investigations did not detect the presence of KDO in the core of *V. cholerae* (37). This can be explained, because this KDO can be released from the LPS only by harsh hydrolytic conditions, unlike the KDO in other species such as *Salmonella* spp., which is more readily released. At least one KDO sugar unit can be detected (11, 77). This finding supports the studies that have shown KDO or KDO-like material in other genera of the *Vibrionaceae* family (3, 36).

The remaining sugars found in the core region of the LPS form a backbone substituted at various points by sugars such as fructose, heptose, ethanolamine, and *N*-acetyl glucosamine. It has been reported that D-fructose is acid labile and may be the linking sugar between the core oligosaccharide and lipid A (42), but it now seems more likely that KDO is the linker. More recent data have shown that the fructose can be destroyed by periodate oxidation and therefore cannot be the linking sugar. Since fructose is released by mild acid hydrolysis, it seems likely that it branches off the core oligosaccharide backbone (49, 77).

Using the cloned *V. cholerae* O1 O-antigen biosynthesis genes, *rfb* (see below), Morona et al. (66) have obtained evidence supporting the proposed existence of glucose in the core oligosaccharide of *V. cholerae*. If the *rfb* genes of *V. cholerae* are introduced into *E. coli* strains lacking glucose in their cores, the production of *V. cholerae* O antigen attached to the *E. coli* core is eliminated. This is consistent with the reported presence of glucose in the *V. cholerae* LPS core oligosaccharide.

Properties and Composition of the O Antigen

The O polysaccharide or O antigen of *V. cholerae* has been examined by both chemical and physical means, including nuclear magnetic resonance spectroscopy. Together with the core oligosaccharide, O antigen can be separated from the lipid A portion of the molecule by mild acid hydrolysis. The lipid A fraction is chloroform soluble, whereas the polysaccharide fraction is water soluble. The polysaccharide moieties can then be separated on a Sephadex G-50 column, yielding two polysaccharide fractions, one of 10,000 molecular weight (MW) (the O antigen) and one of 1,000 MW (51). The 1,000-MW fraction can be refractionated to give a 1,000-MW fraction corresponding to the core oligosaccharide and a fraction containing D-fructose (32).

The same sugars appear to be present in all *V. cholerae* O1 O antigens regardless of serotype (81). The most common sugars found in the O polysaccharide are perosamine and quinovosamine. Earlier reports suggested that the O antigen also contained other sugars such as heptose, fructose, glucosamine, and ethanolamine (58, 81), but these sugars now appear to be present exclusively in the core (33). This understanding is based on the fact that the sugars of the core oligosaccharide are located at the reducing end of the perosamine, which is essentially the backbone of the unit that makes up the O-antigen polymer (33). Furthermore, rough strains that lack O antigen still contain glucose, fructose, heptose, and glucosamine (34, 77).

Perosamine is the major sugar in the O antigen (42, 82) and forms a polymer of approximately 17 or 18 units that has been shown by nuclear magnetic resonance spectroscopy to exist as a homopolymer. The perosamine is acylated with 3-deoxy-L-*glycero*-tetronic acid (52, 84). This substitution is common to both Inaba and Ogawa and has been proposed to correspond to the A antigen common to both serotypes (52). The perosamine can be further substituted at the C-3 position, whereas the tetronate can be substituted at C-2.

The unique sugar quinovosamine has as yet not been localized. Its ratio to perosamine suggests that it may be a "capping sugar" on either the distal or proximal end of the O antigen (42, 77).

The sugar 4-NH$_2$-4-deoxy-L-arabinose has so far been found only in the Ogawa serotype. The original observation by Redmond (83) has been both confirmed and disputed by a number of other workers, leaving some doubt as to the presence of this sugar (35, 48, 83). One explanation for this controversy may be that amino sugars are stable only in acid conditions and require modified thin-layer chromatography stains for visualization (83). The 4-NH$_2$-4-deoxy-L-arabinose of *Salmonella* spp. is released under mild acid hydrolysis, whereas the *V. cholerae* sugar is released only under very harsh hydrolytic conditions. The fact that 4-NH$_2$-4-deoxy-L-arabinose is found only in Ogawa strains suggests that it may be part of the Ogawa-specific epitope. Antibody absorptions using anti-B monoclonal antibody have proved negative; the possibility that 4-NH$_2$-4-deoxy-L-arabinose is involved in the B antigen cannot be ruled out (77).

A number of studies designed to correlate the various O-antigen polysaccharides with particular antigenic specificities have been carried out (28, 59, 92). These oligosaccharides have been generated by partial hydrolysis of either O antigen or O-antigen core oligosaccharides from Ogawa and Inaba strains. It has been shown that various oligosaccharides and monosaccharides can inhibit agglutination of *V. cholerae* cells by antiserum to *V. cholerae* LPS. However, in all cases, the oligosaccharides were better inhibitors than any of the constituent monosaccharides. The data indicate that glucuronic acid followed by glucosamine is the most effective inhibitor of agglutination and that these two are the immunodominant sugars, regardless of serotype. The glucosamine is

present in the core of *V. cholerae* O1, but as yet no glucuronic acid has been found in the LPS of *V. cholerae* O1. It is surprising to find that a sugar found exclusively in the core should be immunodominant, and the claim for the immunodominance of glucuronic acid needs to be looked at critically.

In studies with monoclonal antibodies to LPS, the A antigen appears to be present as multiple determinants; i.e., there is more than one antigenic site per LPS molecule (32, 38). In contrast, the B and C antigens appear to be single determinants. The low epitope density on the LPS molecule has hampered qualitative identification of the antigens. These data may explain why it has not been possible to precisely define the three antigenic determinants. Use of these antibodies in immunoelectron microscopy has also permitted the relative abundances of the A, B, and C antigens on Inaba and Ogawa strains to be determined (97) (Fig. 2).

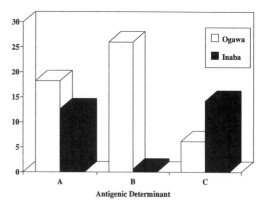

Figure 2. Quantitation of the relative levels of A, B, and C antigenic determinants by immunoelectron microscopy. Whole cells of either CA411 (classical, Ogawa) or 569B (classical, Inaba) were incubated with monoclonal antibodies to the A, B, or C determinants and then with protein A-gold complexes. The cells were then examined in the electron microscope, and the gold particles in random grids were counted. Relative amounts of the determinants are shown by an arbitrary scale on the *y* axis.

Genetics of O-Antigen Biosynthesis

The genes involved in O-antigen biosynthesis in *V. cholerae* O1 strains 569B (Inaba, classical) and O17 (Ogawa, El Tor) have been cloned and expressed in *E. coli* K-12 (60, 106). More recently, the locus associated with O-antigen biosynthesis has been mapped near *ilv* and *arg* on the *V. cholerae* chromosome (104) and corresponds to the locus associated with serotype specificity of the O antigen, *oag*, originally described by Bhaskaran et al. (6–9). Thus, the synthesis of the O antigen and the generation of the serotype-specific determinants are part of the same locus.

The genes responsible for O-antigen biosynthesis have been designated *rfb* according to nomenclature for other gram-negative pathogens (37) and localized to a 20-kb *Sac*I (*Sst*I) fragment of DNA (60, 106).

The complete nucleotide sequence of the 20-kb *Sst*I fragment has been determined for two strains, 569B (classical, Inaba) and O17 (El Tor, Ogawa). The region contains 21 complete open reading frames (ORFs); their organizations are shown in Fig. 3. Although these strains differ markedly in both site and date of isolation as well as in biotype and a number of other markers, the sequences show less than 0.1% difference. The only changes detected were a single base substitution in an intergenic region, a small deletion in *rfbR*, and a single base deletion in *rfbT* (98; see below for further discussion). Thus, it seems that this region is highly conserved or that, in evolutionary terms, it has been relatively recently acquired. Consistent with this latter possibility, the low overall G+C content of the region (about 39.1% compared to the *V. cholerae* average of about 49%) suggests that the region may in fact be derived from another source. This holds particularly true for *rfbT*, which has a G+C content of only 31.7%.

Comparison of the predicted protein sequences with those of known proteins in the various databases has enabled us to make predictions about their possible functions (Table 1).

Perosamine biosynthesis

The first four genes of the *rfb* region are thought to encode a perosamine biosynthesis pathway (50). RfbA is closely related to a series of proteins that have phospho-mannose isomerase or mannose-1-phosphate guanyl transferase activity, and it seems likely that it has both functions and probably uses fructose-6-phosphate as the starting molecule.

RfbB is a putative phospho-manno-mutase required to convert mannose-1-GDP to mannose-6-GDP.

RfbD is a putative oxidoreductase based on weak homology to a variety of proteins. Such a function is likely to be required for the subsequent synthesis of 4-keto-4,6-dideoxymannose-GDP, the putative precursor of perosamine, from mannose-6-GDP.

RfbE is the putative perosamine synthetase.

O-antigen transport

The RfbH and RfbI proteins show a high degree of homology to proteins associated with transport of capsular polysaccharides in *Haemophilus influenzae* and *E. coli* (61). In particular, RfbH is thought to be the inner-membrane transport protein that is energized, via ATP hydrolysis, by RfbI.

Tetronate biosynthesis

The products of *rfbK L M N O* are thought to constitute a pathway that generates 3-deoxy-L-*glycero*-tetronic acid (67), which provides the modification required to convert perosamine to the basic O-antigen subunit.

RfbK has striking similarities with a family of proteins known as acyl carrier proteins (ACPs). The relationship of RfbK to the ACPs involved in fatty acid chain and polyketide synthesis is of interest since the O-antigen component 3-deoxy-L-*glycero*-tetronic

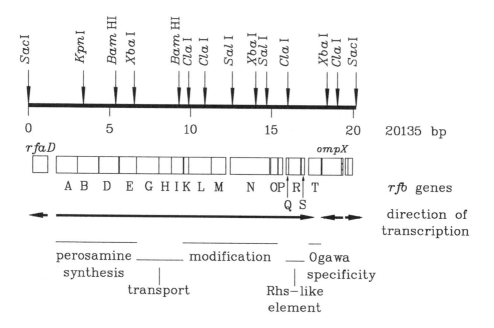

Figure 3. Genetic organization of the 20-kb *Sac*I (*Sst*I) fragment of the chromosome encoding the *rfb* region responsible for O-antigen biosynthesis and assembly. Boxes indicate the extents of the various genes, and horizontal arrows show directions of transcription. The predicted functions of the regions within *rfb* are shown at the bottom of the diagram, and a partial restriction map of the entire fragment is shown at the top. The sequence of the entire region from the El Tor Ogawa strain is available in the EMBL/GenBank/DDBJ database under the accession number X59554 (ID=VCRFBAT).

acid can be considered a molecule with a structure similar to that of the substrates used in those synthetic reactions. Additionally, such ACPs are involved in the synthesis of molecules that are polymerized. The location of an ACP in the *rfb* region encoding the biosynthesis of an O antigen has not been previously reported and emphasizes the unique nature of the *V. cholerae* O1 O antigen.

RfbL has homology with a number of prokaryotic and eukaryotic enzymes (67) that have been recognized as belonging to a superfamily of adenylate-forming enzymes (102). RfbL possesses the conserved sequence elements common to a large number of these enzymes, which suggests that it, too, is an adenylate-forming enzyme that may act on a carboxylic acid substrate to convert it to a coenzyme A derivative that is subsequently used for O-antigen biosynthesis.

RfbM shows homology to a family of iron-containing, long-chain alcohol dehydrogenases (67) but is unrelated to the well-characterized eukaryotic family of zinc-containing alcohol dehydrogenases (46). The strong identity of RfbM with these enzymes indicates that it, too, is an alcohol dehydrogenase that may act on a long-chain aldehyde to convert it to an alcohol that is then a precursor for O-antigen synthesis.

RfbN encodes a large protein that shows marked homology to two proteins, the N terminus to LuxC and the C terminus to LuxE of *Vibrio harveyi* (45, 63, 67), which together with LuxD form a three-component multienzyme complex termed the long-chain aldehyde reductase complex (62). These enzymes act sequentially on a long-chain fatty acid substrate to produce the long-chain aldehyde. The RfbN protein, by homology

Table 1. Possible functions of *V. cholerae* gene products in the vicinity of the *rfb* region

Protein	No. of amino acids	Size (kDa)	Predicted function	Homology to[a]	Accession no. in Swiss-Prot or PIR database	Reference(s)
Orf1 (RfaD)	410	46,836	ADP-*L-glycero-D*-mannoheptose epimerase	*Escherichia coli rfaD*	P17963	18
RfbA	465	51,913	Phospho-mannose isomerase/mannose-1-phosphate guanyltransferase	*Pseudomonas aeruginosa pmi*	A25638	20
				Salmonella typhimurium cpsB	P26340	44
				Salmonella typhimurium rfbM	P26404	96
RfbB	464	51,803	Phospho-manno-mutase	*Escherichia coli cpsG*	P24175	19
RfbD	373	42,051	Dehydratase/oxidoreductase/epimerase	*Salmonella typhimurium rfbBEG*	P26391 P14169 P26397	96, 103
RfbE	367	41,009	Perosamine synthetase	*Saccharopolyspora erythraea erbS*	P14290	21
RfbG	463	53,137				
RfbH	257	29,912	O-antigen transport channel?	*Escherichia coli kpsM*	P23889	95a
				Haemophilus influenzae bexB	P19390	53
RfbI	250	27,561	O-antigen transport energizer?	*Escherichia coli kpsT*	P23888	95a
				Haemophilus influenzae bexA	P10640	53
RfbK	80	8,859	Acyl carrier protein	*Escherichia coli acp*	P02901	19
RfbL	471	52,569	Coenzyme A-binding protein	4-Coumerate-coenzyme A ligase	A39827	5
RfbM	374	41,227	Alcohol dehydrogenase	*Clostridium acetoacetobutylicum* butanol dehydrogenase	P13604	107
RfbN	825	91,857	Fatty acid (acylcoenzyme A) reductase	*Vibrio harveyi luxCE*	P08639 P14286	45, 63
RfbO	188	20,015	Tetronyltransferase	*Bacillus sphaericus cat*	P26840	64
RfbP	73	8,073				
RfbQ	65	7,576	Rhs element	*Escherichia coli rhsB*	P28912	108
RfbR	234	25,845	Rhs element	*Escherichia coli rhsB*	P28912	108
RfbS	68	7,900	Rhs element	*Escherichia coli rhsB*	P28912	108
RfbT	287	32,917	Ogawa determination			
OmpX	410	46,836	Outer membrane porin?			
Orf2	65	7,880				
Orf3	95	10,210				

[a]Only some of the homologies are shown.

with LuxE and LuxC, apparently has both their enzymatic activities in one polypeptide chain. Thus, the aldehyde produced by RfbN may be a substrate for RfbM, the alcohol dehydrogenase. Since in both cases the substrate seems to be a long-chain acyl compound, these enzymes may be involved in the synthesis of the *V. cholerae* O-antigen component 3-deoxy-L-*glycero*-tetronic acid.

RfbO is thought to provide the tetronyltransferase activity that transfers the tetronate onto perosamine (67). This conclusion is based on its homology to a variety of acetyltransferases.

Rhs-related region

The region encoding *rfbQ*, *rfbR*, and *rfbS*, which is adjacent to *rfbT* encoding the Ogawa specificity modification (see below), shows a high degree of similarity to Rhs elements of *E. coli* K-12 (108), in particular RhsB and RhsE. A single peptide, designated ORF-H, is associated with both of these regions. Its function is unknown, but Rhs elements have been associated with large chromosomal rearrangements. ORF-H has 51% identity with RfbQ, RfbR, and RfbS, and it appears that three peptides are required in *V. cholerae* to do what ORF-H does in *E. coli*. It is interesting to speculate that this region plays no role in O-antigen biosynthesis but has basically served to place *rfbT* in juxtaposition to the other *rfb* genes. This possibility is supported by the observation that 8- or 1-bp deletions in *rfbR* have been detected; they lead to severely truncated RfbR proteins in different strains without a concomitant effect on LPS or O-antigen biosynthesis (98, 99).

The functions of the remaining *rfb* genes are unknown, but presumably they encode the O-antigen polymerase and transferases required for the assembly of the O antigen onto the LPS molecule.

Are core and O-antigen genes linked?

Another important homology that was detected is that of Orf1 (RfaD) to the RfaD

protein of *E. coli* (97). The *E. coli* RfaD is an ADP-L-*glycero*-D-mannoheptose epimerase (18) and is one of the critical proteins involved in the synthesis of the core oligosaccharide of the LPS in *E. coli*. In *V. cholerae*, Orf1 (RfaD) is encoded divergent to the *rfb* region (Fig. 3). In addition, Ward and Manning (105) mapped a transposon insertion that leads to a defective LPS phenotype to the *Sac*I fragment immediately to the left of the 20-kb region shown in Fig. 3. Together, these data suggest that at least some core oligosaccharide biosynthesis functions may be linked to those for the O antigen in *V. cholerae*. Such an organization has not been observed in *E. coli* or *Salmonella* spp.

Although homology to some proteins associated with LPS biosynthesis in other organisms, especially *Salmonella* spp., has been found, the most marked homologies are to proteins that are required for the synthesis of capsular polysaccharides. This suggests that perhaps *V. cholerae* O-antigen biosynthesis has more in common with the biosynthesis of capsular polysaccharides than of O antigens of organisms like *S. typhimurium*. Consequently, studies in this system may provide insights into transport of such complex polysaccharides across the membrane that can complement studies of both capsules and LPS. These benefits may also extend to the study of the synthesis of glycoproteins in higher cells where there is a lipid carrier, dolichol, similar to that, bactoprenol or undecaprenol-phosphate, involved in capsule and LPS synthesis.

SEROTYPE CONVERSION

Two serotypes were initially described for *V. cholerae* serogroup O1 (47). These are now designated Inaba and Ogawa but were originally called J and F forms, respectively. A third serotype, called Hikojima, which carries all three antigens (A, B, and C) was subsequently described (27, 71). *V. cholerae* O1 strains are not fixed but can undergo serotype conversion or switching between the Inaba and Ogawa serotypes. This conversion is nonreciprocal and occurs at a frequency of approximately 10^{-5} for the Ogawa-to-Inaba conversion but significantly less for the converse (7, 8).

History of Serotype Conversion

Serotype conversion was first observed by Kabeshima in 1918, although no mechanism was proposed (47). Serotype conversion occurred as a result of exposure of the organism during growth to antiserum directed against its serotype (Ogawa). Since these original studies were carried out prior to a firm definition of Inaba and Ogawa, the experiments have been modified and repeated. In 1947, Shrivastava and White (95) reported the isolation of Inaba strains from Ogawa cells grown in the presence of anti-Ogawa serum. The converse experiment of growing Inaba cells in anti-Inaba serum gave rise only to rough strains. Thus, it appeared that only the serotype conversion from Ogawa to Inaba could be detected.

Until 1966, serotype switching had been reported only in vitro. Any reports of in vivo serotype conversions were dismissed as multiple infections of patients or as reinfections with another serotype. In 1967, Gangarosa et al. reported Ogawa-to-Inaba switching (26). After a patient had been infected for 8 days with V. cholerae biotype El Tor serotype Ogawa, Inaba organisms were isolated from the patient's stool. Isolation of Inaba organisms was followed by a relapse, suggesting that the organism had multiplied in the intestine and that a change in serotype enabled the host immune response to be evaded. In all, the patient excreted all three serotypes within 10 days of contracting cholera, and all putative Hikojima strains isolated from this patient were subsequently typed as Inaba. This supports the highly unstable nature of the Hikojima serotype and suggests that it may be some intermediate form.

Serotype switching from Inaba to Ogawa was first reported in 1966 (94). In a laboratory that used Inaba serotype strains exclusively, a worker acquired an infection and after 3 days began to excrete Ogawa strains. The reason the Inaba-to-Ogawa switch had not been reported earlier as occurring either in vitro or in vivo may be that there is a much lower rate of conversion from Inaba to Ogawa than the reverse. Thus, it may be easier to detect Inaba-to-Ogawa serotype convertants in humans in whom the V. cholerae multiply to large numbers and presumably a selection against the original serotype can occur via the immune response.

Studies using germfree mice confirmed the ability of V. cholerae to undergo serotype conversion (88). In particular, this demonstrated the ability of V. cholerae not only to change serotype from Inaba to Ogawa and vice versa but also to change from smooth to rough strains lacking O antigen and then back to smooth strains, usually of the same original serotype. Thus, it appears that V. cholerae can change serotype and also the presence or absence of O antigen. In addition, the immune response of the mouse host appeared to provide the selective pressure necessary for serotype switch detection: if antibody to one serotype was present, the change to another serotype was detected earlier than in the absence of antibody, and if immunosuppressive drugs such as cyclophosphamide, which inhibits antibody formation, were given, serotype conversion was not detected or was significantly reduced. These data indicate that serotype convertants are selected for by the specific antibodies, a result consistent with those of earlier in vitro studies (72, 95).

Despite our ability to detect Inaba-to-Ogawa switching in vivo, it has remained difficult to demonstrate Inaba-to-Ogawa serotype conversion in vitro. Sakazaki and Tamura (89) attempted to show this change by looking at 13 strains (6 Ogawa, 6 Inaba, and 1 Hikojima strain). The bacteria were grown in the presence of monospecific antibodies to either Inaba or Ogawa. Of the Ogawa strains tested, all showed a number of cells changing to Inaba, Hikojima, or rough strains. The Inaba strains showed no switching, although some rough cells were isolated. Thus, it seems likely that either the rate of Inaba-to-Ogawa conversion is extremely low (89) or that only some Inaba strains can convert to Ogawa.

The significance of serotype conversion to the spread and persistence of cholera epidemics has not been demonstrated. Data to support the notion that serotype conversion is important come from observations of the epidemic that broke out in Latin America in 1991. Extensive biochemical analyses and rRNA restriction fragment length polymorphism (RFLP) analysis has shown that the epidemic strain, an El Tor Inaba, is unique to Latin America (15, 90). However, Ogawa isolates that were identical to the epidemic strain in all other respects began to appear in about the seventh month of the epidemic, suggesting that the epidemic strain had undergone a serotype conversion, possibly because of immune pressure in the population.

Molecular Basis of Serotype Specificity and Serotype Conversion

Ogg et al. have suggested that serotype conversion may be mediated by a lysogenic conversion of *V. cholerae* by the bacteriophage CP-T1 (73, 74). It has been proposed that CP-T1 becomes lysogenic upon infection in a manner similar to that of phages that infect *Shigella flexneri* and *S. typhimurium* (17, 40, 41, 55). Like the phages that modify the LPS of *S. flexneri* and *S. typhimurium*, phage CP-T1 uses the O antigen of the LPS as its receptor and can infect both Inaba and Ogawa strains (29). The proposed CP-T1 lysogens that are resistant to phage killing are presumed to be serotype-converted cells (4, 73, 74). However, this seems unlikely to be the reason for serotype-conversion in *V. cholerae*, since CP-T1 does not form lysogens in either Inaba or Ogawa strains (29–31). It appears to be more likely that any serotype-converted strains detected either were spontaneous Ogawa-to-

Inaba converts or, alternatively, arose because of the ability of CP-T1 to mediate generalized transduction (75). In this case, it is possible that the phage transduce the serotype from one strain to another, although at a low frequency. Furthermore, any rough strains detected are likely to be spontaneous CP-T1-resistant mutants.

Using the cloned DNA, it was not possible to differentiate between the *rfb*-encoding regions of Inaba and Ogawa strains by either heteroduplex analysis by electron microscopy or restriction analysis and Southern DNA hybridizations of the chromosomal DNA (106). These data suggest that the genetic determinant or event associated with serotype conversion is subtle, since this region was all that was required for producing O antigen of the correct serotype specificity when it was cloned into *E. coli* K-12 (60, 66, 68, 106) (Fig. 4).

Genetic complementation studies have suggested that the determinant responsible for Ogawa specificity lies at the distal end of the *rfb* region, an area in which no readily detectable differences could be discerned (60, 68, 106). However, when the entire 20-kb *rfb* region was sequenced from both an Inaba and an Ogawa strain, two significant variations with the potential to affect gene function were identified (98): one was an 8-bp deletion in *rfbR*, and the other was a single base deletion in *rfbT*. When a variety of other strains of both biotypes and serotypes were examined, the only differences that could be correlated with serotype were sequence changes in Inaba strains only. These changes, which would in all cases result in a truncated RfbT protein, were variable and ranged from single base substitutions and 1-bp deletions to larger deletions (15, 98, 99). Construction of a defined *rfbT* mutation in vitro and its reintroduction into the *V. cholerae* chromosome indicated that it was only necessary to mutate *rfbT* to convert a strain from Ogawa to Inaba and that the Ogawa *rfbT* gene was sufficient to convert Inaba strains to Ogawa. Thus, Inaba strains are *rfbT* mutants of an Ogawa strain. Presumably, in the absence of RfbT, a specific epitope is not added to the LPS molecule, and the Inaba serotype is the absence of that epitope. In other words, the C epitope is whatever is exposed when the RfbT modification is removed from the B epitope.

The *rfbT* gene is encoded by a separate transcript within the *rfb* region with its own promoter located about 100 bp upstream from the initiation codon (98). The RfbT protein fractionates with the cytoplasmic membrane, and analysis of the protein sequence suggests that it contains two membrane-spanning domains. Thus, it may be a trans-cytoplasmic-membrane protein that is exposed in both the cytoplasm and the periplasm.

During LPS biosynthesis in organisms like *Salmonella* and *Shigella* spp. and *E. coli*, the O-antigen subunit is thought not to be transferred to the lipid A plus core until it is complete (79). However, additional phage-encoded modifications that are able to alter the O-antigen structure are known (4, 17). These may be added immediately prior to or during LPS export to the outer membrane. How RfbT functions in modifying the basic O-antigen structure (Inaba-like) to generate the Ogawa specificity is unknown, but because of its apparent dispensability without affecting O-antigen biosynthesis, we are tempted to predict that it produces its modification in a manner similar to these phage-encoded modifications, namely, in the terminal stages of LPS synthesis and export.

The various strains resulting from serotype conversion reported in the literature are no longer available, but it is now possible to understand why the interconversion between the Inaba and Ogawa serotypes appears to be nonreciprocal. An Ogawa strain can be converted to an Inaba by any change that leads to a defective RfbT protein, whereas an Inaba can switch to Ogawa only if the specific *rfbT* mutation in that strain is precisely reverted. Simple probability analysis indicates that this is likely to result in at least a 1,000-fold difference in the two switching frequencies. Also, Inaba strains resulting from deletions may not revert at all. Thus, it was of particular interest to assess the nature of the change that occurred in the Latin American Inaba

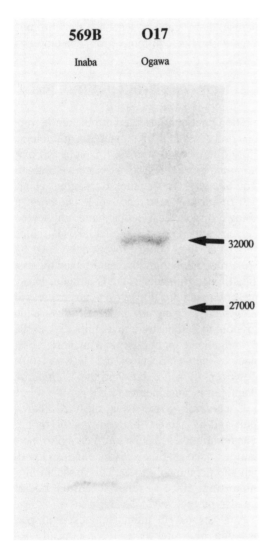

Figure 4. Identification of the RfbT protein in El Tor Ogawa strain O17 and classical Inaba strain 569B. An antiserum to the Ogawa RfbT protein was generated by using purified RfbT$_{Ogawa}$ protein. This antiserum was used to identify the RfbT proteins by Western blot (immunoblot) analysis of sodium dodecyl sulfate-polyacrylamide gels transferred to nitrocellulose filters. The Ogawa protein has an MW of 32,000, and this Inaba protein is truncated to 27,000 as the consequence of a single base deletion that results in a frameshift leading to translation termination.

strains that gave rise to the Ogawa form during the South American cholera epidemic in 1991. As predicted, the Ogawa variant that arose, presumably from an isogenic Inaba epidemic strain, had acquired a single base at the precise position required to convert $rfbT_{Inaba}$ to $rfbT_{Ogawa}$ (15).

The nature of Hikojima strains remains a mystery. There is no genetic evidence to suggest that they differ in any way from Ogawa strains except that they are unstable (7, 8). One possibility is that there are factors that affect *rfbT* expression and consequently could lead to increased or decreased levels of enzyme activity, which in turn could similarly alter the level of expression of the B and C determinants. In this regard, it is worth pointing out that the *rfbT* mRNA has an extended untranslated 5' region (98) that is presumably involved in mRNA stability or translation efficiency and may interact with some other as-yet-undefined regulatory factor. The fact that the region, which encodes *rfbQ*, *rfbR*, and *rfbS* and immediately precedes *rfbT*, is related to the Rhs elements involved in chromosomal rearrangements and duplications (54, 108) may give a clue. Such an element could lead to a duplication producing a merodiploid, as predicted by Bhaskaran in his early studies (6–8).

Serotype as an Epidemiologic Marker?

Serotyping clinical isolates has often been used as a tool for epidemiologic analysis, but the very limited number of subtypes of *V. cholerae* O1 strains permits little differentiation. In addition, the possibility of serotype switching, even though it does not occur with the same frequency in both directions, makes serotype by itself a relatively unreliable marker. However, if the strain of interest is an Inaba with an unusual mutation within *rfbT*, then this particular mutation could be a useful marker, especially when coupled with more complex biochemical, RFLP, or rRNA markers. This would require the particular *rfbT* mutation to be sequenced, but because

of the highly conserved nature of the *rfb* region and newly developed simplified DNA sequencing technology, sequencing is not a difficult task. In fact, oligodeoxynucleotide primers to be synthesized to directly polymerase chain reaction sequence the chromosome without a requirement for gene cloning.

ROLE OF O ANTIGEN IN VIRULENCE

The O antigen has been implicated in various processes in the pathogenesis of cholera. From a number of studies, it is apparent that the O antigen represents a protective antigen (16, 60, 70). The studies by Chitnis et al. (16) of in vitro adherence of *V. cholerae* in isolated rabbit ileal loops have implicated LPS in adherence. Booth et al. (10) also found that they could prevent adherence by using monoclonal antibodies directed against specific determinants of the O antigen. However, by using the cloned *rfb* genes expressed in *E. coli* to generate antisera, it has been possible to demonstrate that antibodies to the O antigen are protective and in part cross-reactive between Inaba and Ogawa (60). However, *E. coli* expressing the *V. cholerae* O antigen is nonadherent (105).

V. cholerae possesses a sheathed flagellum, and by using O-antigen-specific monoclonal antibodies, Fuerst and Perry (24) have demonstrated that the sheath contains O antigen and presumably LPS. This association is interesting, since flagellar sheath antibodies are also protective (2, 22, 39).

Analysis of *rfb* mutants blocked in perosamine biosynthesis and consequently totally defective in O-antigen biosynthesis has revealed a complex phenotype (100). These mutants are poorly motile and by electron microscopy their flagella seem abnormal. They also show few or no toxin-coregulated pili (93, 101) but have instead a pool of unassembled TcpA (TcpA is the major structural subunit or pilin of which toxin-coregulated pili are composed [23]) associated with the cytoplasmic membrane/periplasm. These effects suggest that *rfb* mutants are probably defective in the assembly of a number of membrane components and in membrane integrity. Consequently, it is not surprising that these mutants are attenuated, with 50% lethal doses that are 100- to 1,000-fold higher than those of their O-antigen-producing parent strain.

Studies using the infant mouse cholera model suggest that although the presence of the O antigen is important for a fully virulent organism, the serotype appears to be irrelevant to virulence per se (98). However, one cannot rule out the possibility that the serotype plays a role in avoiding the host immune response, especially because of the ability to select mutations in the gene (*rfbT*) determining Ogawa serotype specificity without losing the ability to produce O antigen.

CONCLUDING REMARKS

The biosynthesis of the LPS of *V. cholerae* O1, and especially the O antigen, is complex and shows some significant differences from other well-characterized systems. It is clear that the O antigen represents an important protective antigen that should be considered when vaccines against cholera are developed. The understanding we now have of serotype specificity and switching simplifies this work. Although there is some cross-protection between the serotypes, primarily owing to the A antigen (60), it is still highly desirable to immunize against both serotypes. Thus, attenuated candidate vaccine strains that have already undergone extensive testing in humans could be simply converted, by allelic exchange, to the alternative serotype. Since it has also been shown that serotype specificity has no effect on virulence, this switch does not require extensive retesting of the strains in order to gain approval for their use in humans.

The fact that all the genes for O-antigen biosynthesis are at a single locus has many benefits for future genetic analysis. This grouping may help us determine the nature of the serotype-specific determinants and elucidate the mechanism of O-antigen synthesis and transport in *V. cholerae* O1.

Acknowledgments. We gratefully acknowledge the support of F. H. Faulding through Enterovax Limited as well as that of the Australian Research Council, the National Medical and Research Council of Australia, and the Clive and Vera Ramaciotti Foundations.

ADDENDUM IN PROOF

In late 1992, apparently for the first time in recorded history, an epidemic of cholera-like disease mediated by a *V. cholerae* non-O1 type began in India and Bangladesh and effectively replaced O1 strains in the epidemic zone (16a). Independent isolates of this non-O1 strain were relatively uniform in their characteristics and antigenically identical but unrelated to the previously defined 138 O serotypes. It was consequently designated O139 and given the synonym Bengal because of the initial sites of its isolation along the coastal regions of the Bay of Bengal. Although these isolates bore the typical biochemical characteristics of *V. cholerae* O1 El Tor strains as well as genetic similarity/identity to a variety of well-characterized O1 virulence determinants, there was clearly a difference in their surface polysaccharides.

Analysis of serogroup O1 and O139 isolates revealed marked differences in their LPSs (Fig. A1). Whereas a distinct ladder corresponding to lipid A plus core oligosaccharide plus O antigen can be detected in O1 strains, no O-antigen-substituted material can be detected in the O139 strains. Instead, LPSs from O139 strains produce a doublet pattern with electrophoretic mobility slightly slower than that of the lipid A plus core oligosaccharide region of O1 strains, although there are lesser amounts of comigrating material. Thus, O139 strains appear to have a modified core structure; whether this altered mobility is due to phosphorylation of core sugars or the substitution of the core with additional sugars has not yet been determined. However, it is clear that O139 strains possess only what is usually referred to as a semi-rough LPS and not a smooth LPS like O1 strains.

Because of the changes in O-antigen expression, it was of interest to assess the changes within the *rfb* region encoding O-antigen biosynthesis. Southern hydridization analysis with *V. cholerae* O1-specific *rfb* gene probes implied that the genes *rfbA*, *-B*, *-D*, *-E*, *-G*, *-H*, *-I*, *-K*, *-L*, *-M*, *-N*, *-O*, *-P*, and *-T* are deleted (see this chapter), whereas *rfaD* involved in core oligosaccharide biosynthesis and *rfhQ*, *-R*, and *-S* were still present. Thus, all the genes required for O1 O antigen and the Ogawa modification were absent. The retention of *rfbQ*, *-R*, and *-S* was particularly interesting because of its relatedness to the Rhs elements associated with chromosomal rearrangements in *E. coli* (see this chapter for discussion). The *rfbQ*, *-R*, and *-S* region is flanked, in an inverted repeat orientation, by a 17-bp sequence which can be used as a simple primer to amplify the entire Rhs-like element by the polymerase chain reaction. Amplification of the *rfdQ, -R,* and *-S* region followed by restriction with *Hind*III and electrophoresis through a 1% agarose gel indicated that this element is approximately 50 bp larger in O139 strains, implicating it in the events which have occurred in the vicinity of the *rfh* region.

However, although LPS changes are evident in the O139 strains, it seems more likely that the actual O139 antigen is a colitose-containing capsular polysaccharide layer which can be visualized by electron microscopy extending out from the cell surface in cell sections (36a, 44a, 106a). Thus, it is clear that the presumed derivation of the O139 clone from an ancestral O1 El Tor strain has required multiple genetic events. Whether this capsule is encoded by a pre-existing pathway in O1 strains which has been activated or whether it has come in via a temperate bacteriophage or some other means remains to be elucidated. In any case, it would seem that the selection pressure for cholera-inducing *V. cholerae* to change serotype is great and that the arrival of O139 heralds the beginning of the eighth pandemic (100a).

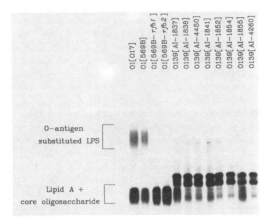

Figure A1. Analysis of LPSs by using proteinase K-treated whole-cell lysates and silver staining following sodium dodecyl sulfate-polyacrylamide gel electrophoresis. The serotypes of the strains are shown and the corresponding strains are shown in square brackets. 569B-*rfh1* and 569B-*rfb2* and two defined *rfb* mutants of the O1 strain 569B and lack the O antigen.

REFERENCES

1. **Armstrong, J. L., and J. W. Redmond.** 1973. The fatty acids present in the lipopolysaccharide of *Vibrio cholerae* 568B (Inaba). *Biochim. Biophys. Acta* **348:**302–305.
2. **Attridge, S. R., and D. Rowley.** 1983. The role of the flagellum in the adherence of *Vibrio cholerae. J. Infect. Dis.* **147:**873–881.
3. **Banoub, J. H., and H. J. Hodder.** 1985. Structural investigation of the lipopolysaccharide core

isolated from a virulent strain of *Vibrio ordalii. Can. J. Biochem. Cell. Biol.* **63:**1199–1205.
4. **Barksdale, L., and S. B. Arden.** 1974. Persisting bacteriophage infections, lysogeny and phage conversions. *Annu. Rev. Microbiol.* **28:**265–299.
5. **Becker-Andre, M., P. Schulze-Lefert, and K. Hahlbrock.** 1991. Structural comparison, modes of expression, and putative cis-acting element of the two 4-coumerate-CoA ligase genes in potato. *J. Biol. Chem.* **266:**8551–8559.
6. **Bhaskaran, K.** 1959. Observations of the nature of genetic recombination in *Vibrio cholerae. Indian J. Med. Res.* **47:**253–260.
7. **Bhaskaran, K.** 1960. Recombination of characters between mutant stocks of *Vibrio cholerae* strain 162. *J. Gen. Microbiol.* **23:**47–54.
8. **Bhaskaran, K., and R. H. Gorrill.** 1957. A study of antigenic variation in *Vibrio cholerae. J. Gen. Microbiol.* **16:**721–729.
9. **Bhaskaran, K., and V. B. Sinha.** 1971. Transmissible plasmid factors and fertility inhibition in *Vibrio cholerae. J. Gen. Microbiol.* **69:**89–97.
10. **Booth, B. A., C. V. Sciortino, and R. A. Finkelstein.** 1985. Adhesins of *Vibrio cholerae,* p. 169–182. *In* D. Mirelman (ed.), *Microbial Lectins and Agglutinins.* John Wiley & Sons, New York.
11. **Brade, H.** 1985. Occurrence of 2-keto-deoxyoctonic 5-phosphate in lipopolysaccharides of *Vibrio cholerae* Ogawa and Inaba. *J. Bacteriol.* **161:**795–798.
12. **Broady, K. W., E. Rietschel, and O. Lüderitz.** 1981. The chemical structure of the lipid A component of lipopolysaccharides from *Vibrio cholerae. Eur. J. Biochem.* **115:**463–468.
13. **Burrows, W., A. N. Mather, V. G. McGann, and S. M. Wagner.** 1946. Studies on immunity to Asiatic cholera. Part I. Introduction. *J. Infect. Dis.* **79:**159–167.
14. **Burrows, W., A. N. Mather, V. G. McGann, and S. M. Wagner.** 1946. Studies on immunity to Asiatic cholera. Part II. The O and H antigenic structure of the cholera and related vibrios. *J. Infect. Dis.* **79:**168–179.
15. **Cameron, D. N., T. Popovic, I. K. Wachsmuth, and P. I. Fields.** Personal communication.
16. **Chitnis, D. S., K. D. Sharma, and R. S. Koynat.** 1982. Role of somatic antigen of *Vibrio cholerae* in adhesion to intestinal mucosa. *J. Med. Microbiol.* **5:**53–61.
16a. **Cholera Working Group, International Centre for Diarrhoeal Diseases Research, Bangladesh.** 1993. Large epidemic of cholera-like disease in Bangladesh caused by *Vibrio cholerae* O139 synonym Bengal. *Lancet* **342:**387–390.
17. **Clark, C. A., J. Beltrame, and P. A. Manning.** 1991. The *oac* gene encoding a lipopolysaccharide O-antigen acetylase maps adjacent to the integrase encoding gene on the genome of *Shigella flexneri* bacteriophage Sf6. *Gene* **107:**43–52.

18. **Coleman, W. G.** 1983. The *rfaD* gene encodes for ADP-L-glycero-D-mannoheptose-6-epimerase. *J. Biol. Chem.* **258:**1985–1990.

19. **Cronan, J. E., Jr., and M. Rawlings.** 1992. The gene encoding Escherichia coli acyl carrier protein lies within a cluster of fatty acid biosynthetic genes. *J. Biol. Chem.* **267:**5751–5754.

20. **Darzins, A., B. Frantz, R. I. Vanags, and A. M. Chakrabarty.** 1986. Nucleotide sequence analysis of the phosphomannose isomerase gene (*pmi*) of *Pseudomonas aeruginosa* and comparison with the corresponding *Escherichia coli* gene *manA. Gene* **42:**293–302.

21. **Dhillon, N., R. S. Hale, J. Cortes, and P. F. Leadlay.** 1989. Molecular characterization of a gene from *Saccharopolyspora erythraea* (*Streptomyces erythraeus*) which is involved in erythromycin biosynthesis. *Mol. Microbiol.* **3:**1405–1414.

22. **Eubanks, E. R., M. N. Guentzel, and L. J. Berry.** 1977. Evaluation of surface components of *Vibrio cholerae* as protective immunogens. *Infect. Immun.* **15:**533–538.

23. **Faast, R., M. A. Ogierman, U. H. Stroeher, and P. A. Manning.** 1989. Nucleotide sequence of the structural gene, *tcpA*, for a major pilin subunit in *Vibrio cholerae. Gene* **85:**229–233.

24. **Fuerst, J. A., and J. W. Perry.** 1988. Demonstration of lipopolysaccharide on sheathed flagella of *Vibrio cholerae* O1 by protein A-gold immunoelectron microscopy. *J. Bacteriol.* **170:**1488–1494.

25. **Gan, K. H., and S. K. Tjia.** 1963. A new method for the differentiation of *Vibrio comma* and *Vibrio* El Tor. *Am. J. Hyg.* **77:**184–186.

26. **Gangarosa, E. J., A. Sonati, H. Saghari, and J. C. Feeley.** 1967. Multiple serotypes of *Vibrio cholerae* from a case of cholera. *Lancet* **i:**646–648.

27. **Gardner, A. D., and K. V. Venkatraman.** 1935. The antigens of the cholera group of vibrios. *J. Hyg.* **25:**262–282.

28. **Guhathakurta, B., M. Majumdar, A. K. Sen, D. Sasmal, A. K. Mukherjee, and A. Datta.** 1986. Immunochemical properties of the lipopolysaccharide O-antigen of *Vibrio cholerae* O1 in relation to its chemical structure. *J. Gen. Microbiol.* **132:**1641–1646.

29. **Guidolin, A., and P. A. Manning.** 1985. *Vibrio cholerae* bacteriophage CP-T1: characterization of the receptor. *Eur. J. Biochem.* **153:**89–94.

30. **Guidolin, A., and P. A. Manning.** 1987. Genetics of *Vibrio cholerae* and its bacteriophages. *Microbiol. Rev.* **51:**285–298.

31. **Guidolin, A., G. Morelli, M. Kamke, and P. A. Manning.** 1984. *Vibrio cholerae* bacteriophage CP-T1: characterization of phage DNA and restriction analysis. *J. Virol.* **51:**163–169.

32. **Gustafsson, B., and T. Holme.** 1985. Immunological characterization of *Vibrio cholerae* O1 lipopolysaccharide, O-side chain, and the core with monoclonal antibodies. *Infect. Immun.* **49:**275–280.

33. **Hisatsune, K., M. Hayashi, Y. Haishima, and S. Kondo.** 1989. Relationship between structure and antigenicity of O1 *Vibrio cholerae* lipopolysaccharide. *J. Gen. Microbiol.* **135:**1901–1907.

34. **Hisatsune, K., and S. Kondo.** 1980. Lipopolysaccharides of R mutants isolated from *Vibrio cholerae. Biochem. J.* **185:**77–81.

35. **Hisatsune, K., S. Kondo, and T. Iguchi.** 1978. Further chemical and immunochemical characterization of lipopolysaccharides of *Vibrio cholerae* including their R-mutants, p. 171–181. In *14th Joint Conference U.S.-Japan Cooperative Medical Science Program*. National Institutes of Health, Bethesda, Md.

36. **Hisatsune, K., S. Kondo, T. Iguchi, M. Muchida, S. Asou, M. Inaguma, and F. Yamamoto.** 1983. Sugar composition of lipopolysaccharides of family *Vibrionaceae*: absence of 2-keto-2-deoxyoctonate (KDO) except in *Vibrio parahaemolyticus* O6 and *Plesiomonas shigelloides*, p. 59–74. *In* S. Kumahara and N. F. Pierce (ed.), *Advances in Research on Cholera and Related Diarrheas*, vol. 1. Martinus Nijhoff Publisher, The Hague.

36a. **Hisatsune, K., S. Kondo, Y. Isshiki, T. Iguchi, Y. Kawamata, and T. Shimada.** 1993. Chemical study of Ogawa antigen factor B of *Vibrio cholerae* and lipopolysaccharide [LPS] isolated from non-O1 *Vibrio cholerae* O139 Bengal, a new epidemic strain for recent cholera endemics in the Indian subcontinent, p. 29–34. *In 29th Joint Conference on Cholera and Related Diarrheal Diseases*. U.S.-Japan Cooperative Medical Science Program.

37. **Hitchcock, P. J., L. Leive, P. H. Mäkelä, E. T. Reitschel, W. Strittmatten, and D. C. Morrison.** 1986. Lipopolysaccharide nomenclature: past, present, and future. *J. Bacteriol.* **166:**699–705.

38. **Holme, T., and B. Gustafsson.** 1985. Monoclonal antibodies against group and type specific antigens of *Vibrio cholerae* O:1, p. 167–189. *In* A. J. L. Macario and E. Conway de Macario (ed.), *Monoclonal Antibodies and Bacteria*, vol. 1. Academic Press, Orlando, Fla.

39. **Hranitzsky, K. W., A. Mulholland, A. D. Larson, E. R. Eubanks, and L. T. Hart.** 1980. Characterization of a flagella sheath protein of *Vibrio cholerae. Infect. Immun.* **27:**597–603.

40. **Iseki, S., and S. Hamano.** 1959. Conversion of type antigen IV in *Shigella flexneri* by bacteriophage. *Proc. Jpn. Acad.* **35:**407–412.

41. **Iseki, S., and K. Kashiwagi.** 1957. Lysogenic conversion and transduction of genetic characters by temperate phage Iota in *Salmonella*. I. Lysogenic conversion with regard to somatic antigen 1 in *Salmonella* groups A, B and D. *Proc. Jpn. Acad.* **33:**481–485.

42. **Jann, B., K. Jann, and G. O. Beyaert.** 1973. 2-Amino-2,6-dideoxy-D-glucose (D-quinavosamine): a constituent of the lipopolysaccharides of *Vibrio cholerae. Eur. J. Biochem.* **37:**531–534.

43. **Jann, K., and O. Westphal.** 1975. Microbial polysaccharides, p. 1–125. *In The Antigen*, vol. 3. Academic Press, New York.

44. **Jiang, X. M., B. Neal, F. Santiago, S. J. Lee, L. K. Romana, and P. R. Reeves.** 1991. Structure and sequence of the *rfb* (O-antigen) gene cluster of *Salmonella* serovar *typhimurium* (strain LT2). *Mol. Microbiol.* **5:**695–713.

44a.**Johnson, J. A., M. J. Albert, P. Panigrahi, A. C. Wright, A. Joseph, L. Comstock, M. Trucksis, J. Michalski, R. J. Johnson, J. B. Kaper, and J. G. Morris, Jr.** 1993. Non-O1 *Vibrio cholerae* (O139 synonym Bengal) from the India/Bangladesh epidemic are encapsulated, p. 35–38. *In 29th Joint Conference on Cholera and Related Diarrheal Diseases*. U.S.-Japan Cooperative Medical Science Program.

45. **Johnston, T. C., K. S. Hruska, and L. F. Adams.** 1989. The nucleotide sequence of the *luxE* gene of *Vibrio harveyi* and a comparison of the amino acid sequences of the acyl-protein synthetases from *V. harveyi* and *V. fischeri. Biochem. Biophys. Res. Commun.* **163:**93–101.

46. **Jörnvall, H., B. Persson, and J. Jeffery.** 1987. Characteristics of alcohol/polyol dehydrogenases. The zinc containing long-chain alcohol dehydrogenases. *Eur. J. Biochem.* **167:**195–201.

47. **Kabeshima, T.** 1918. On the serology of *Vibrio cholerae. Acta Pathol. Microbiol. Scand.* **27:**282.

48. **Kabir, S.** 1982. Characterization of the lipopolysaccharide from *Vibrio cholerae* 395 (Ogawa). *Infect. Immun.* **38:**1263–1272.

49. **Kaca, W., L. Brade, E. T. Rietschel, and H. Brade.** 1986. The effect of removal of D-fructose on the antigenicity of the lipopolysaccharide from a rough mutant of *Vibrio cholerae* Ogawa. *Carbohydr. Res.* **149:**293–298.

50. **Karageorgos, L. E., M. H. Brown, U. H. Stroeher, R. Morona, and P. A. Manning.** A putative pathway for perosamine biosynthesis is the first function encoded within the *rfb* region of *Vibrio cholerae* O1. Submitted for publication.

51. **Kenne, L., B. Lindberg, P. Unger, T. Holme, and J. Holmgren.** 1979. Structural studies of the *Vibrio cholerae* O-antigen. *Carbohydr. Res.* **68:**C16–C17.

52. **Kenne, L., B. Lindberg, P. Unger, T. Holme, and J. Holmgren.** 1982. Structural studies of the *Vibrio cholerae* O-antigen. *Carbohydr. Res.* **100:**341–349.

53. **Kroll, J. S., B. Loynds, L. N. Brophy, and E. R. Moxon.** 1990. The *bex* locus in encapsulated *Haemophilus influenzae*: a chromosomal region involved in capsule polysaccharide export. *Mol. Microbiol.* **4:**1853–1862.

54. **Lin, R.-J., M. Capage, and C. W. Hill.** 1984. A repetitive DNA sequence, *rhs*, responsible for duplications within the *Escherichia coli* K-12 chromosome. *J. Mol. Biol.* **177:**1–8.

55. **Lindberg, A. A., C. G. Hellerquist, G. Bagdian-Motto, and P. H. Mäkelä.** 1978. Lipopolysaccharide modification accompanying antigenic conversion by phage P22. *J. Gen. Microbiol.* **107:**279–287.

56. **Lüderitz, O., K. Jann, and R. Wheat.** 1968. Somatic and capsular antigens of gram negative bacteria, p. 105–227. *In* M. Florkin and E. H. Stotz (ed.), *Comprehensive Biochemistry, Extracellular and Supporting Structures*, vol. 26A. Elsevier, Amsterdam.

57. **Lüderitz, O., O. Westphal, A. M. Staub, and H. Nikaido.** 1971. Isolation and chemical and immunological characterization of bacterial lipopolysaccharides, p. 145–233. *In* G. Weinbaum, S. Kadis, and S. J. Ajl (ed.), *Microbial Toxins*, vol. 4. Academic Press, New York.

58. **Majumdar, M., and A. K. Mukherjee.** 1983. Studies on the partial structure of the O-antigen of *Vibrio cholerae* Ogawa G-2102. *Carbohydr. Res.* **122:**209–216.

59. **Majumdar, M., A. K. Mukherjee, B. Guhathakurta, A. Dutta, and D. Sasmal.** 1983. Structural investigation on the lipopolysaccharide isolated from *Vibrio cholerae* G-2102. *Carbohydr. Res.* **108:**269–278.

60. **Manning, P. A., M. W. Heuzenroeder, J. Yeadon, D. I. Leavesley, P. R. Reeves, and D. Rowley.** 1986. Molecular cloning and expression in *Escherichia coli* K-12 of the O antigens of the Ogawa and Inaba serotypes of the lipopolysaccharides of *Vibrio cholerae* O1 and their potential for vaccine development. *Infect. Immun.* **53:**272–277.

61. **Manning, P. A., L. E. Karageorgos, and R. Morona.** Putative O-antigen transport genes within the *rfb* region of *Vibrio cholerae* O1 are homologous to those for capsule transport. Submitted for publication.

62. **Meighen, E. A.** 1991. Molecular biology of bacterial bioluminescence. *Microbiol. Rev.* **55:**123–142.

63. **Mijamoto, C. M., A. F. Graham, and E. A. Meighen.** 1988. Nucleotide sequence of the *luxC* gene and the downstream DNA from the bioluminescent system of *Vibrio harveyi. Nucleic Acids Res.* **16:**1551–1562.

64. **Monod, M., S. Mohan, and D. Dubnau.** 1987. Cloning and analysis of *ermG*, a new macrolide-lincosamide-streptogramin B resistance element from *Bacillus sphaericus. J. Bacteriol.* **169:**340–350.

65. **Monsur, K. A., S. S. H. Rizvi, M. I. Huq, and A. S. Beneson.** 1965. Effect of Mukerjee's Group IV phage on El Tor vibrios. *Bull. W.H.O.* **32:**211–216.

66. **Morona, R., M. H. Brown, J. Yeadon, M. W. Heuzenroeder, and P. A. Manning.** 1991. Effect of lipopolysaccharide core synthesis mutations on the production of *Vibrio cholerae* O-antigen in *Escherichia coli* K-12. *FEMS Microbiol. Lett.* **82:**279–286.

67. **Morona, R., L. E. Karageorgos, and P. A. Manning.** A putative pathway for biosynthesis of the O-antigen component, 3-deoxy-L-glycero-tetronic acid based on the nucleotide sequence of the *Vibrio cholerae* O1 *rfb* region. Submitted for publication.

68. **Morona, R., M. S. Matthews, J. K. Morona, and M. H. Brown.** 1990. Regions of the cloned *Vibrio cholerae rfb* genes needed to determine the Ogawa form of the O-antigen. *Mol. Gen. Genet.* **224:**405–412.

69. **Mukerjee, S., and G. Roy.** 1961. Evaluation of tests for differentiating *Vibrio cholerae* and El Tor vibrios. *Annu. Biochem. Exp. Med.* **21:**129–132.

70. **Neoh, S. H., and D. Rowley.** 1972. Protection of infant mice against cholera by antibodies to three antigens of *Vibrio cholerae. J. Infect. Dis.* **126:**41–47.

71. **Nobechi, K.** 1923. Immunological studies upon types of *Vibrio cholerae. Sci. Rep. Inst. Infect. Dis. Tokyo Univ.* **2:**43.

72. **Nobechi, K., and E. Nakano.** 1967. Studies on shifting of the serotypes of cholerae vibrios. The first report: studies in vitro, p. 119–121. *In Symposium on Cholera.* National Institutes of Health, Bethesda, Md.

73. **Ogg, J. E., B. J. Ogg, M. B. Shrestha, and L. Poudayl.** 1979. Antigenic changes in *Vibrio cholerae* biotype El Tor serotype Ogawa after bacteriophage infection. *Infect. Immun.* **24:**974–978.

74. **Ogg, J. E., M. B. Shrestha, and L. Poudayl.** 1978. Phage-induced changes in *Vibrio cholerae*: serotype and biotype conversions. *Infect. Immun.* **19:**231–238.

75. **Ogg, J. E., T. L. Timme, and M. M. Alemohammad.** 1981. General transduction in *Vibrio cholerae. Infect. Immun.* **31:**737–741.

76. **Osborn, M. J.** 1979. Biosynthesis and assembly of the lipopolysaccharide of the outer membrane, p. 15–34. *In* M. Inouye (ed.) *Bacterial Outer Membranes.* John Wiley & Sons, Inc., New York.

77. **Packer, N., M. Batley, and J. W. Redmond.** Personal communication.

78. **Pesigan, T. P., C. Z. Gomez, and D. Gaetos.** 1967. Variants of agglutinable *Vibrios* in the Philippines. *Bull. W.H.O.* **37:**795–797.

79. **Raetz, C. R. H.** 1990. Biochemistry of endotoxins. *Annu. Rev. Biochem.* **59:**129–170.

80. **Raziuddin, S.** 1977. Studies of the polysaccharide fraction form the cell wall lipopolysaccharides (O-antigen) of *Vibrio cholerae. Indian J. Biochem. Biophys.* **14:**262–263.

81. **Raziuddin, S.** 1980. Immunochemical studies of the lipopolysaccharides of *Vibrio cholerae*: constitution of O-specific side chain and core polysaccharide. *Infect. Immun.* **27:**211–215.

82. **Redmond, J. W.** 1975. 4-Amino-1,6-dideoxy-D-mannose (D-perosamine): a component of the lipopolysaccharide of *Vibrio cholerae* 569B (Inaba). *FEBS Lett.* **50:**147–149.

83. **Redmond, J. W.** 1978. The 4-amino sugars present in the lipopolysaccharides of *Vibrio cholerae* and related vibrios. *Biochim. Biophys. Acta* **542:**378–384.

84. **Redmond, J. W.** 1979. The structure of the O-antigenic side chain of the lipopolysaccharide of *Vibrio cholerae* 569B (Inaba). *Biochim. Biophys. Acta* **584:**346–352.

85. **Redmond, J. W., M. J. Korsch, and G. D. F. Jackson.** 1973. Immunochemical studies of the O-antigens of *Vibrio cholerae*. Partial characterization of an acid-labile antigenic determinant. *Aust. J. Exp. Biol. Med. Sci.* **51:**229–235.

86. **Rizvi, S., M. I. Huq, and A. S. Beneson.** 1965. Isolation of haemagglutinating non-El Tor vibrios. *J. Bacteriol.* **89:**910–912.

87. **Roy, C., K. Mridha, and S. Mukerjee.** 1965. Action of polymyxin on cholera vibrios. Techniques of determination of polymyxin-sensitivity. *Proc. Soc. Exp. Biol. Med.* **119:**893–896.

88. **Sack, R. B., and C. E. Miller.** 1969. Progressive changes of vibrio serotypes in germ-free mice infected with *Vibrio cholerae. J. Bacteriol.* **99:**688–695.

89. **Sakazaki, R., and K. Tamura.** 1971. Somatic antigen variation in *Vibrio cholerae. Jpn. J. Med. Sci. Biol.* **24:**93–100.

90. **Salazar-Lindo, E., L. Seminario-Carrasco, C. Carrillo-Parodi, and A. Gayoso-Villaflor.** 1991. The cholera epidemic in Peru, p. 9–13. *In The United States-Japan Cooperative Medical Science Program. Twenty-Seventh Joint Conference on Cholera and Related Diarrheal Diseases.* National Institutes of Health, Bethesda, Md.

91. **Sen, A. K., A. K. Mukherjee, B. Guhathakurta, A. Dutta, and D. Sasmal.** 1979. Structural investigation on the lipopolysaccharide isolated from *Vibrio cholerae* Inaba 569B. *Carbohydr. Res.* **72:**191–199.

92. **Sen, A. K., A. K. Mukherjee, B. Guhathakurta, A. Dutta, and D. Sasmal.** 1980.

Studies on the partial structure of the O-antigen of *Vibrio cholerae* 569B. *Carbohydr. Res.* **86**:113–121.

93. **Sharma, D. P., U. H. Stroeher, C. J. Thomas, P. A. Manning, and S. R. Attridge.** 1989. The toxin coregulated pilus TCP of *Vibrio cholerae*: molecular cloning of the genes involved in pilus biosynthesis and evaluation of TCP as a protective antigen in the infant mouse model. *Microb. Pathog.* **7**:437–448.

94. **Sheehy, T. W., H. Sprinz, W. S. Augerson, and S. B. Formal.** 1966. Laboratory *Vibrio cholerae* infection in the United States. *JAMA* **197**:321–325.

95. **Shrivastava, D. L., and P. B. White.** 1947. Note on the relationship of the so-called Ogawa and Inaba types of *V. cholerae. Indian J. Med. Res.* **35**:117–129.

95a. **Smith, A. N., G. J. Bulnois, and I. S. Roberts.** 1990. Molecular analysis of the *Escherichia coli* K5 *kps* locus: identification and characterization of an inner-membrane capsular polysaccharide transport system. *Mol. Microbiol.* **4**:1863–1869.

96. **Stevenson, G., S. J. Lee, L. K. Romana, and P. R. Reeves.** 1991. The *cps* gene cluster of *Salmonella* strain LT2 includes a second mannose pathway: sequence of two genes and relationship to the genes in the *rfb* gene cluster. *Mol. Gen. Genet.* **227**:173–180.

97. **Stroeher, U. H., L. E. Karageorgos, R. Morona, and P. A. Manning.** Divergent *rfa* and *rfb* genes in *Vibrio cholerae* O1: nucleotide sequence of an *rfa* homologue. Submitted for publication.

98. **Stroeher, U. H., L. E. Karageorgos, R. Morona, and P. A. Manning.** 1992. Serotype conversion in *Vibrio cholerae* O1. *Proc. Natl. Acad. Sci. USA* **89**:2566–2570.

99. **Stroeher, U. H., and P. A. Manning.** Unpublished data.

100. **Stroeher, U. H., H. M. Ward, C. J. Thomas, and P. A. Manning.** Virulence of *rfb* mutants, defective in the synthesis of the O-antigen of the lipopolysaccharide of *Vibrio cholerae* O1. Sub-

mitted for publication.

100a. **Swerdlow, D. L., and A. L. Ries.** 1993. *Vibrio cholerae* non-O1—the eighth pandemic? *Lancet* **342**:382–383.

101. **Taylor, R. K., V. L. Miller, D. B. Furlong, and J. J. Mekalanos.** 1987. Use of *phoA* gene fusion to identify a pilus colonization factor coordinately regulated with cholera toxin. *Proc. Natl. Acad. Sci. USA* **84**:2833–2837.

102. **Turgay, K., M. Krause, and M. A. Marhiel.** 1992. Four homologous domains in the primary structure of GrsB are related to domains in a superfamily of adenylate-forming enzymes. *Mol. Microbiol.* **6**:529–546.

103. **Verma, N. K., and P. R. Reeves.** 1989. Identification and sequence of *rfbS* and *rfbE*, which determine antigenic specificity of group A and D salmonellae. *J. Bacteriol.* **171**:5694–5701.

104. **Ward, H. M., and P. A. Manning.** 1989. Mapping of chomosomal loci associated with lipopolysaccharide synthesis and serotype specificity in *Vibrio cholerae* O1 by transposon mutagenesis using Tn*5* and Tn*2680*. *Mol. Gen. Genet.* **218**:367–370.

105. **Ward, H. M., and P. A. Manning.** Unpublished data.

106. **Ward, H. M., G. Morelli, M. Kamke, R. Morona, J. Yeadon, J. A. Hackett, and P. A. Manning.** 1987. A physical map of the chromosomal region determining O-antigen biosynthesis in *Vibrio cholerae* O1. *Gene* **55**:197–204.

106a. **Weintraub, A.** Personal communication.

107. **Youngleson, J. S., J. D. Santangelo, D. T. Jones, and D. R. Woods.** 1988. Cloning and expression of *Clostridium acetobutylicum* dehydrogenase gene in *Escherichia coli. Appl. Environ. Microbiol.* **54**:676–682.

108. **Zhao, S., C. H. Sandt, G. Feulner, D. A. Vlazny, J. A. Gray, and C. W. Hill.** 1993. *Rhs* elements of *Escherichia coli* K-12: complex composites of shared and unique components that have different evolutionary histories. *J. Bacteriol.* **175**:2799–2808.

Vibrio cholerae and Cholera: Molecular to Global Perspectives
Edited by I. Kaye Wachsmuth, Paul A. Blake, and Ørjan Olsvik
© 1994 American Society for Microbiology, Washington, DC 20005

Chapter 7

Vibrio cholerae O139 Bengal

*J. Glenn Morris, Jr., and the Cholera Laboratory Task Force**

Beginning in October 1992, cases of choleralike disease associated with a *Vibrio cholerae* strain that did not agglutinate with O1 antisera were noted in Madras, India (25). Similar cases were subsequently identified in Madurai, Vellore, Calcutta, and, beginning in December, southern Bangladesh (1, 13). During the next several months, this strain (subsequently designated *V. cholerae* O139 Bengal) spread in epidemic form across Bangladesh, with over 100,000 cases reported by the end of March 1993 (13). By the fall of 1993, cases had also been reported in Thailand (5), Malaysia, Pakistan, and Nepal. It has been proposed that these cases reflect the beginning of the eighth cholera pandemic (35).

These cases presage a marked departure from the traditional understanding (dating back to the work of Gardner and Venkatramen and others in the 1930s [9, 22]) that epidemic cholera is caused only by *V. cholerae* within O group 1. Strains of *V. cholerae* in O groups other than 1 (non-O1 *V. cholerae*) are known to cause sporadic illness, probably reflecting periodic introduction of environmental strains into human populations (see chapter 8). However, except for occasional reports of non-O1 *V. cholerae* being isolated from patients during cholera epidemics caused by *V. cholerae* O1 (6, 31), there has been no suggestion that non-O1 strains possess the characteristics necessary to spread in epidemic or pandemic form. As outlined below, *V. cholerae* O139 is closely related to Asian *V. cholerae* O1 El Tor strains associated with the seventh pandemic; at the same time, O139 isolates have certain distinctive features, including the potential for capsule expression (16) and a possible increased capacity for spread and proliferation within the environment (14). The emergence of *V. cholerae* O139 demonstrates that the O1 antigen is not an exclusive marker for epidemic potential and underscores the need to better understand the interplay of genetic and epidemiologic factors that allow strains to spread rapidly through populations.

CLASSIFICATION

O Antigen

V. cholerae O139 Bengal does not agglutinate with monoclonal or polyclonal antisera

J. Glenn Morris, Jr. • Divisions of Geographic Medicine and Infectious Diseases, Department of Medicine, University of Maryland School of Medicine, and Veterans Affairs Medical Center, Baltimore, Maryland 21201. ***Cholera Laboratory Task Force*** • Foodborne and Diarrheal Diseases Branch, National Center for Infectious Diseases, Centers for Disease Control and Prevention, Atlanta, Georgia 30333.

*Tanja Popovic, Patricia I. Fields, Ørjan Olsvik, Joy G. Wells, Gracia M. Evins, Daniel N. Cameron, Kathy Greene, Kaye Wachsmuth, and John C. Feeley.

directed against the O1 antigen. These strains appear to lack portions of at least two genes within the O1 biosynthetic gene cluster (*Vcrfb*), suggesting that failure to express the O antigen represents more than a minor or point mutation in the O1-antigen biosynthetic pathway (16) (see Addendum in Proof to chapter 6 for further discussion of this issue). The designation O group 139 is in the O-group typing system established by Sakazaki and Shimada (28). The relationship of this serovar to O groups in the typing systems established by Smith (33) and Seibling et al. (32) remains to be determined.

Phage Typing/Biotype Characteristics

All isolates appear to be resistant to Mukherjee's phages specific for both biotypes of *V. cholerae* O1 (13, 25). At the same time, strains show high-level polymyxin B resistance and agglutinate chicken erythrocytes, features used in identification of the El Tor biotype of *V. cholerae* O1.

Enzyme Electrophoresis (Zymovar) Analysis

V. cholerae strains can be classified on the basis of enzyme electrophoretic patterns (zymovar analysis). In the zymovar system established by Salles and colleagues (21, 29), O139 strains are grouped in zymovar 14, the zymovar containing the majority of *V. cholerae* O1 El Tor strains, including all epidemic Asian and South American strains (30). To place this in context, other non-O1 *V. cholerae* strains occupy over 100 zymovars, with an intraspecies diversity that approaches that of *Escherichia coli*. In the more sensitive typing system reported by Wachsmuth and colleagues at the Centers for Disease Control and Prevention (37), these strains have a pattern identical to that of electrophoretic type (ET) 3, which encompasses Asian *V. cholerae* El Tor strains. Epidemic South American *V. cholerae* O1 El Tor strains are in ET 4, which differs from ET 3 at a single enzyme locus.

Molecular Typing

DNA sequences of 16S rRNA (positions 330 to 615) from these isolates are typical for *V. cholerae* O1 strains. DNA sequence analyses of both subunits of the *ctx* gene appear to be identical to those of *V. cholerae* O1 El Tor strains from the seventh pandemic (7, 12). In studies conducted at CDC with 20 O139 strains (24), two ribotypes were identified. These two ribotypes differed from each other by a single band of 6 kb. Patterns were very similar (but not identical) to ribotype patterns 3 and 5 of seventh-pandemic O1 El Tor strains (23); in both instances, ribotypes differed from those of seventh-pandemic strains by one 2-kb band (24). Four very similar pulsed-field gel electrophoresis patterns were observed. Two dominant patterns were typical for the majority of strains, while each of the remaining two patterns contained a single strain.

EPIDEMIOLOGY

Epidemic Potential

As noted above, during 1993, *V. cholerae* O139 spread rapidly through India and Bangladesh and into the rest of Asia. Cases were identified in Thailand by April 1993 (5), a little over 6 months after the initial identification of cases in Madras. By way of comparison, during the seventh pandemic it took 2 years for *V. cholerae* O1 El Tor to spread from the Celebes to Thailand (2). This difference may reflect improved transportation and movement of people and products within the region in the 1990s compared with those in the 1960s; at the same time, it suggests that O139 isolates have a unique ability to become established and spread rapidly once introduced into a geographic area. By February 1993, there had already been one known introduction of *V. cholerae* O139 into the United States: the patient in this case was a woman who had returned from Hyderabad, India, with a diarrheal illness (4). As in past instances, it is unlikely that a case of this type will lead to transmission of cholera in the United States because of the extant high level of sanitation. However, its occurrence underscores the ease with which organisms can be transmitted from one continent to another.

As is apparent from the molecular typing studies described above, O139 strains are not strictly clonal in origin; minor differences are noted among strains in the strain collections that were initially examined. There currently are insufficient data to say whether specific clones can be associated with certain geographic areas; however, at some point it may be possible to use these differences to trace the spread of the epidemic. At the same time, the presence of multiple clones raises interesting questions about the origins of these strains. It is possible that observed differences among strains reflect random mutations that have occurred as the epidemic has spread; alternatively (and perhaps less likely), critical genes encoding factors essential for epidemic spread may have been introduced simultaneously into multiple strain backgrounds.

Age- and Sex-Related Incidence

In the 1993 Bangladeshi studies, the majority of cases in the first wave of illness were in adults (74% of patients were < 15 years of age [13]). This differs from the traditional pattern seen in this region, in which the majority of patients are younger than 15 years (10, 13). If, as has been proposed, the concentration of illness in children in areas where the disease is endemic reflects the development of specific immunity over time, the occurrence of large numbers of cases in adults suggests that prior exposure to *V. cholerae* O1 did not protect against infection with *V. cholerae* O139. This pattern of age-related incidence would be analogous to that seen in South America in 1991, in which introduction of cholera into an immunologically naive population resulted in high attack rates in all age groups (34).

Illness due to O139 strains in the 1993 epidemic in Bangladesh was significantly more common among men than among women (13). This again reflects a departure from traditional patterns: the incidence of *V. cholerae* O1 in Bangladesh has been reported to peak a second time among women of childbearing age (10), with a resulting increase in the number of reported cases among adult women compared to the number among men.

Reservoir

Though the data are still limited, the reservoir for *V. cholerae* O139 appears to be the aquatic environment. In studies conducted in Bangladesh in the spring of 1993, O139 strains were isolated from 12% of 92 water samples from ponds, lakes, rivers, and canals in rural Matlab and urban Dhaka (14); *V. cholerae* O1 El Tor was isolated from 1 sample in the same study. These data are particularly striking in that in the past, *V. cholerae* O1 has been isolated from less than 1% of water samples during cholera epidemics (14). While these observations may in part reflect the intensity of the epidemic, they also suggest that O139 strains have an unusual ability to survive and replicate within the environment. At this point, it is unclear whether O139 strains are replacing O1 strains in an ecologic niche or, alternatively, are occupying a larger, as-yet-undefined niche comparable to that normally occupied by other non-O1 *V. cholerae* strains (19, 26).

PATHOPHYSIOLOGY

V. cholerae O139 carries the gene for and expresses cholera toxin. As reported for *V. cholerae* O1 El Tor, these strains have *ctxAB*, *zot* (3), *ace* (36), and RS1 elements (11) organized in typical virulence cassettes (16). In studies at the University of Maryland, seven of eight O139 strains studied had two copies of the cassette, while an eighth strain had more than two copies (16). Strains also carry the *tcpA* gene, encoding the toxin coregulated pilus (TCP) (12). These data suggest that the pathogenesis of illness due to O139 strains parallels that of *V. cholerae* O1, with expression of cholera toxin and adherence factors such as TCP.

In contrast to *V. cholerae* O1 (but analogous to other non-O1 *V. cholerae* strains [17, 18]), *V. cholerae* O139 can also express a polysaccharide capsule (16) (Fig. 1).

Strains are able to undergo phase variation, shifting between encapsulated forms with an opaque colonial morphology after overnight incubation on L agar and minimally encapsulated or unencapsulated forms with a translucent colonial morphology. As previously reported for other non-O1 *V. cholerae* strains (18), encapsulated strains are resistant to killing by healthy human serum and have increased virulence (as measured by a decrease

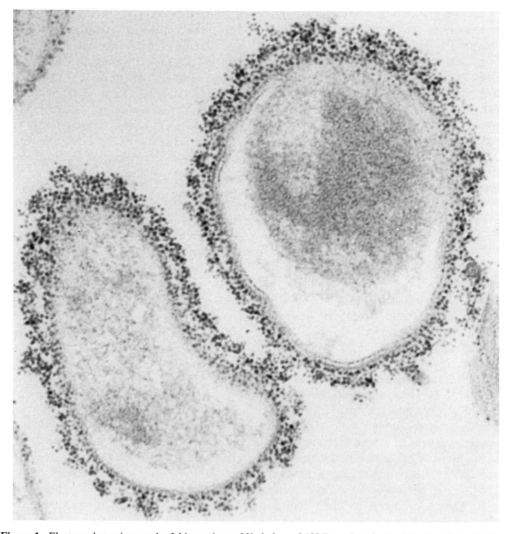

Figure 1. Electron photomicrograph of thin sections of *V. cholerae* O139 Bengal strained with polycationic ferritin, showing capsular polysaccharide.

in the 50% lethal dose) after intradermal inoculation into mice. O139 strains do not appear to be as heavily encapsulated as some non-O1 *V. cholerae* blood isolates (17); nonetheless, the presence of a capsule appears to increase the potential for invasive disease among persons infected with this organism.

CLINICAL SYNDROMES

Illness reported in association with *V. cholerae* O139 strains appears to be virtually identical to that seen with *V. cholerae* O1 (Table 1). In the Bangladeshi studies (13), patients had severe secretory-type watery diarrhea and vomiting, with rapid onset of dehydration. Adult patients were treated with an average of 7 liters of intravenous fluids, 11 liters of oral rehydration solution, and tetracycline. No deaths were reported among persons reaching the treatment facilities.

There has been one case report of septicemia due to *V. cholerae* O139 in a patient with underlying liver disease (15). This is in contrast to findings with *V. cholerae* O1, which virtually never causes extraintestinal disease. It does, however, parallel reports with other non-O1 *V. cholerae* strains: in the United States, 20% of non-O1 *V. cholerae* isolates are from blood (20), with most patients with septicemia having underlying liver disease or immunosuppression (27). As noted above, this may reflect expression of a capsule by O139 strains.

DIAGNOSIS

The methodology for isolation of *V. cholerae* O139 is identical to that for *V. cholerae* O1, with strains producing typical yellow colonies on thiosulfate-citrate-bile salts-sucrose agar and grayish opaque colonies with dark centers on TTG agar (13). However, virtually all isolates are resistant to the vibriostatic compound 2,4-diamino-6,7-diisopropyl-pteridine (O/129) and do not agglutinate with standard O1 antisera. Monoclonal and polyclonal antisera directed against the O139 antigen are available on a limited, experimental basis; such antisera should be available commercially in the near future. In the absence of specific antisera, a strain can be presumptively identified as O139 on the basis of epidemiologic setting, failure to agglutinate in O1 antisera, and presence of the cholera toxin gene (as identified by oligonucleotide [38] or other DNA probes or by polymerase chain reaction).

Table 1. Clinical signs and symptoms of patients with acute diarrhea associated with *V. cholerae* O139[a]

Clinical feature	Southern Bangladesh (n = 87)	Matlab (n = 236)	Dhaka (n = 63)
Age of patients	All ages	All ages	Adults
No. (%) with watery diarrhea	87 (100%)	235 (99.6%)	63 (100%)
Mean (SD) duration of diarrhea before admission	<12 (for 70% of patients)	19.2 (21.3)	10.9 (4.3)
% of patients with severe dehydration on admission	75	55	94
% of patients with vomiting	97	86	98
Mean (SD) stool output after rehydration (ml kg^{-1} h^{-1})[b]		3.6 (3.9)	12.0 (5.4)
Mean (SD) duration of illness (h)[c]		45.4 (27.7)	40.5 (15.1)

[a]Reproduced with permission from reference 13 (copyright held by The Lancet Ltd.).
[b]Measured in Dhaka during first 4 h after rehydration was completed, measured in Matlab during whole hospital stay.
[c]All patients were treated with tetracycline.

THERAPY

As with *V. cholerae* O1, the mainstay of therapy for diarrheal disease is oral rehydration. Isolates have been reported to be susceptible to tetracycline, ampicillin, chloramphenicol, erthromycin, and the quinolones (1, 13). On the basis of prior studies with *V. cholerae* O1, it appears reasonable to treat patients with O139 disease with tetracycline. It should be noted, however, that tetracycline resistance is widespread in areas with endemic cholera (1, 8): 70% of recent Bangladeshi *V. cholerae* O1 isolates are tetracycline resistant (1). Since resistance to antimicrobial agents is in most instances plasmid mediated (8), it may be only a matter of time before tetracycline resistance is acquired by O139 strains.

PREVENTION

Efforts to prevent disease due to O139 strains are best directed toward minimizing exposure to food or water sources that may have been contaminated with the organism. As noted above, the high attack rates in adult populations in Bangladesh suggest that prior exposure to *V. cholerae* O1 does not protect against infection with O139. If this is true, it also suggests that vaccines directed against *V. cholerae* O1 will have minimal efficacy against O139 strains (13).

SUMMARY AND FUTURE DIRECTIONS

At this point, there is every reason to believe that *V. cholerae* O139 Bengal will continue to spread rapidly through areas that are susceptible to cholera. While every effort should be made to limit the spread of the organisms beyond areas that are currently infected, recent history suggests that this may be difficult, with the widespread availability of airline travel and the global market for products from countries in which the disease is already extant. At this point, it may be of more value to concentrate on defining environmental reservoirs and modes of transmission for the organism in an effort to understand why it appears to be able to spread so rapidly once introduced into a geographic area. From a microbiological standpoint, we need to understand the genetic basis for the epidemic potential for this strain; there are clearly critical factors other than the O1 antigen, which has traditionally been used as a surrogate marker for epidemic potential. Finally, there is a need for rapid work on vaccines for this organism. The apparent lack of protection engendered by *V. cholerae* O1 suggests that current vaccines will be of little efficacy; however, the molecular approaches that have been learned during the past decade of work on cholera vaccines should be readily applicable to these new strains.

REFERENCES

1. **Albert, M. J., A. K. Siddique, M. S. Islam, A. S. G. Faruque, M. Ansaruzzaman, S. M. Faruque, and R. B. Sack.** 1993. A large outbreak of clinical cholera due to *Vibrio cholerae* non-O1 in Bangladesh. *Lancet* **341:**704.
2. **Barua, D.** 1992. History of cholera, p. 1–36. *In* D. Barua and W. B. Greenough III (ed.), *Cholera.* Plenum Medical Book Co., New York.
3. **Baudry, B., A. Fasano, J. Ketley, and J. B. Kaper.** 1992. Cloning of a gene (*zot*) encoding a new toxin produced by *Vibrio cholerae*. *Infect. Immun.* **60:**428–434.
4. **Centers for Disease Control and Prevention.** 1993. Imported cholera associated with a newly described toxigenic *Vibrio cholerae* O139 strain—California, 1993. *Morbid. Mortal. Weekly Rep.* **42:**501–503.
5. **Chongsa-Nguan, M., W. Chaicumpa, P. Moolasart, P. Kandhasingha, T. Shimada, H. Kurazono, and Y. Takeda.** 1993. *Vibrio cholerae* O139 in Bangkok. *Lancet* **342:**430–431.
6. **Dutt, A. K., S. Alwi, and T. Velauthan.** 1971. A shellfish-borne cholera outbreak in Malaysia. *Trans. R. Soc. Trop. Med. Hyg.* **65:**815–818.
7. **Fields, P. I., M. Tamplin, and Ø. Olsvik.** 1993. DNA sequence heterogeneity within the genes encoding cholera toxin. Program Abstr. 29th Joint Conf. Cholera Related Diarrheal Dis.
8. **Finch, M. J., J. G. Morris, Jr., J. Kaviti, W. Kagwanja, and M. M. Levine.** 1988. Molecular

epidemiology of antimicrobial-resistant cholera in Kenya and East Africa. *Am. J. Trop. Med. Hyg.* **39:**484–490.

9. **Gardner, A. D., and V. K. Venkatraman.** 1935. The antigens of the cholera group of vibrios. *J. Hyg.* **35:**262–282.

10. **Glass, R. I., S. Becker, M. I. Huq, B. J. Stoll, M. U. Khan, M. H. Merson, J. V. Lee, and R. E. Black.** 1982. Endemic cholera in rural Bangladesh, 1966–1980. *Am. J. Epidemiol.* **116:**959–970.

11. **Goldberg, I., and J. J. Mekalanos.** 1986. Effect of a *recA* mutation on cholera toxin gene amplification and deletion events. *J. Bacteriol.* **165:**723–731.

12. **Hall, R. H., F. M. Khambaty, M. Kothary, and S. P. Keasler.** 1993. Non-O1 *Vibrio cholerae. Lancet* **342:**430.

13. **International Center for Diarrheal Disease Research, Bangladesh, Cholera Working Group.** 1993. Large epidemic of cholera-like disease in Bangladesh caused by *Vibrio cholerae* O139 synonym Bengal. *Lancet* **342:**387–390.

14. **Islam, M. S., M. K. Hasan, M. A. Miah, F. Qadri, M. Yunus, R. B. Sack, and M. J. Albert.** 1993. Isolation of *Vibrio cholerae* O139 Bengal from water in Bangladesh. *Lancet* **342:**430.

15. **Jesudason, M. V., A. M. Cherian, and T. J. John.** 1993. Blood stream invasion by *Vibrio cholerae* O139. *Lancet* **342:**431.

16. **Johnson, J. A., M. J. Albert, P. Panigrahi, A. C. Wright, A. Joseph, L. Comstock, M. Trucksis, J. Michalski, R. J. Johnson, J. B. Kaper, and J. G. Morris, Jr.** 1993. Non-O1 *Vibrio cholerae* (O139 synom. Bengal) from the India/Bangladesh epidemic are encapsulated. Program Abstr. 29th Joint Conf. Cholera Related Diarrheal Dis.

17. **Johnson, J. A., A. Joseph, P. Panigrahi, and J. G. Morris, Jr.** 1992. Frequency of encapsulated versus unencapsulated strains of non-O1 *Vibrio cholerae* isolated from patients with septicemia or diarrhea, or from environmental sources, abstr. B-277, p. 72. Abstr. 92nd Gen. Meet. Am. Soc. Microbiol. 1992.

18. **Johnson, J. A., P. Panigrahi, and J. G. Morris, Jr.** 1992. Non-O1 *Vibrio cholerae* NRT36S produces a polysaccharide capsule that determines colony morphology, serum resistance, and virulence in mice. *Infect. Immun.* **60:**684–689.

19. **Kaper, J. B., H. Lockman, R. R. Colwell, and S. W. Joseph.** 1979. Ecology, serology, and enterotoxin production of *Vibrio cholerae* in Chesapeake Bay. *Appl. Environ. Microbiol.* **37:**91–103.

20. **Kelly, M. T., F. W. Hickman-Brenner, and J. J. Farmer III.** 1991. *Vibrio*, p. 384–395. *In* A. Balows, W. J. Hausler, Jr., K. L. Herrmann, H. D. Isenberg, and H. J. Shadomy (ed.), *Manual of Clinical Microbiology*, 5th ed. American Society for Microbiology, Washington, D.C.

21. **Momen, H., and C. A. Salles.** 1985. Enzyme markers for *Vibrio cholerae:* identification of classical, El Tor, and environmental strains. *Trans. R. Soc. Trop. Med. Hyg.* **79:**773–776.

22. **Pollitzer, R.** 1959. *Cholera.* World Health Organization, Geneva.

23. **Popovic, T., C. Bopp, Ø. Olsvik, and K. Wachsmuth.** 1993. Epidemiologic application of a standardized ribotype scheme for *Vibrio cholerae* O1. *J. Clin. Microbiol.* **31:**2474–2482.

24. **Popovic, T., P. I. Fields, O. Olsvik, J. G. Wells, G. M. Evins, D. N. Carmeron, K. Wachsmuth, R. B. Sack, J. M. Albert, N. Balakrish, and J. C. Feeley.** 1993. Program Abstr. 29th Joint Conf. Cholera and Related Diarrheal Dis.

25. **Ramamurthy, T., S. Garg, R. Sharma, S. K. Bhattacharya, G. B. Nair, T. Shimada, T. Takeda, T. Karasawa, H. Kurazano, A. Pal, and Y. Takeda.** 1993. Emergence of novel strain of *Vibrio cholerae* with epidemic potential in southern and eastern India. *Lancet* **341:**703–704.

26. **Roberts, N. C., R. J. Seibeling, J. B. Kaper, and H. B. Bradford.** 1982. Vibrios in the Louisiana Gulf Coast environment. *Microb. Ecol.* **8:**299–312.

27. **Safrin, S., J. G. Morris, Jr., M. Adams, V. Pons, R. Jacobs, and J. E. Conte, Jr.** 1988. Non-O:1 *Vibrio cholerae* bacteremia: case report and review. *Rev. Infect. Dis.* **10:**1012–1017.

28. **Sakazaki, R., and T. Shimada.** 1977. Serovars of *Vibrio cholerae. Jpn. J. Med. Sci. Biol.* **30:**279–282.

29. **Salles, C. A., A. R. da Silva, and H. Momen.** 1986. Enzyme typing and phenetic relationships in *Vibrio cholerae. Rev. Brasil Genet.* **9:**407–419.

30. **Salles, C. A., and J. G. Morris, Jr.** Unpublished data.

31. **Shehabi, A. A., A. B. Abu Rajab, and A. A. Shaker.** 1980. Observations on the emergence of non-cholera vibrios during an outbreak of cholera. *Jordan Med. J.* **14:**125–127.

32. **Siebeling, R. J., L. B. Adams, Z. Yusof, and A. D. Larson.** 1984. Antigens and serovar-specific antigens of *Vibrio cholerae*, p. 33–58. *In* R. R. Colwell (ed.), *Vibrios in the Environment.* John Wiley & Sons, Inc., New York.

33. **Smith, H. L., Jr.** 1979. Serotyping of non-cholera vibrios. *J. Clin. Microbiol.* **10:**85–90.

34. **Swerdlow, D. L., E. D. Mintz, M. Rodriguez, E. Tejada, C. Ocampo, L. Espejo, K. D. Greene, W. Saldana, L. Seminario, R. V. Tauxe, J. G. Wells, N. H. Bean, A. A. Ries, M. Pollack, B. Vertiz, and P. A. Blake.** 1992. Waterborne transmission of epidemic cholera in Trujillo, Peru: lessons for a continent at risk. *Lancet* **340:**28–32.

35. **Swerdlow, D. L., and A. A. Ries.** 1993. *Vibrio cholerae* non-O1—the eighth pandemic? *Lancet* **342:**382–383.

36. **Trucksis. M., J. E. Galen, J. Michalski, A. Fasano, and J. B. Kaper.** 1993. Accessory cholera enterotoxin (Ace), the third toxin of a *Vibrio cholerae* virulence cassette. *Proc. Natl. Acad. Sci. USA* **90:**5267–5271.

37. **Wachsmuth, I. K., G. M. Evins, P. I. Fields, O. Olsvik, T. Popovic, C. A. Bopp, J. G. Wells, C. Carrillo, and P. A. Blake.** 1993. The molecular epidemiology of cholera in Latin America. *J. Infect. Dis.* **167:**621–626.

38. **Wright, A. C., Y. Guo, J. A. Johnson, J. P. Nataro, and J. G. Morris, Jr.** 1992. Development and testing of a non-radioactive DNA oligonucleotide probe that is specific for *Vibrio cholerae* toxin. *J. Clin. Microbiol.* **30:**2302–2306.

Vibrio cholerae and Cholera: Molecular to Global Perspectives
Edited by I. Kaye Wachsmuth, Paul A. Blake, and Ørjan Olsvik
© 1994 American Society for Microbiology, Washington, DC 20005

Chapter 8

Non-O Group 1 *Vibrio cholerae* Strains Not Associated with Epidemic Disease

J. Glenn Morris, Jr.

A major concern of early cholera investigators was the differentiation of epidemic-associated strains of *Vibrio cholerae* from other, "atypical" vibrios. The work by Gardner and Venkatraman and others in the 1930s (20, 64) led to the concept that *V. cholerae* strains could be divided into two groups: those in O group 1, which agglutinated with antisera directed against antigens present on strains isolated from cholera patients, and other "nonagglutinating" or "noncholera" vibrios, which were regarded primarily as nonpathogenic environmental isolates. Some early writers discussed the possibility that nonagglutinating isolates were responsible for "paracholera" or similar clinical syndromes (64). However, it was not until the 1950s and 1960s that investigators began to identify outbreaks of disease directly attributable to these strains (2, 17, 37, 51, 94).

As more attention was paid to nonagglutinating vibrios, it became increasingly obvious that these isolates were a heterogeneous group, including species besides *V. cholerae*. These observations have led, since the late 1970s, to the designation of a number of new vibrio species. As shown in Table 1,

11 vibrio species other than *V. cholerae* have now been associated with human disease. While most cause gastroenteritis, some, such as *V. vulnificus,* have been implicated primarily as a cause of septicemia and wound infections (8, 54, 62). Recent publications (19, 29, 54) have discussed the clinical and epidemiologic features of pathogenic vibrio species other than *V. cholerae*; these species will not be discussed further.

At the same time, there has been increasing recognition that *V. cholerae* strains in O groups other than 1 (non-O1 *V. cholerae*) can cause both epidemic and endemic disease (1, 15, 28, 55, 56, 66, 81). Epidemic illness has been most strongly associated with *V. cholerae* O139 Bengal; clinical, epidemiologic, and microbiologic characteristics of this serovar are discussed in a separate chapter. Illness associated with most other non-O1 strains is best categorized as sporadic: cases occur at a low rate in a number of geographic areas (53), probably reflecting periodic introduction of what is primarily an environmental organism into human populations. The most common clinical manifestation in these cases is gastroenteritis (Table 2). However, non-O1 *V. cholerae* strains have also been isolated from blood (ca. 20% of isolates submitted to the Centers for Disease Control and Prevention for species confirmation), wound and ear infections (ca. 20% of isolates), and other clinical sites. The factors that determine

J. Glenn Morris, Jr. • Divisions of Geographic Medicine and Infectious Diseases, Department of Medicine, University of Maryland School of Medicine, and Veterans Affairs Medical Center, Baltimore, Maryland 21201.

Table 1. *Vibrio* species implicated as a cause of human disease

Species	Clinical presentations[a]		
	GI	Wound, ear infection	Septicemia
V. cholerae			
O1	++	(+)	
Non-O1	++	+	+
V. parahaemolyticus	++	+	(+)
V. fluvialis	++		
V. mimicus	++	+	
V. hollisae	++		(+)
V. furnissii	++		
V. vulnificus	+	++	++
V. alginolyticus		++	
V. damsela		++	
V. cincinnatiensis			+
V. carchariae		+	
V. metschnikovii	?		?

[a]GI, gastrointestinal; ++, most common presentation; +, other clinical presentations; (+), very rare presentation.

Table 2. Sources of non-O1 *V. cholerae* submitted for species confirmation to the Centers for Disease Control through 1985[a]

Source	No. (%) of isolates submitted
Feces or intestine	94 (46.5)
Spinal fluid	1 (0.5)
Blood	42 (20.8)
Wound	
Hand or arm	2 (1.0)
Foot or leg	9 (4.5)
Other or unknown	3 (1.5)
Ear	24 (11.9)
Gallbladder	1 (0.5)
Urine	4 (2.0)
Respiratory tract	10 (5.0)
Other or unknown	12 (5.9)
Total	202 (100)

[a]Data are from Farmer et al. (18).

whether a strain will spread in epidemic form or simply appear as a sporadic cause of illness remain to be determined.

CLASSIFICATION/TYPING SYSTEMS

O Antigen

O-group 1 and non-O1 *V. cholerae* give identical reactions in standard biochemical assays; differentiation is based on expression of the O1 antigen and agglutination with O1 antisera. The O1 biosynthetic gene cluster in *V. cholerae* (VcRfb) contains ca. 20 kb of DNA (87). In limited studies, probes to internal gene sequences within this cluster do not hybridize with non-O1 strains, suggesting that at least portions of this gene complex are absent in non-O1 *V. cholerae* (36).

Non-O1 strains have been subdivided by several investigators into a series of O groups. The two most commonly used schemes were developed by Smith (84) and Sakazaki and Shimada (76). The Smith typing system uses sera raised to live organisms and contains over 70 O groups. The Sakazaki system, utilizing heat-killed organisms, contains a comparable number of O groups, with O group designations now extending through O139. A third typing system has recently been developed by Siebeling and colleagues (83): while the methodology used is similar to that of Sakazaki, efforts to correlate O types with Sakazaki (or Smith) serotypes have not been completely successful (83).

Zymovar Analysis

V. cholerae strains can also be classified on the basis of enzyme electrophoretic patterns (zymovar analysis). The most extensive work in this area has been done by Salles and colleagues (52, 77, 78). In their system, virtually all epidemic-associated *V. cholerae* strains fall into one of two closely related zymovars: zymovar 13 (O1 classical strains) or zymovar 14 (O1 El Tor and O139 Bengal strains). In contrast, nonepidemic *V. cholerae* strains (O1 and non-O1) can be classified into close to 100 zymovars, with a degree of genetic diversity that approaches that of *Escherichia coli*. These observations emphasize the heterogeneity of *V. cholerae* strains; as is true for *E. coli*, this diversity should prompt caution in making generalizations about nonepidemic *V. cholerae* isolates.

Other Typing Systems

Non-O1 strains can produce a polysaccharide capsule (33), and some preliminary data suggest that a number of different capsular types exist (30); in the future, capsular typing may provide a useful means of classifying strains. It is also possible that typing systems for non-O1 *V. cholerae* could be based on the analysis of outer membrane protein profiles (71) or chromosomal restriction patterns for rRNA or other critical genes (39, 43). However, none of these techniques have been widely used in epidemiologic studies to date.

EPIDEMIOLOGY

Reservoirs

The principal reservoir for *V. cholerae* is the aquatic environment. Non-O1 strains are part of the normal, free-living (autochthonous) bacterial flora in estuarine areas (38, 72). These organisms have been isolated from surface waters at numerous sites in North America (10, 20, 38, 40, 41, 72, 80), Europe (4, 46, 58, 59), Asia (12, 15, 23, 65), and Australia (14), and it is likely that they are present in coastal and estuarine areas throughout the world. Rates of isolation from the environment generally exceed those for *V. cholerae* O1 by several orders of magnitude.

As would be expected with a free-living estuarine organism, environmental studies have generally shown no correlation between isolation of non-O1 *V. cholerae* and the presence of fecal coliforms (4, 38, 46, 72). Growth is influenced by salinity (isolation generally occurs in water with a salinity of 2 to 20 ppt) and water temperature (optimally >17°C) (10, 29, 38). However, isolation from fresh water has been reported (70), and infections have occurred after exposure to freshwater inland lakes (57, 63).

Non-O1 *V. cholerae* are a common isolate from shellfish, particularly from filter feeders such as oysters. In one study conducted by the U.S. Food and Drug Administration, non-O1 *V. cholerae* organisms were isolated from 111 (14%) of 790 samples of freshly harvested oyster shell stock (91). Isolation rates were highest during warm summer months, when counts of the organism would be expected to be highest in harvested waters (Table 3). Isolation of non-O1 *V. cholerae* from other shellfish, including crab (13), shrimp, and cultured prawns (68), has also been reported.

Non-O1 *V. cholerae* have been isolated from a variety of wild and domestic animals, including 6% of seagulls sampled in England (46); 14% of dogs in Calcutta (74); cows, goats, dogs, and chickens in India (79); ducks in Denmark (6); and horses, lambs, and bison in Colorado (69). It is unclear what role animal carriage plays in maintaining the organism in the environment. Asymptomatic carriage of non-O1 *V. cholerae* by humans is also known to occur (49, 97).

Transmission

Virtually all cases of non-O1 *V. cholerae* gastroenteritis acquired in the United States

Table 3. Frequency distribution of oyster samples positive for non-O1 *V. cholerae* by month of harvest: FDA study of oyster shell stock in the United States[a]

Mo. of harvest	No. of samples	
	Total	From which non-O1 *V. cholerae* was isolated
January	73	1 (1.4)
February	68	3 (4.4)
March	68	0
April	64	5 (7.8)
May	53	7 (13.2)
June	62	22 (35.5)
July	104	40 (38.5)
August	85	20 (23.5)
September	93	11 (11.8)
October	53	1 (1.9)
November	40	1 (2.8)
December	27	0
Total	790	111 (14)

[a]Data are from Twedt et al. (91).

are associated with eating raw or undercooked shellfish (particularly raw oysters)
(56). While the linkage is less strong, there is
also a suggestion that septicemia can result
from ingestion of the organism in seafood
(42, 75).

While seafood remains an important vehicle of infection for sporadic non-O1 disease
outside of the United States, it is clear that
transmission can also occur through other
routes (53, 56). Potatoes and chopped egg on
an asparagus salad were implicated as vehicles in outbreaks in Czechoslovakia (2) and
on an airplane flight to Australia (11), respectively. In a case-control study in Cancun,
Mexico (19), infection was significantly associated with eating preprepared gelatin; borderline significance was found with eating
seafood ($P=0.06$) and drinking powdered
milk ($P=0.06$). In the Ban Vinai refugee
camp in Thailand, non-O1 strains were isolated from a number of different foods (4% of
vegetable samples, 39% of meat samples)
and from 16% of drinking water samples
(90). An outbreak in the Sudan in 1968 was
attributed to contamination of surface water
adjacent to a well (37).

Wound and ear infections from which non-
O1 *V. cholerae* have been isolated have almost all been associated with water exposure
(9, 24, 26): in most instances this has been
seawater, although there are reports of cellulitis occurring after exposure to fresh water
(57, 63). Positive sputum cultures have generally been from cases of near drowning (24).
One case of meningitis in a 3-week-old infant
occurred after the child's bottle had been
stored in the same cooler used to keep freshly
caught crabs (73).

Seasonality

Cases of non-O1 *V. cholerae* gastroenteritis
in the United States tend to occur in the late
summer and early fall (56), paralleling the
seasonal increase seen in non-O1 *V. cholerae*
in the environment (water and oysters). In
Bangladesh (7), sporadic cases of non-O1 *V.*
cholerae gastroenteritis have been reported to
increase during the spring and summer (with
a peak in April and May).

Frequency of Isolation

There is wide variability in the reported
frequency of isolation of nonepidemic strains
of non-O1 *V. cholerae* from stool cultures of
persons with diarrhea (16, 19, 24, 44, 45, 51,
54, 89, 90, 92). The highest reported isolation rate is 16% among 134 children and
adults with diarrhea in Cancun, Mexico, seen
during a single 2-month period in July and
August 1983 (19). While matched controls
were not examined, no non-O1 *V. cholerae*
was isolated from a group of 22 well children
in this same study. In a comparable study
conducted in Mexico City (at a high altitude
and at some distance from the coast), no non-
O1 strains were identified from patients with
diarrhea (92). In Bangladesh, isolation rates
of nonepidemic non-O1 strains from persons
with diarrhea at the Cholera Research Laboratory in Dhaka (51) and the Matlab Treatment Center (7) have been reported to be 3
and 7%, respectively. Matched controls were
not selected in these studies; however, in the
Dhaka study (51) it was noted that non-O1 *V.*
cholerae was isolated from only 0.01% of
samples from "healthy village populations."

In passive, laboratory-based studies in the
United States, non-O1 *V. cholerae* has rarely
been isolated. Seven coastal-area hospitals in
four southern states isolated only seven specimens of non-O1 *V. cholerae* (including five
isolates from a single outbreak) from approximately 11,000 stool cultures performed with
use of thiosulfate-citrate-bile salts-sucrose
medium (54). Comparable data were obtained during a 14-year period in a single hospital in Annapolis, Maryland (24). Non-O1
strains were isolated from 13 (2.7%) of 479
persons in a cohort of physicians attending a
convention in New Orleans in late September
(49). Despite this relatively high colonization
rate, only 2 (15%) of the 13 culture-positive
persons were symptomatic, which is compa-

rable to the overall 14% rate of diarrhea reported in the entire cohort.

PATHOPHYSIOLOGY

From the time the strains were first identified, questions have arisen about the pathogenicity of non-O1 *V. cholerae* strains. There is now compelling evidence from volunteer studies that nonepidemic non-O1 strains can indeed cause human disease (55).

At the same time, rates of disease occurrence are relatively low, given the high frequency with which the organism is found in the environment. For example, despite a high frequency of contamination of food and water in a place such as the Ban Vinai refugee camp, non-O1 strains were present in only 0.6% of diarrheal stools (90). These observations lead to the hypothesis that only a minority of strains carry the virulence factors necessary to cause human illness. At this point no single virulence factor that can account for all reported cases has been identified. Analogous to what we now understand of diarrheagenic *E. coli* (47) (and in keeping with the genetic diversity seen among non-O1 strains), there may be several subgroups of non-O1 *V. cholerae* that are able to cause disease, each through a different pathogenic mechanism. Possible mechanisms are summarized below.

Extracellular Toxins

CT

Some non-O1 strains can produce cholera toxin (CT) or choleralike toxins (12, 39, 85, 96). In one study in Bangladesh, CT-producing strains were isolated from 29% of patients with sporadic non-O1 *V. cholerae* gastroenteritis; patients infected with these strains had more severe illness, with greater weight loss, a significantly higher admission specific gravity, and more prolonged and profuse diarrhea (85).

CT-producing strains of non-O1 *V. cholerae* have always been found at relatively high rates in Bangladesh and India, with one Indian study finding CT activity in 9 (26%) of 34 clinical isolates from patients with sporadic cases compared with activity in only 1 of 10 environmental isolates (12). However, the frequency of isolation of CT-positive strains has been much lower outside of the Indian subcontinent. In a study in Thailand, only 5 (2%) of 237 environmental and 0 of 44 clinical isolates carried gene sequences homologous to that of the CT gene (22). In a study of the Louisiana Gulf Coast environment, 7 (0.3%) of ca. 2,500 non-O1 *V. cholerae* carried the gene for CT and produced CT (72) (and in retrospect, some of these CT-positive isolates may have been *V. mimicus*).

NAG-ST

Some non-O1 strains produce a 17-amino-acid heat-stable enterotoxin (designated NAG-ST) that closely resembles the heat-stable toxin produced by enterotoxigenic strains of *E. coli* and *Yersinia enterocolitica* (5, 25, 88). Production of both CT and NAG-ST by a single strain has not been reported. In volunteer studies, a colonizing, NAG-ST-producing strain of non-O1 *V. cholerae* caused diarrhea in healthy North American adults; one volunteer, who received 10^6 CFU, produced over 5 liters of diarrheal stool (55). In epidemiologic studies it has been possible to identify a subgroup of non-O1 strains that are closely related genetically (based on zymovar analysis), are characterized by a common capsular type and production of NAG-ST, and are significantly more likely to be isolated from patients with diarrhea than from the environment (35).

Again, however, the strains within this group account for a small minority of all clinical non-O1 *V. cholerae* isolates. In one study, only 7 (6.8%) of 103 clinical non-O1 *V. cholerae* isolates from Thailand and none of 78 isolates from Mexico and the United States hybridized with a DNA probe for

NAG-ST (23). In a Japanese study, none of 58 clinical isolates were NAG-ST positive (21).

El Tor hemolysin

All non-O1 V. cholerae strains are hemolytic, and all appear to carry the gene for the El Tor hemolysin. The El Tor hemolysin has been shown to have enterotoxic activity, and it has been postulated that the enterotoxigenicity of non-O1 strains is due to this toxin (21, 27, 50, 95). A similar role for the El Tor hemolysin has been proposed in explaining the virulence of CT-negative strains of V. cholerae O1.

At this point there is not a clear consensus on the importance of the hemolysin in either O1 (3, 48) or non-O1 disease. However, several factors suggest that the hemolysin is not a major virulence factor for non-O1 V. cholerae. Since all strains are hemolytic, the presence or absence of hemolytic activity would not help explain the low rate of disease occurrence compared to the frequency of environmental isolation. A colonizing, hemolytic non-O1 strain was fed to volunteers and did not cause disease, even at a dose of 10^9 CFU (55). Finally, no change in virulence in mice was observed when the hemolysin gene of a non-O1 strain was inactivated by transposon mutagenesis (34).

Other putative toxins

A variety of other possible toxins have been identified, including a Kanagawa-like hemolysin (21, 25), a Shiga-like toxin (21, 61), the zonula occludens toxin recently found in V. cholerae O1 (which, however, appears to almost always be linked to CT [32]), and various cell-associated hemagglutinins (82, 96). Further studies, including further volunteer studies and studies with isogenic mutants, are needed to define the importance of these and other factors in the pathogenesis of non-O1 V. cholerae-associated disease.

Colonization Factors

Non-O1 strains differ in their abilities to colonize and cause disease in rabbits (86): isolates from patients with diarrhea are significantly more likely to colonize and cause disease than are environmental isolates. Differences in colonizing ability have been confirmed in volunteer studies (55), and it appears reasonable to hypothesize that colonization is a necessary prerequisite for occurrence of disease. However, colonization alone is not sufficient to cause human illness: in volunteer studies, one strain that was an excellent colonizer failed to cause diarrhea even after administration of 10^9 CFU (55). While some non-O1 strains are fimbriated, colonization can occur in the absence of fimbriae (55, 60).

Encapsulation

Non-O1 strains are able to produce a polysaccharide capsule that confers resistance to serum bactericidal activity and is associated with increased virulence in mice (Fig. 1) (33). Heavily encapsulated strains are significantly more likely to be isolated from patients with septicemia than are strains with minimal or no capsular polysaccharide (31). Strains of V. cholerae O1 are not encapsulated, which may account for their inability to cause septicemia.

Non-O1 V. cholerae strains are able to undergo phase variation, shifting between encapsulated and unencapsulated variants at a rate of ca. 10^{-5} (33). Analogous to findings with V. vulnificus, which also has a polysaccharide capsule (67, 93), encapsulated strains produce opaque colonies, while colonies of unencapsulated strains are translucent. The degree of encapsulation can be reflected in the colonial morphology, with colonies of strains producing a heavy capsule having a denser, more opaque appearance (31).

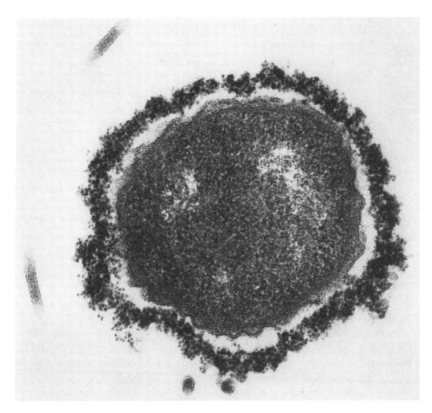

Figure 1. Electron photomicrograph of thin sections of non-O1 *V. cholerae* NRT-36S stained with polycationic ferritin, showing heavy encapsulation.

CLINICAL SYNDROMES

Gastroenteritis

The gastroenteritis associated with non-O1 *V. cholerae* can range from mild illness to profuse, watery diarrhea comparable to that seen in patients with epidemic cholera. In a group of 19 patients with sporadic non-O1 *V. cholerae* gastroenteritis in Dhaka, Bangladesh (51), illness ranged from mild diarrhea, requiring no therapy, to heavy, choleralike purges. The maximum stool volume among these 19 patients was 8 liters (recorded during hospitalization), and the longest duration of diarrhea was 2.5 days; however, the median stool volume during hospitalization was less than 1 liter, and diarrhea lasted for less than 24 h in half of the

cases. Seven of the 16 patients for whom data were available had fever. No mention was made of upper abdominal complaints. Comparable symptoms were seen in a second group of 14 patients from Dhaka (85), although it was also noted that in this group, 11% had vomiting, 71% had abdominal cramps, and 43% had fever; none had bloody diarrhea.

In a group of 14 U.S. cases identified retrospectively on the basis of positive stool cultures (56), symptoms included diarrhea (100% of patients), abdominal cramps (93%), and fever (71%), with nausea and vomiting occurring less frequently (21%). A quarter of the U.S. patients reported bloody diarrhea. Illness was noted in several cases to be quite severe, with 20 to 30 diarrheal stools per day. Median duration of illness was 6.4

days, with 8 of the 14 requiring hospitalization; however, as these patients were identified on the basis of positive cultures, hospitalized patients and patients with more severe, prolonged diarrhea may have been overrepresented.

While the above data provide a general picture of the gastroenteritis that can be seen in association with non-O1 *V. cholerae,* it should be recognized that they probably reflect symptoms from a heterogeneous group of non-O1 strains (with a heterogeneous group of virulence mechanisms). A more accurate assessment of symptoms associated with a single strain or subset of strains can be obtained from volunteer studies and outbreak reports.

One series of volunteer studies has been conducted with non-O1 *V. cholerae* (55). In these studies, three different non-O1 strains were administered to volunteers in doses as high as 10^9 CFU. Two of the strains (an O17 and an O31 strain) caused no illness (55). The third strain, NRT-36S (an O31 strain isolated from a patient with traveler's diarrhea at Narita Airport, Tokyo), caused diarrhea; this strain was an excellent colonizer and produced NAG-ST. The median incubation period before onset of gastroenteritis for volunteers receiving NRT-36S was 10 h (range, 5.5 to 96 h). There was no obvious association between incubation period and inoculum size (55). Diarrheal stool volumes ranged from 140 to 5,397 ml (Table 4). The diarrheal illness tended to be short-lived, with a median duration of 21 h (range, 3.5 to 48 h). Abdominal cramps were prominent, with one volunteer (who received 10^6 CFU) having only abdominal cramps and no diarrhea. While diarrhea was usually mild, two volunteers each produced over 2 liters of diarrheal stool.

Two foodborne outbreaks attributable to non-O1 *V. cholerae* have been reported. In one, on an airplane flight to Australia (11), the mean incubation period was 11.5 h (range, 5.25 to 37.5 h); in an outbreak at a technical school in Czechoslovakia (2), the incubation period was estimated to be 20 to 30 h. Attack rates in both outbreaks exceeded 60%, emphasizing that a virulent non-O1 strain at a high enough inoculum can consistently cause illness in healthy adults. In both outbreaks, diarrhea was generally mild and resolved in the majority of cases in less than 24 h. However, it was also noted that some patients had severe diarrhea (12 or more bowel movements in 24 h; up to 6 bowel movements in half an hour) and that illness was often accompanied by abdominal cramps and vomiting. Unfortunately, the strains responsible for the outbreaks were not well characterized and are no longer available for study.

Septicemia

Non-O1 *V. cholerae* has also been isolated from blood (75). The majority of these cases have involved immunocompromised patients, particularly those with hematologic malignancies or cirrhosis, suggesting that host susceptibility is a critical factor in determining whether septicemia will occur. The case fatality rate among reported cases is 61.5% (75). As noted above, strains responsible for septicemia have uniformly been found to be heavily encapsulated (31).

Infections at Other Sites

Non-O1 *V. cholerae* has been isolated from a variety of other sites, including wounds, ears, sputum, urine, and cerebral spinal fluid (9, 18, 24, 26). The clinical significance of the wound, ear, and sputum isolates is unclear. In many such cases, non-O1 *V. cholerae* has been isolated together with a variety of other potential pathogens (24). In the absence of subsequent bacteremia, it is difficult to say whether the non-O1 strain isolated was a reflection of a significant non-O1 *V. cholerae* infection or was simply a commensal or colonizing strain.

Table 4. Symptoms of volunteers ingesting non-O1 *V. cholerae* NRT-36S[a]

Inoculum size	No. of volunteers	No. of volunteers with:			Stool vol (ml)
		Diarrhea	Fever	Abdominal pain	
10^5	2	0	0	0	
10^6	3	2	0	2	5,397, 140
10^7	2	1	0	0	589
10^9	3	3	0	3	2,091, 276, 253

[a]Reproduced from the *Journal of Clinical Investigation,* 1990, **85:**697–705, by copyright permission of the American Society for Clinical Investigation.

DIAGNOSIS/THERAPY

Methodology for isolation of non-O1 strains is identical to that for *V. cholerae* O1. Non-O1 strains grow well on blood agar and other nonselective media that may be used for wound cultures; they also grow well in standard blood culture media.

As with *V. cholerae* O1, the mainstay of therapy of diarrheal disease is oral rehydration. While efficacy data for non-O1 strains are not available, tetracycline may be beneficial in more severe cases of gastroenteritis. For septicemia and wound infections, tetracycline and/or ciprofloxacin appears to be a reasonable choice for therapy, possibly in combination with an aminoglycoside in severely ill patients.

SUMMARY AND FUTURE DIRECTIONS

Non-O1 *V. cholerae* is ubiquitous in the environment. It is commonly present in shellfish and, in the developing world, is frequently isolated from food and water. Current data suggest that there are subsets of strains that are able to cause sporadic human disease: strains that colonize and produce either CT or NAG-ST have been associated with diarrhea, while heavy encapsulation appears to be a prerequisite for septicemia. However, there are still a number of disease-associated strains for which we do not have a good understanding of pathogenic mechanisms. There is a need for further studies to define other pathogenic subgroups and to evaluate mechanisms by which these subgroups cause disease.

REFERENCES

1. **Albert, M. J., A. K. Siddique, M. S. Islam, A. S. G. Faruque, M. Ansaruzzaman, S. M. Faruque, and R. B. Sack.** 1993. A large outbreak of clinical cholera due to *Vibrio cholerae* non-O1 in Bangladesh. *Lancet* **341:**704.
2. **Aldova, E., K. Laznickova, E. Stepankova, and J. Lietava.** 1968. Isolation of nonagglutinable vibrios from an enteritis outbreak in Czechoslovakia. *J. Infect. Dis.* **118:**25–31.
3. **Alm, R. A., G. Mayrhofer, I. Kotlarski, and P. A. Manning.** 1991. Amino-terminal domain of the El Tor hemolysin of *Vibrio cholerae* O1 is expressed in classical strains and is cytotoxic. *Vaccine* **9:**588–591.
4. **Amaro, C., A. E. Toranzo, E. A. Gonzalez, J. Blanco, M. J. Pujalte, R. Azsnar, and E. Garay.** 1990. Surface and virulence properties of environmental *Vibrio cholerae* non-O1 from Albufera Lake (Valencia, Spain). *Appl. Environ. Microbiol.* **56:**1140–1147.
5. **Arita, M., T. Takeda, T. Honda, and T. Miwatani.** 1986. Purification and characterization of *Vibrio cholerae* non-O1 heat-stable enterotoxin. *Infect. Immun.* **52:**45–49.
6. **Bisgaard, M., R. Sakazaki, and T. Shimada.** 1978. Prevalence of non-cholera vibrios in cavum nasi and pharynx of ducks. *Acta Pathol. Microbiol. Scand. Sect. B* **86:**261–266.
7. **Black, R. E., M. H. Merson, and K. H. Brown.** 1983. Epidemiological aspects of diarrhea associated with known enteropathogens in rural Bangladesh, p. 73–86. *In* L. C. Chen and N. S. Scrimshaw (ed.), *Diarrhea and Malnutrition.* Plenum Publishing Corp., New York.
8. **Bode, R. B., P. R. Brayton, R. R. Colwell, F. M. Russo, and W. E. Bullock.** 1986. A new Vibrio species, *Vibrio cincinnatiensis,* causing meningitis: successful treatment in an adult. *Ann. Intern. Med.* **106:**55–56.
9. **Bonner, J. R., A. S. Coker, C. R. Berryman, and H. M. Pollock.** 1983. Spectrum of Vibrio in-

fections in a Gulf coast community. *Ann. Intern. Med.* **99**:464–469.

10. **Colwell, R. R., P. A. West, D. Maneval, E. F. Remmers, E. L. Elliot, and N. E. Carlson.** 1984. Ecology of pathogenic Vibrios in Chesapeake Bay, p. 367–387. *In* R. R. Colwell (ed.). *Vibrios in the Environment.* John Wiley & Sons, New York.

11. **Dakin, W. P. H., D. J. Howell, R. G. A. Sutton, M. F. O'Keefe, and P. Thomas.** 1974. Gastroenteritis due to non-agglutinable (non-cholera) vibrios. *Med. J. Aust.* **2**:487–490.

12. **Datta-Roy, K., K. Banerjee, S. P. De, and A. C. Ghose.** 1986. Comparative study of expression of hemagglutinins, hemolysins, and enterotoxins by clinical and environmental isolates of non-O1 *Vibrio cholerae* in relation to their enteropathogenicity. *Appl. Environ. Microbiol.* **52**:875–879.

13. **Davis, J. W., and R. K. Sizemore.** 1982. Incidence of Vibrio species associated with blue crabs (*Callinectes sapidus*) collected from Galveston Bay, Texas. *Appl. Environ. Microbiol.* **43**:1092–1097.

14. **Desmarchelier, P. M., and J. L. Reichelt.** 1981. Phenotypic characterization of clinical and environmental isolates of *Vibrio cholerae* from Australia. *Curr. Microbiol.* **5**:123–127.

15. **Dutt, A. K., S. Alwi, and T. Velauthan.** 1971. A shellfish-borne cholera outbreak in Malaysia. *Trans. R. Soc. Trop. Med. Hyg.* **65**:815–818.

16. **Echeverria, P., C. Pitarangsi, B. Eampokalap, S. Vibulbandhitkit, P. Boonthai, and B. Rowe.** 1983. A longitudinal study of the prevalence of bacterial enteric pathogens among adults with diarrhea in Bangkok, Thailand. *Diagn. Microbiol. Infect. Dis.* **1**:193–204.

17. **El-Shawi, N., and A. J. Thewaini.** 1969. Nonagglutinable vibrios isolated in the 1966 epidemic of cholera in Iraq. *Bull. W.H.O.* **40**:163–166.

18. **Farmer, J. J., III, F. W. Hickman-Brenner, and M. T. Kelly.** 1985. *Vibrio,* p. 282–301. *In* E. H. Lennette, A. Balows, W. J. Hausler, Jr., and H. J. Shadomy (ed.). *Manual of Clinical Microbiology,* 4th ed. American Society for Microbiology, Washington, D.C.

19. **Finch, M. J., J. L. Valdespino, J. G. Wells, G. Perez-Perez, F. Arjona, A. Sepulveda, D. Bessudo, and P. A. Blake.** 1987. Non-O1 *Vibrio cholerae* infections in Cancun, Mexico. *Am. J. Trop. Med. Hyg.* **36**:393–397.

20. **Gardner, A. D., and V. K. Venkatraman.** 1935. The antigens of the cholera group of vibrios. *J. Hyg.* **35**:262–282.

21. **Gyobu, Y., H. Kodama, and S. Sato.** 1991. Studies on the enteropathogenic mechanism of non-O1 *Vibrio cholerae.* III. Production of enteroreactive toxins. Kansenshogaku. *Zasshi* J. Jpn. *Assoc. Infect. Dis.* **65**:781–787.

22. **Hanchalay, S., J. Seriwatana, P. Echeverria, J. Holmgren, C. Tirapat, S. L. Moseley, and D. N. Taylor.** 1985. Non-O1 *Vibrio cholerae* in Thailand: homology with cloned cholera toxin genes. *J. Clin. Microbiol.* **21**:288–289.

23. **Hoge, C. W., O. Sethabutr, L. Bodhidatta, P. Echeverria, D. C. Robertson, and J. G. Morris, Jr.** 1990. Use of a synthetic oligonucleotide probe to detect strains of non-O1 *Vibrio cholerae* carrying the gene for heat stable enterotoxin (NAG-ST). *J. Clin. Microbiol.* **28**:1473–1476.

24. **Hoge, C. W., D. Watsky, R. N. Peeler, J. P. Libonati, E. Israel, and J. G. Morris, Jr.** 1989. Epidemiology and spectrum of vibrio infections in a Chesapeake Bay community. *J. Infect. Dis.* **160**:985–993.

25. **Honda, T., M. Arita, T. Takeda, M. Yoh, and T. Miwatani.** 1985. Non-O1 *Vibrio cholerae* produces two newly identified toxins related to *Vibrio parahaemolyticus* hemolysin and *Escherichia coli* heat-stable enterotoxin. *Lancet* **ii**:163–164.

26. **Hughes, J. M., D. G. Hollis, E. J. Gangarosa, and R. E. Weaver.** 1978. Non-cholera vibrio infections in the United States: clinical, epidemiologic, and laboratory features. *Ann. Intern. Med.* **88**:602–606.

27. **Ichinose, Y., K. Yamamoto, N. Nakasone, M. J. Tanabe, T. Takeda, T. Miwatani, and M. Iwanaga.** 1987. Enterotoxicity of El Tor-like hemolysin of non-O1 *Vibrio cholerae.* *Infect. Immun.* **55**:1090–1093.

28. **International Center for Diarrheal Disease Research, Bangladesh, Cholera Working Group.** 1993. Large epidemic of cholera-like disease in Bangladesh caused by *Vibrio cholerae* O139 synonym Bengal. *Lancet* **342**:387–390.

29. **Janda, J. M., C. Powers, R. G. Bryant, and S. L. Abbott.** 1988. Current perspectives on the epidemiology and pathogenesis of clinically significant *Vibrio* species. *Clin. Microbiol. Rev.* **1**:245–267.

30. **Johnson, J. A., C. A. Bush, and J. G. Morris, Jr.** Unpublished data.

31. **Johnson, J. A., A. Joseph, P. Panigrahi, and J. G. Morris, Jr.** 1992. Frequency of encapsulated versus unencapsulated strains of non-O1 *Vibrio cholerae* isolated from patients with septicemia or diarrhea, or from environmental strains, abstr. B-277. *Abstr. 92nd Gen. Meet. Am. Soc. Microbiol. 1992.*

32. **Johnson, J. A., J. G. Morris, Jr., and J. B. Kaper.** 1993. Gene encoding zonula occludens toxin (*zot*) does not occur independently from cholera enterotoxin genes (*ctx*) in *Vibrio cholerae.* *J. Clin. Microbiol.* **31**:732–733.

33. **Johnson, J. A., P. Panigrahi, and J. G. Morris, Jr.** 1992. Non-O1 *Vibrio cholerae* NRT36S produces a polysaccharide capsule that determines colony morphology, serum resistance, and virulence in mice. *Infect. Immun.* **60**:684–689.

34. **Johnson, J. A., P. Panigrahi, R. G. Russell, and J. G. Morris, Jr.** 1991. Role of hemolysin in the pathogenesis of non-serogroup O1 *Vibrio cholerae,* abstr. 1326. *Program Abstr. 31st Intersci. Conf. Antimicrob. Agents Chemother.*

35. **Johnson, J. A., C. A. Salles, and J. G. Morris, Jr.** 1992. Correlation of heat-stable enterotoxin and capsule type of non-O1 *Vibrio cholerae,* Abstr. 613. *Program Abstr. 32nd Intersci. Conf. Antimicrob. Agents Chemother.*

36. **Johnson, J. A., A. C. Wright, and J. G. Morris, Jr.** Unpublished data.

37. **Kamal, A. M.** 1971. Outbreak of gastroenteritis by non-agglutinable (NAG) vibrios in the republic of the Sudan. *J. Egypt. Public Health Assoc.* **46:**125–173.

38. **Kaper, J. B., H. Lockman, R. R. Colwell, and S. W. Joseph.** 1979. Ecology, serology, and enterotoxin production of *Vibrio cholerae* in Chesapeake Bay. *Appl. Environ. Microbiol.* **37:**91–103.

39. **Kaper, J. B., J. P. Nataro, N. C. Roberts, R. Siebeling, and H. B. Bradford.** 1986. Molecular epidemiology of non-O1 *Vibrio cholerae* and *Vibrio mimicus* in the U.S. Gulf Coast region. *J. Clin. Microbiol.* **23:**652–654.

40. **Kaysner, C. A., C. Albeyta, Jr., M. M. Wekell, A. DePaola, Jr., R. F. Scott, and J. M. Leitch.** 1987. Incidence of *Vibrio cholerae* from estuaries of the United States West Coast. *Appl. Environ. Microbiol.* **53:**1344–1348.

41. **Kenyon, J. E., D. C. Gillies, D. R. Piexoto, and B. Austin.** 1983. *Vibrio cholerae* (non-O1) isolated from California coastal waters. *Appl. Environ. Microbiol.* **46:**1232–1233.

42. **Klontz, K. C.** 1990. Fatalities associated with *Vibrio parahaemolyticus* and *Vibrio cholerae* non-O1 infections in Florida (1981–1988). *South. Med. J.* **83:**500–502.

43. **Koblavi, S., F. Gramont, and P. A. Gramont.** 1990. Clonal diversity of *Vibrio cholerae* O1 evidenced by rRNA gene restriction patterns. *Res. Microbiol.* **141:**645–657.

44. **Kotloff, K. L., S. S. Wasserman, J. Y. Steciak, B. D. Tall, G. A. Losonsky, P. Nair, J. G. Morris, Jr., and M. M. Levine.** 1988. Acute diarrhea in Baltimore children attending an outpatient clinic. *Pediatr. Infect. Dis. J.* **7:**753–759.

45. **Lanata, C. F., R. E. Black, R. B. Sack, A. Yi, D. Maurtua, A. Gil, L. Barua, H. Verastegui, E. Miranda, and R. Gilman.** 1989. Epidemiology of acute diarrheal diseases in children under three years of age in Peri-urban Lima, Peru, p. 17. *Program Abstr. 25th Joint Conf. Cholera—U.S.-Jpn. Coop. Med. Sci. Program.*

46. **Lee, J. V., D. J. Bashford, T. J. Donovan, A. L. Furniss, and P. A. West.** 1984. The incidence and distribution of V. cholerae in England, p. 427–450. *In* R. R. Colwell (ed.), *Vibrios in the Environment.* John Wiley & Sons, New York.

47. **Levine, M. M.** 1987. *Escherichia coli* that cause diarrhea: enterotoxigenic, enteropathogenic, enteroinvasive, enterohemorrhagic, and enteroadherent. *J. Infect. Dis.* **166:**377–389.

48. **Levine, M. M., J. B. Kaper, D. Herrington, G. Losonsky, J. G. Morris, Jr., M. L. Clements, R. E. Black, B. Tall, and R. Hall.** 1988. Volunteer studies of deletion mutants of *Vibrio cholerae* O1 prepared by recombinant techniques. *Infect. Immun.* **56:**161–167.

49. **Lowry, P. W., L. M. McFarland, B. H. Peltier, N. C. Roberts, H. B. Bradford, J. L. Herndon, D. F. Stroup, J. B. Mathison, P. A. Blake, and R. A. Gunn.** 1989. Vibrio gastroenteritis in Louisiana: a prospective study among attendees of a scientific congress in New Orleans. *J. Infect. Dis.* **160:**978–984.

50. **McCardell, B. A., J. M. Madden, and D. B. Shah.** 1985. Isolation and characterization of a cytolysin produced by *Vibrio cholerae* serogroup non-O1. *Can. J. Microbiol.* **31:**711–720.

51. **McIntyre, O. R., J. C. Feeley, W. B. Greenough, A. S. Benenson, S. I. Hassan, and A. Saad.** 1965. Diarrhea caused by non-cholera vibrios. *Am. J. Trop. Med. Hyg.* **14:**412–418.

52. **Momen, H., and C. A. Salles.** 1985. Enzyme markers for *Vibrio cholerae:* identification of classical, El Tor, and environmental strains. *Trans. R. Soc. Trop. Med. Hyg.* **79:**773–776.

53. **Morris, J. G., Jr.** 1990. Non-O group 1 *Vibrio cholerae:* a look at the epidemiology of an occasional pathogen. *Epidemiol. Rev.* **12:**179–191.

54. **Morris, J. G., Jr., and R. E. Black.** 1985. Cholera and other vibrioses in the United States. *N. Engl. J. Med.* **312:**343–350.

55. **Morris, J. G., Jr., T. Takeda, B. D. Tall, G. A. Losonsky, S. K. Bhattacharya, B. D. Forrest, B. A. Kay, and M. Nishibuchi.** 1990. Experimental non-O group 1 *Vibrio cholerae* gastroenteritis in humans. *J. Clin. Invest.* **85:**697–705.

56. **Morris, J. G., Jr., R. Wilson, B. R. Davis, I. K. Wachsmuth, C. F. Riddle, H. G. Wathen, R. A. Pollard, and P. A. Blake.** 1981. Non-O Group 1 *Vibrio cholerae* gastroenteritis in the United States. *Ann. Intern. Med.* **94:**656–658.

57. **Mulder, G. D., T. M. Reis, and T. R. Beaver.** 1989. Non-toxigenic *Vibrio cholerae* wound infection after exposure to contaminated lake water. *J. Infect. Dis.* **159:**809–811.

58. **Muller, G.** 1977. Befunde an nicht-agglutinierenden, cholera-ahnlichen Vibrionen (NAGs) in Abwasser, Flubwasser and Meerwasser. [Nonagglutinable cholera vibrios (NAGs) in sewage, riverwater, and seawater.] *Zentralbl. Bakteriol. Hyg. Abt. 1 Orig Reihe B* **165:**487–497.

59. **Muller, H. E.** 1978. Vorkommen und Okologie von NAG-Vibrionen in Oberflachen-gewassern. (Occurrence and ecology of NAG vibrios in surface

waters.) *Zentralbl. Bakteriol. Hyg. Abt. 1 Orig Reihe B* **167**:272–284.

60. Nakasone, N., and M. Iwanaga. 1990. Pili of *Vibrio cholerae* non-O1. *Infect. Immun.* **58**:1640–1646.

61. O'Brien, A. D., M. E. Chen, R. K. Holmes, J. B. Kaper, and M. M. Levine. 1984. Environmental and human isolates of *Vibrio cholerae* and *Vibrio parahaemolyticus* produce a *Shigella dysenteriae* 1 (Shiga-like) cytotoxin. *Lancet* **i**:77–78.

62. Pavia, A. T., J. A. Bryan, K. L. Maher, T. R. Hester, and J. J. Farmer III. 1989. *Vibrio carchariae* infection after a shark bite. *Ann. Intern. Med.* **111**:85–86.

63. Pitrak, D. L., and J. D. Gindorf. 1989. Bacteremia cellulitis caused by non-serogroup O1 *Vibrio cholerae* acquired in a freshwater inland lake. *J. Clin. Microbiol.* **27**:2874–2876.

64. Pollitzer, R. 1959. *Cholera.* Monograph no. 43. World Health Organization, Geneva.

65. Rai, R. N., V. C. Tripathi, and R. D. Joshi. 1991. Persistence of *Vibrio cholerae* in inter-epidemic period—preliminary observations on analysis of water. *J. Communicable Dis.* **23**:44–45.

66. Ramamurthy, T., S. Garg, R. Sharma, S. K. Bhattacharya, G. B. Nair, T. Shimada, T. Takeda, T. Karasawa, H. Kurazano, A. Pal, and Y. Takeda. 1993. Emergence of novel strain of *Vibrio cholerae* with epidemic potential in southern and eastern India. *Lancet* **341**:703–704.

67. Reddy, G. P., U. Hayat, C. Abeygunawardana, C. Fox, A. C. Wright, D. R. Maneval, C. A. Bush, and J. G. Morris, Jr. 1992. Purification and structure determination of *Vibrio vulnificus* capsular polysaccharide. *J. Bacteriol.* **174**:2620–2630.

68. Reilly, P. J., and D. R. Twiddy. 1992. *Salmonella* and *Vibrio cholerae* in brackish water cultured tropical prawns. *Int. J. Food Microbiol.* **16**:293–301.

69. Rhodes, J. B., D. Schweitzer, and J. E. Ogg. 1985. Isolation of non-O1 *Vibrio cholerae* associated with enteric disease of herbivores in western Colorado. *J. Clin. Microbiol.* **22**:572–575.

70. Rhodes, J. B., H. L. Smith, Jr., and J. E. Ogg. 1986. Isolation of non-O1 *Vibrio cholerae* serovars from surface waters in western Colorado. *Appl. Environ. Microbiol.* **51**:1216–1219.

71. Richardson, K., and C. D. Parker. 1985. Identification and characterization of *Vibrio cholerae* surface proteins by radioiodination. *Infect. Immun.* **48**:87–93.

72. Roberts, N. C., R. J. Seibeling, J. B. Kaper, and H. B. Bradford. 1982. Vibrios in the Louisiana Gulf Coast environment. *Microb. Ecol.* **8**:299–312.

73. Rubin, L. G., J. Altman, L. K. Epple, and R. H. Yolken. 1981. *Vibrio cholerae* meningitis in a neonate. *J. Pediatr.* **98**:940–942.

74. Sack, R. B. 1973. A search for canine carriers of Vibrio. *J. Infect. Dis.* **127**:709–712.

75. Safrin, S., J. G. Morris, Jr., M. Adams, V. Pons, R. Jacobs, and J. E. Conte, Jr. 1988. Non-O:1 *Vibrio cholerae* bacteremia: case report and review. *Rev. Infect. Dis.* **10**:1012–1017.

76. Sakazaki, R., and T. Shimada. 1977. Serovars of *Vibrio cholerae*. *Jpn. J. Med. Sci. Biol.* **30**:279–282.

77. Salles, C. A., A. R. da Silva, and H. Momen. 1986. Enzyme typing and phenetic relationships in *Vibrio cholerae*. *Rev. Bras. Genet.* **9**:407–419.

78. Salles, C. A., and J. G. Morris, Jr. Unpublished data.

79. Sanyal, S. C., S. J. Singh, I. C. Tiwari, P. C. Sen, S. M. Marwah, U. R. Hazarika, A. Singh, T. Shimada, and R. Sakazaki. 1974. Role of household animals in maintenance of cholera infection in a community. *J. Infect. Dis.* **130**:575–579.

80. Seidler, R. J., and T. M. Evans. 1984. Computer-assisted analysis of Vibrio field data: four coastal areas, p. 411–425. *In* R. R. Colwell (ed.). *Vibrios in the Environment.* John Wiley & Sons, New York.

81. Shehabi, A. A., A. B. Abu Rajab, and A. A. Shaker. 1980. Observations on the emergence of non-cholera Vibrios during an outbreak of cholera. *Jordan Med. J.* **14**:125–127.

82. Shehabi, A. A., H. Drexler, and S. H. Richardson. 1986. Virulence mechanisms associated with clinical isolates of non-O1 *Vibrio cholerae*. *Zentralbl. Bakteriol Hyg. Reihe A* **261**:232–239.

83. Siebeling, R. J., L. B. Adams, Z. Yusof, and A. D. Larson. 1984. Antigens and serovar-specific antigens of *Vibrio cholerae*, p. 33–58. *In* R. R. Colwell (ed.), *Vibrios in the Environment.* John Wiley & Sons, New York.

84. Smith, H. L., Jr. 1979. Serotyping of non-cholera vibrios. *J. Clin. Microbiol.* **10**:85–90.

85. Spira, W. M., R. R. Daniel, Q. S. Ahmed, A. Huq, A. Yusuf, and D. A. Sack. 1978. Clinical features and pathogenicity of O group 1 non-agglutinating *Vibrio cholerae* and other vibrios isolated from cases of diarrhea in Dacca, Bangladesh, pp. 137–153. *In* J. Takeya and Y. Zinnaka (ed.), *Proceedings of the 14th Joint Conference U.S.-Japan Cooperative Medical Science Program Cholera Panel.* Toho University, Tokyo.

86. Spira, W. M., P. J. Fedorka-Cray, and P. Pettebone. 1983. Colonization of the rabbit small intestine by clinical and environmental isolates of non-O1 *Vibrio cholerae* and *Vibrio mimicus*. *Infect. Immun.* **41**:1175–1183.

87. Stroeher, U. H., L. E. Karageorgos, R. Morona, and P. A. Manning. 1992. Serotype conversion in *Vibrio cholerae* O1. *Proc. Natl. Acad. Sci. USA* **89**:2566–2570.

88. Takeo, T., Y. Shimonishi, M. Kobayashi, O. Nishimura, M. Arita, T. Takeda, T. Honda, and T. Miwatani. 1985. Amino acid sequence of heat

stable enterotoxin produced by *Vibrio cholerae* non-O1. *FEBS Lett.* **193:**250–254.

89. **Taylor, D. N., and P. Echeverria.** 1986. Etiology and epidemiology of traveler's diarrhea in Asia. *Rev. Infect. Dis.* **8:**S136–S141.

90. **Taylor, D. N., P. Echeverria, C. Pitarangsi, J. Seriwatana, O. Sethabutr, L. Bodhidatta, C. Brown, J. E. Herrmann, and N. R. Blacklow.** 1988. Application of DNA hybridization techniques in the assessment of diarrheal disease among refugees in Thailand. *Am. J. Epidemiol.* **127:**179–187.

91. **Twedt, R. M., J. M. Madden, J. M. Hunt, D. W. Francis, J. T. Peeler, A. P. Duran, W. O. Hebert, S. G. McKay, C. N. Roderick, G. T. Spite, and T. J. Wazenski.** 1981. Characterization of *Vibrio cholerae* isolated from oysters. *Appl. Environ. Microbiol.* **41:**1475–1478.

92. **Varlea, G., J. Olarte, A. Perez-Miravete, and L. Filloy.** 1971. Failure to find cholera and non-cholera Vibrios in diarrheal disease in Mexico City, 1966-7. *Am. J. Trop. Med. Hyg.* **20:**925–926.

93. **Wright, A. C., L. M. Simpson, J. D. Oliver, and J. G. Morris, Jr.** 1990. Phenotypic evaluation of acapsular transposon mutants of *Vibrio vulnificus*. *Infect. Immun.* **58:**1769–1773.

94. **Yajnik, B. S., and B. G. Prasad.** 1954. A note on vibrios isolated in Kumbh Fair, Allahbad, 1954. *Indian Med. Gazette* **89:**341– 349.

95. **Yamamoto, K., Y. Ichinose, N. Nakasone, M. Tanabe, M. Nagahama, J. Sakurai, and M. Iwanaga.** 1986. Identity of hemolysins produced by *Vibrio cholerae* non-O1 and *V. cholerae* O1, biotype El Tor. *Infect. Immun.* **51:**927–931.

96. **Yamamoto, K., Y. Takeda, T. Miwatani, and J. P. Craig.** 1983. Evidence that a non-O1 *Vibrio cholerae* produces enterotoxin that is similar but not identical to cholera enterotoxin. *Infect. Immun.* **41:**896–901.

97. **Zafari, Y., A. Z. Zarifi, S. Rahmanzadeh, and N. Fakhar.** 1973. Diarrhea caused by non-agglutinable Vibrio cholerae (non-cholera Vibrio). *Lancet* **ii:**429–430.

Vibrio cholerae and Cholera: Molecular to Global Perspectives
Edited by I. Kaye Wachsmuth, Paul A. Blake, and Ørjan Olsvik
© 1994 American Society for Microbiology, Washington, DC 20005

Chapter 9

Vibrios in the Environment: Viable but Nonculturable *Vibrio cholerae*

Rita R. Colwell and Anwarul Huq

When bacteria are transported from one environment to another, the environmental changes with which they are confronted include changes in temperature, nutrient concentration, salinity, osmotic pressure, pH, and many other factors. However, bacterial cells dynamically adapt to shifts in environmental parameters by employing a variety of genotypic and phenotypic mechanisms. With constitutive and inducible enzyme synthesis, bacteria can accommodate to growth-limiting nutrients and adjust or reroute metabolic pathways to avoid metabolic and/or structural disruption caused by specific nutrient limitations. Furthermore, they are able to coordinate rates of synthesis to maintain cellular structure and function. These adaptive capabilities, in fact, provide bacterial cells with the mechanisms by which they are able to respond to the surrounding environment and survive.

Until recently, the ability to culture microorganisms on artificial media in the laboratory constituted proof of viability. Depending on the efficiency and/or selectivity of the media employed, interpretation of the viability of bacteria in environmental samples varied, with various terms such as "live" cells, "dead" cells, "vegetative" cells, "viable" cells, "nonviable" cells, "stressed" cells, "injured" cells, "moribund" cells, and "static" cells being applied, sometimes quite ambiguously. Additional terms used to describe the metabolic and reproductive states for the bacteria can be found in the literature. Roszak et al. (76) first coined the term "viable but nonculturable" for bacteria that demonstrated detectable metabolic function but were not culturable by any available methods.

It has been recognized by microbial ecologists that one of the major limitations to research in microbial ecology is the inability to isolate and grow in culture the vast majority of the bacteria that occur in nature. Thus, the "viable but nonculturable" phenomenon represents a state of dormancy, survival, and persistence in the environment. The discovery of this phenomenon has revitalized the interest of microbial ecologists in microbial survival, persistence, and distribution in the natural environment. Bacterial cells that are "intact and alive" according to selected metabolic criteria but are not recoverable by culture in or on routinely employed bacteriologic media have been observed in both terrestrial and aquatic environments and, most recently, in the clinical laboratory. Food microbiolo-

Rita R. Colwell • Center of Marine Biotechnology, University of Maryland Biotechnology Institute, 600 East Lombard Street, Baltimore, Maryland, 21202, and Department of Microbiology, University of Maryland at College Park, College Park, Maryland 20742. *Anwarul Huq* • Department of Microbiology, University of Maryland at College Park, College Park, Maryland 20742.

gists have addressed "injury" and "resuscitation" of bacteria in a variety of foods (73) but have not been able to develop satisfactory hypotheses for these phenomena.

The problem of determining the metabolic states of organisms observed in direct counts but not recoverable by plating or most-probable-number counts was recognized in the late 1800s and reported by Knaysi (53) and Janinson (48), who did not propose specific terminology to define that stage of bacterial growth and survival. The phenomenon appears to reflect various mechanisms of bacterial survival, and individual strains within a species respond differently to environmental conditions as well as to the length of exposure to those conditions.

HISTORY OF THE VIABLE BUT NONCULTURABLE PHENOMENON OF BACTERIAL CELLS

Valentine and Bradfield (87) proposed the term "viable" to describe cells capable of multiplying and forming colonies, but the term "live" was suggested for respiring cells unable to divide under the same conditions. Postgate and Hunter (71) described "dead" bacteria as those that did not divide; however, they considered that nondividing bacteria may be in some sense "alive," because they retain their osmotic barrier even after "death." By this definition, cells are dead if they do not multiply.

Another term, "moribund," was suggested by Postgate (70) for the transient state between viability and death that bacterial cells enter when they are exposed to starvation and when they are incapable of multiplication but maintain other metabolic functions.

The word "dormancy," a term frequently employed among microbiologists, has been used to describe sporeformers or cyst formers but not organisms that do not produce spores or cysts. Spores and cysts produced by some soil microorganisms are mechanisms of survival. "Resting" forms are found in physi-

ologically and genetically diverse groups of organisms, such as *Myxococcus, Azotobacter, Streptomyces, Bacillus,* and *Clostridium* spp., notably when they are exposed to altered nutritional conditions. Both soil and aquatic bacteria are able to function at very low metabolic rates when environmental conditions are extreme. Spores develop when only a limited source of carbon, nitrogen, or phosphorous is available, and the spore state is considered highly resistant to extreme environments, e.g., very high or low temperatures, desiccation, exposure to toxic chemicals, etc.

Kurath and Morita (57) supported the definition of Valentine and Bradford (87) for viable cells but described cells as "nonviable" once the cells lost the ability to form a colony. They also acknowledged that the number of "nonviable" respiring cells was 10-fold greater than the number of those that were capable of multiplying.

After comparing various available methods for isolation and detection of cells, Roszak et al. (76) used the term "viable but nonculturable" to describe *Salmonella enteritidis* during the stage when it is detectable by direct viable count (DVC) (56) but is not culturable. They were able to "resuscitate" cells assumed to have "died off" in seawater, defining the nonculturable cells as viable but nonculturable.

Three years later, Roszak and Colwell (75) proposed a resting or "somnicell" stage for gram-negative bacteria that is analogous to spore formation in gram-positive bacteria. Later, in an attempt to show the genetic basis of nonculturability, Vadivelu et al. (86) observed an interesting level of homology between described fragments of *Vibrio* genome and *spo* genes when they used probes prepared from cloned *spo* genes from *Bacillus* spp.

Bacteria that have the greatest capacity to survive starvation are small and have lower metabolic rates than other bacteria. Dawes and Ribbons (19) have shown that this characteristic low metabolic rate is shared with spores, which exhibit a capacity for survival by shutting off almost all metabolic activity and yet remain viable. The "rounding up" phenomenon, with concomitant reduction in cell volume, was reported for *Vibrio* spp. under low nutrient conditions, i.e., in an environment relatively free of organic nutrients, at which time bacterial cells are observed to adjust to the substrate-limited stress condition by adopting necessary physiologic changes (4). *Campylobacter jejuni* becomes a typical coccoidlike cell with intact cell membranes and retains its viability after entering the nonculturable state (74).

When morphologic changes take place, the rounding up and reduction in size of the cells may be as much as 15- to 300-fold, depending on the level of nutritional stress (29, 65). Anderson and Hofferman (3) were able to filter organisms from seawater that passed through a 0.45-μ-pore-size filter membrane but were retained on 0.22-μm-pore-size membranes. These bacteria were later identified as *Spirillum*, *Leucothrix*, *Flavobacterium*, and *Vibrio* spp. This size reduction has also been observed for viable but nonculturable *V. cholerae*, as membranes with pores no larger than 0.22 or 0.10 μm must be used to collect the cells (88).

The three O-antigenic forms or subtypes of *Vibrio cholerae* O1 (Ogawa, Inaba, and Hikojima) are based on structure and response to various tests that determine an organism's role in causing disease. For purposes of discussion, epidemic strains of *V. cholerae* O1, i.e., the "cholera vibrio," are defined as possessing the O1 antigen, being capable of producing cholera toxin (CT$^+$), and therefore causing cholera in susceptible hosts. Both classical and El Tor biovars are included in *V. cholerae* O1. Nontoxigenic *V. cholerae* O1 strains comprise those isolates that agglutinate in O1 antiserum but do not possess the cholera toxin gene. Nevertheless, these strains may produce diarrheal disease and are designated nonepidemic *V. cholerae* O1. Recently, a non-O1 *V. cholerae* strain, *V. cholerae* O139, which does not agglutinate with O1 antiserum, has caused epidemics in India, Bangladesh, and elsewhere (1, 72).

Environmental and clinical isolates are identical by 5S rRNA sequencing (59). Interestingly, both the 5S rRNA and the 16S rRNA data indicate a deep branching of *V. cholerae* from other vibrios (reported by MacDonell and Colwell [60]). Huq et al. (40) recently reported little difference in the major outer membrane proteins of clinical and environmental isolates of *V. cholerae* O1 as determined by sodium dodecyl sulfate-polyacrylamide gel electrophoresis. A band of 22-kDa protein was observed for all the environmental strains but not for the clinical strains. A significant decrease in concentration of several of the protein bands, including the 22-kDa band, was detected for environmental strains passaged through rabbit ileal loops, demonstrating a possible environmentally induced change. The new cholera-causing *V. cholerae* O139 (72) should prove interesting when comparative studies can be done.

The anecdotal information that persisted for ca. 100 years maintained that the only reservoir of toxigenic *V. cholerae* O1, the epidemic strain of cholera, was the human intestinal tract and that this strain was taxonomically separate from *V. cholerae* non-O1, also known as "nonagglutinable vibrios" and found in aquatic, predominantly estuarine environments. An important advance was made when it was shown by Citarella and Colwell, using DNA-DNA hybridization, that O1 and non-O1 *V. cholerae* constitute a single species (10). The distinction between "cholera vibrios" and other *V. cholerae* was vigorously championed at that time (36), but a mounting volume of evidence, including results of numerical taxonomy, (85), DNA-DNA hybridization (10), and nucleic acid sequencing (59, 60), provided solid underpinning for the conclusion reached by Citarella and Colwell. The recent epidemic of cholera occurring in India and Bangladesh, in which *V. cholerae* O139 was isolated and concluded to be responsible for the epidemic, suggests the need to finally put to rest the notion of cholera vibrios and *V. cholerae* non-O1 strains as separate species (1, 72).

CURRENT CONCEPTS OF *V. CHOLERAE* ECOLOGY

Until the late 1970s and early 1980s, *V. cholerae* was believed to be highly host adapted and incapable of surviving longer than a few hours or days outside the human intestine. The view, wrote Felsenfeld (21), was that "some authors claimed that cholera vibrios may survive in water, particularly seawater, for as long as 2 months. This is, however, scarcely possible under natural conditions, if reinfection of the water does not take place." This perspective of cholera ecology has dominated the literature since the organism was first identified by Robert Koch in 1884 (54).

For *V. cholerae,* the term "survival" had been viewed to reflect a high degree of host adaptation, i.e., cholera vibrios could exist for only very short periods outside the human intestine. Koch's speculation that multiplication takes place in river water without any assistance was proven to be prescient, however, since the most recent data show that toxigenic *V. cholerae* O1 exists for long periods in laboratory microcosm water (46). In fact, the evidence accumulated over the past decade shows that *V. cholerae* is an autochthonous inhabitant of brackish water and estuarine systems (15). The autochthonous nature of *V. cholerae* O1 is an important factor in the epidemiology of cholera and is significantly so in endemic areas. Thus, the very early studies of *V. cholerae,* prior to 1970, were aimed at identifying environmental conditions associated with unusual delays in the inevitable "death" of the cholera vibrios in order to establish the length of time after which an environment could be considered cholera free, unless recontaminated with infected stool.

The remarkable discoveries of the past decade have revealed the existence of the dormant or somnambulant, i.e., viable but nonculturable, state into which *V. cholerae* O1 and *V. cholerae* non-O1 enter in response to nutrient deprivation and other environmental conditions (11). This phenomenon represents a new perspective and imparts a dynamic meaning to the term "survival," with the evidence showing that *V. cholerae* cells do not necessarily die when discharged into aquatic environments but instead remain viable and are capable of transforming into a culturable state if environmental conditions again become favorable (38). The implications of dormancy of *V. cholerae* O1 are both significant and relevant, considering that dormant, i.e., viable but nonculturable, forms are not recoverable on conventional bacteriologic media routinely used to isolate and maintain cells of *V. cholerae* in culture. However, when inoculated in rabbit ileal loops and tested in human volunteers, nonculturable forms of *V. cholerae* O1 cause accumulation of a large amount of fluid and diarrhea, respectively (11).

The evidence, primarily based on physiologic studies of *V. cholerae* in aquatic environments, is such that the dormant or viable but nonculturable state transcends mere survival and addresses the more fundamental questions of adaptation and response of microorganisms to environmental conditions (76). Studies of temperature and salinity relationships, adherence, and colonization of chitinaceous and mucilaginous macrobiota in microcosms simulating saline, estuarine, brackish, and freshwater environments provided new and important information on *V. cholerae* physiology.

Studies prior to 1970 were all based on methods of isolation and characterization of *V. cholerae* originally developed for clinical diagnosis of cholera in hospital laboratories (24). The many difficulties associated with isolation of *V. cholerae* O1 from the aquatic environment can be related to the simple fact that methods for isolating *V. cholerae* were developed for clinical specimens containing large numbers of actively growing cells, and such methods do not work for environmental samples that are likely to contain cells exposed to and thereby adapted to a variety of environmental conditions including, most commonly, low nutrient concentration, pH in the range of 7 to 8 (seawater), fluctuating temperatures and pH, variations in oxygen tension, exposure to UV via sunlight, etc. (58). The findings of Colwell et al. (11) that *V. cholerae* O1 in environmental samples may not grow on laboratory media routinely used for isolation was a pivotal point in the debate concerning the ecology of *V. cholerae*, i.e., that viable that nonculturable cells may go undetected unless appropriate methods for detection, e.g., molecular, immunologic, or direct microscopic, are employed. With modern molecular biology techniques, combined with immunologic direct detection methods, it can be convincingly demonstrated that viable but nonculturable *V. cholerae* O1 occurs even in clinical specimens from patients with the disease (11, 12, 16, 39).

It has long been known that cholera can be transmitted via water, wastewater, and food (26). However, a consistent theme of reports on this subject is that when cells of *V. cholerae* are suspended in fresh or saline water, over time they show a steady decrease in the number of cells enumerable by plate counts. To address this issue, the conversion of *V. cholerae* from the culturable to the viable but nonculturable state was demonstrated in laboratory microcosms (6, 88). Subsequent studies showed that survival is enhanced in saline waters but that the organism maintains itself less well in potable water. In any case, there is an absolute requirement of Na$^+$ for growth of *V. cholerae* in water (79). The traditional argument in support of the view that *V. cholerae* O1 does not survive in environmental waters has been the failure to isolate the organism from environmental water sources unless existing cases of cholera occurred in close proximity. For example, epidemiologic surveillance in the intensively monitored

Matlab area in Bangladesh in 1969 failed to isolate *V. cholerae* O1 from water sources between cholera seasons or from water not subject to contamination by patients with active cholera (62). In contrast, Huq et al. (38) showed that *V. cholerae* O1 could be detected by direct immunofluorescence microscopy throughout the year in these waters even when the organism could not be isolated, i.e., cultured, from water.

By using the direct detection technique, Islam et al. (46) demonstrated the persistence of culturable *V. cholerae* O1 under laboratory conditions for over 15 months in association with blue-green bacteria (cyanobacteria). However, Huq et al. (38) had reported, for the first time, the presence of *V. cholerae* O1 throughout the year by using a fluorescent-antibody detection method to detect organisms in >63% of plankton specimens. Results for samples collected in Matlab, Bangladesh, were positive for *V. cholerae* O1 when <1% of the specimens collected were positive for culturable *V. cholerae* O1. Earlier, Tamplin et al. (84) reported attachment of *V. cholerae* O1 to zooplankton and phytoplankton from natural waters. Thus, in light of the most recent findings, the existence of a viable but nonculturable state for *V. cholerae* O1 was indicated. Clearly, the microorganism is able to adapt and maintain itself in the aquatic environment in a form that will be missed by surveillance methodologies other than direct detection by fluorescent antibody or gene probes.

Probably the most profound challenge to dogma concerning *V. cholerae* O1 ecology derives from studies by Colwell et al. (11) that suggest that *V. cholerae* O1 possesses the ability to enter, or approximate, a state of dormancy in response to nutrient deprivation, elevated salinity, and/or reduced temperature. Exposure to low-nutrient conditions has recently been recognized as one important stimulus to entering the viable but nonculturable state that represents a common strategy for survival among bacteria in nutrient-poor environments (76). Novitsky and Morita (66–

68) demonstrated that cultures of the marine psychrophilic *Vibrio* sp. strain ANT-300 responded to starvation in either natural or artificial seawater by increasing the number of cells producing progeny cells significantly decreased in volume. Morphology also was altered from the typical bacillus, i.e., rod, shape to a coccoid shape. The coccoid cells exhibited an endogenous respiration less than 1% of the original as well as a 40% decrease in cellular DNA but remained culturable. Novitsky and Morita (68) did not postulate a viable but noncultural state but, rather, presented excellent evidence of adaptation of cells to a low-nutrient environment, since they studied only those cells remaining culturable, not recognizing cells that were viable but nonculturable.

BACTERIAL ADAPTATION TO CONDITIONS NOT OPTIMAL FOR REPRODUCTION

Colwell et al. (11), Huq et al. (38), Roszak et al. (76), Xu et al. (88), and Brayton et al. (6) demonstrated by both field and laboratory studies that *V. cholerae* indeed undergoes conversion to a viable but nonculturable state whereby the cells are reduced in size and become ovoid but, in contrast to starved cells, do not grow at all on standard laboratory media yet remain responsive to nalidixic acid and continue to assimilate radiolabeled substrate (76, 82). In other experiments, cells of *V. cholerae* O1 (strain CA401) were inoculated into microcosms of chemically defined sea salt solution adjusted to 0.5% salinity and incubated at 10 or 25°C for 24 or 96 h. The DVC represented in all cases ca. 98% of the acridine orange direct count (AODC [31a]), while the percentage of culturable cells ranged from 13% (10°C, 2.5% salinity) to 87% (25°C, 0.5% salinity) of AODC.

Hood and Ness (33) reported results similar to those obtained by Xu et al. (88) but used a nontoxigenic *V. cholerae* O1 strain isolated from oysters and five nontoxigenic *V.*

cholerae non-O1 environmental isolates as well as a clinical isolate of *V. cholerae* O1. Survival was monitored in sterile and nonsterile estuarine waters and sediments at 20°C by AODC and plate count on Trypticase soy–1.5% NaCl agar. All of the strains showed similar patterns in sterile estuarine water, where the culturable counts remained relatively constant for the duration of monitoring (up to 15 days) and were consistently 0.5 to 1 $\log_{10}$ less than the AODC counts.

Later, Baker et al. (4) provided evidence for the formation of small coccoid cells by both an environmental strain (WF110) from shellfish and a clinical strain of *V. cholerae* CA401 exposed to nutrient-free artificial seawater and filter-sterilized natural seawater microcosms. Total counts were determined by direct epifluorescence microscopy on preparations stained with 4', 6-diamino-2-phenylinodole (DAPI). Viability was determined by DVC and plate counts on a seawater-based complete medium incubated for 72 h at 30°C. Microcosms were maintained at 21°C for up to 330 days following inoculation, yielding approximately 5×10^3 cells by culturable count. The results showed no significant difference between the two seawater solutions or between the two strains of *V. cholerae* with respect to starvation survival. Upon exposure to starvation conditions, the clinical as well as the environmental isolates underwent reductive cell division, increased their cell numbers, and formed coccoid cells while greatly reducing individual cell volume. By 2 days after inoculation, the culturable counts increased approximately 2.5 $\log_{10}$. From that peak, all counts decreased steadily and gradually (<0.5 $\log_{10}$ overall) until the experiment was terminated at 55 or 75 days. At the time of inoculation, the total count (AODC) was approximately 0.5 $\log_{10}$ greater than the DVC and 1 $\log_{10}$ greater than the culturable count. By 20 days and thereafter, this difference increased to 1 and 2 $\log_{10}$, respectively. The relationship between the slopes for total count, DVC, and culturable count remained essentially constant throughout; there was no sudden or rapid decrease in viability or increase in viable but nonrecoverable cells in any of the periods monitored. Since incubation was at 21°C, in contrast to the much lower temperature employed by Xu et al. (88), these data indicate the importance of temperature in inducing the viable but nonculturable state in *V. cholerae*. The morphologic response to starvation, monitored by electron microscopy, appeared to be uniform for the two strains studied.

At 24 h, cells of both strains were uniformly electron dense, vibrioid, and of normal size (1.5 by 0.38 μm). At 48 h, small regions of electron transparency appeared at the peripheries of cells that were unchanged in size or shape. These regions were observed by Felter et al. (22) and Kennedy et al. (51), who reported "round-body" formation by marine vibrios. At day 18, cells were spherical, with an electron-dense region (0.56-μm diameter) in the center surrounded by a remnant cell wall, as observed by Felter et al. (22) for marine and estuarine vibrios.

The coccoid-shaped cells studied by Hood and Ness (33) regained normal size and vibrioid shape within 2 h after nutrient supplementation with a dilute solution of peptone and yeast extract. The beginning of the reversion process was observed within 10 min after nutrient addition, and initiation of cell division occurred within 4 to 5 h. Thus, Hood and Ness (33) did not induce the viable but nonculturable state in *V. cholerae* but instead provided some useful information concerning nutrient depletion in this species, showing that generation times during exponential growth were significantly longer than those of control, nutrient-conditioned cells and suggesting the use of nutrient depletion to induce the starvation response as opposed to complete dormancy in *V. cholerae*.

Hood et al. (32) and Guckert et al. (28) further characterized the starvation response of *V. cholerae* CA401 at the macromolecular level. Nutrient deprivation included a rapid decline in total lipids and carbohydrates within the first 7 days and a constant decline in protein and DNA over 30 days (the period of monitoring). After 7 days, 88.7% of carbohydrates and 99.8% of total lipids remained, with little change thereafter. After 30 days, a reduction to 70% of protein and 75% of DNA content was noted. RNA concentrations, on the other hand, remained relatively constant. Profiles of lipids and carbohydrates changed during the time of maximum loss; i.e., neutral lipids declined much more slowly than did phospholipids, poly-β-hydroxybutyrate disappeared completely, and a relative decline in ribose and N-acetylglucosamine and a relative increase in six-carbon sugars, especially glucose, were observed.

The morphologic response of *V. cholerae* CA401 to starvation shared some but not all characteristics of that of the marine vibrio ANT-300 (2, 66). Common important features included conservation of ribosomal structure, disappearance of granules, and compression of the nuclear region into the center of the cell, where it was surrounded by a denser cytoplasm. On the other hand, ANT-300 showed little distortion of the cell wall, while *V. cholerae* CA401 exhibited a great deal of convolution, like marine vibrios (22, 51).

The ancillary characterization of fatty acids carried out by Guckert et al. (28) demonstrated that lipid utilization by *V. cholerae* CA401 was preferential for fatty acids, which were most easily and rapidly metabolized (*cis*-monoenoic fatty acids), but conservation of cyclopropyl and *trans*-monoenoic fatty acids was noted, with the latter believed to be important in the maintenance of membrane integrity and fluidity. The strikingly reproducible pattern of lipid degradation during starvation in the clinical isolate *V. cholerae* CA401 paralleled in several respects the pattern of starvation survival of other organisms studied and demonstrates clearly that *V. cholerae* O1 possesses a well-regulated set of starvation responses.

In a recently completed study, Chowdhury et al. (9) observed that the uptake of [^{3}H]thymidine, [^{14}C]glucose, and [^{14}C]acetate by nonculturable *V. cholerae* dramatically increased with temperature upshift, leading to recovery of culturable cells of *V. cholerae*. Uptake was 10- to 100-fold higher than initial levels, i.e., compared to uptake in cells that were normal in size and culturable.

The annual distribution of *V. cholerae* detected by the fluorescent-antibody method has been reported by Huq et al. (38). However, the seasonality of this organism coupled with the starvation response and nonculturable phenomenon was earlier reported by Colwell et al. (11) and reflects the origin of *V. cholerae* as an autochthonous estuary dweller. The capacity of *V. cholerae* to undergo the starvation response as well as to enter the viable but nonculturable state makes it clear that long-term survival of this organism in the environment, perhaps for years, must be considered a source of the organism in cholera epidemics. When the cells are subjected to nutrient depletion or addition or to reduction or elevation in salinity and temperature, the cells rapidly go nonculturable but remain viable and potentially pathogenic, as demonstrated by Colwell et al. (11), who employed rabbit loop assays to recover viable but not culturable cells of *V. cholerae* from cold, full-strength seawater. Furthermore, nonculturable cells of *V. cholerae* ingested by volunteers in a feeding experiment produced diarrhea even though the cholera toxin gene had been removed from the strain to reduce risk to the volunteers (13). Clearly cholera toxin-positive strains of *V. cholerae* O1 do not dramatically alter their toxigenic character even after they enter the nonculturable state in the aquatic environment.

INFLUENCE OF SALINITY AND TEMPERATURE

Isolation of *V. cholerae* from the natural aquatic environment has been patchy, and the concentration of cells varies, depending on the culture methods used for detection. More-frequent isolations in the summer months

may be because of higher concentrations of culturable cells. A linear correlation with salinity was observed, with greater frequency of isolations at sites with salinities between 0.2 and 2.0‰. The effect of temperature was more strongly correlated with the frequency of isolations when the water temperature was higher than 17°C. From these data, Colwell et al. (17) concluded that *V. cholerae* is autochthonous in brackish water and estuarine environments. This finding was supported by Hood and coworkers (34, 35), who reported strong linear correlations between *V. cholerae* non-O1 and temperature and salinity in two estuaries in Florida.

The constantly changing conditions in tidal estuaries suggest an association of the ability of *V. cholerae* to adapt to a wide range of salinity and temperature conditions with the fact that its natural ecological habitat is brackish, riverine, and estuarine waters.

To obtain an understanding of salinity and temperature relationships for *V. cholerae*, laboratory microcosms simulating environmental conditions that could be controlled and replicated as required were prepared. This approach provided the advantage of allowing us to measure quantitatively the environmental effects on *V. cholerae* O1 as well as the effects of environmental parameters on pathogenic properties during exposure to water of various salinities, temperatures, etc. Cells of *V. cholerae* O1 in natural waters are not consistently detectable by culture methods, as discussed above. In fact, detection, enumeration, and monitoring of *V. cholerae* in the environment are not yet feasible unless a method that allows detection of nonculturable organisms is used. The fluorescent-antibody direct-staining approach has proven to be very effective for samples collected from the natural environment (38).

The first extensive studies of salinity and temperature relationships using microcosms were reported by Singleton et al. (79, 80), who used *V. cholerae* O1 strain LA4808, isolated during the 1978 El Tor cholera outbreak in Louisiana, and nine other clinical or environmental strains, including both O1 and non-O1 *V. cholerae*. The number of bacterial cells in the microcosms over time were calculated from viable plate and microscopic AODC of samples taken from microcosms prepared with a chemically defined sea salt solution in which the salinity, organic nutrient concentration, or temperature was varied. Following initial studies by Singleton et al. (79), survival and growth of *V. cholerae* under various conditions of salinity, pH, temperature, and presence of cations was reported by Miller et al. (63). A total of 59 strains representing a variety of biovars were examined, using viable plate counts on Trypticase soy agar. Six *V. cholerae* O1 isolates were extensively studied. Results from these studies showed that strains of *V. cholerae* varied greatly in survival in the culturable state under conditions of low salinity (0.05%) and that survival was unrelated to serotype (O1 or non-O1), source (clinical or environmental), or country of origin (e.g., Tanzania or Bangladesh). At 25°C, 18 of the 20 strains tested remained viable, with less variation at optimal salinity.

It has been demonstrated in in situ studies that vibrios make up a notable portion of the normal flora in many aquatic environments, with the presence of *V. cholerae* strongly influenced by salinity and temperature, as discussed above. Parallel laboratory microcosm studies showed that the strict requirement for Na$^+$ is not limiting in natural waters but that salinity, as the key determinant of osmolarity, normally would restrict the distribution of *V. cholerae* to water of more than potable salinity. Studying the influence of water temperature, salinity, and pH on survival and growth of toxigenic *V. cholerae* O1 associated with live copepods, Huq et al. (42) concluded that 15‰ salinity, 30°C water temperature, and pH 8.5 supported increased attachment and multiplication of *V. cholerae* on the copepod. Results provided by Colwell et al. (15, 17) and Kaper et al. (49) also suggest that growth of *V. cholerae* in the environment is limited to waters with temperatures

above 10°C. Survival through cold seasons, however, depends on the capacity of the organism to enter a dormant state, as discussed above.

Tamplin and Colwell (83) demonstrated that toxin production in microcosm cultures was related to salinity, demonstrating a salinity optimum between 2.0 and 2.5% for toxin production that was independent of cell concentration and toxin stability. A study reported by Miller et al. (64) showed that cells do not lose enterotoxigenicity after long-term exposure (64 days) in microcosms under a variety of conditions, nor did a selection for either hyper- or hypotoxigenic mutants occur. However, there is increased toxin production by toxigenic *V. cholerae* O1 when the organism is associated with *Rhizoclonium fontanum*, a green alga commonly found in natural water (43).

ASSOCIATION OF *V. CHOLERAE* WITH ZOOPLANKTON (COPEPODS)

The phenomenon of the nonculturable stage of *V. cholerae* and its association with phyto- and zooplankton and survival in the aquatic environment are very important in the ecology of this organism. Huq et al. (41, 42) demonstrated that *V. cholerae* associated with planktonic live copepods survived longer, i.e., remained culturable, in laboratory microcosms than *V. cholerae* exposed to dead copepods or to *Pseudoisochrysis* sp., a blue-green alga used to feed live copepods. In this study, cells of *V. cholerae* were enumerated by plate count, and it was found that the organisms could be cultured on plates for up to 336 h. However, no attempts were made to detect viable cells of *V. cholerae* that entered the nonculturable state. Also, the fluorescent-antibody method had not yet been fully developed nor was an established method for detection of toxigenic *V. cholerae*, such as polymerase chain reaction (PCR), available at that time. In a subsequent study, Huq et al. (38) were able to detect cells of *V. cholerae*

O1 from 63% of plankton samples collected in Bangladesh during a 3-year study. In this same study, they were able to culture *V. cholerae* O1 on plates from <1% of the water samples and from none of the plankton samples. These findings suggested that cells of *V. cholerae* attached to plankton enter the nonculturable stage as a part of their mechanism for survival in the environment until proliferation, or "blooms" of copepods occur. Further evidence on the survival of this organism in the environment will be reported elsewhere (14).

DETECTION OF VIABLE BUT NONCULTURABLE *V. CHOLERAE* O1

Because the standard plate count method cannot be used to enumerate viable but nonculturable organisms, alternative methods based on direct counting were developed to enumerate these bacteria. The most common of these methods are epifluorescence microscopy and acridine orange staining (18, 25). The AODC is considered a presumptive count of viable organisms, since in general, intact, double-stranded DNA binds monomers of the dye that fluoresce green, but broken, single-stranded DNA binds as dimers that fluoresce red-orange (Fig. 1). At the very least, the AODC permits enumeration of unlysed cells containing intact DNA. However, subsequent studies showed that the color shift cannot be used to distinguish between viable and nonviable cells.

A more persuasive assay for testing viability is the DVC developed by Kogure et al. (56), in which active cells are identified by growth without multiplication after incubation for 6 h at 25°C in response to the addition of yeast extract in the presence of nalidixic acid. Under these conditions, active cells carry out protein synthesis in the absence of DNA replication or cell division and produce elongated cells that are easily identified (Fig. 2). In general, the AODC is higher than the DVC for a given bacterial suspen-

sion, suggesting that non-substrate-responsive cells may not be viable. However, this observation requires further study, because the substrate employed for the DVC can influence the end result.

Xu et al. (88) first reported that *V. cholerae* O1 could be induced to enter a nonculturable state under certain laboratory conditions; i.e., after suspension in saline solution for 9 days, cells of *V. cholerae* O1 could not be recovered by direct plating or alkaline peptone enrichment. In contrast, AODC and fluorescent-antibody direct counting indicated the continued presence of intact cells at the same concentration as when the experiment was initiated.

Viable but nonculturable cells will not form colonies on agar media or grow in broth culture, a response similar to the response to nutritional deprivation by *Escherichia coli* (88), *C. jejuni* (74), *S. enteritidis* (76), and other gram-negative and gram-positive bacteria (8). The starvation response, leading to formation of small coccoid cells, may be a preliminary step in the induction of the viable but nonculturable state in *V. cholerae*.

In a series of microcosm experiments, it was observed that *V. cholerae* O1 and related human pathogens enter a viable but nonculturable state under particular nutritional depreciation, especially in the environment (6, 88). Thus, DVC epifluorescence microscopy will consistently remain higher than corresponding plate counts, and the assumption that *V. cholerae* O1 dies off or decays in the marine environment is not valid. Because indirect immunofluorescence microscopy is significantly more sensitive than DVC in detecting *V. cholerae* O1 in environmental samples, the organism can now be readily detected when it had been assumed to have died off.

Testing of the viability and pathogenicity of *E. coli* in a stage when it could not be cultured by conventional culture methods was accomplished by using membrane chambers submerged in semitropical waters at Bimini, Bahamas (27). Ligated ileal loop assays (81) were employed, with nonculturable cells of *V. cholerae* O1 strain CA401 harvested by centrifugation and suspended to a concentration of 10^8 cells per ml by DVC. Ligated loops separated by interloops were inoculated, with 10^6 culturable organisms serving as positive controls (7). Results showed that *V. cholerae* could be recovered from the positive loops injected with nonculturable *V. cholerae*. Bacterial cells that cannot be cultured are conventionally presumed to be dead, but fluorescence microscopy reveals that nonculturable cells continue to be viable and remain intact and metabolically active for weeks, months, and even years (61). In a volunteer feeding experiment, it was found that nonculturable vibrios can produce clinical symptoms of cholera in human volunteers, confirming that nonculturable *V. cholerae* maintains pathogenic potential in the environment, even though the cells are nonculturable, with human passage effectively triggering outgrowth of cells to the viable and culturable stage (16).

PCR has been used by various investigators to detect *V. cholerae* toxin genes in food and in pure cultures (23, 55). This method is especially useful when organisms are in the nonculturable stage. In environmental specimens, the concentration of bacterial cells is usually very low, making it difficult to detect them. Koch et al. (55) claimed to detect as few as 1 *V. cholerae* cell per 10 g of food with amplification reaction from crude bacterial lysates. Earlier, Shirai et al. (78) reported the use of PCR for detection of the cholera enterotoxin operon of *V. cholerae* O1 in stool specimens. They concluded that the cells of *V. cholerae* were dead when they did not grow on plates, a premature conclusion in the absence of corroborating evidence of cell death. In a later report, cholera toxin genes were detected in viable but nonculturable *V. cholerae* O1 in laboratory microcosms, employing PCR (30a). The capability of bacterial cells to adapt to environmental conditions adverse to cell multiplication needs to be assessed.

It is routine practice in clinical laboratories to culture for enteric pathogens within 2 h of collection of a specimen for optimal recovery of the pathogen under study (50). Delay in culture may result in overgrowth by other fast-growing flora normally present in the intestine, or the target organism may enter a nonculturable stage and thus be easily confused with dead organisms. Islam et al. (47) reported the detection by PCR of viable but nonculturable *Shigella dysenteriae* type 1 in laboratory microcosms; the report of detection of viable but nonculturable cells of *S. dysenteriae* type 1 for up to 6 weeks after suspension in water is extremely valuable for diagnostic purposes. However, further study is needed to determine the disease-producing capacity of these organisms, the knowledge of which will be extremely valuable in understanding transmission and virulence (20).

A rapid-detection test kit for detection of *V. cholerae,* CholeraScreen, has proven useful and was concluded to be highly satisfactory for field use (12), since detection is possible within minutes and the test requires no culture procedures. An interesting finding that was particularly important was that up to 20% of the clinically confirmed cholera patients harbored nonculturable *V. cholerae;* i.e., the stools failed to yield growth of *V. cholerae* on culture plates but gave positive results by direct detection methods. In another study, PCR test results were in agreement with results of the CholeraScreen coagglutination test (5). Thus, a significant number of cholera cases will go undiagnosed if there is reliance solely on culture methods.

Yet another test, Cholera DFA, a modified fluorescent-antibody method originally described by Brayton et al. (6), has been optimized to kit form, reducing the time required to complete the test to less than 30 min (31) (Fig. 3). A colorimetric immunoassay, Cholera SMART (30), recently optimized for clinical specimens has also proven promising. All three of these test kits require no culture, making the tests valuable for detection of *V. cholerae* even in the nonculturable stage.

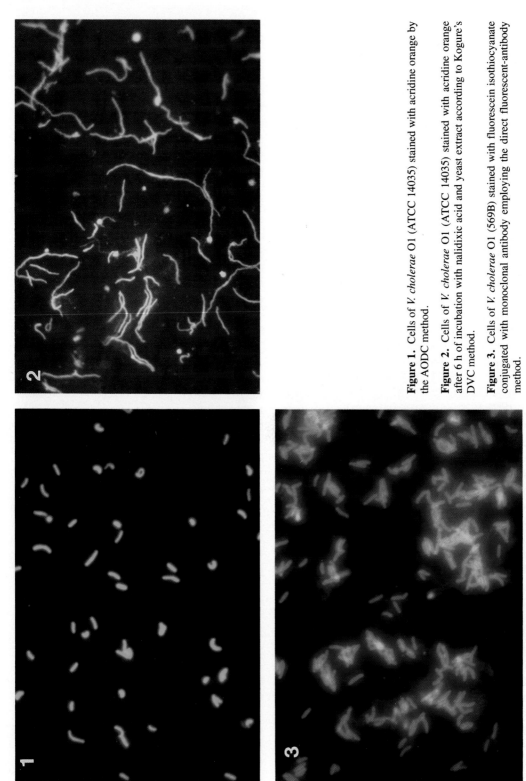

Figure 1. Cells of *V. cholerae* O1 (ATCC 14035) stained with acridine orange by the AODC method.

Figure 2. Cells of *V. cholerae* O1 (ATCC 14035) stained with acridine orange after 6 h of incubation with nalidixic acid and yeast extract according to Kogure's DVC method.

Figure 3. Cells of *V. cholerae* O1 (569B) stained with fluorescein isothiocyanate conjugated with monoclonal antibody employing the direct fluorescent-antibody method.

Field use of these tests in developing countries should provide improved diagnostic capability where complete laboratory facilities are absent.

Most recently, detection and isolation of toxigenic *V. cholerae* O1 in a nonepidemic area were reported by Huq et al. (39). The study was carried out with divers in the Chesapeake Bay region of the United States and at the Black Sea off the coasts of Russia and Ukraine. *V. cholerae* O1 was isolated by culture and also was detected in seawater and from swabs of the ears, noses, and throats of divers by the fluorescent-antibody method when culture was unsuccessful. Blood samples drawn from the divers before and 2 weeks after their dives demonstrated a clinically significant rise in vibriocidal antibody titer, even when the divers did not manifest clinical symptoms. It is interesting to note that in June 1991, *V. cholerae* O1, in water and plankton specimens collected in the Black Sea was detected by direct detection kits (39) and fluorescent-antibody direct

staining. Except for a few cases, these same samples, however, were not culture positive. About 4 weeks after this field work was completed, a local newspaper, *Rechernya Odessa* (11 September 1991), reported an outbreak of cholera in the same coastal area of swimming and bathing beaches (39).

The nonculturability demonstrated by *V. cholerae* and other pathogens, e.g., *Campylobacter* (74), *Helicobacter* (77), and *Legionella* (69) spp., appears to be a mechanism of adaptation to natural, low-nutrient aquatic environments. Starvation-induced changes in bacterial surface topography and hydrophobicity, electrostatic interaction, and irreversible binding to glass surfaces have been reported for several strains of marine bacteria by Kjelleberg and Hermansson (52). The cells of *V. cholerae* demonstrate a predilection for association with chitinaceous surfaces and the mucilaginous sheath of algae that can be interpreted as an ecologic advantage. For example, enhanced attachment to inert suspended matter can provide a mecha-

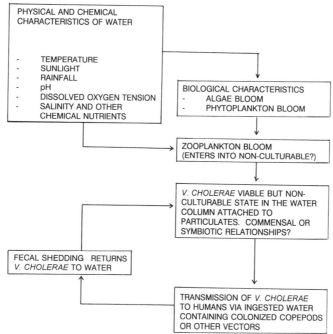

Figure 4. Hypothetical model for transmission of *V. cholerae* in nature. Modified with permission (Huq et al., p. 262, *in* M. Yasuno and B. A. Whitton, ed., *Biological Monitoring of Environment Pollution*, Tokai University Press, Tokyo, 1988).

nism for transport of the bacteria to sediment during periods of nutrient depletion in the overlying water column. In the sediment, cells are in close contact with a higher concentration of nutrient than in the water column; hence they are better able to survive and reproduce for extended periods (4).

A few examples of adherence of *V. cholerae* to nonchitinaceous surfaces include adherence to roots of water hyacinth by *V. cholerae* O1 (81) in fresh water in areas of cholera endemicity and to green algae (44, 45). An earlier review by Colwell and Spira summarizes much of the ecological work (15a).

In summary, in Bangladesh, thousands of water samples can be analyzed without a single isolation of *V. cholerae* O1 in culture. At the same time, millions of people ingest water several times daily directly from these water sources, creating a much greater "sampling" intensity that leads to a finite number of primary cases, i.e., cases arising from ingestion of autochthonous *V. cholerae* O1. This suggests that these organisms enter a stage during which they cannot be cultured. At the same time, some factors, either in the environment or associated with human or animal passage, must trigger conversion of the organism to the culturable state. A possible hypothetical cycle is suggested (37) in Fig. 4. What the actual sequence of events is in the life cycle of *V. cholerae* remains to be definitely proven, but our hypothesis offers the most workable explanation to date.

Acknowledgments. The work reported here was supported in part by NOAA grant NA16RU0264-02, National Science Foundation grant BSR-9020268, and Environmental Protection Agency cooperative agreement CR817791-01.

We gratefully acknowledge Sam Joseph for careful review of the manuscript and Jafrul Hasan for kind assistance in preparation of the photomicrograph.

REFERENCES

1. **Albert, M. J., A. K. Siddique, M. S. Islam, A. S. G. Faruque, M. Ansaruzzaman, S. M. Faruque, and R. B. Sack.** 1993. Large outbreak of clinical cholera due to *V. cholerae* non O1 in Bangladesh. *Lancet* **341:**704.

2. **Amy, P. S., and R. Y. Morita.** 1983. Starvation-survival patterns of sixteen freshly isolated open-ocean bacteria. *Appl. Environ. Microbiol.* **45:**1109–1115.

3. **Anderson, J. I., and W. P. Hofferman.** 1965. Isolation and characterization of filterable marine bacteria. *J. Bacteriol.* **90:**1713–1718.

4. **Baker, R. M., F. L. Singleton, and M. A. Hood.** 1983. Effects of nutrient deprivation on *Vibrio cholerae. Appl. Environ. Microbiol.* **46:**930–940.

5. **Bonilla, C. G., G. S. Villegas, R. R. Robles, A. V. Ramudio, C. W. Agambula, G. H. Palomares, L. G. Cogeo, P. F. Arredondo, J. L. Gomez, J. A. K. Hasan, and R. R. Colwell.** Unpublished data.

6. **Brayton, P. R., and R. R. Colwell.** 1987. Fluorescent antibody staining method for enumeration of viable environmental *Vibrio cholerae* O1. *J. Microb. Methods* **6:**309–314.

7. **Brayton, P. R., D. B. Roszak, L. M. Palmer, S. A. Huq, D. J. Grimes, and R. R. Colwell.** 1986. Fluorescent antibody enumeration of *Vibrio cholerae* in the marine environment. *IFREMER* **3:**507–514.

8. **Byrd, J., and R. R. Colwell.** 1990. Maintenance of plasmids pBR322 and pUC8 in nonculturable *Escherichia coli* in the marine environment. *Appl. Environ. Microbiol.* **56:**2104–2107.

9. **Chowdhury, M. A., R. T. Hill, and R. R. Colwell.** Unpublished data.

10. **Citarella, R. V., and R. R. Colwell.** 1970. Polyphasic taxonomy of the genus *Vibrio:* polynucleotide sequence relationships among selected *Vibrio* species. *J. Bacteriol.* **104:**434–442.

11. **Colwell, R. R., P. R. Brayton, D. J. Grimes, D. R. Roszak, S. A. Huq, and L. M. Palmer.** 1985. Viable, but non-culturable *Vibrio cholerae* and related pathogens in the environment: implication for release of genetically engineered microorganisms. *Bio/Technology* **3:**817–820.

12. **Colwell, R. R., J. A. K. Hasan, A. Huq, L. Loomis, R. J. Siebling, M. Torres, S. Galvez, S. Islam, and D. Bernstein.** 1992. Development and evaluation of a rapid, simple sensitive monoclonal antibody-based co-agglutination test for direct detection of *V. cholerae* O1. *FEMS Microbiol. Lett.* **97:**215–220.

13. **Colwell, R. R., A. Huq, and P. R. Brayton.** Unpublished data.

14. **Colwell, R. R., A. Huq, E. Russek-Cohen, and D. Jacobs.** Unpublished data.

15. **Colwell, R. R., J. Kaper, and S. W. Joseph.** 1977. *Vibrio cholerae, Vibrio parahaemolyticus* and other vibrios: occurrence and distribution in Chesapeake Bay. *Science* **198:**394–396.

15a. **Colwell, R. R., and W. M. Spira.** 1992. The ecology of *Vibrio cholerae*, p. 107–127. *In* D. Barua and W. B. Greenough III (ed.), *Cholera.* Plenum Medical Book Company, New York.

16. **Colwell, R. R., M. L. Tamplin, P. R. Brayton, A. L. Gauzens, B. D. Tall, D. Harrington, M. M. Levine, S. Hall, A. Huq, and D. A. Sack.** 1990. Environmental aspects of *V. cholerae* in transmission of cholera, p. 327–343. *In* R. B. Sack and Y. Zinnaka (ed.), *Advances in Research on Cholera and Related Diarrhoeas,* 7th ed. KTK Scientific Publishers, Tokyo.

17. **Colwell, R. R., P. A. West, D. Maneval, E. F. Remmers, E. L. Elliot, and N. E. Carlson.** 1984. Ecology of pathogenic vibrios in Chesapeake Bay, p. 367–387. *In* R. R. Colwell (ed.), *Vibrios in the Environment.* John Wiley & Sons, New York.

18. **Daley, R. J., and J. E. Hobbie.** 1975. Direct counts of aquatic bacteria by a modified epi-fluorescent technique. *Limnol. Oceanogr.* **20:**875–882.

19. **Dawes, E. A., and D. W. Ribbons.** 1965. Studies on the endogenous metabolism of *Escherichia coli. Biochem. J.* **95:**332–343.

20. **Ewald, P. W.** 1993. The evolution of virulence. *Sci. Am.* **268:**(4):86–93.

21. **Felsenfeld, O.** 1974. The survival of cholera vibrios, p. 359–366. *In* D. Barua and W. Burrows (ed.), *Cholera.* W. B. Saunders, Philadelphia.

22. **Felter, R. A., R. R. Colwell, and G. B. Chapman.** 1969. Morphology and round body formation of *Vibrio marinus. J. Bacteriol.* **99:**326–335.

23. **Field, P. I., T. Popovic, K. Wachsmuth, and Ø. Olsvik.** 1992. Use of the polymerase chain reaction for detection of toxigenic *Vibrio cholerae* O1 strains from the Latin American cholera epidemic. *J. Clin. Microbiol.* **30:**2118–2121.

24. **Finkelstein, R. A.** 1973. Cholera. *Crit. Rev. Microbiol.* **2:**553–623.

25. **Floodgate, G. D.** 1979. The assessment of marine microbial biomass and activity, p. 217–252. *In* R. R. Colwell and J. Foster (ed), *Aquatic Microbial Ecology.* Maryland Sea Grant publication no. UM-SG-TS-80-03. University of Maryland, College Park.

26. **Glass, R. I., and R. Black.** 1992. The epidemiology of cholera, p. 129–154. *In* D. Barua and W. B. Greenough III (ed.), *Cholera.* Plenum Book Co., New York.

27. **Grimes, D. J., and R. R. Colwell.** 1986. Viability and virulence of *Escherichia coli* suspended by membrane chamber in semi-tropical ocean water. *FEMS Microbiol. Lett.* **34:**161–165.

28. **Guckert, J. B., M. A. Hood, and D. C. White.** 1986. Phospholipid ester-linked fatty acid profile changes during nutrient deprivation of *Vibrio cholerae:* increases in the *trans/cis* ratio and proportions of cyclopropyl fatty acids. *Appl. Environ. Microbiol.* **52:**794–801.

29. **Guetin, A. M., I. E. Mishustina, L. V. Andreev, M. A. Bobyk, and V. A. Lambina.** 1979. Some problems of the ecology and taxonomy of marine microvibrios. *Biol. Bull. Acad. Sci. USSR.* **5:**336–340.

30. **Hasan, J. A. K., L. Loomis, A. Huq, D. Bernstein, M. L. Tamplin, M. Wier, H. Bodenheimer, R. J. Siebling, and R. R. Colwell.** 1992. Development of a fast and sensitive immunoassay to detect *V. cholerae* O1 from clinical samples using SMART™ technology, abstr. C-118, p. 440. *Abstr. 92nd Gen. Meet. Am. Soc. Microbiol. 1992.*

30a. **Hasan, I. A. K., M. Shahabuddin, A. Huq, L. Loomis, and R. R. Colwell.** 1992. Polymerase chain reaction for detection of cholera toxin genes in viable but non-culturable *Vibrio cholerae* O1, abstr. D-138, p. 119. Abstr. 92nd Gen. Meet. Am. Soc. Microbiol. 1992.

31. **Hasan, J. A. K., M. Wier, A. Huq, D. Bernstein, L. Loomis, and R. R. Colwell.** Unpublished data.

31a. **Hobbie, J. E., R. J. Daley, and S. Jasper.** 1977. Use of nucleopore filters for counting bacteria by fluorescence microscopy. *Appl. Environ. Microbiol.* **33:**1225–1228.

32. **Hood, M. A., J. B. Guckert, D. C. White, and F. Deck.** 1986. Effect of nutrient deprivation on lipid, carbohydrate, DNA, RNA, and protein levels in *Vibrio cholerae. Appl. Environ. Microbiol.* **52:**788–793.

33. **Hood, M. A., and G. E. Ness.** 1982. Survival of *Vibrio cholerae* and *Escherichia coli* in estuarine waters and sediments. *Appl. Environ. Microbiol.* **43:**578–584.

34. **Hood, M. A., G. E. Ness, G. E. Rodrick, and N. J. Blake.** 1983. Distribution of *Vibrio cholerae* in two Florida estuaries. *Microb. Ecol.* **9:**65–75.

35. **Hood, M. A., G. E. Ness, G. E. Rodrick, and N. J. Blake.** 1984. The ecology of *Vibrio cholerae* in two Florida estuaries, p. 399–409. *In* R. R. Colwell (ed.), *Vibrios in the Environment.* John Wiley & Sons, New York.

36. **Hugh, R., and J. C. Feeley.** 1972. Report (1966–1970) of the subcommittee on taxonomy of vibrios to the International Committee on Nomenclature of Bacteria. *Int. J. Syst. Bacteriol.* **22:**123.

37. **Huq, A., M. A. R. Chowdhury, A. Felsenstein, R. R. Colwell, R. Rahman, and K. M. B. Hossain.** 1988. Detection of *V. cholerae* from aquatic environments in Bangladesh, p. 259–264. *In* M. Yasuno and B. A. Whitton (ed.), *Biological Monitoring of Environment Pollution.* Tokai University Press, Tokyo.

38. **Huq, A., R. R. Colwell, R. Rahman, A. Ali, M. A. R. Chowdhury, S. Parveen, D. A. Sack, and R. Russek-Cohen.** 1990. Detection of *Vibrio cholerae* O1 in the aquatic environment by fluorescent-monoclonal antibody and culture methods. *Appl. Environ. Microbiol.* **56:**2370–2373.

39. **Huq, A., J. A. K. Hasan, G. Losonsky, and R. R. Colwell.** Unpublished data.

40. **Huq, A., S. Parveen, F. Qadri, and R. R. Colwell.** 1993. Comparison of *V. cholerae* serotype O1

isolated from patient and aquatic environment. *J. Trop. Med. Hyg.* **96:**86–92.

41. **Huq, A., E. B. Small, P. A. West, M. I. Huq, R. Rahman, and R. R. Colwell.** 1983. Ecology of *Vibrio cholerae* O1 with special reference to planktonic crustacean copepods. *Appl. Environ. Microbiol.* **45:**275–283.

42. **Huq, A., P. A. West, E. B. Small, M. I. Huq, and R. R. Colwell.** 1984. Influence of water temperature, salinity, and pH on survival and growth of toxigenic *Vibrio cholerae* serovar O1 associated with live copepods in laboratory microcosms. *Appl. Environ. Microbiol.* **48:**420–424.

43. **Islam, M. S.** 1990. Increased toxin production by *V. cholerae* O1 during survival with a green algae, *Rhizoclonium fontanam,* in an artificial aquatic environment. *Microbiol. Immunol.* **34:**557–563.

44. **Islam, M. S.** 1992. Seasonality and toxigenicity of *Vibrio cholerae* non-O1 isolated from different components of ponds ecosystem of Dhake City, Bangladesh. *World J. Microbiotechnol.* **8:**160–163.

45. **Islam, M. S., B. S. Drasar, and D. J. Bradley.** 1989. Attachment of toxigenic *V. cholerae* O1 to various fresh water plants and survival with a filamentous green algae, *Rhizoclonium fontanam J. Trop. Med. Hyg.* **92:**396–401.

46. **Islam, M. S., B. S. Draser, and D. J. Bradley.** 1990. Survival of toxigenic *Vibrio cholerae* O1 with common duckweed, *Lema minor* in artificial aquatic ecosystem. *Trans. R. Soc. Trop. Med. Hyg.* **84:**422–424.

47. **Islam, M. S., M. K. Hasan, M. A. Miah, G. C. Sur, A. Felsenstein, M. Venkatesan, R. B. Sack, and M. J. Albert.** 1993. Use of polymerase chain reaction and fluorescent antibody methods for detecting viable but nonculturable *Shigella dysentery* type 1 in laboratory microcosm. *Appl. Environ. Microbiol.* **59:**536–540.

48. **Janinson, M. W.** 1937. Relation between plate counts and direct microscopic counts of *Escherichia coli* during logarithmic period. *J. Bacteriol.* **33:**461–469.

49. **Kaper, J. B., H. Lockman, R. R. Colwell, and S. W. Joseph.** 1979. Ecology, serology, and enterotoxin production of *Vibrio cholerae* O1 in Chesapeake Bay. *Appl. Environ. Microbiol.* **37:**91–103.

50. **Kelly, M. T., D. Brenner, and J. J. Farmer III.** 1985. *Enterobacteriaceae,* p. 263–277. *In* E. H. Lennette, A. Balows, W. J. Hausler, Jr., and H. J. Shadomy (ed.), *Manual of Clinical Microbiology,* 4th ed. American Society for Microbiology, Washington, D.C.

51. **Kennedy, S. F., R. R. Colwell, and G. B. Chapman.** 1970. Ultrastructure of a psychrophilic marine vibrio. *Can. J. Microbiol.* **16:**1027–1032.

52. **Kjelleberg, S., and M. Hermansson.** 1984. Starvation-induced effects on bacterial surface characteristics. *Appl. Environ. Microbiol.* **48:**497–503.

53. **Knaysi, G.** 1935. A microscopic method of distinguishing dead from living cells. *J. Bacteriol.* **30:**193–206.

54. **Koch, R.** 1884. An address on cholera and its bacillus. *Br. Med. J.* **2:**403–407, 453–459.

55. **Koch, W. H., W. L. Payne, B. A. Wentz, and T. A. Cebula.** 1993. Rapid polymerase chain reaction method for detection of *Vibrio cholerae* in Foods. *Appl. Environ. Microbiol.* **59:**556–560.

56. **Kogure, K., U. Simidu, and N. Taga.** 1979. A tentative direct microscopic method for counting living marine bacteria. *Can. J. Microbiol.* **25:**415–420.

57. **Kurath, G., and Y. Morita.** 1983. Starvation-survival physiological studies of a marine *Pseudomonas* sp. *Appl. Environ. Microbiol.* **45:**1206–1211.

58. **Litsky, W.** 1979. Gut critters are stressed in the environment, more stressed by isolation procedures, p. 345–347. *In* R. R. Colwell and J. Foster (ed.), *Aquatic Microbial Ecology.* Maryland Sea Grant publication no. UM-SG-TS-80-03. University of Maryland, College Park.

59. **MacDonell, M. T., and R. R. Colwell.** 1984. Identical 5S rRNA nucleotide sequence of *Vibrio cholerae* strains representing temporal, geographical, and ecological diversity. *Appl. Environ. Microbiol.* **48:**119–121.

60. **MacDonell, M. T., and R. R. Colwell.** 1984. Nucleotide base sequence of vibrionaceae 5S rRNA. *FEMS Microbiol. Lett.* **175:**183–188.

61. **Mai, U. E. H., M. Shahamat, and R. R. Colwell.** 1990. Survival of *Helicobacter pylori* in the aquatic environment, p. 91–96. *In* H. Menge (ed.), *Helicobactor pylori—1990.* Springer-Verlag, New York.

62. **McCormack, W. M., M. S. Islam, M. Fahimuddin, and W. H. Mosley.** 1969. Endemic cholera in rural East Pakistan. *Am. J. Epidemiol.* **89:**393–404.

63. **Miller, C. J., B. S. Drasar, and R. G. Feachem.** 1984. Response to toxigenic *Vibrio cholerae* O1 to physico-chemical stresses in aquatic environments. *J. Hyg.* (Cambridge) **93:**475–495.

64. **Miller, C. J., B. S. Drasar, R. G. Feacham, and R. J. Hayes.** 1986. The impact of physico-chemical stress on the toxigenicity of *Vibrio cholerae. J. Hyg.* (Cambridge) **96:**49–57.

65. **Mishustina, I. E., and T. G. Kameneva.** 1981. Bacterial cells of minimal size in the Barents Sea during the polar night. *Microbiologiya* **50:**360–363.

66. **Novitsky, J. S., and R. Y. Morita.** 1976. Morphological characterization of small cells resulting from nutrient starvation of a psychrophilic marine vibrio. *Appl. Environ. Microbiol.* **32:**617–622.

67. **Novitsky, J. S., and R. Y. Morita.** 1977. Survival of a psychrophilic marine vibrio under long-term nutrient starvation. *Appl. Environ. Microbiol.* **33:**635–641.

68. Novitsky, J. S., and R. Y. Morita. 1978. Possible strategy for the survival of marine bacteria under starvation conditions. *Mar. Biol.* **48:**289–295.

69. Paszko-Kolva, C., M. Shahamat, H. Yamamoto, T. K. Sawyer, J. Vives-Rego, and R. R. Colwell. 1991. Survival of *Legionella pneumophila* in the aquatic environment. *Microb. Ecol.* **22:**75–83.

70. Postgate, J. R. 1967. Viability, measurements and survival of microbes under minimum stress. *Adv. Microb. Physiol.* **1:**1–24.

71. Postgate, J. R., and J. R. Hunter. 1962. The survival of starved bacteria. *J. Gen. Microbiol.* **29:**233–263.

72. Ramamurthy, T., S. Garg, R. Sharma, S. Bhattacharya, G. B. Nair, T. Shimada, T. Takeda, T. Karasawa, H. Kurazano, A. Pal, and Y. Takeda. 1993. Emergence of novel strains of *V. cholerae* with epidemic potential in southern and eastern India. *Lancet* **341:**703–704.

73. Ray, B. 1979. Methods to detect stressed microorganisms. *J. Food Prot.* **42:**346–355.

74. Rollins, D. M., and R. R. Colwell. 1986. Viable but nonculturable stage of *Campylobacter jejuni* and its role in survival in the natural aquatic environment. *Appl. Environ. Microbiol.* **52:**531–538.

75. Roszak, D. B., and R. R. Colwell. 1987. Survival strategies of bacteria in the natural environment. *Microbiol. Rev.* **51:**365–379.

76. Roszak, D. B., D. J. Grimes, and R. R. Colwell. 1984. Viable but non-recoverable stage of *Salmonella enteritidis* in aquatic systems. *Can. J. Microbiol.* **30:**334–338.

77. Shahamat, M., C. Paszko-Kolva, J. Keiser, and R. R. Colwell. 1991. Sequential culturing method improves recovery of *Legionella* spp. from contaminated environmental samples. *Zentralbl. Bakteriol.* **275:**312–319.

78. Shirai, H., M. Nishibuchi, T. Ramamurthy, S. K. Bhattacharya, S. C. Pal, and Y. Takeda. 1991. Polymerase chain reaction for detection of the cholera enterotoxin operon of *Vibrio cholerae*. *J. Clin. Microbiol.* **29:**2517–2521.

79. Singleton, F. L., R. W. Atwell, M. S. Jangi, and R. R. Colwell. 1982. Influence of salinity and nutrient concentration on survival and growth of *Vibrio cholerae* in aquatic microcosms. *Appl. Environ. Microbiol.* **43:**1080–1085.

80. Singleton, F. L., R. W. Atwell, M. S. Jangi, and R. R. Colwell. 1984. Effects of temperature and salinity on *Vibrio cholerae* growth. *Appl. Environ. Microbiol.* **44:**1047–1058.

81. Spira, W. M., A. Huq, Q. S. Ahmad, and Y. A. Sayeed. 1981. Uptake of *Vibrio cholerae* biotype El Tor from contaminated water by water hyacinth (*Eichornia crassipis*). *Appl. Environ. Microbiol.* **42:**550–553.

82. Stevenson, L. H. 1978. A case for bacterial dormancy in aquatic systems. *Microb. Ecol.* **4:**127–133.

83. Tamplin, M. L., and R. R. Colwell. 1986. Effects of microcosm salinity and organic substrate concentration on production of *Vibrio cholerae* enterotoxin. *Appl. Environ. Microbiol.* **52:**297–301.

84. Tamplin, M. L., A. L. Gauzens, A. Huq, D. A. Sack, and R. R. Colwell. 1990. Attachment of *V. cholerae* serogroup O1 to zooplankton and phytoplankton of Bangladesh waters. *Appl. Environ. Microbiol.* **56:**1977–1980.

85. West, P. A., and R. R. Colwell. 1984. Identification and classification of *Vibrionaceae*: an overview, p. 285–363. *In* R. R. Colwell (ed.), *Vibrios in the Environment*. John Wiley & Sons, New York.

86. Vadivelu, J., K. S. Gobius, and R. R. Colwell. 1989. Evidence for nucleotide sequence similarity between *Bacillus subtilis spoO* genes and *Aeromonas caviae*, *Escherichia coli* and *V. cholerae* restriction fragments, abstr. P-188, p. 187. *Abstr. 5th Int. Symp. Microb. Ecol.*

87. Valentine, R. C., and J. R. G. Bradfield. 1954. The urea method for bacterial viability counts with the electron microscope and its relation to other viability counting methods. *J. Gen. Microbiol.* **11:**349–357.

88. Xu, H.-S., N. Roberts, F. L. Singleton, R. W. Atwell, D. J. Grimes, and R. R. Colwell. 1982. Survival and viability of non-culturable *Escherichia coli* and *Vibrio cholerae* in the estuarine and marine environment. *Microb. Ecol.* **8:**313–323.

Vibrio cholerae and Cholera: Molecular to Global Perspectives
Edited by I. Kaye Wachsmuth, Paul A. Blake, and Ørjan Olsvik
© 1994 American Society for Microbiology, Washington, DC 20005

Chapter 10

Serologic Diagnosis of *Vibrio cholerae* O1 Infections

Timothy J. Barrett and John C. Feeley

Isolation of *Vibrio cholerae* O1 from the stools of persons with diarrheal illness remains the definitive diagnostic method for cholera. In some cases, however, isolation of the organism is not possible. Stool specimens may not be obtained or may be obtained after the person stops excreting the organism (because of antimicrobial treatment or the passage of time). Specimens may also be mishandled so that the organism dies before being cultured. In these situations, measurement of serum antibodies may be the best method for determining whether an individual was infected with *V. cholerae* O1. Serum antibody assays may be particularly valuable in identifying control subjects for case-control studies who have not had recent *V. cholerae* O1 infections, in identifying suspect cases of cholera in persons who may not have had appropriate cultures taken, and in using stored serum collections to determine whether cholera previously existed undetected in an area.

Several methods have been used to measure the human immune response to *V. cholerae* O1 infection. Pollitzer reviewed the early literature on the subject in 1959 (29). More recent reviews have been written by Barua (1) and

Holmgren and Svennerholm (22). The most commonly performed methods for serodiagnosis of *V. cholerae* O1 infection involve measurement of antibacterial antibodies by agglutination or vibriocidal antibody assays. The measurement of antibodies directed against the heat-labile enterotoxin (cholera toxin [CT]) has also proven useful. These two approaches are discussed at length below. Detailed methods for performing these assays are described elsewhere (39).

TESTS BASED ON ANTIBACTERIAL ANTIBODIES

A variety of methods have been used to measure the antibacterial immune responses of humans to infection with *V. cholerae*. Older literature on the subject has been extensively reviewed by Pollitzer (29). Subsequent observations have been summarized by Wachsmuth et al. (39), Young et al. (41), and, most recently, Pal (27). Perhaps the simplest and most easily performed procedures are those based on the measurement of antibodies against lipopolysaccharide somatic O antigens of *V. cholerae* by agglutination or vibriocidal antibody assays.

Agglutination Tests

Older literature on the topic records conflicting results based on use of a variety of

Timothy J. Barrett and John C. Feeley • Foodborne and Diarrheal Diseases Branch, Division of Bacterial and Mycotic Diseases, National Center for Infectious Diseases, Centers for Disease Control and Prevention, Atlanta, Georgia 30333.

procedures and antigens (29). A number of studies have shown that O antibodies are best measured in agglutination tests with agglutinogens consisting of a suspension of living vibrios (15, 16, 19, 38). Live antigens are far more sensitive than boiled antigens or antigens killed by a number of methods (38). In contrast to experience with *Salmonella* spp., O agglutination develops rapidly, while H agglutination develops slowly and is not observed under recommended test conditions (39). This is of considerable diagnostic importance, since H antigens lack the specificity of O antigens and are shared by serogroups of *V. cholerae* other than the cholera-causing O1 and O139 serogroups and by other closely related *Vibrio* species (7, 33).

When live antigens are used, a fourfold or greater rise in titer between acute- and convalescent-phase serum specimens from patients with bacteriologically confirmed cholera has been observed 90 to 95% of the time (2, 5, 6, 12, 16, 32, 34).

Adaptation of the agglutination test to the microtiter system favored these days by many laboratories has been described elsewhere (5). Our laboratory and others favor the use of vibriocidal antibody assays (discussed below) rather than agglutination tests with the microtiter system because the results are easy to read.

Vibriocidal Tests

V. cholerae is exceedingly susceptible to the bactericidal effect of O antibody in the presence of complement. A number of procedures have been employed to take advantage of this phenomenon. In principle, they depend on exposure of dilutions of serum to a carefully standardized constant inoculum of *V. cholerae* in the presence of excess guinea pig complement, an incubation period to permit killing of the bacteria, and subculture of the dilutions of serum and appropriate controls on agar (13) or in broth (3, 12, 41). While the procedure requires very careful

standardization, it can be made very sensitive by the use of a small inoculum. The procedure that we have the most experience with at Centers for Disease Control and Prevention (CDC) (41) is similar to the microtiter procedure described by Benenson et al. (3) and is based on a tube test described earlier (25). With either procedure, it is imperative to use fresh active complement, smooth cultures, and a carefully standardized inoculum. At CDC we use Ogawa and Inaba strains VC12 and VC13, respectively, but other smooth strains may be used.

Like the agglutination test, the vibriocidal test can be expected to show a rise in titer in 90 to 95% of sera from patients with bacteriologically confirmed cholera due to *V. cholerae* O1. A high, in fact nearly perfect, correlation exists between the results of living-vibrio agglutination tests and those of vibriocidal tests in detecting diagnostic rises in antibody levels with paired acute- and convalescent-phase sera. The vibriocidal test is significantly more sensitive than the agglutination test and therefore gives a greater background of antibody levels in healthy or acute-phase sera and higher titers with convalescent-phase sera (39).

The vibriocidal antibody test has been used extensively for serologic surveys, since it is amenable to the titration of large numbers of sera by microtiter methodology, and the acquisition of antibody through natural exposure or vaccination shows an apparent correlation to immunity on a population basis and in volunteers challenged with the organism.

As of this writing, data establishing a similar usefulness of either vibriocidal or agglutinating antibody with *V. cholerae* O139 have not been published.

Other Tests with Somatic Antigens of *V. cholerae*

Other procedures such as the indirect hemagglutination test and enzyme-linked immunosorbent assays (ELISAs) have been investigated by a number of workers. Generally, these tests have not shown any advantage over simpler agglutination or vibriocidal antibody tests. We have confirmed (unpublished data) the observations of Barua and Sack (2) that the hemagglutination test is less sensitive than the agglutination test with living antigens. This is contrary to most experience with enterobacterial antigens (26).

The ELISA has been used by Holmgren and Svennerholm (21) to measure antibody to lipopolysaccharide in human sera, jejunal fluids, and colostrum. While convenient for such studies, an anti-lipopolysaccharide ELISA was much less sensitive than the vibriocidal test with human sera (unpublished data). While O antibodies are clearly reactive in the vibriocidal test, it is possible that other antigens such as outer membrane proteins and cell-bound hemagglutinins participate in the vibriocidal reaction.

ANTITOXIN ANTIBODY

Most strains of *V. cholerae* O1 produce a potent heat-labile enterotoxin. CT is strongly antigenic and generally elicits an antibody response in persons recovering from clinical illness. CT is functionally and antigenically similar to the LT of *Escherichia coli,* creating problems in interpreting antitoxin titers, as discussed below.

The titration of antitoxic antibodies was reported as early as 1966, when Kasai and Burrows detected antitoxic antibodies in human serum by neutralization of toxin in the rabbit ileal loop assay (23). The ileal loop assay is not a suitable method for most laboratories. It requires animal surgery and prolonged survival, is sensitive to differences in surgical technique, and is too laborious for testing large numbers of specimens.

Other well-characterized in vivo assays for antitoxic antibodies include the infant rabbit model (14) and the rabbit skin permeability factor (PF) test (10). Craig has proposed a standard antitoxin unit based on the PF test

(10). Standard antitoxins calibrated according to PF units are available from the Bureau of Biologics, Food and Drug Administration, and the Microbiology and Infectious Diseases Programs, National Institute of Allergy and Infectious Diseases. These antitoxins have been used to standardize the cell culture and enzyme assays described below.

The development of cell culture models for detecting CT also provided marked improvement in the ability to measure antitoxic antibodies. Both the Y1 strain of mouse adrenal cells (11) and Chinese hamster ovary (CHO) cells (17) are exquisitely sensitive to CT. Neutralization of CT activity in cell culture provides major advantages over animal models. Besides the obvious advantage of eliminating the use of live animals, cell cultures provide a constant source of assay material and can be done in microtiter format, allowing much larger numbers of serum specimens to be tested. Neutralization of CT in both these cell culture models has been well described elsewhere (11, 17).

Antitoxin levels have also been measured by passive hemagglutination (20). This test can be done in microtiter format with stable, preserved, sensitized erythrocytes. Titers obtained with this test correlate well with those obtained by toxin neutralization.

Perhaps the most broadly useful assays for antitoxic antibodies are ELISAs. The ELISA can be based on inhibition of toxin binding by antitoxic antibodies or on direct detection of antitoxic antibodies bound to fixed CT (21, 36, 40). These methods have the added advantage of being performed with commercially available reagents (CT, GM_1, and enzyme-conjugated anti-human immunoglobulin). The ELISA has also been carefully validated by comparison with the rabbit skin PF technique (40). The ELISA has the potential for greater reliability than cell culture assays because it removes the subjective interpretation of cell morphology required by the Y1 and CHO cell assays (11, 17). On the other hand, ELISA detects only immunologic reactivity with CT and does not demonstrate the presence of toxin-neutralizing antibody.

Immunologic response to CT in persons with *V. cholerae* O1 infection is not universal. Using the rabbit ileal loop assay, Pierce et al. found the antitoxin response to be highly variable (28). Of 16 patients, 7 showed a titer rise of threefold or less, and 1 of these 7 showed no rise. Conversely, convalescent-phase titers in two patients increased by 100-fold or more. The geometric mean titer increased ninefold, peaking at 15 days postonset. The mean titer fell slowly thereafter but remained six-fold above initial titers after 18 months. This study also demonstrated the close relationship between toxin-neutralizing antibodies as measured by the ileal loop assay and anti-PF antibodies.

Benenson et al. reported a diagnostic rise in antitoxin titer by PF neutralization in 73% of culture-confirmed cases (4). Using a hemagglutination test, Hochstein et al. demonstrated a response in 60% (9 of 15) of confirmed cases (20). More recent publications have generally reported a higher percentage of patients showing an antibody response. The ability to demonstrate a diagnostic rise in titer appears to have been improved by the use of immunoassays rather than biologic assays.

Snyder et al. reported that all 11 persons with culture-confirmed *V. cholerae* O1 infection during a 1978 outbreak in the United States had elevated titers to CT (35). Four persons had serum samples tested 8 to 30 days, 3 months, and 9 to 11 months after they had eaten contaminated crabs. All showed elevated titers by 20 days, and titers remained high at 3 months. Titers in two samples declined to insignificant levels by 10 months after infection. When the same criterion is used for a positive value (ELISA net optical density [OD] greater than the mean net OD of controls plus 3 standard deviations), 3 (4%) of 75 serum samples from a community survey were also positive. These persons showed no other evidence of *V. cholerae* infection, and the serologic reactions would have to be considered false positives.

Levine et al. used the same assay to examine serum samples from 31 volunteers who had ingested *V. cholerae* O1 (24). Using a positive cutoff value as the control mean OD plus 3 standard deviations, all 31 volunteers had elevated antitoxin levels; 7 had elevated levels before challenge. If a more stringent positive cutoff was used (a positive/negative OD ratio of ≥4), 30 of 31 were positive postchallenge and none were positive prechallenge. The development of clinical illness played a significant role in the immune response. The only person who did not seroconvert was not ill, and the mean titer for 24 persons who developed clinical illness was significantly higher than the mean for the 7 with subclinical infection. Twenty of 24 ill persons still had elevated serum antitoxin titers at 19 to 25 months postinfection, while titers from 3 of seven persons without illness had declined below significance after 4 months. Antitoxin titers would thus appear to be more useful in making a retrospective diagnosis of cholera in persons with a recognized diarrheal illness than in serum surveys of areas where diarrheal disease is not known to be occurring.

As discussed briefly above, CT is structurally, biologically, and immunologically similar to *E. coli* LT (30, 31). Immunologic cross-reactivity complicates serodiagnosis of cholera by measurement of antitoxic antibodies (8, 37). Fortunately, patients usually develop higher titers to the homologous toxin (CT in cholera patients and LT in *E. coli* patients). Svennerholm et al. found that 16 of 17 volunteers infected with *V. cholerae* O1 developed a fourfold or greater rise in titer to CT, and all titers were at least 1.5 times the corresponding serum titer to LT (37). Likewise, 11 of 14 patients infected with LT-producing *E. coli* showed a fourfold or greater rise in titer to LT, and all titers were at least 1.5 times the corresponding titer to CT.

When Svennerholm et al. tested sera from persons with *V. cholerae* O1 infection for antibodies to purified CT subunits A and B, most sera showed high titers of antibody to subunit B but not to subunit A (37). Serum samples from patients infected with LT-producing *E. coli* showed little reaction with either individual subunit. Titration of antibody against purified B subunit may thus be a means of distinguishing persons with *V. cholerae* infection from those with LT-producing-*E. coli* infection.

Harris et al. used the CT/LT titer ratio to identify possible cholera cases during and after a cholera outbreak in a previously uninfected island population (18). During investigation of a cholera outbreak on Truk in 1982, postoutbreak serum was collected from both ill and well persons. Results of antitoxin antibody tests were compared with those obtained with sera collected during a rotavirus outbreak in 1964, before cholera was known to have affected the island. The prevalences of elevated anti-CT and anti-LT titers were similar in both years, presumably owing to the cross-reaction of anti-LT antibodies with CT. In contrast, anti-CT/anti-LT ratios were elevated in 42% of serum samples collected in 1982 versus 4% of serum samples collected in 1964. The anti-CT/anti-LT ratio thus appeared capable of distinguishing naturally acquired infection with *V. cholerae* O1 from infection with LT-producing *E. coli* even in areas where both are prevalent.

The studies discussed here clearly illustrate both the value and the limitations of antitoxin antibody testing. As with any serodiagnostic test, specimen collection must be appropriately timed, and a rising titer is stronger evidence of recent infection than is a single elevated titer. Cross-reactivity between CT and LT remains the biggest problem in interpretation of antitoxin titers. Cross-reactivity may not present a significant problem in the United States or other areas where toxigenic *E. coli* is uncommon, but the possibility of cross-reacting antibodies must always be considered.

APPLICATION OF ANTIBODY TESTING TO SERODIAGNOSIS

Whenever possible, paired acute- and convalescent-phase serum specimens should be collected from each individual being tested. Since the vibriocidal antibody titer normally rises rapidly after infection, acute-phase serum should be collected during the first 3 days of illness. Vibriocidal antibody titers in volunteers ingesting *V. cholerae* O1 usually peaked within 10 days after ingestion, began to decline within 1 month, fell markedly by 6 to 12 months, and returned to near baseline after 1 year (9). The rapid rise and gradual decline in vibriocidal titers occurred in most infected persons regardless of the presence or severity of symptoms (9).

A fourfold or greater rise in vibriocidal antibody titer may be considered diagnostic of recent infection with *V. cholerae* O1. Since low titers (20 or 40) may not be reproducible (unpublished data) and since infected persons typically have very high vibriocidal titers, the convalescent-phase titer should be at least 320. There is no one titer that can be considered diagnostic of recent infection. In one study, 79% of volunteers had titers of $\geq 1,280$ at 1 month postinfection, and 99% had titers of ≥ 20 (9). A titer of < 20 thus implies a very low likelihood of recent *V. cholerae* O1 infection. Since titers of > 640 were found in only 0.9% of healthy persons in one U.S. study (41), it can be assumed that titers of 1,280 or greater imply recent infection. Intermediate titers are more problematic. Titers of 160, 320, or even 640 may represent true infection with *V. cholerae* O1 but may also be due to cross-reacting antibodies elicited by infection with other organisms. Organisms known to elicit cross-reacting antibodies include *Yersinia enterocolitica* serotype O9 and *Brucella* and *Citrobacter* species. These complications to interpreting vibriocidal titers reinforce the desirability of obtaining paired serum samples.

The interpretation of antitoxin antibody titers was discussed at length above. As with vibriocidal antibody titers, interpretation is easier if paired sera can be obtained. The antitoxin antibody response peaks somewhat later (21 to 28 days postinfection) than the vibriocidal titer (10 to 14 days). Convalescent-phase sera obtained relatively early may thus show a rise in vibriocidal titer without showing a rise in antitoxin titer. In contrast to vibriocidal titers, antitoxin titers may remain high for as long as 2 years (24, 28). The longevity of the antitoxin response may be helpful in epidemiologic investigations to determine whether cholera might have been occurring before an outbreak was recognized and serum samples were collected.

Finally, the utility of using both antibacterial and antitoxin antibody assays must be emphasized. While cross-reactions with other antigens are known to occur with both types of assays, the likelihood of a rise in titer to two independent antigens of *V. cholerae* O1 occurring by chance in an uninfected person is remote.

REFERENCES

1. **Barua, D.** 1974. Laboratory diagnosis of cholera, p. 85–126. *In* D. Barua and W. Burrows (ed.), *Cholera*. The W. B. Saunders Co., Philadelphia.
2. **Barua, D., and R. B. Sack.** 1964. Serological studies in cholera. *Indian J. Med. Res.* **52:**855–866.
3. **Benenson, A. S., A. Saad, and W. H. Mosley.** 1968. Serological studies in cholera. 2. The vibriocidal antibody response of cholera patients determined by a microtechnique. *Bull. W.H.O.* **38:**277–285.
4. **Benenson, A. S., A. Saad, W. H. Mosley, and A. Ahmed.** 1968. Serological studies in cholera. 3. Serum toxin neutralization—rise in titre in response to infection with *Vibrio cholerae*, and the level in the "normal" population of East Pakistan. *Bull. W.H.O.* **38:**287–295.
5. **Benenson, A. S., A. Saad, and M. Paul.** 1968. Serological studies in cholera. 1. Vibrio agglutinin response of cholera patients determined by a microtechnique. *Bull. W.H.O.* **38:**267–276.
6. **Beran, G. W.** 1964. Serological studies on cholera. 1. Antibody levels observed in vaccinated and convalescent persons. *Am. J. Trop. Med. Hyg.* **13:**698–707.
7. **Bhattacharya, F. K., and S. Mukerjee.** 1974. Serological analysis of the flagellar or H agglutinating antigens of cholera and NAG vibrios. *Ann. Microbiol.* (Paris) **125A:**167–181.
8. **Clements, J. D., and R. A. Finkelstein.** 1978. Immunological cross-reactivity between a heat-labile enterotoxin(s) of *Escherichia coli* and subunits of *Vibrio cholerae* enterotoxin. *Infect. Immun.* **21:**1036–1039.
9. **Clements, M. L., M. M. Levine, C. R. Young, R. E. Black, Y. L. Lim, R. M. Robins-Browne, and J. P. Craig.** 1982. Magnitude, kinetics, and duration of vibriocidal antibody responses in North Americans after ingestion of *Vibrio cholerae*. *J. Infect. Dis.* **145:**465–473.
10. **Craig, J. P.** 1971. Cholera toxins, p. 189–254. *In* S. Kadis, T. C. Montie, and S. J. Ajl (ed.), *Microbial Toxins*, vol. IIA. Academic Press, Inc., New York.
11. **Donta, S. T., and D. M. Smith.** 1974. Stimulation of steroidogenesis in tissue culture by enterotoxigenic *Escherichia coli* and its neutralization by specific antiserum. *Infect. Immun.* **9:**500–505.
12. **Feeley, J. C.** 1965. Comparison of vibriocidal and agglutinating antibody responses in cholera patients, p. 220–222. *In* D. A. Bushell and C. S. Brookhyer (ed.), *Proceedings of the Cholera Research Symposium*, U.S. Public Health Service Publication no. 1328.
13. **Finkelstein, R. A.** 1962. Vibriocidal antibody inhibition (VAI) analysis: a technique for the identification of the predominant vibriocidal antibodies in serum and for the detection and identification of *Vibrio cholerae* antigens. *J. Immunol.* **89:**264–271.
14. **Finkelstein, R. A., and P. Atthasampunna.** 1967. Immunity against experimental cholera. *Proc. Soc. Exp. Biol. Med.* **125:**465–469.
15. **Gallut, J., and G. Brounst.** 1949. Sur la mise en evidence des agglutines choleriques. *Ann. Inst. Pasteur* (Paris) **76:**557–559.
16. **Goodner, K., H. L. Smith, Jr., and H. Stempen.** 1960. Serologic diagnosis of cholera. *J. Albert Einstein Med. Cent.* **8:**143–147.
17. **Guerrant, R. L., L. L. Brunton, T. C. Schnaitman, L. I. Rebhun, and A. G. Gilman.** 1974. Cyclic adenosine monophosphate and alteration of Chinese hamster ovary cell morphology: a rapid, sensitive *in vitro* assay for the enterotoxins of *Vibrio cholerae* and *Escherichia coli*. *Infect. Immun.* **10:**320–327.
18. **Harris, J. R., S. D. Holmberg, R. D. R. Parker, D. E. Kay, T. J. Barrett, C. R. Young, M. M. Levine, and P. A. Blake.** 1986. Impact of epidemic cholera in a previously uninfected island population: evaluation of a new seroepidemiologic method. *Am. J. Epidemiol.* **123:**424–430.
19. **Heiberg, B.** 1935. *On the Classification of Vibrio cholerae and the Cholera-Like Vibrios.* Nyt Nordisk Forlag-Arnold Busck, Copenhagen.
20. **Hochstein, H. D., J. C. Feeley, and W. E. DeWitt.** 1970. Titration of cholera antitoxin in human sera by microhemagglutination with formalinized erythrocytes. *Appl. Microbiol.* **19:**742–745.
21. **Holmgren, J., and A. M. Svennerholm.** 1973. Enzyme-linked immunosorbent assays for cholera serology. *Infect. Immun.* **7:**759–763.

22. **Holmgren, J., and A. M. Svennerholm.** 1974. Mechanisms of disease and immunity in cholera: a review. *J. Infect. Dis.* **136**(Suppl.):S105–S112.

23. **Kasai, G. J., and W. Burrows.** 1966. The titration of cholera toxin and antitoxin in the rabbit ileal loop. *J. Infect. Dis.* **116**:606–614.

24. **Levine, M. M., C. R. Young, T. P. Hughes, S. O'Donnell, R. B. Black, M. L. Clements, R. Robins-Browne, and Y.-L. Lim.** 1981. Duration of serum antitoxin response following *Vibrio cholerae* infection in North Americans: relevance for seroepidemiology. *Am. J. Epidemiol.* **114**:348–354.

25. **McIntyre, O. R., and J. C. Feeley.** 1964. Passive serum protection of the infant rabbit against experimental cholera. *J. Infect. Dis.* **114**:468–475.

26. **Neter, E.** 1956. Bacterial hemagglutination and hemolysis. *Bacteriol. Rev.* **20**:166–188.

27. **Pal, S. C.** 1992. Laboratory diagnosis, p. 229–251. *In* D. Barua and W. B. Greenough III (Ed.), *Cholera.* Plenum Publishing Corp., New York.

28. **Pierce, N. F., J. G. Banwell, R. B. Sack, R. C. Mitra, and A. Mondal.** 1970. Magnitude and duration of antitoxic response to human infection with Vibrio cholerae. *J. Infect. Dis.* **121**(Suppl.):S31–S35.

29. **Pollitzer, R.** 1959. *Cholera,* p. 266–270. W.H.O. Monograph no. 43. World Health Organization, Geneva.

30. **Richardson, K. L., and S. D. Douglas.** 1978. Pathophysiological effects of *Vibrio cholerae* and enterotoxigenic *Escherichia coli* and their exotoxins in eucaryotic cells. *Microbiol. Rev.* **42**:592–613.

31. **Sack, R. B.** 1975. Human diarrheal disease caused by enterotoxigenic *Escherichia coli. Annu. Rev. Microbiol.* **29**:333–353.

32. **Sack, R. B., D. Barua, R. Saxena, and C. C. J. Carpenter.** 1966. Vibriocidal and agglutinating antibody patterns in cholera patients. *J. Infect. Dis.* **116**:630–640.

33. **Sakazaki, R., K. Tamura, C. Z. Gomez, and R. Sen.** 1970. Serological studies on the cholera group of vibrios. *Jpn. J. Med. Sci. Biol.* **23**:13–20.

34. **Smith, H. L., Jr., and K. Goodner.** 1965. Antibody patterns in cholera, p. 215–219. *In* O. A. Bushnell and C. S. Brookhyer (ed.), *Proceedings of the Cholera Research Symposium.* U.S. Public Health Service Publication no. 1328. U.S. Government Printing Office, Washington, D.C.

35. **Snyder, J. D., D. T. Allegra, M. M. Levine, J. P. Craig, J. C. Feeley, W. E. DeWitt, and P. A. Blake.** 1981. Serologic studies of naturally acquired infection with *Vibrio cholerae* serogroup O1 in the United States. *J. Infect. Dis.* **143**:182–187.

36. **Svennerholm, A. M., and J. Holmgren.** 1978. Identification of *Escherichia coli* heat-labile enterotoxin by means of a ganglioside immunosorbent assay (GM₁-ELISA) procedure. *Curr. Microbiol.* **1**:19–23.

37. **Svennerholm, A.-M., J. Holmgren, R. Black, M. Levine, and M. Merson.** 1983. Serologic differentiation between antitoxin responses to infection with *Vibrio cholerae* and enterotoxin-producing *Escherichia coli. J. Infect Dis.* **147**:514–522.

38. **Vella, E. E., and P. Fielding.** 1963. Note on the agglutinability of cholera suspensions. *Trans. R. Soc. Trop. Med. Hyg.* **57**:112–114.

39. **Wachsmuth, I. K., J. C. Feeley, W. E. DeWitt, and C. R. Young.** 1980. Immune response to *Vibrio cholerae*, p. 464–473. *In* N. R. Rose and H. Friedman (ed.), *Manual of Clinical Immunology,* 2nd ed. American Society for Microbiology, Washington, D.C.

40. **Young, C. R., M. M. Levine, J. P. Craig, and R. Robins-Browne.** 1980. Microtiter enzyme-linked immunosorbent assay for immunoglobulin G cholera antitoxin in humans: method and correlation with rabbit skin vascular permeability factor technique. *Infect. Immun.* **27**:492–496.

41. **Young, C. R., I. K. Wachsmuth, Ø. Olsvik, and J. C. Feeley.** 1986. Immune response to *Vibrio cholerae*, p. 363–370. *In* N. R. Rose, H. Friedman, and J. L. Fahey (ed.), *Manual of Clinical Immunology,* 3rd ed. American Society for Microbiology, Washington, D.C.

II.   VIRULENCE FACTORS

Vibrio cholerae and Cholera: Molecular to Global Perspectives
Edited by I. Kaye Wachsmuth, Paul A. Blake, and Ørjan Olsvik
© 1994 American Society for Microbiology, Washington, DC 20005

Chapter 11

Toxins of *Vibrio cholerae*

James B. Kaper, Alessio Fasano, and Michele Trucksis

Vibrio cholerae produces a variety of extracellular products that have deleterious effects on eukaryotic cells. The massive diarrhea induced by *V. cholerae* is caused by cholera enterotoxin (CT), but strains deleted of genes encoding CT can still cause mild to moderate diarrhea in many individuals. The search for a cause of the diarrhea seen after ingestion of CT⁻ strains has resulted in the discovery of additional toxins produced by this species. Furthermore, recent reports about CT itself have raised questions about the long-standing model of how CT causes diarrhea. In this chapter, we review new information about CT and about recently discovered toxins produced by *V. cholerae*.

CHOLERA ENTEROTOXIN

History

In 1884, Robert Koch proposed that the agent responsible for cholera produces "a special poison" acting on the epithelium and that the symptoms of cholera could be "regarded essentially as a poisoning" (96). Considerable time elapsed before the existence of

this hypothesized toxin was demonstrated by two separate groups of investigators working in India. In 1959, S. N. De, working in Calcutta, injected sterile filtrates of *V. cholerae* culture into ligated rabbit ileal loops, which produced an outflowing of fluid, an effect he described as "enterotoxicity" (26). In the same year, N. K. Dutta and colleagues, working in Bombay, demonstrated that ingestion of sterile lysates of *V. cholerae* 569B produced fatal diarrhea in infant rabbits (36). Ten years later, purification of the toxin to homogeneity was accomplished by Finkelstein and LoSpalluto (48). Availability of purified toxin allowed a number of investigators to discover fundamental properties of the toxin such as structure, receptor binding, and mode of action. These discoveries have been extensively reviewed in a historical context (44, 192). Perhaps the ultimate proof of the role of CT in the disease was an experiment reported in 1983 by Levine et al. (104) in which purified CT was fed to volunteers. These investigators found that ingestion of 25 μg of pure CT (administered with cimetidine and NaHCO$_3$ to diminish gastric acidity) caused over 20 liters of rice water stool, and ingestion of as little as 5 μg of pure CT resulted in between 1 and 6 liters of diarrheal stool in five of six volunteers. These results closed the circle on the role of the "special poison" postulated by Koch nearly a century earlier.

James B. Kaper, Alessio Fasano, and Michele Trucksis • Center for Vaccine Development, University of Maryland School of Medicine, 10 South Pine Street, Baltimore, Maryland 21201-1116.

Structure

CT (also called choleragen) is the proto-type of the A-B subunit toxins. As with other toxins of this group, the function of the sub-units are specific: the B subunit serves to bind the holotoxin to the eukaryotic cell receptor, and the A subunit possesses a specific enzymatic function that acts intracellularly. Neither of the subunits individually has significant secretogenic activity in animal or intact cell systems. The mature B subunit contains 103 amino acids and has a subunit weight of 11.6 kDa. The mature A subunit has a mass of 27.2 kDa and is proteolytically cleaved (see below) to yield two polypeptide chains, a 195-residue A_1 peptide of 21.8 kDa and a 45-residue A_2 peptide of 5.4 kDa. After proteolytic cleavage, the A_1 and A_2 peptides are still linked by a disulfide bond before internalization (58). Cross-linking experiments by Gill (56) determined that the ratio of B subunits to A subunits is 5:1. A recent review focuses extensively on the structure and function of CT (175).

The crystal structures for CT and the highly similar heat-labile enterotoxin of the enterotoxigenic *Escherichia coli* (LT) have recently been determined (172, 194). The most extensively characterized structure is that of LT, shown in Fig. 1. The structure for CT is essentially identical in the major features (194). The major features of the LT B subunit include a small N-terminal α-helical structure and two three-stranded antiparallel β sheets (called sheets I and II) (Fig. 2A). A disulfide bond between residues 9 and 86 connects the N-terminal helix with strand $\beta5$ located in the C-terminal half of the monomer. The B subunits form a pentamer via interactions between β sheet I of one monomer and sheet II of an adjacent monomer. Each monomer contains a large central α helix (residues 59 to 78), and in the pentameric form, these helices form a "barrel" in the center of the pentamer, creating a pore with a diameter of 11 to 15 Å (1.1 to 1.5 nm). In the pore formed by the B pentamer sits the C-terminal end of the A_2 peptide, which binds to the B pentamer by multiple interactions. The crystal structure confirms previous cross-linking studies indicating that the A_2 peptide links the A_1 peptide to the B pentamer (56). The N-terminal half of the A_2 is a long α helix that extends outside the B pentamer and interacts with the A_1 peptide (Fig. 2B). The A_1 peptide itself is a mixture of β strands, α helices, and extended chains. The shape of the A_1 subunit roughly resembles a triangle with a base of 57 Å (5.7 nm) and sides of ~ 35 Å (3.5 nm). The A_1 peptide has structural homology with the catalytic region of *Pseudomonas aeruginosa* exotoxin A and diphtheria toxin (172).

Receptor Binding and Translocation

The interaction between CT and the receptor occurs via the B subunit. Addition of purified B subunit (also known as choleragenoid) to rabbit ileal loops before addition of CT inhibits fluid secretion (158). Antibodies directed against B, the immunodominant subunit, are much more efficient at neutralizing toxin activity than are antibodies directed against the A subunit (155, 179). King and van Heyningen (95) identified the CT receptor as the ganglioside GM_1. As with B subunit, addition of GM_1 to rabbit ileal loops before addition of CT inhibits fluid secretion (95, 158). The structure of GM_1 is Gal$\beta1 \rightarrow$ 3GalNAc$\beta1$(NeuAc$\alpha2 \rightarrow$ 3) $\rightarrow$ Glc-ceramide, where Gal is galactose, GalNAc is *N*-acetylgalactosamine, Glc is glucose, and NeuAc is *N*-acetylneuraminic acid (sialic acid) (175). GM_1 is ubiquitous in the body, being present on such diverse cell types as ovarian and neural cells as well as intestinal cells. Both CT and LT bind the terminal sugar sequence Gal$\beta1 \rightarrow$ 3GalNAc$\beta1$(NeuAc$\alpha2 \rightarrow$ 3) $\rightarrow$ 4Gal (51, 73); a recent crystallography study shows that LT binds specifically to the terminal galactose (171). Unlike CT, LT can also bind to GM_2 and asialo-GM_1 in addition to GM_1 (51). The neuraminidase produced by *V. cholerae* can increase the number of receptors by acting upon higher-order gangliosides to

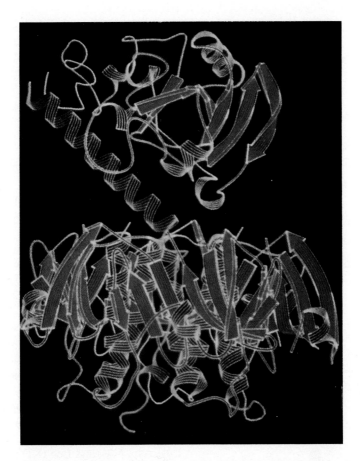

Figure 1. Crystal structure of the LT holotoxin shown as a ribbon plot. The A_1 subunit shown at the top is connected to the B pentamer via the long helical structure of the A_2 peptide. Reprinted from *Nature* with permission (171).

convert them to GM_1 (see below). Binding of the toxin appears to require that at least two of the five B subunits interact with GM_1, but it is not clear that all five subunits need to bind (30, 172).

Several key residues in the B subunit have been identified as being necessary for receptor binding. The single tryptophan residue (Trp-88) in each subunit is essential for binding, as shown by chemical modification (29, 30, 114) and site-directed mutagenesis studies (79). Substitution of Gly-33 by negatively charged or large hydrophobic residues also abolishes binding and toxicity (79). The two cysteine residues at positions 9 and 86 are also essential for proper functioning of the B subunit.

Crystallographic studies by Sixma et al. (171) have located the GM_1-binding site of the B subunit on the "bottom" of the CT, i.e., on the side opposite the A_1 subunit (Fig. 1). These data strongly imply that the B pentamer binds to the GM_1-containing membranes with the A subunit pointing away from the membrane. It is not clear how the A_1 subunit makes its way past the subunit B and into the cell. Simple insertion of the A_1 peptide through the pore in the B pentamer and into the cell is not supported by the crystallography studies, which show a pore size of only 11 to 15 Å. Biochemical studies show that the B pentamer does not enter the lipid bilayer (175, 187, 195). Although it has recently been suggested that the B pentamer

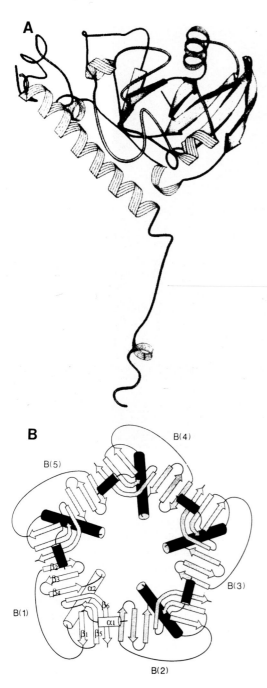

resembles an iris and that sliding actions between B subunits could open a large pore (113), the available evidence indicates that the B pentamer does not undergo major conformational changes (162, 172). Reduction of the disulfide bond between A_1 and A_2 on the external surface of the membrane is necessary for penetration of the A_1 into the cell (187). The fate of the A_2 peptide is not known, but there is little evidence that it actually enters the cell.

Several studies indicate that CT undergoes receptor-mediated endocytosis with exposure to acidic pH (77, 78, 98). Most of these studies utilized nonpolarized cells, but a recent study shows that in T_{84} cells, a polarized cell line of intestinal origin, CT is also found within endocytic vesicles (102). In an endocytic event, the entire toxin, both A and B subunits, could be internalized into the cell, with subsequent translocation of the A subunit across the endosomal membrane. It has recently been proposed that after translocation and while still associated with the endosomal membrane, the A_1 peptide ADP-ribosylates the G protein (78). Interestingly, the final four residues of the A subunit are Lys, Asp, Glu, and Leu (KDEL), a sequence that retains newly formed proteins within the endoplasmic reticulum (142). A similar sequence (RDEL) is found at the C-terminal ends of LT and the exotoxin A of *Pseudomonas* spp. (22). Perhaps the role of the B pentamer is merely to present the A subunit close enough to the membrane so that the intrinsic properties of the A subunit allow it to be translocated. Because only a few molecules of A_1 are enough to ADP-ribosylate all the G proteins in a cell, the entry of the A_1 peptide does not have to be highly efficient (191).

Figure 2. (A) Ribbon plot of the structure of the LT A subunit showing the triangular shape of the A_1 peptide (top right) and the long α helix (left) of A_2, terminated by a single small helix and tail. (B) Schematic secondary structure diagram of the B pentamer. A single B subunit is shown in light gray. Reprinted from *Nature* with permission (171, 172).

Enzymatic Activity

CT affects one of the most important regulatory factors of the eukaryotic cell, namely, adenylate cyclase. Adenylate cyclase mediates the transformation of ATP to cyclic AMP (cAMP), a crucial intracellular messenger for a variety of cellular pathways. Normally, adenylate cyclase is activated or inactivated in response to a variety of stimuli. Regulation of adenylate cyclase is mediated by G proteins, which serve to link many cell surface receptors to effector proteins at the plasma membrane. The specific G protein involved is the G_s protein, activation of which leads to increased adenylate cyclase activity. CT catalyzes the transfer of the ADP-ribose moiety of NAD to a specific arginine residue in the α subunit of the G_s protein. Gill and King (57) showed that the activity was specifically associated with the A_1 peptide. ADP-ribosylation of the α subunit constitutively activates adenylate cyclase, thereby leading to significant increases in intracellular levels of cAMP. cAMP activates a cAMP-dependent protein kinase, leading to protein phosphorylation, alteration of ion transport, and ultimately diarrhea (Fig. 3).

After CT binds to intact cells, there is a lag of 15 to 60 min before adenylate cyclase is activated (57). However, in lysed pigeon erythrocytes, the activation of adenylate cyclase begins immediately after addition of toxin to the lysate (57). The lag time is necessary to allow the A_1 peptide to translocate through the membrane and come into contact with the G proteins.

Since adenylate cyclase is located on the basolateral membrane of polarized intestinal epithelial cells and CT attaches to the apical membrane, a fundamental question concerns how these two proteins interact. This question has not been definitely answered, but there are multiple possibilities. One possibility is that the A_1 peptide diffuses freely through the cytoplasm, but there is no convincing evidence that this actually occurs (43). A second possibility is that the A_1 peptide interacts with an α subunit of the G protein in the brush border membrane. G proteins are heterotrimers composed of three distinct subunits: α (M_r = 39 to 46 kDa), β (M_r = 37 kDa), and γ (M_r = 8 kDa) (67).

$G_{s\alpha}$ can dissociate from the other subunits and interact with adenylate cyclase, thereby activating it. In the rabbit small intestine, most of the G_s is localized to the brush border membrane, and CT can ADP-ribosylate the α subunit at this site. Dominguez et al. (32) propose that the activated α subunit dissociates from the apical membrane, thus becoming able to traverse the cell and attach to the adenylate cyclase located in the basolateral membrane. However, the ability of $G_{s\alpha}$ to move and activate effectors at the opposite side of the cell is by no means universally accepted. A third model proposes that neither the A_1 subunit nor the activated $G_{s\alpha}$ moves freely through the cytosol to reach the basolateral adenylate cyclase. Instead, Janicot et al. (78) propose that generation of the A_1 peptide and activation of adenylate cyclase are functionally linked to toxin endocytosis. In this model, whole CT is internalized by endocytosis, and the low pH of the endosomal compartment facilitates translocation of the A subunit across the endosomal membrane. The endosome then travels through the cell with the A_1 peptide still associated with the endosomal membrane. Finally, the A_1 peptide ADP-ribosylates $G_{s\alpha}$ located in the basolateral membrane, perhaps after an endosome-plasma membrane fusion, which in turn activates adenylate cyclase. Recent support for this model comes from work by Lencer et al. (102), who used polarized T_{84} cells. (Unlike CHO or Y1 adrenal cells, polarized cells such as T_{84} and Caco-2 have clearly differentiated apical and basolateral membranes.) These investigators suggest that the mechanism of CT action involves vesicular transport of toxin-containing membranes beyond the apical endosomal compartment. These investigators have also recently demonstrated that Cl^- secretion by CT could be completely inhibited by addition of brefeldin A, a fungal metabolite known to interfere with vesicular transport in endosomal and transcytotic pathways of many eukaryotic cells (101). Nambiar et al. (143) have also shown that entry of CT into Y1 and CHO

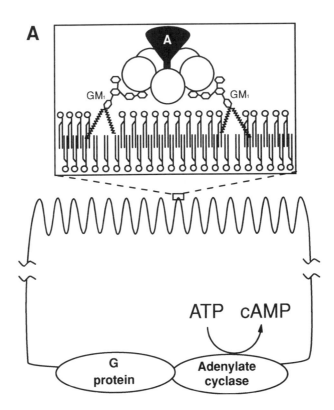

Figure 3. Mode of action of CT; see text for details. (A) Adenylate cyclase, located in the basolateral membrane of intestinal epithelial cells, is regulated by G proteins. CT binds via the B-subunit pentamer to the GM_1 ganglioside receptor inserted into the lipid bilayer. (B) The A subunit enters the cell, perhaps via endosomes, and is proteolytically cleaved, with subsequent reduction of the disulfide bond to yield A_1 and A_2 peptides. The A_1 peptide is activated (at least in vitro) by ARF and transfers an ADP-ribose moiety (ADPR) and NAD to the α subunit of the G_s protein. The ADP-ribosylated α subunit dissociates from the other subunits of $G_{s\alpha}$ and activates adenylate cyclase, thereby increasing the intracellular cAMP concentration. Three possible scenarios have been proposed to explain entry of the toxin and activation of adenylate cyclase. Possibility 1 proposes that the A_1 subunit translocates through the apical membrane, leaving the B pentamer on the apical membrane. The A_1 diffuses freely through the cytoplasm to the basolateral membrane, where it ADP-ribosylates $G_{s\alpha}$. Possibility 2 is that the A_1 peptide ADP-ribosylates an α subunit in the apical membrane. The ADP-ribosylated α subunit traverses the cell to attach to the adenylate cyclase located in the basolateral membrane. Possibility 3 is that the entire toxin enters the cell via endosomes and the A subunit translocates through the endosomal membrane. The endosome travels through the cell with the A_1 peptide still associated with the endosomal membrane. The A_1 peptide ADP-ribosylates the $G_{s\alpha}$ located in the basolateral membrane, perhaps after an endosome-plasma membrane fusion. (C) Increased cAMP activates protein kinase A, leading to protein phosphorylation. Protein phosphorylation leads to increased Cl^- secretion in crypt cells and decreased NaCl-coupled absorption in villus cells.

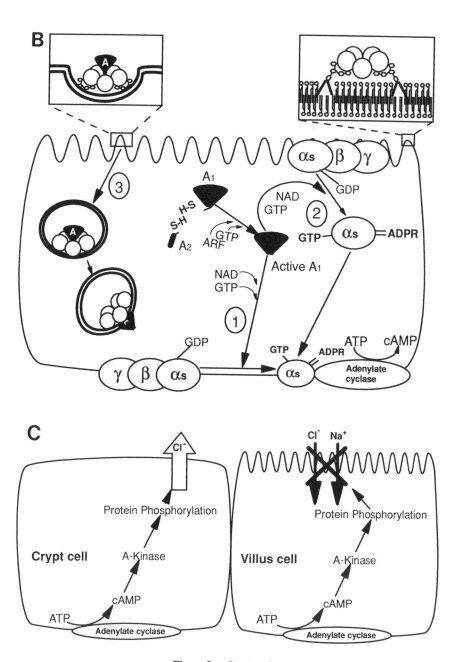

Figure 3.—*Continued.*

cells is inhibited by brefeldin A, thereby indicating that an intact Golgi region is required for intracellular trafficking of CT. Thus, at least three groups of investigators using four different cell types have demonstrated the involvement of toxin endocytosis in the entry of CT into cells.

By whatever means A_1 reaches adenylate cyclase, the A_1 peptide catalyzes the transfer of ADP-ribose from NAD to an arginine residue on $G_{s\alpha}$. Moss and Vaughn (137) first demonstrated the ADP-ribosyltransferase activity of CT, which catalyzes the reaction $NAD + G_{s\alpha} \rightarrow [ADP\text{-ribosyl } G_{s\alpha}] + $ nicotinamide $+ H^+$. The α subunit of G_s contains a GTP-binding site and an intrinsic GTPase activity (reviewed in reference 67). Binding of GTP to the α subunit leads to dissociation of the α and the β-γ subunits and subsequent increased affinity of α for adenylate cyclase. The resulting activation of adenylate cyclase continues until the intrinsic GTPase activity hydrolyzes GTP to GDP, thereby inactivating the G protein and adenylate cyclase. ADP-ribosylation of the α subunit by the A_1 peptide of CT inhibits the hydrolysis of GTP to GDP, thus leaving adenylate cyclase constitutively activated, probably for the life of the cell (16, 82).

The ADP-ribosylation activity of A_1 is stimulated in vitro by a family of proteins termed ADP-ribosylation factors (ARFs; reviewed in reference 138). The ARF proteins are ca. 20-kDa GTP-binding proteins found in both soluble and membrane fractions. The ARF proteins constitute a distinct family within the larger group of ca. 20-kDa guanine nucleotide-binding proteins that include the *ras* oncogene protein. In the presence of GTP, ARF appears to act directly on the A_1 peptide, independently of B subunits or $G_{s\alpha}$. ARFs serve as allosteric activators of A_1 and increase the ADP-ribosyl transferase activity of the proteins. The stimulatory effects of ARF proteins on A_1 have been shown only in vitro, and the presumed importance of ARFs in the pathogenesis of cholera has not yet been shown in vivo.

Cellular Response

The activation of adenylate cyclase by CT leads to a substantial increase in intracellular cAMP concentrations. The increased cAMP levels lead to increased Cl^- secretion by intestinal crypt cells and decreased NaCl-coupled absorption by villus cells (40) (Fig. 3). The net movement of electrolytes into the lumen results in a transepithelial osmotic gradient that causes water flow into the lumen. The massive volume of water overwhelms the absorptive capacity of the intestine, resulting in diarrhea (see chapter 15). The steps between increased levels of cAMP and secretory diarrhea are not known in their entirety. Pioneering work in this area by Field and colleagues (39, 42) demonstrated that CT could concomitantly increase cAMP formation and ion transport in isolated intestinal epithelium mounted in Ussing chambers. These early observations have subsequently been confirmed and expanded by many investigators (reviewed in reference 44). One crucial step resulting from increased cAMP levels is activation of protein kinase A. The inactive form of this kinase consists of two regulatory and two catalytic subunits (185). Increased cAMP levels lead to binding of the nucleotide to the regulatory subunits, leading to dissociation of the regulatory and catalytic subunits and activation of the catalytic subunit. The activated kinase then phosphorylates numerous substrates in the cell (21).

Although it is clear that chloride channels can be regulated by cAMP-dependent protein kinases, the actual ion channel or channels that are affected by the action of CT are not known with certainty. There are multiple types of Cl^- channels in apical membranes with different modes of activation and subcellular distributions (27). It is also not certain whether protein kinase A directly phosphorylates the ion channel or whether it phosphorylates an intermediate protein that then phosphorylates other proteins in a cascade. One attractive candidate for a relevant target protein is the cystic fibrosis (CF) gene product, CFTR. The CFTR protein is a Cl$^-$ channel (7) and has multiple potential substrate sequences for kinase A. Unlike healthy intestinal tissue, tissues obtained from patients with CF (CF homozygotes) do not respond to either cAMP- or Ca-mediated secretagogs (9). Interestingly, it has been hypothesized that individuals who are CF heterozygotes may have a selective advantage over "normal" homozygotes in surviving cholera (164). Heterozygotes presumably have only half of the normal number of chloride channels responsive to kinase. After infection with *V. cholerae,* the CF heterozygote may have less intestinal chloride secretion and therefore less diarrhea, owing to the smaller number of chloride channels.

Alternative Mechanisms of Action

The mechanism of action for CT outlined above is by no means complete. Key questions, such as the identity of the ion channel activated by kinase A, need to be answered. Although the great majority of the biochemical steps detailed above have been reproduced by numerous investigators, the importance of these reactions in the disease cholera is not universally accepted. Data from several groups suggest that the increased levels of cAMP and subsequent protein kinase A activation may not explain all of the secretory effects of CT. The majority of these dissenting reports, which are summarized below, do not suggest that the above scenario is wrong; rather, they suggest that this scenario simply does not explain all of the secretion due to CT. Further investigations into alternative explanations for the mechanism of CT in causing diarrhea are clearly warranted.

Prostaglandins

The possible involvement of prostaglandins in the pathogenesis of cholera was first suggested in 1971 (8) along with initial reports that the prostaglandins E$_1$ and E$_2$ (PGE$_1$ and PGE$_2$) could stimulate adenylate cyclase and increase short circuit current (Isc) in Ussing chambers (94). Cholera patients in the active secretory disease stage have elevated jejunal concentrations of PGE$_2$ compared to patients in the convalescent stage (176). Furthermore, high jejunal PGE$_2$ concentrations correlated with high stool volumes. The role of prostaglandins, leukotrienes, and other metabolites of arachidonic acid in causing intestinal secretion have been well documented (52), although the exact mechanisms are unclear. Drugs that affect prostaglandin synthesis, such as aspirin and indomethacin, have been shown to inhibit secretory effects of CT in animal studies but do not reduce fluid loss in patients (160). Peterson and Ochoa (156) have reported that addition of cAMP induced only a small, transient fluid accumulation in rabbit intestinal loops, whereas addition of PGE$_2$ caused a much larger fluid accumulation in rabbit loops. Addition of CT led to increases in both cAMP and PGE in rabbit loops and resulted in the release of arachidonic acid from membrane phospholipids in Chinese hampster ovary (CHO) cells (161). Release of PGE$_2$ into the intestinal lumen after addition of CT has also been reported in a rat model (10). Other reports have suggested that CT activates phospholipase A$_2$ (14) and that inhibitors of transcription and translation can block the turnover of phospholipids in response to CT (157). A model has been suggested in which cAMP levels increased by CT serve not only to activate protein kinase A but also to regulate transcription of a phospholipase or a phospholipase-activating protein. The activated phospholipase could act on membrane phospholipids to produce arachidonic acid, a precursor of prostaglandins and leukotrienes (157).

Other investigators have also reported the importance of arachidonic acid metabolites in fluid secretion due to CT. De Jonge and colleagues (27), using the human intestinal cell line HT29.c1.19A mounted in Ussing chambers, report that 40 to 60% of the Isc response to CT is inhibited by relatively low concentrations of the phospholipase A_2 inhibitor mepacrine. The effect of CT on arachidonic acid metabolism might occur through ADP-ribosylation of G_s in the apical membrane, which would then directly activate a phospholipase, or the effect may be mediated through protein kinase A, which would activate the phospholipase.

ENS

The vast majority of data concerning the mode of action of CT has been obtained with tissue culture cells, erythrocytes, or intestinal tissue stripped of muscles and nerves mounted in Ussing chambers. Such studies would underestimate the role of the enteric nervous system (ENS), which, as part of the autonomic nervous system, plays an important role in intestinal secretion and absorption (24). The intestine also contains a variety of cells that can produce hormones and neuropeptides, products that can affect secretion. Two such substances with well-established roles in causing secretion are vasoactive intestinal peptide (VIP) and serotonin (5-hydroxytryptamine [5-HT]) (24). These components of the intestine could play a significant role in the secretory response to CT. Lundgren and colleagues (115) have reported several studies that support the concept that the ENS plays a crucial role in secretion due to CT. The overall hypothesis proposed by these investigators is that CT binds to "receptor cells," namely, enterochromaffin cells, that release a substance such as 5-HT that activates dendritelike structures located beneath the intestinal epithelium. This leads to the release of VIP, resulting in electrolyte and fluid secretion.

This model is supported by three major types of experiments, conducted largely in cat and rat systems.

(i) CT-induced secretion is inhibited by a variety of ganglionic or neurotransmitter blockers placed on the mucosal or serosal side of the intestinal tissue. The secretion induced by CT is significantly inhibited by tetrodotoxin (a sodium gate blocker), hexamethonium (a cholinergic nicotinic receptor antagonist), and lidocaine (a local anesthetic) (17, 18, 20). It should be noted, however, that each of these agents has multiple effects, including, potentially, a direct effect on epithelial cell function (71).

(ii) CT stimulates release of 5-HT from enterochromaffin cells into the intestinal lumen (144). Furthermore, fluid secretion induced by CT is markedly diminished by 5-HT receptor antagonists (173). It is interesting that chlorpromazine, a compound used in the treatment of cholera (159), is also a 5-HT receptor blocker (144).

(iii) CT-induced secretion is accompanied by an increased release of VIP from the small bowel (144). The increased release of VIP is blocked by tetrodotoxin. Cholera patients also show increased levels of VIP in blood (71). VIP, normally found only in the nerves of the gut (115), binds to receptors on epithelial cells to inhibit absorption or induce secretion (24). (It has also been found that very low concentrations of VIP can increase the concentration of cAMP in intestinal cells [3]).

The majority of the experiments described above have been performed in vivo, and the results often differ from those obtained in in vitro experiments examining the role of the ENS. Using muscle-stripped guinea pig ileum, Carey and Cooke (15) could not block the effect of CT on Isc by using tetrodotoxin. Tetrodotoxin also failed to block the in vitro effect of CT on rabbit ileal tissue mounted in Ussing chambers (134). The denervation of the extrinsic nervous supply in the in vitro experiments may account for the observed differences.

Cassuto et al. (19) estimate that ca. 60% of the effect of CT on intestinal fluid transport could be attributed to nervous mechanisms. The release of 5-HT from enterochromaffin cells in response to binding of CT may cause secretion by two pathways: (i) the 5-HT could directly stimulate the ENS, leading to release of VIP, or (ii) the 5-HT might release prostaglandins that could alter transport function directly (see above) or could activate the ENS. Besides the direct secretory effect, CT may also increase intestinal motility and thus contribute to diarrhea. Mathias et al. (124) found that pure CT and live *V. cholerae* organisms evoke a strong myoelectrical activity in the intestine, producing a pattern known as migrating action potential complex. This activity is not seen with genetically engineered *V. cholerae* lacking CT, which nonetheless still produces diarrhea in volunteers (109).

Additional support for the role of the ENS in secretion due to CT may be found in the experiments of Levine et al. (104) in which volunteers ingested pure CT. The very small amounts of CT used in these experiments might reasonably be expected to be largely bound to the most proximal part of the small intestine. Yet the massive amounts of diarrheal stool seen after ingestion of only 25 μg of CT suggests that a much larger part of the gut is involved in secretion. Activation of the ENS by CT could explain the widespread effect throughout the gut.

Interactions with the Immune System

CT is one of the most potent oral immunogens ever studied (116). Vigorous immune responses are engendered not only against CT delivered orally by itself but also against unrelated antigens delivered orally with CT. Lycke and colleagues have shown that CT can stimulate immune responses to protein antigens such as keyhole limpet hemocyanin up to 50-fold greater than those observed when antigen is delivered alone (120). This adjuvant effect is seen not only with purified protein antigens but also with whole virus particles (108). Chemical conjugation or genetic fusion of antigens to the B subunit of CT or LT has yielded enhanced responses to streptococcal antigens (25, 28), hepatitis B antigen (170), and enterotoxigenic *E. coli* heat-stable enterotoxin (ST) (167).

Both subunits of CT contribute to the adjuvant effect. The adjuvancy of B subunit

alone is probably due to the avid binding of B subunit to GM_1 (25, 49). However, the most potent adjuvant effect of CT and LT is due to the ADP-ribosyltransferase activity of the A subunit. Lycke et al. (119) have shown that alteration of a single residue in the active site of the A subunit greatly diminishes the adjuvant effect of LT. CT has a variety of effects on cells of the immune system, including stimulating interleukin 1 proliferation and enhancing antigen presentation by macrophages (12), promoting B-cell isotype differentiation (118), and inhibiting Th1 cells (141). It has also been suggested that the adjuvant action of CT is due to increased intestinal permeability in response to CT, perhaps providing increased access of antigens to the gut mucosal immune system (117).

A variety of monoclonal antibodies have been generated against CT, primarily against the immunodominant B subunit (70, 91, 92, 110, 183). These have been used to map B-subunit epitopes and to compare CT to LT I and LT II of enterotoxigenic *E. coli*. The majority of these monoclonal antibodies recognize conformation epitopes (91). A few monoclonal antibodies that recognize the A subunit have been isolated, and the majority of these reacted only with denatured A subunit in Western blots (immunoblots) (47). Finkelstein and colleagues have used monoclonal antibody reactivity and sequence information to construct a "family tree" showing the possible evolution of the LT-CT family of toxins (47). At least three variants of the B subunit (CT-1, CT-2, CT-3) are found, and convalescent-phase sera from cholera patients react slightly differently to the different variants (91). It has been suggested that differences in B-subunit sequences may be significant in designing cholera vaccines (91). However, given the overwhelming importance of antibacterial versus antitoxic immunity in protection against cholera (see chapter 26), these differences are probably of minimal importance for vaccines.

Genetics

The genes encoding CT (*ctxAB*) were initially cloned by exploiting their homology to the genes encoding the LT of enterotoxigenic *E. coli* (55, 84, 152). DNA sequence analysis shows 78% overall nucleotide homology and 80% predicted protein homology between the two genes (111, 112, 128, 200, 201). A study of the evolutionary origin of the toxins suggests that the toxin genes originated in *V. cholerae* O1 and were acquired by *E. coli* some 130 million years ago, i.e., before the presumed appearance of humans (199).

The A and B subunits are encoded on two separate but overlapping open reading frames; the first two bases of the *ctxA* translation termination signal (TGA) are the last two of the *ctxB* translation initiation codon (ATG) (Fig. 4) (11, 111, 128). The *ctxA* DNA sequence predicts a translation product of 258 amino acids, of which the first 18 residues are a putative signal peptide, and the *ctxB* sequence predicts a translation product of 124 amino acids, including a putative 21-residue signal peptide. The A and B cistrons possess ribosomal-binding sites immediately upstream of their start codons, with the site for *ctxB* being located in the 3' end of the *ctxA* sequence. The higher expression of the B

subunit, leading to the 5:1 ratio of B:A subunits in the holotoxin, results from more-efficient translation due to a stronger ribosomal-binding site for the B subunit (128). Transcriptional regulation of the *ctx* operon is quite complex and is discussed in chapter 12.

Many strains of *V. cholerae* O1 contain multiple copies of the *ctx* operon. Classical strains contain two copies, which are separated on the *V. cholerae* O1 chromosome by an unknown distance (126, 136). Most El Tor strains contain only a single copy of the *ctx* operon, but about 30% of El Tor strains contain two or more adjacent gene copies (126, 127). The duplicated *ctx* genes within each classical or El Tor strain appear to be identical (150).

The *ctx* operon, along with genes encoding the Zot and Ace toxins (see below), is located on a 4.5-kb region called the "core region" (59). Flanking the 4.5-kb region are one or more copies of a 2.7-kb sequence called RS1. Recombination between RS1 sequences can lead to tandem duplication and amplification of the core region as well as deletion of the core region (59). Mekalanos (126) demonstrated that amplification of the core region can occur in vivo by serially passing an El Tor strain through three sets of rabbit intestinal loops. Isolates recovered after the third

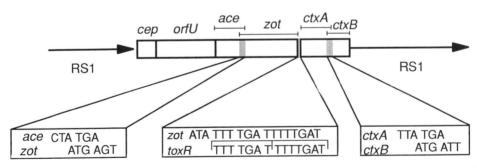

Figure 4. Arrangement of *ctxAB*, *zot*, and *ace* genes in the *V. cholerae* core region. RS1 elements are shown on both sides of the core region, a common arrangement in El Tor but not classical strains. The enlarged regions depict the overlapping open reading frames of *ctxA* and *ctxB* and of *ace* and *zot*. In addition, the region where the stop codon of *zot* overlaps the first ToxR binding repeat upstream of the *ctx* operon is also enlarged. Also shown are *orfU*, an open reading frame of unknown function, and *cep* (core-encoded pilus), the gene for a recently described colonization factor (153).

passage expressed higher levels of CT and contained more *ctx* copies than the strain injected into the first loop. This in vivo amplification may be enhanced by the presence of an intestinal colonization factor on the 4.5-kb core region (153) (Fig. 4). Classical strains and some El Tor strains possess RS1 sequences only on one side of the core region, and therefore this gene duplication and amplification are not seen (126, 127). The RS1 sequence encodes a site-specific transposable element that can insert into a specific 18-bp sequence called attRS1 (153). Naturally occurring isolates of *V. cholerae* that do not produce CT lack sequences homologous to *ctx* and the rest of the core region (88, 131) but do not contain attRS1 sequences (153). Pearson et al. (153) recently demonstrated that when the core region and RS1 sequences are cloned into a suicide plasmid incapable of replicating in *V. cholerae* and transferred from a conjugation-proficient *E. coli*, the RS1 and *ctx* sequences can insert into the attRS1 site in a *recA*-independent manner.

Sequence conservation among *ctx* genes from different strains of *V. cholerae* O1 is quite high. Olsvik et al. (150) have recently determined the *ctxB* sequences from 45 strains of *V. cholerae* O1 by using the polymerase chain reaction technique and automated sequencing. The *V. cholerae* O1 strains studied were isolated in 29 countries over a period of 70 years. Among the 300 bp sequenced for each strain, variation was found only at three nucleotide positions, and each base change resulted in a change in the predicted amino acid sequence. Variation at the three nucleotide positions divided the 45 strains into three genotypes of *ctxB*. Genotype 1 contained all classical strains and El Tor strains from the U.S. Gulf Coast, genotype 2 contained only El Tor strains from Australia, and genotype 3 contained El Tor strains from the seventh pandemic and from the recent Latin American epidemic. A potential fourth genotype, from the sequence originally reported by Mekalanos et al. (128),

was concluded to represent misprints or sequencing errors (150).

Accessory *V. cholerae* Enzymes Affecting CT

Neuraminidase

V. cholerae produces a neuraminidase (EC 3.2.1.18; NANase) that can enhance the effect of CT by catalyzing the conversion of higher-order gangliosides to GM_1 (72). It has been hypothesized that NANase may produce locally high concentrations of the GM_1 receptor for CT in vivo, thereby enhancing the binding of CT and leading to greater fluid secretion (72, 95). Experiments using purified NANase and ^{125}I-labeled CT showed that isolated rabbit small-intestine brush borders or cultured mouse neuroblastoma cells exposed to NANase bound two to seven times more CT than did untreated cells (61, 132). The enhanced binding can also lead to increased internalization of CT; exposure of isolated rat adrenal cells to NANase prior to incubation with CT caused up to a sevenfold increase in intracellular cAMP levels (63). Furthermore, secretion in ligated canine ileal loops perfused with CT increased fourfold in loops pretreated with NANase compared to untreated control loops (178). It has been reported that El Tor strains produce less NANase with lower specific activity than that produced by classical strains (81), leading to the suggestion that the decreased NANase activity by El Tor strains may result in a higher ratio of asymptomatic or mild cases to symptomatic infections than with classical strains (53).

The gene encoding *V. cholerae* NANase *(nanH)* has been cloned by Galen et al. (53), and the nucleotide sequence predicts a molecular mass of 83.0 kDa for the extracellular enzyme. These investigators constructed isogenic strains mutated in the *nanH* gene and examined the effect of culture filtrates of these strains. In using flow cytometry, the fluorescence due to binding of fluorescein-conjugated CT to mouse fibroblasts exposed to NANase⁺ culture filtrates increased five- to eightfold relative to binding to cells exposed to NANase⁻ filtrates. Furthermore, the Isc measured in Ussing chambers increased 65% with NANase⁺ filtrates compared to NANase⁻ filtrates. With growth conditions under which CT expression was low, the effect of NANase was greater than with conditions under which toxin expression was high. These investigators concluded that while NANase was not a primary virulence factor of *V. cholerae*, it played a subtle but significant role in binding and uptake of CT.

Disulfide isomerase

The assembly and secretion of CT across the bacterial outer membrane into the extracellular environment is a multistep process. Like other proteins secreted into the periplasm of gram-negative bacteria, the A and B subunits are secreted into the periplasm after cleavage of 18- and 21-residue signal peptides, respectively. This process is presumably mediated by the *V. cholerae* counterpart of the *E. coli* Sec system (68). However, assembly of the subunits and excretion of them into the extracellular environment represent additional and as yet poorly characterized steps.

One enzyme that plays a crucial role in the biogenesis of CT is a disulfide isomerase that catalyzes the formation of disulfide bonds between Cys-187 and Cys-199 of the A subunit and Cys-9 and Cys-86 of the B subunit (202). Mutations in the gene encoding this 20.5-kDa protein (*dsbA*) lead to rapid degradation of B-subunit monomers in the periplasm and prevention of holotoxin assembly. Oxidation of the cysteine residues in the B subunit had previously been shown to be required for formation of B-subunit pentamers (64). This same enzyme is essential for assembly of the toxin-coregulated colonization pilus of *V. cholerae*, which contains an intrachain disulfide bond (154).

The other steps of CT assembly and secretion into the extracellular environment are not as well characterized as disulfide bond formation (reviewed in reference 68). It is known that the subunits differ in their abilities to reach the extracellular environment; the B-subunit pentamer can be assembled and excreted in the absence of the A subunit, but the A subunit is not found extracellularly without first being assembled into the holotoxin (69). The recent cloning of genes involved in secretion of *E. coli* LT toxin through the outer membrane of *V. cholerae* should offer insights into the assembly and secretion of CT (168).

Protease

The proteolytic cleavage of the A subunit to the A_1 and A_2 peptides is a crucial step in the activation of CT (58). Multiple enzymes including trypsin (58) can mediate this cleavage, but an endogenous enzyme of *V. cholerae* has specifically been shown to process or "nick" the A subunit to the A_1 and A_2 peptides. The *V. cholerae* soluble hemagglutinin-protease (HA/protease) is a zinc metalloenzyme that nicks CT and also cleaves fibronectin, mucin, and lactoferrin (46). El Tor strains produce higher levels of HA/protease than do classical strains (34, 180), and excessive protease production by some strains results in degradation rather than just nicking of the A subunit (34). The gene encoding the HA/protease (*hap*) has been cloned (65), and the predicted amino acid sequence of the 46.7-kDa protein shows 61.5% identity with the *P. aeruginosa* elastase. Although the HA/protease is a very attractive virulence factor, an isogenic strain of *V. cholerae* specifically mutated in the gene encoding this protein was no less virulent in infant rabbits than the parent strain (45). Thus, there are additional proteases, either produced by *V. cholerae* or found within the intestinal environment, that can nick the A subunit. Interestingly, the HA/protease has a "detachase" activity that allows the vibrios to detach from cultured human intestinal epithelial cells (45). The significance of this activity is not known, but the activity could conceivably aid the spread of the organism through the environment (45).

OTHER TOXINS PRODUCED BY *V. CHOLERAE*

Volunteer Experiments with Δ*ctx V. cholerae* Strains

The study in which volunteers ingested purified CT and experienced the severe diarrhea characteristic of cholera suggested that all of the symptoms of disease due to *V. cholerae* were due to a single toxin (104). Thus, it was a great surprise when the first genetically engineered *V. cholerae* strains specifically deleted for genes encoding either the A_1 peptide or both A and B subunits were found to cause diarrhea in volunteers (106).

Levine et al. (106) tested a number of attenuated *V. cholerae* strains from which genes encoding the A subunit (Δ*ctxA*) or both the A and B subunits (Δ*ctxAB*) had been specifically deleted by recombinant DNA techniques. Two such strains were JBK70 (Δ*ctxAB*) (86), a derivative of El Tor Inaba strain N16961, and CVD101 (Δ*ctxA*) (85), a derivative of classical Ogawa strain 395. When volunteers ingested doses of these organisms ranging from 10^4 to 10^{10} CFU, 20 of 38 subjects experienced mild to moderate diarrhea. The diarrhea lasted from 1 to 3 days, and the total diarrheal stool volumes ranged from 0.3 to 2.1 liters. The severity of diarrhea was not cholera-like; ingestion of the CT^+ parent strain of CVD101, Ogawa 395, resulted in over 40 liters of diarrheal stool (103). However, the stool volumes engendered by CT^- strains are comparable to those seen with volunteers ingesting many strains of enterotoxigenic *E. coli* (107). In addition, many recipients of the attenuated *V. cholerae* strains experienced other symptoms, such as headache, fever, nausea, intestinal cramps, and vomiting. Two hypotheses were advanced to explain the residual diarrhea observed in ca. 50% of vaccines (106). One hypothesis suggested that the act of colonization of the proximal small intestine by adherent vibrios itself results in a secretory response leading to diarrhea. The alternative hypothesis holds that *V. cholerae* possesses an additional enterotoxin(s) distinct from CT and that this secondary toxin(s) is responsible for the diarrhea observed after ingestion of the Δ*ctx* strains. Several additional toxins produced by *V. cholerae* are discussed below.

Hemolysin-Cytolysin

Hemolysis of sheep erythrocytes was traditionally used to distinguish between the El Tor and classical biotypes of *V. cholerae*, although more recent El Tor isolates are only poorly hemolytic on sheep erythrocytes (54). Genes encoding this hemolysin, *hlyA*, are present in classical, El Tor, and non-O1 strains of *V. cholerae* (13). The hemolysin was initially purified by Honda and Finkelstein (74) and was shown to be cytolytic for a variety of erythrocytes and mammalian cells in culture and rapidly lethal for mice. The hemolysin produced by *V. cholerae* non-O1 is indistinguishable biologically, physicochemically, and antigenically from hemolysin produced by El Tor strains (196) and is capable of causing fluid accumulation in ligated rabbit ileal loops (76). In contrast to the watery fluid produced in response to CT, the accumulated fluid produced in response to hemolysin was invariably bloody with mucus (76). McCardell et al. (125) and Spira et al. (177) have described a cytolysin that is cytotoxic for Y1 adrenal and CHO cells and causes fluid accumulation in rabbit ileal loops. This cytolysin was proposed to be identical to the El Tor hemolysin. The cytolysin has recently been shown to form anion-selective channels in planar lipid bilayers (99).

To test the hypothesis that this hemolysin was responsible for the diarrhea seen with Δ*ctx* strains of *V. cholerae*, Kaper et al. (87) constructed derivatives of JBK70 and

CVD101 that were mutated in the *hlyA* gene by deletion of an internal 400-bp *Hpa*I fragment. (The *hlyA* gene was mutated in the classical strain CVD101 because *E. coli* containing cloned *hlyA* genes from classical 395 were capable of lysing chicken and rabbit but not sheep erythrocytes [163].) When tested in volunteers, the Δ*hlyA* strains CVD104 and CVD105 still caused diarrhea in 33% of the subjects (106). These results indicate that the hemolysin-cytolysin is probably not the cause of diarrhea seen in recipients of Δ*ctx V. cholerae* strains.

The hemolysin-cytolysin is initially made as a 82-kDa protein and processed in two steps to a 65-kDa active cytolysin (197). In classical strain 569B, an 11-bp deletion leads to a truncated *hlyA* gene product of 27 kDa (2). Alm et al. (1) suggest that the ability to lyse sheep erythrocytes is present in the C-terminal end of the hemolysin-cytolysin and that the enterotoxic activity lies in the N-terminal end, within the truncated 27-kDa product termed "HlyA*." Because the *hlyA* deletion present in CVD104 and CVD105 removes only 2.5 kDa of the 27-kDa HlyA* product, these authors conclude that the enterotoxic activity in these strains is not inactivated. However, interpretation of the results in the study by Alm et al. (1) is difficult, because the only strain tested in animal models that expressed the truncated HlyA* also produces CT.

ZO Toxin (Zot)

In 1991, Fasano et al. (38) reported that *V. cholerae* produced a toxin that increases the permeability of the small intestinal mucosa by affecting the structure of the intercellular tight junction, or zonula occludens (ZO). This activity was discovered by testing culture supernatants of *V. cholerae*, both wild type and Δ*ctx*, in Ussing chambers, a classic technique for measuring transepithelial transport of electrolytes across intestinal tissue. The parameters that can be measured or calculated with Ussing chambers are (i) poten-tial difference (PD), or the difference in voltage measured on the mucosal side versus the serosal side of the tissue; (ii) Isc, or the amount of current needed to nullify the PD; and (iii) tissue resistance (Rt), or the resistance of a membrane to ion movements, which is the inverse of tissue conductivity (Gt). These parameters are related by Ohm's law: $Isc = PD/Rt$.

When culture supernatants of *V. cholerae* CVD101 (Δ*ctxA*) were added to rabbit ileal tissue mounted in Ussing chambers, an immediate increase in tissue conductivity was observed (38). Unlike the increase in PD observed with CT, which reflects ion transport across the membrane (41), i.e., the transcellular pathway, variation in transepithelial conductance reflects modification of tissue permeability through the intercellular space, i.e., the paracellular pathway (31). Since tight junctions (ZO) represent the major barrier in this paracellular pathway (31, 121), the integrity of the tight junctions was examined by electron microscopy. As shown in Fig. 5, exposure of ileal tissue to culture supernatant of CVD101 resulted in "loosening" of the tight junction so that an electron-dense marker could permeate into the paracellular space. In contrast, tissue treated with uninoculated broth control was not permeable to this marker. In freeze-fracture electron microscopy studies, tight junctions appear as an anastomosing network of strands (Fig. 6). In general, as the tissue resistance decreases, the number of ZO strands also decreases (62). When tissue treated with *V. cholerae* CVD101 or 395 culture supernatants was examined by freeze-fracture electron microscopy, a striking decrease in strand complexity and the density of strand intersections compared to those of the negative control was observed (Fig. 6) (38). The toxin responsible for this striking effect on ZO was named Zot, for ZO toxin.

Intestinal tight junctions restrict or prevent the diffusion of water-soluble molecules through the intercellular space (the paracellular pathway) into the lumen (121). The permeability of the tight junction can be affected by a variety of factors, including Ca^{2+} ions (60), protein kinase C (139), cAMP (35), glucose (4), and tumor necrosis factor (140). Recent evidence from Madara and Atisook (5, 122) shows that small peptides can transit through the tight junction from the lumen to the submucosa, which suggests that the paracellular pathway is a major route for uptake of nutrients. Fasano and colleagues (38) speculate that Zot may cause the diarrhea and other symptoms seen with CT^- *V. cholerae* strains in two possible ways. Since Zot decreases tissue resistance, thereby increasing intestinal permeability, water and electrolytes may leak into the lumen under the force of hydrostatic pressure, resulting in diarrhea. Alternatively, similar to the mechanism of nutrient uptake proposed by Madara and Atisook (5, 122), loosening of the tight junctions may allow uptake of bacterial endotoxin or other products and thereby cause symptoms such as abdominal cramps, vomiting, and headache. There is recent evidence in cholera patients of Zot-like activity in vivo. Mathan and Mathan (123), working in Vellore, India, have found alterations of the intercellular tight junctions in ileal-biopsy samples obtained from cholera patients. The patients had been treated with intravenous rehydration only, so that alteration of the tight junctions due to glucose in oral rehydration solution is not responsible for this effect.

The mechanism by which Zot affects tight junctions is not known, but a possible mechanism can be proposed. Since the cytoskeleton has an important role in the regulation of tight junctions (62), loosening of the tight junctions may reflect broader changes affecting the cytoskeleton such as alterations of actin microfilaments. Intracellular messengers such as Ca^{2+}, cAMP, and protein kinase C can affect tight junctions (35, 60, 139). The action of Zot is reversible, and preliminary evidence does not implicate cAMP in the mode of action (38). Binding of Zot to a membrane receptor could result in signal transduction involving an intracellular mes-

senger such as Ca^{2+}, with subsequent alterations of the cytoskeleton and tight junctions.

The gene encoding Zot was cloned and found to be located immediately upstream of the *ctx* locus (6). The *zot* gene consists of a 1.3-kb open reading frame that could potentially encode a 44.8-kDa polypeptide. The termination codon of *zot* (TGA) overlaps the initial TTTTGAT repeat of the ToxR binding site for the *ctx* operon (Fig. 4). The predicted amino acid sequence of the Zot protein shows no homology to any other bacterial toxin including the toxin A of *Clostridium difficile*, a large 300-kDa protein that also alters tight junctions (66). However, toxin A differs from Zot in size (33), cytopathicity (190), and attenuation of glucose-stimulated Na^+ cotransport (133). A recent search of the GenBank sequence data revealed an interesting homology of unknown significance (97). Zot shares three conserved regions with the product of gene I of filamentous bacteriophages, which is believed to be an ATPase. However, the putative nucleotide triphosphate-binding domain of Zot contains a drastic substitution of tyrosine for glycine, and the possibility that Zot contains ATPase activity has not yet been shown experimentally.

The distribution of *zot* gene sequences has been studied by several investigators (37, 80, 90), and homologous sequences have been found in both O1 and non-O1 strains of *V. cholerae*. These studies show that strains that contain *ctx* sequences almost always contain *zot* sequences and vice versa. Furthermore, the limited DNA sequence data that are available show a high conservation of *zot* sequences. Two strains of the classical biotype, strains 395 and 569B, share 100% homology in the *zot* gene, and an El Tor strain shares 99% homology (23).

Accessory Cholera Enterotoxin

Another potential enterotoxin of *V. cholerae* has recently been identified by Trucksis et al. (188). These investigators demon-

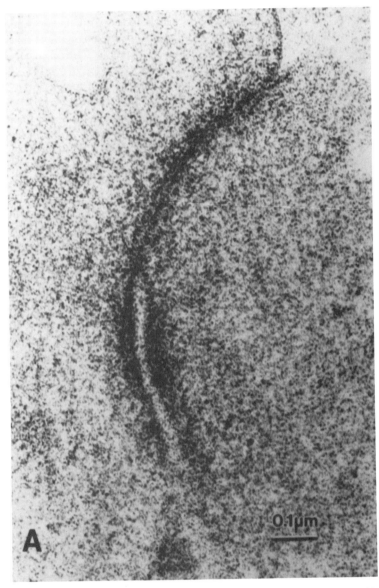

Figure 5. Electron micrograph showing the entry of wheat germ agglutinin-horseradish peroxidase into the paracellular space of rabbit ileal cells exposed for 60 min to medium (A) or to culture supernatants of Zot+ *V. cholerae* CVD101 (B). Taken from Fasano et al. (38).

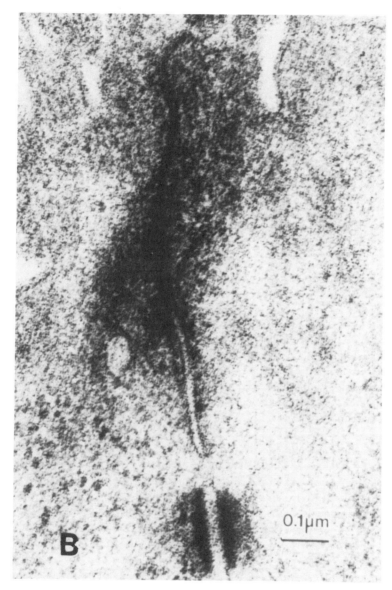

Figure 5.—*Continued.*

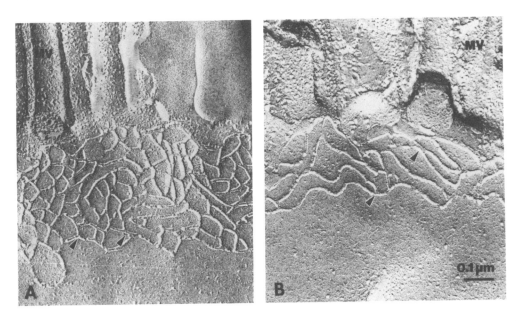

Figure 6. Freeze-fracture studies of rabbit ileal tissue showing effect of culture supernatants of *V. cholerae* on ZO. (A) An intact ZO with numerous intersections (arrowheads) between junctional strands. MV, microvilli. (B) An affected ZO from ileal tissue exposed to culture supernatants of *V. cholerae* 395. Taken from Fasano et al. (38).

strated that the gene product of an open reading frame located immediately upstream of *zot* can increase Isc in Ussing chambers. Like CT, and in contrast to Zot, this new toxin increases PD rather than tissue conductivity. This toxin was called Ace, for accessory cholera enterotoxin.

The *ctx, zot,* and *ace* sequences are located on a 4.5-kb region of DNA called the core region (59) (Fig. 4). As noted above, recombination between RS1 elements flanking the core region can lead to deletion of the entire 4.5-kb region encoding CT, Zot, and Ace (59, 129). A derivative of pathogenic El Tor strain E7946 was constructed in this manner to yield *V. cholerae* CVD110 (129). In addition to the deletion of *ctx, zot,* and *ace,* this strain also contained a deletion-insertion mutation of the *hlyA* gene. Culture supernatants of CVD110 produced no change in Isc in Ussing chambers. When a 289-bp open reading frame located immediately upstream of *zot* was cloned into CVD110, culture supernatants of the result-

ing strain gave a significant increase in PD and Isc in Ussing chambers. In ligated rabbit ileal loops, CVD110 containing the cloned *ace* gene causes significant fluid accumulation compared to CVD110 without *ace* (188).

The *ace* open reading frame could potentially encode a 96-residue peptide with a predicated M_r of 11.3 kDa (188). When the predicted amino acid sequence of Ace was compared to the Swiss-Prot database, a striking similarity was found between Ace and a family of eukaryotic ion-transporting ATPases, including the human plasma membrane calcium pump, the calcium-transporting ATPase from rat brain, and the product of the CF gene, the CF transmembrane regulator (CFTR) (Fig. 7). The first two proteins are involved in transport of calcium ions across the membrane, and CFTR functions as a chloride ion channel (7). In addition, Ace also shows sequence similarity to a virulence protein of *Salmonella dublin*, SpvB, which is essential for virulence in mice (100).

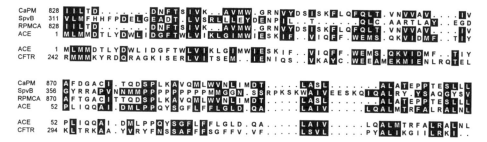

Figure 7. Amino acid sequence comparison of the predicted *ace* gene product and predicted sequence of two ion-transporting ATPases, the human plasma membrane Ca^{2+} pump (CaPM) and the Ca^{2+}-transporting ATPase from rat brain (RPMCA); a virulence protein in *Salmonella dublin* (SpvB); and the CF transmembrane conductance regulator (CFTR). Boxes indicate identical or similar amino acid residues in Ace and the other sequences. Ace was aligned separately with CFTR to maximize sequence similarity. Taken from Trucksis et al. (188).

Analysis of the predicted Ace protein sequence revealed an α-helical region with a high hydrophobic moment at the C-terminal end of Ace (188). This 20-residue region has an amphipathic nature, in which the majority of hydrophobic residues are on one side of the helix (Fig. 8). This structure is consistent with a model in which Ace multimers insert into the eukaryotic membrane with hydrophobic surfaces facing the lipid bilayer and hydrophilic sides facing the interior of a transmembrane pore. Such a structure is found for the 26-residue δ toxin of *Staphylococcus aureus* (50), which has a variety of effects on eukaryotic tissue, including increasing vascular permeability in guinea pig skin and inhibiting water absorption and increasing cAMP concentration in the ileum (89, 148). The C-terminal region of Ace shows 47% amino acid similarity with residues 2 to 20 of δ toxin (188), thus lending some support to a model in which Ace acts by aggregating and inserting into the eukaryotic membrane to form an ion channel.

The last codon of the *ace* open reading frame overlaps with the start codon of *zot* (Fig. 4), suggesting translational coupling between *ace* and *zot*. Interestingly, this same structure is seen with the two subunit genes of *ctx* (see above). It is intriguing that the genes encoding these three toxins are located in tandem on a dynamic region of the *V. cholerae*

chromosome. The arrangement of the 4.5-kb core region flanked by RS1 elements can, in some strains, lead to amplification or deletion of all three toxin genes as a unit (59). This region was recently shown to also contain a gene encoding an intestinal colonization fac-

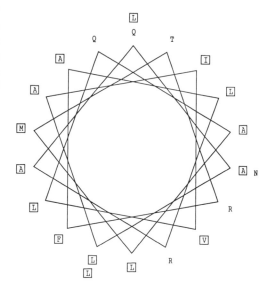

Figure 8. Potential amphipathic region of Ace. The C-terminal end of the Ace protein containing a predicted α helix is arranged in the form of a wheel to show the predominance of hydrophobic (boxed) residues on one side of the helix. The glutamine (Q) at the top of the wheel is residue 76 of the predicted amino acid sequence of Ace. Taken from Trucksis et al. (188).

tor, *cep* (core-encoded pilin) (153). Thus, this region may be perceived as a "virulence cassette" of *V. cholerae* (188).

Miscellaneous Toxins

Shiga-like toxin

In 1984, O'Brien et al. (147) reported that *V. cholerae* O1 produced a Shiga-like toxin. This toxin was identified on the basis of cytotoxicity in HeLa cells that was neutralized by antibody raised against Shiga toxin purified from *Shigella dysenteriae* 1. It was hypothesized that this toxin was responsible for diarrhea in volunteers who ingested genetically engineered CT$^-$ strains. Consistent with this hypothesis, *V. cholerae* CVD103-HgR (Δctx), which does not produce detectable Shiga-like activity, causes little or no reactogenicity in volunteers (105). However, Morris et al. (135) showed that a strain of *V. cholerae* Non-O1 producing a Shiga-like toxin did not cause diarrhea in volunteers. Genes encoding a Shiga-like toxin activity have not been cloned from *V. cholerae*, despite repeated efforts (130). Although *V. cholerae* genomic fragments hybridizing under reduced stringency to the *E. coli slt-1* gene (encoding Shiga toxin) have been cloned, the DNA sequence derived from these fragments showed no significant homology to the *slt-1* coding sequence (151).

ST

Some strains of *V. cholerae* non-O1 produce a toxin that shares 50% protein sequence homology to the ST of enterotoxigenic *E. coli* (149). Production of this toxin has been associated with diarrhea in volunteers (135; also see chapter 8). It is exceedingly uncommon for this toxin to be found in *V. cholerae* O1, but one study did identify this toxin in 1 of 197 isolates of *V. cholerae* O1 (182).

NCT

In 1983, Sanyal and colleagues reported that environmental strains of *V. cholerae* O1 that lack genes encoding CT could cause fluid accumulation in ligated rabbit ileal loops (169). Fluid accumulation was observed when either whole cells or culture filtrates were used. The filtrates also increased the capillary permeability of rabbit skin, but unlike CT, they also caused blanching or necrosis along with the bluing reaction (169). The enterotoxic activity was subsequently found in CT$^+$ strains such as classical Inaba 569B (165). The toxin was termed "new cholera toxin" (NCT) and proposed as the cause of diarrhea in volunteers fed genetically engineered CT$^-$ *V. cholerae* strains (166). The enterotoxic activity was not neutralized by antisera raised against CT (169) and was also found in strains that naturally lack *zot* and *ace* sequences (6). Antisera raised against a crude preparation of NCT neutralized the enterotoxic activity, and it was suggested that attenuated *V. cholerae* vaccine strains should also confer immunity against NCT as well as CT (166). No additional biochemical or genetic characterization of this toxin has been reported.

Sodium channel inhibitor

Tamplin et al. (184) reported that strains of *V. cholerae* O1, including CT$^-$ CVD101, produced a factor that inhibited sodium channels. The initial characterization of this factor suggested that the active compound is tetrodotoxin or a related toxin. This report is interesting in light of studies showing that tetrodotoxin is produced by some marine *Vibrio* species (146), but no further characterization of this sodium channel inhibitor in *V. cholerae* O1 has been reported.

TDH

V. parahaemolyticus produces a hemolysin called thermostable direct hemolysin (TDH), which is responsible for the enterotoxic activity of this species (145). Genes encoding a TDH-like toxin have been found on a plasmid in some strains of *V. cholerae* non-O1 (75), but homologous sequences have not yet been found in *V. cholerae* O1 (186).

Role of Additional Toxins in Disease

The role of toxins other than CT in the pathogenesis of disease due to *V. cholerae* is unknown. These toxins clearly cannot cause cholera gravis, because the diarrhea seen with Δctx strains presumably still producing these toxins is not the severe purging seen with wild-type *V. cholerae* strains. However, toxins other than CT may contribute to the diarrhea and other symptoms seen with wild-type *V. cholerae* strains and may be responsible for the symptoms seen with CT⁻ strains. Such toxins may serve as a secondary secretogenic mechanism when conditions for producing CT are not optimal. Two of the toxins reviewed above, ST and TDH, have well-established mechanisms of enterotoxicity but are rarely produced by *V. cholerae*. The Shiga-like toxin of *V. cholerae* may cause diarrhea, but the expression of this toxin by a *V. cholerae* non-O1 strain that caused no symptoms in volunteers does not support a role for this toxin. Two of the toxins, the sodium channel inhibitor and the NCT of Sanyal and colleagues, are insufficiently characterized to access their potential roles in disease.

The three non-CT *V. cholerae* toxins that are the best characterized are the hemolysin-cytolysin, Zot, and Ace. Genes encoding the hemolysin-cytolysin are found in nearly all pathogenic and nonpathogenic strains of *V. cholerae* O1 and non-O1, with no correlation seen with the presence of *ctx* sequences. In contrast, *ace* and *zot* are almost always found in strains containing *ctx* but are rarely found in strains lacking *ctx*. Thus, there is a strong epidemiologic correlation of the presence of *zot* and *ace* with disease. The facts that *ctx*, *zot*, and *ace* are located in tandem on a dynamic chromosomal region of virulent *V. cholerae* and that *ace* and *zot* sequences are highly conserved (23, 189) suggest that all three toxins may have a role in disease. It is also intriguing that the three toxins have distinctly different modes of action. CT is the classic enterotoxin, increasing Isc and PD by increasing intracellular cAMP. Zot also increases Isc but primarily affects tissue resistance, possibly leading to diarrhea by the paracellular rather than the transcellular pathway. Ace has a classic enterotoxic effect in Ussing chambers in that it increases PD and Isc and also has a fascinating similarity to eukaryotic ion-transporting proteins.

Sequences encoding Zot, Ace, and hemolysin-cytolysin as well as the CT A subunit have been deleted from *V. cholerae* El Tor CVD110 (see above). Recently, CVD110 was fed to volunteers (181) to test the hypothesis that these toxins were responsible for diarrhea and other symptoms observed with CVD101 and JBK70. However, this strain caused mild to moderate diarrhea in 7 of 10 volunteers (mean stool volume of 861 ml) as well as fever and abdominal cramps. While these volunteer studies do not preclude a role for Ace, Zot, and hemolysin-cytolysin in the pathogenesis of cholera, they clearly indicate that there are additional features of *V. cholerae* that result in diarrhea.

The assumption that removal of *ctx*, *zot*, and *ace* genes would result in a nondiarrheagenic *V. cholerae* strain was based on Ussing chamber studies showing that culture supernatants of CVD110 did not increase Isc (129). If there is another enterotoxin produced by *V. cholerae*, then there are a number of potential reasons why culture supernatants of CVD110 did not alter Ussing chamber activity (181). First, as with CT, expression of the hypothesized toxin may be very sensitive to environmental conditions, and the culture conditions used to grow CVD110 for the Ussing chamber assays may not favor expression. Second, the Ussing chamber studies employed rabbit intestinal tissue, which may lack receptors for the toxin. Third, the toxin may be cell associated rather than secreted into the supernatant. Fourth, the toxin may be electrogenically silent, meaning that although net Isc may not be altered, secretion of Na⁺ and CL⁻ ions that could be detected with radiolabeled ions can still occur. An example of this situation is Shiga toxin produced by *S. dysenteriae,* which selectively kills the absorptive villus tip cells while sparing the secreting crypt cells, thereby changing the balance toward net secretion (83).

The alternative hypothesis to explain the diarrhea produced by CT⁻ strains is that mere colonization of the proximal small bowel by *V. cholerae* may result in diarrhea. The classic experiments of Smith and Linggood (174) demonstrated that in piglets, enterotoxigenic *E. coli* strains that express both a colonization factor and an enterotoxin caused diarrhea; however, strains that had lost the enterotoxin but still expressed the colonization factor were able to cause diarrhea in about one-third of infected animals. Subsequent experiments by Wanke and Guerrant (193) showed that a human strain of enterotoxigenic *E. coli* that had lost the enterotoxin but still possessed the colonization factor antigen II caused diarrhea in the rabbit reversible ileal tie model. These authors suggested that normal metabolic or enzymatic products of bacteria colonizing the small bowel might act in the gut as secretagogs or as inhibitors of absorption, perhaps via the ENS. The attenuated *V. cholerae* vaccine strain CVD103-HgR is well tolerated in volunteers and does not cause diarrhea and other symptoms seen with CVD101 and JBK70 (see chapter 26). Interestingly, this strain is diminished in its intestinal colonization ability relative to these earlier attenuated *V. cholerae* strains (93, 105).

CONCLUSIONS

The study of CT has produced profound insights into the pathogenesis, treatment, and prevention of cholera. Furthermore, it has yielded knowledge on basic cellular functions

and structures such as intracellular messengers and neurological pathways. The study of *V. cholerae* toxins other than CT has demonstrated new potential mechanisms of diarrhea and may also produce new insights into basic cellular functions such as the assembly and regulation of tight junctions. One definite conclusion that can be drawn from studies with Δ*ctx* strains is that despite more than 100 years of study, there is still much to be learned about the ways which *V. cholerae* can cause disease.

Acknowledgments. Investigations in our laboratories were supported by NIH grant AI 19716.

We thank Carol Tacket for careful review of the manuscript and Michael Field, Joel Moss, Martha Vaughan, and Randall Holmes for their review of Fig. 3.

REFERENCES

1. **Alm, R. A., G. Mayrhofer, I. Kotlarski, and P. A. Manning.** 1991. Amino-terminal domain of the El Tor haemolysin of *Vibrio cholerae* O1 is expressed in classical strains and is cytotoxic. *Vaccine* 9:588–594.

2. **Alm, R. A., U. H. Stroeher, and P. A. Manning.** 1988. Extracellular proteins of *Vibrio cholerae:* nucleotide sequence of the structural gene (*hlyA*) for the haemolysin of the haemolytic El Tor strain 017 and characterization of the *hlyA* mutation in the non-haemolytic classical strain 569B. *Mol. Microbiol.* 2:481–488.

3. **Amiranoff, B., M. Laburthe, and G. Rosselin.** 1980. Characterization of specific binding sites for vasoactive intestinal peptide in intestinal epithelial cell membranes. *Biochim. Biophys. Acta* 627:215–224.

4. **Atisook, K., S. Carson, and J. L. Madara.** 1990. Effects of phlorizin and sodium on glucose-elicited alterations of cell junctions in intestinal epithelia. *Am. J. Physiol.* 258:C77–C85.

5. **Atisook, K., and J. L. Madara.** 1991. An oligopeptide permeates intestinal tight junctions at glucose-elicited dilatations. *Gastroenterology* 100:719–724.

6. **Baudry, B., A. Fasano, J. Ketley, and J. B. Kaper.** 1992. Cloning of a gene (*zot*) encoding a new toxin produced by *Vibrio cholerae. Infect. Immun.* 60:428–434.

7. **Bear, C. E., C. Li, N. Kartner, R. J. Bridges, T. J. Jensen, M. Ramjeesingh, and J. R. Riordan.** 1992. Purification and functional reconstitution of the cystic fibrosis transmembrane conductance regulator (CFTR). *Cell* 68:809–818.

8. **Bennett, A.** 1971. Cholera and prostaglandins. *Nature* (London) **231:**536.

9. **Berschneider, H. M., M. R. Knowles, R. G. Azizkhan, R. C. Boucher, N. A. Tobey, R. C. Orlando, and D. W. Powell.** 1988. Altered intestinal chloride transport in cystic fibrosis. *FASEB J.* **2:**2625–2629.

10. **Beubler, E., G. Kollar, A. Saria, K. Bukhave, and J. Rask-Madsen.** 1989. Involvement of 5-hydroxytryptamine, prostaglandin E_2, and cyclic adenosine monophosphate in cholera toxin-induced fluid secretion in the small intestine of the rat in vivo. *Gastroenterology* **96:**368–376.

11. **Booth, I. W., M. M. Levine, and J. T. Harries.** 1984. Oral rehydration therapy in acute diarrhoea in childhood. *J. Pediatr. Gastroenterol. Nutr.* **3:**491–499.

12. **Bromander, A., J. Holmgren, and N. Lycke.** 1991. Cholera toxin stimulates IL-1 production and enhances antigen presentation by macrophages in vitro. *J. Immunol.* **146:**2908–2914.

13. **Brown, M. H., and P. A. Manning.** 1985. Haemolysin genes of *Vibrio cholerae:* presence of homologous DNA in non-haemolytic O1 and haemolytic non-O1 strains. *FEMS Microbiol. Lett.* **30:**197–201.

14. **Burch, R. M., C. Jelsema, and J. Axelrod.** 1988. Cholera toxin and pertussis toxin stimulate prostaglandin E_2 synthesis in a murine macrophage cell line. *J. Pharmacol. Exp. Ther.* **244:**765–773.

15. **Carey, H. V., and H. J. Cooke.** 1986. Submucosal nerves and cholera toxin-induced secretion in guinea pig ileum in vitro. *Digest. Dis. Sci.* **31:**732–736.

16. **Cassel, D., and Z. Selinger.** 1977. Mechanism of adenylate cyclase activation by cholera toxin: inhibition of GTP hydrolysis at the regulatory site. *Proc. Natl. Acad. Sci. USA* **74:**3307–3311.

17. **Cassuto, J., J. Fahrenkrug, M. Jodal, R. Tuttle, and O. Lundgren.** 1981. The release of vasoactive intestinal polypeptide from the cat small intestine exposed to cholera toxin. *Gut* **22:**958–963.

18. **Cassuto, J., M. Jodal, and O. Lundgren.** 1982. The effect of nicotinic and muscarinic receptor blockade on cholera toxin induced intestinal secretion in rats and cats. *Acta Physiol. Scand.* **114:**573–577.

19. **Cassuto, J., M. Jodal, R. Tuttle, and O. Lundgren.** 1981. On the role of intramural nerves in the pathogenesis of cholera toxin-induced intestinal secretion. *Scand. J. Gastroenterol.* **16:**377–384.

20. **Cassuto, J., A. Siewert, M. Jodal, and O. Lundgren.** 1983. The involvement of intramural nerves in cholera toxin induced intestinal secretion. *Acta Physiol. Scand.* **117:**195–202.

21. **Chang, E. B., and M. C. Rao.** 1991. Intracellular mediators of intestinal electrolyte transport, p. 49–72. *In* M. Field (ed.), *Diarrheal Diseases.* Elsevier, New York.

22. **Chaudhary, V. K., Y. Jinno, D. Fitzgerald, and I. Pastan.** 1990. *Pseudomonas* exotoxin contains a specific sequence at the carboxyl terminus that is required for cytotoxicity. *Proc. Natl. Acad. Sci. USA* **87:**308–312.

23. **Comstock, L., and J. B. Kaper.** Unpublished data.

24. **Cooke, H. J.** 1991. Hormones and neurotransmitters regulating intestinal ion transport, p. 23–48. *In* M. Field (ed.), *Diarrheal Diseases.* Elsevier, New York.

25. **Czerkinsky, C., M. W. Russell, N. Lycke, M. Lindblad, and J. Holmgren.** 1989. Oral administration of a streptococcal antigen coupled to cholera toxin B subunit evokes strong antibody responses in salivary glands and extramuscosal tissues. *Infect. Immun.* **57:**1072–1077.

26. **De, S. N.** 1959. Enterotoxicity of bacteria-free culture filtrate of *Vibrio cholerae. Nature* (London) **183:**1533–1534.

27. **de Jonge, H. R.** 1991. Intracellular mechanisms regulating intestinal secretion, p. 107–114. *In* T. Wadström, P. H. Mäkelä, A.-M. Svennerholm, and H. Wolf-Waltz (ed.), *Molecular Pathogenesis of Gastrointestinal Infections.* Plenum Press, New York.

28. **Dertzbaugh, M. T., D. L. Peterson, and F. L. Macrina.** 1990. Cholera toxin B-subunit gene fusion: structural and functional analysis of the chimeric protein. *Infect. Immun.* **58:**70–79.

29. **De Wolf, M. J. S., M. Fridkin, M. M. Epstein, and L. D. Kohn.** 1981. Structure-function studies of cholera toxin and its A protomers and B protomers—modification of tryptophan residues. *J. Biol. Chem.* **256:**5481–5488.

30. **De Wolf, M. S. J., M. Fridkin, and L. D. Kohn.** 1981. Tryptophan residues of cholera toxin and its A protomers and B protomers. Intrinsic fluorescence and solute quenching upon interacting with the ganglioside GM_1, oligo-GM_1, or dansylated oligo-GM_1. *J. Biol. Chem.* **256:**5489–5496.

31. **Diamond, J. M.** 1977. The epithelial junction: bridge, gate and fence. *Physiologist* **20:**10–18.

32. **Dominguez, P., G. Velasco, F. Barros, and P. S. Lozo.** 1987. Intestinal brush border membranes contain regulatory subunits of adenylyl cyclase. *Proc. Natl. Acad. Sci. USA* **84:**6965–6969.

33. **Dove, C. H., S. Z. Wang, S. B. Price, C. J. Phelps, D. M. Lyerly, T. D. Wilkens, and J. L. Johnson.** 1990. Molecular characterization of the *Clostridium difficile* toxin A gene. *Infect. Immun.* **58:**480–488.

34. **Dubey, R. S., M. Lindblad, and J. Holmgren.** 1990. Purification of El Tor cholera enterotoxins

and comparisons with classical toxin. *J. Gen. Microbiol.* **136:**1839–1847.

35. **Duffey, M. E., B. Hainou, S. Ho, and C. J. Bentzel.** 1981. Regulation of epithelial tight junction permeability by cyclic AMP. *Nature* (London) **294:**451–453.

36. **Dutta, N. K., M. W. Panse, and D. R. Kulkarni.** 1959. Role of cholera toxin in experimental cholera. *J. Bacteriol.* **78:**594–595.

37. **Farques, S. M., J. B. Kaper, and M. J. Albert.** Unpublished data.

38. **Fasano, A., B. Baudry, D. W. Pumplin, S. S. Wasserman, B. D. Tall, J. M. Ketley, and J. B. Kaper.** 1991. *Vibrio cholerae* produces a second enterotoxin, which affects intestinal tight junctions. *Proc. Natl. Acad. Sci. USA* **88:**5242–5246.

39. **Field, M.** 1971. Intestinal secretion: effect of cyclic AMP and its role in cholera. *N. Engl. J. Med.* **284:**1137–1144.

40. **Field, M.** 1981. Secretion of electrolytes and water by mammalian small intestine, p. 963–982. *In* L. R. Johnson (ed.), *Physiology of the Gastrointestinal Tract.* Raven Press, New York.

41. **Field, M., D. Fromm, Q. Al-Awqati, and W. B. Greenough III.** 1972. Effect of cholera enterotoxin on ion transport across isolated ileal mucosa. *J. Clin. Invest.* **51:**796–804.

42. **Field, M., D. Fromm, C. K. Wallace, and W. B. Greenough III.** 1969. Stimulation of active chloride secretion in small intestine by cholera exotoxin. *J. Clin. Invest.* **48:**24a.

43. **Field, M., M. C. Rao, and E. B. Chang.** 1989. Intestinal electrolyte transport and diarrheal disease. Part 2. *N. Engl. J. Med.* **321:**879–883.

44. **Finkelstein, R. A.** 1992. Cholera enterotoxin (choleragen)—a historical perspective, p. 155–187. *In* D. Barua and W. B. Greenough III (eds.), *Cholera.* Plenum Medical Book Co., New York.

45. **Finkelstein, R. A., M. Boesman-Finkelstein, Y. Chang, and C. C. Häse.** 1992. *Vibrio cholerae* hemagglutinin/protease, colonial variation, virulence, and detachment. *Infect. Immun.* **60:**472–478.

46. **Finkelstein, R. A., M. Boesman-Finkelstein, and P. Holt.** 1983. *Vibrio cholerae* hemagglutinin/lectin/protease hydrolyzes fibronectin and ovomucin: F.M. Burnet revisited. *Proc. Natl. Acad. Sci. USA* **80:**1092–1095.

47. **Finkelstein, R. A., M. F. Burks, A. Zupan, W. S. Dallas, C. O. Jacob, and D. S. Ludwig.** 1987. Epitopes of the cholera family of enterotoxins. *Rev. Infec. Dis.* **9:**544–561.

48. **Finkelstein, R. A., and J. J. LoSpalluto.** 1969. Pathogenesis of experimental cholera: preparation and isolation of choleragen and choleragenoid. *J. Exp. Med.* **130:**185–202.

49. **Francis, M. L., J. Ryan, M. G. Jobling, R. K. Holmes, J. Moss, and J. J. Mond.** 1992. Cyclic AMP-independent effects of cholera toxin on B cell activation. II. Binding of ganglioside G_{M1} induces B cell activation. *J. Immunol.* **148:**1999–2005.

50. **Freer, J. H., and T. H. Birkbeck.** 1982. Possible conformation of delta-lysin, a membrane-damaging peptide of *Staphylococcus aureus. J. Theor. Biol.* **94:**535–540.

51. **Fukuta, S., E. M. Twiddy, J. L. Magnani, V. Ginsburg, and R. K. Holmes.** 1988. Comparison of the carbohydrate binding specificities of cholera toxin and *Escherichia coli* heat-labile enterotoxins LTH-1, LT-1a, and LT-1b. *Infect. Immun.* **56:**1748–1753.

52. **Gaginella, T. S.** 1990. Eicosanoid-mediated intestinal secretion, p. 15–30. *In* E. Lebenthal and M. E. Duffey (ed.), *Textbook of Secretory Diarrhea.* Raven Press, New York.

53. **Galen, J. E., J. M. Ketley, A. Fasano, S. H. Richardson, S. S. Wasserman, and J. B. Kaper.** 1992. Role of *Vibrio cholerae* neuraminidase in the function of cholera toxin. *Infect. Immun.* **60:**406–415.

54. **Gallut, J.** 1974. Bacteriology, p. 17–40. *In* D. Barua and W. Burrows (ed.), *Cholera.* The W. B. Saunders Co., Philadelphia.

55. **Gennaro, M. L., P. J. Greenaway, and D. A. Broadbent.** 1982. The expression of biologically active cholera toxin in *Escherichia coli. Nucleic Acids Res.* **10:**4883–4890.

56. **Gill, D. M.** 1976. The arrangement of subunits in cholera toxin. *Biochemistry* **15:**1242–1248.

57. **Gill, D. M., and C. A. King.** 1975. The mechanism of action of cholera toxin in pigeon erythrocyte lysates. *J. Biol. Chem.* **250:**6424–6432.

58. **Gill, D. M., and R. S. Rappaport.** 1979. Origin of the enzymatically active A1 fragment of cholera toxin. *J. Infect. Dis.* **139:**674–680.

59. **Goldberg, I., and J. J. Mekalanos.** 1986. Effect of a *recA* mutation on cholera toxin gene amplification and deletion events. *J. Bacteriol.* **165:**723–731.

60. **Gonzalez-Marischal, L., R. G. Contreras, J. J. Bolívar, A. Ponce, B. Chávez de Ramirez, and M. Cereijido.** 1990. Role of calcium in tight junction formation between epithelial cells. *Am. J. Physiol.* **259:**C978–C986.

61. **Griffiths, S. L., R. A. Finkelstein, and D. R. Critchley.** 1986. Characterization of the receptor for cholera toxin and *Escherichia coli* heat-labile toxin in rabbit intestinal brush borders. *Biochem. J.* **238:**313–322.

62. **Gumbiner, B.** 1987. Structure, biochemistry, and assembly of epithelial tight junctions. *Am. J. Physiol.* **253:**C749–C758.

63. **Haksar, A., D. V. Maudsley, and F. G. Peron.** 1974. Neuraminidase treatment of adrenal cells increases their response to cholera enterotoxin. *Nature* (London) **251:**514–515.

64. **Hardy, S. J., J. Holmgren, S. Johansson, J. Sanchez, and T. R. Hirst.** 1988. Coordinated assembly of multisubunit proteins: oligomerization of bacterial enterotoxins in vivo and in vitro. *Proc. Natl. Acad. Sci. USA* **85**:7109–7113.

65. **Häse, C. C., and R. A. Finkelstein.** 1991. Cloning and nucleotide sequence of the *Vibrio cholerae* hemagglutinin/protease (HA/protease) gene and construction of an HA/protease-negative strain. *J. Bacteriol.* **173**:3311–3317.

66. **Hecht, G., C. Pothoulakis, J. T. LaMont, and J. L. Madara.** 1988. *Clostridium difficile* toxin A perturbs cytoskeletal structure and tight junction permeability of the cultured human. *J. Clin. Invest.* **82**:1516–1524.

67. **Hepler, J. R., and A. G. Gilman.** 1992. G proteins. *Trends Biochem. Sci.* **17**:383–387.

68. **Hirst, T. R.** 1991. Assembly and secretion of oligomeric toxins by Gram-negative bacteria, p. 75–100. *In* J. E. Alouf and J. H. Freer (ed.), *Sourcebook of Bacterial Protein Toxins.* Academic Press, Ltd., London.

69. **Hirst, T. R., J. Sanchez, J. B. Kaper, S. J. Hardy, and J. Holmgren.** 1984. Mechanism of toxin secretion by *Vibrio cholerae* investigated in strains harboring plasmids that encode heat-labile enterotoxins of *Escherichia coli*. *Proc. Natl. Acad. Sci. USA* **81**:7752–7756.

70. **Holmes, R. K., and E. M. Twiddy.** 1983. Characterization of monoclonal antibodies that react with unique and cross-reacting determinants of cholera enterotoxin and its subunits. *Infec. Immun.* **42**:914–923.

71. **Holmgren, J.** 1992. Pathogenesis, p. 199–208. *In* D. Barua and W. B. Greenough III (ed.), *Cholera.* Plenum Medical Book Co., New York.

72. **Holmgren, J., I. Lonnroth, J. Mansson, and L. Svennerholm.** 1975. Interaction of cholera toxin and membrane G_{M1} ganglioside of small intestine. *Proc. Natl. Acad. Sci. USA* **72**:2520–2524.

73. **Holmgren, J., I. Lonnroth, and L. Svennerholm.** 1973. Tissue receptor for cholera exotoxin: postulated structure studies with GM_1 ganglioside and related glycoproteins. *Infec. Immun.* **8**:208–214.

74. **Honda, T., and R. A. Finkelstein.** 1979. Purification and characterization of a hemolysin produced by *Vibrio cholerae* biotype El Tor: another toxic substance produced by cholera vibrios. *Infect. Immun.* **26**:1020–1027.

75. **Honda, T., M. Nishibuchi, T. Miwatani, and J. B. Kaper.** 1986. Demonstration of a plasmid-borne gene encoding a thermostable direct hemolysin in *Vibrio cholerae* non-O1 strains. *Appl. Environ. Microbiol.* **52**:1218–1220.

76. **Ichinose, Y., K. Yamamoto, N. Nakasone, M. J. Tanabe, T. Takeda, T. Miwatani, and M. Iwanaga.** 1987. Enterotoxicity of El Tor-like hemolysin of non-O1 *Vibrio cholerae. Infect. Immun.* **55**:1090–1093.

77. **Janicot, M., and B. Desbuquois.** 1987. Fate of injected ^{125}I-labeled cholera toxin taken up by rat liver in vivo. Generation of the active A1 peptide in the endosomal compartment. *Eur. J. Biochem.* **161**:433–442.

78. **Janicot, M., F. Fouque, and B. Desbuquois.** 1991. Activation of rat liver adenylate cyclase by cholera toxin requires toxin internalization and processing in endosome. *J. Biol. Chem.* **266**:12858–12865.

79. **Jobling, M. G., and R. K. Holmes.** 1991. Analysis of structure and function of the B subunit of cholera toxin by the use of site-directed mutagenesis. *Mol. Microbiol.* **5**:1755–1767.

80. **Johnson, J. A., J. G. Morris, Jr., and J. B. Kaper.** 1993. Gene encoding zonula occludens toxin (*zot*) does not occur independently from cholera enterotoxin genes (*ctx*) in *Vibrio cholerae. J. Clin. Microbiol.* **31**:732–733.

81. **Kabir, S., N. Ahmad, and S. Ali.** 1984. Neuraminidase production by *Vibrio cholerae* O1 and other diarrheagenic bacteria. *Infect. Immun.* **44**:747–749.

82. **Kahn, R. A., and A. G. Gilman.** 1984. ADP-ribosylation of G_s promotes dissociation of its a and b subunits. *J. Biol. Chem.* **259**:6235–6240.

83. **Kandel, G., A. Donohue-Rolfe, M. Donowitz, and G. T. Keusch.** 1989. Pathogenesis of shigella diarrhea. XVI. Selective targetting of Shiga toxin to villus cells of rabbit jejunum explains the effect of the toxin on intestinal electrolyte transport. *J. Clin. Invest.* **84**:1509–1517.

84. **Kaper, J. B., and M. M. Levine.** 1981. Cloned cholera enterotoxin genes in study and prevention of cholera. *Lancet* **ii**:1162–1163.

85. **Kaper, J. B., H. Lockman, M. M. Baldini, and M. M. Levine.** 1984. A recombinant live oral cholera vaccine. *Bio/Technology* **2**:345–349.

86. **Kaper, J. B., H. Lockman, M. M. Baldini, and M. M. Levine.** 1984. Recombinant nontoxinogenic *Vibrio cholerae* strains as attenuated cholera vaccine candidates. *Nature* (London) **308**:655–658.

87. **Kaper, J. B., H. L. T. Mobley, J. M. Michalski, D. A. Herrington, and M. M. Levine.** 1988. Recent advances in developing a safe and effective live oral attenuated *Vibrio cholerae* vaccine, p. 161–167. *In* N. Ohtomo and R. B. Sack (ed.), *Advances in Research on Cholera and Related Diarrheas,* vol. 6. KTK Scientific Publishers, Tokyo.

88. **Kaper, J. B., S. L. Moseley, and S. Falkow.** 1981. Molecular characterization of environmental and nontoxigenic strains of *Vibrio cholerae. Infect. Immun.* **32**:661–667.

89. **Kapral, F. A., A. D. O'Brien, P. D. Ruff, and W. J. Drugan, Jr.** 1976. Inhibition of water ab-

sorption in the intestine by *Staphylococcus aureus* delta-toxin. *Infect. Immun.* **13:**140–145.

90. **Karasawa, T., T. Mihara, H. Kurazono, G. B. Nair, S. Garg, T. Ramamurthy, and Y. Takeda.** 1993. Distribution of the *zot* (zonula occludens toxin) gene among strains of *Vibrio cholerae* O1 and non-O1. *FEMS Microbiol. Lett.* **106:**143–146.

91. **Kazemi, M., and R. A. Finkelstein.** 1990. Study of epitopes of cholera enterotoxin-related enterotoxins by checkerboard immunoblotting. *Infect. Immun.* **58:**2352–2360.

92. **Kazemi, M., and R. A. Finkelstein.** 1991. Mapping epitopic regions of cholera toxin B-subunit protein. *Mol. Immunol.* **28:**865–876.

93. **Ketley, J. M., J. Michalski, J. Galen, M. M. Levine, and J. B. Kaper.** 1993. Construction of genetically-marked *Vibrio cholerae* O1 vaccine strains. *FEMS Microbiol. Lett.* **111:**15–22.

94. **Kimberg, D. K., M. Field, J. Johnson, E. Henderson, and E. Gershon.** 1971. Stimulation of intestinal mucosal adenyl cyclase by cholera enterotoxin and prostaglandins. *J. Clin. Invest.* **50:**1218–1230.

95. **King, C. A., and W. E. van Heyningen.** 1973. Deactivation of cholera toxin by a sialidase-resistant monosialosylganglioside. *J. Infect. Dis.* **127:**639–647.

96. **Koch, R.** 1884. An address on cholera and its bacillus. *Br. Med. J.* **2:**403–407.

97. **Koonin, E. V.** 1992. The second cholera toxin, Zot, and its plasmid-encoded and phage-encoded homologues constitute a group of putative ATPases with an altered purine NTP-binding motif. *FEBS Lett.* **312:**3–6.

98. **Krasilnikov, O. V., J. N. Muratkhodjaev, A. E. Voronov, and Y. V. Yezepchuk.** 1991. The ionic channels formed by cholera toxin planar bilayer lipid membranes are entirely attributable to its B-subunit. *Biochim. Biophys. Acta* **1067:**166–170.

99. **Krasilnikov, O. V., J. N. Muratkhodjaev, and A. O. Zitzer.** 1992. The mode of action of *Vibrio cholerae* cytolysin. The influences on both erythrocytes and planar lipid bilayers. *Biochim. Biophys. Acta* **1111:**7–16.

100. **Krause, M., C. Roudier, J. Fierer, J. Harwood, and D. Guiney.** 1991. Molecular analysis of the virulence locus of the *Salmonella dublin* plasmid pSDL2. *Mol. Microbiol.* **5:**307–316.

101. **Lencer, W. I., J. B. de Almeida, S. Moe, J. L. Stow, D. A. Ausiello, and J. L. Madara.** 1993. Entry of cholera toxin into polarized human epithelial cells: identification of an early brefeldin-A sensitive event required for A₁ peptide generation. *J. Clin. Invest.* **92:**2941–2951.

102. **Lencer, W. I., C. Delp, M. R. Neutra, and J. L. Madara.** 1992. Mechanism of cholera toxin

action on a polarized human intestinal epithelial cell line: role of vesicular traffic. *J. Cell Biol.* **117:**1197–1209.

103. **Levine, M. M.** 1980. Immunity to cholera as evaluated in volunteers, p. 195–203. *In* O. Ouchterlony and J. Holmgren (ed.), *Cholera and Related Diarrheas.* Karger, Basel.

104. **Levine, M. M., J. B. Kaper, R. E. Black, and M. L. Clements.** 1983. New knowledge on pathogenesis of bacterial enteric infections as applied to vaccine development. *Microbiol. Rev.* **47:**510–550.

105. **Levine, M. M., J. B. Kaper, D. Herrington, J. Ketley, G. Losonsky, C. O. Tacket, B. Tall, and S. Cryz.** 1988. Safety, immunogenicity, and efficacy of recombinant live oral cholera vaccines, CVD 103 and CVD 103-HgR. *Lancet* **ii:**467–470.

106. **Levine, M. M., J. B. Kaper, D. Herrington, G. Losonsky, J. G. Morris, M. L. Clements, R. E. Black, B. Tall, and R. Hall.** 1988. Volunteer studies of deletion mutants of *Vibrio cholerae* O1 prepared by recombinant techniques. *Infect. Immun.* **56:**161–167.

107. **Levine, M. M., D. R. Nalin, D. L. Hoover, E. J. Bergquist, R. B. Hornick, and C. R. Young.** 1979. Immunity to enterotoxigenic *Escherichia coli*. *Infect. Immun.* **23:**729–736.

108. **Liang, X., M. E. Lamm, and J. G. Nedrud.** 1988. Oral administration of cholera toxin-Sendai virus conjugate potentiates gut and respiratory immunity against Sendai virus. *J. Immunol.* **141:**1495–1501.

109. **Lind, C. D., R. H. Davis, R. L. Guerrant, J. B. Kaper, and J. R. Mathias.** 1991. Effects of *Vibrio cholerae* recombinant strains on rabbit ileum in vivo. Enterotoxin production and myoelectric activity. *Gastroenterology* **101:**319–324.

110. **Lindholm, L., J. Holmgren, M. Wikstrom, U. Karlsson, K. Andersson, and N. Lycke.** 1983. Monoclonal antibodies to cholera toxin with special reference to cross-reactions with *Escherichia coli* heat-labile enterotoxin. *Infect. Immun.* **40:**570–576.

111. **Lockman, H., and J. B. Kaper.** 1983. Nucleotide sequence analysis of the A2 and B subunits of *Vibrio cholerae* enterotoxin. *J. Biol. Chem.* **258:**13722–13726.

112. **Lockman, H. A., J. E. Galen, and J. B. Kaper.** 1984. *Vibrio cholerae* enterotoxin genes: nucleotide sequence analysis of DNA encoding ADP-ribosyltransferase. *J. Bacteriol.* **159:**1086–1089.

113. **London, E.** 1992. How bacterial protein toxins enter cells: the role of partial unfolding in membrane translocation. *Mol. Microbiol.* **6:**3277–3282.

114. **Ludwig, D. S., R. K. Holmes, and G. K. Schoolnik.** 1985. Chemical and immunochemical studies on the receptor binding domain of cholera

toxin B subunit. *J. Biol. Chem.* **260:**12528–12534.

115. **Lundgren, O.** 1988. Factors controlling absorption and secretion in the small intestine, p. 97–112. In W. Donachie, E. Griffiths, and J. Stephen (ed.), *Bacterial Infections of Respiratory and Gastrointestinal Mucosae.* IRL Press, Oxford.

116. **Lycke, N., and J. Holmgren.** 1986. Strong adjuvant properties of cholera toxin on gut mucosal immune responses to orally presented antigens. *Immunology* **59:**301–308.

117. **Lycke, N., U. Karlsson, A. Sjölander, and K.-E. Magnusson.** 1991. The adjuvant action of cholera toxin is associated with an increased intestinal permeability for luminal antigens. *Scand. J. Immunol.* **33:**691–698.

118. **Lycke, N., and W. Strober.** 1989. Cholera toxin promotes B cell isotype differentiation. *J. Immunol.* **142:**3781–3787.

119. **Lycke, N., T. Tsuji, and J. Holmgren.** 1992. The adjuvant effect of *Vibrio cholerae* and *Escherichia coli* heat-labile enterotoxins is linked to their ADP-ribosyltransferase activity. *Eur. J. Immunol.* **22:**2277–2281.

120. **Lycke, N. Y., and A.-M. Svennerholm.** 1990. Presentation of immunogens at the gut and other mucosal surfaces, p. 207–227. In G. C. Woodrow and M. M. Levine (ed.), *New Generation Vaccines.* Marcel Dekker, Inc., New York.

121. **Madara, J. L.** 1989. Loosening tight junctions—lessons from the intestine. *J. Clin. Invest.* **83:**1089–1094.

122. **Madara, J. L.** 1990. Contributions of the paracellular pathway to secretion, absorption, and barrier function in the epithelium of the small intestine, p. 125–138. In E. Lebenthal and M. Duffey (ed.), *Textbook of Secretory Diarrhea.* Raven Press, Ltd., New York.

123. **Mathan, V. I., and M. Mathan.** Unpublished data.

124. **Mathias, J. R., G. M. Carlson, A. J. DiMarino, G. Bertiger, H. E. Morton, and S. Cohen.** 1976. Intestinal myoelectric activity in response to live *Vibrio cholerae* and cholera enterotoxin. *J. Clin. Invest.* **58:**91–96.

125. **McCardell, B., J. M. Madden, and D. B. Shah.** 1985. Isolation and characterization of a cytolysin produced by *Vibrio cholerae* serogroup non-O1. *Can. J. Microbiol.* **31:**711–720.

126. **Mekalanos, J. J.** 1983. Duplication and amplification of toxin genes in Vibrio cholerae. *Cell* **35:**253–263.

127. **Mekalanos, J. J.** 1985. Cholera toxin: genetic analysis, regulation, and role in pathogenesis. *Curr. Top. Microbiol. Immunol.* **118:**97–118.

128. **Mekalanos, J. J., D. J. Swartz, G. D. Pearson, N. Harford, F. Groyne, and M. de Wilde.** 1983. Cholera toxin genes: nucleotide sequence, dele-

tion analysis and vaccine development. *Nature* (London) **306:**551–557.

129. **Michalski, J., J. E. Galen, A. Fasano, and J. B. Kaper.** 1993. CVD110, an attenuated *Vibrio cholerae* O1 El Tor live oral vaccine strain. *Infect. Immun.* **61:**4462–4468.

130. **Michalski, J. M., and J. B. Kaper.** Unpublished data.

131. **Miller, V. L., and J. J. Mekalanos.** 1984. Synthesis of cholera toxin is positively regulated at the transcriptional level by toxR. *Proc. Natl. Acad. Sci. USA* **81:**3471–3475.

132. **Miller-Podraza, H., R. M. Bradley, and P. H. Fishman.** 1982. Biosynthesis and localization of gangliosides in cultured cells. *Biochemistry* **21:**3260–3265.

133. **Moore, R., C. Pothoulakis, J. T. LaMont, S. Carlson, and J. Madara.** 1990. *C. difficile* toxin A increases intestinal permeability and induces Cl- secretion. *Am. J. Physiol.* **259:**G165–G172.

134. **Moriarty, K. J., N. B. Higgs, M. Woodford, and L. A. Turnberg.** 1989. An investigation of the role of possible neural mechanisms in cholera toxin-induced secretion in rabbit ileal mucosa in vitro. *Clin. Sci.* **77:**161–166.

135. **Morris, J. G., Jr., T. Takeda, B. D. Tall, G. A. Losonsky, S. K. Bhattacharya, B. D. Forrest, B. A. Kay, and M. Nishibuchi.** 1990. Experimental non-O group 1 *Vibrio cholerae* gastroenteritis in humans. *J. Clin. Invest.* **85:**697–705.

136. **Moseley, S. L., and S. Falkow.** 1980. Nucleotide sequence homology between the heat-labile enterotoxin gene of *Escherichia coli* and *Vibrio cholerae.* *J. Bacteriol.* **144:**444–446.

137. **Moss, J., and M. Vaughan.** 1977. Mechanism of activation of choleragen. Evidence for ADP-ribosyltransferase activity with arginine as an acceptor. *J. Biol. Chem.* **252:**2455–2457.

138. **Moss, J., and M. Vaughan.** 1991. Activation of cholera toxin and *Escherichia coli* heat-labile enterotoxins by ADP-ribosylation factors, a family of 20-kDa guanine nucleotide-binding proteins. *Mol. Microbiol.* **5:**2621–2627.

139. **Mullin, J. M., and T. O'Brien.** 1986. Effects of tumor promoters on LLC-pk, renal epithelial tight junctions and transepithelial fluxes. *Am. J. Physiol.* **251:**C597–C602.

140. **Mullin, J. M., and K. V. Snock.** 1990. Effect of tumor necrosis factor on epithelial tight junctions and transepithelial permeability. *Cancer Res.* **50:**2172–2176.

141. **Munoz, E., A. M. Zubiaga, M. Merrow, N. P. Sauter, and B. T. Huber.** 1990. Cholera toxin discriminates between T helper 1 and T helper 2 cells in T cell receptor mediated activation: role of cAMP in T cell proliferation. *J. Exp. Med.* **172:**95–103.

142. **Muron, S., and H. R. B. Pelham.** 1987. A C-terminal signal prevents secretion of luminal ER proteins. *Cell* **48:**899–907.

143. **Nambiar, M. P., T. Oda, C. Chen, Y. Kuwazuru, and H. C. Wu.** 1993. Involvement of the Golgi region in the intracellular trafficking of cholera toxin. *J. Cell. Physiol.* **154:**222–228.

144. **Nilsson, O., J. Cassuto, P.-A. Larsson, M. Jodal, P. Lidberg, H. Ahlman, A. Dahlström, and O. Lundgren.** 1983. 5-Hydroxytryptamine and cholera secretion: a histochemical and physiological study in cats. *Gut* **24:**542–548.

145. **Nishibuchi, M., A. Fasano, R. G. Russell, and J. B. Kaper.** 1992. Enterotoxigenicity of *Vibrio parahaemolyticus* with and without genes encoding thermostable direct hemolysin. *Infect. Immun.* **60:**3539–3545.

146. **Noguchi, T., J. K. Jeon, O. Arakawa, H. Sugita, Y. Deguchi, Y. Shida, and K. Hashimoto.** 1986. Occurrence of tetrodotoxin and anhydrotetrodotoxin in *Vibrio* sp. isolated from the intestines of a xanthid crab, *Atergatis floridus. J. Biochem.* (Tokyo) **99:**311–314.

147. **O'Brien, A. D., M. E. Chen, R. K. Holmes, J. Kaper, and M. M. Levine.** 1984. Environmental and human isolates of *Vibrio cholerae* and *Vibrio parahaemolyticus* produce a *Shigella dysenteriae* 1 (Shiga)-like cytotoxin. *Lancet* **i:**77–78.

148. **O'Brien, A. D., and F. A. Kapral.** 1976. Increased cyclic adenosine 3',5'-monophosphate content in guinea pig ileum after exposure to *Staphylococcus aureus* delta-toxin. *Infect. Immun.* **13:**152–162.

149. **Ogawa, A., J.-I. Kato, H. Watanabe, B. G. Nair, and T. Takeda.** 1990. Cloning and nucleotide sequence of a heat-stable enterotoxin gene from *Vibrio cholerae* non-O1 isolated from a patient with traveler's diarrhea. *Infect. Immun.* **58:**3325–3329.

150. **Olsvik, Ø., J. Wahlberg, B. Petterson, M. Uhlén, T. Popovic, I. K. Wachsmuth, and P. I. Fields.** 1993. Use of automated sequencing of polymerase chain reaction-generated amplicons to identify three types of cholera toxin subunit B in *Vibrio cholerae* O1 strains. *J. Clin. Microbiol.* **31:**22–25.

151. **Pearson, G. D., V. J. DiRita, M. B. Goldberg, S. A. Boyko, S. B. Calderwood, and J. J. Mekalanos.** 1990. New attenuated derivatives of *Vibrio cholerae. Res. Microbiol.* **141:**893–899.

152. **Pearson, G. D., and J. J. Mekalanos.** 1982. Molecular cloning of *Vibrio cholerae* enterotoxin genes in *Escherichia coli* K-12. *Proc. Natl. Acad. Sci. USA* **79:**2976–2980.

153. **Pearson, G. D. N., A. Woods, S. L. Chiang, and J. J. Mekalanos.** 1993. CTX genetic element encodes a site-specific recombination system and an intestinal colonization factor. *Proc. Natl. Acad. Sci. USA* **90:**3750–3754.

154. **Peek, J. A., and R. K. Taylor.** 1992. Characterization of periplasmic thiol:disulfide interchange protein required for the functional maturation of secreted virulence factors of *Vibrio cholerae. Proc. Natl. Acad. Sci. USA* **89:**6210–6214.

155. **Peterson, J. W., K. E. Hejtmancik, D. E. Markel, J. P. Craig, and A. Kurosky.** 1979. Antigenic specificity of neutralizing antibody to cholera toxin. *Infect. Immun.* **24:**774–779.

156. **Peterson, J. W., and L. G. Ochoa.** 1989. Role of prostaglandins and cAMP in the secretory effects of cholera toxin. *Science* **245:**857–859.

157. **Peterson, J. W., J. C. Reitmeyer, C. A. Jackson, and G. A. S. Ansari.** 1991. Protein synthesis is required for cholera toxin-induced stimulation of arachidonic acid metabolism. *Biochim. Biophys. Acta* **1092:**79–84.

158. **Pierce, N. F.** 1973. Differential inhibitory effects of cholera toxoids and ganglioside on the enterotoxin of *Vibrio cholerae* and *Escherichia coli. J. Exp. Med.* **137:**1009–1023.

159. **Rabbani, G. H., W. Greenough III, J. Holmgren, and I. Lonnroth.** 1979. Chlorpromazine reduces fluid-loss in cholera. *Lancet* **i:**410–412.

160. **Rabbani, G. H., and W. B. Greenough III.** 1992. Pathophysiology and clinical aspects of cholera, p. 209–228. *In* D. Barua and W. B. Greenough III (ed.), *Cholera.* Plenum Medical Book Co., New York.

161. **Reitmeyer, J. C., and J. W. Peterson.** 1990. Stimulatory effects of cholera toxin on arachidonic acid metabolism in Chinese hamster ovary cells. *Proc. Soc. Exp. Biol. Med.* **193:**181–184.

162. **Ribi, H. O., D. S. Ludwig, K. L. Mercer, G. K. Schoolnik, and R. D. Kornberg.** 1988. Three-dimensional structure of cholera toxin penetrating a lipid membrane. *Science* **239:**1272–1276.

163. **Richardson, K., J. Michalski, and J. B. Kaper.** 1986. Hemolysin production and cloning of two hemolysin determinants from classical *Vibrio cholerae. Infect. Immun.* **54:** 415–420.

164. **Rodman, D. M., and S. Zamudio.** 1991. The cystic fibrosis heterozygote—advantage in surviving cholera? *Med. Hypoth.* **36:**253–258.

165. **Saha, S., and S. C. Sanyal.** 1988. Cholera toxin gene-positive *Vibrio cholerae* O1 Ogawa and Inaba strains produce the new cholera toxin. *FEMS Microbiol. Lett.* **50:**113–116.

166. **Saha, S., and S. C. Sanyal.** 1990. Immunobiological relationships of the enterotoxins produced by cholera toxin gene-positive (CT⁺) and -negative (CT⁻) strains of *Vibrio cholerae* O1. *J. Med. Microbiol.* **32:**33–37.

167. **Sanchez, J., S. Johansson, B. Lowenadler, A. M. Svennerholm, and J. Holmgren.** 1990. Recombinant cholera toxin B subunit and gene fu-

sion proteins for oral vaccination. *Res. Microbiol.* **141**:971–979.

168. **Sandkvist, M., V. Morales, and M. Bagdasarian.** 1993. A protein required for secretion of cholera toxin through the outer membrane of *Vibrio cholerae. Gene* **123**:81–86.

169. **Sanyal, S. C., K. Alam, P. K. B. Neogi, M. I. Huq, and K. A. Al-Mahmud.** 1983. A new cholera toxin. *Lancet* **i**:1337.

170. **Schödel, F., H. Will, S. Johansson, J. Sanchez, and J. Holmgren.** 1991. Synthesis in *Vibrio cholerae* and secretion of hepatitis B virus antigens fused to *Escherichia coli* heat-labile enterotoxin subunit B. *Gene* **99**:255–259.

171. **Sixma, T. K., S. E. Pronk, K. H. Kalk, B. A. M. van Zanten, A. M. Berghuis, and W. G. J. Hol.** 1992. Lactose binding to heat-labile enterotoxin revealed by X-ray crystallography. *Nature* (London) **355**:561–564.

172. **Sixma, T. K., S. E. Pronk, K. H. Kalk, E. S. Wartna, B. A. M. van Zanten, B. Witholt, and W. G. J. Hol.** 1991. Crystal structure of a cholera toxin-related heat-labile enterotoxin from *E. coli. Nature* (London) **351**:371–377.

173. **Sjöqvist, A., J. Cassuto, M. Jodal, and O. Lundgren.** 1992. Actions of serotonin antagonists on cholera-toxin-induced intestinal fluid secretion. *Acta Physiol. Scand.* **145**:229–237.

174. **Smith, H. W., and M. A. Linggood.** 1971. Observations on the pathogenic properties of the K88, Hly and Ent plasmids of *Escherichia coli* with particular reference to porcine diarrhoea. *J. Med. Microbiol.* **4**:467–485.

175. **Spangler, B. D.** 1992. Structure and function of cholera toxin and the related *Escherichia coli* heat-labile enterotoxin. *Microbiol. Rev.* **56**:622–647.

176. **Speelman, P., G. H. Rabbani, K. Bukhave, and J. Rask-Madsen.** 1985. Increased jejunal prostaglandin E$_2$ concentrations in patients with acute cholera. *Gut* **26**:188–193.

177. **Spira, W. M., D. A. Sack, P. J. Fedorka-Cray, S. Sanyal, J. Madden, and B. McCardell.** 1986. Description of a possible new extracellular virulence factor in nontoxigenic *Vibrio cholerae* O1, p. 263–270. *In* S. Kuwahara and N. F. Pierce (ed.), *Advances in Research in Cholera and Related Diarrheas,* vol. 3. KTK Scientific Publishers, Tokyo.

178. **Staerk, J., H. J. Ronneberger, and H. Wiegandt.** 1974. Neuraminidase, a virulence factor in *Vibrio cholerae* infection? *Behring Inst. Mitt.* **55**:145–146.

179. **Svennerholm, A.-M.** 1980. The nature of protective immunity in cholera, p. 171–184. *In* O. Ouchterlony and J. Holmgren (ed.), *Cholera and Related Diarrheas.* 43rd Nobel Symposium, Stockholm, 1978. S. Karger, Basel.

180. **Svennerholm, A. M., G. J. Stromberg, and J. Holmgren.** 1983. Purification of *Vibrio cholerae* soluble hemagglutinin and development of enzyme-linked immunosorbent assays for antigen and antibody quantitations. *Infect. Immun.* **41**:237–243.

181. **Tacket, C. O., G. Losonsky, J. P. Nataro, S. J. Cryz, R. Edelman, A. Fasano, J. Michalski, J. B. Kaper, and M. M. Levine.** Safety, immunogenicity, and transmissibility of live oral cholera vaccine candidate CVD110, a Δ*ctxA* Δ*zot* Δ*ace* derivative of El Tor Ogawa *Vibrio cholerae. J. Infect. Dis.,* in press.

182. **Takeda, T., Y. Peina, A. Ogawa, S. Dohi, H. Abe, G. B. Nair, and S. C. Pal.** 1991. Detection of heat-stable enterotoxin in a cholera toxin gene-positive strain of *Vibrio cholerae* O1. *FEMS Microbiol. Lett.* **80**:23–28.

183. **Tamplin, M. L., M. K. Ahmed, R. Jalali, and R. R. Colwell.** 1989. Variation in epitopes of the B subunit of El Tor and classical biotype *Vibrio cholerae* O1 cholera toxin. *J. Gen. Microbiol.* **135**:1195–1200.

184. **Tamplin, M. L., R. R. Colwell, S. Hall, K. Kogure, and G. R. Strichartz.** 1987. Sodium-channel inhibitors produced by enteropathogenic *Vibrio cholerae* and *Aeromonas hydrophila. Lancet* **i**:975.

185. **Taylor, S. S., J. Bubis, J. Toner Webb, L. D. Saraswat, E. A. First, J. A. Buechler, D. R. Knighton, and J. Sowadski.** 1988. CAMP-dependent protein kinase: prototype for a family of enzymes. *FASEB J.* **2**:2677–2685.

186. **Terai, A., K. Baba, H. Shirai, O. Yoshida, Y. Takeda, and M. Nishibuchi.** 1991. Evidence for insertion-sequence-mediated spread of the thermostable direct hemolysin gene among *Vibrio* species. *J. Bacteriol.* **173**:5036–5046.

187. **Tomasi, M., and C. Montecucco.** 1981. Lipid insertion of cholera toxin after binding to GM$_1$-containing liposomes. *J. Biol. Chem.* **256**:11177–11181.

188. **Trucksis, M., J. E. Galen, J. Michalski, A. Fasano, and J. B. Kaper.** 1993. Accessory cholera enterotoxin (Ace), the third toxin of a *Vibrio cholerae* virulence cassette. *Proc. Natl. Acad. Sci. USA* **90**:5267–5271.

189. **Trucksis, M., and J. B. Kaper.** Unpublished data.

190. **Tucker, K. D., P. E. Carrig, and T. D. Wilkins.** 1990. Toxin A of *Clostridium difficile* is a potent cytotoxin. *J. Clin. Microbiol.* **28**:869–871.

191. **van Heyningen, S.** 1991. The ring on a finger. *Nature* (London) **351**:351.

192. **van Heyningen, W. E., and J. R. Seal.** 1983. *Cholera—The American Scientific Experience, 1947–1980.* Westview Press, Boulder, Colo.

193. **Wanke, C. A., and R. L. Guerrant.** 1987. Small-bowel colonization alone is a cause of diarrhea. *Infect. Immun.* **55**:1924–1926.

194. **Westbrook, E.** Personal communication.
195. **Wisnieski, B. J., and J. S. Bramhall.** 1981. Photolabelling of cholera toxin subunits during membrane penetration. *Nature* (London) **289**:319–321.
196. **Yamamoto, K., Y. Ichinose, N. Nakasone, M. Tanabe, M. Nagahama, J. Sakurai, and M. Iwanaga.** 1986. Identity of hemolysins produced by *Vibrio cholerae* non-O1 and *V. cholerae* O1, biotype El Tor. *Infect. Immun.* **51**:927–931.
197. **Yamamoto, K., Y. Ichinose, H. Shinagawa, K. Makino, A. Nakata, M. Iwanaga, T. Honda, and T. Miwatani.** 1990. Two-step processing for activation of the cytolysin/hemolysin of *Vibrio cholerae* O1 biotype El Tor: nucleotide sequence of the structural gene (*hlyA*) and characterization of the processed products. *Infect. Immun.* **58**:4106–4116.
198. **Yamamoto, K., A. C. Wright, J. B. Kaper, and J. G. Morris.** 1990. The cytolysin gene of *Vibrio*

vulnificus: sequence and relationship to the *Vibrio cholerae* El Tor hemolysin gene. *Infect. Immun.* **58**:2706–2709.
199. **Yamamoto, T., T. Gojobori, and T. Yokota.** 1987. Evolutionary origin of pathogenic determinants in enterotoxigenic *Escherichia coli* and *Vibrio cholerae* O1. *J. Bacteriol.* **169**:1352–1357.
200. **Yamamoto, T., T. Tamura, and T. Yokota.** 1984. Primary structure of heat-labile enterotoxin produced by *Escherichia coli* pathogenic for humans. *J. Biol. Chem.* **259**:5037–5044.
201. **Yamamoto, T., and T. Yokota.** 1983. Sequence of heat-labile enterotoxin of *Escherichia coli* pathogenic for humans. *J. Bacteriol.* **155**:728–733.
202. **Yu, J., H. Webb, and T. R. Hirst.** 1992. A homologue of the *Escherichia coli* DsbA protein involved in disulphide bond formation is required for enterotoxin biogenesis in *Vibrio cholerae.* *Mol. Microbiol.* **6**:1949–1958.

Vibrio cholerae and Cholera: Molecular to Global Perspectives
Edited by I. Kaye Wachsmuth, Paul A. Blake, and Ørjan Olsvik
© 1994 American Society for Microbiology, Washington, DC 20005

Chapter 12

Regulation of Cholera Toxin Expression

Karen M. Ottemann and John J. Mekalanos

Vibrio cholerae resides in two distinct environments: the human intestine and water reservoirs (7, 15, 31). These two environments differ in many respects: temperature, osmolarity, and type and level of nutrients. Residing in two such different environments requires many different gene products for growth. Indeed, *V. cholerae* contains a regulon that is specifically required for growth and survival in the human host and that responds to specific environmental signals. This regulon is under the master control of the ToxR protein and is referred to as the ToxR regulon. The composition and molecular mechanism of control of this regulon are the subjects of this chapter.

ENVIRONMENTAL CONDITIONS THAT REGULATE TOXIN EXPRESSION

Early studies of the in vitro production of cholera toxin defined certain conditions that increased overall levels of cholera toxin (5, 8, 12, 16, 53). The earliest defined variable was growth temperature. Surprisingly, cultures grown at lower temperatures (30°C) produce more toxin than cultures grown at 37°C, the

Karen M. Ottemann and John J. Mekalanos • Department of Microbiology and Molecular Genetics, Harvard Medical School, 200 Longwood Avenue, Boston, Massachusetts 02115.

body temperature of the human host. In addition, Evans and Richardson (12) found that a starting pH of 6.6 favored toxin production, and they later expanded these studies to determine that adequate aeration was also an important parameter.

Further characterization of these regulatory signals has occurred since the identification of the ToxR regulon, when it became apparent that conditions that promote production of cholera toxin also promote expression of other genes in the ToxR regulon (39, 50, 59). *V. cholerae* requires a specific ion concentration in the growth medium for optimal toxin expression, suggesting that osmolarity is a regulatory signal (39). Amino acids also appear to be important, as cholera toxin production occurs in minimal medium only with the addition of amino acids such as asparagine, arginine, glutamate, and serine (5, 6, 16, 39, 50). Miller and Mekalanos hypothesized that these conditions have potential physiologic relevance, since serum osmolarity is equivalent to 150 mM NaCl and the action of *V. cholerae* proteases would release amino acids from mucus during colonization of the intestinal epithelium (39). Both of these signals may act through the same mechanism, because amino acids that promote toxin production also act as osmoprotectants to facilitate growth at high salt concentrations (39).

Other parameters that affect toxin production include the presence of 5% CO_2, which has been shown to increase toxin production of cultures grown at 37°C (55). Bile salts–0.1% sodium deoxycholate also increases the production of toxin from cultures grown aerobically at 37°C (14). Interestingly, laboratory conditions for expression of cholera toxin differ greatly between the classical and El Tor biotypes of *V. cholerae* (27).

THE ToxR REGULON

Early work on the genetics of cholera toxin expression relied heavily on the use of chemical mutagenesis to produce mutants altered in toxinogenesis. Several groups isolated hypotoxinogenic mutants (4, 17, 23–25, 32, 54) and hypertoxinogenic mutants (32, 34). Subsequent mapping studies (3, 34, 57) provided evidence that loci distinct from the *ctx* structural genes influenced toxin expression in a variety of different *V. cholerae* mutants, suggesting that a regulatory system controlled toxin production.

The suspicion that *V. cholerae* had mechanisms for regulating the expression of cholera toxin found further support in the early 1980s when the genes for cholera toxin *(ctxAB)* were cloned and expressed in *Escherichia coli* (19, 28, 48). Despite the use of a variety of culture conditions, *ctxAB* expression in *E. coli* was relatively poor (35). This suggested that the genes were no longer regulated as in *V. cholerae* and prompted Miller and Mekalanos to set up a genetic screen to identify *V. cholerae* regulatory factors (37). This was done by fusing the cholera toxin promoter to the gene for β-galactosidase *(lacZ)* and integrating this into the *E. coli* chromosome to create the strain VM2 (37). In *E. coli* VM2, *lacZ* expression from this fusion was low, and colonies were white on plates containing the β-galactosidase chromogenic substrate X-Gal (5-bromo-4-chloro-3-indolyl-β-D-galactopyranoside). By transforming *E. coli* VM2 with a genomic library from *V. cholerae* and screening for blue colonies on X-Gal, a *V. cholerae* gene that enhanced transcription of *lacZ* from the *ctx* promoter 10- to 12-fold was identified (37). This gene was called *toxR* and was shown to complement a variety of different hypotoxinogenic mutants of *V. cholerae* (37). The ToxR protein was shown to be a transmembrane protein with both a cytoplasmic domain and a periplasmic domain (40) (see below).

Subsequent characterization of the *toxR* gene product indicated that it regulated expression of other gene products in addition to cholera toxin (39, 50, 59). By screening *V. cholerae* mutagenized by Tn*PhoA*, 17 *phoA* gene fusions (called *tags*, for ToxR-activated genes) that were regulated by culture conditions in a manner identical to cholera toxin expression were identified (29, 50). Several of these gene fusions were located in the *tcp* operon, which consists of 8 to 10 genes that are required for biogenesis of the toxin-coregulated pilus, an essential intestinal colonization factor (see chapter 13 for more detail) (21, 50, 59). Also identified was the accessory colonization locus *(acf)*, a cluster of four genes whose products enhance colonization in an infant mouse model (50). ToxR was further shown to regulate the expression of two lipoproteins encoded by the *acfD* and *tagA* genes (47) as well as production of aldehyde dehydrogenase encoded by the *aldA* gene (46). In addition, two major outer membrane proteins, OmpU and OmpT, are regulated by ToxR (39).

In addition to regulating these specific proteins, ToxR regulates the serum resistance phenotype (47). Mutations in the *tcp* or *toxR* locus render *V. cholerae* 10^4 to 10^6 times more sensitive to complement-mediated killing. Recent evidence suggests that this phenotype is associated with the *tcpC* locus, which encodes a major lipoprotein distinct from AcfD or TagA (18, 47).

ToxR may also regulate motility (18). Hypermotile mutants of *V. cholerae* produce less cholera toxin and fewer toxin-coregulated pili, suggesting that motility and ToxR-regulated gene expression are connected by some form of regulatory "cross-talk." Motility has been shown to be involved in virulence but may be expressed at a stage in the infection that is distinct from the stage at which ToxR plays its most important role (see below) (20, 52).

With the exception of *ctxAB*, most of the virulence factors listed above are not directly controlled by ToxR. *V. cholerae* strains that contain *toxR* null mutations do not express these virulence factors, but ToxR alone is not able to activate transcription of these genes when both are present in the heterologous organism *E. coli*. Direct activation has been shown to be accomplished by ToxT, another member of the ToxR regulon, which was also identified as a locus that could mediate increased transcription from the *ctx* promoter in *E. coli* (10). Characterization of the *toxT* gene product indicated that the ToxT protein is a 32-kDa cytoplasmic protein belonging to the AraC-like family of transcriptional activators (22). ToxT is able to activate transcription of many of the genes in the ToxR regulon directly. This has been demonstrated for *ctxAB, aldA, tagA, tcpA, tcpI,* and *tcpC* (10).

In addition to ToxR and ToxT, the ToxS protein is also required for correct expression of the ToxR regulon (9, 36). The *toxS* gene is transcribed downstream from and in an operon with the *toxR* gene, and its gene product is a 19-kDa protein that is located in the inner membrane and periplasm (9, 36). Evidence from experiments analyzing the behavior of different ToxR-alkaline phosphatase fusion proteins indicates that the ToxR and ToxS proteins interact directly (9). This interaction probably occurs in the periplasm and may act to stabilize ToxR by driving ToxR into a favored conformation (9, 42). Mutations in *toxR* that affect the ToxR-ToxS interaction have been identified (9). These mutations lie in the cytoplasmic portion of ToxR in the region of the OmpR-homologous domain (see below) that is most carboxy terminal. This region of ToxR has been designated the "linker," a terminology used with chemotaxis proteins because the region appears to link events in the periplasm (the ToxR-ToxS interaction) to a conformational change in the cytoplasm (2, 9).

The requirement of three proteins for proper expression of the ToxR regulon led to the proposal that a regulatory cascade control virulence factor expression in *V. cholerae* (Fig. 1) (10). In this scheme, ToxR in conjunction with ToxS activates transcription of *ctxAB* and *toxT,* and it is ToxT that activates transcription of the many virulence factors of *V. cholerae*. ToxT is regulated by ToxR, since expression of *toxT* mRNA is abolished in a *toxR* mutant background as well as under environmental conditions that eliminate expression of the ToxR regulon (10). Interestingly, recent analysis of the *toxT* gene indicates that it resides within the *tcp* gene cluster (22).

MOLECULAR ANALYSIS OF ToxR

The *toxR* gene has been cloned and characterized quite extensively (37, 38, 40). The gene product is a 32-kDa integral inner membrane protein. The amino-terminal two-thirds of the protein is cytoplasmic, and the carboxy-terminal one-third is periplasmic. The cytoplasmic portion of ToxR is homologous to the "two-component" family of transcriptional regulators (reviewed in references 1 and 58; 40). This family is involved in the regulation of the expression of a variety of bacterial proteins. The prototype member of this family consists of two proteins, the sensor and the effector. The effector protein is modified by the sensor protein (usually by phosphorylation or dephosphorylation) and then carries out the response to the perceived signal, usually by altering gene expression. Proteins belonging to the effector family contain sequence homology in two domains, the domain that is modified by the sensor and the domain that is required for transcriptional activation. ToxR contains only the domain that is required for transcriptional activation, which is referred to as the "OmpR-homologous domain" (42). Evidence from OmpR indicates that this family of proteins activates transcription by binding DNA and then directly interacting with the alpha subunit of RNA polymerase (56).

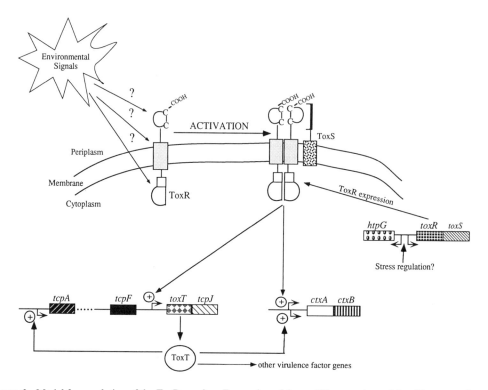

Figure 1. Model for regulation of the ToxR regulon. Expression of the *toxRS* operon is modulated by expression of the *htpG* gene in response to stress. The ToxR protein is depicted as a monomer that senses environmental signals and then undergoes a conformational change (here shown as dimerization) to become competent to activate transcription. This active ToxR is stabilized by the ToxS protein and activates transcription of both the *ctxAB* operon and the *toxT* gene. The ToxT protein then activates transcription of other virulence factors, including the *tcp* and *ctxAB* operons.

ToxR binds directly to the cholera toxin promoter region, and deletion analysis has identified a series of three to eight copies of a TTTTGAT repeat motif that is necessary for this binding. This motif is found immediately upstream of the −35 region of the *ctx* promoter (40). In keeping with the indirect activation of the remainder of the ToxR regulon, this repeat motif has not been found in promoter regions of the other members of the regulon (46).

The periplasmic portion of ToxR is ideally situated for sensing environmental signals, and indeed, early evidence suggested that this might be the function of this domain. ToxR-alkaline phosphatase fusions that replaced the majority of the periplasmic domain no longer responded to high osmolarity (40). This could be due to the absence of sequences required for sensing the environment or because these fusion proteins adopt an activated conformation (e.g., dimerized; see below) that bypasses the need for the ToxR periplasmic domain. Curiously, ToxR-alkaline phosphatase fusion proteins lack the majority of the periplasmic portion of ToxR yet are still capable of regulating toxin expression in response to pH, temperature, and amino acids (40). This result suggests that the ability of the ToxR protein to respond to environmental signals exists at several levels, i.e., that the cytoplasmic portion of ToxR as well as the periplasmic portion responds to various environmental signals.

REGULATION OF ToxR ACTIVITY AT THE CONFORMATIONAL LEVEL

The possibility that a protein conformational change was responsible for ToxR function was suggested by ToxR-alkaline phosphatase fusions that replaced the majority of the periplasmic region of ToxR but still retained the ability to activate transcription. This led Miller et al. (40) to propose that the alkaline phosphatase moiety was substituting for a normal function of this periplasmic domain. Since alkaline phosphatase is active only as a dimer, it was proposed that the periplasmic region of ToxR was a dimerization domain and that the active state of ToxR was a dimer. This provided a model for ToxR activation in which ToxR is inactive as a monomer but in the presence of appropriate environmental conditions dimerizes and becomes competent to activate transcription.

In order to test this model, a set of fusion proteins composed of the amino-terminal portion of ToxR and known dimeric and monomeric proteins has been constructed (Fig. 2) (43). ToxR fusion proteins with the periplasmic region replaced by either the leucine zipper from GCN4 (a dimerization domain) or β-lactamase (a monomeric domain) were constructed (ToxR-GCN4-M and ToxR-Bla, respectively). Both fusion proteins were fully competent to activate transcription and responded like wild-type ToxR to environmental conditions. When the fusion junction of ToxR-GCN4 was moved amino terminally to eliminate the membrane-spanning segment, this protein (ToxR-GCN4-C) no longer activated transcription, indicating that membrane localization is required for correct ToxR function. It appears from these studies that the periplasmic portion of ToxR does not need to be dimerized and that it may be dispensable for correct functioning under laboratory conditions. Interestingly, ToxR proteins truncated just beyond the transmembrane region or 50 amino acids from the carboxy terminus are not able to activate transcription even though they contain more ToxR than the fusion proteins described above (9). The addition of any periplasmic domain to the transmembrane domain of ToxR may act to stabilize ToxR or to more firmly anchor it into the membrane (43).

Although current evidence suggests that the periplasmic domain of ToxR does not require dimerization for activity, two additional lines of evidence suggest that ToxR may be dimerized when active. First, site-directed mutagenesis has been used to remove one of two cysteine residues in the periplasmic domain of ToxR. The mutated form of ToxR (*toxRC236S*) was able to activate transcription and formed disulfide-cross-linked ToxR

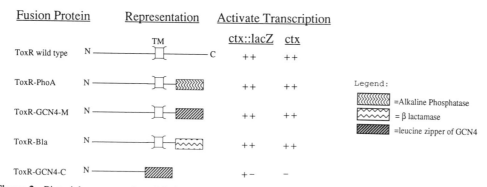

Figure 2. Pictorial representation of fusion protein analysis of ToxR. Wild-type ToxR is shown at the top as a plain line, with the transmembrane region depicted as a box with TM above it. At the right is indicated the ability of ToxR to activate transcription in two different organisms: *E. coli* from a *ctx::lacZ* construct and *V. cholerae* from the *ctx* promoter. The different moieties that were fused to ToxR are depicted below, with their identities indicated in the legend at the bottom. The ability of each of these fusion proteins to activate transcription from the two constructs is indicated at the right.

homodimers on nonreducing sodium dodecyl sulfate-polyacrylamide gels (41). The second line of evidence that ToxR is dimerized comes from experiments in which the amino-terminal domain of lambda repressor (cI) was used as a reporter for dimerization. Fusion of the cI N-terminal domain to the N-terminal domain of ToxR produced a fusion protein (cI-ToxR) that has both ToxR activity and cI activity as assayed by repression of a P_R-lacZ reporter construct (11). Because the N-terminal domain of cI dimerizes inefficiently unless carboxy-terminal sequences are able to propagate dimerization, these results suggest that ToxR provides the dimerization function in this fusion protein (26, 44, 63). This dimerization domain of ToxR may well be located in its cytoplasmic domain.

TRANSCRIPTIONAL REGULATION OF toxR

In addition to conformational regulation, the amount of toxR mRNA is regulated. Studies examining the toxR promoter found another gene directly upstream of toxR that was divergently transcribed (45). Sequence analysis revealed that this gene was homologous to htpG, a member of the Hsp90 family of heat shock proteins. Indeed, transcription from this gene is increased in response to higher temperatures, and this increase is accompanied by a concomitant decrease in levels of toxR mRNA. This regulation appears to be via direct competition for promoter binding between the heat shock sigma factor (σ^{32}) and σ^{70}. Parsot and Mekalanos (45) proposed that V. cholerae entering the intestine via the stomach would encounter harsh conditions like low pH, anoxia, starvation, proteases, and bile salts that would induce a stress response that would induce the htpG gene and thus inhibit toxR transcription. This type of temporal control may be beneficial to the organism in preventing expression of ToxR-activated genes too early in the infection process and allowing expression of ToxR-repressed genes and other phenotypes such as motility, which are regulated in a fashion opposite to that of ToxR-activated genes.

ROLE OF THE OmpR HOMOLOGOUS DOMAIN

The cytoplasmic portion of ToxR contains a region of homology to OmpR. The function of this region has been characterized by taking advantage of this homology to indicate which amino acid residues may be important for function (42). Changes of several of these amino acid residues resulted in proteins that no longer bound ctx promoter DNA and thus were unable to activate transcription. Two substitutions had more complex phenotypes. Changes at arginine 96 universally abolished the ability to activate transcription, but changes to leucine preserved the ability to recognize promoter DNA. This may localize a region of ToxR that interacts with RNA polymerase, as has been shown for OmpR (56). Amino acid substitution of glutamic acid 51 with lysine resulted in inactive protein that was dominant over wild type. This protein may exert its effect by forming defective heterodimers between the mutant and the wild-type proteins, in keeping with the evidence that the ToxR is able to dimerize.

ADDITIONAL REGULATION OF CHOLERA TOXIN AND VIRULENCE FACTORS

Control of the synthesis of virulence factors in the ToxR regulon is largely at the level of transcription; factors identified to date require toxR and toxT for maximal expression, although at least one gene product (OmpT) is negatively regulated. In addition to being subject to transcriptional control, the cholera toxin genes are also regulated by gene amplification. The ctxAB operon resides on a segment of DNA called the core region that is flanked by directly repeated segments called RS1 sequences. Collectively, the core region and the RS1 sequences make up the CTX genetic element, a type of site-specific transposon (30, 49). The core region contains the genes for several other virulence factors, including the zona occludans toxin (encoded by the *zot* genes) (13) (see also chapter 11), a pilin-like factor called Cep that enhances colonization (49), and a newly discovered enterotoxin called Ace (60). Different biotypes of *V. cholerae* contain different arrangements of the CTX element: classical strains contain two widely separated copies, while El Tor strains that contain more than one copy have them tandemly arranged (30). The CTX element can undergo *recA*-independent amplification in response to intestinal passage, resulting in increased copy number and increased toxin production (30). It is also of interest that recently isolated non-O1 strains of *V. cholerae* O139, which is responsible for a major cholera epidemic in India (51), contain duplications and further amplifications of the CTX genetic element (62). Thus, transposition of the CTX element into antigenically new strains of *V. cholerae* may be involved in the molecular evolution of new epidemic strains of *V. cholerae*. The presence of the ToxR regulatory system in many non-O1 *V. cholerae* strains, including O139 (61), makes this entirely feasible (37).

Other than inducing diarrhea, additional roles for cholera toxin in *V. cholerae* infection are not understood. While the toxin clearly causes the bulk of the secretory response, it may also provide benefit to the bacterial cells in an as-yet-undefined way. This may be by conferring a selective advantage, since strains that do not express toxin grow more slowly in vivo (4, 23, 25, 33). Mekalanos and colleagues hypothesized two possible roles for cholera toxin: alteration of epithelial cell metabolism so as to release a growth-limiting nutrient, or undefined intoxication of an intestinal cell that is responsible for a bactericidal activity that may be related to complement in its mode of action (31, 47).

CONCLUSIONS

V. cholerae possesses an array of virulence factors that are highly regulated. Among the factors necessary for a successful infection are cholera toxin, the toxin-coregulated pilus, and other accessory colonization factors. These virulence factors are organized into a regulon that is under the control of the ToxR protein (Fig. 1). Heat shock can modulate the expression of *toxR*. ToxR is at the top of a regulatory cascade in which the ToxR and ToxS proteins interact and then activate transcription of the *ctx* genes and the *toxT* gene. It is the ToxT protein that directly activates transcription of most of the virulence regulon. The entire ToxR regulon is sensitive to several environmental conditions, including temperature, pH, and osmolarity. The ToxR protein may have a role in sensing these various parameters directly and, in the presence of the correct conditions, assumes an active state. Responsiveness to a variety of conditions as well as multiple levels of control provides *V. cholerae* with the ability to fine-tune its virulence response.

Acknowledgments. We thank M. Dziejman and C. Gardel for permission to cite unpublished data.

REFERENCES

1. **Albright, L. M., E. Huala, and F. M. Ausubel.** 1989. Prokaryotic signal transduction mediated by sensor and regulator protein pairs. *Annu. Rev. Genet.* **23:**311–336.
2. **Ames, P., and J. S. Parkinson.** 1988. Transmembrane signaling by bacterial chemoreceptors: *E. coli* transducers with locked signal output. *Cell.* **55:**817–826.
3. **Baine, W. B., M. L. Vasil, and R. K. Holmes.** 1978. Genetic mapping of mutations in independently isolated nontoxinogenic mutants of *Vibrio cholerae. Infect. Immun.* **21:**194–200.
4. **Baselski, V. S., R. A. Medina, and C. D. Parker.** 1979. In vivo and in vitro characterization of virulence-deficient mutants of *Vibrio cholerae. Infect. Immun.* **24:**111–116.
5. **Callahan, L. T., and S. H. Richardson.** 1973. Biochemistry of *Vibrio cholerae* virulence. III. Nutritional requirements for toxin production and the effects of pH on toxin elaboration in chemically defined media. *Infect. Immun.* **7:**567–574.

6. **Callahan, L. T., R. C. Ryder, and R. C. Richardson.** 1971. Biochemistry of *Vibrio cholerae* virulence. II. Skin permeability factor/cholera enterotoxin production in chemically defined media. *Infect. Immun.* **4**:611–618.

7. **Colwell, R. R. (ed.).** 1984. *Vibrios in the Environment.* John Wiley & Sons, Inc., New York.

8. **Craig, J. P.** 1966. Preparation of the vascular permeability factor of *Vibrio cholerae*. *J. Bacteriol.* **92**:793–795.

9. **DiRita, V. J., and J. J. Mekalanos.** 1991. Periplasmic interaction between two membrane regulatory proteins, ToxR and ToxS, results in signal transduction and transcriptional activation. *Cell* **64**:29–37.

10. **DiRita, V. J., C. Parsot, G. Jander, and J. J. Mekalanos.** 1991. Regulatory cascade controls virulence in *Vibrio cholerae*. *Proc. Natl. Acad. Sci. USA* **88**:5403–5407.

11. **Dziejman, M., and J. J. Mekalanos.** ToxR dimerization as assayed by λ repressor fusions in *E. coli*. Submitted for publication.

12. **Evans, D. J., and S. H. Richardson.** 1968. In vitro production of choleragen and vascular permeability factor by *Vibrio cholerae*. *J. Bacteriol.* **96**:126–130.

13. **Fasano, A., B. Baudry, D. W. Pumplin, S. S. Wasserman, B. D. Tall, J. N. Ketley, and J. B. Kaper.** 1991. *Vibrio cholerae* produces a second enterotoxin, which affects intestinal tight junctions. *Proc. Natl. Acad. Sci. USA* **88**:5242–5246.

14. **Fernandes, P. B., and J. H. L. Smith.** 1977. The effects of anaerobiosis and bile salts on the growth and toxin production by *Vibrio cholerae*. *J. Gen. Microbiol.* **98**:77–86.

15. **Finkelstein, R. A.** 1973. Cholera. *Crit. Rev. Microbiol.* **2**:553–623.

16. **Finkelstein, R. A., and J. J. LoSpalluto.** 1970. Production, purification and assay of cholera toxin. Production of highly purified choleragen and choleragenoid. *J. Infect. Dis.* **121S**:S63–S72.

17. **Finkelstein, R. A., M. L. Vasil, and R. K. Holmes.** 1974. Studies on toxinogenesis in *Vibrio cholerae*. I. Isolation of mutants with altered toxinogenicity. *J. Infect. Dis.* **129**:117–123.

18. **Gardel, C., and J. J. Mekalanos.** Unpublished data.

19. **Gennaro, M. L., P. J. Greenaway, and D. A. Broadbent.** 1982. The expression of biologically active cholera toxin in *Escherichia coli*. *Nucleic Acids Res.* **10**:4883–4890.

20. **Guentzel, M. N., and L. J. Berry.** 1975. Motility as a virulence factor of *Vibrio cholerae*. *Infect. Immun.* **11**:890–897.

21. **Herrington, D. A., R. H. Hall, G. Losonsky, J. J. Mekalanos, R. K. Taylor, and M. M. Levine.** 1988. Toxin, toxin-coregulated pili, and the *toxR* regulon are essential for *Vibrio cholerae*

pathogenesis in humans. *J. Exp. Med.* **168**:1487–1492.

22. **Higgins, D. E., E. Nazareno, and V. J. DiRita.** 1992. The virulence gene activator ToxT from *Vibrio cholerae* is a member of the AraC family of transcriptional activators. *J. Bacteriol.* **174**:6874–6980.

23. **Holmes, R. K., W. B. Baine, and M. L. Vasil.** 1978. Quantitative measurements of cholera enterotoxin in cultures of toxinogenic wild-type and nontoxigenic mutant strains of *Vibrio cholerae* by using a sensitive and specific reversed passive hemmagglutination assay for cholera enterotoxin. *Infect. Immun.* **19**:101–106.

24. **Holmes, R. K., M. L. Vasil, and R. A. Finkelstein.** 1975. Studies on toxinogenesis in *Vibrio cholerae*. I. Isolation of mutants with altered toxinogenic mutants in vitro and in experimental animals. *J. Clin. Invest.* **55**:551–560.

25. **Howard, B. D.** 1971. A prototype live oral cholera vaccine. *Nature* (London) **230**:97–99.

26. **Hu, J. C., E. K. O'Shea, P. S. Kim, and R. T. Sauer.** 1990. Sequence requirements for coiled-coils: analysis with λ repressor-GCN4 leucine zipper fusions. *Science* **250**:1400–1403.

27. **Iwanaga, M., K. Yamamoto, N. Hiha, Y. Ichinose, N. Nakasone, and M. Tanabe.** 1986. Culture conditions for stimulating cholera toxin production by *Vibrio cholerae* O1 El Tor. *Microbiol. Immunol.* **30**:1075–1083.

28. **Kaper, J. B., and M. M. Levine.** 1981. Cloned cholera enterotoxin genes in study and prevention of cholera. *Lancet* **ii**:1162–1163.

29. **Manoil, C., and J. Beckwith.** 1985. TnphoA: a transposon probe for protein export signals. *Proc. Natl. Acad. Sci. USA* **82**:8129–8133.

30. **Mekalanos, J. J.** 1983. Duplication and amplification of toxin genes in *Vibrio cholerae*. *Cell* **35**:253–263.

31. **Mekalanos, J. J.** 1985. Cholera toxin: genetic analysis, regulation, and role in pathogenesis. *Curr. Top. Microbiol. Immunol.* **118**:97–118.

32. **Mekalanos, J. J., R. J. Collier, and W. R. Romig.** 1978. Affinity filters, a new approach to the isolation of *tox* mutants of *Vibrio cholerae*. *Proc. Natl. Acad. Sci. USA* **75**:941–945.

33. **Mekalanos, J. J., S. L. Moseley, J. R. Murphy, and S. Falkow.** 1982. Isolation of enterotoxin structural gene deletion mutations in *Vibrio cholerae* induced by two mutagenic vibriophages. *Proc. Natl. Acad. Sci. USA* **79**:151–155.

34. **Mekalanos, J. J., R. Sublett, and W. R. Romig.** 1979. Genetic mapping of toxin regulatory mutations in *Vibrio cholera*. *J. Bacteriol.* **139**:859–865.

35. **Mekalanos, J. J., D. J. Swartz, G. D. N. Pearson, N. Harford, F. Groyne, and M. de Wilde.** 1983. Cholera toxin genes: nucleotide sequence, deletion analysis and vaccine development. *Nature* (London) **306**:551–557.

36. **Miller, V. L., V. J. DiRita, and J. J. Mekalanos.** 1989. Identification of *toxS,* a regulatory gene whose product enhances ToxR-mediated activation of the cholera toxin promoter. *J. Bacteriol.* **171**:1288–1293.

37. **Miller, V. L., and J. J. Mekalanos.** 1984. Synthesis of cholera toxin is positively regulated at the transcriptional level by *toxR. Proc. Natl. Acad. Sci. USA* **81**:3471–3475.

38. **Miller, V. L., and J. J. Mekalanos.** 1985. Genetic analysis of the cholera toxin-positive regulatory gene *toxR. J. Bacteriol.* **163**:580–585.

39. **Miller, V. L., and J. J. Mekalanos.** 1988. A novel suicide vector and its use in construction of insertion mutations: osmoregulation of outer membrane proteins and virulence determinants in *Vibrio cholerae* requires *toxR. J. Bacteriol.* **170**:2575–2583.

40. **Miller, V. L., R. K. Taylor, and J. J. Mekalanos.** 1987. Cholera toxin transcriptional activator ToxR is a transmembrane DNA binding protein. *Cell* **48**:271–279.

41. **Ottemann, K. M., V. J. DiRita, and J. J. Mekalanos.** Unpublished data.

42. **Ottemann, K. M., V. J. DiRita, and J. J. Mekalanos.** 1992. ToxR proteins with substitutions in residues conserved with OmpR fail to activate transcription from the cholera toxin promoter. *J. Bacteriol.* **174**:6807–6814.

43. **Ottemann, K. M., and J. J. Mekalanos.** Unpublished data.

44. **Pabo, C. O., R. T. Sauer, J. M. Strurtevant, and M. Ptashne.** 1979. The λ repressor contains two domains. *Proc. Natl. Acad. Sci. USA* **76**:1608–1612.

45. **Parsot, C., and J. J. Mekalanos.** 1990. Expression of ToxR, the transcriptional activator of virulence factors of *Vibrio cholerae,* is modulated by the heat shock response. *Proc. Natl. Acad. Sci. USA* **87**:9898–9902.

46. **Parsot, C., and J. J. Mekalanos.** 1991. Expression of *Vibrio cholerae* gene encoding aldehyde dehydrogenase is under control of ToxR, the cholera toxin transcriptional activator. *J. Bacteriol.* **173**:2842–2851.

47. **Parsot, C., E. Taxman, and J. J. Mekalanos.** 1991. ToxR regulates the production of lipoproteins and the expression of serum resistance in *Vibrio cholerae. Proc. Natl. Acad. Sci. USA* **88**:1641–1645.

48. **Pearson, G. D. N., and J. J. Mekalanos.** 1982. Molecular cloning of *Vibrio cholerae* enterotoxin genes in *Escherichia coli* K-12. *Proc. Natl. Acad. Sci. USA* **79**:2976–2980.

49. **Pearson, G. D. N., A. Woods, S. L. Chiang, and J. J. Mekalanos.** 1993. CTX genetic element encodes a site-specific recombination system and an intestinal colonization factor. *Proc. Natl. Acad. Sci. USA* **90**:3750–3754.

50. **Peterson, K. M., and J. J. Mekalanos.** 1988. Characterization of the *Vibrio cholerae* ToxR regulon: identification of novel genes involved in intestinal colonization. *Infect. Immun.* **56**:2822–2829.

51. **Ramamurthy, T., S. Garg, R. Sharma, S. K. Bhattacharya, G. B. Nair, T. Shimada, T. Takeda, T. Karasawa, H. Kurazano, A. Pal, and Y. Takeda.** 1993. Emergence of a novel strain of *Vibrio cholerae* with epidemic potential in southern and eastern India. *Lancet* **341**:703–704.

52. **Richardson, K.** 1991. Roles of motility and flagellar structure in pathogenicity of *Vibrio cholerae:* analysis of motility mutants in three animal models. *Infect. Immun.* **59**:2727–2736.

53. **Richardson, S. H.** 1969. Factors influencing in vitro skin permeability factor production by *Vibrio cholerae. J. Bacteriol.* **100**:27–34.

54. **Ruch, R. E., J. R. Murphy, L. H. Graf, and M. Field.** 1978. Isolation of nontoxinogenic mutants of *Vibrio cholerae* in a colorimetric assay for cholera toxin using the S49 mouse lymphosarcoma cell line. *J. Infect. Dis.* **137**:747–755.

55. **Shimamura, T., S. Watanabe, and S. Sasaki.** 1985. Enhancement of enterotoxin production by carbon dioxide in *Vibrio cholerae. Infect. Immun.* **49**:455–456.

56. **Slauch, J. M., F. D. Russo, and T. J. Silhavy.** 1991. Suppressor mutations in *rpoA* suggest that OmpR controls transcription by direct interaction with the alpha subunit of RNA polymerase. *J. Bacteriol.* **173**:7501–7510.

57. **Sporeke, I., D. Castro, and J. J. Mekalanos.** 1984. Genetic mapping of *Vibrio cholerae* enterotoxin structural genes. *J. Bacteriol.* **157**:253–261.

58. **Stock, J. B., A. J. Ninfa, and A. M. Stock.** 1989. Protein phosphorylation and regulation of adaptive responses in bacteria. *Microbiol. Rev.* **53**:450–490.

59. **Taylor, R. K., V. L. Miller, D. B. Furlong, and J. J. Mekalanos.** 1987. Use of *phoA* gene fusions to identify a pilus colonization factor coordinately regulated with cholera toxin. *Proc. Natl. Acad. Sci. USA* **84**:2833–2837.

60. **Trucksis, M., J. E. Galen, J. Michalski, A. Fasano, and J. B. Kaper.** 1993. Accessory cholera enterotoxin (Ace), the third toxin of a *Vibrio cholerae* virulence cassette. *Proc. Natl. Acad. Sci. USA* **90**:5267–5271.

61. **Waldor, M. K., and J. J. Mekalanos.** 1994. ToxR regulates virulence gene expression in non-O1 strains of *Vibrio cholerae* that cause epidemic cholera. *Infect. Immun.* **62**:72–78.

62. **Waldor, M. K., A. Woods, G. D. Pearson, J. C. Sadoff, and J. J. Mekalanos.** Molecular analysis of novel *Vibrio cholerae* strains and development of live vaccine prototypes. Submitted for publication.

63. **Weiss, M. A., C. O. Pabo, M. Karplus, and R. T. Sauer.** 1987. Dimerization of the operator binding domain of phage λ repressor. *Biochemistry* **26**:897–904.

Vibrio cholerae and Cholera: Molecular to Global Perspectives
Edited by I. Kaye Wachsmuth, Paul A. Blake, and Ørjan Olsvik
© 1994 American Society for Microbiology, Washington, DC 20005

Chapter 13

The Toxin-Coregulated Pilus: Biogenesis and Function

Melissa R. Kaufman and Ronald K. Taylor

Although production of cholera toxin directly results in the manifestation of diarrhea, cholera pathogenesis relies on a variety of carefully orchestrated factors for virulence. After surviving passage through the acidic environment of the stomach, cholera vibrios must resist the potent cleansing mechanisms of the small bowel in order to proliferate on the intestinal epithelium and establish infection. In addition to enterotoxin, *Vibrio cholerae* carefully regulates synthesis of (i) a polar flagellum for chemotaxis to and penetration of the mucous barrier (22), (ii) a protease to dissolve the mucous gel (19), (iii) cell-associated and soluble hemagglutinins (29), (iv) at least three types of pili for association with and colonization of microvilli (14, 28, 81), (v) a neuraminidase to convert gangliosides to toxin receptors (23, 35), (vi) a Shiga-like toxin (50), (vii) a zonal occlusion toxin (Zot) (3, 18); and (viii) iron-regulated proteins for survival in the iron-limited environment of the gut (25). Many of the virulence factors utilized by *V. cholerae* reflect common themes in microbial pathogenesis. For example, pili (fimbriae), the filamentous surface organelles that mediate specific binding and colonization of host tissue, are essential components of the infection strategy for a variety of pathogenic bacteria (4, 54).

IDENTIFICATION OF TCP AND *tcpA*

The best characterized pilus of *V. cholerae* is the toxin-coregulated pilus (TCP) described by Taylor et al. (81). Electron micrographs of *V. cholerae* cultured to stationary phase in pH 6.5 Luria broth containing 60 mM salt at 25 to 30°C with moderate aeration reveal abundant pili 5 to 7 nm in width and 10 to 15 μm in length. These pili aggregate to form lateral bundles (Fig. 1), probably because of the hydrophobic nature of the pilus (discussed later in this chapter). Expression of TCP bundles results in bacterial autoagglutination and sedimentation in liquid media. Elaboration of this pilus confers upon the bacteria the ability to hemagglutinate erythrocytes from CD-1 mice in the presence of fucose or mannose, sugars that inhibit the majority of *V. cholerae* hemagglutinins. Partial purification of pili shearates by differential centrifugation followed by sodium dodecyl sulfate-polyacrylamide gel electrophoresis (SDS-PAGE) resolves the pilin subunit as a 20.5-kDa protein. Production of 20.5-kDa pilin parallels expression of cholera toxin under a variety of growth conditions. At the

Melissa R. Kaufman • Hopkins Marine Station, Stanford University, Ocean View Boulevard, Pacific Groves, California 93950. *Ronald K. Taylor* • Department of Microbiology, Dartmouth Medical School, Hanover, New Hampshire 03755.

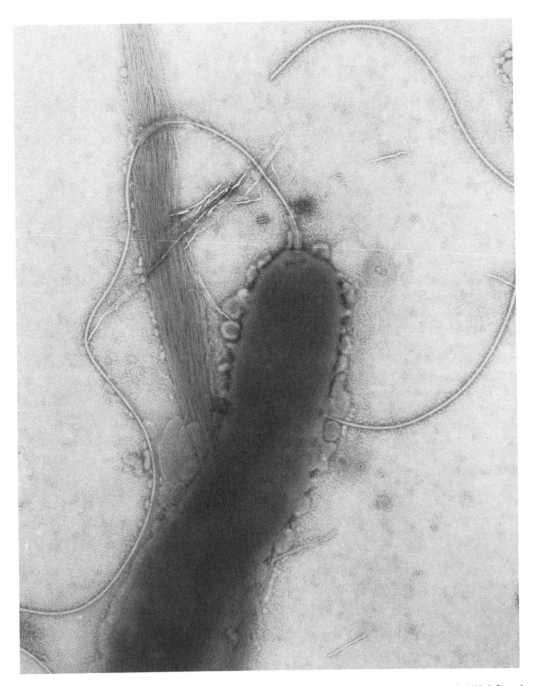

Figure 1. Electron micrograph of the TCP expressed by *V. cholerae* cultured at 30°C in Luria broth (pH 6.5) and negatively stained. Note the bundling of the pilus into large, hydrophobic masses.

molecular level, pilin synthesis is coordinately expressed within a virulence regulon by the transmembrane, DNA-binding regulatory protein ToxR and by the recently identified global regulator ToxT (see below). Therefore, the pilus is designated TCP, for toxin-coregulated pilus, and the pilin subunit is designated TcpA.

Random insertion mutagenesis of the *V. cholerae* chromosome identified the gene encoding the TCP pilin subunit, *tcpA*. Tn*phoA*, a Tn*5* derivative, was employed for this mutagenesis because of the unique properties of this transposon that enrich for identification of genes encoding secreted proteins such as a pilin (40, 80). Gene fusions created by Tn*phoA* code for hybrid proteins composed of the carboxyl-terminal portion of alkaline phosphatase (PhoA) fused in frame to an amino-terminal portion of the target. Since alkaline phosphatase is enzymatically active only when secreted beyond the reducing environment of the cytoplasm and since the *phoA* gene in the transposon is devoid of a signal sequence, only fusions with target genes carrying information able to compensate for the secretory defect will appear PhoA$^+$ on indicator plates. Cloning of the disrupted gene is facilitated by kanamycin resistance of the transposon, selected in conjunction with scoring for a PhoA$^+$ phenotype. Taylor et al. (81) employed this strategy to isolate mutants of *V. cholerae* Ogawa 395 deficient in production of the 20.5-kDa pilin. Correlation of the N-terminal pilin sequence with the coding region of the cloned fusion gene confirmed that the insertions were indeed within the *tcpA* structural gene.

ROLE OF TCP IN COLONIZATION

Mutants carrying Tn*phoA* fusions to the *tcpA* gene showed a 50% lethal dose 5 logs higher than that of the wild-type parent strain in the infant mouse model. This same model demonstrated that the mutant was outcompeted 500 to 1 when administered in equal amounts with the wild-type parent (81). Human volunteer studies utilizing a *tcpA ctxA* (cholera toxin gene) double mutant confirmed the role of TCP in colonization. These double mutants were deficient for colonization at doses that readily allowed colonization and seroconversion by the *tcp*$^+$ parent strain (31). Passive immunization experiments with infant mice further supported these results. Antibodies to TcpA provided a high level of protection when administered with virulent vibrios to infant mice (76). This protection was found to be more pronounced for classical than for El Tor biotype strains (70, 76). Because of these results and the differences in growth conditions that elicit TCP expression by El Tor strains, there has been some controversy regarding the role for TCP in El Tor biotype colonization. Recent cloning of the El Tor biotype *tcpA* gene and the construction of a defined *tcpA* knockout mutation have demonstrated that TCP expression is required for colonization to a degree similar to that previously established for classical biotype strains, thus establishing TCP's role in both biotypes (61a). For these experiments, El Tor biotype strains E7946 and N16961 were seen to express TCP bundles using the described culture conditions (76). The predicted amino acid sequence of TcpA from El Tor strain E7946 shows 82% identity with the classical biotype sequence. An insertion in this gene was crossed into El Tor strain N16961, which is virulent in infant mice. An in vivo competition between the parent strain and the mutant in infant mice showed a 42-fold advantage for the parent; for a similar pair of classical strains, the parent strain had a 120-fold advantage. Immunoblotting has also shown that TcpA is expressed in vivo by both biotypes, although for many El Tor strains, it has not been detectable on the cell surface (27, 34).

TCP GENE CLUSTER

Previously characterized pilus systems in *Escherichia coli* reveal that numerous genes, often located adjacent to the major subunit, are required for pilus formation and function (6). Primary evidence that linked genes were required for TCP formation came from studies by Taylor et al. (73, 79) in which a *V. cholerae* genomic library in *E. coli* was tested for TcpA expression by Western blot (immunoblot) analysis with anti-TcpA sera. Two classes of cosmid clones were distinguished: those that expressed a TcpA product of the expected 20.5 kDa and those that produced a 23-kDa pilin precursor. The gene necessary for processing the precursor to the mature pilin has been identified and characterized as *tcpJ* (36) (see below). Since proteins instrumental to pilus formation are often secreted, these cosmid clones were further analyzed by Tn*phoA* mutagenesis. An additional mutagenesis of the chromosome was included to elucidate genes unlinked to *tcpA* that lost properties associated with TCP expression by screening for ToxR-regulated alkaline phosphatase expression and loss of bacterial autoagglutination (59). Using Southern analysis and Western blotting with anti-PhoA antibodies, fusions were physically mapped both upstream and downstream of *tcpA*, revealing coding regions for *tcpBCDEFHI* (71, 79). One unlinked gene fusion exhibiting the desired properties, *tcpG,* was identified by Tn*phoA* insertion. To evaluate the role of each gene in TCP assembly and function, fusions were subcloned and recombined into the *V. cholerae* chromosome by utilizing the incompatible plasmid method of Ruvkun and Ausubel (66). Characterization of the resulting mutant strains with regard to TcpA expression, pilus assembly, autoagglutination, and fucose[r] hemagglutination allowed tentative functions to be assigned to each TCP gene product (Fig. 2). Fusions to genes *tcpBCDEF* show reduced levels of pilin and lack pili on the cell surface. Therefore, these genes probably encode assembly and transport proteins required for pilus biogenesis. Further analyses of TcpE, TcpF, and TcpC function are discussed below. Fusions in *tcpH* result in an agglutination-defective or-

ganism that assembles low levels of normal pili, suggesting a positive regulatory function for *tcpH*. The *tcpI* mutant produces an overabundance of pili even under nonexpressing growth conditions and therefore is possibly involved in negative regulation of pilin production (72). Nucleotide sequencing of the TCP region has identified additional open reading frames (ORFs) (52). Although the roles of these ORFs await genetic analysis, it is likely that they encode products with functions similar to those outlined above. The *tcpG* mutant displays pili that are visually indistinguishable from wild type; however, the strain fails to autoagglutinate or mediate fucose[r] hemagglutination. This protein is involved in disulfide bond formation within TcpA (see below).

TcpA: HOMOLOGY TO TYPE IV PILINS AND FUNCTIONAL DOMAINS

Amino acid sequencing of gel-purified TcpA reveals an amino-terminal region that bears striking homology to type IV pilins, formerly referred to as *N*-methylphenylalanine pilins because of the modification present on their amino-terminal residues. The sequencing of mature pilin from *V. cholerae* has identified the N-terminal modified amino acid residue as *N*-methylmethionine (36). The recently recognized type IV pilin from enteropathogenic *E. coli* also shows N-terminal methionine in Edman degradation sequencing (24); however, the DNA sequence from two isolates predicts a leucine in this position (10). This discrepancy might suggest a modified leucine at the N terminus. Thus, type IV pilins with at least two different N-terminal residues and at least one type of modification have now been identified. A number of bacterial pathogens elaborate type IV pili, thought to be critical for virulence. These include *Pseudomonas aeruginosa* (67), *Neisseria gonorrhoeae* (44), *Moraxella bovis* (41), and *Dicholobacter nodosus* (15) (Fig. 3).

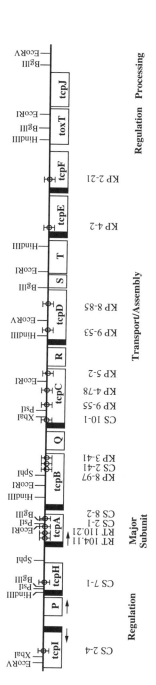

Figure 2. Organization of the TCP gene cluster. Tn*phoA* insertions and relevant restriction sites are indicated at their approximate locations. Dark bars at the 5' ends of genes indicate the presence of a signal sequence as determined by PhoA fusion protein activity and nucleotide sequencing. Functions as determined by phenotypic analysis of mutants are listed under the genes. ORFs revealed by nucleotide sequencing are indicated by capital letters (52). Arrows denote directions of transcription.

Figure 3. Comparison of TcpA and other type IV pilins. Symbols: ■, residues in the majority of type IV pilins that are identical to TcpA; ○, conserved hydrophobic residues; arrow, pilin processing site. The hydrophobicity plot is a Kyte and Doolittle (38) analysis for TcpA. The pattern is similar for all the pilins shown.

An exceptionally hydrophobic domain is created by the first 24 residues of all type IV mature pilins. This domain is essential for translocation of *P. aeruginosa* pilin beyond the cytoplasmic membrane (74) and likely serves as an internal signal sequence for export for all the type IV pilins. Preceding this hydrophobic stretch is a short, hydrophilic leader peptide that differs significantly from the typical proteolytically cleaved signal sequences found on most exported proteins (53). TcpA and the *E. coli* type IV pilins contain a leader sequence up to 18 amino acids longer than the other type IV pilins, but considerable homology is retained in the region surrounding the processing site. The cleavage at this site requires specialized processing machinery, as will be discussed below. The general motifs described here extend beyond type IV pili to proteins involved in the process of extracellular secretion in many bacterial species. Homology is retained in the areas of the unusual leader, the processing site, and an N-terminal Phe or Met on the mature protein (for a recent review, see reference 60).

The mechanism by which TCP and other type IV pili mediate colonization has yet to be elucidated in the molecular detail that has been accomplished for some *E. coli* fimbrial types. Studies on the binding domains of type IV pili are complicated by the lack of identified cellular receptors, with the possible exception of one for the *P. aeruginosa* pilus, which binds specific proteins from buccal epithelial cells after solubilization, SDS-PAGE separation, and transfer to a nitrocellulose membrane (33).

Evidence exists that at least part of the colonizing function of type IV pili is mediated through the pilin subunit and not an ancillary adhesin molecule. For example, regions within the *Neisseria* type IV pilin have been identified by monoclonal antibodies and by antibodies to synthetic peptides that inhibit binding activity in vitro (62, 69). Studies utilizing a combination of peptide binding and competitive inhibition as well as antibodies

directed against synthetic peptides have demonstrated that the carboxyl disulfide loop region of the *Pseudomonas* PAK pilin (PilA) mediates binding to the buccal epithelial cell (BEC) receptors (9, 33).

In the case of TCP, both polyclonal antibodies and a subset of monoclonal antibodies directed against the major subunit of the pilus are able to provide passive immunization in infant mice challenged with high doses of virulent strains (76). The epitopes recognized by the protective monoclonal antibodies have been mapped by using synthetic peptides to the carboxyl region of the pilin subunit (77, 78). This region contains the disulfide loop common among type IV pilins. Additional evidence regarding the functionality of this region is based on the phenotype of mutant strains that lack the function of the TcpG periplasmic disulfide isomerase. Such strains elaborate nonfunctional TCP and are altered in disulfide bonding properties in general, leading to the hypothesis that the functional domain near the carboxyl end of TcpA is misfolded in such mutants (57) (see below). It should be noted that the significant hydrophobic nature of the type IV pili may, in and of itself, also contribute to colonization by nonspecific target cell interactions or perhaps through microcolony interactions. This property is exemplified by the autoagglutination phenomenon attributable to TCP expression (81) and might provide an attractive hypothesis for the colonization role of type IV pili, since receptors have not generally been conclusively identified and no carbohydrates that inhibit any of the hemagglutination characteristics of these pili have yet been described. In this regard, autoagglutination and hydrophobicity are concomitantly lost with the ability of *V. cholerae* to colonize *tcpA* or *tcpG* mutants.

For consideration of potential inclusion of pilus antigens in vaccine formulations, we have determined whether there is any extensive antigenic diversity found between isolates. The *tcpA* gene has been sequenced from three classical biotype isolates with only two variant bases, neither of which results in an amino acid change (17, 73). Sequence analysis of El Tor strain E7946 reveals divergence from the classical strain sequence. The primary structure of E7946 TcpA shows 82% identity to classical TcpA (61a). Much of the diversity is within the carboxyl region and cystine loop implicated in type IV pilus function. Determination of the *tcpA* sequence from the Peru El Tor isolate (currently under way) will help determine whether there is any variation within this biotype.

TcpG: CATALYSIS OF DISULFIDE BOND FORMATION IS COUPLED TO THE ELABORATION OF FUNCTIONAL PILI

The *tcpG::phoA* fusion strain KP8-96 elaborates TCP colonization pili that appear morphologically normal by transmission electron microscopy yet fail to mediate colonization in vivo and are functionally deficient by several in vitro assays (59, 71). Consistent with the identification of *tcpG* by *phoA* fusion, TcpG is a periplasmic protein. The predicted amino acid sequence of TcpG initially revealed homology to both thioredoxin and protein disulfide isomerase (PDI), centering around the thiol:disulfide interchange sites of these molecules. Purified TcpG can function in vitro with a level of activity comparable to that of *E. coli* thioredoxin, suggesting that TcpG might interact with target proteins that contain disulfide bonds. The major TcpA pilin subunit contains an intrachain disulfide bond that creates a domain contributing to the hydrophobic nature of the pilus (77). This hydrophobic character of TcpA is lost in TcpG null strains (unpublished data). Recent studies (57a) using site-directed mutagenesis have shown that the homologous redox site is indeed required for TcpG function in vivo. The properties of TcpG, coupled with the ToxR regulation of its expression, prompted examination of the effect of *tcpG*

null mutations on other disulfide-containing, ToxR-regulated molecules such as cholera exotoxin (42). A combination of gene fusion and Western analysis demonstrated that the toxin genes are expressed normally in a *tcpG* mutant background, but toxin is no longer detectable in the culture supernatant. In the periplasm of these strains, the A subunit is present but the B subunit is not, suggesting that it is quickly degraded. Since the A subunit must associate with the B pentamer to be exported (32), it seems likely that the conformations required for this association are not efficiently achieved in the *tcpG* mutant, thus explaining the loss of secreted toxin. A similar finding has been reported for the secretion of *E. coli* exotoxin by *V. cholerae* and its dependence on TcpG (86). The pleiotropic nature of a *tcpG* mutation also includes a significant decrease in the amount of active protease secreted by such mutants (57). A likely protease to be affected is the soluble hemagglutinin-protease that contains four cysteine residues and shares extensive homology with the elastase of *P. aeruginosa* (30), in which these residues have been shown to be disulfide bonded (82).

The discovery of TcpG and its role in disulfide bond formation was surprising, because it was previously believed that disulfides could form spontaneously in exported proteins. In fact, disulfides can form spontaneously, but not at a physiologic rate. This has been established through studies of the DsbA protein of *E. coli* (2). Using pulse-chase experiments, it was shown that the disulfides of β-lactamase, OmpA, and alkaline phosphatase all formed at a slower rate in *dsbA* mutant strains (2). This resulted in the instability of some of these proteins, similar to what is seen with cholera toxin. It is now realized that DsbA and TcpG are representatives of a newly identified global class of periplasmic thiol:disulfide interchange proteins. Comparison of the *E. coli dsbA* gene product to TcpG reveals 40% identity throughout the entire sequence, extending beyond the active site, and *tcpG* can complement a *dsbA* mutation. Another member of this class has recently been identified in *Haemophilus influenzae* (83) (Fig. 4), and there are unpublished reports of numerous others.

While the exact roles and mechanisms of TcpG function are not yet defined, insight may be provided from the functions attributable to related thiol:disulfide interchange molecules. For example, a protein with a homologous redox site, PDI, has been shown to catalyze thiol:disulfide interchange reactions in protein substrates and has also been closely correlated to the processes of secretory protein synthesis and secretion (20, 21). Thus, proteins involved with disulfide bond formation may be important in mediating protein conformation during the export process.

Given the properties of the TcpA pilin in the *tcpG* mutant, TcpG may have two roles in polypeptide maturation, one as a thiol oxidant and possibly one as an isomerase/chaperone similar to Bip (26), whereby TcpG may act to locate and guide portions of polypeptide chains into a state in which complex surfaces can form. These surfaces, such as the externalized hydrophobic domain of TcpA, might otherwise be energetically unfavorable and would rarely form under physiologic conditions.

TcpJ: NOVEL SECRETORY APPARATUS FOR TCP EXPORT

Secretion of bacterial cell surface proteins composing macromolecular complexes occurs by a mechanism only recently elucidated in comparison to the initial stages of general prokaryotic protein export (68, 85). The pathway for TCP biogenesis provides a model system for investigating these novel secretion processes. A primary step in pilus formation is the processing of precursor TcpA pilin into its mature, export-competent form. Prepilin does not contain the consensus signal sequence common to most exported proteins and fails to be processed in *E. coli* (71). Instead, precursor TcpA possesses a hydrophilic N-terminal domain that shares homology with type IV pilins and other recently defined proteins involved with extracellular export from gram-negative bacteria (49, 60). Processing of this leader is accomplished in *E. coli* when one of the linked *V. cholerae* pilus biogenesis genes, *tcpJ*, is supplied in *trans*.

The deduced primary structure of TcpJ indicates an exceptionally hydrophobic protein that is likely to span the cytoplasmic membrane by analogy to the highly homologous PilD protein of *P. aeruginosa* (49). A *V. cholerae* mutant of *tcpJ* shows a significant pilin-processing defect, fails to produce assembled

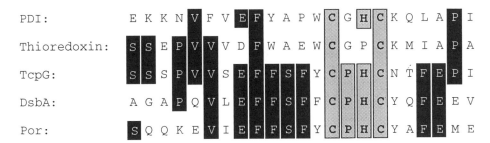

Figure 4. Active-site homologies between TcpG, PDI, thioredoxin, DsbA, and Por. TcpG residues 36 to 58 are aligned with the region surrounding the catalytic site (gray boxes) of rat liver PDI (13, 16), thioredoxin from *Corynebacterium nephridii* (43), the highly related DsbA protein of *E. coli* (2), and the highly related Por protein of *H. influenzae* (83).

pili, and can be complemented in *trans* by plasmid-encoded TcpJ. When cloned into *E. coli,* TcpJ is able to catalyze pilin processing in the absence of leader peptidases involved in the classical export pathway, suggesting that TcpJ functions independently to cleave the leader peptide from precursor TcpA. This is consistent with the in vitro demonstration of peptidase function for *P. aeruginosa* PilD, defining a common class of prepilin peptidases and homologs involved in the biogenesis of certain extracellular proteins. An additional peptidase homolog has recently been identified in the type IV pilus system in *N. gonorrhoeae* (39). TcpA processing is also independent of at least one major component of the classical export pathway, SecA (36).

During general export, a complex including SecA, the signal sequence of the targeted protein, and in some cases the SecB chaperone interacts with an integral membrane complex containing the SecY and SecE proteins (5, 68). The signal sequence that promotes these events lies at the N terminus of the precursor form of the secreted protein and is characterized by one or two basic residues followed by a hydrophobic core and a consensus proteolytic processing site (84). Following the initiation of export through the membrane, processing of the signal sequence occurs on the periplasmic side of the cytoplasmic membrane. In the case of type IV pilins, the processed N-terminal region is not itself sufficient for secretion; the hydrophobic portion that is retained by the mature protein and functions as an internal signal sequence is also necessary (74). The cleaved leader of these pilins is positively charged, reminiscent of the first residue generally found on a typical leader sequence. Thus, for type IV pilin-like leaders, it is likely to be a combination of this charged region and the hydrophobic domain that functions as a signal for membrane translocation (Fig. 5). The N-terminal leader is cleaved by prepilin peptidase after membrane association that, at least for TcpA, occurs in a SecA-independent fashion. The cleavage event most likely takes place on the cytoplasmic side of the membrane, judging by the deduced orientation of the protein and the position of N methylation of the mature pilin (12, 36). This mechanism likely imparts to the protein the ability to remain in the membrane until released to the outside of the cell, perhaps through association with membrane junctions. Supporting evidence for membrane association during export is derived from fractionation and electron microscopy studies, in which it was observed that both TcpA and PilA are localized exclusively to the membrane (36, 56, 75).

A specialized apparatus has evolved for both type IV pilin secretion and some other types of extracellular export by gram-negative bacteria (60). Homology of the pilin leader region, cleavage site, and often the leader peptidase extends to proteins involved in extracellular secretion in other systems such as pullulanase of *Klebsiella pneumoniae,* extracellular products of *Erwinia* spp. such as pectate lyase, cellulase secretion in *Xanthomonas campestris,* and transformation proteins of *Bacillus subtilis* (1). This homology extends through additional corresponding secretion molecules associated with these systems, as discussed in greater detail below.

TCP BIOGENESIS GENES: ANALOGIES TO OTHER VIRULENCE FACTOR SECRETORY SYSTEMS

A number of the *phoA* fusions initially isolated on the basis of loss of TCP expression still express the TcpA pilin but fail to assemble or secrete pili. Thus, these genes are postulated to encode proteins with biogenesis functions (79). The first two of these genes identified by fusion analysis to be extensively characterized were *tcpE* and *tcpF* (37). As discussed above, homology between TcpJ and the PulO protein involved in pullulanase secretion in *K. pneumoniae* (36) led us to investigate whether TcpE or TcpF showed homology to other pullulanase secretion fac-

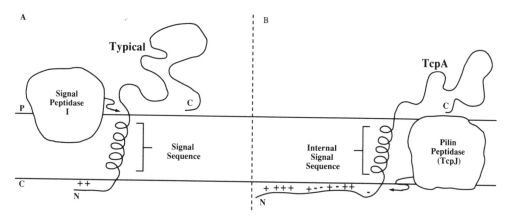

Figure 5. Comparison of TcpJ function with leader peptidase. (A) During general export in *E. coli*, the N-terminal hydrophobic signal sequence of proteins to be translocated has a transmembrane orientation and is cleaved by signal peptidase I in the periplasm. (B) For secretion of TcpA, the hydrophilic leader peptide is likely to interact with the inner membrane and be cleaved on the cytoplasmic face. The membrane-embedded signal sequence for export is located in the mature pilin molecule. Cleavage of the basic leader could then allow release of pilin from the membrane for surface assembly, which is likely mediated by the products of additional TCP biogenesis genes. This figure has been reprinted with permission (36).

tors. Computer-generated alignments indicate that TcpE shows 47% similarity and 21% identity to the PulF molecule and is additionally related to *X. campestris* extracellular enzyme secretory factor XpsF (11) and *B. subtilis* genetic transformation protein ComG-ORF2 (1). As noted above, another *B. subtilis* competence factor, ComC, also shows substantial homology with the TcpJ peptidase (36). Consistent with this observation are the data indicating that additional *B. subtilis* ComG ORFs show significant similarity to type IV pilins (1, 60). TcpE also shows 43% similarity and 19% identity to the *K. pneumoniae* PulD protein. PulD shares regions of homology with the YscC protein required for secretion of the *Yersinia enterocolitica* extracellular virulence factors designated Yops (45). Alignments of TcpE and YscC indicate 43% similarity and 21% identity. Yops, analogous to TcpA, do not contain a classic signal sequence for secretion, instead relying on an internal secretion signal encoded within the mature N terminus of the molecule (36, 45). YscC and PulD also share domains with the filamentous phage biogen-

esis protein pIV, postulated to provide an outer membrane channel for phage translocation (46, 65). Considering the possible evolutionary relatedness between biogenesis of filamentous bacteriophages and fimbrial structures, this observation reinforces our hypothesis that TcpE belongs to this recently recognized class of virulence factor secretion molecules. Additionally, TcpF displayed limited homology to several pullulanase secretion factors, although rigorous alignment resulted in numerous large gaps being introduced. Other secretion machinery homologies with molecules encoded within the TCP cluster have recently been reported for TcpT. The deduced amino acid sequence of TcpT is most closely related to the pilus biogenesis protein PilB of *P. aeruginosa*, the pullulanase secretion protein PulE of *Klebsiella oxytoca*, and ComG ORF1 of *B. subtilis* involved with transformation competence (52). Components of a similar apparatus for protein translocation have also been identified for *N. gonorrhoeae* (39). The *N. gonorrhoeae* proteins PilF, PilT, and PilD, presumed to play a role in pilus biogenesis, share substantial homology to pilus assembly proteins of the same name from *P. aeruginosa.*

A database search found a C-terminal portion of TcpE that displayed 46% identity to the highly conserved transmembrane domain of the integrin beta-1 subunit from numerous eukaryotic organisms (37, 63). Exceptional primary structure conservation exists between integrins in this transmembrane domain, suggesting that very precise constraints operate for the structure of this region (64). The transmembrane domain has been implicated in signal transduction across the cell membrane, alpha-chain binding, and lipid binding. Interestingly, integrins codistribute and copurify with certain sialylated gangliosides such as the cholera toxin receptor GM_1 (64). TcpE homology to the intergrin transmembrane region could thus possibly represent a domain involved in GM_1 interaction to bring the bacteria close enough to the ganglioside for efficient toxin delivery. Thus, the *tcp* locus appears to encode multiple factors that enhance colonization of the small intestine.

TcpC: EVASION OF BACTERICIDAL ACTIVITY IN THE GUT

As mentioned above, it is postulated that the *tcp* locus encodes numerous functions involved in *V. cholerae* adherence to and colonization of eukaryotic mucosal surfaces. Recent data suggest that the TcpC protein may contain novel functions independent of its hypothesized role in pilus biogenesis. ToxR regulates expression of TcpC, an abundant outer membrane protein discovered to be a lipoprotein by DNA sequence analysis and [^{3}H]palmitate labeling of hybrid proteins encoded by Tn*phoA* gene fusions (55). Production of bacterial lipoproteins has been shown to be associated with resistance to the bactericidal effects of complement (51, 58, 61). Because of the noninvasive nature of vibrios, results of serum sensitivity assays with TcpC mutants were especially intriguing (55). *V.*

cholerae mutants of *tcpC* were approximately 10^5 times more sensitive to the bactericidal effect of anti-whole-cell antibody and guinea pig complement than isogenic strains carrying insertion mutations in unrelated genes. These observations may imply the presence of a novel bactericidal activity in the intestinal mucosa that mimics the activities of complement and antibody in serum. Perhaps evasion of this complement-like activity via production of a lipoprotein resembling TcpC may play an essential role in the colonization process of numerous bacterial pathogens.

ToxR AND ToxT: A REGULATORY CASCADE FOR TCP EXPRESSION

Many of the *tcp* genes were identified on the basis of their regulation properties as well as on loss of bacterial autoagglutination following Tn*phoA* mutagenesis. Central in this regulation is the ToxR protein (for a recent review, see reference 7). ToxR is an integral membrane protein that activates the expression of a number of virulence genes in response to environmental conditions. This regulation appears to differ between the classical and El Tor biotypes. For example, the original condition used to elicit high-level TCP expression in classical strains, pH 6.5 Luria broth at 30°C with aeration (81), does not enhance TCP expression by El Tor strains. However, conditions developed by K. Peterson (AKI medium, pH 7.0, 30°C in 5% CO_2) elicit maximal TCP expression by both biotypes (76). These non-biotype-specific expression conditions are possibly those that more closely mimic the in vivo environment.

In at least the case of the cholera toxin gene operon, *ctxAB*, activation is accomplished by direct binding of ToxR to *cis*-acting elements upstream of the RNA polymerase binding site (48). The *toxR* gene was initially isolated by screening a *V. cholerae* gene library for clones that activated a *ctx-lacZ* gene fusion (47). This same screen identified a second gene, designated *toxT,* which encodes a product also capable of activating *ctx* expression but to a lesser extent than ToxR (8). The level of *toxT* mRNA is regulated by ToxR such that ToxT is part of a regulatory cascade. When transcription of the *tcp* genes in *E. coli* was analyzed, it was found that, contrary to the case of the *ctx* genes, *tcp* expression was not activated when *toxR* was present in *trans.* However, when *toxT* alone was provided in *trans, tcp* expression was activated (5a, 8). Thus, the regulation of the *tcp* genes by ToxR is indirect, being mediated through ToxT. Interestingly, the *tcpI* gene, which acts in a negative manner to influence other *tcp* gene expression (71), is also activated by ToxT. It is currently hypothesized that the presence of TcpI would allow additional modulation of *tcp* gene expression after initial activation by ToxT. The *tcpI* gene was identified by *phoA* fusion, and sequence analysis supports that it is likely to code for a transmembrane protein (71).

SUMMARY

As mentioned above, although the causative agent of cholera was first isolated over a century ago, the molecular mechanisms utilized by *V. cholerae* to provoke disease are only now being elucidated. One essential component of the infection process is colonization of the small bowel, mediated by expression of the TCP. Investigation of the TCP, in particular work on the major pilin subunit, has revealed novel strategies for vaccine development for *V. cholerae* and other bacteria that utilize type IV pili for adherence. Additionally, our general knowledge of protein export and protein folding has been expanded by studies of the TCP secretory machinery used for biogenesis and membrane translocation. In fact, an entirely new pathway for protein secretion in gram-negative bacteria has been discovered, in part owing to results obtained from studies on TcpG and TcpJ. Thus, genetic and biochemical analyses of one

member of the type IV pilus family not only provide findings of clinical relevance for prevention of cholera but also impact significantly on our understanding of fundamental biological processes.

REFERENCES

1. **Albano, M., R. Briettling, and D. A. Dubnau.** 1989. Nucleotide sequence and genetic organization of the *Bacillus subtilis comG* operon. *J. Bacteriol.* **171:**5386–5404.
2. **Bardwell, J. C. A., K. McGovern, and J. Beckwith.** 1991. Identification of a protein required for disulfide bond formation in vivo. *Cell* **67:**581–589.
3. **Baudry, B., A. Fasano, J. Ketley, and J. B. Kaper.** 1992. Cloning of a gene (*zot*) encoding a new toxin produced by *Vibrio cholerae. Infect. Immun.* **60:**428–434.
4. **Beachey, E. H.** 1981. Bacterial adherence: adhesin-receptor interactions mediating the attachment of bacteria to mucosal surfaces. *J. Infect. Dis.* **143:**325–345.
5. **Bieker, K. L., G. J. Phillips, and T. Silhavy.** 1990. The *sec* and *prl* genes of *Escherichia coli. J. Bioenerg. Biomembr.* **22:**291–310.
5a. **Brown, R. C., and R. K. Taylor.** Unpublished data.
6. **DeGraaf, F. K.** 1990. Genetics of adhesive fimbriae of intestinal *Escherichia coli. Curr. Top. Microbiol. Immunol.* **151:**29–53.
7. **DiRita, V. J.** 1992. Coordinate expression of virulence genes by ToxR in *Vibrio cholerae. Mol. Microbiol.* **6:**451–458.
8. **DiRita, V. J., C. Parsot, G. Jander, and J. J. Mekalanos.** 1991. Regulatory cascade controls virulence in Vibrio cholerae. *Proc. Natl. Acad. Sci. USA* **88:**5403–5407.
9. **Doig, P., P. A. Sastry, R. S. Hodges, K. K. Lee, W. Paranchych, and R. T. Irvin.** 1990. Inhibition of pilus-mediated adhesion of *Pseudomonas aeruginosa* to human buccal epithelial cells by monoclonal antibodies directed against pili. *Infect. Immun.* **58:**124–130.
10. **Donnenberg, M. S., J. A. Giron, J. P. Nataro, and J. B. Kaper.** 1992. A plasmid-encoded type IV fimbrial gene of enteropathogenic *Escherichia coli* associated with localized adherence. *Mol. Microbiol.* **6:**3427–3437.
11. **Dums, F., J. M. Dow, and M. J. Daniels.** 1991. Structural characterization of protein secretion genes of the bacterial phytopathogen Xanthomonas campestris pathovar campestris: relatedness to secretion systems of other gram-negative bacteria. *Mol. Gen. Genet.* **229:**357–364.
12. **Dupuy, B., M.-K. Taha, A. P. Pugsley, and C. Marchal.** 1991. *Neisseria gonorrhoeae* prepilin

export studied in *Escherichia coli. J. Bacteriol.* **173**:7589–7598.

13. **Edman, J. C., L. Ellis, R. W. Blacher, R. A. Roth, and W. J. Rutter.** 1985. Sequence of protein disulfide isomerase and implications of its relationship to thioredoxin. *Nature* (London) **317**:267–270.

14. **Ehara, M., M. Ishibashi, Y. Ichinose, M. Iwanaga, S. Shimodori, and T. Naito.** 1987. Purification and partial characterization of *Vibrio cholerae* O1 fimbriae. *Vaccine* **5**:283–288.

15. **Elleman, T. C., and P. A. Hoyne.** 1984. Nucleotide sequence of the gene encoding pilin of *Bacteroides nodosus*, the causal organism of ovine footrot. *J. Bacteriol.* **160**:1184–1187.

16. **Ellis, R. J.** 1990. The molecular chaperone concept. *Semin. Cell Biol.* **1**:1–19.

17. **Faast, R., M. A. Ogierman, U. H. Stroeher, and P. A. Manning.** 1989. Nucleotide sequence of the structural gene, *tcpA*, for a major pilin subunit of *Vibrio cholerae. Gene* **85**:227–231.

18. **Fasano, A., B. Baudry, D. W. Pumplin, S. S. Wasserman, B. D. Tall, J. M. Ketley, and J. B. Kaper.** 1991. Vibrio cholerae produces a second enterotoxin which affects intestinal tight junctions. *Proc. Natl. Acad. Sci. USA.* **88**:5242–5246.

19. **Finkelstein, R. A., M. Boesman-Finkelstein, and P. Holt.** 1983. *Vibrio cholerae* hemagglutinin/lectin/protease hydrolyzes fibronectin and ovomucin: F. M. revisited. *Proc. Natl. Acad. Sci. USA* **80**:1092–1095.

20. **Freedman, R. B.** 1984. Native disulfide bond formation in protein biosynthesis: evidence for the role of protein disulfide isomerase. *Trends Biochem. Sci.* **9**:438–441.

21. **Freedman, R. B., N. J. Bulleid, H. C. Hawkins, and J. L. Paver.** 1989. Role of protein disulfide isomerase in the expression of native proteins. *Biochem. Soc. Symp.* **55**:167–192.

22. **Freter, R., P. C. M. O'Brien, and M. S. Macsai.** 1981. Role of chemotaxis in the association of motile bacteria with intestinal mucosa: in vivo studies. *Infect. Immun.* **34**:234–240.

23. **Galen, J. E., J. M. Ketley, A. Fasano, S. H. Richardson, S. S. Wasserman, and J. B. Kaper.** 1992. Role of *Vibrio cholerae* neuraminidase in the function of cholera toxin. *Infect. Immun.* **60**:406–415.

24. **Giron, J. A., A. S. Y. Ho, and G. K. Schoolnik.** 1991. An inducible bundle-forming pilus of enteropathogenic *Escherichia coli. Science* **254**:710–713.

25. **Goldberg, M. B., V. J. DiRita, and S. B. Calderwood.** 1990. Identification of an iron-regulated virulence determinant in *Vibrio cholerae*, using Tn*phoA* mutagenesis. *Infect. Immun.* **58**:55–60.

26. **Haas, I. G., and M. Wabl.** 1983. Immunoglobulin heavy chain binding protein. *Nature* (London) **306**:387–389.

27. **Hall, R. H., G. Losonsky, A. P. D. Silveira, R. K. Taylor, J. J. Mekalanos, N. D. Witham, and M. M. Levine.** 1991. Immunogenicity of *Vibrio cholerae* O1 toxin coregulated pili in experimental and clinical cholera. *Infect. Immun.* **59**:2508–2512.

28. **Hall, R. H., P. A. Vial, J. B. Kaper, J. J. Mekalanos, and M. M. Levine.** 1988. Morphological studies on fimbriae expressed by *Vibrio cholerae* O1. *Microb. Pathog.* **4**:257–265.

29. **Hanne, L. F., and R. A. Finkelstein.** 1982. Characterization and distribution of the hemagglutinins produced by *Vibrio cholerae. Infect. Immun.* **36**:209–214.

30. **Hase, C. C., and R. A. Finkelstein.** 1991. Cloning and nucleotide sequence of the *Vibrio cholerae* hemagglutination protease (HA/protease) gene and construction of an HA/protease-negative strain. *J. Bacteriol.* **173**:3311–3317.

31. **Herrington, D. A., R. H. Hall, G. Losonsky, J. J. Mekalanos, R. K. Taylor, and M. M. Levine.** 1988. Toxin, toxin-coregulated pili and the *toxR* regulon are essential for *Vibrio cholerae* pathogenesis in humans. *J. Exp. Med.* **168**:1487–1492.

32. **Hirst, T. R., J. Sanchez, J. B. Kaper, S. J. S. Hardy, and J. Holmgren.** 1984. Mechanism of toxin secretion by *Vibrio cholerae* investigated in strains harboring plasmids that encode heat-labile enterotoxins of *Escherichia coli. Proc. Natl. Acad. Sci. USA* **81**:7752–7756.

33. **Irvin, R. T., P. Doig, K. K. Lee, P. A. Sastry, W. Paranchych, T. Todd, and R. S. Hodges.** 1989. Characterization of the *Pseudomonas aeruginosa* pilus adhesin: confirmation that the pilin structural protein subunit contains a hyman epithelial cell-binding domain. *Infect. Immun.* **57**:3720–3726.

34. **Jonson, G., A. M. Svennerholm, and J. Holmgren.** 1992. Analysis of expression of toxin-coregulated pili in classical and El Tor *Vibrio cholerae* O1 in vitro and in vivo. *Infect. Immun.* **60**:4278–4284.

35. **Kabir, S., N. Ahmad, and S. Ali.** 1984. Neuraminidase production by *Vibrio cholerae* O1 and other diarrheagenic bacteria. *Infect. Immun.* **44**:747–749.

36. **Kaufman, M. R., J. M. Seyer, and R. K. Taylor.** 1991. Processing of TCP pilin by TcpJ typifies a common step intrinsic to a newly recognized pathway of extracellular protein secretion by gram-negative bacteria. *Genes Dev.* **5**:1834–1846.

37. **Kaufman, M. R., C. E. Shaw, I. D. Jones, and R. K. Taylor.** 1993. Biogenesis and regulation of the Vibrio cholerae toxin-coregulated pilus: analogies to other virulence factor secretory systems. *Gene* **126**:43–49.

38. **Kyte, J., and R. F. Doolittle.** 1982. A method for displaying the hydropathic character of a protein. *J. Mol. Biol.* **157**:105–132.

39. Lauer, P., N. H. Albertson, and M. Koomey. 1993. Conservation of genes encoding components of a type IV pilus assembly-two-step protein export pathway in *Neisseria gonorrhoeae*. *Mol. Microbiol.* **8:**357–368.

40. Manoil, C., and J. Beckwith. 1985. Tn*phoA*: a transposon probe for protein export signals. *Proc. Natl. Acad. Sci. USA* **82:**8129–8133.

41. Marrs, C., G. Schoolnik, J. M. Koomey, J. Hardy, J. Rothbard, and S. Falkow. 1985. Cloning and sequencing of a *Moraxella bovis* pilin gene. *J. Bacteriol.* **163:**132–139.

42. Mekalanos, J. J., R. J. Collier, and W. R. Romig. 1977. Simple method for purifying choleragenoid, the natural toxoid of *Vibrio cholerae*. *Infect. Immun.* **16:**789–795.

43. Meng, M., and H. P. C. Hogenkamp. 1981. Purification, characterization, and amino acid sequence of thioredoxin from Corynebacterium nephridii. *J. Biol. Chem.* **256:**9174–9182.

44. Meyer, T. F., E. Billyard, R. Haas, S. Storzbach, and M. So. 1984. Pilus genes of *Neisseria gonorrheae*: chromosomal organization and DNA sequence. *Proc. Natl. Acad. Sci. USA* **81:**6110–6114.

45. Michiels, T., and G. R. Cornelis. 1991. Secretion of hybrid proteins by the *Yersinia* Yop export system. *J. Bacteriol.* **173:**1677–1685.

46. Michiels, T., J.-C. Vanooteghem, C. Lambert De Rouvriot, B. China, A. Gustin, P. Boudry, and G. R. Cornelis. 1991. Analysis of *virC*, an operon involved in the secretion of Yop proteins by *Yersinia entercolitica*. *J. Bacteriol.* **173:**4994–5009.

47. Miller, V. L., and J. J. Mekalanos. 1984. Synthesis of cholera toxin is positively regulated at the transcriptional level by *toxR*. *Proc. Natl. Acad. Sci. USA* **81:**3471–3475.

48. Miller, V. L., R. K. Taylor, and J. J. Mekalanos. 1987. Cholera toxin transcriptional activator ToxR is a transmembrane DNA binding protein. *Cell* **48:**271–279.

49. Nunn, D. N., and S. Lory. 1991. Product of the *Pseudomonas aeruginosa* gene *pilD* is a prepilin leader peptidase. *Proc. Natl. Acad. Sci. USA* **88:**3281–3285.

50. O'Brien, A. D., M. E. Chen, R. K. Holmes, J. Kaper, and M. M. Levine. 1984. Environmental and human isolates of *Vibrio cholerae* and *Vibrio parahaemolyticus* produce a *Shigella dysenteriae 1* (Shiga)-like cytotoxin. *Lancet.* **1:**77–78.

51. Ogata, R. T., C. Winters, and R. P. Levine. 1982. Nucleotide sequence analysis of the complement resistance gene from plasmid R100. *J. Bacteriol.* **151:**819–827.

52. Ogierman, M. A., S. Zabihi, L. Mourtzios, and P. A. Manning. 1993. Genetic organization and sequence of the promoter-distal region of the *tcp* gene cluster of *Vibrio cholerae*. *Gene* **126:**51–60.

53. Oliver, D. 1985. Protein secretion in *Escherichia coli*. *Annu. Rev. Microbiol.* **39:**615–648.

54. Paranchych, W., and L. S. Frost. 1988. The physiology and biochemistry of pili. *Adv. Microb. Physiol.* **29:**53–114.

55. Parsot, C., E. Taxman, and J. J. Mekalanos. 1991. ToxR regulates the production of lipoproteins and the expression of serum resistance in Vibrio cholerae. *Proc. Natl. Acad. Sci. USA* **88:**1641–1645.

56. Pasloske, B. L., and W. Parancych. 1988. The expression of mutant pilins in Pseudomonas aeruginosa: fifth position glutamate affects pilin methylation. *Mol. Microbiol.* **2:**489–495.

57. Peek, J. A., and R. K. Taylor. 1992. Characterization of a periplasmic thiol:disulfide interchange protein required for the functional maturation of secreted virulence factors of *Vibrio cholerae*. *Proc. Natl. Acad. Sci. USA* **89:**6210–6214.

57a. Peek, J. A., and R. K. Taylor. Submitted for publication.

58. Perumal, N. B., and E. G. J. Minkley. 1984. The product of the F sex factor *traT* surface exclusion gene is a lipoprotein. *J. Biol. Chem.* **259:**5357–5360.

59. Peterson, K. M., and J. J. Mekalanos. 1988. Characterization of the *Vibrio cholerae* ToxR regulon: identification of novel genes involved in intestinal colonization. *Infect. Immun.* **56:**2822–2829.

60. Pugsley, A. P. 1993. The complete general secretory pathway in gram-negative bacteria. *Microbiol. Rev.* **57:**50–108.

61. Rhen, M., and S. Sukupolvi. 1988. The role of the *traT* gene of the *Salmonella typhimurium* virulence plasmid for serum resistance and growth within liver macrophages. *Microb. Pathog.* **5:**275–285.

61a. Rhine, J. A., and R. K. Taylor. Submitted for publication.

62. Rothbard, J. B., R. Fernandez, L. Wang, N. N. H. Teng, and G. K. Schoolnik. 1985. Antibodies to peptides corresponding to a conserved sequence of gonococcal pilins block bacterial adhesion. *Proc. Natl. Acad. Sci. USA* **82:**915–919.

63. Ruoslahti, E. 1988. Fibronectin and its receptors. *Annu. Rev. Biochem.* **57:**375–413.

64. Ruoslahti, E. and M. D. Pierschbacher. 1987. New perspectives in cell adhesion: RGD and integrins. *Science* **238:**491–497.

65. Russel, M. 1991. Filamentous phage assembly. *Mol. Microbiol.* **5:**1607–1613.

66. Ruvkun, G. B., and F. M. Ausubel. 1981. A general method for site-directed mutagenesis in prokaryotes. *Nature* (London) **289:**85–88.

67. Sastry, P. A., J. R. Pearlstone, L. B. Smillie, and W. Paranchych. 1983. Amino acid sequence of pilin isolated for *Pseudomonas aeruginosa* PAK. *FEBS Lett.* **151:**253–256.

68. **Schatz, P., and J. Beckwith.** 1990. Genetic analysis of protein export in *Escherichia coli. Annu. Rev. Genet.* **24:**215–248.

69. **Schoolnik, G. K., R. Fernandez, J. Y. Tai, J. Rothbard, and E. C. Gotschlich.** 1984. Gonococcal pili: primary structure and receptor binding domain. *J. Exp. Med.* **159:**1351–1370.

70. **Sharma, D. P., C. Thomas, R. H. Hall, M. M. Levine, and S. R. Attridge.** 1989. Significance of toxin-coregulated pili as protective antigens of *Vibrio cholerae* on the infant mouse model. *Vaccine* **7:**451–456.

71. **Shaw, C. E., K. M. Peterson, J. J. Mekalanos, and R. K. Taylor.** 1990. Genetic studies of *Vibrio cholerae* TCP pilus biogenesis, p. 51–58. *In* R. B. Sack and Y. Zinnaka (ed.), *Advances in Research on Cholera and Related Diarrheas*, vol. 7. KTK Scientific Publishers, Tokyo.

72. **Shaw, C. E., K. M. Peterson, D. Sun, J. J. Mekalanos, and R. K. Taylor.** 1988. TCP pilus expression and biogenesis by classical and El Tor biotypes of *Vibrio cholerae* O1, p. 23–35. *In* Switalski, Hook, and Beachey (ed.), *Molecular Mechanisms of Microbial Adhesion.* Springer-Verlag, New York.

73. **Shaw, C. E., and R. K. Taylor.** 1990. *Vibrio cholerae* O395 *tcpA* pilin gene sequence and comparison of predicted protein structural features to those of type 4 pilins. *Infect. Immun.* **58:**3042–3049.

74. **Strom, M. A., and S. Lory.** 1987. Mapping of export signals of *Pseudomonas aeruginosa* pilin with alkaline phosphatase fusions. *J. Bacteriol.* **169:**3181–3188.

75. **Strom, M. S., and S. Lory.** 1991. Amino acid substitutions in pilin of *Pseudomonas aeruginosa*: effect on leader peptide cleavage, amino-terminal methylation, and pilus assembly. *J. Biol. Chem.* **266:**1656–1664.

76. **Sun, D., J. J. Mekalanos, and R. K. Taylor.** 1990. Antibodies directed against the toxin-coregulated pilus from *Vibrio cholerae* provide protection in the infant mouse experimental cholera model. *J. Infect. Dis.* **161:**1231–1236.

77. **Sun, D., J. M. Seyer, I. Kovari, R. A. Sumrada, and R. K. Taylor.** 1991. Localization of protective epitopes within the pilin subunit of the *Vibrio cholerae* toxin-coregulated pilus. *Infect. Immun.* **59:**114–118.

78. **Sun, D., D. M. Tillman, T. N. Marion, and R. K. Taylor.** 1990. Production and characterization of monoclonal antibodies to the toxin coregulated pilus (TCP) of Vibrio cholerae that protect against experimental cholera in infant mice. *Serodiagn. Immunother. Infect. Dis.* **4:**73–81.

79. **Taylor, R., C. Shaw, K. Peterson, P. Spears, and J. Mekalanos.** 1988. Safe live *Vibrio cholerae* vaccines? *Vaccine* **6:**151–154.

80. **Taylor, R. K., C. Manoil, and J. J. Mekalanos.** 1989. Broad host-range vectors for delivery of Tn*phoA*: use in genetic analysis of secreted virulence determinants of *Vibrio cholerae. J. Bacteriol.* **171:**1870–1878.

81. **Taylor, R. K., V. L. Miller, D. B. Furlong, and J. J. Mekalanos.** 1987. The use of *phoA* gene fusions to identify a pilus colonization factor coordinately regulated with cholera toxin. *Proc. Natl. Acad. Sci. USA* **84:**2833–2837.

82. **Thayer, M. M., K. M. Flaherty, and D. B. McKay.** 1991. Three dimensional structure of the elastase of Pseudomonas aeruginosa at 1.5Å resolution. *J. Biol. Chem.* **266:**2864–2871.

83. **Tomb, J. F.** 1992. A periplasmic protein disulfide oxidoreductase is required for transformation of *Haemophilus influenzae* Rd. *Proc. Natl. Acad. Sci. USA* **89:**10252–10256.

84. **von Heijne, G.** 1985. Signal sequences: the limits of variation. *J. Mol. Biol.* **184:**99–105.

85. **Wickner, W., A. J. M. Driessen, and F.-U. Hartl.** 1991. The enzymology of protein translocation across the *Escherichia coli* plasma membrane. *Annu. Rev. Biochem.* **60:**101–124.

86. **Yu, J., H. Webb, and T. R. Hirst.** 1992. A homologue of the *Escherichia coli* DsbA protein involved in disulfide bond formation is required for enterotoxin biogenesis in *Vibrio cholerae. Mol. Microbiol.* **6:**1949–1958.

Vibrio cholerae and Cholera: Molecular to Global Perspectives
Edited by I. Kaye Wachsmuth, Paul A. Blake, and Ørjan Olsvik
© 1994 American Society for Microbiology, Washington, DC 20005

Chapter 14

Animal Models in Cholera Research

Stephen H. Richardson

It is probably fair to say that there are few mammals, birds, tissues, or cells that have not been inoculated with *Vibrio cholerae* or intoxicated with its classical exo-enterotoxin (CT). Our current knowledge of the disease and the highly effective oral treatment of cholera could not have been attained without the use of animal models and human volunteers. Research on cholera has largely been driven by a constant search for effective immunologic and/or pharmacologic interventions in the disease cycle and by the universal academic question: What is the real (normal) function(s) of CT? The in vivo verification eventually required of cholera investigators (even when using the most sophisticated molecular genetics techniques to produce microbial mutations) is affirmative answers to the questions (i) do the altered *V. cholerae* survive, colonize, and or replicate; (ii) do their modified virulence factors incite or inhibit fluid accumulation (i.e., diarrhea); and (iii) do the bacteria or their products stimulate or prevent a long-lasting protective immune response in a suitable animal model. In spite of the inevitable requirement for a test animal, cholera research has served as a prime example of how alternative in vitro methods and models can largely supplant the use of animals. These examples have been abundantly documented in several recent books and reviews covering the subject (2, 6, 22, 36–39, 88, 89, 114, 137, 150).

In this chapter the basic starting point will be 1965, shortly after I entered the cholera arena (8a, 122). The mid-1960s were about the midpoint of a decade of intensive animal experimentation and a period in which there was still uncertainty about how *V. cholerae* caused cholera. Selected key contributions of animal experimentation from that period will be highlighted and then correlated with our current knowledge. Several prophetic statements from that era will be quoted so that we will remember that "what goes around, comes around," or WGACA.

ANIMAL MODELS

Ever since the discovery of the etiologic agent of cholera, there have been attempts to develop suitable animal models for studying the interactions of *V. cholerae* with its accidental human niche. Up to the present, no animal model replicates all of the features of the human disease. However, a few models have overcome the difficulties inherent in studying a noninvasive human intestinal disease, and these have provided illuminating answers to some complex questions.

Stephen H. Richardson • Department of Microbiology and Immunology, Bowman Gray School of Medicine, Wake Forest University, Medical Center Boulevard, Winston-Salem, North Carolina 27157.

Except for humans and healthy adult animals modified by prior nutritional, chemical, antibiotic, or surgical manipulations, *V. cholerae* can successfully colonize and multiply only in suckling animals. This limitation is usually considered to be due to immature host defenses. Physiologic factors include low stomach acidity, a transient paucity of digestive enzymes, and increased permeability of the neonatal intestinal tract related to the infant's need to take up the mother's protective gift of colostrum. Within hours after birth, the intestinal tracts of uninfected suckling animals begin to be colonized by microorganisms that evolve in 7 to 10 days into a protective carpet of normal flora.

Initially, lactobacilli and streptococci from mother's milk are predominant. These microaerophilic organisms are soon displaced by facultative gram-negative rods in the small bowel, obligate anaerobes in the large bowel, and a variety of host-specific yeasts, spirilla, spirochetes, atypical bacteria, and other microorganisms in the stomach and the small bowel. These normal flora, many of which have never been cultivated in vitro, have been shown by a number of workers, particularly Freter (48) and Savage (132, 133), to provide the first line of protection against colonization and invasion by even the most virulent pathogens.

The two animals most frequently used in cholera research, rabbits and mice, are both coprophagic, maintaining a constant renewal of indigenous bacteria in their small bowel at all times. This is not true of humans, in whom the small intestine is only transiently colonized by bacteria from swallowed oral flora and other microorganisms taken in with food. One answer to the colonization problem appears to be the use of gnotobiotic animals, which have no natural flora. However, in addition to purchase and maintenance expenses, germfree animals are deficient in peristalsis and often exhibit abnormal immune responses to exogenous microbial infections. Germfree animals have been used only occasionally by investigators who have appropri-

ate facilities (131). Young animals, as pointed out above, can be readily colonized and infected with *V. cholerae*. However, because their intestines are immunologically and physiologically immature, the use of young animals is primarily restricted to short-term colonization, pathogenesis, and passive-immunity experiments. In spite of these handicaps, much progress has been made through the employment of infant mice and infant rabbits, as will be described in later sections of the chapter.

Adult animals, especially rabbits, have been utilized extensively in cholera research through surgical modifications of their intestinal tracts. Preparation involves tying off segments of the small intestine and then installing the test materials in the ligated loops. This procedure traps viable microorganisms and/or toxins in a restricted area of the intestinal lumen and artificially exposes the absorptive surface to supranormal concentrations of bacteria and antigens. False-positive loops are a common problem, as is the highly artificial environment created in the closed system. Ligated loops have also lost favor with investigators because of the recent regulations regarding postoperative observation and care of animals, which requires that the animals be kept under constant postsurgery observation for up to the 12 to 18 h required by the procedure. Dogs have been employed extensively in cholera research, but in most settings they are too expensive and require more-stringent standards of animal care.

Table 1 catalogs the major animal models that have been employed in cholera research and offers some examples of the valuable information provided by these models. For the most part, the models described are relevant to the disease caused when *V. cholerae* toxins or antigens are delivered into the lumen of the intestine. However, several exceptions that do not meet that criterion are included: (i) Ussing chamber experiments, in which freshly prepared animal tissues are used to measure ion fluxes, thus providing part of the basis for Phillips oral rehydration therapy

Table 1. Major animal models in cholera research[a]

Model (reference)	Preparation and use	Key observation or data
Fasted guinea pig (47)	Food is withheld for 4–5 days; oral challenge is with morphine or opium postfeeding; animals die in 18–24 h with cholera-like disease.	*E. coli* administered with *V. cholerae* increased animal's resistance to disease; oral passive immunity protects; i.p. passive and active immunity with high titer circulating Ab is not protective.

"It may thus be possible that other methods for active prophylactic immunization of human beings directed at maintaining high levels of coproantibody would give more favorable results." (47)

Mouse protection (31)	Adult mice are vaccinated i.p. with graded dose of test material; challenge is with 1,000 LD_{50} std *V. cholerae* in 5% mucin i.p. at 14 days; ED_{50} = dead or alive.	ED_{50} of experimental vaccines quantitated and compared with that of U.S. vaccine of known potency; vaccine strain NIH 41 was lysogenic for phage VcA1.
Ussing chamber (63)	Tissue mounted between fluid-filled, oxygenated chambers; microammeter measures ion flux (Na) between chambers; test material put on either surface; flux = short circuit current; later show that ammonium ion contaminant was the heat-stable element.	Cholera stool from patients inhibited Na transport; fluid from *Vibrio*-infected ileal loops contained similar activity; low- MW heat-stable agent in fluids, theophylline and cAMP have same effect (32); basis for oral rehydration therapy.

"Thus it is evident that studies on man must be made to determine the relative role of sodium transport inhibition and increased filtration or secretion to determine what actually transpires in severe, acute cholera in *man*. When such studies have been completed it will then be possible, we hope, to devise an animal model which will mimic or duplicate the state of affairs that occurs in the cholera victim. Until such studies have been undertaken and delineated, conclusions drawn from any models, while interesting in their own right, cannot be used to interpret the state of affairs which exist in cholera in man." (109)

Infant rabbit (25, 26)	10-day-old suckling rabbits are fed orally after gastric lavage or injected by laparotomy directly in gut; diarrhea, dehydration, and death in 18–24 h.	Live *V. cholerae* and bacterial lysates provoke cholera-like disease; 569B culture conditions were peptone, pH 8, 18 h, 37°C; lysates are prepared by sonication.
Ligated rabbit ileal loop (18–21, 46)	Adult rabbit ileum is divided into 5- to 10-cm segments by ligatures and injected with live vibrios or test materials; false positives recognized as chronic problem; after 12–24 h animal is sacrificed; loop length and fluid volume = ml/cm readout.	Live vibrios and culture filtrates of 5% peptone; medium, pH, oxygen, and temp are varied; optimum is shallow medium, pH 7.1–7.5, no CT at 6.5 or 8.1 titer (maximum at 37.5°C; less at 32°C); peak at 12 h.
Guinea pig (73)	Animals are starved for 4 days before *V. cholerae* is given orally followed with opium; bacteria increase up to 6 $\log_{10}$; dehydration or death is end point.	Corroborated human study; i.e., pathology of lethal cholera limited to lumen; vibrios carpeted intact villi per fluorescent-Ab stain.

Continued on following page

Table 1. *Continued.*

Model (reference)	Preparation and use	Key observation or data
Rabbit/guinea pig PF skin test (15, 16)	Shaved backs are injected with test materials in buffer for PF test or are mixed with std amt of antitoxin for Lb; at 18–24 h, Evans blue dye is injected i.v.; at 1 h, edema or induration is measured; toxin = BD 4–8-mm endpoint; Lb = 4-mm endpoint; BD4 = 30 pg.	Detected PF activity in stools of cholera patients that was neutralized by convalescent-phase sera; PF was found in *V. cholerae* peptone broth filtrates; Ab titrations were done using mixtures of test Ab and std toxin; toxin can be detected at picogram levels.
Canine model (130)	Adult dogs are fasted 2–5 days; challenge is with $NaHCO_3$ bolus with *V. cholerae* in Syncase medium per os; later variations include jejunal Thiry-Vella loops and cannulated loops with external ports; strains 569B, 0395, and B1307 are used at 10^9 CFU total.	Dogs with severe dehydration died at 4–14 h postfeed; 7/22 had typical cholera; 17/47 had clinical cholera; dogs fed suboptimal dose were partially or fully protected vs 2nd challenge; 4-fold titer increase in agglutination seen in all animals fed.
Rat loop model (1)	Rats are fed water and chow for 72 h and then 5% glucose for 72 h; two 10-cm loops/rat; 0.5 ml of test organism or toxin/loop without glucose feed; inhibitor of CT found in gut lumen.	12-h culture filtrates of 569B–2% peptone; avg in 48 rats was 3.0 ml/12.8 cm of loop; feces or gut wash fluid from untreated rats inhibited CT; rats used by Pierce and Gowens to define anti-CT-Ab-containing cell trafficking after gut challenge (112); CT incited fat cell lipolysis; cAMP was implicated in CT mechanism of action (56).
Infant mouse model (147)	Mice are inoculated per os with *V. cholerae* or test materials; at sacrifice stomach and entire gut are removed and weighed; gut wt/carcass wt = fluid accumulation ratio; overnight fast at 30°C improves reproducibility.	*V. cholerae* multiplied in intestines with fluid accumulation; mother's milk in stomach protected infants vs challenge; injection in full stomach used as std assay for *E. coli* heat-stable toxin.
Germfree adult mice (131)	Ogawa 395 or Inaba 569B was grown in peptone for 6–18 h; 10 μl is given per os (ca. 10^6 CFU); food and water are given ad libitum up to 14 wk with no illness; serum Ab monitored with time.	All mice colon cultures were positive; inverse relation between rising agglutinin titers toward Ogawa or Inaba serotype and switch to opposite serotype; high serum Ab had no effect on vibrio viability in lumen; prior Inaba vaccination hastened switch to Ogawa; immunosuppression prevented switch.

Continued on following page

Table 1. *Continued.*

Model (reference)	Preparation and use	Key observation or data
i.v. mouse assay for lethal toxin (17)	Mice are inoculated i.v.; no change at 24 h; increasing wt loss and death occur at 2 LD$_{50}$ (ca. 5 μg); wt loss is not due to diarrhea but likely to diuresis.	Lethal factor was converted to toxoid with formaldehyde given i.p. and protected mice vs 16 LD$_{50}$ at 6 wk (98) 0.02–2 μg of CT is adjuvant for SRBC response; >2 μg of toxin is selective for RES (12, 13).
Adult mouse (51)	Mice are fed 0.5 ml of test materials in 10% NaHCO$_3$; challenge is with 4 LD$_{50}$ i.v. or 10^8 569B inoculated into 6- to 12-cm loops, one loop/mouse; sacrifice is at 12–18 h; readout in mg/cm of fluid; footpad edema and Ab titers are also indicators of immune status; parenteral route gave variable results.	Demonstrated utility of immunocompetent adult mice for study of Ab response to CT, toxoid procholeragenoid, choleragenoid, and conventional whole-cell vaccine; 4 μg for detectable diarrhea is equivalent to human dose for purified toxin response.
Nonhuman primates (97)	30 rhesus monkeys are inoculated s.c. with 1, 2, and 50 μg of formalin toxoid from 569B and with similar toxin; toxoid assay by PF, lipolysis (i.v. in mice, i.v. and s.c. in dogs); reaction = skin thickness and erythema; antitoxin titration by Lb.	No reaction at 1 and 2 μg; 50 μg = 10- to 25-mm skin reaction (normal = 2 mm) by day 2; erythema = 35 mm; marked histologic changes like that with CT control; no reaction in monkeys immunized passively or actively; this toxoid later assayed in vitro; 50 μg of toxoid is equivalent to 200–400 ng of active CT; high antitoxin titers in toxoid revertants may be due to adjuvant effect of active toxin.
Reversible intestinal tie adult rabbit diarrhea (RITARD) rabbit (139)	Adult rabbit cecum is ligated through abdominal incision and removable tie is placed at ile-cecum junction at time of challenge; later modifications were cimetidine with NaHCO$_3$ at challenge, ip opium plus morphine at 30 min, and repeat opiates.	Challenge with 569B and 0395 titered; 10^3 CFU caused diarrhea; >10^3 CFU caused diarrhea and death; key to model appears to be retaining vibrios in small intestine long enough for colonization; Pierce et al. quantitated vibrio colonization with RITARD (113).
SAM (124–126)	Adult mice are fasted overnight, anorectal canal is occluded with cyanoacrylamide surgical adhesive; toxin or bacteria are fed in NaHCO$_3$ per os; paregoric is given ip; for toxins sacrifice is at 2–6 h; predose of 25 mg of Sm and Sm mutants and longer incubation for bacteria; (wt of fluid/carcass wt) × 1,000 = fluid accumulation.	Dose curve with CT = 3 to 4 μg minimum dose for response, 10 μg for full response; CT and LT reveal *H*-2-linked ETS phenotype also expressed in i.v. lethal toxin model; Tox$^-$ *V. cholerae* O1 strains cause fluid accumulation in SAM.

In 1982, the late W. E. van Heyningen, on hearing about the SAM model from me, replied, ". . . I like your mouse test, and I understand why you find it aesthetic [with respect to oral administration of test materials], but I doubt if a layman would find plugging a mouse arse with superglue and blowing up its gut as pleasing as you and I would" (148).

Continued on following page

Table 1. *Continued.*

Model (reference)	Preparation and use	Key observation or data
Backpack mouse model (154)	Adult mice are fed *V. cholerae* repeatedly; Peyer's patch cells are fused for MAb source; antivirulence factor MAb-producing clones are injected into mice to form tumors; 1-day-old mice are inoculated with hybrid cells and then challenged with 10^7 CFU of bacteria orally.	*V. cholerae* were found in or on M cells; anti-LPS MAb clone was used to induce tumors in adult mice; Ig secreted in gut binds to *V. cholerae* surface and cross-links; neonates inoculated with same hybrid cells were protected from lethal oral challenge with same serotype.
In vivo expression technology (87)	Auxotrophs are complemented with promoterless operon fusion to defective gene; promoters are supplied from chromosome library; growth in vivo selects for promoters and genes that function in vivo and can be selected in vitro for cloning.	Used *Salmonella typhimurium* as model; recovers genes that are required for in vivo survival; knockout of transduced *ivi* gene increased oral $LD_{50} > 4$ logs; model could be used for *V. cholerae* in gut or for M-cell or Peyer's patch enrichment.

[a]Abbreviations: Ab, antibody; BD, bluing dose; ED_{50}, 50% effective dose; i.p., intraperitoneally; Lb, limit of bluing; MAb, monoclonal antibody; MW, molecular weight; RES, reticuloendothelial system; s.c., subcutaneously; Sm, streptomycin; SRBC, sheep erythrocytes.

regimen and suggesting that cyclic (cAMP) might play a role in CT intoxication as well as reveal the newly discovered Zot and Ace toxins; (ii) the intradermal permeability factor (PF) assay, which is a simple yet sensitive assay for CT anti-CT titrations facilitating the biochemical manipulations involved in CT purification; (iii) the intravenous "lethal factor" mouse test, demonstrating the dual adjuvant and suppressive immunologic properties of CT and verifying the *H-2* linked ETS (enterotoxin sensitivity) phenotype in mice; and (iv) the rat fat cell assay, which supports the concept that cAMP is the signal sent by CT intoxication. The models, experiments, and my comments or quotes are presented in Table 1 in semichronological order and reflect my bias as to the importance of the information they provide.

IN VITRO TOXIN PRODUCTION AND REGULATION

By the mid-1960s, there was compelling evidence that *V. cholerae* produced a true exo-enterotoxin, but there was confusion about the importance of other factors (endotoxin, intracellular toxins, low-molecular-weight ion transport inhibitors, etc.) that also appeared to influence ion fluxes and fluid accumulation in the intestine (8a). The animal models in vogue at the time were the infant rabbit model, the ileal loop model, and the intradermal skin test (PF test). The PF test was useful because of its picogram sensitivity, but it was often believed to be measuring factors unrelated to CT (16).

Finkelstein clearly demonstrated the presence of CT (choleragen) and proceeded to work on purification of the material from Syncase medium, using infant rabbits as his biological assay (2, 34, 35, 41–43). Conflicting data concerning growth conditions for CT production had appeared in the literature. Intuitively, investigators believed that a rich medium incubated at 37°C at slightly alkaline pH would maximize CT production, and because of the strong interest in purifying CT, there was a powerful impetus to increase CT yields and further simplify growth media.

My group began our investigations with the goal in mind of maximizing CT production (9, 10, 28, 29, 82, 119–123, 128). Monitoring the effects of variations in growth conditions on toxin yields, assaying column fractions, and quantitating other purification maneuvers were greatly facilitated by the use of the PF test, which allowed titration of as many as 100 samples on the back of a single rabbit. Initially, these tests confirmed earlier studies demonstrating that lower temperatures were optimum for toxin production from a number of strains of *V. cholerae*. It was also found that strong aeration and an initial pH of 6.5 enhanced CT production and release, which were correlated with a rise in pH of the growth medium (9). These seemingly odd requirements for producing high CT titers remained unexplained until the discovery of the two-component sensory-regulatory system 20 years later by Mekalanos and his colleagues (89, 93, 94) and Colwell and Spira's ecological investigations (14), which reminded many of us working in cholera research that *V. cholerae* is really out of its niche in the human intestine.

The Bowman Gray group then showed that the nutritional requirements for toxigenicity could be reduced to four amino acids plus a carbon source and salts and ultimately to a medium that contained only glucose and asparagine as sources of carbon and nitrogen. The same laboratory also investigated other factors such as strain differences (e.g., 569B, which carries multiple copies of the *ctxAB* operon, made high levels of CT under all growth conditions), mucinase (rediscovered by them in the guise of "ion translocase inhibitor"), ATPase, neuraminidase (NANase), and protease production. They showed that synthesis of these "contaminants" was repressed in the simpler culture medium and that CT denaturation was minimal (121). In retrospect, decreasing the level of "contaminating substances" is counterproductive if stimulating a complete protective library of mucosal antibodies against *V. cholerae* and its antigens is the goal. These studies are an example of the "CT and anti-toxic immunity is the complete answer" philosophy extant at the time. By the late 1960s, most laboratories

involved in CT purification were employing immunologic assays; e.g., Finkelstein used monospecific horse serum (34), and Richardson et al. used passive hemagglutination inhibition (60) in place of animals, which were reserved for monitoring biological activity after CT had been biochemically purified.

In 1969, CT was purified by Finkelstein and LoSpalluto (42, 43). Their protocols, combined with those developed by Spyrides and Feeley (140), and Richardson and Noftie (128), were then used by a group headed by Ruth Rappaport at Wyeth to produce the first large-scale, semipurified, endotoxin-free toxin for use in vaccine trials as a toxoid vaccine (115–117). The toxoid was tested in all the standard animal models prior to administration to human volunteers. As indicated above, only the primate model (and human volunteers) revealed evidence of toxoid reversion to toxin (97).

Purified CT was made available through Finkelstein, the National Institutes of Health, and commercial sources to a wide circle of scientists who began investigating the physicochemical and immunologic properties of the toxin. An immediate observation was that CT appeared to be made up of subunits of distinct molecular weights. Holotoxin was active in the PF, ileal loop, and mouse models, but the subunits were not. Holmgren in Sweden labeled the subunits H and L for heavy and light and showed that only the L subunits were immunogenic (85). In the United States the subunits were labeled A for active and B for binding. The terminology melded into A and B, and it became apparent that A was the biologically active subunit of the toxin and that B was involved in binding (40, 54, 99).

At the same time, Parker et al., working in Romig's lab at the University of California Los Angeles (UCLA), constructed a detailed genetic linkage map of *V. cholerae* (102). This basic information provided the foundation for targeted genetic manipulation of cholera vibrios to create CT⁻ mutants or to obtain higher yields of toxin. For example,

Mekalanos, also working with Romig and Collier, devised an affinity filter screening method to identify hypo- and hypertoxigenic mutants on a large scale. Nitrosoguanidine (NTG)-induced hypertoxigenic mutants producing 3.5- to 4.5-fold more toxin than the parent 569B where isolated and tested in ileal loops, where they multiplied and evoked differential levels of fluid accumulation. The authors pointed out that even though in vitro toxigenicity was high, other properties of the mutants or the in vivo environment might influence overall secretion (was this early evidence of global regulation thinking?) (90, 91).

Concomitant with toxin purification, attempts were made to derive mutants of *V. cholerae* that would produce altered toxins or natural toxoids. Initially, these studies involved the use of potent mutagens like NTG that often result in multiple mutations. Also, because of the logistics of selecting isolates from thousands of colonies, various in vitro assays for toxinogenicity were employed for screening. Using these techniques, Vasil et al. identified hypotoxigenic mutants that appeared to be *tox* mutants and that were able to colonize rabbit ileal loops and infant rabbits without eliciting fluid accumulation. Prototype strain M13 showed no evidence of reversion to toxigenicity on multiple passage in infant rabbits and was able to survive in the gut for up to 48 h (151). Later, this strain was found to be a regulatory mutant that produced low levels of toxin when assayed by the more sensitive PF and CHO cell assays (127) and reverted to full toxigenicity when it was tested in volunteers (39).

Honda and Finkelstein isolated an $A^- B^+$ mutant (Texas Star-SR) of El Tor Ogawa 3083 by using NTG mutagenesis. The mutant was nontoxic in infant rabbits, rabbit ileal loops, and the PF test; after sequential infant mouse passages; and in a variety of in vitro assays. Vaccination of chinchillas (4) with 4.6×10^9 CFU of Texas Star protected them against fluid accumulation and colonization with 6.8×10^8 CFU of wild-type parent

3083 challenge at 20 days after immunization (62).

Mekalanos et al. used derivatives of bacteriophage VcA1, originally isolated from the NIH41 standard vaccine strain (153), to construct enterotoxin structural gene deletion mutants in *V. cholerae* RV79. They reasoned that deletion mutants would not be able to revert in vivo. CT^- mutants were screened using the homology between *Escherichia coli* heat-labile enterotoxin gene fragments and CT. After eliminating auxotrophic mutants, which do not colonize or survive well in vivo (e.g., the wild-type *Salmonella* strain used to construct the in vivo expression technology model), three candidate isolates were tested in ileal loops. The candidates produced no fluid accumulation but grew to cell densities ranging from 2×10^9 to 8×10^9 CFU/ml of intestine. The authors also disclosed that they had cloned from 569B a 5.1-kb piece of DNA containing the genes for CT, inactivated the A subunit, and expressed the B subunit free of A in *E. coil* (92).

Building on these preliminary successes, several groups, notably those at Harvard and at the University of Maryland Center for Vaccine Development (CVD), have engineered a number of $A^- B^+$ and $A^- B^-$ mutants as putative vaccine strains. M13, Texas Star, the $A^- B^+$ strain described above, and a number of other putative vaccine strains have been tested in humans in the CVD volunteer facility and in field trials (65, 105). Even though vaccine development relies heavily on animal assays, these investigations will not be discussed further in this chapter, because they will be covered in detail in other sections of this book.

In an ironic turnabout, two new toxins, Zot (zonula occludens) and Ace (accessory cholera enterotoxin), revealed themselves initially by causing adverse reactions in CVD human volunteers following administration of two different $A^- B^+$ vaccine candidate strains. Like the CT toxoid cited above, both strains had already gone through exhaustive animal testing (78). Unlike filtrates of the wild-type

O395 parent, overnight culture filtrates from the offending CVD101 and 395N1 mutants showed no evidence of fluid accumulation in rabbit ileal loop or suckling mouse assays. The filtrates did, however, cause increased tissue conductance in Ussing chamber preparations of rabbit ileal tissue. In spite of temporal and quantitative differences in the electrical responses, O395 and mutant filtrates were shown by light and electron microscopy to increase conductance by loosening the intracellular tight junctions (zonula occludens) in the test strips. The activity was heat labile, reversible, and not due to loss of tissue viability or to cAMP-mediated effects. The *zot* gene was cloned and localized to a 1.3-kb open reading frame whose 3′ end overlaps the 5′ end of the *ctx* genes coding for CT (5, 30).

The CVD group then investigated the possibility that the two additional open reading frames in the "core region" coding for *zot* and *ctx* might contain genes for other toxins responsible for the adverse side effects seen in the volunteers. The region upstream from *zot* was cloned and tested in animals and Ussing chambers. Culture supernatants from constructs devoid of the *ctx* and *zot* genes elicited a significant increase in short circuit current secondary to an increase in potential difference in Ussing preparations. CVD110, containing a plasmid with an insert carrying the region upstream of *zot* and *ctx*, also caused fluid accumulation in adult rabbit ileal loops and in a sealed infant mouse model (52a). Sequence analysis placed the *ace* gene upstream from *zot* and *ctx* (146). *zot* and *ace* are discussed extensively in chapter 11 by Kaper et al., who suggest that this region of the chromosome, which is flanked by insertion sequences, represents a "virulence cassette" in *V. cholerae*.

Of related interest were data showing that the sequences of both *zot* and *ace* have homology with bacteriophage, bacterial, and eukaryotic sequences known to code for ATP binding, pore forming, calcium transport, or ATPase activities. Hydrophobicity plots suggested that they may be localized in the membrane of *V. cholerae* (69). In 1965 it was reported that filtrates from 569B and several other standard strains grown under the conditions employed in the CVD studies of *zot* and *ace* possessed ATPase activity and caused loss of Na from Na-loaded erythrocytes against a salt gradient (122). The *V. cholerae* strains employed in that study are now all known to be lysogenic for defective bacteriophage, and one, NIH41, was the source of VcA1 used to make the first A$^-$B$^+$ deletion mutant (92). The ATPase was localized to the membrane of the vibrios by electron microscopic histochemistry and was shown to be maximally released under growth conditions that result in lysis of *V. cholerae*. Release of the ATPase appeared to be unrelated to CT biosynthesis and release (i.e., not coregulated) (66). In the original publication (122) it was suggested that "the unique ATP-ADP-AMPase activity demonstrated in filtrates is part of a constellation of previously demonstrated enzymes such as neuraminidase, lethinase, mucinase, phospholipase, etc., which together in a concerted effort could (theoretically) exhibit choleragenic properties—this hypothetical situation has no basis in experimental fact and is strictly conjecture. We are now studying various aspects of this model in the hope of demonstrating whether or not the ATPase of *V. cholerae* plays a direct or indirect role in the virulence of supposedly choleragenic strains of the organism." (WGACA).

Sixteen of 32 clinical and environmental isolates of CT$^-$ *V. cholerae* O1 were shown by Moyenuddin et al. (95) to elicit strong fluid accumulation responses in sealed adult mice (SAM). The SAM activity, which was maximum at 5 h postfeeding, was not consistently related to cytotoxicity in CHO, Vero, or HeLa cells. Because the bacteria are probe negative for heat-labile, heat-stable, and Shiga-like toxins as well, their pathogenesis may involve new mechanisms. It is possible that these O1 serotypes represent cholera vibrios whose virulence cassette has been lost. This could be tested for by probing for the residual "signature" insertion sequence that is left behind when the excision event occurs spontaneously. The nontoxigenic O1 *V. cholerae* is treated in full in chapter 5.

Several other related "new" toxins are on the horizon. Three groups of environmental vibrios have been implicated in occasional episodes of diarrhea following accidental oral intake by humans. The spectrum of toxic activities of these vibrios is unrelated to their source, whether clinical or environment (84). Noncytotoxic enterotoxins from the halophilic opportunistic pathogens *V. fluvialis* and *V. hollisae* elongated CHO cells and caused fluid accumulation in the infant mouse model. Both of these species lack *ctx* and CT-like gene sequences (70, 71, 83, 84). Their enterotoxins have been purified and do not react with antibodies against CT, heat-labile enterotoxins, or heat-stable enterotoxins (91a). Purified toxins do not bind to GM_1 and do not ribosylate or activate either adenylate or guanylate cyclase. These CHO cell elongation factors are the first examples of a CHO cell elongation agent whose activity is not cAMP mediated. It appears that these enterotoxins represent a class of toxins with a different mechanism of action. Enterotoxin activity has also been reported from *V. mimicus* (138). In studying these toxins, the estaurine origin of the vibrios must be remembered (14).

The current exciting studies on the regulation of *V. cholerae* virulence are part of a long series of investigations begun at UCLA by Mekalanos and continued in the Murphy laboratory at Harvard. It was demonstrated that CT production is under the regulatory control of a cascade of two-component sensory-transcription factors that also coregulate a wide array of other potential virulence factors (88, 93, 94). The coregulated genes constitute a regulon. This regulation is the subject of chapter 12. In light of the regulon concept, the observations by Dutta et al. (26), De et al. (21), Finkelstein (39), Craig (16), and Richardson et al. (9, 121) that more CT was produced at lower temperatures and lower pHs and that the nutrient requirements were quite specific finally have an explanation.

For example, after 25 years, the temperature effect has recently been explained at the genetic level. ToxR, the key transcriptional activator of the regulon, has been found to be in a regulatory link with a heat shock (stress) protein. In its simplest form, heat shock gene expression increases with increasing temperature and ToxR expression decreases, which in turn shuts off transcription of the *ctx* and other virulence genes (89, 103). The authors speculate that other signals (stomach acid, alkaline pH of the duodenum) might also regulate the stress protein in a manner that might delay "turn on" of CT and pilus synthesis until the vibrios reach the surface of the enterocytes or M cells. In retrospect, it is interesting that most cholera researchers fell in love with toxins and forgot that the vibrio that was producing them is an environmental organism and does not inhabit the intestinal tracts of humans simply to make them miserable.

It should be pointed out that in vaccine development, the final test of the ingenuity of the genetic engineers is whether the putative vaccine strains penetrate, survive, colonize, replicate, and immunize the host with no overt pathogenesis. These parameters can be assessed only in vivo in suitable animal models or in humans. With that in mind, we begin the next section with the satisfaction of knowing that *zot, ace*, and the last two additions to the preceding list of animal models were developed in the laboratories of two dedicated microbial geneticists (or maybe *Vibrio* biologists), Kaper and Mekalanos, who are well aware that the answers are not all on gels or in Eppendorf tubes.

PENETRATION AND ATTACHMENT

In order to be pathogenic, *V. cholerae* needs to establish itself in the intestinal tract. Even in humans, where there is little indigenous normal flora in the upper intestinal tract, a number of protective elements keep the cholera vibrios from colonizing the intestine. Enteric pathogens must first survive the acid pH of the stomach and the alkaline pH and high levels of digestive enzymes of the duodenum, which they often can do in a bolus of food. Once they reach the small intestinal tract, they have to deal with a rapid peristaltic activity that tends to expel them toward the exit. The normal mucosa is blanketed with a mucous layer of glycoprotein that is constantly being renewed by the goblet cells of the gut while the older layer is being swept downward by peristalsis. That this layer had to be penetrated before vibrios could be virulent was realized early on by Jenkin and Rowley (64) and Burnet and Stone (8), who demonstrated that cholera vibrios produce NANase and mucinase. Lankford and Legsomburana (75, 76) found that mucinase was produced at a high titer only in complex media, and Evans found that the enzyme (which we thought was an ion translocase inhibitor) was inducible by high-molecular-weight substances found in peptone and in extracts of glycoprotein from rabbit intestinal mucosa (28, 29, 119, 120).

Using an ileal loop model, Schrank and Verwey proposed that agglutination of vibrios by anti-somatic-antigen antibodies prevented their penetration of the mucous layer and hastened subsequent removal by peristalsis (134). Agglutination by somatic antibodies would be expected to be more efficient in *V. cholerae* than in some other bacteria, because the lipopolysaccharide (LPS) layer extends into the sheath that surrounds the polar flagellum (-a) and might even result in impairment of motility. Nelson et al. (96) compared penetration and adherence in infant and adult rabbits. *V. cholerae* carpeted the infant intestine but appeared to localize in patches of M cells in the ligated adult loop. Bacterial numbers reached a peak 5 to 7 h after inoculation but were largely cleared by 12 to 16 h (96). Most studies agreed that motility was important for

reaching the enterocyte surface, but no mechanisms were defined.

The role of chemotaxis in the ability of *V. cholerae* to penetrate the mucous gel was investigated in germfree and infant mice and in rabbit ileal loops by Freter et al. Spontaneous chemotaxis-negative mutants were isolated from the streptomycin-resistant CA401 parent and differentiated and quantitated on special media by colonial morphology. The main positive chemoattractants were amino acids and mucosal extracts. In infant mice, the non-chemotactic mutants survived longer but ultimately disappeared. It was postulated that penetration of the mucus might be detrimental to the bacteria because of an increasing gradient of host protective factors near the epithelial-cell surface. In rabbit and germfree-mouse loops, the chemotactic parent displayed a distinct advantage and outgrew the mutant. Penetration is nonspecific and guided by chemotaxis; therefore, bacteria may follow "channels" in the thick gel formed as it is released by the crypt cells (49, 50). Motile chemotactic vibrios are clearly superior in being able to penetrate the mucous barrier, but few details concerning chemotaxis in *V. cholerae* are available. Buried vibrios would have a distinct advantage in terms of "hanging on" until their more-specific attachment mechanisms are armed and ready to function. Factors present in the mucous layer may play an important role in signaling the ToxR regulon to prepare for adhesion to the underlying cell surface.

For example, Leitch (77) demonstrated with ileal loops that mucous secretion reached a peak at 3.5 h postinstillation of CT and then rapidly declined. Dibutryl cAMP or theophylline also increased mucous secretion, but secretion was delayed. The mucous blanket thickened and remained thick for the duration of the experiment. A new acidic glycoprotein appeared in the mucous and continued to be secreted in parallel with CT-induced fluid accumulation. Perhaps the acidic glycoprotein should be tested for its effects on ToxR.

Richardson (118) recently assessed the contributions of flagellar structure and motility to *V. cholerae* pathogenesis in rabbit loops, RITARD, and infant mice. Nonmotile mutants were constructed by transposon insertion mutagenesis. Three types of mutants were obtained: nonmotile flagellate, nonmotile aflagellate, and nonmotile aflagellate with vestigial sheath. In all three models, the motile parents outperformed the mutants in terms of pathogenesis measured by fluid accumulation, CFU per milliliter of fluid, and adherence. Competition experiments pitting the mutants against the parent strains and against each other revealed that motility was the main contributor to enhanced virulence. In the RITARD, the nonmotile mutant bearing residual sheath material incited stronger diarrhea that persisted longer than that incited by the sheathless mutant, indicating that flagellar structure (or LPS) may play a subtle role in virulence.

COLONIZATION

Once the vibrios have penetrated the mucous layers and the glycocalyx, they still have to approach and adhere to the surface of the epithelial cells. Colonization studies are largely dependent on animal models, because the complex interactions between the two dynamic living systems are impossible to duplicate in vitro. Most early microscopy studies of *V. cholerae* did not reveal known attachment elements such as pili, so many workers studied cell-bound adhesins. Lankford and Legsomburana pioneered biochemical and physiologic studies on vibrio hemagglutinins, working out conditions for maximum expression and partial purification (75, 76). Their efforts were extended by Finkelstein and Mukerjee, who demonstrated that El Tor vibrios and some classical vibrios produced hemagglutinins that were specific to particular animals cells. For example, El Tor vibrios agglutinate chicken erythrocytes under specific conditions when classical vibrios do not.

This is a useful test, because some El Tor strains are no longer hemolytic, hemolysis being the original definition of the biotype (44).

Baselski and her colleagues employed the infant mouse model to select for NTG-induced colonization and pathogenesis mutants of *V. cholerae*. Their approach was to screen mutants for decreased survival, replication, or colonization in vivo and then identify the phenotype. One of the major mutations appears to be a protease deficiency that might be linked to effective secretion of exoenzymes. It was also found that protease-negative strains were NANase defective. One revertant simultaneously regained both activities, suggesting that protease and NANase activities were somehow linked. The protease mutants grew to lower cell densities in vivo but not in vitro. Other mutations that decreased virulence were rough colony type; absence of *tox*, chemotaxis, mucinase, or NANase; and nutritional auxotrophy. A unique mutant, designated Shr, was deficient not only in Tox but in several other extracellular products as well. In retrospect, this isolate might have carried a regulatory mutation in ToxR, ToxT, ToxS, or one of the other components of the ToxR regulon (3).

Pierce et al. (113) assessed the role of CT itself in colonization by using the opium-treated version of the RITARD rabbit model. It has been speculated ever since the identification of the CT receptor that the high affinity of CT B for GM_1 on the enterocyte surface might enhance *V. cholerae*-GM_1 attachment during the actual disease, when the microorganisms are tightly interfaced with the cell surface. Wild type and A^-B^+ and A^-B^- mutants were fed to rabbits, and the extent of colonization for each was determined by viable colony counts. Colonization increased up to 30-fold in the A-subunit-containing strains compared to the A^-B^- or A^-B^+ mutants. These results indicate that the B subunit has no major effect on vibrio adhesion. CT added concomitantly with the latter mutants increased their colonization to parental levels. Pierce et al. suggest that hypertoxinogenic strains might be selected in vivo because of their enhanced abilities to attach. It is also possible that toxin attachment selects for M-cell sites for colonization because viability is a requirement for M-cell interaction. Additional studies with rabbits demonstrated that the only factor consistently linked to effective immunogenicity was strong colonization (113). An important observation by Owen et al. (101), who used rabbit ileal loops, was that viable but not dead *V. cholerae* were phagocytosed by M cells over the Peyer's patches and introduced into the underlying lymphoid tissue. It was also noted that CT is not involved in uptake (101). Yamamoto and Yokota (155) corroborated these findings in fixed human tissue by demonstrating that *V. cholerae* grown for 3 h at 37°C showed heavy attachment to the microfolds of M cells. The same bacteria grown at 37°C for 20 h or heat treated did not adhere well (155). These studies have major implications for strategies of vaccine development, but to date, little research has addressed this important area.

Using the powerful newly developed tool of transposon inactivation with Tn*phoA*, Taylor et al. (145) created a library of gene fusions between alkaline phosphatase and genes of *V. cholerae* that code for secreted proteins (i.e., putative virulence factors or surface antigens). On preliminary screening by polyacrylamide gel electrophoresis, several mutants were identified that had lost bands that corresponded to proteins in the wild-type O395 strain that were coregulated with CT (i.e., on at pH 6.5 and 30°C; off at pH 8.0 and 37°C). One of the missing bands corresponds to the major subunit of *V. cholerae* pilus protein. This pilus mutant, an outer membrane protein (OMP) mutant, and a variety of *tox* mutants were tested in infant mice for colonization competence by using the Freter competition assay in which mutants are pitted against the genetically marked wild-type parent by simultaneous feeding of various ratios of the two test organisms.

Pilus, OMP, and *tox* mutants had low competitive indexes and 4- to 5-log-higher 50% lethal doses (LD_{50}s). These pili were called TCP, or toxin-coregulated pili. Recently, antibodies against some of these OMP proteins were shown by other workers (136) to protect infant mice against colonization. Taylor et al. stressed the importance of proper growth conditions for the expression of essential colonization factors in the preparation of vaccine components. These studies could not have been done without the aid of an animal model in combination with the advancing technology of molecular genetics.

The same mouse competition system was employed by Peterson and Mekalanos (108) to screen another Tn*phoA* insertion library for a battery of uncharacterized genes essential for colonization. Appropriate isolates were screened for expression under conditions promoting high or low CT production. One of the on-off conditions employed was minimal medium with and without asparagine, which was shown earlier to be the one essential amino acid required for CT production (WGACA) (81). In addition to identifying new TCP mutants, a new class of minimal medium mutations called accessory colonization factors *(acf)* that caused a marked drop in the mouse competitive index was found. The mouse competitive index was also used by Goldberg et al. to identify iron-regulated membrane proteins of *V. cholerae* that result in a colonization defect and a higher LD_{50} in mutants missing the IrgA protein (55).

Sun et al. (141, 142) purified the major pilin subunit TcpA from O395, produced a polyclonal antibody directed against it, and used the antibody to protect infant mice against challenge with a classical Inaba and an El Tor Ogawa *V. cholerae*. In more recent studies, the protective epitope within the pilin subunit of TCP has been identified by using monoclonal antibodies and antigenic cross-reactivity with other type IV pili (104). The authors assert that in addition to eliciting protective antibodies, TCP is ubiquitous and may represent a common antigen that should be included in any proposed vaccine preparation. They employ these data to defend their position that TCP protects against El Tor infection, a position that has been questioned by other investigators.

Osek et al. are among those who have challenged the universal role of TCP in colonization, especially with regard to the El Tor biotype. This group described the mannose-binding pilus on El Tor earlier and showed that even though the El Tor biotype carries the *tcpA* gene, the gene is not functionally expressed in vivo (100). Polyclonal and monoclonal antibodies were raised against a purified mannose-sensitive hemagglutinin (MSHA). Specificity was analyzed by immunoblotting, and it was shown that anti-LPS recognition was low or absent. The polyclonal antibody protected 10 of 12 infant mice against El Tor challenge but only 2 of 6 against challenge with a classical strain. Results with the monoclonal antibody were 8 of 12 and 1 of 6, respectively. The protective effect for Fab fragments was calculated to be 64% of that for total anti-MSHA. The protective effect of the antibodies against El Tor challenge in rabbit ileal loops was found to be 14 to 80 times greater than that of control sera against specific antisera. The same antibodies did not protect any better than controls when classical strains were employed in the challenge. This question will be explored further in chapter 17. The TCP versus MSHA question may go a long way toward explaining why the untoward susceptibility of blood group O individuals to severe cholera is manifested only in persons infected with El Tor *V. cholerae*.

Thus, it appears that colonization by *V. cholerae* is much more complex than previously imagined, and recent findings only confirm the comment by Lankford and Legsomburana (76) during their attempts to purify the hemagglutinin: "the demonstration that preparations of the slime envelope material (bound hemagglutinin) and of the culture supernatant hemagglutinin produce certain toxic effects in mice and rabbits raises the question as to whether they play active and possibly multiple roles in tissue virulence, in certain experimental infections and in man." (WGACA).

Interaction of CT with Enterocytes and Other Cells

Many attempts to block the activity of crude CT with a variety of natural and synthetic materials have been made, but none gave conclusive results. Once CT was available in a purified form, rapid progress was made in attempts to block its activity. The initial significant finding was that in 1971 by van Heyningen et al. (149), who demonstrated that crude mixtures of gangliosides were able to block the effects of multiple doses of CT in the PF skin test, rabbit loops, and rat epididymal fat cells (the first direct biochemical assay for CT-induced cAMP accumulation). King and van Heyningen (68) then showed that the principal factor in the mixed gangliosides was a sialidase-resistant monosialoganglioside, GM_1. GM_1 inhibited CT in the PF skin test and decreased the production of cAMP in guinea pig mucosa to basal levels. Treatment of di- and tri-sialogangliosides with NANase increased their CT-inactivating capacity. King and van Heyningen conjectured that "if the reaction between mucosal cells and CT is relevant to the diarrhoeagenic action of the toxin, then it is possible that the sialidase produced by *V. cholerae* may affect the action of the toxin" (68).

Holmgren et al. (61) purified GM_1 to homogeneity and showed that GM_1 and CT formed precipitin complexes in Ouchterlony gels. GM_1-CT mixtures were titrated for activity in the PF and rabbit ileal loop assays. A volume of 100 ng or 100 pg of GM_1 was sufficient to neutralize the action of 3 μg of CT in the ileal loop or PF test, respectively. The same investigators measured the GM_1 contents of intestinal mucosae from humans, cattle, pigs, and rabbits and then looked at the effects of preincubating the tissues with

GM$_1$ prior to exposure to CT. In each case, the amount of CT bound was proportional to the amount of GM$_1$ added. Ileal loops pretreated with GM$_1$ showed enhanced fluid accumulation responses to CT, but when the loops were pretreated with choleragenoid (a naturally formed toxoid of CT) (42, 106, 110), fluid accumulation responses were markedly inhibited. To retest an old theory, ileal loops were preincubated for up to 200 min with NANase prior to CT challenge. However, sialidase treatment did not alter the fluid accumulation response or the inhibition of CT binding by choleragenoid, nor was more GM$_1$ released in wash fluids. It was concluded that NANase had no effect on the fluid accumulation response when purified CT was used to intoxicate rabbit loops.

In 1983, Richardson et al. developed the SAM model for CT and showed that strains of mice possessing increased amounts of GM$_1$ are more sensitive than other strains to CT. It also appeared that non-GM$_1$ gangliosides inhibit the effects of CT in strains of mice that produce increased amounts of these CT-binding materials. They also showed that this ETS phenotype is *H-2* linked, as are the genes for ganglioside biosynthesis (72, 124, 126). The implications of these studies are considered in detail in chapter 17.

Galen and coworkers tested wild-type and isogenic NANase-negative strains of *V. cholerae* and their culture filtrates in Ussing chambers and suckling mice (52, 53). In contrast to the results of Holmgren et al. (61) they found a 65% increase in short circuit current between the NANase$^+$ and NANase$^-$ mutants and an 18% increase in fluid accumulation in suckling mice fed 10^9 CFU of the NANase$^+$ wild-type bacteria. Fibroblasts derived from C57BL/6 and C3H mice showed eight- and fivefold increases in fluorescein-labeled CT binding when analyzed by flow cytometry. It may be that NANase presented in the context of a *V. cholerae* infection functions differently from a bolus of CT in a closed ileal loop.

The simultaneous discoveries that CT bound to GM$_1$ and stimulated cAMP responses spurred the development of numerous in vitro assays for CT. These two monumental findings probably did more to decrease dependence on animal studies in cholera research than any other event before or since.

In Vivo Activities of CT

In the late 1960s, Field et al. (32, 33) demonstrated that purified CT had activities in Ussing chamber preparations that resembled those of the newly discovered second messenger cAMP and that of theophylline, an inhibitor that prevents cAMP breakdown. This observation prompted Greenough et al. to test the ability of CT to release glycerol from isolated rat fat cells, which was a standard assay used to measure cAMP biological function (56). The lipolysis assay was positive and had a sensitivity equal to or better than that of the rabbit loop test and about 10-fold higher than that of the PF test. These experiments confirmed the key role of cAMP in the action of CT and started an avalanche of reports on the effects of CT on various biological systems. Evidence for increased levels of cAMP in dogs, rabbits, and humans intoxicated with CT were presented (23, 35, 57).

At the same time, other evidence suggesting that CT-stimulated prostaglandin synthesis was the important factor in the active secretory process was accumulating. This hypothesis was based on the observations that indomethacin and aspirin inhibited fluid accumulation. For example, in a study of rat intestinal loops, Roomi et al. (129) demonstrated that CT-induced fluid secretion and mucous secretion are activated by different mechanisms. They suggested that attention should be given to calcium and calmodulin as secretory modulators in addition to cAMP. Peterson and his colleagues have been the major champions of alternative CT-induced secretory mechanisms (107). In support of their cause, rabbit loops were injected with cAMP, prostaglandin E (PGE), and CT, and overnight fluid accumulation was measured. CT elicited fluid accumulation, but dibutryl cAMP and forskolin did not; however, two PGEs evoked almost as much fluid as CT. When stimulated by CT, PGE released in the fluid increased in concentration in parallel with the fluid. The CT B subunit did not stimulate PGE accumulation. The authors point out that the intestinal cAMP response is relatively short-lived compared to the long-term secretory activity. They suggest that PGE may be an accessory after the fact in terms of promoting continued secretory activity during cholera and that this might explain the decrease in secretion following the administration of drugs that inhibit prostaglandin synthesis. A recent report showed that CT acted in synergy with LPS to incite interleukin 1β (IL-1β)-induced PGE$_2$ release (see further reference to cytokines below), suggesting that, in addition to cAMP, LPS and IL-1β might be involved in pathogenesis (59). With the discovery of Zot and Ace, whose mechanisms of action are unknown, a more open-minded approach should be taken in considering additional mechanisms for virulence factors of *V. cholerae.*

Immune Responses to *V. cholerae* Infection

Of all the areas of research in which animal models are of paramount importance, immunology ranks first. In this section I summarize data from animal studies that have guided the approaches being used to actively immunize against cholera. Several later chapters will deal with vaccines. The standard measure of vaccine efficacy was usually determined by the mouse protection assay described above. A recent modification of this assay applied to the newest genetically engineered live attenuated vaccine strains uses pure CT as the challenge. Values for 50% effective doses were obtained by comparing test vaccines with standard vaccine. It should be noted that Margaret Pittman, another chol-

era pioneer and developer of this vaccine assay, was acknowledged by the authors of that study for helpful comments (24) (WGACA).

From the beginning, the primary controversy about immunity to cholera has been the relative importance of parenteral versus oral immunization (see quote by Freter below). Countless attempts to immunize parenterally have been made, and even the most successful efforts have been less than encouraging (80). While the debate has gone on, the answer was always available: recovered victims of cholera, whether the disease is natural or obtained voluntarily and whether the victims are already primed or are strangers to *V. cholerae*, are immune to the disease for years and maybe for life. This fact strongly suggests that delivery of *V. cholerae* to the immune system through the gut is the way to promote solid long-lasting immunity.

A staunch proponent of this idea was Rolf Freter (Table 1), a champion of coproantibodies (immunoglobulin A[IgA] and other antibodies secreted into the intestinal tract) and the use of oral vaccines. Even though he went largely unheard for many years, Freter offered a prophetic statement at the 1965 Cholera Research Symposium: "In view of the firm experimental basis of this approach (i.e., oral vaccination) the chances are good that such a trial of an oral vaccination would yield useful data, which would allow one to answer the most basic question in the immunity to cholera: the location of the protective antibody. Once this problem has been solved, other protective antigens might easily be included in the cholera vaccine, if and when they are discovered." (8a). In light of the recent discovery of new toxins and a better understanding of the adjuvant activities of CT, this statement represents remarkable foresight.

With the availability of purified CT, Fujita and Finkelstein (51) carried out an elegant series of studies with adult mice to address the question of the route of antigen administration. The antigens were killed whole cells, choleragenoid, procholeragenoid, and a formalized toxoid (which reverted to toxin; see Northrup and primate model [97] in Table 1 above) later used in a large-scale field trial. Mice were vaccinated subcutaneously or orally. The challenges were 20 μg CT given intravenously (i.v.) and live *V. cholerae* in ligated loops. End points were loss of weight and/or death in i.v. inoculated mice and fluid accumulation at 16 to 18 h in loops. Choleragenoid protected 75% of the mice against i.v. challenge and 50% against loop challenge. The formalinized toxoid protected 50% of the mice against both routes of challenge. A commercial vaccine containing killed organisms failed to protect any mice against i.v. challenge with CT, but when these immunized animals were challenged intestinally with live organisms, up to 90% survived or were positive for fluid accumulation. Because parenteral immunization gave erratic protection, the authors suggested that it might prove more feasible to immunize humans against cholera by the oral route (51).

Northrup and Fauci (98) (see Table 1) discovered in a study employing infant mice that CT was an adjuvant when administered simultaneously with sheep erythrocytes and an immunosuppressant when given before or after the antigen. Their studies confirmed preliminary investigations that had demonstrated that CT is an adjuvant in low doses and an immunosuppressant at higher doses. Chisari et al. (12, 13) then showed that high i.v. doses of CT caused targeted atrophy of the thymus and depletion of lymphoid cells from the spleen accompanied by massive weight loss, increased serum calcium levels, and death in the mice. All of the lethal effects appear to be associated with prolonged increases in cAMP. In chapter 17, I present data derived from this i.v. model that demonstrate that susceptibility or resistance to i.v. CT is genetically linked to the murine histocompatibility gene complex (126).

Using the rat model, Pierce and Gowans (112) demonstrated that orally administered CT first stimulated the local immune system

via the Peyer's patches. Using fluorescent and iodinated anti-CT antibodies, trafficking of antibody-containing cells was traced through the immune system. Although it could be demonstrated that anti-CT IgA- and IgG-producing cells were present on all mucosal surfaces, the majority of the strongly responsive antibody-containing cells homed backed into the intestine.

In the same model, Pierce compared the immune responses to purified CT, crude CT filtrate, purified CT B, and purified toxoid. The materials were administered directly into the intestine after the rats were fasted overnight. The abilities of the toxin-toxoid preparations to prime, cause a sustained response, and boost the primary response were evaluated. Combinations of crude toxoid and toxin were most effective in mucosal priming. The major positive effects appeared to be linked to the abilities of these antigens to bind to the mucosal surface and stimulate adenyl cyclase activity (111).

A continuing series of important investigations from the Holmgren laboratory, using rabbit ileal loops and mice, showed that the immune responses in these animals were dependent on how much antigen reached the appropriate cells and not so much on the route by which it was administered. Protection was 100-fold better against intestinal challenge when CT-LPS combinations were used than when either of the agents was used alone (143).

In the first of many elegant papers dissecting the immune response to CT in mice, Elson and Ealding (27) showed that CT administered into the intestinal tract generates both a systemic IgG and a mucosal IgA response to CT. They concluded that both of these responses originate in the gut-associated lymphoid tissue and then are disseminated to other tissues. Keyhole limpet hemocyanin at higher doses evoked no response unless CT was also fed. It was assumed that the disparate responses were caused by the effects of CT on the regulatory environment in the intestine. This is interesting in light of earlier studies in which it was demonstrated that parenteral immunization with CT results in intestinal immune antibody responses. These investigations showed that CT is a valuable probe for understanding the regulation of both immunity and tolerance to intestinal antigens. In continuing this work, Elson and his coworkers found that the murine immune response to CT was controlled by the I-A region of the *H-2* complex. These results will be discussed in chapter 17.

Lange et al. (74), using the mouse model, showed that mice orally immunized with whole CT were resistant to subsequent intestinal hypersecretion when challenged with CT, but when previously immunized with CT B, they responded with strong fluid accumulation. These authors suggest that intestinal resistance to CT is due to desensitization of adenylate cyclase in the hypersecreting cells of the intestine. These data may correlate with some of the evidence for additional non-cAMP secretory mechanisms discussed above.

A partial explanation for some of the earlier observations of Chisari et al. regarding CT regulation of the immune response might be offered by studies carried out by Tannebaum and Hamilton (144) and Chedid and Mizel (11). From these studies it is clear that elevation of cAMP (by CT and other known and unknown activators) is involved in the signal transduction pathway for IL-1 biosynthesis. Supersecretion of IL-1 (e.g., in the i.v. mouse model) could trigger a cytokine cascade of excess tumor necrosis factor and other tissue-damaging reactions, leading to the gross pathology observed in the mice. Lycke et al. (86) demonstrated beautifully that CT stimulates IL-1 production and enhances antigen presentation by macrophages in vitro. The authors postulated that oral administration of CT to mice may stimulate antigen-processing cells and that the potentiating effect of IL-1 may partially explain the adjuvant effect of CT on the immune response. Their study suggested that the underlying mechanism was stimulation of IL-1 production in the antigen-presenting cells,

with its attendant signaling responses up-regulating the local immune system (7).

Another approach to passive protection against cholera was employed by Sciortino (135). Monoclonal antibodies against LPS and a panel of vibrio outer membrane proteins were mixed with suspensions of *V. cholerae* and administered to infant rabbits. Six of the monoclonal antibodies, each recognizing a different outer membrane antigen, offered no protection against fatal *V. cholerae* infection. Four antibodies against an 18-kDa OMP antigen on the surface were protective. The protected animals had $> 10^8$ viable vibrios in their intestines, suggesting that the monoclonal antibodies were not vibriocidal but that they might block either adherence or toxin expression.

A new animal model for studying the effects of monoclonal IgA in the intestine was promoted by Winner et al. (154). In this unique model, hybridoma cells producing dimeric monoclonal IgA antibodies directed against Ogawa-specific LPS carbohydrate antigens on the bacterial surface were injected into syngeneic mice to produce backpack tumors. This resulted in the secretion of monoclonal secretory IgA onto mucosal surfaces. Newborn mice bearing these backpack tumors were specifically protected against oral challenge with the homologous organism. The authors suggest that this approach could be used to define effective mucosal immunogens or to identify protective IgA antibodies and that monoclonal IgAs could be used passively to provide passive mucosal protection against the disease. This model seems to be an ideal vehicle for investigating TCP versus MSHA immune responses against El Tor infections.

In order to determine how serum vibriocidal antibodies might prevent cholera, Guptka et al. (58) prepared conjugates of detoxified LPS with CT and a variety of other proteins to elicit serum LPS antibodies having vibriocidal activity. The conjugates incited vibriocidal titers in mice similar to those seen with cellular vaccines.

WGACA

Thirty years ago, there was debate about the nature of the mechanisms of cholera pathogenesis. There were ion translocase inhibitors, a variety of heat-labile and heat-stable toxins, and low-molecular-weight toxic materials. Pili were not yet proven to be involved in attachment of *V. cholerae*, and several hemagglutinins were invoked as important virulence factors. Today, the debate about mechanisms of pathogenesis continues. It is clear that cAMP plays a major role, but maybe it plays several roles. A strong case has been made for PGE and its derivatives as potent secretory mediators. Zot, Ace, and IL-1 are involved as well, but how? TCP and MSHA both facilitate attachment but differ in expression depending on the environment. For all of the remarkable advances that have been made in cholera research at the bench, the real answers will have to be provided by *V. cholerae* in its normal environmental niche and in its transient home away from home, the intestines of humans and experimental animals.

Acknowledgments. I dedicate this chapter to my mentor, Sydney C. Rittenburg, who encountered coproantibodies early in his career and who personifies the highest standards for any scientist and teacher. I am indebted to John Feeley for introducing me to cholera and the late John Seal and the U.S.-Japan Cooperative Medical Sciences Program for support and encouragement through the early days. I am fortunate to have known, admired, and had many fruitful discussions with cholera pioneers Charles Lankford, Margaret Pittman, Willard Verwey, William Burrows, Robert Phillips, and William E. van Heyningen, who, in their time, "wrote the book" on cholera.

REFERENCES

1. **Aziz, K. M. S., A. K. M. Mohsin, W. K. Hare, and R. A. Phillips.** 1968. Using the rat as a cholera "model." *Nature* (London) **220:**814–815.
2. **Barua, D., and W. B. Greenough III (ed.).** 1992. *Cholera.* Plenum Medical Book Co., New York.
3. **Baselski, V. S., S. Upchurch, and C. D. Parker.** 1978. Isolation and phenotypic characterization of virulence deficient mutants of *Vibrio cholerae. Infect. Immun.* **22:**181–188.
4. **Basu, S., R. L. Robinson, and M. J. Pickett.** 1970. Preliminary studies on the bioassay of anti-

cholera vaccines in chinchillas. *J. Infect. Dis.* **121:**S56–S57.
5. **Baudry, B., A. Fasano, J. Ketley, and J. B. Kaper.** 1992. Cloning of a gene (*zot*) encoding a new toxin produced by *Vibrio cholerae. Infect Immun.* **60:**428–434.
6. **Betley, M. J., V. L. Miller, and J. J. Mekalanos.** 1986. Genetics of bacterial enterotoxins. *Annu. Rev. Microbiol.* **40:**577–605.
7. **Bromander, A., J. Holmgren, and N. Lycke.** 1991. Cholera toxin stimulates IL-1 production and enhances antigen presentation by macrophages *in vitro. J. Immunol.* **146:**2908–2914.
8. **Burnet, F. M., and J. D. Stone.** 1947. Desquamation of intestinal epithelium by *V. cholerae* filtrates. Characterization of mucinase and tissue disintegrating enzymes. *Aust. J. Exp. Biol. Med. Sci.* **25:**219–226.
8a. **Bushnell, O. A., and C. S. Brookhyer (ed.).** 1965. *Proceedings of the Cholera Research Symposium, Honolulu, 1965.* U.S. Public Health Service publication no. 1328. U.S. Government Printing Office, Washington, D.C.
9. **Callahan, L. T., III, and S. H. Richardson.** 1973. Biochemistry of *Vibrio cholerae* virulence, III. Nutritional requirements for toxin production and the effects of pH on toxin elaboration in chemically defined media. *Infect. Immun.* **7:**567–572.
10. **Callahan, L. T., III, R. C. Ryder, and S. H. Richardson.** 1971. Biochemistry of *Vibrio cholerae* virulence. II. Skin permeability factor-cholera enterotoxin production in a chemically defined medium. *Infect. Immun.* **4:**611–618.
11. **Chedid, M., and S. B. Mizel.** 1990. Involvement of cyclic AMP-dependent protein kinases in the signal transduction pathway for interleukin-1. *Mol. Cell. Biol.* **10:**3824–3827.
12. **Chisari, F. V., and R. S. Northrup.** 1974. Pathophysiologic effects of lethal and immunoregulatory doses of cholera enterotoxin in the mouse. *J. Immunol.* **113:**740–749.
13. **Chisari, F. V., R. S. Northrup, and L. C. Chen.** 1974. The modulating effect of cholera enterotoxin on the immune response. *J. Immunol.* **113:**729–739.
14. **Colwell, R. R., and W. M. Spira.** 1992. The ecology of *Vibrio cholerae*, p. 107–127. *In* D. Barua and W. B. Greenough III, (ed.), *Cholera.* Plenum Medical Book Co., New York.
15. **Craig, J. P.** 1965. A permeability factor (toxin) found in cholera stools and culture filtrates and its neutralization by convalescent cholera sera. *Nature* (London) **207:**614–616.
16. **Craig, J. P.** 1965. The effect of cholera stool and culture filtrates on the skin of guinea pigs and rabbits, p. 153–158. *In* O. A. Bushnell and C. S. Brookhyer (ed.), *Proceedings of the Cholera Research Symposium, Honolulu, 1965.* U.S. Public

Health Service publication no. 1328. U.S. Government Printing Office, Washington D.C.

17. **Craig, J. P.** 1971. Cholera toxins, p. 189–254. *In* S. Kadis, T. C. Montie, and S. J. Ajl (ed.), *Microbial Toxins*, vol. IIA. Academic Press, New York.

18. **De, S. N.** 1959. Enterotoxicity of bacteria-free culture-filtrate of *Vibrio cholerae*. *Nature* (London) **183**:1533–1534.

19. **De, S. N.** 1959. False reaction in ligated loop of rabbit intestine. *Indian J. Pathol. Bacteriol.* **2**:121–128.

20. **De, S. N., M. L. Ghose, and J. Chandra.** 1962. Further observations on cholera enterotoxin. *Trans. R. Soc. Trop. Med. Hyg.* **3**:241–245.

21. **De, S. N., M. L. Ghose, and A. Sen.** 1960. Activities of bacteria-free preparations from *Vibrio cholerae*. *J. Pathol. Bacteriol.* **79**:373–380.

22. **DiRita, V. J.** 1992. Co-ordinate expression of virulence genes by ToxR in *Vibrio cholerae*. *Mol. Microbiol.* **6**:451–458.

23. **Donta, S. T., and M. King.** 1973. Induction of steroidogenesis in tissue culture by cholera enterotoxin. *Nature* (London) **243**:246–247.

24. **Dragunsky, E. M., E. Rivera, W. Aaronson, T. M. Dolgaya, H. D. Hochstein, W. H. Habig, and I. S. Levenbook.** 1992. Experimental evaluation of antitoxic protective effect of new cholera vaccines in mice. *Vaccine* **10**:735–736.

25. **Dutta, N. K., and M. K. Habbu.** 1955. Experimental cholera in infant rabbits: a method for chemotherapeutic investigation. *Br. J. Pharmacol. Chemother.* **10**:153–159.

26. **Dutta, N. K., M. V. Panse, and D. R. Kulkarni.** 1959. Role of cholera toxin in experimental cholera. *J. Bacteriol.* **78**:594–595.

27. **Elson, C. O., and W. Ealding.** 1984. Generalized systemic and mucosal immunity in mice after mucosal stimulation with cholera toxin. *J. Immunol.* **132**:2736–2741.

28. **Evans, D. J.** 1968. Characterization of an ion translocase inhibitor from *Vibrio cholerae*, p. 1–172. Ph.D. dissertation, Wake Forest University, Winston-Salem, N.C.

29. **Evans, D. J., Jr., and S. H. Richardson.** 1968. In vitro production of choleragen and vascular permeability factor by *Vibrio cholerae*. *J. Bacteriol.* **96**:126–130.

30. **Fasano, A., B. Baudry, D. W. Pumplin, S. S. Wasserman, B. D. Tall, J. M. Ketley, and J. B. Kaper.** 1991. *Vibrio cholerae* produces a second enterotoxin, which affects intestinal tight junctions. *Proc. Natl. Acad. Sci. USA* **88**:5242–5246.

31. **Feeley, J. C., and M. Pittman.** 1962. A mouse protection test for assay of cholera vaccine, p. 92–94. *In SEATO Conference on Cholera*. Post Publishing Co., Bangkok, Thailand.

32. **Field, M.** 1971. Intestinal secretion: effect of cyclic AMP and its role in cholera. *N. Engl. J. Med.* **284**:1137–1144.

33. **Field, M., G. R. Plotkin, and W. Silen.** 1968. Effects of vasopressin, theophylline and cyclic adenosine monophosphate on short-circuit current across isolated rabbit ileal mucosa. *Nature* (London) **217**:469–471.

34. **Finkelstein, R. A.** 1970. Monospecific equine antiserum against cholera exo-enterotoxin. *Infect. Immun.* **2**:691–697.

35. **Finkelstein, R. A.** 1973. Cholera. *Crit. Rev. Microbiol.* **2**:553–623.

36. **Finkelstein, R. A.** 1976. Progress in the study of cholera and related enterotoxins, p. 53–84. *In* A. Bernheimer (ed.), *Mechanisms in Bacterial Toxicology*. John Wiley & Sons, New York.

37. **Finkelstein, R. A.** 1988. Structure of the cholera enterotoxin (choleragen) and the immunologically related ADP-ribosylating heat-labile enterotoxins, p. 1–38. *In* M. C. Hardegree, W. H. Habig, and A. Tu (ed.), *Handbook of Natural Toxins*, vol. II. *Bacterial Toxins*. Marcel Dekker, Inc., New York.

38. **Finkelstein, R. A.** 1988. Cholera, the cholera enterotoxins, and the cholera enterotoxin-related enterotoxin family, p. 85–101. *In* P. Owen and T. J. Foster (ed.), *Immunochemical and Molecular Genetic Analysis of Bacterial Pathogens*. Elsevier/North Holland Science Publishing, Amsterdam.

39. **Finkelstein, R. A.** 1992. Cholera enterotoxin (choleragen): a historical perspective, p. 155–187. *In* D. Barua and W. G. Greenough III (ed.), *Cholera*. Plenum Medical Book Co., New York.

40. **Finkelstein, R. A., M. Boesman, S. H. Neoh, M. K. LaRue, and R. Delaney.** 1974. Dissociation and recombinations of the subunits of the cholera enterotoxin (choleragen). *J. Immunol.* **113**:145–150.

41. **Finkelstein, R. A., and C. E. Lankford.** 1955. Nutrient requirements of *Vibrio cholerae*, p. 49. *Bacteriol. Proc. 1955.*

42. **Finkelstein, R. A., and J. J. LoSpalluto.** 1969. Pathogenesis of experimental cholera: preparation and isolation of choleragen and choleragenoid. *J. Exp. Med.* **130**:185–202.

43. **Finkelstein, R. A., J. J. LoSpalluto.** 1970. Production of highly purified choleragen and choleragenoid. *J. Infect. Dis.* **121**:(Suppl):S63–S72.

44. **Finkelstein, R. A., and S. Mukerjee.** 1963. Hemagglutination: a rapid method for differentiating *Vibrio cholerae* and El Tor vibrios. *Proc. Soc. Exp. Biol. Med.* **112**:355–359.

45. **Finkelstein, R. A., M. L. Vasil, and R. K. Holmes.** 1974. Studies on toxinogenesis in *Vibrio cholerae*. I. Isolation of mutants with altered toxinogenicity. *J. Infect. Dis.* **129**:117–123.

46. **Formal, S. B., D. Kundel, H. Schneider, N. Kunev, and H. Sprinz.** 1961. Studies with *Vibrio cholerae* in the ligated loop of the rabbit intestine. *Br. J. Exp. Pathol.* **42**:504–510.

47. **Freter, R.** 1956. Coproantibody and bacterial antagonism as protective factors in experimental enteric cholera. *J. Exp. Med.* **104:**419–426.

48. **Freter, R.** 1974. Interactions between mechanisms controlling the intestinal flora. *Am. J. Clin. Nutr.* **27:**1409–1416.

49. **Freter, R., and P. C. M. O'Brien.** 1981. Role of chemotaxis in the association of motile bacteria with intestinal mucosa: fitness and virulence of nonchemotactic *Vibrio cholerae* mutants in infant mice. *Infect. Immun.* **34:**222–233.

50. **Freter, R., P. C. M. O'Brien, and M. S. Macsal.** 1981. Role of chemotaxis in the association of motile bacteria with intestinal mucosa: in vivo studies. *Infect. Immun.* **34:**234–240.

51. **Fujita, K., and R. A. Finkelstein.** 1972. Antitoxic immunity in experimental cholera: comparison of immunity induced perorally and parenterally in mice. *J. Infect. Dis.* **125:**647–655.

52. **Galen, J., A. Fasano, J. Ketley, and S. Richardson.** 1990. The role of neuraminidase in *Vibrio cholerae* pathogenesis, p. 41. *Abstr. Annu. Meet. Am. Soc. Microbiol. 1990.*

52a. **Galen, J., and J. Kaper.** Personal communication.

53. **Galen, J. E., J. M. Ketley, A. Fasano, S. H. Richardson, S. S. Wasserman, and J. B. Kaper.** 1992. Role of *Vibrio cholerae* neuraminidase in the function of cholera toxin. *Infect. Immun.* **60:**406–415.

54. **Gill, D. M.** 1976. The arrangement of subunits in cholera toxin. *Biochemistry* **15:**1242–1248.

55. **Goldberg, M. B., V. J. DiRita, and S. B. Calderwood.** 1990. Identification of an iron-regulated virulence determinant in *Vibrio cholerae,* using Tn*phoA* mutagenesis. *Infect. Immun.* **58:**55–60.

56. **Greenough, W. B., III, N. F. Pierce, and M. Vaughn.** 1970. Titration of cholera enterotoxin and antitoxin in isolated fat cells. *J. Infect. Dis.* **121:**S111–S113.

57. **Guerrant, R. L., L. L. Brunton, T. C. Schnaitman, L. I. Rebbun, and A. G. Gilman.** 1974. Cyclic adenosine monophosphate and alteration of Chinese hamster ovary cell morphology: a rapid, sensitive in vitro assay for the enterotoxins of *Vibrio cholerae* and *Escherichia coli. Infect. Immun.* **10:**320–327.

58. **Guptka, R. K., S. C. Szu, R. A. Finkelstein, and J. B. Robbins.** 1992. Synthesis characterization, and some immunological properties of conjugates composed of the detoxified lipopolysaccharide of *Vibrio cholera* O1 serotype Inaba bound to cholera toxin. *Infect. Immun.* **60:**3201–3208.

59. **Hill, S. J., and J. L. Ebersole.** 1991. Cholera toxin synergizes LPS and IL-1 beta-induced PGE2 release: potential amplification systems in cholera. *Lymphokine Cytokine Res.* **10:**445–449.

60. **Hochstein, H. D., J. C. Feeley, and S. H. Richardson.** 1970. Titration of cholera antitoxin levels

61. **Holmgren, J., I. Lonnroth, and L. Svennerholm.** 1973. Tissue receptor for cholera exotoxin: postulated structure from studies with GM_1 ganglioside and related glycolipids. *Infect. Immun.* **8:**208–214.

62. **Honda, T., and R. A. Finkelstein.** 1979. Selection and characteristics of a novel *Vibrio cholerae* mutant lacking the A (ADP-ribosylating) portion of the cholera enterotoxin. *Proc. Natl. Acad. Sci. USA* **76:**2052–2056.

63. **Huber, G. S., and R. A. Phillips.** 1962. Cholera and the sodium pump, p. 37–40. *In SEATO Conference on Cholera.* Post Publishing Co., Bangkok, Thailand.

64. **Jenkin, C. R., and D. Rowley.** 1959. Possible factors in the pathogenesis of cholera. *Br. J. Exp. Pathol.* **40:**474–482.

65. **Kaper, J. B., and M. M. Levine.** 1990. Recombinant attenuated *Vibrio cholerae* strains used as live oral vaccines. *Res. Microbiol.* **141:**901–906.

66. **Kennedy, J. R., and S. H. Richardson.** 1969. Fine structure of *Vibrio cholerae* during toxin production. *J. Bacteriol.* **100:**1393–1401.

67. **Kennedy, J. R., and S. H. Richardson.** 1972. Effects of cholera permeability factor on guinea pig skin. An electron microscopic study. *Lab. Invest.* **26:**409–418.

68. **King, C. A., and W. E. van Heyningen.** 1973. Deactivation of cholera toxin by a sialidase-resistant monosialosyl ganglioside. *J. Infect. Dis.* **127:**639–647.

69. **Koonin, E. V.** 1992. The second cholera toxin, Zot, and its plasmid-encoded and phage-encoded homologues constitute a group of putative ATPases with an altered purine NTP-binding motif. *FEBS Lett.* **312:**3–6.

70. **Kothary, M. H., E. F. Claverie, J. M. Madden, and S. H. Richardson.** 1991. Purification and characterization of a *Vibrio hollisae* toxin which elongates chinese hamster ovary cells, p. 53–58. *Proc. 27th Joint Conf. Cholera Related Diarrheal Dis.*

71. **Kothary, M. H., and S. H. Richardson.** 1987. Fluid accumulation in infant mice caused by *Vibrio hollisae* and its extracellular enterotoxin. *Infect. Immun.* **55:**626–630.

72. **Kuhn, R. E., and S. H. Richardson.** 1984. A study on the genetic and cellular basis of sensitivity to cholera toxin in the SAM model, p. 221. *Abstr. Joint Meet. R. Am. Soc. Trop. Med. Hyg.*

73. **LaBrec, E. H., H. Sprinz, H. Schneider, and S. Formal.** 1965. Localization of vibrios in experimental cholera: a fluorescent antibody study in guinea pigs, p. 272–276. *In* O. A. Bushnell and C. S. Brookhyer (ed.), *Proceedings of the Cholera*

by passive hemagglutination tests using fresh and formalinized sheep erythrocytes. *Proc. Soc. Exp. Biol. Med.* **133:**120–124.

Research Symposium, Honolulu, 1965. U.S. Public Health Service publication no. 1328. U.S. Government Printing Office, Washington, D.C.

74. **Lange, S., I. Lonnroth, and H. Nygren.** 1984. Intestinal resistance to cholera toxin in mouse. Antitoxic antibodies and desensitization of adenylate cyclase. *Int. Arch. Allergy Appl. Immunol.* **74:**221–225.

75. **Lankford, C. E.** 1960. Factors of virulence of *Vibrio cholerae. Ann. N.Y. Acad. Sci.* **88:**1203–1212.

76. **Lankford, C. E., and U. Legsomburana.** 1965. Virulence factors of choleragenic vibrios, p. 109–120. *In* O. A. Bushnell and C. S. Brookhyer (ed.), *Proceedings of the Cholera Research Symposium,* Honolulu, 1965. U.S. Public Health Service publication no. 1328. U.S. Government Printing Office, Washington, D.C.

77. **Leitch, G. J.** 1988. Cholera enterotoxin-induced mucus secretion and increase in the mucus blanket of the rabbit ileum in vivo. *Infect. Immun.* **56:**2871–2875.

78. **Levine, M. M., and J. B. Kaper.** 1993. Live oral vaccines against cholera: an update. *Vaccine* **11:**207–212.

79. **Levine, M. M., J. B. Kaper, R. E. Black, and M. L. Clements.** 1983. New knowledge on pathogenesis of bacterial enteric infection as applied to vaccine development. *Microbiol. Rev.* **47:**510–550.

80. **Levine, M. M., and N. F. Pierce.** 1992. Immunity and vaccine development, p. 285–327. *In* D. Barua and W. B. Greenough III (ed.), *Cholera.* Plenum Medical Book Co., New York.

81. **Lewis, A. C., and S. H. Richardson.** 1974. Properties of cholera enterotoxin from chemically defined media, p. 82–92. *In Proceedings of the Ninth Joint Cholera Conference.* Department of State publication no. 8762. U.S. Government Printing Office, Washington, D.C.

82. **Lewis, A. C., S. H. Richardson, and B. Sheridan.** 1976. Biochemistry of *Vibrio cholerae* virulence: purification of cholera enterotoxin by preparative disc electrophoresis. *Appl. Environ. Microbiol.* **32:**288–293.

83. **Lockwood, D., S. H. Richardson, and M. Aiken.** 1982. In vivo biologic activities of *Vibrio fluvialis* and its toxic products, p. 31. *Abstr. Annu. Meet. Am. Soc. Microbiol. 1982.*

84. **Lockwood, D. E., A. S. Kreger, and S. H. Richardson.** 1982. Detection of toxins produced by *Vibrio fluvialis. Infect. Immun.* **35:**702–708.

85. **Lönnroth, I., and J. Holmgren.** 1973. Subunit structure of cholera toxin. *J. Gen. Microbiol.* **76:**417–427.

86. **Lycke, N., A. K. Bromander, L. Ekman, U. Karlsson, and J. Holmgren.** 1989. Cellular basis of immunomodulation by cholera toxin *in vitro* with possible association to the adjuvant function *in vivo. J. Immunol.* **142:**20–27.

87. **Mahan, M. J., J. M. Slauch, and J. J. Mekalanos.** 1993. Selection of bacterial virulence genes that are specifically induced in host tissues. *Science* **259:**686–688.

88. **Mekalanos, J. J.** 1985. Cholera toxin: genetic analysis, regulation and role in pathogenesis. *Curr. Top. Microbiol. Immunol.* **118:**97–118.

89. **Mekalanos, J. J.** 1992. Environmental signals controlling expression of virulence determinants in bacteria. *J. Bacteriol.* **174:**1–7.

90. **Mekalanos, J. J., R. J. Collier, and W. R. Romig.** 1978. Purification of cholera toxin and its subunits: new methods of preparation and the use of hypertoxinogenic mutants. *Infect. Immun.* **20:**552–558.

91. **Mekalanos, J. J., R. J. Collier, and W. R. Romig.** 1978. Affinity filters, a new approach to the isolation of tox mutants of *Vibrio cholerae. Proc. Natl. Acad. Sci. USA* **75:**941–945.

91a. **Mekalanos, J. J., and M. H. Kothary.** Personal communication.

92. **Mekalanos, J. J., S. L. Moseley, J. R. Murphy, and S. Falkow.** 1982. Isolation of enterotoxin structural gene deletion mutations in *Vibrio cholerae* induced by two mutagenic vibriophages. *Proc. Natl. Acad. Sci. USA.* **79:**151–155.

93. **Miller, J. F., J. J. Mekalanos, and S. Falkow.** 1989. Coordinate regulation and sensory transduction in the control of bacterial virulence. *Science* **243:**916–922.

94. **Miller, V. L., R. K. Taylor, and J. J. Mekalanos.** 1987. Cholera toxin transcriptional activator ToxR is a transmembrane DNA binding protein. *Cell* **48:**271–279.

95. **Moyenuddin, M., K. Wachsmuth, S. H. Richardson, and W. L. Cook.** 1992. Enteropathogenicity of non-toxigenic *Vibrio cholerae* O1 for adult mice. *Microb. Pathog.* **12:**451–458.

96. **Nelson, E. T., J. D. Clements, and R. A. Finkelstein.** 1976. *Vibrio cholerae* adherence and colonization in experimental cholera: electron microscopic studies. *Infect. Immun.* **14:**527–547.

97. **Northrup, R. S., and F. V. Chisari.** 1972. Response of monkeys to immunization with cholera toxoid, toxin, and vaccine: reversion of cholera toxoid. *J. Infect. Dis.* **125:**471–479.

98. **Northrup, R. S., and A. S. Fauci.** 1972. Adjuvant effect of cholera enterotoxin on the immune response of the mouse to sheep red blood cells. *J. Infect. Dis.* **125:**672–673.

99. **Ohtomo, N., T. Muraoka, A. Tashiro, Y. Zinnaka, and K. Amako.** 1976. Size and structure of the cholera toxin molecule and its subunits. *J. Infect. Dis.* **133**(Suppl):S31–S40.

100. **Osek, J., A. M. Svennerholm, and J. Holmgren.** 1992. Protection against *Vibrio cholerae* El Tor infection by specific antibodies against mannose-binding hemagglutinin pili. *Infect. Immun.* **60:**4961–4964.

101. **Owen, R. L., N. F. Pierce, R. T. Apple, and W. C. Cray, Jr.** 1986. M cell transport of *Vibrio cholerae* from the intestinal lumen into Peyer's patches: a mechanism for antigen sampling and for microbial transepithelial migration. *J. Infect. Dis.* **153:**1108–1118.

102. **Parker, C., D. Gauthier, A. Tate, K. Richardson, and W. R. Romig.** 1979. Extended linkage map of *Vibrio cholerae. Genetics* **91:**191–214.

103. **Parsot, C., and J. J. Mekalanos.** 1991. Expression of ToxR, the transcriptional activator of the virulence factors in *Vibrio cholerae,* is modulated by the heat shock response. *Proc. Natl. Acad. Sci. USA* **87:**9898–9902.

104. **Patel, P., C. F. Marrs, J. S. Mattick, W. M. Ruehl, R. K. Taylor, and M. Koomey.** 1991. Shared antigenicity and immunogenicity of type 4 pilins expressed by *Pseudomonas aeruginosa, Moraxella bovis, Neisseria gonorrhoeae, Dichelobacter nodosus,* and *Vibrio cholerae. Infect. Immun.* **59:**4674–4676.

105. **Pearson, G. D. N., V. J. DiRita, M. B. Goldberg, S. A. Boyko, S. B. Calderwood, and J. J. Mekalanos.** 1990. New attenuated derivatives of *Vibrio cholerae. Res. Microbiol.* **14:**893–899.

106. **Peterson, J. W., J. J. LoSpalluto, and R. A. Finkelstein.** 1972. Localization of cholera toxin *in vivo. J. Infect. Dis.* **126:**617–628.

107. **Peterson, J. W., and L. G. Ochoa.** 1989. Role of prostaglandins and cAMP in the secretory effects of cholera toxin. *Science* **245:**857–859.

108. **Peterson, K. M., and J. J. Mekalanos.** 1988. Characterization of the *Vibrio cholerae* ToxR regulon: identification of novel genes involved in intestinal colonization. *Infect. Immun.* **56:**2822–2829.

109. **Phillips, R. A.** 1965. Pathophysiology of cholera, p. 82–84. *In* O. A. Bushnell and C. S. Brookhyer (ed.), *Proceedings of the Cholera Research Symposium,* Honolulu, 1965. U.S. Public Health Service publication no. 1328. U.S. Government Printing Office, Washington, D.C.

110. **Pierce, N. F.** 1973. Differential inhibitory effects of cholera toxoids and ganglioside on the enterotoxins of *Vibrio cholerae* and *Escherichia coli. J. Exp. Med.* **137:**1009–1023.

111. **Pierce, N. F.** 1978. The role of antigen form and function in the primary and secondary intestinal immune responses to cholera toxin and toxoid in rats. *J. Exp. Med.* **145:**195–206.

112. **Pierce, N. F., and J. L. Gowans.** 1975. Cellular kinetics of the intestinal immune response to cholera toxoid in rats. *J. Exp. Med.* **142:**1550–1563.

113. **Pierce, N. F., J. B. Kaper, J. J. Mekalanos, and W. C. Cray, Jr.** 1985. Role of cholera toxin in enteric colonization by *Vibrio cholerae* O1 in rabbits. *Infect. Immun.* **50:**813–816.

114. **Pollitzer, R.** 1959. *Cholera.* Monograph no. 43. World Health Organization, Geneva.

115. **Rappaport, R. S., and G. Bonde.** 1981. Development of a vaccine against experimental cholera and *Escherichia coli* diarrheal disease. *Infect. Immun.* **32:**534–542.

116. **Rappaport, R. S., G., Bonde, T. McCann, B. A. Rubin, and H. Tint.** 1974. Development of a purified cholera toxoid. II. Preparation of a stable, antigenic toxoid by reaction of purified toxin with glutaraldehyde. *Infect. Immun.* **9:**304–317.

117. **Rappaport, R. S., B. A. Rubin, and H. Tint.** 1974. Development of a purified cholera toxoid. I Purification of toxin. *Infect. Immun.* **9:**294–303.

118. **Richardson, K.** 1991. Roles of motility and flagellar structure in pathogenity of *Vibrio cholerae:* analysis of motility mutants in three animal models. *Infect. Immun.* **59:**2727–2736.

119. **Richardson, S. H.** 1966. Inhibition of intestinal ion translocase enzymes by culture filtrates of *Vibrio cholerae. J. Bacteriol.* **91:**1384–1386.

120. **Richardson, S. H.** 1968. An ion translocase system from rabbit intestinal mucosa. Preparation and properties of the (Na+K+)-activated ATPase. *Biochim. Biophys. Acta* **150:**572–577.

121. **Richardson, S. H.** 1969. Factors influencing in vitro skin permeability factor production by *Vibrio cholerae J. Bacteriol.* **100:**27–34.

122. **Richardson, S. H., and D. J. Evans, Jr.** 1965. The effects of *Vibrio cholerae* extracts on membrane transport mechanisms, p. 139–144. *In* O.A. Bushnell and C. S. Brookhyer (ed.), *Proceedings of the Cholera Research Symposium,* Honolulu, 1965. U.S. Public Health Service publication no. 1328. U.S. Government Printing Office, Washington, D.C.

123. **Richardson, S. H., and D. J. Evans, Jr.** 1968. Isolation of cholera toxins by dextran sulfate precipitation. *J. Bacteriol.* **96:**1443–1445.

124. **Richardson, S. H., J. Giles, and K. S. Kruger.** 1983. H-2 modulation of cholera enterotoxin effects in SAM (sealed adult mice), p. 28 *Abstr. Annu. Meet. Am. Soc. Microbiol. 1983.*

125. **Richardson, S. H., J. C. Giles, and K. S. Kruger.** 1984. Sealed adult mice: new model for enterotoxin evaluation. *Infect. Immun.* **43:**482–486.

126. **Richardson, S. H., and R. E. Kuhn.** 1986. Studies on the genetic and cellular control of sensitivity to enterotoxins in the sealed adult mouse model. *Infect. Immun.* **54:**522–528.

127. **Richardson, S. H., W. McKenzie, E. Daves, D. J. Evans, and D. G. Evans.** 1975. Transmissibility of the plasmid controlling vascular permeability factor (PF) and heat labile enterotoxin (LT) in E. coli from patients with "cholera-like" diarrhea. p. 37–44. *In* H. Fukumi and M. Ohashi (ed.), *Symposium on Cholera Kyoto 1974.* Japanese Cholera Panel, Tokyo.

128. **Richardson, S. H., and K. A. Noftie.** 1970. Purification and properties of permeability factor/chol-

era enterotoxin from complex and synthetic media. *J. Infect. Dis.* **121**:S73–S79.

129. **Roomi, N., M. Laburthe, N. Fleming, R. Crowther, and J. Forstner.** 1984. Cholera-induced mucin secretion from rat intestine: lack of effect of cAMP, cycloheximide, VIP and colchicine. *Am. J. Physiol.* **247**:G140–G148.

130. **Sack, R. B., C. C. J. Carpenter, R. W. Steenburg, and N. F. Pierce.** 1966. Experimental cholera: a canine model. *Lancet* **7456**:206–207.

131. **Sack, R. B., and C. E. Miller.** 1969. Progressive changes of vibrio serotypes in germ-free mice infected with *Vibrio cholerae. J. Bacteriol.* **99**:688–695.

132. **Savage, D. C.** 1978. Factors involved in colonization of the gut epithelial surface. *Am. J. Clin. Nutr.* **31**:S131–S135.

133. **Savage, D. C.** 1980. Adherence of normal flora to mucosal surfaces, p. 33–59. *In* E. H. Beachey (ed.), *Bacterial Adherence.* Chapman & Hall, London.

134. **Schrank, G. D., and W. F. Verwey.** 1976. Distribution of cholera organisms in experimental *Vibrio cholerae* infections: proposed mechanisms of pathogenesis and antibacterial immunity. *Infect. Immun.* **13**:195–203.

135. **Sciortino, C. V.** 1989. Protection against infection with *Vibrio cholerae* by passive transfer of monoclonal antibodies to outer membrane antigens. *J. Infect. Dis.* **160**:248–252.

136. **Sengupta, D. K., T. K. Sengupta, and S. C. Ghose.** 1992. Major outer membrane proteins of *Vibrio cholerae* and their role in induction of protective immunity through inhibition of intestinal colonization. *Infect. Immun.* **60**:4848–4855.

137. **Spangler, B. D.** 1992. Structure and function of cholera toxin and the related *Escherichia coli* heat-labile enterotoxin. *Microbiol. Rev.* **56**:622–647.

138. **Spira, W. M., and P. J. Fedorka-Cray.** 1984. Purification of enterotoxins from *Vibrio mimicus* that appear to be identical to cholera toxin. *Infect. Immun.* **45**:679–684.

139. **Spira, W. M., R. B. Sack, and J. L. Froehlich.** 1981. Simple adult rabbit model for *Vibrio cholerae* and enterotoxigenic *Escherichia coli* diarrhea. *Infect. Immun.* **32**:739–747.

140. **Spyrides, G. J., and J. C. Feeley.** 1970. Concentration and purification of cholera exotoxin by absorption on aluminum compound gels. *J. Infect. Dis.* **121**(Suppl):S96–S99.

141. **Sun, D., J. J. Mekalanos, and R. K. Taylor.** 1990. Antibodies directed against the toxin-coregulated pilus isolated from *Vibrio cholerae* provide protection in the infant mouse experimental cholera model. *J. Infect. Dis.* **161**:1231–1236.

142. **Sun, D., J. M. Seyer, J. Kovari, R. A. Sumrada, and R. K. Taylor.** 1991. Localization of protective epitopes within the pilin subunit of the *Vibrio cholerae* toxin-coregulated pilus. *Infect. Immun.* **59**:114–118.

143. **Svennerholm, A. M., and J. Holmgren.** 1976. Synergistic protective effect in rabbits of immunization with *Vibrio cholerae* lipopolysaccharide and toxin/toxoid. *Infect. Immun.* **13**:735–740.

144. **Tannenbaum, C. S., and T. A. Hamilton.** 1989. Lipopolysaccharide-induced gene expression in murine peritoneal macrophages is selectively suppressed by agents that elevate intracellular cAMP. *J. Immunol.* **142**:1274–1280.

145. **Taylor, R. K., V. L. Miller, D. B. Furlong, and J. J. Mekalanos.** 1987. Use of phoA gene fusions to identify a pilus colonization factor coordinately regulated with cholera toxin. *Proc. Natl. Acad. Sci. USA* **84**:2833–2837.

146. **Trucksis, M., J. E. Galen, J. Michalski, A. Fasano, and J. B. Kaper.** 1993. Accessory cholera enterotoxin (Ace), the third member of a *Vibrio cholerae* virulence cassette. *Proc. Natl. Acad. Sci. USA* **90**: 5267–5271.

147. **Ujiiye, A., A. Nakatomi, K. Utsunomiya, K. Mitsui, S. Sogame, M. Iwanaga, and K. Kobari.** 1968. Experimental cholera in mice. I. First report on the oral infection. *Trop. Med.* (Nagasaki University) **10**:65–71.

148. **van Heyningen, W. E.** Personal communication.

149. **van Heyningen, W. E., C. C. J. Carpenter, N. F. Pierce, and W. B. Greenough III.** 1971. Deactivation of cholera toxin by ganglioside. *J. Infect. Dis.* **124**:415–418.

150. **van Heyningen, W. E., and J. R. Seal.** 1983. *Cholera. The American Scientific Experience 1947–80.* Westview Press, Boulder, Colo.

151. **Vasil, J. L., R. K. Holmes, and R. A. Finkelstein.** 1975. Conjugal transfer of a chromosomal gene determining production of enterotoxin in *Vibrio cholerae. Science* **187**:849–850.

152. **Vaughan, M., N. F. Pierce, and W. B. Greenough III.** 1970. Stimulation of glycerol production in fat cells by cholera toxin. *Nature* (London) **226**:658–659.

153. **Weston, L., H. Drexler, and S. H. Richardson.** 1973. Characterization of vibriophage VA-1 *J. Gen. Virol.* **21**:155–158.

154. **Winner, L. III, J. Mack, R. Weltzin, J. J. Mekalanos, J. Kraehenbuhl, and M. Neutra.** 1991. New model for analysis of mucosal immunity: intestinal secretion of specific monoclonal immunoglobulin A from hybridoma tumors protects against *Vibrio cholerae* infection. *Infect. Immun.* **59**:977–982.

155. **Yamamoto, T., and T. Yokota.** 1989. *Vibrio cholerae* O1 adherence to human small intestinal cells in vitro. *J. Infect. Dis.* **160**:168–169.

III. CHOLERA: THE DISEASE

Vibrio cholerae and Cholera: Molecular to Global Perspectives
Edited by I. Kaye Wachsmuth, Paul A. Blake, and Ørjan Olsvik
© 1994 American Society for Microbiology, Washington, DC 20005

Chapter 15

Cholera: Pathophysiology, Clinical Features, and Treatment

Michael L. Bennish

There are few if any diseases that make as striking an impression on a physician (or a layperson) as cholera. Infection with *Vibrio cholerae* serogroup O1 (or with a recently described *V. cholerae* serotype, O139, that, like serogroup O1, produces large amounts of cholera toxin [3, 11, 66, 125, 136]) can cause a watery diarrhea of such prodigious volume that hypotensive shock and death can occur within 12 h of the first symptom. Because *V. cholerae* O1 is capable of epidemic spread, there are usually numerous cases within affected communities. The ability of *V. cholerae* O1 infection to rapidly kill otherwise healthy persons (including adults) combined with the large number of persons affected during *V. cholerae* O1 epidemics has historically instilled within the public psyche a fear that few other diseases elicit. This impact on the public imagination is evident from the frequent references to cholera in fiction, with allusions to cholera often being used to convey feelings of foreboding or doom. Cholera was also commonly invoked in traditional curses. "A cholerie zol der treppen" (may the cholera strike you) is a Yiddish curse that dates from the 19th century, when cholera was epidemic in Eastern Europe, and was intended for one's most implacable enemies.

The dread that cholera instilled in persons who witnessed those afflicted with it is evident from the description in the *Lancet* of two patients Dr. William O'Shaughnessy encountered in Sunderland, England, in 1831, during the second cholera pandemic: "On the bed lay an expiring woman . . . presenting an attitude of death which . . . I never saw paralleled in terror. . . . On the floor . . . lay a girl of slender make and juvenile height, but with the face of a superannuated hag. She uttered no moan, gave no expression of pain, but she languidly flung herself from side to side . . . her eyes were sunk deep into her sockets, as though they had been driven an inch behind their natural position; her mouth was squared; her features flattened; her eyelids black; her fingers shrunk, bent and inky in their hue. All pulse was gone at the wrist, and a tenacious sweat moistened her bosom. In short, Sir, that face and form I can never forget, were I to live beyond the period of man's natural age" (27).

O'Shaughnessy (and other physicians of his time, such as Thomas Latta) not only described the clinical features of cholera but also discerned the need for replacement of the water and electrolytes lost in the stool. In additional correspondence with the *Lancet*, O'Shaughnessy wrote that ". . . the indications of cure . . . are two in a number—viz.

Michael L. Bennish • Division of Geographic Medicine and Infectious Diseases, Departments of Pediatrics and Medicine, New England Medical Center, Tufts University School of Medicine, Boston, Massachusetts 02111.

1st to restore the blood to its natural specific gravity; 2nd to restore its deficient saline matters . . ." (27). Unfortunately, these therapeutic recommendations went largely unheeded during the next 100 years, and the mortality rate in severely ill cholera patients remained 20% or more throughout the 19th century and the first half of the 20th century (18, 108). It is only in the last 40 years that the pathophysiology of fluid loss in cholera has been elucidated and that treatment regimens providing adequate replacement of water and electrolytes lost in the stool have become standard. Currently, no patient with cholera who reaches even the most modest medical facility should die, as there is no therapy that is simpler in either concept or execution than the correction of dehydration caused by cholera (47).

This chapter will review the pathophysiology of cholera, focusing on the most common and important complication, dehydration; describe the clinical features of patients with cholera; and outline treatment for patients with this disease.

PATHOPHYSIOLOGY

Diarrhea and Dehydration

In humans, *V. cholerae* O1 infection results from ingestion of the organism, usually in contaminated food or water. Depending on the size of the inoculum and the susceptibility of the person who has been exposed, the incubation period for *V. cholerae* O1 infection can be as short as 12 h or as long as 72 h (20). In comparison to that required for most other enteric pathogens, the inoculum required to establish *V. cholerae* O1 infection is relatively large. This may be due in part to the fact that *V. cholerae* O1 is highly acid labile and that most ingested *V. cholerae* O1 organisms are killed in the acidic environment of the stomach.

Studies conducted in volunteers experimentally infected with *V. cholerae* O1 and epidemiologic studies of cholera outbreaks

have found that the infectious dose is lower and the risk of illness is higher in persons who are hypochlorhydric (20, 41, 100, 101). In one volunteer study, none of 10 otherwise healthy American prison inmates developed either infection or disease after ingestion of inocula ranging from 10^4 to 10^7 *V. cholerae* O1 organisms (20). Inocula of 10^8 to 10^{11} organisms produced diarrhea in five of nine volunteers, only one of whom had disease severe enough to warrant intravenous fluid therapy (20). When gastric acid was neutralized by administering sodium bicarbonate, the infectious dose of *V. cholerae* O1 in these prison volunteers was much lower: 35 of 44 volunteers who received inocula ranging from 10^4 to 10^6 organisms became ill, including 7 patients who had severe diarrhea requiring intravenous therapy (20).

In another volunteer study in the United States, persons who smoked cannabis more than twice a week were at increased risk of developing diarrhea following ingestion of *V. cholerae* O1 compared with persons who did not smoke cannabis or who smoked cannabis less frequently (100). The increased risk of infection in heavy users of cannabis was attributable to the effect of cannabis on stomach acid production; heavy cannabis users had lower mean basal and histamine-induced concentrations of gastric acid than did other volunteers (100). This finding may have implications for patterns of infection in the Ganges delta, where cholera is endemic and where cannabis use is common among certain segments of the population.

The small intestine is the primary site of infection with *V. cholerae* O1 and is the source of the secretory diarrhea that is the major pathologic disturbance during cholera. The rate of fluid loss is greatest in the jejunum, with fluid losses in the jejunum of up to 11 ml/cm/h being recorded during perfusion studies of patients with acute cholera (5). *V. cholerae* O1 colonizes the intestinal epithelium but does not invade epithelial cells (other than antigen-processing M cells [103]) or cause any structural alteration to the epithelium (39). The major effects of *V. cholerae* O1 infection are to increase active chloride and bicarbonate secretion into the intestinal lumen by crypt cells and to decrease villous absorption of sodium chloride (34, 35). Both of these effects are mediated by cholera toxin, the B subunit of which binds to receptors on the mucosal surface of the intestinal epithelium that contain the glycolipid GM_1 ganglioside, and the A subunit of which enzymatically activates adenylate cyclase and increases intracellular concentrations of cyclic AMP (cAMP) (34, 71, 135). cAMP then acts as an intracellular second messenger to inhibit active sodium chloride absorption and increase chloride and bicarbonate secretion (see chapter 11 for details).

Mechanisms other than increased cellular concentrations of cAMP have been proposed as contributing to intestinal fluid secretion during cholera. Prostaglandins increase intestinal fluid secretion in vitro, and increased concentrations of prostaglandins are present in the stools of cholera patients (71, 112, 139). It is uncertain, however, whether the observed increase in stool prostaglandin concentrations is a primary event (70) or whether it occurs in response to the dehydration and hypovolemia. Another postulated mechanism for the secretory diarrhea of cholera is that cholera toxin induces the release of amines and peptides that activate enteric intramural neurotransmitters, such as vasointestinal peptide, which then stimulate fluid secretion. Despite the increase in research activity exploring mechanisms other than an increase in intracellular cAMP as the cause for fluid secretion during cholera, the extensive in vivo and in vitro data supporting increased cAMP as the primary mechanism of increased intestinal fluid production during cholera remain convincing.

The increased chloride and bicarbonate secretion in crypt cells during cholera is associated with an increase in the passive transfer of water across the mucosal surface into the gut lumen. Water absorption, which normally occurs when sodium chloride is absorbed by villous cells, is also inhibited by the action of cholera toxin. The net effect of these changes is to markedly increase the amounts of water and electrolytes in the lumen of the small intestine (35). In patients with severe *V. cholerae* O1 infection, the volume of small-intestine fluid reaching the colon far exceeds the maximum resorptive capacity of the colon, which is approximately 6 liters a day (31, 109, 130). The result is the profuse watery diarrhea that characterizes cholera. In a few patients, most commonly children with ileus, hypotension can occur with little or no diarrhea, as the intestinal secretions induced by *V. cholera* infection remain contained within a markedly distended small bowel and colon. This has been termed "cholera sicca." In most cholera patients, however, colonic transit time is very rapid, with a mean of 15 min being reported in one study (138).

In severely affected patients, the volume of diarrheal stool can exceed 200 ml/kg of body weight per day, and if antimicrobial treatment is not provided, some patients can purge a volume of stool in excess of their body weight in the course of the disease (107). Unless the water is replaced, its loss in cholera stools results in proportional losses from first the extracellular and then the intracellular body compartments. Loss of water from the extracellular compartment results in a marked decrease in intravascular volume (52, 53, 78, 108), and loss of water from the intracellular compartment affects cell function. If intravascular volume decreases by more than 20%, hypotension and hypoperfusion of critical organs occur, and death ensues.

Electrolyte and Acid-Base Derangements

Cholera stool is isotonic with plasma. As a consequence, the plasma of dehydrated cholera patients is almost always iso-osmolar or slightly hypo-osmolar in comparison with normal plasma (Tables 1 and 2). The osmotically active solutes in cholera stool are electrolytes, primarily sodium chloride, that are normally found in plasma (50, 81, 94,

Table 1. Electrolyte composition of cholera stool and of fluids used for hydration of patients with cholera

Fluid	Electrolyte and glucose concn (mmol/liter)					Osmolality (mosmol/liter)[a]
	Na$^+$	Cl$^-$	K$^+$	HCO$_3^-$	Glucose or dextrose	
Cholera stool						
Adults	130	100	20	44	0	300[b]
Children	100	90	33	30	0	300[b]
WHO ORT	90	80	20	30[c]	111	331
Intravenous solutions						
Lactated Ringer's	130	109	4	28[d]	0[e]	271
Dhaka	133	98	13	48	0	292
Peru polyelectrolyte	90	80	20	30	111	331
Normal saline	154	154	0	0	0[e]	308

[a]For rehydration solutions, the value listed is the calculated osmolality. Actual osmolality is slightly less, as the salts do not completely disassociate.
[b]Osmolality includes unmeasured osmotically active molecules (primarily organic acids) in addition to electrolytes.
[c]Now most often replaced with 10 mmol of trisodium citrate per liter, which has a longer shelf life than sodium bicarbonate.
[d]As lactate.
[e]Both lactated Ringer's solution and normal saline solution are available as solutions containing 50 g of dextrose per liter (5%; 277 mmol/liter). Because of the risk of hypoglycemia in patients with cholera, such solutions are preferred to the non-dextrose-containing preparations.

Table 2. Changes in plasma electrolytes, creatinine, and total protein following rapid, 4-h rehydration of children with cholera[a]

Time after initiation of fluid therapy	Concn				
	Na$^+$ (mmol/liter)	K$^+$ (mmol/liter)	HCO$_3^-$ (mmol/liter)	Creatinine (mmol/liter)	Protein (g/liter)
0 h (admission)	132	4.4	13.3	138	102
4 h	135	3.8	19.4	106	72
24 h	134	3.7	21.9	83	73

[a]Data are from reference 123 and are based on 32 children with moderate to severe dehydration due to *V. cholerae* O1 infection who received a median of 100 ml of Dhaka Solution per kg of body weight intravenously in the 4 h after admission (see Table 1 for constituents of Dhaka Solution). During the subsequent 20 h the patients primarily received oral rehydration solution.

107, 148) (Table 1). In this respect, cholera stool differs from the stool produced in some other infectious diarrheas (such as rotavirus-caused diarrhea), in which organic acids contribute a larger proportion of stool solutes and hypertonic dehydration is more common.

The sodium concentration in cholera stool ranges from approximately 90 to 140 mmol/liter, with adult cholera stools having a higher sodium concentration than those of children (50). Because the stool concentrations of sodium and chloride are roughly proportional to their concentrations in blood, concentrations of sodium and chloride in plasma are usually normal or slightly decreased in the untreated cholera patient (7, 50, 51, 81, 107, 123, 148) (Table 2).

The concentrations of potassium and bicarbonate in cholera stool are approximately two- and four-fold higher, respectively, than they are in plasma (5, 7, 94, 149) (Table 1). This is a result of the active secretion by the ileum and colon of potassium and bicarbonate during cholera (5, 138). The colonic secretion of potassium and bicarbonate has been attributed to a direct effect of cholera toxin on colonic epithelial cells and to an increase in circulating concentrations of aldosterone elicited by hypovolemia (5, 138). Aldosterone acts directly on colonic epithelial cells to increase bicarbonate and potassium secretion and sodium absorption (138). Despite the large potassium losses in the stool, plasma potassium concentrations are usually normal (or modestly elevated) when cholera patients first present for care (16, 19, 50, 51, 77, 79, 81, 123, 147–149). This is a consequence of the accompanying acidosis, since in a homeostatic response to the acidosis, extracellular H^+ ions are exchanged for intracellular K^+. When the acidosis is corrected by the administration of bicarbonate, plasma K^+ moves back into cells and plasma K^+ concentrations can fall precipitously. The resulting hypokalemia can be severe enough to cause ileus and characteristic electrocardiogram changes (16, 72).

The most important electrolyte derangement in patients with cholera is the metabolic acidosis that results from the large loss of bicarbonate in cholera stool (7). The metabolic acidosis caused by bicarbonate loss in the stool is exacerbated by lactic acid accumulation secondary to tissue hypoperfusion and anaerobic glycolysis (147). An increase in the relative concentrations of anionic plasma proteins and phosphate also contributes to the metabolic acidosis (147).

Derangements in Glucose Homeostasis

Patients with cholera and severe dehydration are usually hyperglycemic (121). The hyperglycemia is caused by elevations in the concentrations of epinephrine, glucagon, and cortisol in serum; production of these compounds is stimulated by the profound hypotension that occurs in patients with severe dehydration. All three of these hormones have an anti-insulin effect.

Some patients with cholera, most commonly children, are hypoglycemic (9, 58, 91). In a study in Bangladesh, hypoglycemia was identified in only 0.5% of all cholera patients admitted to the hospital (9). The mortality rate in patients with hypoglycemia, however, was 15%, compared with less than 1% in patients who were not hypoglycemic. Patients who were hypoglycemic accounted for approximately 25% of all deaths (9).

Hypoglycemia in children with cholera is most often a result of inadequate gluconeo-genesis (9). Children with cholera and hypoglycemia have usually been sick for 12 or more hours before they come to a medical facility for care and usually have not eaten during that time (9). These children have exhausted their glycogen stores (14) and are dependent on gluconeogenesis for maintaining their blood glucose concentrations. The reasons for impaired gluconeogenesis are not known with any certainty but probably involve either insufficient mobilization of substrates (primarily fatty acids) for gluconeogenesis or impairment in the metabolic pathways of gluconeogenesis (9). The response of glucoregulatory (insulin) and counterregulatory (epinephrine, glucagon, and cortisol) hormones in these patients is appropriate for their degree of hypoglycemia (9).

The majority of children with cholera and hypoglycemia have neither frank marasmus nor kwashiorkor and resemble other cholera patients in their general clinical features (9). Neither hyper- nor hypoglycemia is specific to cholera diarrhea, as both may be seen in children with diarrhea caused by other enteric pathogens and in children who have nonenteric infections (9, 68).

Derangements in Calcium, Phosphate, and Magnesium Homeostasis

Serum calcium and magnesium concentrations are elevated in dehydrated patients with cholera. The increased concentrations of these two cations, both of which are largely bound to serum proteins, are a result of the hemoconcentration and resultant increase in serum protein concentration that is found in dehydrated cholera patients (147). When the volume depletion and the acidosis of severely dehydrated cholera patients are corrected, the concentration of ionized calcium (which is the physiologically important component) usually decreases, sometimes to abnormally low levels (129, 147). This is especially true in patients who have had excessive alkali administered during intravenous rehydration, as alkalosis increases calcium binding to serum proteins and thus decreases the proportion of serum calcium that is ionized (129).

Renal Insufficiency

Cholera patients with severe dehydration and volume depletion have a marked decrease in renal perfusion and glomerular filtration, with a resulting decrease in (or absence of) urine production (123). The diminished glomerular filtration further exacerbates electrolyte and acid-base disturbances, as the kidney is unable to excrete excess hydrogen ion. This "prerenal" failure also results in increased concentrations of creatinine and blood urea nitrogen in serum; both of these are normally filtered and excreted by the kidney.

Renal failure has been reported in patients with cholera and has been attributed to acute tubular necrosis caused by hypotension and renal hypoperfusion and to hypokalemic nephropathy (10). Although rarely seen now, renal failure was more commonly reported in the period before aggressive rehydration regimens were routinely employed in treating cholera patients (10, 30). In most reports of cholera patients with renal failure, the volume of fluids administered to patients was insufficient to restore normal intravascular volume. As a consequence patients were hypovolemic for prolonged periods (18).

Pulmonary Edema

Pulmonary edema has been reported in cholera patients with severe acidosis who received fluids lacking bicarbonate (49). The persistent acidosis in those patients was thought to have resulted in sustained peripheral vasoconstriction (52, 53). Because of the peripheral vasoconstriction, a disproportionate share of administered fluid remained in the large vessels, with a resultant increase in pulmonary blood flow and central venous pressure (53).

Pulmonary edema can also occur if the volume of fluids administered exceeds require-ments. Patients at greatest risk of developing pulmonary edema are the elderly, who often have diminished cardiac and renal function and are unable to handle the excess fluid load. Pulmonary edema can occur even in patients with normal cardiac reserve if the rapid rates of fluid infusion used to rehydrate cholera patients are continued after rehydration is complete.

Sepsis

V. cholerae serogroup O1 rarely invades the bloodstream. Only a few cases of *V. cholerae* O1 bacteremia have been reported (64), with the best-documented case being that of an alcoholic woman with rheumatoid arthritis who was the index case for an outbreak of cholera in Queensland, Australia, in 1977 (128).

When bacteremia does occur in cholera patients, it is most often the result of nosocomial infection caused by bacterial contamination of the intravenous infusate or colonization of the catheter used for infusion. The organisms responsible are those that usually cause intravenous-infusion-related sepsis, including non-glucose fermenting gram-negative rods and *Staphylococcus aureus*.

Cholera in Pregnant Women

For reasons that are not entirely clear, cholera is a more severe disease in pregnant women (55). Although pregnant women have historically had a much higher mortality from cholera than women who were not pregnant, current fluid therapy regimens have reduced mortality in pregnant women to approximately the same level as that of the general population. Fetal loss, however, remains high, even in women who receive adequate rehydration. In one report, the fetal death rate was 50% when women in their third trimester of pregnancy developed severe dehydration from cholera (55).

CLINICAL FEATURES

Signs and Symptoms

The most distinctive features of cholera are the production of a voluminous watery stool and the dehydration that results if water lost in the stool is not replaced. Although the stools of cholera patients may initially contain fecal matter or bile, in patients with high purging rates the stool quickly becomes white and opalescent, the characteristic rice water stool of cholera (so termed because it resembles water that rice has been cooked or washed in (Fig. 1). Not frankly malodorous, cholera stool can have a fishy odor. Cholera stool contains few if any inflammatory cells or erythrocytes and almost no protein. The absence of leukocytes, erythrocytes, and protein reflects the noninflammatory and noninvasive character of *V. cholerae* O1 infection of the intestinal lumen. The diarrhea is painless and is not accompanied by straining. Some patients do, however, complain of intestinal cramping (146). When cramping occurs, it is presumably a result of fluid distension of the bowel or is associated with more-

generalized muscle cramping resulting from disturbances in calcium metabolism.

Vomiting commonly accompanies the diarrhea, especially early in the illness. In some patients the vomiting can be profuse (99, 146). The cause of the vomiting is not well established. Because vomiting usually abates with adequate replacement of fluid and electrolytes (99), some investigators have attributed the vomiting to electrolyte disturbances, especially disturbances in acid-base balance. In persons experimentally infected with *V. cholerae* O1, however, vomiting precedes alterations in acid-base status and indeed precedes the development of signs and symptoms of dehydration (99). The vomiting has also been attributed to a direct effect of *V. cholerae* O1 infection or intestinal motility. Although both live *V. cholerae* O1 and cholera toxin alter small intestinal myoelectric activity when injected into rabbit ileal loops (83), a study of 30 actively purging cholera patients did not find disturbances in gastric emptying (25). That study, however, was conducted after the patients had been fully rehydrated.

In patients who have mild diarrhea or in patients who have more-marked fluid loss but

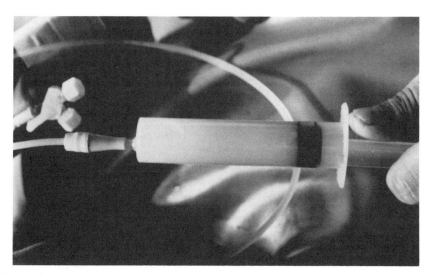

Figure 1. A syringe containing the characteristic rice water stool of cholera. Reprinted with permission (G. T. Keusch and M. L. Bennish, p. 601, *in* R. D. Feigin and J. D. Cherry, ed., *Textbook of Pediatric Infectious Diseases*, 1992).

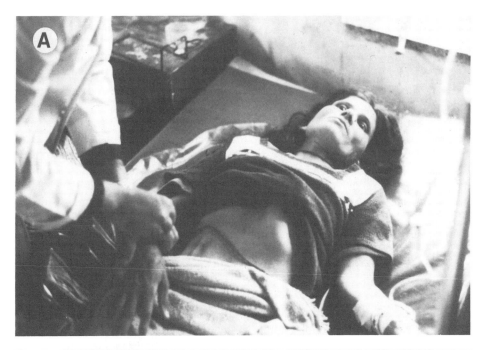

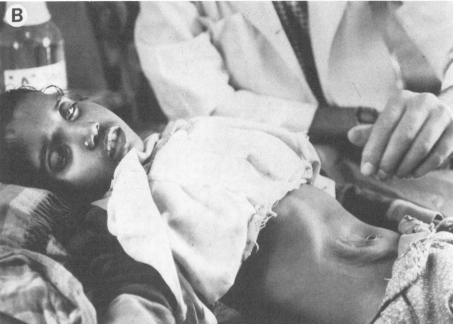

Figure 2. Characteristic clinical features of cholera. (A) A Bangladeshi woman in her early twenties with severe dehydration due to cholera who exhibits many of the features described by O'Shaughnessy 150 years ago. She appears "superannuated," she "uttered no moan, gave no expression of pain," "her eyes were sunk deep into her sockets," "all pulse was gone at the wrist," and "a tenacious sweat moistened her bosom." (B) A young girl with cholera. This young girl is severely dehydrated, as can be seen from her sunken eyes, somnolence, and increased skin tenting in response to pinching of the skin. Panel B is reprinted with permission (G. T. Keusch and M. L. Bennish, p. 601, *in* R. D. Feigin and J. D. Cherry, ed., *Textbook of Pediatric Infectious Diseases,* 1992).

have received sufficient fluids to prevent dehydration, diarrhea and vomiting are usually the only manifestations of the disease. Most patients who come to the attention of physicians will, however, have some degree of dehydration, and the appearance of patients with cholera is dominated by the features of dehydration (Fig. 2).

Severe dehydration produces such distinctive and pronounced features that cholera is one of the few diseases in adults that can be accurately diagnosed at first sight. Characteristic signs and symptoms of severely dehydrated patients include an increase in pulse rate and a decrease in pulse volume; hypotension; an increase in respiratory rate accompanied by a deep, Kussmaul pattern of respiration; sunken eyes and cheeks; dry mucous membranes; a decrease in skin turgor (determined by forcefully pinching the abdominal skin and subcutaneous tissues and determining how long it takes for the subcutaneous tissues to retract ["the skin pinch"; Fig. 2B]); wrinkled "washerwoman" hands; a scaphoid abdomen; a decrease in urine output; and thirst. Altered mentation, including restlessness and irritability, weakness, lethargy, stupor, and more rarely coma, is a common finding in severely dehydrated patients. Cholera patients with hypoglycemia, especially those with severe hypoglycemia (serum glucose concentration of less than 1 mmol/liter) may have seizures in addition to other alterations in mentation (9, 58).

In addition to intestinal cramping, cramping of the extremities also occurs in patients with cholera (146), usually when patients are acutely dehydrated. Although a source of considerable discomfort to patients, cramps have no serious sequelae and usually abate when the patient is rehydrated. Tetany can occur in patients with cholera as a result of the decrease in ionized calcium brought on by administration of alkali.

Because *V. cholerae* O1 does not invade the intestinal mucosa and does not elicit a host inflammatory response, marked fever does not occur in cholera patients unless superinfection (usually from sepsis due to contaminated intravenous fluids or catheters) is present or unless pyrogen-containing fluids have been administered. Lesser degrees of temperature elevation may occur, especially in young children, presumably because of peripheral vasoconstriction and a resulting inability to dissipate heat (77).

To help facilitate management of patients with cholera, dehydration has been classified on the basis of clinical signs and symptoms into three grades of severity: none or mild, some or moderate, and severe. These three categories of dehydration have in turn been correlated with the volume of body water that has been lost (Table 3). The distinction between the three categories is relatively simple: patients with no or mild dehydration have thirst as their only symptom and no clinical signs of dehydration, patients with some or moderate dehydration have signs of dehydration but no clinical evidence of diminished cardiac output, and patients with severe dehydration have evidence of diminished cardiac output (an absent or low-volume pulse or hypotension) in addition to other signs of dehydration. Of all the signs of dehydration, decreased skin turgor is probably the most consistent (and reproducible) finding in patients with cholera who have been ill for at least a few hours.

Diagnosis of dehydration based on disease signs and symptoms is less reliable in malnourished children, who often have decreased skin turgor, sunken eyes, and lethargy in the absence of dehydration.

Laboratory Features

The major laboratory abnormalities in patients with cholera are alterations in the concentration of serum electrolytes and creatinine and the effects of hemoconcentration on other blood constituents. The pathogenesis of the electrolyte disturbances is described above; typical serum electrolyte concentrations in children with cholera are shown in Table 2, and values in adults are similar.

Table 3. Classification of dehydration and fluid deficit based on clinical findings

Parameter	Sign, symptom, or laboratory finding with:		
	Mild or no dehydration	Some or moderate dehydration	Severe dehydration
Mentation	Alert	Restless or lethargic	Apprehensive, lethargic, stuporous, or comatose
Thirst	Present	Present	Marked
Radial pulse rate and character	Normal	Rapid	Rapid and feeble or impalpable
Respiratory rate and character	Normal	Tachypneic	Tachypneic, deep, labored
Elasticity of subcutaneous tissues as determined by skin pinch	Pinch retracts immediately	Pinch retracts slowly	Pinch retracts very slowly
Eyes	Normal	Sunken	Dramatically sunken
Urine flow	Normal	Scant and dark	Scant or absent
Serum specific gravity[a]	< 1.027	1.028–1.034	> 1.034
Approximate fluid deficit (mg/kg of body wt)	≤ 50	≥ 51–90	> 90
Preferred method of fluid replacement	ORT	Oral or intravenous, depending on presence of vomiting and rate of continued stool loss	Intravenous

[a]Patients with malnutrition may have a lower baseline specific gravity; thus, the values listed may not be applicable to these patients.

Total serum protein, serum specific gravity, and the hematocrit are increased as a result of the loss of much of the water component of the blood. Serum protein and serum specific gravity (which are closely correlated) are the most reliable indicators of the extent of dehydration, and both can be estimated at the bedside by using a simple and relatively inexpensive hand-held refractometer (78, 107). Most patients without dehydration have a serum specific gravity of 1.024 to 1.027, those with moderate dehydration have a serum specific gravity of 1.028 to 1.034, and those with severe dehydration have a serum specific gravity of > 1.034 (19, 78, 148). Malnourished persons often are hypoproteinemic in the absence of dehydration. Specific gravity measurements are therefore not a good indicator of initial volume status in malnourished patients. Measurement of specific gravity remains a reliable indicator of the adequacy of fluid replacement in malnourished patients, however. Once a baseline concentration is determined, the impact of volume replacement in lowering serum specific gravity will be proportionately similar in malnourished and well-nourished individuals.

Blood leukocyte concentrations in patients with cholera are often elevated, with the mean value reported in one study of children and adults with cholera being 21,000/mm³ (48). This increase in leukocyte concentration is presumably caused by the outpouring of stress hormones in response to hypovolemic shock and the resulting demargination of leukocytes adhering to vascular epithelium. Concentrations of the acute-phase reactant C-reactive protein (CRP) in serum are also elevated in patients with cholera. This increase in CRP is less easily explained, as CRP production by the liver is initiated by mediators of inflammation. The increase in CRP concentrations in patients with cholera suggests that there may be more alteration (and inflammation) of the intestinal mucosa than is commonly believed.

DIAGNOSIS

Cholera is a clinical diagnosis. Any patient with watery diarrhea and severe dehydration in an area where *V. cholerae* O1 is known to be endemic or epidemic should be suspected of having cholera and treated accordingly. Laboratory confirmation of *V. cholerae* infection is not required for the treatment of individual patients, as fluid replacement in patients with watery diarrhea is the same regardless of the infecting pathogen.

Although isolation of the organism from stool is the definitive means of diagnosing cholera (see chapter 11 for a detailed description of microbiologic methods for isolating *V. cholerae* O1), reliable and inexpensive rapid diagnostic methods are available. The most commonly used of these has been dark-field microscopy, which relies on identifying motile vibrios under dark-field illumination and immobilizing the motile vibrios with specific antisera (8). Newer rapid diagnostic methods based on detection of bacterial antigens or cholera toxin in stool samples have also been employed (26, 124, 134).

THERAPY

Fluid and Electrolyte Therapy

Background

The administration of fluid and electrolytes either intravenously or orally is the cornerstone of therapy for cholera (15, 108) (Table 4). When fluids are administered early in the course of illness, they can prevent dehydration from occurring. When administered later in the illness, after dehydration has developed, they are essential for restoring fluid balance and preventing death. The principles of fluid therapy remain those enumerated by O'Shaughnessy more than 160 years ago: ". . . the ingredients deficient in the blood were detected in the dejections, or in other words, the addition of the dejection to the

Table 4. Principles of fluid therapy in cholera patients

1. Fluid therapy is divided into two components: rehydration of dehydrated patients and maintenance of continuing stool losses.
2. All severely dehydrated patients should be rehydrated with intravenous fluids.
3. The intravenous fluid used should have a composition similar to that of cholera stool (see Table 1).
4. Rehydration can be accomplished rapidly within 2 to 4 h of admission.
5. Oral rehydration solution should be used for maintenance fluid therapy in all patients except those with continued high purging rates (> 10 ml/kg/h) or persistent vomiting. ORT can also be used to rehydrate patients with mild or moderate dehydration.
6. Oral rehydration solution should be administered in a volume sufficient to match ongoing stool losses and insensible losses. Stool losses should be monitored, and a record should be kept of stool losses and fluids provided.
7. Patients can be discharged from the hospital (or clinic) once the rate of stool volume has moderated and they have demonstrated that they can drink sufficient oral rehydration solution to keep pace with stool water loss. For most severely dehydrated patients, this will require a minimum of 24 h from the time of admission and will rarely require more than 72 h.

blood, in due proportion, would have restored the latter to its normal constitution" (27).

Although O'Shaughnessy and Latta had administered intravenous fluids to patients with cholera in the 1830s (using a goose trachea as intravenous tubing) and this therapy had been lifesaving for a number of the patients with whom it was tried (27), intravenous fluid therapy fell out of favor during the remainder of the 19th century. Although intravenous rehydration was more commonly practiced in the first half of this century, the fluids used often had inappropriate concentrations of electrolytes, and the volumes administered were predetermined and often did not keep pace with stool losses (18). It was only in the 1950s and 1960s, after carefully conducted studies by Phillips, Watten, Carpenter, and others (Table 5), that an intravenous fluid containing the proper concentration of electrolytes for use in cholera patients came into routine use (15, 19, 108).

Oral rehydration therapy (ORT) was developed in the 1960s and was based on the discovery that glucose-mediated cotransport of sodium and water across the mucosal surface of the small-intestine epithelium remains intact during cholera despite the poisoning by cholera toxin of other channels of intestinal sodium and water absorption (107, 131). The small-intestine glucose-sodium cotransport mechanism had been previously identified,

first in animal models and subsequently in human experiments, by a number of different investigators (35, 54). The potential importance of this cotransport mechanism for patients with toxin-mediated diarrhea was demonstrated by studies conducted with cholera patients in the Philippines, Bangladesh, and India (57, 107, 110). In those studies, perfusion of the small intestine with a glucose-containing polyelectrolyte solution (administered orally or with a nasogastric tube) enhanced net water absorption in the gut (57, 110). Following these studies, the utility of ORT in the treatment of diarrhea was quickly demonstrated in clinical and field trials conducted in Bangladesh and India (21, 22, 80, 96, 113, 132) and subsequently in trials conducted in other developing and developed countries (114).

Intravenous fluid therapy

Fluid therapy is divided into two phases: the rehydration phase, during which the water and electrolytes that have been lost by dehydrated patients are replaced, and the maintenance phase, during which ongoing stool losses are replaced. Intravenous fluid therapy remains the treatment of choice for the rehydration of patients who are severely dehydrated and for maintenance fluid replacement in patients with

Table 5. Research accomplishments from 1947 to 1970 crucial to developing effective intravenous and oral fluid therapies for cholera

Accomplishment	When accomplished	Where accomplished
Characterization of electrolyte content of cholera stools	Late 1940s	Egypt
	Late 1950s	Thailand and Philippines
Use of serum specific gravity to determine extent of	Late 1940s	Egypt
dehydration and vol of replacement fluid required	Late 1950s	Thailand and Philippines
Discovery that *V. cholerae* O1 secretes a toxin that causes intestinal fluid secretion	1959	India
Development of intravenous solutions containing appropriate concn of sodium, potassium, and bicarbonate	Late 1950s and early 1960s	Thailand, Bangladesh, and India
Demonstration that cholera does not cause morphologic insult to the gut	1960	Thailand
Development of the cholera cot for closely monitoring stool losses	Late 1950s	Thailand and Philippines
Demonstrations that antimicrobial therapy significantly decreases stool volume and duration of illness	Mid-1960s	Bangladesh and India
Demonstration of glucose-mediated sodium chloride and	1964	Philippines
water absorption in actively purging cholera patients	1968	Bangladesh and India
Clinical and field trials of oral rehydration solution	1968–1970	Bangladesh and India

persistent vomiting or high rates of stool losses. ORT is the preferred mode of therapy for rehydration of patients who are mildly dehydrated or who are moderately dehydrated but do not have excessive purging and vomiting and for maintenance of hydration following correction of dehydration.

There have been reports of patients with severe dehydration successfully rehydrated by oral therapy. An anecdotal report from a refugee camp with very limited supplies of intravenous fluids noted a cholera case fatality rate of 3.6% during an epidemic (80). On the other hand, a hospital-based study in which ORT was initially used to rehydrate all patients found that ORT failed to adequately rehydrate 29 of 60 patients who were initially severely dehydrated (28). In general, unless necessity absolutely demands otherwise, all severely dehydrated patients should be rehydrated with intravenous fluids.

Although there is no absolute cutoff for the rate of stool losses that require intravenous therapy during the maintenance phase, most patients with rates of stool loss greater than 10 ml/kg of body weight per h require intravenous therapy. In one study, all 24 cholera patients who were initially rehydrated intravenously and then had diarrhea rates of less than 10 ml/kg/h had hydration successfully maintained with ORT compared to 11 of 15 patients with diarrhea rates of 11 to 20 ml/kg/h and only 3 of 16 patients with rates greater than 20 ml/kg/h (104).

Intravenous rehydration of severely dehydrated patients can be accomplished within 2 to 4 h of the initiation of treatment (123). Although the standard practice in developed countries has been to rehydrate dehydrated patients with hypotonic solutions over a period of 24 h or more (54), there is no empiric evidence to support the use of such a slow rehydration regimen or the use of hypotonic solutions (which may actually be dangerous) (54, 75). There is, on the other hand, ample experimental evidence that rapid rehydration is safe and effective in both children and adults (54, 123). The practical advantages of a rapid rehydration regimen are conspicuous. During a cholera epidemic, the capacity of treatment facilities to care for patients would be quickly overwhelmed if a slow rehydration regimen were used and patients received intravenous solutions for 24 h or more.

The volume of fluid required for rehydration depends on the severity of dehydration and the rate of ongoing fluid losses (Table 3). In general, patients with severe dehydration require 90 to 120 ml (and occasionally even more) of intravenous fluid per kg of body weight plus replacement of ongoing water loss, those with moderate dehydration require 50 to 90 ml per kg of body weight plus replacement of ongoing water loss, and those with mild dehydration require 20 to 50 ml/kg of body weight plus replacement of ongoing water loss. For practical purposes, during the 2- to 4-h rehydration period, ongoing water losses can be assumed to consist of continuing losses from cholera stool, as insensible and urine water losses will make a minimal contribution to total water loss during this period.

Establishing venous access in severely dehydrated patients with vascular collapse can be difficult, especially in facilities with little previous experience in treating cholera patients. The antecubital or external jugular veins are probably the most accessible vessels for inserting an intravenous needle or catheter if hand veins cannot be accessed. In small children, scalp veins (and scalp needles) can be used. If venous access cannot be readily established by percutaneous venipuncture, an alternative means of providing fluids should be established as quickly as possible, while efforts at venipuncture continue. One alternative method for supplying fluid to severely dehydrated patients, especially those with diminished consciousness, is insertion of a nasogastric tube and instillation of oral rehydration solution (97), which will at least initiate the rehydration process. Rehydration has also been performed by the intraosseous and intraperitoneal routes (45, 127). The effectiveness of these routes for fluid administration in patients with diarrhea, especially cholera patients with severe dehydration, has not been well established, and some complications have been associated with their use (38). If all efforts at venipuncture fail, the best alternative for administering intravenous fluid is probably the surgical exposure of a vein (a "cut-down"). With a trained nursing and physical staff (and persistence), the need for venous cut-downs can be kept to a minimum. At the Treatment Centre of the International Centre for Diarrhoeal Disease Research in Bangladesh, over 100,000 patients with cholera have been treated without resort to a venous cut-down.

A number of intravenous fluids can be used for treatment of patients with cholera (Table 1). The ideal intravenous solution is one with an electrolyte composition similar to that of cholera stool (15, 19, 50, 51, 79, 81, 107, 108, 121, 123). Such a solution (Dhaka Solution) (123) is manufactured in Bangladesh, where cholera is a continuing public health problem. Polyelectrolyte solutions for use in rehydrating patients with diarrhea, including cholera diarrhea, have been manufactured in some other countries, with the intravenous solution used for pediatric cholera patients in Peru (the epicenter of the 1991 South American cholera epidemic) being identical to that of the standard World Health Organization (WHO) ORT solution (Table 1).

Lactated Ringer's solution is the most widely available commercially produced solution that resembles Dhaka Solution. Lactated Ringer's solution contains adequate amounts of sodium, chloride, and bicarbonate for rehydrating cholera patients but contains only 4 mmol of potassium per liter compared to 13 mmol of potassium per liter in Dhaka Solution (79, 123). The lower concentration of potassium in lactated Ringer's solution in comparison to that in Dhaka Solution can lead to hypokalemia, especially in malnourished children, who usually have a total body potassium deficit before contracting cholera. Other than this limitation, the two solutions (Dhaka Solution and lactated Ringer's solution) have equivalent efficacies when used to treat patients with cholera (79).

Normal saline solution has been used to treat patients with cholera (148), but it contains neither bicarbonate nor potassium. It should be used for treatment of cholera patients only when polyelectrolyte solutions are not available. Routinely adding electrolytes to stock solutions such as normal saline or glucose in water is not advisable. Cholera patients are often treated in field facilities during epidemics, and in such circumstances, adding electrolytes individually is not only not practicable but is also likely to lead to contamination of solutions or the inadvertant addition of inappropriate amounts of electrolytes.

Although each intravenous solution used for the rehydration of cholera patients has advantages and disadvantages, the critical determinant of outcome in patients with cholera is the administration of adequate fluid volume rather than the precise electrolyte concentration of the fluid that is administered. Within limits, once intravascular fluid volume is restored and glomerular filtration resumes, the kidney can regulate electrolyte concentrations in plasma.

If an adequate volume of fluid and appropriate electrolytes are provided, patients at the end of the rehydration phase of fluid therapy will have normal serum specific gravity and total protein concentration, near-normal sodium and chloride concentrations, and a potassium concentration lower than that on admission (123) (Table 2). The decrease in serum potassium concentrations occurs despite the administration during rehydration of large amounts of potassium and the absence of urine production in most patients. The absence of urine production is not a contraindication to including potassium as part of the initial rehydration fluid, since without the administration of potassium, serum potassium concentrations would fall even more precipitously. Creatinine concentrations decrease following the initial rehydration, but because not all excess creatinine is filtered by the kidneys immediately upon restoration of intravascular volume, creatinine concentrations can remain elevated for 24 h or more even in the absence of renal failure (123).

Oral fluid therapy

Once a severely dehydrated patient has been rehydrated with an intravenous solution,

administration of an oral rehydration solution is usually sufficient to maintain hydration. Using ORT during the maintenance phase of fluid therapy has at least three important advantages: it reduces the requirement for intravenous fluids, which are far more expensive than ORT, by up to 80% (96, 97); it reduces the risk of sepsis from contaminated intravenous solutions or apparatus; and it reduces the need for trained medical staff. Despite these advantages, many physicians and other medical personnel will persist in using intravenous fluids during the maintenance phase of therapy unless a conscious effort is made to promote the use of ORT (140). The use of ORT should be encouraged for all patients, and administration of ORT can start as soon as a severely dehydrated patient is alert and strong enough to ingest the solution.

The volume of oral therapy solution ingested during the maintenance phase of therapy should be sufficient to replace normal insensible fluid losses plus continuing losses from diarrhea and vomiting. Although there are a number of ways to calculate insensible fluid losses, the most practical determination is based on body weight. By this method, insensible fluid loss is calculated as being 100 ml/kg for the first 10 kg of body weight, 50 ml/kg of body weight for the next 10 kg (i.e., from 11 to 20 kg of body weight), and 10 ml/kg for every kilogram of body weight over 20 kg.

Monitoring ongoing fluid losses in actively purging patients is best done by having patients defecate into a calibrated bucket. This may be done by using a traditional cholera cot, which is a simple plastic-covered cot with a hole in the middle through which stool drains into a bucket (107) (Fig. 3), or by using a commode with a removable bucket, as has been done during the recent cholera epidemic in Peru. No matter which method is used, it is crucial to frequently monitor stool loss and fluid intake. In rapidly purging patients, this may have to be done every hour or even more often if a recurrence of dehydration is to be prevented. A sheet on which to

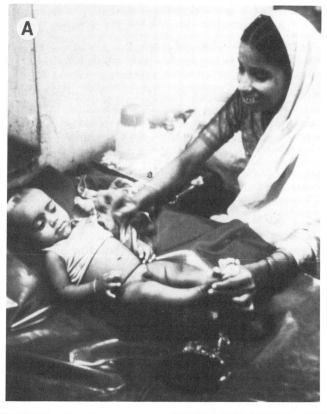

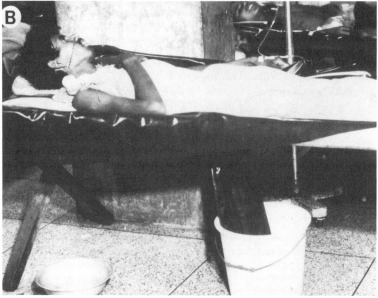

Figure 3. Cholera cots used for children and adults with diarrhea. This plastic-lined cot contains a hole in the middle (A) that allows stool to drain into a calibrated bucket (B) and facilitates recording of the volume of stool losses. Panel B is reprinted with permission (G. T. Keusch and M. L. Bennish, p. 603, *in* R. D. Feigin and J. D. Cherry, ed., *Textbook of Pediatric Infectious Diseases*, 1992).

record the volume of stool and the volume of fluids received should be kept at every patients' bedside. If stool loss exceeds intake, the patient should be encouraged to increase oral intake. If the patient is not capable of increasing the ingestion of oral rehydration solution, intravenous fluids should be restarted.

Repeated vomiting precludes the use of ORT (99) and is an additional indication for continued provision of intravenous fluids. Occasional vomiting (less than two or three times an hour), however, should not impede efforts to provide ORT. Caretakers should continue to monitor such patients very closely to ensure that they do not again become dehydrated. In general, once a patient has been adequately rehydrated, the frequency of vomiting markedly decreases.

As with intravenous solutions, a number of different formulations of oral fluid therapy solution have been used. The electrolyte content of the most commonly used oral rehydration solution, that recommended by the WHO and widely available in most countries (but not generally available in the United States), contains 90 mmol of sodium, 80 mmol of chloride, 20 mmol of potassium, and 30 mmol of bicarbonate or an equivalent base per liter (Table 1). The sodium concentration of WHO ORT is lower than the sodium concentration of intravenous solutions used for rehydration of patients with cholera. This is because ORT is routinely used to treat patients with watery diarrhea caused by rotavirus and other organisms in which the stool sodium concentration is lower than in cholera (94) and because once the rehydration phase of fluid replacement is completed in cholera patients, insensible water loss, which contains negligible amounts of sodium, accounts for a greater proportion of total water loss.

WHO ORT mix is widely available in developing countries as prepackaged sachets of salts and glucose that are dissolved in a predetermined volume of water (usually 500 ml or 1 liter) before being administered (Fig. 4). The ORT solution prepared from these sa-

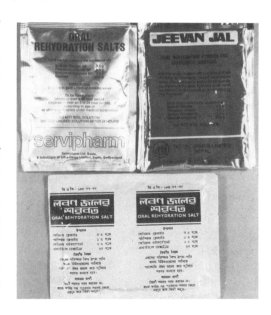

Figure 4. Prepackaged sachets of standard oral rehydration solution from different countries. These packages are available for sale from pharmacists. The contents are dissolved in water and administered to patients.

chets usually has an appropriate electrolyte concentration, especially if a container of standard size is provided for mixing the solution. The sachets of oral rehydration salts have a relatively long shelf life and are easily transportable from clinic (or pharmacy) to home. The sachets are inexpensive, usually costing no more than 25¢ (and often much less) in most developing countries. In the United States, sachets of WHO ORT may be ordered from Jianis Brothers Packaging Co. (2533 Southwest Blvd., Kansas City, MO 64108; telephone 816-421-2880). A travelers' packet containing 20 1-liter packets currently costs $25.00.

Alternative means of administering ORT are preparation of a sugar-salt solution from household ingredients or home preparation of a cereal-based salt-containing suspension (89, 92, 93, 106), the oligosaccharides of which are hydrolyzed to glucose within the gut lumen. A cereal-based suspension has the advantage, compared to either prepackaged sachets or homemade sugar-salt solutions, of tasting better and providing a limited number of additional calories. In some studies, cereal-based oral therapy has also decreased the duration of diarrhea and the volume of diarrheal stool compared to results with glucose-based solution (89, 92, 93, 106). The effect of cereal-based suspensions on stool volume has been greatest in children with cholera (46). In at least two studies (only one of which included patients with cholera), feeding cereal and other foods with standard ORT reduced stool volume to the level achieved with the use of a cereal-based suspension (2, 133). Both the home-prepared sugar-salt solution and the cereal-based suspension have the disadvantages of not containing appreciable quantities of potassium and bicarbonate and, because of often inappropriate preparation, are subject to considerable variation in sodium concentration (6, 24, 33). With more-intensive training and education of mothers on proper preparation, this variability can be decreased (122). When prepared in a hospital or clinic setting, the electrolyte concentration can be more closely monitored, and bicarbonate and potassium can be added. Bottles of premixed oral rehydration solution, using either rice or glucose as a substrate, are available in developed countries (114); because of cost and difficulties in transportation, these premixed solutions are not practical for use in developing countries.

ORT preparations containing organic acids in addition to glucose have also been evaluated (17, 69, 98). These preparations have the theoretic advantage of increasing sodium and water absorption, as the intestinal cotransport mechanism for organic acids such as glycine is distinct from that for glucose. In some studies, these enhanced ORT preparations have increased fluid absorption from the small intestine when compared with solutions containing only glucose. The enhanced solutions cost more to prepare, and none of the enhanced ORT preparations have had sufficient advantages in trials conducted to date to warrant their routine use.

ORT can be administered with a cup or, for small children, a spoon. Using bottles with nipples to provide ORT to small children should be discouraged, as the bottles used are often contaminated and as the use of bottles detracts from programs emphasizing the importance of breast feeding of infants. ORT should be continued for as long as a patient continues to have diarrhea, but patients need not be kept in a hospital or clinic for all this time. When the diarrhea volume abates and a patient is capable of ingesting sufficient ORT to maintain hydration, the patient can be discharged. This usually requires 24 to 72 h. On discharge, patients should be provided with packets of ORT mix and told to return to the clinic if the diarrhea worsens or if signs of dehydration reappear. Relapses are uncommon in patients who have been treated with an effective antimicrobial agent.

Antimicrobial Therapy

Although replacement of fluid lost in cholera stool remains the crucial element of treatment of cholera patients, antimicrobial therapy is an important adjunctive therapy, especially in patients with high purging rates. Antimicrobial therapy can reduce the volume of cholera stool purged during illness by half, as well as shorten the duration of symptoms and the duration of excretion of *V. cholerae* by about the same amount (40, 48, 61, 76, 111). Cholera patients who do not receive antimicrobial therapy will usually have diarrhea that continues for 4 to 6 days. This is reduced to 2 to 3 days in patients who have received antimicrobial therapy. If an inexpensive antimicrobial agent is used, antimicrobial treatment of cholera patients is extremely cost-effective, as it reduces the duration of hospital stay in addition to reducing the volume of intravenous and oral fluids required for hydration.

Options for antimicrobial therapy of *V. cholerae* O1 infections are shown in Table 6. Tetracycline was the first antimicrobial agent systematically evaluated for the treatment of

Table 6. Options for antimicrobial therapy of persons infected with *V. cholerae* serogroup O1[a]

Drug	Dose	
	Adult	Pediatric
Tetracycline	500 mg four times daily for 3 days or 1 g as a single dose	50 mg/kg of body wt/day divided into four doses for 3 days
Doxycycline	300 mg as a single dose	Not evaluated
Furazolidone	100 mg four times daily for 3 days	5 mg/kg of body wt/day divided into four doses and given for 3 days, or 7 mg/kg of body wt as a single dose
Trimethoprim-sulfamethoxazole	320 mg of trimethoprim-1,600 mg of sulfamethoxazole twice daily for 3 days	8 mg of trimethoprim-40 mg of sulfamethoxazole/kg of body wt/ day divided into two doses for 3 days.
Norfloxacin[b]	400 mg twice daily for 3 days	Not recommended for use in children

[a] Drug therapy should be based on a knowledge of the susceptibility pattern of isolates of *V. cholerae* O1 in the community. *V. cholerae* O1 isolates resistant to all of the listed antimicrobial agents (with the exception of norfloxacin) have been identified.
[b] Other second-generation quinolone agents are likely to be as effective as norfloxacin but have not been evaluated.

cholera (48), and it (or its congener, doxycycline) remains the most commonly used agent for the treatment of cholera. A single 1-g dose of tetracycline (61) or a single 300-mg dose of doxycycline (1) is as effective as longer courses of therapy in reducing stool volume in cholera patients. Short-course regimens, however, have consistently been less efficacious than 3- or 5-day courses of therapy in eradicating *V. cholerae* O1 from the stool (29, 61, 118, 120, 145). Although tetracycline is generally contraindicated in children less than 8 years old because it causes staining of permanent teeth, this is not thought to be a common problem when tetracycline is used for short courses of therapy such as that for treatment of cholera (74, 76). Antimicrobial agents other than tetracycline that have been evaluated and found effective in the treatment of cholera include erythromycin (13), furazolidone (67, 73, 111, 118, 120), trimethoprim-sulfamethoxazole (23, 40, 105), and the newer quinolone agent norfloxacin (12).

When compared to 3- or 5-day treatment regimens, the lower cost and greater ease of administration of short-course antimicrobial regimens outweigh the disadvantage of slower eradication of *V. cholerae* O1 from the stool. Although providing antimicrobial therapy to community contacts of patients with cholera has been shown to reduce the inci-

dence of disease (85), simply reducing the duration of excretion of *V. cholerae* O1 in patients who come to medical facilities for care is unlikely to have much impact on transmission of the disease within the community. Even if all patients with cholera who came to a medical facility for care were treated with an antimicrobial agent, there would remain a large environmental reservoir of *V. cholerae* O1 as well as many asymptomatic or mildly symptomatic persons who are excreting the organism and never come to medical attention. Although it is uncommon, some infected individuals continue to shed *V. cholerae* O1 for months without symptoms (4). Shortening the duration of excretion of *V. cholerae* O1 in patients in the hospital may affect transmission within hospitals, however, as nosocomial outbreaks of cholera have been reported (86, 142).

Antimicrobial therapy of cholera is complicated by the marked increase in multiply antibiotic-resistant strains during the last 20 years (102). Strains resistant to all of the first-line antimicrobial agents have been identified in disparate geographic regions (37, 43, 95). Currently in Bangladesh, more than 90% of *V. cholerae* O1 isolates are resistant to tetracycline, ampicillin, and trimethoprim-sulfamethoxazole (132a). Most isolates of *V. cholerae* O139, however, remain susceptible to these agents. Tetracycline-resistant strains have also been reported from India (65, 126), Thailand (143), and a number of African countries (37). Such multiply resistant strains are best treated with norfloxacin or another of the newer quinolones, all of which are likely to have equivalent efficacy in the treatment of *V. cholerae* O1 infection. The newer quinolone agents, however, cost substantially more than older antimicrobial agents, such as tetracycline, that have been the drugs of choice for treating cholera. If the new quinolones come into common use for treating *V. cholerae* infections, strains resistant to these agents will likely also appear. Because of the increase in multiply resistant strains of *V. cholerae* O1, it is important in areas where cholera is endemic or epidemic to periodically monitor the antimicrobial susceptibility pattern of *V. cholerae* O1.

When patients develop a high fever, nosocomial infection from a contaminated intravenous infusion or, alternatively, a fluid that is sterile but contains pyrogens should be suspected. The principal therapy for this complication is replacement of the intravenous infusion as well as the intravascular needle or catheter used for the infusion. If fever persists, the initiation of antimicrobial therapy directed at the gram-negative organisms and *S. aureus* that commonly cause intravenous therapy-related sepsis is indicated. Sepsis can largely be prevented if intravenous solutions are used for only a single patient (in an effort to conserve supplies and limit costs, some treatment centers unwisely save partially used intravenous solutions for use in another patient) and if intravenous catheters are removed as soon as they are no longer needed.

Treatment with Drugs Other Than Antimicrobial Agents

In vitro studies and animal model studies of cholera have identified a number of pharmaceutical agents that diminish cholera toxin-induced gut secretion (63). On the basis of these studies, a number of agents have been evaluated for their efficacies in reducing stool volume in cholera patients. Agents evaluated include the anti-inflammatory drugs aspirin (60) and indomethacin (144), the phenothiazine agent chlorpromazine (62, 119), nicotinic acid (115), the plant alkaloid berberine (116) (which has been used by traditional healers in India and China for the treatment of diarrhea), the antihypertensive agent clonidine (117), and somatostatin (90). Although some of these agents have reduced stool output in clinical trials, none has had a substantial enough effect to warrant inclusion as part of the routine therapy of cholera patients.

Passively acquired breast milk antibodies have been shown to protect infants from clinically apparent illness following exposure to *V. cholerae* O1 (44); administering hyperimmune bovine colostrum did not decrease stool volume in adults with cholera, however (84). Two separate trials evaluated the effectiveness of the B subunit of cholera toxin and GM_1 ganglioside (the cholera toxin receptor). The B subunit of toxin was administered to contacts of patients with cholera, and GM_1 ganglioside was given to actively purging cholera patients (42, 141). The rationale for evaluating B subunit was that it would block gut receptors for holotoxin, and the rationale for administering GM_1 ganglioside was that it would bind cholera toxin and thereby decrease binding of cholera toxin to receptors on the gut mucosal surface. Despite the seemingly convincing logic for using these two agents, neither had a marked effect in reducing stool volume, and B subunit did not appreciably reduce the incidence of disease (42, 141). Other biological agents tried unsuccessfully in the treatment of cholera include cholera phages (82) and *Streptococcus faecium* SF68 (sold commercially in Europe as Bioflorin) (87). Treatment with bismuth subsalicylate has had modest effects on stool volume and illness duration in travelers (32) and in children in developing countries (36) with noncholera diarrhea. It has not been evaluated in patients with cholera, however. Antiperistaltic agents, such as loperamide

(Imodium) and diphenoxylate hydrochloride (Lomotil), which are often promoted for treatment of patients with diarrhea, have no place in the treatment of patients with cholera. Similarly, antiemetic agents are ineffective in patients with cholera and may cause severe dystonic reactions.

Management of Hypoglycemia

It is impossible to determine on the basis of signs and symptoms at the time of admission to the hospital which patients with cholera are hypoglycemic. Most patients with severe dehydration have altered consciousness even if they are not hypoglycemic. On the other hand, some patients with hypoglycemia but without severe dehydration are alert. Because of the difficulty in detecting hypoglycemia, it is best to include dextrose in all intravenous solutions administered to cholera patients. Dextrose concentrations of 2 to 5% (2 to 5 g/liter of intravenous solution) are sufficient to prevent hypoglycemia and will not cause an osmotic diuresis in dehydrated patients even if a rapid rehydration regimen is used (123). Lactated Ringer's solution with 5% dextrose is commercially available and is pre-ferred to lactated Ringer's solution without dextrose for the treatment of cholera.

Adding dextrose to intravenous solutions for selected patients is an ineffective method of managing the problem of hypoglycemia. Because of the difficulty in ascertaining which patients are hypoglycemic, many such patients go untreated. Using a bedside glucometer to measure blood glucose will increase the detection rate of patients with hypoglycemia, but supplies for this determination are often unavailable in field facilities. Selectively adding dextrose to pre-packaged intravenous solutions also substantially increases the risk of contamination of the intravenous solution.

Nutritional Support

V. cholerae O1 infection does not substantially affect the ability of the intestinal mucosa to absorb nutrients (59, 88), and once the initial vomiting subsides, nausea and anorexia are not common problems. Therefore, patients with cholera can be fed when rehydration is complete, which is usually within 4 h of admission. Breast feeding should be continued in children who are being breast-fed.

Table 7. Supplies required for the treatment of cholera in field facilities

Supplies required	Supplies not required
1. Intravenous solution (5–10 liters/patient) 2. Needles: 16–20 gauge for adults; 21–25 gauge for children 3. Tape, arm splints, poles to hang intravenous drips 4. Sachets of oral rehydration salts (10–15 1-liter packages/patient) 5. 500-ml or 1-liter bottles for oral rehydration solution 6. Clean water for mixing oral rehydration solution 7. Antimicrobial agent effective against *V. cholerae* O1 8. Stethoscopes 9. Cholera cots 10. Plastic sheets 11. Calibrated buckets to collect stool and vomit 12. Bleach to disinfect stool 13. Hand soap 14. Nasogastric tubes 15. Paper and pencil to record intake and output.	1. Sphygmomanometers 2. Radiographs 3. Blood gas or electrolyte analyzers 4. Microbiologic culture facilities 5. Drugs other than antimicrobial agents for cholera

The effects of cholera on nutrition are not great unless food is mistakenly withheld from a patient, as is still all too commonly recommended by many physicians. Withholding food not only limits caloric intake but may affect intestinal mucosal enzymatic activity, which appears to be in part dependent on directly absorbed nutrients (56).

Supplies Required for Field Facilities

Because of the rapidity with which dehydration may occur in patients infected with *V. cholerae* O1, it is important to make care easily accessible to ill patients. This is a crucial issue in rural areas, where the time required to travel to existing health facilities may be too great to allow timely rehydration of patients with severe disease. When epidemics occur in rural areas, it is often necessary to establish temporary cholera treatment centers (137). The supplies required for such temporary facilities are quite straightforward (Table 7), and the costs not great.

REFERENCES

1. **Alam, A. N., N. H. Alam, T. Ahmed, and D. A. Sack.** 1990. Randomised double blind trial of single dose doxycycline for treating cholera in adults. *Br. Med. J.* **300:**1619–1621.
2. **Alam N. H., T. Ahmed, M. Khatun, and A. M. Molla.** 1992. Effects of food with two oral rehydration therapies: a randomized controlled clinical trial. *Gut* **33:**560–562.
3. **Albert, M. J., A. K. Siddique, M. S. Islam, A. S. G. Faruque, M. Ansaruzzaman, S. M. Faruque, and R. B. Sack.** 1993. Large outbreak of clinical cholera due to *Vibrio cholerae* non-O1 in Bangladesh. *Lancet* **341:**704. (Letter.)
4. **Azurin, J. C., K. Kobari, D. Barua, M. Alvero, C. Z. Gomez, J. J. Dizon, E.-I. Nakano, R. Suplido, and L. Ledesma.** 1967. A long-term carrier of cholera. Cholera Delores. *Bull. W.H.O.* **37:**745–749.
5. **Banwell, J. G., N. F. Pierce, R. C. Mitra, K. L. Brigham, G. J. Caranasos, R. J. Keimowitz, D. S. Fedson, J. Thomas, S. L. Gorbach, R. B. Sack, and A. Mondal.** 1970. Intestinal fluid and electrolyte transport in human cholera. *J. Clin. Invest.* **49:**183–195.
6. **Barros, F. C., C. G. Victoria, B. Forsberg, A. G. Maranhao, M. Stegeman, A. Gonzalez-** Richmond, R. M. Martins, Z. A. Neuman, J. McAullife, and J. A. Branco, Jr. 1991. Management of childhood diarrhoea at the household level: a population-based survey in north-east Brazil. *Bull. W.H.O.* **69:**59–65.
7. **Beisel, W. R., R. H. Watten, R. Q. Blackwell, C. Benyajati, and R. A. Phillips.** 1963. The role of bicarbonate pathophysiology and therapy in Asiatic cholera. *Am. J. Med.* **35:**58–66.
8. **Benenson, A. S., M. R. Islam, and W. B. Greenough III.** 1964. Rapid identification of *Vibrio cholerae* by darkfield microscopy. *Bull. W.H.O.* **30:**827–831.
9. **Bennish, M. L., A. K. Azad, O. Rahman, and R. E. Phillips.** 1990. Hypoglycemia during diarrhea. Prevalence, pathophysiology, and outcome. *N. Engl. J. Med.* **322:**1357–1363.
10. **Benyajati, C., M. Keoplug, W. R. Beisel, E. J. Gangarosa, H. Sprinz, and V. Sitprija.** 1960. Acute renal failure in Asiatic cholera: clinicopathologic correlations with acute tubular necrosis and hypokalemic nephropathy. *Ann. Intern. Med.* **52:**960–975.
11. **Bhattacharya, M. K., S. K. Bhattacharya, S. Garg, P. K. Saha, D. Dutta, G. B. Nair, B. C. Deb, and K. P. Das.** 1993. Outbreak of *Vibrio cholerae* non-O1 in India and Bangladesh. *Lancet* **341:**1346–1347. (Letter.)
12. **Bhattacharya, S. K., M. K. Bhattacharya, P. Dutta, D. Dutta, S. P. De, S. N. Sikdar, A. Maitra, A. Dutta, and S. C. Pal.** 1990. Double-blind, randomized, controlled clinical trial of norfloxacin for cholera. *Antimicrob. Agents Chemother.* **34:**939–940.
13. **Burans, J. P., J. Podgore, M. M. Mansour, A. H. Farah, S. Abbas, R. Abu-Elyazeed, and J. N. Woody.** 1989. Comparative trial of erythromycin and sulphatrimethoprim in the treatment of tetracycline-resistant *Vibrio cholerae* O1. *Trans. R. Soc. Trop. Med. Hyg.* **83:**836–838.
14. **Butler, T., M. Arnold, and M. Islam.** 1989. Depletion of hepatic glycogen in the hypoglycaemia of fatal childhood diarrhoeal illnesses. *Trans. R. Soc. Trop. Med. Hyg.* **83:**839–843.
15. **Carpenter, C. C. J.** 1992. The treatment of cholera: clinical science at the bedside. *J. Infect. Dis.* **166:**2–14.
16. **Carpenter, C. C. J., R. O. Biern, P. P. Mitra, R. B. Sack, P. E. Dans, S. A. Wells, and S. S. Khanra.** 1967. Electrocardiogram in Asiatic cholera. Separated studies of the effects of hypovolemia, acidosis, and potassium loss. *Br. Heart J.* **29:**103–111.
17. **Carpenter, C. C. J., W. B. Greenough, and N. F. Pierce.** 1988. Oral-rehydration therapy—the role of polymeric substrates. *N. Engl. J. Med.* **319:**1346–1348.
18. **Carpenter, C. C. J., P. P. Mitra, and R. B. Sack.** 1966. Clinical studies in Asiatic cholera. I.

Preliminary observations, November, 1962–March 1963. *Bull. Johns Hopkins Hosp.* **118**:165–173.

19. **Carpenter, C. C. J., A. Mondal, R. B. Sack, P. P. Mitra, P. E. Dans, S. A. Wells, E. J. Hinman, and R. N. Chaudhuri.** 1966. Clinical studies in Asiatic cholera. II. Development of 2:1 saline:lactate regimen. Comparison of this regimen with traditional modes of treatment, April and May, 1963. *Bull. Johns Hopkins Hosp.* **118**:174–196.

20. **Cash, R. A., S. I. Music, J. P. Libonati, M. J. Snyder, R. P. Wenzel, and R. B. Hornick.** 1974. Response of man to infection with *Vibrio cholerae*. I. Clinical, serologic, and bacteriologic responses to a known inoculum. *J. Infect. Dis.* **129**:45–52.

21. **Cash, R. A., D. R. Nalin, J. N. Forrest, and E. Abrutyn.** 1970. Rapid correction of acidosis and dehydration of cholera with oral electrolyte and glucose solution. *Lancet* **ii**:549–550.

22. **Cash, R. A., D. R. Nalin, R. Rochat, L. B. Reller, Z. A. Haque, and A. S. M. M. Rahman.** 1970. A clinical trial of oral therapy in a rural cholera-treatment center. *Am. J. Trop. Med. Hyg.* **19**:653–656.

23. **Cash, R. A., R. S. Northrup, and A. S. M. M. Rahman.** 1973. Trimethoprim and sulfamethoxazole in clinical cholera: comparison with tetracycline. *J. Infect. Dis.* **128**:S749–S753.

24. **Chowdhury, A. M. R., J. P. Vaughn, and F. H. Abed.** 1988. Use and safety of home-made oral rehydration solutions: an epidemiological evaluation from Bangladesh. *Int. J. Epidemiol.* **17**:655–665.

25. **Collins, B. J., F. P. L. Van Loon, A. Molla., A. M. Molla, and N. H. Alam.** 1989. Gastric emptying of oral rehydration solutions in acute cholera. *J. Trop. Med. hyg.* **92**:290–294.

26. **Colwell, R. R., J. A. Hasan, A. Huq, L. Loomis, R. J. Siebeling, M. Torres, S. Galvez, S. Islam, and D. Bernstein.** 1992. Development and evaluation of a rapid, simple, sensitive, monoclonal antibody-based co-agglutination test for direct detection of *Vibrio cholerae* O1. *FEMS Microbiol. Lett.* **76**:215–219.

27. **Cosnett, J. E.** 1989. The origins of intravenous fluid therapy. *Lancet* **i**:768–771.

28. **De, S.** 1975. Oral glucose-electrolyte solution in the treatment of cholera and other diarrhoeas. *J. Indian Med. Assoc.* **65**:230–232.

29. **De, S., A. Chaudhuri, P. Dutta, S. P. De, and S. C. Pal.** 1976. Doxycycline in the treatment of cholera. *Bull. W.H.O.* **54**:177–179.

30. **De, S. N., K. P. Sengupta, and N. N. Chandra.** 1954. Renal changes including total cortical necrosis in cholera. *Arch. Pathol.* **37**:505–515.

31. **Debongnie, J. C., and S. F. Phillips.** 1978. Capacity of the human colon to absorb fluid. *Gastroenterology* **74**:698–703.

32. **Dupont, H. L., P. Sullivan, L. K. Pickering, G. Haynes, and P. B. Ackerman.** 1977. Symptomatic treatment of diarrhea with bismuth subsalicylate among students attending a Mexican university. *Gastroenterology* **73**:715–718.

33. **Ellerbrock, T. V.** 1981. Oral replacement therapy in rural Bangladesh with home ingredients. *Trop. Doct.* **11**:179–183.

34. **Field, M., D. Fromm, Q. Al-Awqati, and W. B. Greenough III.** 1972. Effects of cholera enterotoxin on ion transport across isolated ileal mucosa. *J. Clin. Invest.* **51**:796–804.

35. **Field, M., M. C. Rao, and E. B. Chan.** 1989. Intestinal electrolyte transport and diarrheal diseases. *N. Engl. J. Med.* **321**:800–806, 879–883.

36. **Figueroa-Quintanilla, D., E. Salazar-Lindo, R. B. Sack, R. León-Barúa, S. Sarabia-Arce, M. Campos-Sanchez, and E. Eyzaguirre-Maccan.** 1993. A controlled trial of bismuth subsalicylate in infants with acute watery diarrheal disease. *N. Engl. J. Med.* **328**:1653–1658.

37. **Finch, M. J., J. G. Morris, Jr., J. Kavati, W. Kagwanja, and M. M. Levine.** 1988. Epidemiology of antimicrobial resistant cholera in Kenya and East Africa. *Am. J. Trop. Med. Hyg.* **39**:484–490.

38. **Galpin, R. D., J. B. Kronick, R. B. Willis, and T. C. Frewen.** 1991. Bilateral lower extremity compartment syndromes secondary to intraosseous fluid resuscitation. *J. Pediatr. Orthop.* **11**:773–776.

39. **Gangarosa, E. J., W. R. Beisel, C. Benyajati, H. Sprinz, and P. Piyaratn.** 1960. The nature of the gastrointestinal lesion in Asiatic cholera and its relation to pathogenesis. A biopsy study. *Am. J. Trop. Med. Hyg.* **9**:125–135.

40. **Gharagozloo, R. A., K. Naficy, M. Mouin, M. H. Nassirzadeh, and R. Yalda.** 1970. Comparative trial of tetracycline, chloramphenicol, and trimethoprim/sulphamethoxazole in eradication of *Vibrio cholerae* El Tor. *Br. Med. J.* **4**:281–282.

41. **Gitelson, S.** 1971. Gastrectomy, achlorhydria and cholera. *Isr. J. Med. Sci.* **7**:663–667.

42. **Glass, R.I., J. Holmgren, M. R. Khan, K. M. B. Hossain, M. I. Huq, and W. B. Greenough.** 1984. A randomized, controlled trial of the toxin-blocking effects of B subunit in family members of patients with cholera. *J. Infect. Dis.* **149**:495–500.

43. **Glass, R. I., I. Huq, A. R. M. A. Alim, and M. Yunus.** 1980. Emergence of multiply antibiotic resistant *V. cholerae* in Bangladesh. *J. Infect. Dis.* **142**:939–942.

44. **Glass, R. I., A.-M. Svennerholm, B. J. Stoll, M. R. Khan, K. M. B. Hossain, M. I. Huq, and**

J. Holmgren. 1983. Protection against cholera in breast-fed children by antibodies in breast milk. *N. Engl. J. Med.* **308:**1389–1392.

45. Goldstein, B., D. Doody, and S. Briggs. 1990. Emergency intraosseous infusion in severely burned children. *Pediatr. Emerg. Care* **6:**195–197.

46. Gore, S. M., O. Fontaine, and N. F. Pierce. 1992. Impact of rice based oral rehydration solution on stool output and duration of diarrhoea: meta-analysis of 13 clinical trials. *Br. Med. J.* **304:**287–291.

47. Greenough, W. B., III. 1990. *Vibrio cholerae,* p. 1640. In G. L. Mandell, R. G. Douglas, and J. E. Bennett (ed.), *Principles and Practice of Infectious Diseases.* Churchill Livingstone, New York.

48. Greenough, W. B., III, R. S. Gordon, I. S. Rosenberg, and B. I. Davies. 1964. Tetracycline in the treatment of cholera. *Lancet* **i:**355–357.

49. Greenough, W. B., III, N. Hirschhorn, R. S. Gordon, Jr., J. Lindenbaum, and K. M. Kelly. 1976. Pulmonary oedema associated with acidosis in patients with cholera. *Trop. Geogr. Med.* **28:**86–90.

50. Griffith, L. S. C., J. W. Fresh, R. H. Watten, and M. P. Villaroman. 1967. Electrolyte replacement in paediatric cholera. *Lancet* **i:**1197–1199.

51. Gutman, R. A., D. J. Drutz, G. E. Whalen, and R. H. Watten. 1969. Double blind fluid therapy evaluation in pediatric cholera. *Pediatrics* **44:**922–931.

52. Harvey, R. M., Y. Enson, M. L. Lewis, W. B. Greenough, K. M. Ally, and R. A. Panno. 1966. Hemodynamic effects of dehydration and metabolic acidosis in Asiatic cholera. *Trans. Assoc. Am. Phys.* **29:**177–186.

53. Harvey, R. M., Y. Enson, M. L. Lewis, W. B. Greenough, K. M. Ally, and R. A. Panno. 1968. Hemodynamic studies on cholera. Effects of hypovolemia and acidosis. *Circulation* **37:**709–728.

54. Hirschhorn, N. 1980. The treatment of acute diarrhea in children. An historical and physiological perspective. *Am. J. Clin. Nutr.* **33:**637–663.

55. Hirschhorn, N., A. K. M. A. Chowdhury, and J. Lindenbaum. 1969. Cholera in pregnant women. *Lancet* **i:**1230–1232.

56. Hirshhorn, N., and K. M. Denny. 1975. Oral glucose-electrolyte therapy for diarrhea: a means to maintain or improve nutrition? *Am. J. Clin. Nutr.* **28:**189–192.

57. Hirschhorn, N., J. L. Kinzie, D. B. Sachar, R. S. Northrup, J. O. Taylor, S. Z. Ahmad, and R. A. Phillips. 1968. Decrease in net stool output in cholera during intestinal perfusion with glucose-containing solutions. *N. Engl. J. Med.* **279:**176–181.

58. Hirschhorn, N., J. Lindenbaum, W. B. Greenough III, and S. M. Alam. 1966. Hypoglycemia in children with acute diarrhea. *Lancet* **ii:**128–133.

59. Hirschhorn, N., and A. Molla. 1969. Reversible jejunal disaccharidase deficiency in cholera and other acute diarrheal diseases. *Johns Hopkins Med. J.* **125:**291–300.

60. Islam, A., P. K. Bardhan, M. R. Islam, and M. Rahman. 1986. A randomized double blind trial of aspirin versus placebo in cholera and non-cholera diarrhoea. *Trop. Geogr. Med.* **38:**221–225.

61. Islam, M. R. 1987. Single dose tetracycline in cholera. *Gut* **28:**1029–1032.

62. Islam, M. R., D. A. Sack, J. Holmgren, P. K. Bardhan, and G. H. Rabbani. 1982. Use of chlorpromazine in the treatment of cholera and other severe acute watery diarrheal diseases. *Gastroenterology* **82:**1335–1340.

63. Jacoby, H. I., and C. H. Marshall. 1972. Antagonism of cholera enterotoxin by antiinflammatory agents in the rat. *Nature* (London) **235:**163–165.

64. Jamil, B., A. Ahmed, and A. W. Sturm. 1992. *Vibrio cholerae* O1 septicaemia. *Lancet* **340:**910–911. (Letter.)

65. Jesudason, M. V., and T. J. John. 1990. Transferable trimethoprim resistance of *Vibrio cholerae* O1 encountered in southern India. *Trans. R. Soc. Trop. Med. Hyg.* **84:**136–137.

66. Jesudason, M. V., and T. J. John. 1993. Major shift in prevalence of non-O1 and El Tor *Vibrio cholerae. Lancet* **341:**1090–1091. (Letter.)

67. Karchmer, A. W., G. T. Curlin, M. I. Huq, and N. Hirschhorn. 1970. Furazolidone in pediatric cholera. *Bull. W.H.O.* **43:**373–378.

68. Kawo, N. G., A. E. Msengi, A. B. M. Swai, L. M. Chuwa, K. G. M. M. Alberti, and D. G. McLarty. 1990. Specificity of hypoglycaemia for cerebral malaria in children. *Lancet* **336:**454–457.

69. Khin-Maung-U, M.-K., and Nyunt-Nyunt-Wai. 1991. Comparison of glucose/electrolyte and maltodextrin/glycine/glycyl-glycine/electrolyte oral rehydration solutions in cholera and watery diarrhoea in adults. *Ann. Trop. Med. Parasitol.* **85:**645–650.

70. Kimberg, D. V., M. Field, E. Gershon, and A. Henderson. 1974. Effects of prostaglandins and cholera enterotoxin on intestinal mucosal cyclic AMP accumulation. Evidence against an essential role for prostaglandins in the action of the toxin. *J. Clin. Invest.* **53:**941–949.

71. Kimberg, D. V., M. Field, J. Johnson, A. Henderson, and E. Gershon. 1971. Stimulation of intestinal mucosal adenyl cyclase by cholera enterotoxin and prostaglandins. *J. Clin. Invest.* **50:**1218–1230.

72. Kitamoto, O., K. Hiraishi, T. Takigami, M. Saito, and N. Shimizu. 1967. Electrocardiographic findings in cholera El Tor patients. *Bull. W.H.O.* **37:**787–793

73. Kobari, K., I. Takakura, M. Nakatomi, S. Sogame, and C. Uylangco. 1970. Antibiotic-resistant strains of El Tor Vibrio in the Philippines and the use of furalazine for chemotherapy. *Bull W.H.O.* **43**:365–371.

74. Kucers, A., and N. M. Bennett. 1988. The use of antibiotics. A comprehensive review with clinical emphasis, 4th ed., p. 1004–1007. J. B. Lippincott Co., Philadelphia.

75. Levin, M., J. Brown, and D. Mabey. 1987. Danger of hypotonic half-Darrow's solution. *Lancet* **i**:1204. (Letter.)

76. Lindenbaum, J., W. B. Greenough, and M. R. Islam. 1967. Antibiotic therapy of cholera in children. *Bull. W.H.O.* **37**:529–538.

77. Lindenbaum, J. L., R. S. Gordon, Jr., N. Hirschhorn, R. Akbar, W. B. Greenough III, and M. R. Islam. 1966. Cholera in children. *Lancet* **i**:1066–1068.

78. Love, A. H. G., and R. A. Phillips. 1969. Measurement of dehydration in cholera. *J. Infect. Dis.* **119**:39–42.

79. Mahalanabis D., J. B. Brayton, A. Mondal, and N. F. Pierce. 1972. The use of Ringer's lactate in the treatment of children with cholera and acute noncholera diarrhoea. *Bull. W.H.O.* **46**:311–319.

80. Mahalanabis, D., A. B. Choudhuri, N. C. Bagchi, A. K. Bhattacharya, and T. W. Simpson. 1973. Oral fluid therapy of cholera among Bangladesh refugees. *Johns Hopkins Med. J.* **132**:197–204.

81. Mahalanabis, D., C. K. Wallace, R. J. Kallen, A. Mondal, and N. F. Pierce. 1970. Water and electrolyte losses due to cholera in infants and small children: a recovery balance study. *Pediatrics* **45**:374–385.

82. Marcuk L. M., V. N. Nikiforov, J. F. Scerbak, T. A. Levitov, R. I. Kotljarova, M. S. Naumsina, S. U. Davydov, K. A. Monsur, M. A. Rahman, M. A. Latif, R. S. Northrup, R. A. Cash, I. Huq, C. R. Dey, and R. A. Phillips. 1971. Clinical studies of the use of bacteriophage in the treatment of cholera. *Bull. W.H.O.* **45**:77–83.

83. Mathias, J. R., G. M. Carlson, A. J. DiMarino, G. Bertiger, H. E. Morton, and S. Cohen. 1976. Intestinal myoelectric activity in response to live *Vibrio cholerae* and cholera enterotoxin. *J. Clin. Invest.* **58**:91–96.

84. McClead, R. E., T. Butler, and G. H. Rabbani. 1988. Orally administered bovine colostral anticholera toxin antibodies: results of two clinical trials. *Am. J. Med.* **85**:811–816.

85. McCormack, W. M., A. M. Chowdhury, N. Jahangir, A. B. F. Ahmed, and W. H. Mosley. Tetracycline prophylaxis in families of cholera patients. *Bull. W.H.O.* **38**:787–792.

86. Mhalu, F. S., F. D. E. Mtango, and A. E. Msengi. 1984. Hospital outbreaks of cholera transmitted through close person to person contact. *Lancet* **ii**:82–84.

87. Mitra, A. K., and G. H. Rabbani. 1990. A double-blind, controlled trial of bioflorin (*Streptococcus faecium* SF68) in adults with acute diarrhea due to *Vibrio cholerae* and enterotoxigenic *Escherichia coli*. *Gastroenterology* **99**:1149–1152.

88. Molla, A., A. M. Molla, A. Rahim, S. A. Sarker, Z. Mozaffar, and M. Rahman. 1982. Intake and absorption of nutrients in children with cholera and rotavirus infection during acute diarrhea and after recovery. *Nutr. Res.* **2**:233–242.

89. Molla, A. M., S. M. Ahmed, and W. B. Greenough III. 1985. Rice-based oral rehydration solution decreases the stool volume in acute diarrhoea. *Bull. W.H.O.* **63**:751–756.

90. Molla, A. M., K. Gyr, P. K. Bardhan, and A. Molla. 1984. Effect of intravenous somatostatin on stool output in diarrhea due to *Vibrio cholerae*. *Gastroenterology* **87**:845–847.

91. Molla, A. M., M. Hossain, R. Islam, P. K. Bardhan, and S. A. Sarker. 1981. Hypoglycemia: a complication of diarrhea in childhood. *Indian Pediatr.* **18**:181–185.

92. Molla, A. M., M. Hossain, S. A. Sarker, A. Molla, and W. B. Greenough III. 1982. Rice-powder electrolyte solution as oral therapy in diarrhoea due to *Vibrio cholerae* and *Escherichia coli*. *Lancet* **i**:1317–1319.

93. Molla, A. M., A. Molla, S. K. Nath, and M. Khatun. 1989. Food-based oral rehydration salt solution for acute childhood diarrhoea. *Lancet* **ii**:429–431.

94. Molla, A. M., M. Rahman, S. A. Sarker, D. A. Sack, and A. Molla. 1981. Stool electrolyte content and purging rates in diarrhea caused by rotavirus, enterotoxigenic *E. coli*, and *V. cholerae* in children. *J. Pediatr.* **98**:825–838.

95. Morris, J. G., J. H. Tenney, and G. L. Drusano. 1985. In vitro susceptibility of pathogenic *Vibrio* species to norfloxacin and six other antimicrobial agents. *Antimicrob. Agents Chemother.* **28**:442–445.

96. Nalin, D. R., R. A. Cash, R. Islam, M. Molla, and R. A. Phillips. 1968. Oral maintenance therapy for cholera in adults. *Lancet* **ii**:370–373.

97. Nalin, D. R., R. A. Cash, and M. Rahaman. 1970. Oral (or nasogastric) maintenance therapy for cholera patients in all age groups. *Bull. W.H.O.* **43**:361–36.

98. Nalin, D. R., R. A. Cash, M. Rahman, and M. Yunus. 1970. Effect of glycine and glucose on sodium and water absorption in patients with cholera. *Gut* **11**:768–772.

99. Nalin, D. R., M. M. Levine, R. B. Hornick, E. J. Berquist, D. Hoover, H. P. Holley, D. Water-

man, J. VanBlerk, S. Matheny, S. Sotman, and M. Rennels. 1978. The problem of emesis during oral glucose-electrolyte therapy given from the onset of severe cholera. *Trans. R. Soc. Trop. Med. Hyg.* **73**:10–14.

100. Nalin, D. R., M. M. Levine, J. Rhead, E. Bergquist, M. Rennels, T. Hughes, S. O'Donnell, and R. B. Hornick. 1978. Cannabis, hypochlorhydria, and cholera. *Lancet* ii:859–862.

101. Nalin, D. R., R. J. Levine, M. M. Levine, D. Hoover, E. Bergquist, J. McLaughlin, J. Libonati, J. Alam, and R. B. Hornick. 1978. Cholera, non-vibrio cholera and stomach acid. *Lancet* ii:856–859.

102. O'Grady, F., M. J. Lewis, and N. J. Pearson. 1976. Global surveillance of antibiotic sensitivity of *Vibrio cholerae*. *Bull. W.H.O.* **54**:181–185.

103. Owen, R. L., N. F. Pierce, R. T. Apple, and W. C. Cray, Jr. 1986. M cell transport of *Vibrio cholerae* from the intestinal lumen into Peyer's patches: a mechanism for antigen sampling and for microbial transepithelial migration. *J. Infect. Dis.* **153**:1108–1118.

104. Palmer, D.L., F. T. Koster, A. F. M. R. Islam, A. S. M. M. Rahman, and R. B. Sack. 1977. Comparison of sucrose and glucose in the oral electrolyte therapy of cholera and other severe diarrheas. *N. Engl. J. Med.* **297**:1107–1110.

105. Pastore, G., G. Rizzo, and G. Fera. 1977. Trimethoprim-sulfamethoxazole in the treatment of cholera. Comparison with tetracycline and chloramphenicol. *Chemotherapy* **23**:121–128.

106. Patra, F. C., D. Mahalanabis, K. N. Jalan, A. Sen, and P. Banerjee. 1982. Is oral rice-electrolyte solution superior to glucose electrolyte solution in infantile diarrhoea? *Arch. Dis. Child.* **57**:910–912.

107. Phillips, R. A. 1964. Water and electrolyte losses in cholera. *Fed. Proc.* **23**:705–718.

108. Phillips, R. A. 1967. Twenty years of cholera research. *JAMA* **202**:610–613.

109. Phillips, S. F., and J. Giller. 1973. The contribution of the colon to electrolyte and water conservation in man. *J. Lab. Clin. Med.* **81**:733–746.

110. Pierce, N. F., J. Banwell, R. C. Mitra, G. J. Caranasos, R. I. Keimowitz, A. Mondal, and P. M. Manji. 1968. Effect of intragastric glucose-electrolyte infusion upon water and electrolyte balance in Asiatic cholera. *Gastroenterology* **55**:333–343.

111. Pierce, N. F., J. G. Banwell, R. C. Mitra, G. J. Caranasos, J. R. I. Keimowitz, J. Thomas, and A. Mondal. 1968. Controlled comparison of tetracycline and furazolidone in cholera. *Br. Med. J.* **3**:277–280.

112. Pierce, N. F., C. C. J. Carpenter, H. L. Elliott, and W. B. Greenough. 1971. Effects of prostaglandins and theophylline on net transmural water and electrolyte movement in the canine jejunum. *Gastroenterology* **60**:22–32.

113. Pierce, N. F., R. B. Sack, R. C. Mitra, J. G. Banwell, K. L. Brigham, D. S. Fedson, and A. Mondal. 1969. Replacement of water and electrolyte losses in cholera by an oral glucose-electrolyte solution. *Ann. Intern. Med.* **70**:1173–1181.

114. Pizzaro, D., G. Posada, L. Sandi, and J. R. Moran. 1991. Rice-based oral electrolyte solutions for the management of infantile diarrhea. *N. Engl. J. Med.* **324**:517–521.

115. Rabbani, G. H., T. Butler, P. K. Bardhan, and A. Islam. 1983. Reduction of fluid-loss in cholera by nicotinic acid: a randomized controlled clinical trial. *Lancet* ii:1439–1442.

116. Rabbani, G. H., T. Butler, J. Knight, S. C. Sanyal, and K. Alam. 1987. Randomized, controlled trial of berberine sulfate therapy for diarrhea due to enterotoxigenic *Escherichia coli* and *Vibrio cholerae*. *J. Infect. Dis.* **155**:979–984.

117. Rabbani, G. H., T. Butler, D. Patte, and R. L. Abud. 1989. Clinical trial of clonidine hydrochloride as an antisecretory agent in cholera. *Gastroenterology* **97**:321–325.

118. Rabbani, G. H., T. Butler, M. Shahrier, R. Mazumdar, and M. R. Islam. 1991. Efficacy of a single dose of furazolidone for treatment of cholera in children. *Antimicrob. Agents Chemother.* **35**:1864–1867.

119. Rabbani, G. H., W. B. Greenough III, J. Holmgren, and B. Kirkwood. 1982. Controlled trial of chlorpromazine as antisecretory agent in patients with cholera hydrated intravenously. *Br. Med. J.* **284**:1361–1364.

120. Rabbani, G. H., M. R. Islam, T. Butler, M. Shahrier, and K. Alam. 1989. Single-dose treatment of cholera with furazolidone or tetracycline in a double-blind randomized trial. *Antimicrob. Agents Chemother.* **33**:1447–1450.

121. Rahaman, M. M., M. A. Majid, and K. A. Monsur. 1979. Evaluation of two intravenous rehydration solutions in cholera and non-cholera diarrhoea. *Bull. W.H.O.* **57**:977–981.

122. Rahman, A. S. M., A. Bari, A. M. Molla, and W. B. Greenough III. 1985. Mothers can prepare and use rice-salt oral rehydration solution in rural Bangladesh. *Lancet* ii:539–540.

123. Rahman, O., M. L. Bennish, A. N. Alam, and M. A. Salam. 1988. Rapid intravenous rehydration by means of a single polyelectrolyte solution with or without dextrose. *J. Pediatr.* **113**:654–660.

124. Ramamurthy, T., S. K. Bhattacharya, Y. Uesaka, K. Horigome, M. Paul, D. Sen, S. C. Pal, T. Takeda, Y. Takeda, and G. B. Nair. 1992. Evaluation of the bead enzyme-linked immunosorbent assay for detection of cholera toxin directly from stool specimens. *J. Clin. Microbiol.* **30**:1783–1786.

125. Ramamurthy, T., S. Garg, R. Sharma, S. K. Bhattacharya, G. B. Nair, T. Shimada, T. Takeda, T. Karasawa, H. Kurazano, A. Pal, and Y. Takeda. 1993. Emergence of novel strain of *Vibrio cholerae* with epidemic potential in southern and eastern India. *Lancet* **341**:703–704. (Letter.)

126. Ramamurthy, T., A. Pal, M. K. Bhattacharya, S. K. Bhattacharya, A. S. Chowdhury, Y. Takeda, T. Takeda, S. C. Pal, and G. B. Nair. 1992. Serovar, biotype, phage type, toxigenicity, and antibiotic susceptibility patterns of *Vibrio cholerae* isolated during two consecutive cholera seasons (1989–90) in Calcutta. *Indian J. Med. Res.* **95**:125–129.

127. Ransome-Kuti, O., O. Elebute, T. Agusto-Odutola, and S. Ransome-Kuti. 1969. Intraperitoneal fluid infusion in children with gastroenteritis. *Br. Med. J.* **3**:500–503.

128. Rao, A., and B. A. Stockwell. 1980. The Queensland cholera incident of 1977. *Bull. W.H.O.* **58**:661–664.

129. Rapoport, S., M. Dodd, M. Clark, and I. Sylem. 1947. Postacidotic state of infantile diarrhea: symptoms and chemical data. Postacidotic hypocalcemia and associated decreases in levels of potassium, phosphorus and phosphatase in the plasma. *Am. J. Dis. Child.* **73**:391–441.

130. Read, N. W. 1982. Diarrhea: the failure of colonic salvage. *Lancet* **ii**:481–483.

131. Rohde, J. E., and R. A. Cash. 1973. Transport of glucose and amino acids in human jejunum during Asiatic cholera. *J. Infect. Dis.* **127**:190–192.

132. Sack, R. B., J. Cassells, R. Mitra, C. Merritt, T. Butler, J. Thomas, B. Jacobs, A. Chaudhuri, and A. Mondal. 1970. The use of oral replacement solutions in the treatment of cholera and other severe diarrhoeal disorders. *Bull. W.H.O.* **43**:351–360.

132a. Salam, M. A. Personal communication.

133. Santosham, M., I. M. Fayad, M. Hashem, J. G. Goepp, M. Rafat, and R. B. Sack. 1990. A comparison of rice-based oral rehydration solution and "early feeding" for the treatment of acute diarrhea in infants. *J. Pediatr.* **116**:868–875.

134. Shaffer, N., E. Silva do Santos, P. Andreason, and J. J. Farmer III. 1989. Rapid laboratory diagnosis of cholera in the field. *Trans. R. Soc. Trop. Med. Hyg.* **83**:119–120.

135. Sharp, G. W. G., and S. Hynie. 1971. Stimulation of intestinal adenyl cyclase by cholera toxin. *Nature* (London) **229**:266–269.

136. Shimada, T., G. B. Nair, B. C. Deb, M. J. Albert, R. B. Sack, and Y. Takeda. 1993. Outbreak of *Vibrio cholerae* non-O1 in India and Bangladesh. *Lancet* **341**:1347. (Letter.)

137. Siddique, A. K., P. Mutsuddy, Q. Islam, Y. Mazumder, A. Mitra, and A. Eusof. 1990.

Makeshift treatment centre during a cholera epidemic in Bangladesh. *Trop. Doct.* **20**:83–85.

138. Speelman P., T. Butler, I. Kabir, A. Ali, and J. Banwell. 1986. Colonic dysfunction during cholera infection. *Gastroenterology* **91**:1164–1170.

139. Speelman, P., G. H. Rabbani, K. Bukhave, and J. Rask-Madsen. 1985. Increased jejunal prostaglandin E_2 concentrations in patients with acute cholera. *Gut* **26**:188–193.

140. Stoll, B. J., R. I. Glass, M. I. Huq, M. U. Khan, J. E. Holt, and H. Banu. 1982. Surveillance of patients attending a diarrhoeal disease hospital in Bangladesh. *Br. Med. J.* **285**:1185–1189.

141. Stoll, B. J., J. Holmgren, P. K. Bardhan, I. Huq, W. B. Greenough III, P. Fredman, and L. Svennerholm. 1980. Binding of intraluminal toxin in cholera: trial of GM1 ganglioside charcoal. *Lancet* **ii**:888–891.

142. Swaddiwudhipong, W., and P. Kunasol. 1989. An outbreak of nosocomial cholera in a 755-bed hospital. *Trans. R. Soc. Trop. Med. Hyg.* **83**:279–281.

143. Tabtieng, R., S. Wattanasri, P. Echeverria, J. Serwitana, L. Bodhidatta, A. Chatkaeomorakot, and B. Rowe. 1989. An epidemic of *Vibrio cholerae* El Tor Inaba resistant to several antibiotics with a conjugative group C plasmid coding for type II dihydrofolate reductase. *Am. J. Trop. Med. Hyg.* **41**:680–686.

144. Van Loon, F. P., G. H. Rabbani, K. Bukhave, and J. Rask-Madsen. 1992. Indomethacin decreases jejunal fluid secretion in addition to luminal release of prostaglandin E_2 in patients with acute cholera. *Gut* **33**:643–645.

145. Wallace, C. K, P. N. Anderson, T. C. Brown, S. R. Khanra, G. W. Lewis, N. F. Pierce, S. N. Sanyal, G. V. Segre, and R. H. Waldman. 1968. Optimal antibiotic therapy in cholera. *Bull. W.H.O.* **39**:239–245.

146. Wallace, C. K., A. E. Fabie, O. Mangubat, E. Velasco, C. Juinio, and R. A. Phillips. 1964. The 1961 cholera epidemic in Manila, Republic of the Phillippines. *Bull. W.H.O.* **30**:795–810.

147. Wang, F., T. Butler, G. H. Rabbani, and P. K. Jones. 1986. The acidosis of cholera. Contributions of hyperproteinemia, lactic acidemia, and hyperphosphatemia to an increased anion gap. *N. Engl. J. Med.* **315**:1591–1595.

148. Watten, R. H., F. M. Morgan, Y. N. Songkhla, B. Vanikiati, and R. A. Phillips. 1959. Water and electrolyte studies in cholera. *J. Clin. Invest.* **38**:1879–1889.

149. Watten, R. H., and R. A. Phillips. 1960. Potassium in the treatment of cholera. *Lancet* **ii**:999–1001.

Vibrio cholerae and Cholera: Molecular to Global Perspectives
Edited by I. Kaye Wachsmuth, Paul A. Blake, and Ørjan Olsvik
© 1994 American Society for Microbiology, Washington, DC 20005

Chapter 16

Immunity to *Vibrio cholerae* Infection

Ann-Mari Svennerholm, Gunhild Jonson, and Jan Holmgren

Cholera infection evokes significant antibacterial as well as antitoxin antibody responses both in intestine and in serum. In animals, both of these antibody specificities protect against experimental cholera and cooperate synergistically in mediating protective immunity against experimental infection. Since neither the bacteria nor cholera toxin (CT) penetrates beyond the intestinal epithelium during infection, protective immunity against cholera is thought to depend solely on preventive mechanisms operating against colonization or toxin action in the intestine, i.e., in the gut lumen, in the mucous layer, or on the epithelium surface. In this review we outline present knowledge on the immune mechanisms operating against cholera based on studies of experimental animals infected with *Vibrio cholerae* as well as of humans living in areas where cholera is endemic or convalescing from cholera disease.

SPECIFICITY OF PROTECTIVE IMMUNITY

The relative roles of antibacterial and antitoxin immunities in cholera remain incom-

Ann-Mari Svennerholm, Gunhild Jonson, and Jan Holmgren • Department of Medical Microbiology and Immunology, University of Göteborg, Guldhedsgatan 10, S-413 46 Göteborg, Sweden.

pletely defined. Owing to the critical role of CT in the pathogenesis of *V. cholerae* O1, much interest has been focused on the importance of antitoxic immunity for protection against cholera. However, studies both with human volunteers and with animals infected with live *V. cholerae* have provided evidence of strong antibacterial immunity that inhibits and often completely prevents bacterial colonization upon rechallenge.

Antitoxic Immune Responses in Animals

The classic animal model for studying enterotoxin-induced fluid secretion in the gut is the ligated small bowel loop technique (10). This method has been used in different species, e.g., rabbits, rats, and mice, to assess the protective effect of antibacterial and antitoxic antibodies either in active or passive immunization studies. Using this model, we could show that immunization with CT or its B subunits (CTB) as well as with whole vibrios or isolated lipopolysaccharide (LPS) resulted in significant protection against experimental cholera in rabbits (21, 63). Furthermore, combined immunization with toxin and bacterial antigens induced a protective effect that was considerably better than the sum of the effects induced by each antigen component alone; i.e., antibacterial and antitoxic immunities cooperated synergistically in protecting against cholera.

Studies with rats and mice have shown an indirect correlation between protection against CT-induced fluid secretion and intestinal synthesis of secretory immunoglobulin A (sIgA) antibodies and also between protection and the number of antitoxin-producing cells in the intestine (40, 64). These results, together with the nature of cholera disease, suggest that locally formed sIgA antibodies are of major importance for providing antitoxic immunity in the gut.

Immune responses against CT are mainly directed against the CTB portion of the toxin molecule. Furthermore, neutralization of CT is primarily provided by anti-CTB, whereas antibodies against the A-subunit portion of CT (CTA) have very poor neutralizing capacity. Thus, antibodies against CTB are equally effective as antibodies against holotoxin in protecting against CT-induced fluid accumulation in the gut (20, 61). The neutralization does not necessarily require direct steric blocking of the GM_1-binding site of the B subunit but may result from antibody-mediated confirmational changes induced via other epitopes.

CTA, on the other hand, seems to have strong immunomodulating properties in various species of animals. Thus, whereas peroral administration of isolated CTB is very inefficient in inducing antienterotoxin immune responses in mice, rats, or rabbits, feeding of CT has resulted in strong local antitoxic antibody formation (40, 51). Indeed, CT may also potentiate immune responses to nonrelated protein antigens up to 50-fold, as determined by the increase in antibody-producing cells in the intestine when the antigen was given perorally together with CT (40). This strong adjuvant action has a multifactorial background; e.g., CT has been found to increase uptake of antigens across intestinal epithelial cells, stimulate antigen-presenting cells, induce isotype switching of IgA in B cells, and strongly increase antigen-specific CD4 T-cell priming (39, 40).

The findings that the two biotypes of *V. cholerae* O1, classical and El Tor, produce CTs that differ in a few amino acids in the CTB monomer and that it is possible to induce biotype-specific monoclonal antibodies (MAbs) against CT (33) suggest the existence of biotype-specific antitoxin immunity. However, rabbit polyclonal antisera raised against classical CTB seemed to react equally well with purified B subunits from classical and El Tor strains in immunoblotting analyses (33). Furthermore, polyclonal antisera against classical CT are equally effective in neutralizing classical and El Tor toxins, and immunization with classical or El Tor CTB has resulted in equally high titers against both toxins as determined by different enzyme-linked immunosorbent assays (ELISAs) and toxin neutralization tests (unpublished data).

Antibacterial Immunity

To evaluate anticolonization cholera immunity in animals, nonligated-intestine models are preferable. By using the reversible intestinal tie adult rabbit diarrhea (RITARD) model (58), strong protection against disease caused by rechallenge with live vibrios of homologous as well as heterologous biotypes and serotypes has been demonstrated (23, 41, 71). The protection induced was associated with marked inhibition of vibrio colonization and large rises in titer of vibriocidal and anti-LPS antibodies in serum (41), supporting a protective role for antibacterial antibodies. Another animal model that has been extensively used to study antibacterial cholera immunity is the suckling mouse assay (4). Using this model, Attridge and Rowley (3), for instance, could demonstrate strong protection against experimental cholera by antibacterial antibodies of different specificities.

Results from studies during the 1960s and early 1970s suggested that *V. cholerae* LPS is the most important antigen in affording antibacterial cholera immunity. Thus, immunization with isolated LPS induced significantly increased resistance to experimental cholera in animals (64). Furthermore, purified LPS was capable of inducing significant protection against clinical cholera in a field trial in Pakistan (46). We were also able to show that elimination of anti-LPS antibodies from polyclonal antisera raised against whole killed vibrios (that had been cultured to stationary phase) resulted in complete loss of the passive protective effect of such sera against experimental cholera (61).

The capacity of LPS from the different serotypes of *V. cholerae* O1, i.e., Inaba and Ogawa, to induce cross-protective immunity has also been studied. In a number of cholera field trials that used monovalent cholera vaccines, vaccines made from Inaba serotypes protected against homologous as well as Ogawa serotypes but not vice versa (11). This lack of cross-protection by Ogawa vaccines could have been due either to formation of very low titers of antibodies to the shared *V. cholerae* O1 LPS determinant A or to different protective effects of antibodies against the type-specific determinants, i.e., B for Ogawa and C for Inaba. The low efficiency of Ogawa vaccines may also be due to possible serotype conversion from Ogawa to Inaba during infection, a conversion that has been described on the genetic level (59); conversion from Inaba to Ogawa may also occur if a functional *rfbT* gene is present in the infecting strain. The finding of serotype conversion in all cholera epidemics since 1959, including the ongoing one in Latin America, suggests that serotype-specific immunity may develop during natural infection (see chapter 23).

Although LPS apparently is responsible for much of the antibacterial cholera immunity induced, both by cholera vaccines and by *V. cholerae* infection, additional antigens, some of which may be expressed only during certain culture conditions or during growth in the intestine, may contribute to antibacterial protective immunity. Thus, evidence for non-LPS protective antigens is our observation that an initial intestinal infection-immunization of rabbits with rough *V. cholerae* O1 mutants or non-O1 *V. cholerae* bacteria conferred significant protection against subsequent challenge with fully virulent smooth cholera vibrios in the absence of any anti-O1 (anti-LPS) antibody response. The immunity induced seemed to be directed against colonization, since reinfected animals excreted the strain used for rechallenge for less than 2 days, i.e., during a significantly shorter period than this strain was excreted in nonimmunized animals. Furthermore, antibacterial antibodies against non-LPS structures afforded significant protection against experimental cholera in mice (3). Some of these protective antibodies seem to be directed against outer membrane proteins, i.e., both porin-like proteins and other surface-located proteins (54, 56). In particular, an 18-kDa protein that promotes colonization in the infant rabbit model (54) and is present in both biotypes has been suggested as an important protective antigen.

During recent years, interest has been focused on the role of different putative colonization factors in *V. cholerae* O1 in inducing protective immunity. Previously, the toxin-coregulated pilus (TCP) has aroused the most interest in this regard. This pilus is a colonization factor of classical vibrios (70), and antibodies against TCP afford anticolonization immunity (47a, 60) that may enhance anti-LPS in protecting against cholera caused by classical vibrios. However, TCP is poorly immunogenic in humans. Thus, immune responses against TCP could not be recorded either in serum or in intestinal specimens collected from human volunteers after challenge with classical vibrios (18), and only low responses have been observed in a few instances in patients convalescing from clinical cholera (18; see Table 2). There is also some evidence that the possession of TCP may not be a prerequisite for colonization and virulence of *V. cholerae,* since bacteria harvested from human cholera (El Tor) stools have been TcpA negative in four of five specimens tested (29). Furthermore, despite the facts that the genes for TCP are present in *V. cholerae* O1 of both biotypes and that the subunit protein TcpA may be detected by sodium dodecyl sulfate-polyacrylamide gel electrophoresis (SDS-PAGE) immunoblotting of classical as well as El Tor vibrios, assembled pili have been detected only on the surfaces of vibrios of the classical biotype (26). Thus, neither in vitro culture conditions yielding high levels of TcpA and CT nor growth of bacteria in vivo, e.g., in ileal loops, resulted in expression of TCP on the surface of El Tor vibrios, as could be detected by antibodies against classical TCP (29). These results may explain the poor protection afforded by anti-TCP antibodies against El Tor infection (60; chapter 27). Moreover, although classical vibrios express TCP at a high concentration in vivo, e.g., in the RITARD model, no significant anti-TCP antibody response could be detected in the infected animals (unpublished results).

Another putative colonization factor in *V. cholerae* O1 that may induce anticolonization immunity is the mannose-sensitive hemagglutinin (MSHA) that has been identified on all vibrio strains of El Tor biotype but rarely on classical vibrios (30). Recently we demonstrated that MSHA is a pilus that is expressed on the surfaces of El Tor but not classical vibrios after culture in vivo and in vitro (27, 28). By analogy with TCP, the MSHA subunit protein is present in vibrios of both biotypes, whereas MSHA activity is associated with El Tor bacteria. Experimental infection with El Tor vibrios in the RITARD model induces anti-MSHA antibody responses in serum (see Table 1), although the responses were of relatively low magnitude. In accordance with the biotype-associated expression of MSHA, recent studies have shown that MAbs as well as specific polyclonal antisera raised against MSHA confer significant protection against experimental cholera caused by El Tor but not classical vibrios (49).

Other antigens in *V. cholerae* that may promote colonization and induce specific immune responses are the fucose-sensitive hemagglutinin associated with classical vibrio strains (30) and the mannose-fucose-resistant hemagglutinin, which is a 25-kDa protein

present in both biotypes (13). In addition, various outer membrane proteins (OMPs) including iron-regulated OMPs and also nonspecified hemagglutinins have been implicated as virulence factors and important protective antigens (12, 17, 43, 54, 56). Thus, antibodies against non-LPS components of the outer membranes may block the adhesion of vibrios to the intestinal mucosa, and such antibodies have been shown to protect against challenge with vibrios of both biotypes.

The observation that motile vibrios are more virulent than nonmotile strains has been regarded as evidence that the polar flagellum of *V. cholerae* is a virulence factor. Accordingly, it has been suggested that antiflagellar antibodies that may be induced by experimental cholera infection can contribute to protection by preventing colonization of vibrios in the gut (2, 73). Recently, however, it was claimed that active immunization of rabbits with purified surface-located flagellar antigen (40- and 38-kDa proteins) did not afford protection in the ileal loop model. This was in contrast to the effect of immunization with a 33-kDa nonspecified antigen (57). This discrepant result may be explained by the fact that *V. cholerae* O1 flagellae normally are covered by a sheath that extends from the outer membrane and that may prevent specific antibodies against purified flagellin from binding to the flagellar structure.

Secreted Antigens

Other factors that may facilitate the colonization process of *V. cholerae* are mucinases, proteases, and neuraminidase (34, 74). One of these factors is the soluble hemagglutinin that is an extracellular protease that can proteolytically cleave (nick) and thereby activate CTA (5). MAbs as well as polyclonal antisera against soluble hemagglutinin have also provided some protection against experimental cholera in animals (72). The protection afforded by the polyclonal antiserum may have been due to antinicking as well as antimucinase activities (32, 72), whereas it was clearly shown that the effect obtained with the MAbs was not due to inhibition of mucinase activity (72). Instead, the protective effect of the MAbs may have been due to their capacity to inhibit proteolytic cleavage and thus activation of CTA. During infection, however, proteolytic cleavage of the toxin may also be provided by the action of proteases, e.g., trypsin, in the gut fluid. Thus, human intestinal fluids have been found to nick the A subunit of the related *Escherichia coli* heat-labile enterotoxin (LT) molecule (unpublished results). However, although soluble hemagglutinin is a strong immunogen when given parenterally and is produced (although in very variable concentrations) by most *V. cholerae* O1 strains in vitro, it is poorly immunogenic in vivo. Thus, analyses of numerous serum and intestinal specimens collected from cholera patients in areas of endemicity as well as from American volunteers convalescing from clinical cholera revealed significant immune responses to soluble hemagglutinin in only a few instances (68).

In Vivo Antigens

Recent studies have shown that *V. cholerae* may express a number of antigens during growth in vivo that cannot be identified on corresponding bacteria grown under various conditions in vitro (31, 32). Experimental cholera infection may also induce "in vivo-specific" antibody formation. Thus, antisera collected from cholera-infected rabbits reacted with several antigens that were present in vibrios derived from the intestine but not in vibrios grown under different conditions in vitro as shown by immunoblotting (25a, 32). Also, antisera from rabbits given one immunizing dose of either rough *V. cholerae* O1 or non-O1 *V. cholerae* reacted with some of these low-molecular-weight in vivo-induced antigens (as shown in Fig. 1).

To further evaluate the extent of immune responses to bacterial antigens, including in vivo-specific antigens, the frequency and

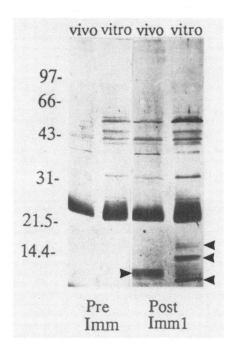

Figure 1. Immunoblot showing the antibody response after infection with rough *V. cholerae* O1 (strain 57) against in vivo-grown (vivo) and in vitro-grown (vitro) classical vibrios (Cairo 48). Antigens (one of which was in vivo specific) evoking the strongest responses had molecular masses of 10 to 18 kDa (indicated by arrowheads). The strong reactivity with the 22-kDa area reflects the presence of antibodies cross-reacting with core LPS in both pre- and postimmune sera. Low-molecular-mass references from Bio-Rad are indicated (in kilodaltons).

specificities of the antibody responses induced by *V. cholerae* O1 in rabbits infected according to the RITARD model were studied (Table 1). All of these animals were protected against challenge infection with a 1,000-fold-higher dose of homologous or heterologous *V. cholerae* O1 given 14 days later. The 10 animals infected with either classical or El Tor vibrios responded with vibriocidal antibody formation after the initial infection. In most cases, the vibriocidal antibody responses were accompanied by significant anti-LPS antibodies.

Table 1. Serum antibody responses in rabbits after intestinal infection with smooth and rough
V. cholerae O1 and non-O1 *V. cholerae*

Immunizing infection[a] (strain)	Frequency of responders[b]							
	Vibriocidal antibodies[c]	Antibody response against:						
		LPS[c]	CT	TCP	MSHA	≤ 18-kDa antigens	≤ 18-kDa antigens in vivo	62 to 150-kDa antigens in vivo
Classical Inaba (Cairo 48)	4/4	2/4	2/4	1/4	0/4	4/4	2/4	4/4
El Tor Inaba (T19479)	6/6	5/6	6/6	0/6	4/6	6/6	4/6	3/6
Rough *V. cholerae* O1 (381)	3/10	2/10[d]	1/10	0/2	0/2	0/2	2/2	0/2
Rough *V. cholerae* O1 (57)	3/6	3/6[d]	1/6	NT[e]	NT	2/4	2/4	1/4
Non-O1 *V. cholerae* O1 (WBDH 712)	0/5	0/5	1/5	NT	1/1	2/3	2/3	0/3

[a]Rabbits were infected with smooth (10^3 to 10^5 bacteria) or rough (10^6 to 10^{11} bacteria) *V. cholerae* O1 or with non-O1 *V. cholerae* (10^9 to 10^{11} bacteria) in the RITARD model, and sera were collected 14 days later.
[b]Number of rabbits with significant (24) antibody response/total number studied.
[c]Against *V. cholerae* O1 bacteria or LPS.
[d]Antibody response against core LPS as determined by immunoblotting analyses.
[e]NT, not tested.

Antibody responses against TCP and MSHA as well as against the fimbrial subunits TcpA and MshA (as studied by ELISA or immunoblotting) were variable in animals. Thus, only one of four rabbits infected with classical vibrios in the RITARD model developed a low antibody response against TCP, and this result correlated with the result from immunoblotting analyses (not shown). However, four of five rabbit preimmune serum samples tested reacted strongly with TcpA in immunoblotting, although these antibodies could not be absorbed with TCP-positive whole bacteria. This finding may suggest the presence of cross-reactive antibodies in preimmune sera, since a recent report (50) has shown that antibodies against subunits of type 4 pili other than TCP may cross-react immunologically with TcpA, making evaluation of anti-TcpA antibody responses by immunoblotting difficult. Four of 6 rabbits intestinally infected with El Tor vibrios responded with anti-MSHA antibodies, but none showed a response against TCP. These immune sera also reacted significantly better with MshA than did preimmune sera in immunoblotting. Interestingly, a rabbit immunized with non-O1 vibrios also responded with anti-MSHA (and anti-MshA) antibodies (Table 1).

In addition to these defined antibody responses after infection, we observed immune responses against various antigens present in whole bacteria grown in vitro and in vivo. Immunoblotting analyses revealed the presence of antibodies to strong antigens in the range of 10 to 18 kDa in the majority of rabbits immunized either with smooth or rough *V. cholerae* O1 or with non-O1 *V. cholerae*. Some of these antibodies reacted only with in vivo-grown bacteria. Immune responses to in vivo-induced antigens in the range of 62 to 150 kDa (31) were also evoked (Table 1). In view of these results, the additional role of non-LPS antigens, e.g., various fimbriae, hemagglutinins, OMPs, flagellar antigens, etc., in affording protective immunity should be further evaluated.

IMMUNE MECHANISMS AGAINST CHOLERA

Since neither the vibrios nor CT penetrate beyond the intestinal epithelium during infection, protective immunity is thought to depend solely on preventive mechanisms operating against colonization or toxin action in the gut.

Nonspecific Factors

A number of nonspecific defense mechanisms may act as a first line of defense to prevent *V. cholerae* O1 from colonizing the small intestine (20, 64). These factors include the acid milieu in the stomach, to which vibrios are very sensitive; intestinal peristalsis, which is effective in preventing bacterial colonization; and intestinal secretions, which continuously wash the intestinal surface and contain a number of proteolytic enzymes. For example, neutralization of gastric acidity before ingestion of *V. cholerae* results in a drastic decrease in the infectious dose required to induce classical cholera in human volunteers (7). Other mechanisms possibly include host-derived antibiotic factors (35); the normal flora, which competes with more pathogenic microorganisms for essential metabolites; and the mucous layer, which functions as a mechanical and chemical barrier.

sIgA

The best-known entity providing specific immune protection for the gut is the sIgA system. In noninvasive enteric infections like cholera, sIgA appears to be the main, although perhaps not the only, protective molecule (44). Systemic immunity may play only a limited role, although antibodies may diffuse from the circulation and act on the mucosal surface (20). Such circulating antibodies are, however, predominantly of the IgG isotype and are consequently very sensitive to proteolytic degradation by intestinal enzymes. The resistance of sIgA to intestinal proteases (44) makes antibodies of this isotype uniquely well suited to protect intestinal mucosal surfaces.

sIgA antibodies against different *V. cholerae* O1 antigens may appear within a week after onset of infection and even more rapidly after repeated exposure to cholera antigens (62). The protective effect of such antibodies may be to prevent the vibrios or the secreted CT from binding to and further acting on the intestinal epithelial cells (20, 64). sIgA may have additional functions in protecting against disease. Thus, such antibodies have been shown to mediate antibody-dependent

T-cell-mediated cytotoxicity in other experimental enteric infections; to interfere with the utilization of necessary growth factors, e.g., iron, for bacterial pathogens in the intestinal milieu; and to facilitate antigen uptake by Peyer's patches (6, 69).

Immunologic Memory for sIgA Responses

Although the intestinal sIgA response to antigen exposure is of relatively short duration, lasting for only weeks to a few months, this system has been shown to exhibit for more than a year a potent immunologic memory that could rapidly and efficiently be stimulated by repeated antigen exposure. Lycke and Holmgren (38) have shown a practically life-long immunologic memory in mice after peroral immunization, since they could adoptively transfer an anamnestic immune response in intestine by isolated B cells 1 year after oral priming with CT. Such a long-lasting local immunologic memory in the intestine has also been shown in humans. Thus, a single oral dose of CTB given to adult Bangladeshi volunteers 15 months after initial oral immunization with CTB resulted in antibody responses in serum and intestine that were significantly earlier appearing and stronger than corresponding antitoxin responses seen in adult Bangladeshis previously not given any peroral cholera vaccine (62). Furthermore, Swedish adult volunteers who had been given peroral cholera vaccine containing CTB 5 years earlier responded to a single oral dose of vaccine with an IgA antitoxin response considerably stronger than that of concurrently immunized Swedes who had not been given CTB previously (25). Moreover, memory antitoxin-producing cells have been isolated from peripheral blood lymphocytes of Swedish volunteers after peroral vaccination with CTB-containing cholera vaccine (52).

Cell-Mediated Immune Responses

Cell-mediated immune reactions such as major histocompatibility complex-restricted cellular cytotoxicity, natural killer cell activity, and antibody-dependent cytotoxicity are probably of little importance in preventing a noninvasive infection like cholera. Recently, however, evidence has accumulated that gamma-interferon-producing T cells in the gut mucosa might also play a role in intestinal immune defense. Such cells are present in very large numbers in the human duodenal mucosa, and there they may undergo a significant increase in numbers following intestinal antigenic exposure (52). Gamma interferon has several properties that may contribute to protection against cholera. For example, it may enhance IgA production, increase the expression of major histocompatibility complex antigen as well as secretory component receptors on the enterocyte surface, and probably allow antigen presentation by different mucosal cells including enterocytes. It may also interfere with tight junction permeability between intestinal epithelial cells (42) as well as with active electrolyte secretion by enterocytes, thereby preventing enterotoxin-induced fluid secretion (19).

ANTIBODY RESPONSES IN HUMANS

The immune responses induced by clinical cholera in humans appear to be very efficient in providing prolonged protective immunity against subsequent infection with *V. cholerae* O1 bacteria. This prolonged protection seems to be due to the capacity of the infection to induce high levels of specific sIgA antibodies as well as immunologic memory for such antibody formation locally in the intestine (44, 52). In addition to containing locally produced sIgA antibodies, intestinal secretions may contain specific antibodies that have diffused from the circulation, i.e., predominantly antibodies of the IgG and IgM classes. However, antibodies of these isotypes are very sensitive to proteolytic degradation and, e.g., IgM antibodies have very poor toxin-neutralizing capacities. Hence, evaluation of mucosal immune responses to cholera should

include determination of sIgA antibodies or intestinally derived immunocytes producing such antibodies against the most important protective antigens.

Methods for Determination of sIgA Antibodies against *V. cholerae* O1

Determination of intestinal antibody responses in humans has been subject to considerable problems with regard both to collection of clinical specimens and to lack of suitable antibody detection methods. Analyses of stool specimens for specific antibodies may present difficulties in extracting the coproantibodies and also in comparing antibody levels before and after immunization because of variable sIgA concentrations in specimens collected on different days (40, 64). Other approaches frequently used to assess intestinal immune responses include determination of sIgA antibodies in intestinal aspirates or of antibody-producing cells in intestinal biopsy specimens (37, 52). However, these methods allow determination of specific antibody formation only in the duodenum and maybe the proximal jejunum or rectum.

Another approach that has been used by us and others (47, 66) is to determine specific antibodies in intestinal washings collected by the lavage method. By this method, the entire intestine is rinsed by letting the volunteer drink several liters of an isotonic salt solution until a watery diarrhea ensues. The rice water stools seen in patients with fulminant clinical cholera may represent natural lavage. The specific sIgA antibodies formed after either artificial lavage or natural disease may then be determined from the clear stools by different ELISAs. By introducing a number of modifications of the lavage method, including prevention of gastric enzyme activity through addition of multiple enzyme inhibitors (47), addition of high concentrations of nonspecific proteins, and freeze-drying of specimens (1a), the intestinal lavage method allows collection of intestinal specimens with a sIgA content that is very stable for at least 2 years after collection (1). Although laborious, this method allows collection of specimens containing antibodies produced in the entire gastrointestinal tract.

On the basis of our knowledge of a common mucosal immune system (6, 20), the correlation of immune responses to cholera in intestinal secretions with those in other more easily available secretions such as milk, saliva, and serum has been evaluated. Unfortunately, no nonintestinal secretion that satisfactorily allows such indirect determinations has been identified. However, serum IgG and particularly IgA antitoxin responses induced by oral antigen stimulation reflect intestinal sIgA responses against CT relatively well (24).

Although clinical cholera and peroral administration of whole-cell cholera vaccines usually result in substantial IgA antibody responses against *V. cholerae* LPS locally in intestine, only modest and irregular rises in anti-LPS titers have been observed in serum (15, 66). However, antibacterial antibodies as determined by different vibriocidal antibody techniques are usually observed in serum both after clinical disease and after peroral cholera vaccination. Indeed, it has been suggested that the levels of vibriocidal antibodies may correlate with protection and reflect successful stimulation of antibacterial immunity in the intestine (15, 24, 45), although parenteral cholera vaccines, which confer only limited protection of short duration, are usually very efficient in inducing high vibriocidal antibody responses in serum (21, 66). The discrepancy between anti-LPS and vibriocidal antibody responses is surprising, since LPS is probably the prime contributory antigen in the vibriocidal assay and is strongly immunogenic in animal studies. Possible explanations may be that the vibriocidal test detects anti-LPS antibodies of broader specificity and different isotypes (mainly IgM) than anti-LPS ELISAs or that the vibriocidal test better discriminates between antibodies of different affinities (24). Another possibility may be

that vibriocidal antibodies are directed against other surface antigens in addition to LPS, although absorption of immune sera with purified LPS regularly has resulted in elimination of their vibriocidal antibody activity.

As an alternative to determining sIgA antibodies in intestinal fluids, the number of antibody-secreting cells found either in intestinal biopsy samples or circulating in the blood may be assayed by using the so-called enzyme-linked immunospot method (9, 52). By this approach, suspensions of lymphoid cells are incubated in tissue culture medium in antigen-coated wells, and the specific antibodies formed are determined by conventional ELISA. Single antibody-secreting cells of different isotypes could then be recognized as colored spots appearing on the antigen-coated surface. By using this method, human lymphoid cells secreting specific antibodies of different isotypes and with different specificities can be detected.

Antibody Responses Induced by Clinical Cholera or Asymptomatic Infection

In patients convalescing from clinical cholera in areas of endemicity, the disease gives rise to high levels of intestinal sIgA antibodies not only against CT but also against the cell wall LPS (66). Thus, a majority of Bangladeshi patients convalescing from moderate to severe cholera produced high levels of antitoxin as well as anti-LPS IgA antibodies in intestinal lavage fluid 1 to 2 weeks after onset of disease. Elevated sIgA levels often appear also in milk and saliva after clinical disease, and most convalescents develop significant antitoxic IgG and IgA titer rises as well as vibriocidal antibody responses in serum. These antibody levels have usually been significantly higher in convalescents than in healthy persons living in the same area who have been subjected to repeated natural exposure to *V. cholerae* organisms (15, 66).

The observation of increased vibriocidal antibody titers in serum with age in persons living in areas where cholera is endemic suggests that significant serum antibody responses to cholera may also develop in response to repeated natural exposure (45). Antitoxic antibody levels, on the other hand, do not show the same increase with age, probably because of high infection rates with LT-producing *E. coli* in early childhood. Furthermore, preexisting antibacterial antibodies may prevent colonization by enterotoxin-producing bacteria and thus boosting with toxin antigen.

Studies with human volunteers have also demonstrated that adult Americans experimentally infected with *V. cholerae* O1 responded with high levels of specific antitoxic and antibacterial (vibriocidal) antibodies in serum as well as with antitoxin IgA antibodies in jejunal fluids (36). Comparison of antitoxic immune responses in patients convalescing from cholera and from disease caused by LT-producing *E. coli* showed that antitoxin titers in serum developed both against CT and LT but that the titer against the homologous toxin was always higher than that against the heterologous toxin (65).

In recent studies we have also evaluated whether *V. cholerae* O1 infection can result in significant antibody responses to antigens other than CT and LPS. Thus, sera from adult Peruvians infected with *V. cholerae* El Tor bacteria during the ongoing epidemic in Latin America were analyzed for antibody responses against TCP and MSHA. As shown in Table 2, ca. 60% of the cholera convalescents responded with significant antibody responses against El Tor-associated MSHA pili, whereas none of them responded to TCP (68a). Significant vibriocidal and antitoxin immune responses in serum were, however, observed in all of the convalescents. Whereas the magnitude of the antibody responses against MSHA was very modest in most cases, i.e., two- to four-fold in the responders, vibriocidal and antitoxic titer rises were considerably higher. Indeed, the antitoxin responses were comparable to or even higher than corresponding responses in vol-

Table 2. Serum antibody responses to different *V. cholerae* antigens in Peruvians convalescing from clinical cholera (68a)

Antibody	Frequency of response (no. responding/ no. tested)	Median magnitude of response (fold)
Anti-MSHA[a]	16/27	3
Anti-TCP[b]	0/16	
Vibriocidal (Inaba)	28/29	>10
Anti-CT[c]	29/29	25

[a]As determined by ELISA using a specific MAb against MSHA followed by purified MSHA for coating.
[b]As determined by ELISA using a specific MAb against TCP (48) followed by purified TCP for coating.
[c]As determined by GM$_1$-ELISA (65).

unteers living in areas where cholera is endemic (15).

Antibody responses to *V. cholerae* O1 OMPs after experimental challenge with different *V. cholerae* strains have also been studied (53, 55). These analyses revealed that approximately half of the volunteers recovering from experimentally induced cholera responded to several of the OMPs (22 to 38 kDa) of *V. cholerae* O1 in serum. Also, antibodies reacting with in vivo-induced antigens of high molecular mass (84 to 152 kDa) were detected (53). In jejunal fluids of these volunteers, however, the non-LPS antibacterial sIgA antibodies were directed only against antigens of low molecular mass, i.e., of <25 kDa. Some of these low-molecular-mass antigens were present only in in vivo-grown bacteria.

Immune sera from Bangladeshi cholera convalescents, cholera-infected human volunteers, and Swedish vaccinees were also studied for reactivity with in vivo-induced proteins (31; unpublished data). Sera from all these groups of volunteers contained antibodies against in vivo-specific antigen proteins of molecular masses ranging from 62 to 200 kDa. When healthy human sera were examined, however, seven of nine serum samples from adults not exposed to cholera but none of nine serum samples from 18-month-old children contained antibodies reacting with several of the high-molecular-mass proteins (140 to 200 kDa) seen in SDS-PAGE-separated *V. cholerae* O1. These proteins may therefore not be restricted to *V. cholerae*, and it is possible that they are important for adaptation to the intestinal milieu.

We have also evaluated the presence of different *V. cholerae* O1 antigens, e.g., TCP, MSHA, and other fimbriae, on El Tor vibrios collected from stools from Bangladeshi cholera patients without subculturing (29; unpublished data) and the presence of specific antibodies against these antigens in the stool specimens. Fimbriae were seen on the vibrios in most of the specimens; in addition to fimbriae reacting with an MSHA-specific MAb (67), at least three different kinds of aggregating fimbriae were seen (28, 29). None of the six specimens studied contained bacteria expressing TCP or antibodies against this pilus. However, IgA antibody titers against CT, LPS, whole bacteria, and MSHA were seen in most of the stools that were collected from patients a couple of days after onset of disease (unpublished data).

In addition to eliciting an sIgA antibody response in intestine that is usually of short duration (lasting for one or a few months after onset of disease), cholera infection may also induce long-lasting immunologic memory. Induction of such a memory may explain why protective immunity after cholera infection can be seen in the absence of significant mucosal levels of sIgA antibodies. As support for this assumption, we have found that oral administration of relatively low doses of cholera antigens to cholera convalescents results in considerably higher and earlier appearing sIgA antibody responses in intestinal secretions than are observed in similarly immunized healthy controls (62, 66). Furthermore, infection with *E. coli* that produces LT that cross-reacts immunogically with CT could prime the intestine to respond with an anamnestic IgA response to a subsequent oral vaccination with CTB (22).

Symptomatic or asymptomatic infections with *V. cholerae* in early childhood may result in partial or complete immunity to cholera disease or even infection. This is supported by the finding of a reverse disease-to-infection rate with age for cholera in children in areas where cholera is endemic, e.g., in Bangladesh (45). Thus, the age group between 2 and 9 years is affected at higher frequencies than older children, adults, and younger children. The relatively low cholera disease rate in the youngest age groups may at least to some extent be ascribed to breast feeding.

Further support for acquired immunity to cholera is provided by previous observations by Glass et al. (14) in Bangladesh of a very low reinfection rate after symptomatic cholera infections, suggesting approximately 90% protection against a second episode of cholera in an area where cholera is endemic. However, subsequent studies have suggested that such protection is biotype associated (8). Thus, in a population in rural Bangladesh that was monitored between 1985 and 1988 for recurrent cholera episodes, initial infection with classical vibrios was associated with solid protection against reinfection with either biotype, whereas several cases of cholera were seen in those with initial episodes of El Tor cholera. Whether this difference could be ascribed to induction of different antibody specificities by the two biotypes, to induction of stronger immune responses by the classical than the El Tor biotype, to the low number of recurrent cholera episodes observed (7 of 2,214 with an initial cholera episode), or even to different host genetic predispositions remains to be elucidated.

Studies of the incidence of diarrhea in breast-fed children also suggest a protective effect of antitoxic as well as antibacterial antibodies in milk against cholera. Specific sIgA antibodies against the most prevalent enteropathogens are usually found in the milk of lactating mothers. The protective role of these antibodies has been difficult to assess, e.g., by comparing diarrhea morbidity in breast-fed and bottle-fed children, owing to several confounding factors. However, in a study in Bangladesh, strong evidence for the protective effect of sIgA antibodies against cholera was achieved (16). Thus, symptomatic cholera was seen in significantly lower numbers in children who drank milk that contained high levels of anti-CT or anti-bacterial sIgA antibodies or both than in children who drank milk with low levels of such antibodies.

Acknowledgments. All of our coworkers at the University of Göteborg and at the International Center for Diarrhoeal Disease Research, Bangladesh (ICDDR,B) are greatly acknowledged for their contributions to these studies.

Financial support was obtained from the Swedish Medical Research Council, the Swedish Agency for Research Cooperation with Developing Countries, and the World Health Organization.

REFERENCES

1. Åhrén, C., et al. Unpublished data.
1a. Åhrén, C., C. Wennerås, J. Holmgren, and A.-M. Svennerholm. Intestinal antibody response after oral immunization with a prototype enterotoxigenic *Escherichia coli* vaccine. *Vaccine* 11:929–934.
2. Attridge, S. R., and D. Rowley. 1983. The role of the flagellum in adherence of *Vibrio cholerae. J. Infect. Dis.* 147:864–872.
3. Attridge, S. R., and D. Rowley. 1983. Prophylactic significance of nonlipopolysaccharide antigens of *Vibrio cholerae. J. Infect. Dis.* 148:931–939.
4. Baselski, V., R. Briggs, and C. Parker. 1977. Intestinal fluid accumulation induced by oral challenge with *Vibrio cholerae* or cholera toxin in infant mice. *Infect. Immun.* 15:704–706.
5. Booth, B. A., M. Boesman-Finkelstein, and R. A. Finkelstein. 1983. *Vibrio cholerae* soluble haemagglutinin/protease is a metalloenzyme. *Infect. Immun.* 42:639–644.
6. Brandtzaeg, P., K. Bjerke, T. S. Haltsensen, M. Hvatum, K. Kett, P. Krajci, D. Kvale, R. Müller, D. Wilsson, R. O. Roynum, H. Scott, L. M. Sollid, P. Thrane, and K. Valnes. 1990. Local immunity; the human mucosa in health and disease, p. 1–12. *In* T. MacDonald, J. Challacombe, and W. Bland (ed.), *Advances in Mucosal Immunology.* Kluwer, London.
7. Cash, R., S. Music, J. Libonati, M. Snyder, R. Wenzel, and B. Hornick. 1974. Response of man to infection with *V. cholerae*. I. Clinical, serologic and bacteriologic responses to a known inoculum. *J. Infect. Dis.* 129:45–52.
8. Clemens, J. D., F. Van Loon, D. A. Sack, M. R. Rao, F. Ahmed, J. Chakraborty, B. A. Kay, M. R. Khan, M. D. Yunus, J. R. Harris, A.-M. Svennerholm, and J. Holmgren. 1991. Biotype as a determinant of the natural immunizing effect of *V. cholerae* O1: implications for vaccine development. *J. Infect. Dis.* 23:473–479.
9. Czerkinsky, C., A.-M. Svennerholm, M. Quiding, R. Jonsson, and J. Holmgren. 1991. Antibody-producing cells in peripheral blood and salivary glands after oral cholera vaccination in humans. *Infect. Immun.* 59:996–1001.
10. De, S. N., and D. N. Chatterje. 1953. An experimental study of the mechanism of action of *Vibrio cholerae* on the intestinal mucous membrane. *J. Pathol. Bacteriol.* 46:559–562.
11. Feeley, J. C., and E. J. Gangarosa. 1980. Field trials of cholera toxin, p. 204–210. *In* Ö. Ouchterlony, and J. Holmgren (ed.), *Cholera and related diarrhoeal diseases.* 43rd Nobel Symposium, Stockholm, 1978. Karger, Basel.
12. Foo, E. S. A., and W. Chaicumpa. 1981. Protection against *Vibrio cholerae* infection afforded by fragments of anti-haemagglutinin. *Southeast Asian J. Trop. Med. Public Health* 12:506–512.
13. Franzon, V. L., A. Barker, and P. A. Manning. 1993. Nucleotide sequence encoding the mannose-fucose-resistant hemagglutinin of *Vibrio cholerae* O1 and construction of a mutant. *Infect. Immun.* 61:3032–3037.
14. Glass, R. I., S. Becker, M. I. Huq, B. J. Stoll, M. U. Khan, M. H. Merson, J. V. Lee, and R. E. Black. 1982. Endemic cholera in rural Bangladesh, 1966–1980. *Am. J. Epidemiol.* 116:959–970.
15. Glass, R. I., A.-M. Svennerholm, M. R. Khan, S. Huda, I. Huq, and J. Holmgren. 1985. Seroepidemiological studies of El Tor cholera in Bangladesh: association of serum antibody levels with protection. *J. Infect. Dis.* 151:236–242.
16. Glass, R., A.-M. Svennerholm, B. J. Stoll, M. R. Khan, K. M. B. Hossain, M. I. Huq, and J. Holmgren. 1983. Protection against cholera in breast-fed children by antibodies in breast milk. *N. Engl. J. Med.* 308:1389–1392.
17. Goldberg, M. B., V. J. DiRita, and S. B. Calderwood. 1990. Identification of an iron-regulated virulence determinant in *Vibrio cholerae*, using Tn*phoA* mutagenesis. *Infect. Immun.* 58:55–60.
18. Hall, R. H., G. Losonsky, A. P. D. Silveira, R. K. Taylor, J. J. Mekalanos, N. D. Witham, and M. M. Levine. 1991. Immunogenicity of *Vibrio cholerae* O1 toxin-coregulated pili in experimental and clinical cholera. *Infect. Immun.* 59:2508–2512.
19. Holmgren, J., J. Fryklund, and H. Larsson. 1989. Gamma-interferon-mediated down-regulation of electrolyte secretion by intestinal epithelial cells: a local immune mechanism? *Scand. J. Immunol.* 30:499–503.

20. **Holmgren, J., and N. Lycke.** 1986. Immune mechanisms of enteric infections, p. 9–22. *In* J. Holmgren, A. Lindberg, and R. Möllby (ed.), *Development of Vaccines and Drugs against Diarrhea.* 11th Nobel Conference, Stockholm. Studentlitteratur, Lund, Sweden.

21. **Holmgren, J., and A.-M. Svennerholm.** 1983. Cholera and the immune response. *Prog. Allergy* **33:**106–112.

22. **Holmgren, J., A.-M. Svennerholm, L. Gothefors, M. Jertborn, and B. Stoll.** 1988. Enterotoxigenic *Escherichia coli* diarrhea in an endemic area prepares the intestine for an anamestic immunoglobulin antitoxin response to oral cholera B subunit vaccination. *Infect. Immun.* **56:**230–233.

23. **Jansen, W. H., H. Gielen, S. G. T. Rijpkema, and P. A. M. Guinée.** 1988. Priming and boosting of the rabbit intestinal immune system with live and killed, smooth and rough *Vibrio cholerae* cells. *Microb. Pathog.* **4:**21–26.

24. **Jertborn, M., A.-M. Svennerholm, and J. Holmgren.** 1986. Saliva, breast milk and serum antibody responses as indirect measures of intestinal immunity after oral cholera vaccination or natural disease. *J. Clin. Microbiol.* **24:**203–209.

25. **Jertborn, M., A.-M. Svennerholm, and J. Holmgren.** 1988. Five-year immunologic memory in Swedish volunteers after oral cholera vaccination. *J. Infect. Dis.* **157:**374–377.

25a. **Jonson, G., et al.** Unpublished data.

26. **Jonson, G., J. Holmgren, and A.-M. Svennerholm.** 1991. Epitope differences in toxin-coregulated pili produced by classical and El Tor *Vibrio cholerae* O1. *Microb. Pathog.* **11:**179–188.

27. **Jonson, G., J. Holmgren, and A.-M. Svennerholm.** 1991. Identification of a mannose-binding pilus on *Vibrio cholerae* El Tor. *Microb. Pathog.* **11:**433–441.

28. **Jonson, G., J. Holmgren, and A.-M. Svennerholm.** 1992. Expression of toxin-coregulated pilus in *Vibrio cholerae* O1 *in vitro* and *in vivo,* abstr. B-400, p. 92. *Abstr. 92nd Gen. Meet. Am. Soc. Microbiol. 1992.*

29. **Jonson, G., J. Holmgren, and A.-M. Svennerholm.** 1992. Analysis of expression of toxin-coregulated pili in classical and El Tor *Vibrio cholerae* O1 in vitro and in vivo. *Infect. Immun.* **60:**4278–4284.

30. **Jonson, G., J. Sanchez, and A.-M. Svennerholm.** 1989. Expression and detection of different biotype-associated cell-bound haemoagglutinins of *Vibrio cholerae* O1. *J. Gen. Microbiol.* **135:**111–120.

31. **Jonson, G., A.-M. Svennerholm, and J. Holmgren.** 1989. *Vibrio cholerae* expresses cell surface antigens during intestinal infection which are not expressed during in vitro culture. *Infect. Immun.* **57:**1809–1815.

32. **Jonson, G., A.-M. Svennerholm, and J. Holmgren.** 1990. Expression of virulence factors by classical and El Tor *Vibrio cholerae in vivo* and *in vitro*. *FEMS Microbiol. Ecol.* **74:**221–228.

33. **Kazemi, M., and R. Finkelstein.** 1990. Study of epitopes of cholera enterotoxin related enterotoxins by checkerboard immunoblotting. *Infect. Immun.* **58:**2352–2360.

34. **Kusama, H., and J. P. Craig.** 1970. Production of biologically active substances by two strains of *Vibrio cholerae.* *Infect. Immun.* **1:**80–87.

35. **Lee, J.-Y., A. Boman, S. Chuanxin, M. Andersson, H. Jörnvall, V. Mutt, and H. G. Boman.** 1989. Antibacterial peptides from pig intestine: isolation of a mammalian cecropin. *Proc. Natl. Acad. Sci. USA* **86:**9159–9162.

36. **Levine, M. M., J. B. Kaper, R. E. Black, and M. L. Clements.** 1983. New knowledge on pathogenesis of bacterial enteric infections as applied to vaccine development. *Microbiol. Rev.* **47:**510–550.

37. **Levine, M. M., J. G. Morris, G. Losonsky, E. Boedeker, and B. Rowe.** 1986. Fimbriae (pili) adhesins as vaccines, p. 143–145. *In* D. Lark (ed.). *Protein-Carbohydrate Interactions in Biological Systems.* Academic Press, London.

38. **Lycke, N., and J. Holmgren.** 1989. Adoptive transfer of gut mucosal antitoxin memory by isolated B cells 1 year after oral immunization with cholera toxin. *Infect. Immun.* **57:**1137–1141.

39. **Lycke, N., U. Karlsson, A. Sjölander, and K.-E. Magnusson.** 1991. The adjuvant action of cholera toxin is associated with an increased intestinal permeability for luminal antigens. *Scand. J. Immunol.* **33:**691–698.

40. **Lycke, N., and A.-M. Svennerholm.** 1990. Presentation of immunogens at the gut and other mucosal surfaces, p. 207–227. *In* G. C. Woodrow and M. M. Levine (ed.), *New Generation Vaccines.* Marcel Dekker, Basel.

41. **Lycke, N., A.-M. Svennerholm, and J. Holmgren.** 1986. Strong biotype and serotype cross-protective antibacterial and antitoxic immunity in rabbits after cholera infection. *Microb. Pathog.* **1:**361–371.

42. **Madara, J. L., and J. Stafford.** 1989. Interferon-gamma directly affects barrier function of cultured intestinal epithelial monolayers. *J. Clin. Invest.* **83:**724.

43. **Manning, P. A., A. F. Imbesi, and D. R. Haynes.** 1982. Cell envelope proteins in *Vibrio cholerae.* *FEBS Microbiol. Lett.* **14:**159–166.

44. **Mestecky, J., and J. R. McGhee.** 1987. Immunoglobulin A (IgA): molecular and cellular interactions involved in IgA biosynthesis and immune response. *Adv. Immunol.* **40:**153–245.

45. **Mosley, W. H.** 1969. Vaccines and somatic antigens. The role of immunity in cholera. A review of

epidemiological and serological studies. *Tex. Rep. Biol. Med.* **27**(Suppl. 1):227–241.

46. **Mosley, W. H., W. E. Woodward, K. M. A. Aziz, M. A. S. M. Rhaman, A. K. M. A. Chowdhury, A. Ahmed, and J. C. Feeley.** 1970. The 1968–69 cholera vaccine field trial in rural East Pakistan. Effectiveness of monovalent Ogawa and Inaba vaccines and a purified Inaba antigen, with comparative results of serological and animal protection tests. *J. Infect. Dis.* **121**:471–479.

47. **O'Mahony, S., E. Arranz, J. R. Barton, and A. Ferguson.** 1990. Appraisal of gut lavage in the study of intestinal humoral immunity. *Gut* **31**:1341–1344.

47a. **Osek, J., et al.** Unpublished data.

48. **Osek, J., G. Jonson, A.-M. Svennerholm, and J. Holmgren.** 1992. Enzyme-linked immunosorbent assay for determination of antibodies to *Vibrio cholerae* toxin-coregulated pili. *APMIS* **100**:1027–1032.

49. **Osek, J., A.-M. Svennerholm, and J. Holmgren.** 1992. Protection against *Vibrio cholerae* El Tor infection by specific antibodies against mannose-binding hemagglutinin pili. *Infect. Immun.* **60**:4961–4964.

50. **Patel, P. P., C. F. Marrs, J. S. Mattick, W. W. Ruehl, R. K. Taylor, and M. Koomey.** 1991. Shared antigenicity and immunogenicity of type 4 pilins expressed by *Pseudomonas aeruginosa, Moraxella bovis, Neisseria gonorrhoeae, Dichelobacter nodosus,* and *Vibrio cholerae. Infect. Immun.* **59**:4674–4676.

51. **Pierce, N. F.** 1978. The role of antigen form and function in the primary and secondary intestinal immune responses to cholera toxoid and toxin in rats. *J. Exp. Med.* **148**:195–206.

52. **Quiding, M., I. Nordström, A. Kilander, G. Andersson, L.-Å. Hanson, J. Holmgren, and C. Czerkinsky.** 1991. Intestinal immune responses in humans. Oral cholera vaccination induces strong intestinal antibody responses, gamma-interferon production, and evokes local immunological memory. *J. Clin. Invest.* **88**:143–148.

53. **Richardson, K., J. B. Kaper, and M. M. Levine.** 1989. Human immune response to *V. cholerae* O1 whole cells and isolated outer membrane antigens. *Infect. Immun.* **57**:495–501.

54. **Sciortino, C. V.** 1989. Protection against infection with *V. cholerae* by passive transfer of monoclonal antibodies to outer membrane antigens. *J. Infect. Dis.* **160**:248–252.

55. **Sears, S. D., K. Richardson, C. Young, C. D. Parker, and M. M. Levine.** 1984. Evaluation of the human immune response to outer membrane proteins of *Vibrio cholerae. Infect. Immun.* **44**:439–444.

56. **Sengupta, D. K., T. K. Sengupta, and A. C. Ghose.** 1992. Major outer membrane proteins of

Vibrio cholerae and their role in induction of protective immunity through inhibition of intestinal colonization. *Infect. Immun.* **60**:4848–4855.

57. **Sinha, V. B., A. Jacob, R. Srivastava, J. B. Kaper, and B. S. Srivastava.** 1993. Identification of the flagellar antigens of *Vibrio cholerae* El Tor and their role in protection. *Vaccine* **11**:372–375.

58. **Spira, W. M., R. B. Sack, and J. L. Froelich.** 1981. Simple adult rabbit model for *Vibrio cholerae* and enterotoxigenic *Escherichia coli* diarrhea. *Infect. Immun.* **32**:739–747.

59. **Stroeher, U. W., L. E. Karageorgos, R. Morona, and P. Manning.** 1992. Serotype conversion in *Vibrio cholerae* O1. *Proc. Natl. Acad. Sci. USA* **89**:2566–2570.

60. **Sun, D., D. M. Tillman, T. N. Marion, and R. K. Taylor.** 1990. Production and characterization of monoclonal antibodies to the toxin coregulated pilus (TCP) of *Vibrio cholerae* that protect against experimental cholera in infant mice. *Serodiagn. Immunother. Infect. Dis.* **4**:73–81.

61. **Svennerholm, A.-M.** 1980. The nature of protective immunity in cholera, p. 171–184. *In* Ö. Ouchterlony and J. Holmgren (ed.), *Cholera and Related Diarrheal Diseases.* 43rd Nobel Symposium, Stockholm, 1978, Karger, Basel.

62. **Svennerholm, A.-M., L. Gothefors, D. A. Sack, P. K., Bardhan, and J. Holmgren.** 1984. Local and systemic antibody responses and immunological memory in humans after immunization with cholera B subunit by different routes. *Bull. W.H.O.* **62**:909–918.

63. **Svennerholm, A.-M., and J. Holmgren.** 1976. Synergistic protective effect in rabbits of immunization with *Vibrio cholerae* lipopolysaccharide and toxin/toxoid. *Infect. Immun.* **13**:735–740.

64. **Svennerholm, A.-M., and J. Holmgren.** 1992. Immunity to enterotoxin-producing bacteria, p. 227–246. *In* T. T. MacDonald (ed.), *Immunology of Gastrointestinal Disease,* vol. 19. Kluwer Academic Publishers, London.

65. **Svennerholm, A.-M., J. Holmgren, R. Black, and M. Levine.** 1983. Serologic differentiation between antitoxin responses to infection with *V. cholerae* and enterotoxin-producing *Escherichia coli. J. Infect. Dis.* **147**:514–521.

66. **Svennerholm, A.-M., M. Jertborn, L. Gothefors, A. M. M. M. Karim, D. A. Sack, and J. Holmgren.** 1984. Mucosal antitoxic and antibacterial immunity after cholera disease and after immunization with a combined B subunit-whole cell vaccine. *J. Infect. Dis.* **149**:884–893.

67. **Svennerholm, A.-M., G. Jonson, and Y. Chang.** 1991. A method for studies of an El Tor-associated antigen of *Vibrio cholerae* O1. *FEMS Microbiol. Lett.* **79**:179–186.

68. **Svennerholm, A.-M., M. M. Levine, and J. Holmgren.** 1984. Weak serum and intestinal anti-

body responses to *Vibrio cholerae* soluble hemagglutinin in cholera patients. *Infect. Immun.* **45:**792–794.

68a. **Svennerholm, A.-M., D. Taylor, G. Johnson, and C. K. English.** Unpublished data.

69. **Tagliabue, A., D. Boraschi, D. F. Villa, D. R. Keren, G. H. Lowell, R. Rappuoli, and L. Nencioni.** 1984. IgA-dependent cell-mediated activity against enteropathogenic bacteria: distribution, specificity and characterization of the effector cells. *J. Immunol.* **133:**988–992.

70. **Taylor, R. K., V. L. Miller, D. B. Furlong, and J. J. Mekalanos.** 1987. Use of *phoA* gene fusions to identify a pilus colonization factor coordinately regulated with cholera toxin. *Proc. Natl. Acad. Sci. USA* **84:**2833–2837.

71. **Tokunaga, E., W. C., Cray, Jr., and N. F. Pierce.** 1984. Compared colonizing and immunizing efficiency of toxinogenic (A+ B+) *Vibrio cholerae* and an A⁻ B+ mutant (Texas Star-SR) studied in adult rabbits. *Infect. Immun.* **44:**364–369.

72. **Wikström, M., G. Jonson, and A.-M. Svennerholm.** 1991. Production and characterization of monoclonal antibodies to *Vibrio cholerae* soluble hemagglutinin. *APMIS* **99:**249–256.

73. **Yancey, R. J., D. L. Willis, and L. J. Berry.** 1979. Flagella-induced immunity against experimental cholera in adult rabbits. *Infect. Immun.* **25:**220–228.

74. **Young, D. B., and D. A. Broadbent.** 1982. Biochemical characterization of extracellular proteases from *Vibrio cholerae*. *Infect. Immun.* **37:**875–883.

Vibrio cholerae and Cholera: Molecular to Global Perspectives
Edited by I. Kaye Wachsmuth, Paul A. Blake, and Ørjan Olsvik
© 1994 American Society for Microbiology, Washington, DC 20005

Chapter 17

Host Susceptibility

Stephen H. Richardson

Diarrheal diseases, especially cholera, are among the leading causes of human morbidity and mortality and have always played a prominent role in historical events. Because of this prominence, an extensive literature, both scientific and nonscientific, has evolved. A largely undocumented theme in this literature, however, has been innate (i.e., nonimmunologic) genetic factors controlling susceptibility to cholera and disease severity: in other words, the unexplained pattern in which different individuals at essentially equal risk are infected and develop different intensities of disease during epidemics and pandemics.

For example, contacts sharing food, water, sanitary facilities, a dwelling, and/or a bed with a cholera gravis index case often present with no signs or only minor signs and symptoms like seroconversion or subclinical infections. Historically, fate or a reward for living a good life was invoked as an explanation for this hit-or-miss phenomenon. In modern times, fate was replaced by a better (but still not complete) understanding of the roles of passive and active immunity as the immunology of cholera was elucidated. The current state of knowledge regarding the immunol-

ogy is summarized in chapter 27. Even though immune status is clearly important in protection against overt cholera disease, it is now clear that serum lipopolysaccharide (LPS) or antitoxin antibody titers are of little or no value in predicting the extent of colonization or disease intensity. There still are no good explanations for the variable degrees of severity exhibited by cholera patients and contacts who have similar antibody titers. In this chapter, the term cholera gravis will denote the most severe life-threatening form of the disease, which can be rapidly fatal without treatment. Much of the discussion presented is speculative. Speculation was part of the author's charge when he agreed to prepare this chapter.

ENVIRONMENTAL AND SOCIAL FACTORS

Outside of that on immunologic relationships, the most extensive literature dealing with susceptibility to cholera is probably that addressing sanitation and nutrition. One of the clear messages about cholera is that if the environment is made safer by sound sanitary practices and if good personal hygiene is put in place, the disease largely goes away. Besides the cost, overcoming generations of built-in day-to-day habits and traditions is the most difficult problem faced by public health

Stephen H. Richardson • Department of Microbiology and Immunology, Bowman Gray School of Medicine, Wake Forest University, Medical Center Boulevard, Winston-Salem, North Carolina 27157.

officials. One anecdotal example will suffice. On entering a rural village in Bangladesh, it was noticed that the handle on a recently installed tube well had been cut off with a saw to render it inoperable. When the village elder was quizzed about the missing handle, he explained that a cow had bumped into the pump and broken the handle off. In a later private conversation, the visitor was told by the investigating epidemiologist that other village people had told him that the pump water "did not taste good," so they sabotaged the pump in order to go back to using water from the village tank, which was already beginning to get low (personal observations).

The literature is rich with population studies that offer some clues to innate factors involved in cholera susceptibility, but it is the recent family studies that have provided data backed up by numbers rather than anecdotes. Starting in 1980, cases of confirmed cholera arriving at the International Center for Diarrheal Disease Research, Bangladesh Matlab field area hospital were selected as "index" cases. Within 24 h after cholera was diagnosed, the families (in some instances, large extended groups of individuals) of the index cases were visited. The family members were enrolled in a variety of studies, some aimed at intervention in the possible spread of disease from the contact (18). In the 2 years of the study, 370 index cases and 1,992 contacts were surveyed. Of the contacts, 29% became colonized, and 36% of these got diarrhea. Of the persons with diarrhea, 91% had cases varying from mild to moderate, and 9% exhibited severe diarrhea. This translates to 12 of 1,992 contacts with cholera gravis. In this study, the binding (CTB) subunit of cholera toxin (CT) was administered in a double-blind protocol to attempt to block infection of the contacts. The authors commented on the known variances in intestinal CT receptors (i.e., GM_1 or other glycolipids) as a possible risk factor (18).

In a continuation of the same survey, the contacts were screened for vibriocidal, antitoxin immunoglobulin A (IgA) and IgG, and anti-LPS IgG antibodies on days 1 and 21. Of a subgroup of 370 index cases 1,658 contacts were monitored. Of these, 29% were colonized and 10% got cholera. There was no correlation between anti-LPS antibody titers and protection, but rises in antitoxin titers were deemed useful in identifying recently ill contacts. Of culture-negative contacts, 15 to 20% exhibited a twofold or greater rise in vibriocidal or antitoxin titers. It was concluded that IgG antitoxin titers were independent of colonization or disease and that the vibriocidal and anti-LPS antibody levels were of no value in predicting either event (19).

There is a long-standing impression that nutritional factors must play a role in susceptibility to cholera and in disease severity. The abundant literature on this subject is confusing, but one feature that seems to be agreed on is that chronic undernutrition or malnutrition has major effects on the humoral and cell-mediated immune (CMI) responses, which in turn may influence host susceptibility. There is clear evidence from studies in animals that cell-mediated events are involved in the development of immunity to cholera, but data relating to humans is not as straightforward (11). Palmer and Zaman (37) looked at selected measures of CMI in cholera patients. Twenty-four hours after rehydration, patients responded only half as well as controls to a battery of standard CMI tests. Malnourished children or children with cholera were significantly less able to respond to the test antigens than the village control population or refugees. The authors were unable to determine whether the CMI depression was a physiological response to elevated cyclic AMP (cAMP) during acute disease (as demonstrated in mice) or anergy to the test antigens (37).

Anti-CTB antibody titers in contact children's saliva were used to correlate their immune responses to CTB vaccination with nutritional status as part of the intervention arm of the family study mentioned above (18). The good news was that undernutrition (70% of normal parameters) was not associated with increased risk of colonization, disease, duration or severity of illness, or levels of total IgA, initial serum antitoxin, or vibriocidal antibodies. This is in contrast to the situation in which malnutrition has been implicated in increased risk for reasons that are not clearly understood. In a related observation, Clemens et al. found in a comparison of breast-fed children and nonnursed children less than 3 years old that breast feeding reduced the risk of severe cholera 70% (7).

There is ample evidence that individuals with low stomach acid due to any cause are at increased risk of cholera. The first of the U.S. Gulf Coast cases of cholera in 1973 falls into the low-stomach-acid-indicator group. The same factor was evident in an outbreak on Guam, when a shared contaminated meal was hardest on the individuals who had predisposing gastric problems. Hard data documenting the risk factors of hypochlorhydria were reported in 1978 by Nalin et al. (33). Acid production was low in 16 of 37 Bengalees recovering from cholera. Low acid production predisposed to cholera severity in American volunteers as well. In the volunteer group, the correlation between mean diarrhea volume and basal stomach acid before challenge was strong ($r = 0.99$).

Maryland investigators looked at beer drinking and cannabis use as cholera risk factors in their volunteer study populations (34). The 137 subjects were tested for basal acid levels before and after administration of an acid stimulator. High beer intake correlated with increased peak acid production, and high cannabis use correlated with low acid. The cholera attack rates were 88% in the heavy users of cannabis and 37% in light users. Because of the extensive use of cannabis-containing products in many areas where cholera is endemic, habitual usage may be a factor in host susceptibility where the risk of intake of contaminated food and water is high. A recently described noninvasive test for measuring stomach acid (breath hydrogen post-stimulation) was used to measure the acid contents of matched cases (recovered cholera victims) and controls (contacts of the index case). Acid was more than twice as high in 12 of 15 controls, confirming previous studies. This test should prove to be a valuable epidemiologic tool for predicting one of the known risk factors for cholera (51).

Helicobacter pylori is associated with active gastritis and plays a major role in duodenal ulcer pathogenesis. The epidemiology of *H. pylori* infection and its rate of colonization are just now being examined in controlled studies (28, 38), but available data suggest that the incidence of *H. pylori* carriage is indirectly proportional to the general level of sanitation (i.e., the same as cholera). Cater (4) has hypothesized that *H. pylori* may be a major factor in chronic hypochlorhydria. On initial contact, *H. pylori* colonizes the surface of the gastric epithelium and presumably protects itself from the acidic mileau by secreting a potent urease that surrounds the bacteria with a neutralized microenvironment. *H. pylori* also produces a vacuolating toxin that has an adverse effect on surrounding cells. Other studies (8) have shown that the toxin's activity is potentiated by nicotine, suggesting that heavy smoking (common in areas where cholera is endemic) might contribute to increased damage to the stomach lining and higher stomach pH. In addition, long-term colonization with *H. pylori* is known to cause irreversible damage to the secretory cells of the gastric mucosa, resulting in permanent hypochlorhydria. Thus, an organism that is probably transmitted person to person by the same fecal-oral route as *Vibrio cholerae* (indeed, microaerophilic vibrios were renamed *Campylobacter* spp., and some of these were later put in the genus *Helicobacter*) may increase the risk of cholera infection by lowering the pH barrier so that fewer vibrios are required for infection and disease.

An interesting speculation that cystic fibrosis (CF) heterozygotes have a survival advantage in cholera and that they have been selected out by cholera over the ages has recently been advanced (47). CF heterozygotes have only half the normal number of protein kinase A gatable chloride channels. When infected with cholera, these individuals would be expected to secrete less chloride and have less diarrhea. Studies measuring chloride levels after induced sweating in CF heterozygotes and controls support this idea. This hypothesis has not been tested directly, because the population in areas where cholera is endemic has a low prevalence of overt CF disease. When animals transgenic for CF become available, this idea could be looked at experimentally.

Another potentially exciting connection of CF with cholera is the recent discovery of a new cholera toxin, accessory cholera enterotoxin (ACE). This toxin, which may be responsible for some of the secretory problems seen in volunteers given CTA$^-$ vaccine strains, has 42% amino acid sequence similarity to the CF transmembrane conductance regulator. The implications of this confluence of an infectious disease virulence factor with a genetic defect will have to await purification of ACE and more information about its mechanism(s) of action (27).

CORRELATIONS OF CHOLERA SUSCEPTIBILITY WITH BLOOD TYPE

The first nonanecdotal evidence of a potential genetic basis for individual differences in susceptibility to cholera was reported in 1977 by Barua and Paguio (2), who noted that the incidence of cholera gravis in hospital patients with blood type A was lower than that represented in the general population, and incidence with blood type O was significantly higher. No biotype breakdown of the causative etiologic agents was offered, but at the time, most cases of cholera were due to the El Tor biotype. These authors wrote: "a plea is made for further studies on these lines to explain the host susceptibility in cholera and other acute enterotoxigenic diarrheas."

Table 1 summarizes responses to that plea. Results very similar to Barua's observations were seen in India, where the difference between the controls and cholera patients was even more striking (61.5% type O among cholera gravis patients versus 33% type O in the control population) (49). The investigators further suggested that blood group secretor status might be involved (i.e., substances secreted in saliva and/or digestive juices could block the action of CT), but they offered no experimental evidence to support their suggestion.

Oral challenge of 66 volunteers at the University of Maryland Center for Vaccine Development with Ogawa or Inaba *V. cholerae* resulted in 55 cases of diarrhea. Analysis of the clinical responses of the volunteers demonstrated that disease severity but not susceptibility to infection was associated with type O blood. No data regarding how the classical and El Tor biotypes were distributed among the heavily purging patients were presented. Using a limited battery of reagents, no correlation between human lymphocyte antigen type and risk of infection or severity was observed. The precedent of B27 haplotype and predilection for reactive arthritis subsequent to enteric infections with *Yersinia, Salmonella,* and *Shigella* spp. was discussed as an example of the involvement of genetic factors in enteric infections (30).

A double-blind investigation by Sircar and his colleagues (49) confirmed the increased preponderance of type O blood among hospitalized cholera patients, but no such association was seen in a control group with diarrhea due to *V. parahaemolyticus* or in control individuals from whom no enteric pathogen was isolated.

Glass and his coworkers in Bangladesh found no connection between blood type and diarrhea due to *Escherichia coli* enterotoxigenic strains, non-O1 *V. cholerae, Shigella* spp., or rotavirus. In a survey of 682 patients and 664 controls, blood type O was strongly correlated with severe cholera episodes (57% in cholera patients versus 30% in the general population). On the other side, group AB blood appeared least often among patients (1 versus 9%), and appearance of group B blood was in between (18 versus 35%). Among

Table 1. Evidence linking cholera severity to ABO blood type

Location, date, and reference	% type O in:		Comment[a]
	Controls	Cholera gravis patients	
Philippines, 1977 (2)	45.5	64.4	El Tor biotype predominant
India, 1977 (5)	33.0	61.5	Secretor status possible factor?
United States, 1979 (30)	37.0	64.0	HLA not involved; severity O linked
India, 1981 (49)	32.4	44.3	V. parahemolyticus severity no link
Bangladesh, 1985 (19)	36.0	68.0	NAG vibrio, Shiga E. coli severity not O linked; presence or absence of secretor no link
Bangladesh, 1989 (6)	32.4	53.4	Only El Tor severity O linked; vaccine efficacy versus both biotypes lower with type O

[a]HLA, human lymphocyte antigen; NAG, non-O1 vibrios.

contacts, no association with blood group O was detected in infection (i.e., colonization) but blood group O influenced subsequent development of disease (49% type O in severe cholera patients versus 36% type O in the control population). Increased risk of being infected with V. cholerae (i.e., susceptibility) was not linked to blood type. Disease severity and susceptibility were shown to be independent of group O antigen secretor status. The E. coli data are intriguing because the heat-labile enterotoxin (LT) shares receptors with and is functionally and immunologically closely related to CT. These results suggest that factors other than toxin binding and/or enhanced function are implicated in the blood group O association with severe disease. The authors speculated that over a long period, cholera may have been a factor in lowering the percentage of type O in the Bengali population to one of the lowest in the world (17).

In analyzing results of field trials of a combined CTB–killed-whole-cell vaccine, Clemens and his colleagues corroborated the data from prior studies vis-à-vis the heightened risk of severe cholera for type O patients (6). A total of 3,647 persons were bled and typed. Among 208 cholera patients, 55% carried classical Ogawa, and 32% carried El Tor Ogawa. For El Tor cholera, the risk factors compared to those for controls increased from 1.1 for type AB blood to 2.8 for type A-plus-

B blood to 6.2 for type O blood. Comparable numbers for classical cholera were 7.6 for the placebo, 6.0 for type A-plus-B blood, and 6.1 for type O blood. Surprisingly, the O severity link appeared to be operative only in disease due to the El Tor biotype and was not seen among patients infected by classical strains.

In contrast to these results, type O persons vaccinated with either whole-cell or combined vaccine demonstrated lower protection against severe disease than did other blood groups (40% for O versus 76% for other types). The blood group-related difference in protective efficacy was not limited to El Tor disease but was also seen with severe classical cholera (44% type O versus 79% other). In older and younger vaccinees, protection against disease due to either biotype was 43% for type O versus 85% for other types. Classical strains are now reemerging in areas where cholera is endemic after a 35- to 40-year domination by the El Tor biotype. The increased incidence of classical cholera in this study population might explain the relatively lower percentage of type O blood among cholera gravis patients than was noted in the earlier reports, when El Tor was predominant.

Fifteen years after Barua's plea for expanded investigation, the results of this study once again emphasized the importance of research aimed at elucidating the mechanisms underlying the risks or benefits associated with being in the O or AB blood group sub-population in an area where cholera is endemic. The emphasis is even more urgent now that the 1991 South American epidemic is spreading. That epidemic is due to *V. cholerae* of the El Tor biotype, which would seem to place more of the population at risk of severe cholera if the above results are verified in this experiment of nature. In addition, the proportion of type O individuals in the South American population is much higher than it is in Bangladesh.

Not much is known about binding of CT to the gangliosides of the human intestine or about the expression of gangliosides in situ. Analysis of intestinal epithelial cell and mucosal scrapings from the large bowels of elective surgery patients revealed gangliosides and blood group ABH antigens in the non-epithelial fractions. The Le blood substances were detected in epithelial cells, but only traces of GM_1 were found in those fractions. In contrast, the cells of the small intestine displayed larger amounts of all blood group antigens as well as the genes for their synthesis and expression, which were distributed nonuniformly in those tissues (23).

Bennun et al. (3) and Barra et al. (1) used iodinated CT and LT to analyze the blood group glycosphingolipids of pig intestines, which are similar to those of humans. Initially, they found that blood group A^+ pig extracts contained six to eight CT-binding components. A^- pig extracts showed a different CT-binding pattern, with the components running below ganglioside GD_{1b} missing. Their conclusions were that type O individuals are missing components present on A, B, or AB cells that can bind CT and prevent it from interacting with its receptor. In the more recent report (1), the group used ABH blood group glycolipids isolated from human erythrocytes in addition to the pig intestinal fractions to compare *E. coli* human heat-labile enterotoxin (LT_H) and CT binding. Both toxins bound better to A^+ glycolipids than to A^- material. LT but not CT interacted with ABH antigens from human erythrocytes. LT bound better to A, B,

and AB glycolipids than to O glycolipids. Competition binding experiments corroborated many other investigations indicating that LT and CT have different carbohydrate structural requirements for binding.

The molecular genetic basis for the group ABO system was elucidated by Yamamoto et al. (53), who suggested that the ABO markers be designated a histo-blood system, since the genes and their products are also found in many other cell types, e.g., epithelial cells. Three carbohydrate antigens are coded for by the ABH genes. The H antigen is converted by specific glycosyltransferases into A, B, or both on cells expressing the A, B, or AB phenotype. Type O persons lack the transferase activity to convert H further and do not express any surface antigen. Deletions of only a few base pairs result in different gene expression patterns. This imparts flexibility and polymorphism to the system, so that a single individual may express a different array of antigens on different tissues or in a different location within the same tissue (e.g., the intestine).

These data fit well with the hypothesis drawn from the porcine experiments; i.e., type O cells are better able to bind CT (or to bind more CT) because their surfaces are not covered with A and B glycolipids. This theory is yet to be proven, but as a working model, it fits the epidemiologic data (and parallels data from the murine studies to be described in a later section of this chapter). A precedent for genetic control of an infectious disease by expression (or secretion) of blood group substances is found in the predilection of women nonsecretors of the Lewis ($Le_a^+b^-$ and $Le_a^-b^-$) blood group to have recurrent urinary tract infections due to uropathogenic strains of *E. coli* that employ the antigens as receptors in attachment (48).

COLONIZATION MECHANISMS OF *V. CHOLERAE*

At present, two major attachment systems are known to be involved in *V. cholerae* adherence and colonization of animals and humans. CT itself may play a significant role in attachment to epithelial cells. Toxin-coregulated pili (TCP) are part of an armamentarium of virulence factors including CT, outer membrane proteins, iron-gathering proteins, etc., that are regulated in concert in response to environmental signals such as pH and temperature. Together, they form a regulon and are under the transcriptional control of a master gene called *toxR*. The TCP system has been beautifully dissected by Taylor and his colleagues (22) and is discussed in chapter 13. The other adhesin was discovered 35 years ago by Lankford and is called the mannose-sensitive hemagglutinin (MSHA) because its ability to adhere is blocked in vitro by high levels of mannose and mannose derivatives. In general, it is believed that classical *V. cholerae* attach mainly by TCP and that El Tor biotypes primarily use the MSHA mechanism (see below). Some classical strains have a fucose-sensitive hemagglutinin in addition to or instead of TCP (24).

There is disagreement about the relative importance of these colonization factors in cholera. From the field data cited above it is clear that classical and El Tor *V. cholerae* display different patterns of colonization during disease and that the O blood group propensity for severe disease is more associated with El Tor than with classical *V. cholerae*. One possibility that might explain this important difference is that the El Tor CT (long thought to be produced in minimal amounts compared to classical CT) might be slightly altered structurally so that it binds in a different configuration or to additional receptors not recognized by LT and classical CT. Holotoxin (CTA plus CTB) has been shown to be necessary for effective colonization in animal models, so it is quite possible that toxin, in addition to adhesins, influences attachment and subsequent hypersecretion in clinical disease. Also, as will be discussed below, the most extensive evidence for host-controlled susceptibility differences to cholera comes from work with the murine system, where

humoral and CMI responses and response to CT are all linked to the murine major histocompatibility gene complex (MHC) called *H-2* (45).

Dubey et al. (10) purified both CT and El Tor enterotoxins to address the possibility of major structural or functional differences. They found that growth conditions for the maximum expression of the two toxins were disparate and that El Tor toxin was produced best under a two-phase (anaerobic followed by aerobic) growth procedure. Elimination of the soluble hemagglutinin-protease coproduced by El Tor vibrios by incorporation of antisera eliminated nicking of the holotoxin and destruction of CTA during synthesis and purification. The hemagglutinin-protease has recently been implicated in detachment of the growing vibrios from the intestine during clinical disease (15). With the technical problems averted, the classical and El Tor CTs were compared immunologically and immunochemically and found to be identical except for minor biotype-specific epitope differences that remain undefined. In spite of the in vitro evidence that the two toxins are nearly identical, this does not mean that the toxins are necessarily comparable when produced in vivo. In vivo structure and function of the two toxins is a problem that should be addressed by using some of the animal models currently available. It should also not be assumed that the "minor" immunochemical differences seen in vitro do not contribute strongly to the observed host susceptibility and severity patterns, even though the major effects appear now to be related to colonization factors. The role of their toxins such as ACE and ZOT also need to be added to the overall virulence equation.

To evaluate the TCPs produced by classical and El Tor vibrios, Jonson et al. (24) compared enzyme-linked, immunosorbent assay (ELISA) and immunoblotting patterns of the antigens with those of a panel of polyclonal and monoclonal antibodies. The TCPs share 80% amino acid homology at the amino acid level. It was discovered by monoclonal mapping that a major pilin (TcpA) subunit epitope of classical TcpA is not present in El Tor TcpA and that one of the shared epitopes is more strongly expressed in classical TcpA. Polyclonal antibodies reacted strongly with both TcpAs in immunoblotting, indicating that El Tor vibrios produce as much TcpA subunit as the classical strains. However, intact El Tor strains were not inhibitory in a competition ELISA, and no TCP bundles were detected on them by immunoelectron microscopy. The investigators concluded that functional TCPs were poorly expressed on El Tor vibrios under experimental conditions that supported maximum CT production.

Because TCP did not appear to be expressed well by El Tor strains and because the MSHA had been shown previously to be a colonization factor for El Tor, Jonson et al. (25) purified and characterized mannose-binding pili from *V. cholerae* El Tor. This hemagglutinin agglutinates chicken erythrocytes and is the basis for differentiating El Tor from classical vibrios in the laboratory. A 17-kDa protein that fit all of the criteria of a pilin subunit was identified. Immunochemical data show that these pili are expressed only by El Tor vibrios grown under appropriate conditions in vitro and in vivo, including fresh isolates from cholera patients. The MSHA is proposed by Jonson et al. (25) to be the major adhesin of El Tor vibrios and has been identified on all isolates tested so far. These results are exciting, because if verified, they show a clear mechanistic difference between classical and El Tor *V. cholerae* attachment and colonization that might help explain the O blood group response.

Jonson et al. continued their investigations of TCP in El Tor and classical vibrios by looking for expression of TCP under laboratory conditions and in rabbit intestine (26). Even though the genes for TcpA have been found in both biotypes, regulation of expression is quite different. In rabbit ileal loops, 10 of 11 classical vibrio isolates expressed TCP, as monitored by inhibition ELISA. On the other hand, only one of nine El Tor vi-

brios expressed TCP in loops, and none of the El Tor vibrios from cholera stool had detectable TCP reactive material on their surface. These data support the Swedish group's contention that MSHA, not TCP, is the relevant in vivo adhesin molecule on El Tor vibrios (26). Those authors make a strong case for inclusion of MSHA in any future vaccine formulations (i.e., either El Tor strains in killed-whole-cell vaccines or MSHA genes in a genetically engineered live-vaccine strain).

Tox$^-$ mutant *V. cholerae,* TCP, LPS, CT, and synthetic TcpA peptide epitopes were given to human volunteers at the University of Maryland Center for Vaccine Development (22). In addition, serologic responses to the same antigens were measured in acute- and convalescent-phase sera from Indonesian cholera patients from whom *V. cholerae* of the same serotype and biotype had been isolated. The volunteers responded to CT, whole-cell preparations, and LPS. The group that had never been exposed to cholera did not mount a detectable immune response to purified TCP or to the TcpA peptide. TCP seroconversion occurred in three of six Thais, and one of six responded to the peptide antigen with seroconversion. Vibriocidal antibody titers were equivalent in sera from TCP$^+$ and TCP$^-$ responders. Thus, it appears that disease can cause TCP seroconversion but that long-term protection can occur in the absence of a detectable TCP response.

Putting these data together, it is clear that humans can vary greatly in their expression of CT and attachment factor receptors, even to the extent that the receptors are presented at different degrees of saturation within one individual's intestinal tract. The CTs of classical and El Tor *V. cholerae* are similar but not identical, and TCP and MSHA are distinct and expressed in very different patterns. It appears that the human and bacterial genetic systems are jockeying for an advantage, and we are trying to find out what is going on. To complicate the situation, it should not be forgotten that humans are just a temporary ecological niche for *V. cholerae* of either biotype, so we are really dealing with an estaurine-based bacterium trying to survive in two quite different worlds. It is clear that multiple genetic factors operate to affect host susceptibility, and if these factors can be deciphered, they may be exploited to our benefit. Since opportunities to test hypotheses related to host susceptibility to cholera in humans are limited, a few investigators have used animal models to try to address some of the basic questions.

MURINE MODELS OF HOST SUSCEPTIBILITY TO CHOLERA

Like humans, many animals provide their young with natural protection against the agents of diarrheal disease. An example of one such factor is a nonimmunoglobulin substance released by lactating rats that protects the infants against CT challenge (29). Female rats were fed CT on several occasions prior to and during gestation. Their milk contained an antisecretory factor (not an immunoglobulin) that was shown to protect neonates against diarrhea after CT oral challenge. Milk and bile from the vaccinated females also passively protected adult rats against intravenous (i.v.) and oral challenge. This phenomenon is not limited to rats or to cholera. The investigators point out that the factor is active against prostaglandin E-stimulated secretion (which they claim is a major contributor to cholera diarrhea) and not against that due to cAMP elevation. These observations are being followed up in humans to see whether a similar nonimmunologic protective effect can be demonstrated. In populations in an area where cholera is endemic, the oral CT challenge continuously goes on, beginning at birth in females, so a phenomenon like that demonstrated in the rats might be involved in innate host resistance. The question of passively and actively acquired immunities is addressed in chapters 26 and 27.

Charles Elson and his collaborators have long been interested in various aspects of in-

testinal immunity and have utilized CT extensively in their studies. In the course of their experimentation, they noted that certain strains of mice were more responsive to immunization with CT than others. They postulated that the immune response gene (part of the *H-2* murine MHC) might be involved. Mice were immunized intraperitoneally with alum-bound CT. Serum IgG and IgA anti-CT titers were measured by ELISA. $H-2^b$ and $H-2^q$ haplotypes were high responders, while $H-2^k$, $H-2^s$, and $H-2^d$ haplotypes were low responders. The IgG response was mapped to the I-A region. Serum IgA responses were low and ambiguous and did not parallel those of IgG. The investigators concluded that the anti-CT IgG response was under immune response gene control but that the IgA control mechanisms were quite different. The authors discussed their observations in light of the known immunosuppressive-adjuvant activities of CT and speculated on the possible implications for regulation of local immune responses (12).

In a continuation of the investigation, inbred mice, congenic mice, and mice with recombinations within the *H-2* immune response region were immunized orally with 10 μg of CT and boosted at 14 days (13). Intestinal secretions and plasma were collected. Again, the $H-2^b$ and $H-2^q$ haplotypes responded well, while the $H-2^k$, $H-2^s$, and $H-2^d$ haplotypes responded poorly. The recombinant-inbred-mouse responses localized the intestinal IgA response to the I-A subregion. Plasma IgG and secretory IgA responses were coincident in high or low titers, and both appeared to be controlled by the I-A region.

In a recent contribution, Elson (11) studied the adjuvanticity of CT and the contribution of LPS to CT's effects on the immune systems of mice. Previous studies (cited in chapter 14) showed that CT can act as a mucosal adjuvant for other proteins under appropriate conditions. Mice used in Elson's studies were high and low responders for CT plus other strains showing high and low responses to

LPS. Keyhole limpet hemocyanin (KLH), which except in the presence of CT does not stimulate an immune response orally, was used as the mucosal antigen. Plasma IgG and secretory IgA KLH responses were higher in H-2^b mice and lower in H-2^k mice, and they paralleled anti-CT responses. The same effects were noted in LPS-responsive and nonresponsive mice. Tolerance to KLH broken by CT was not H-2 restricted. The investigator concluded that CT adjuvanticity is controlled in mice by the H-2 and lps genes and that good immune responses depend on a strong initial response to CT. These data have many important implications for studies of human host resistance to cholera and disease severity and for future vaccine development.

SAM MODEL AND RESPONSE TO CT

In 1981, our laboratory developed the sealed adult mouse (SAM) model for studying virulence factors of enterotoxigenic bacteria. CT was used to standardize SAM because CT was readily available and its mechanism of action was well understood. To create the SAM model, mice were deprived of food overnight, anally occluded with cyanoacrylamide surgical adhesive, fed toxins or bacteria per os, and sacrificed at 6 h, when gut and body weights were determined. The gut weight/remaining body weight ratio measures fluid accumulation, which we called FA. Among several thousand mice assayed, the FA scale ran from a control-corrected low value of 11 to a high of 136 (39).

We discovered that inbred SAM expressing the H-2^b haplotype (alleles) of the murine MHC accumulated two to three (average FA, 100 versus 35) times as much fluid as H-2^k mice fed an identical dose of CT. The brevity of CT exposure and the unlikely possibility that only H-2^k strains had prior experience with CT ruled out an immunologic explanation for the increased FA. Following the precedent of diphtheria toxin sensitivity no-

menclature, we designated the enterotoxin-sensitive phenotype ETS$^+$ and the "normal" phenotype ETS$^-$. We use the terms ETS$^-$ and "normal" interchangeably here (44).

RESPONSE OF ETS$^+$ AND ETS$^-$ MICE TO i.v. CT

The possibility that the ETS phenotypes were an artifact of the SAM model or were restricted to intestinal CT responses was tested by employing an i.v. assay for toxin potency developed by Craig and his associates (see chapter 14 for details). CT was titrated in groups of ETS$^+$ and ETS$^-$ mice using death subsequent to i.v. injection of lethal doses or the reversible nondiarrhetic 3- to 5-day weight loss incited by sublethal doses as end points. Mice were weighed and checked for viability daily. Normal (ETS$^-$) 20-g C3H, CBA, and B10.BR mice did not lose any weight with a CT dose of 2.5 or 5 µg. The same doses caused ETS$^+$ strains B10 and B6 to die or lose up to 4.0 g of weight during the course of the experiment. These data offered supportive evidence that the ETS phenotype demonstrated in SAM was associated with the H-2^b haplotype and that the differential responses to CT extended to extraintestinal tissues. Encouraged by these preliminary observations, we pursued the study of the ETS phenotype in mice as a model for the human type O subpopulation at risk for cholera gravis (39–46).

CHARACTERISTICS AND MAPPING OF THE ETS PHENOTYPE

All ETS$^+$ mice had the H-2^b haplotype (e.g., C57BL/6, C57BL/10, C57L, and LP mice). ETS$^-$ inbred mice expressed other haplotypes (e.g., C3H, C57BR, CBA, DBA/1). ETS$^-$ appeared to be dominant, because F$_1$ progeny of ETS$^+$ × ETS$^-$ matings were invariably ETS$^-$ (e.g., C3H × B6, C3H × C3H.SW, B10 × B10.BR, B6 × DBA, and B6 × WB).

To map the ETS trait, inbred recombinant mice expressing different combinations of alleles within the *H-2* complex were employed. ETS⁻ mice isogenic for the *H-2ᵇ* complex (e.g., C3H.SW) expressed the ETS⁺ phenotype; conversely, ETS⁺ mice congenic for the *H-2ᵏ* allele (e.g., B6-H-2ᵏ, B10.BR) were ETS⁻. The most revealing data were derived from the recombinant inbred strains in the C57BL/10 background. B10.A, B10.AH4ᵇ, and B10.A(4R) mice all had genetic crossovers in the *H-2* region that converted the *Kᵇ* end of the B10 MHC to *Kᵏ*. These animals all expressed the ETS⁻ phenotype. In contrast, B10.A(5R) and B10.MBR mice had crossovers that preserved the *b* allele on the *K* end; these strains remained ETS⁺. Substitutions of other alleles throughout the I or D region of the *H-2* complex did not alter the ETS phenotype. Our conclusions from these studies were that the ETS⁺ phenotype was associated with the *b* haplotype, was *H-2* linked, and mapped centromeric to the *K* end of the *H-2* complex (43).

DEMONSTRATION OF THE ETS PHENOTYPE IN CULTURED CELLS

The results of the i.v. experiments (i.e., that the ETS phenotype was not limited to enterocytes) described above and our earlier studies on the effects of CT and LT on cultured cells prompted us to try to demonstrate the ETS phenotype at the cellular level by measuring CT-induced increases in cAMP (42). Simian virus 40-transformed mouse fibroblasts were challenged with a standard dose of CT in the presence of phosphodiesterase inhibitors, which block conversion of cAMP to 5′-AMP, and the total cAMP accrued was determined by radioimmunoassay (specific responses were expressed as picomoles of cAMP/2 h/10⁶ cells). B6 cells accumulated fourfold more cAMP (2,870 pmol) than C3H cells (750 pmol), and C3H.SW (*H-2ᵇ*) cells accumulated threefold more (2,018 pmol). The same differential response was seen when the cells were exposed to *E. coli* LT, which is similar to CT in biological activity and molecular structure (875 versus 154 pmol).

Because transformation is known to alter membrane ganglioside composition and CT responses, nontransformed fibroblasts from B10, B10.BR, and B10 × B10.BR F₁ mice were included as controls. In nontransformed ETS⁺ cells, cAMP levels were only twofold greater than in comparable ETS⁻ cells (8,018 versus 4,702 pmol). It should be noted that the dominance of the ETS⁻ trait was also expressed by cultured fibroblasts, as shown by the low cAMP responses of the B10 × BR F1 cells (4,864 pmol), which was equivalent to the cAMP produced by ETS⁻ cells. Isoproterenol, which uses cAMP as a second messenger, and forskolin, a reversible nonspecific adenylate cyclase activator with no specific receptor, also caused greater cAMP accumulation in the ETS⁺ cells. These results implied that differences in adenylate cyclase activity could account in part for the ETS⁺ and ETS⁻ phenotypes.

ADENYLATE CYCLASE ACTIVITY IN ETS⁺/ETS⁻ CELL MEMBRANES

To test the possibility that differences in enzyme levels and/or activity were involved in the ETS phenotype, adenylate cyclase and CT ribosylation activities were assayed in enterocyte membranes isolated from ETS⁺ and ETS⁻ prototype mice. Basal cyclase levels in B6 (ETS⁺) and BR (ETS⁻) membranes were identical. Exposure of the membranes to CT (which ribosylates a cyclase regulatory subunit, locking it in the "on" position) prior to the assay increased cyclase activity comparably for both phenotypes. Stimulation of untreated or CT-pretreated membranes with forskolin or NaF caused comparable two- to fourfold increases in cyclase activity in both phenotypes. ETS⁺ and ETS⁻ membranes from liver, kidney, and brain with and without added NaF exhibited equivalent cyclase activities as well, and if anything, ETS⁻ responses were slightly higher.

ANALYSIS OF CT BINDING TO ETS⁺ AND ETS⁻ CELLS

Because there were no apparent differences in adenylate cyclase or ribosylation activities between the phenotypes, we searched for alternative explanations for the observed phenotypic variations. Our preliminary feeding experiments in SAM had indicated that ETS⁺ mice were able to bind more radiolabeled CT to their intestines than their normal counterparts (42). To better quantitate the observed differences in CT binding, we employed prototype ETS⁺ and ETS⁻ fibroblasts grown in tissue culture. Cultured cells were exposed to ^{125}I-labeled CT with and without excess competing unlabeled CT, and the number of receptors and their affinities were estimated by Scatchard analysis. As predicted by the in vivo results, simian virus 40-transformed B6 cells expressed nearly threefold more receptors (1.1×10^6 versus 0.4×10^6) than comparable C3H ETS⁻ cells.

The nontransformed cells displayed about eight times as many receptors (8.5×10^6) as the transformed cells in all cases, but receptor density on the B6 cells was still 2.2 to 2.5 times greater than that on ETS⁻ cells. CT binding in both phenotypes, primary and transformed, had K_d values near 10^{-9} making it likely that receptor number, not difference in avidity, was a major factor contributing to the ETS phenotype. The observation that transformed cell lines exhibited essentially the same phenotypes as primary cells made us comfortable in using the more stable transformed lines for further studies.

Our hypothesis at this point was that ETS⁺ cells expressed more GM₁ than ETS⁻ cells, thereby binding more CT and resulting in a more vigorous biological response. To test this hypothesis, we carried out concomitant binding and cAMP assays using prototype ETS⁺ and ETS⁻ transformed fibroblasts plus

a ganglioside-deficient null cell line. As shown in Table 2, the prototype B6 cells expressed 4.6 times as many high-affinity receptors and accumulated double the amount of cAMP as their C3H counterparts. The HTI cells, derived from a recombinant inbred mouse with an $H\text{-}2^b/H\text{-}2^d$ haplotype, had 3.7-fold more receptors and produced twice as much cAMP as the $H\text{-}2^d/H\text{-}2^b$ HTG line. The NCTC 2071 null cells expressed virtually no GM₁ ganglioside on their surfaces, which resulted in an inability to bind or respond to CT. These experiments showed clearly that biological responses and CT binding were linked to the $H\text{-}2$ haplotype at the cellular level.

CT INTERACTIONS WITH ETS⁺ AND ETS⁻ ENTEROCYTE GANGLIOSIDE EXTRACTS (46)

Because CT binding appeared to be a key element in ETS expression, we isolated enterocytes from prototype ETS⁺ and ETS⁻ mice, prepared ganglioside extracts from them, and analyzed the extracts for ganglioside composition and CT binding by using a combination of high-pressure liquid chromatography (HPLC) and ELISA. The gradient used in the HPLC separation resolved the glycolipids into mono-, di-, and trisialogangliosides on a Mono Q anion-ex-

Table 2. CT binding and cAMP accumulation in ETS⁺ and ETS⁻ cell lines[a]

Cell line	ETS	No. of receptors/ cell (10^5)	pmol of cAMP/ 10^6 cells/2 h
B6	+	18.1	1,240 ± 62
C3H.SW	+	14.9	1,475 ± 100
HTI	+	12.8	1,528 ± 56
C3H	−	3.9	612 ± 50
HTG	−	3.5	767 ± 34
NCTC 2071	−	0.005	143 ± 5

[a]Assays were done as described in the text. HTI has the $H\text{-}2^b/H\text{-}2^d$ haplotype. HTG is $H\text{-}2^d/H\text{-}2^b$. NCTC 2071 is a chemically transformed ganglioside-deficient C3H derivative. C3H.SW is a congenic mouse with the $H\text{-}2^b$ haplotype in the C3H genetic background.

change column. Portions of individual fractions were then immobilized on microtiter plates, purified CT was added, and the bound CT was quantified by using peroxidase-conjugated anti-CT. Cumulative ELISA titers revealed that B6 (ETS$^+$) extracts contained 77% GM$_1$ and 23% non-GM$_1$ glycolipids, whereas C3H, BR, and WBB6 (ETS$^-$) preparations displayed 38, 54, and 100% non-GM$_1$ CT-reactive components, respectively. These results suggested that the ETS$^+$ phenotype was due not only to the availability of more GM$_1$ receptors but also to a decrease in the level of expression of nonfunctional CT-binding glycolipids that limited CT access to GM$_1$ in healthy mice.

To further resolve differences in CT-binding elements, ETS$^+$ and ETS$^-$ enterocyte extract components were quantitated after resolution by two-dimensional chromatography employing HPLC in the first dimension and thin-layer chromatography in the second, labeling with ^{125}I-CT, and autoradiography. These data confirmed the ELISA results by showing that ETS$^-$ glycolipids contained 40 to 60% non-GM$_1$ CT-binding moieties. Similar results were reported recently by Bennun et al. (3), who found six to eight non-GM$_1$ CT-binding components in glycolipids isolated from pig gastric mucosa.

If nonfunctional glycolipids hinder CT access to the biologically relevant GM$_1$ receptor, elimination or modification of these alternative sites should "elevate" the ETS$^-$ phenotype to near-ETS$^+$ status. Conversion could occur, because existing GM$_1$ becomes more accessible, because new GM$_1$ sites are created, because more free CT is available to saturate GM$_1$ receptors, or because of a combination of these reactions. To test this hypothesis, B6 and C3H fibroblasts were preincubated with *V. cholerae* neuraminidase or with *E. coli* LT IIa or LT IIb (kindly provided by Randall Holmes), which bind to a variety of gangliosides in addition to GM$_1$ (9, 50). After preincubation with the modifying reagents, the cells were exposed to 100 ng of CT and 1 μg of LT to assay for cAMP accu-

mulation or to 100 ng of fluorescein isothiocyanate (FITC; List Biological Laboratories, Campbell, Calif.)-labeled CTB subunit to quantitate binding by flow cytometry. These experiments were not done in the presence of competing unlabeled CT, so all binding, not just high-affinity binding, was registered. Table 3 contains a summary of the cAMP experiments.

Neuraminidase, which converts polysialylated gangliosides to GM$_1$ (16), effected a small decrease in the ETS$^+$ cAMP response but caused a slight elevation in the ETS$^-$ response. LT IIa (bound by GM$_1$, GD$_{1a}$, GD$_{1b}$, GM$_2$, and other gangliosides) effected a 3.3-fold-greater cAMP response in the B6 cells than in the C3H cells. LT IIb elicited no detectable response in the B6 cells but evoked a response equivalent to that seen with LT IIa and LT$_H$ in ETS$^-$ cells. These are the expected results, because LT IIb does not bind to GM$_1$ but substitutes GD$_{1a}$ as functional receptors. LT$_H$, the prototype *E. coli* enterotoxin whose major functional receptor is GM$_1$ (9), exhibited the same phenotype as CT (i.e., a 5.7-fold-greater cAMP response in ETS$^+$ cells). These data confirmed the observations that non-GM$_1$ receptors were dominant on

Table 3. Effects of preincubation with neuraminidase, LT IIa, and LT IIb on CT-induced cAMP accumulation in ETS$^+$ and ETS$^-$ transformed cells[a]

Expt	pmol of cAMP accumulated in 10^6 cells/2 h	
	B6 ETS$^+$ cells	C3H ETS$^-$ cells
Cell control	0	0
Cell control plus:		
Neuraminidase	0	0
100 ng of CT	732	153
Neuraminidase + CT	646	172
LT IIb	0	138
LT IIa	403	130
1 μg of LT	875	154

[a]Where indicated, 3.5 × 10^5 cells were preincubated with 0.2 U of *V. cholerae* neuraminidase or buffer for 30 min at 25°C, washed, and then incubated with 100 ng of CT, LT IIa, or LT IIb or 1 μg of LT for 2 h at 37°C. Cells were disrupted by sonication, and total cAMP accumulation was quantitated by radioimmunoassay.

ETS⁻ cells and interfered with access to GM$_1$ by CT in that phenotype.

ANALYSIS OF CT BINDING TO ETS⁺ AND ETS⁻ CELLS BY FLOW CYTOMETRY

To verify the effects of neuraminidase and LT IIb on CT binding, cells were treated as described above, stained with FITC-labeled CT, and analyzed by flow cytometry. Neuraminidase had no effect on CT binding to B6 cells but enhanced CT binding to C3H cells, in agreement with the cAMP responses noted in Table 3. Preincubation with LT IIb markedly decreased CT binding to both B6 and C3H cells. These results are consistent with the model that ETS⁻ cell surfaces are coated with decoy nonfunctional receptors for CT, making them less susceptible to its biological effects.

The reverse of the above experimental approach was undertaken, "reconstituting" NCTC 2071 murine fibroblast null cells, which express no GM$_1$ or other gangliosides and are able to respond to CT only after preincubation with exogenously supplied gangliosides. NCTC cells were preincubated with ganglioside extracts from ETS⁻ and ETS⁺ enterocytes, washed, labeled with FITC-CT as described above, and analyzed by flow cytometry. Control NCTC cells bind very low levels of CT (ca. 0.5 that of C3H and 0.3 that of B6). When the cells were preincubated with C3H extracts, CT binding increased to a level equivalent to that in C3H cells themselves. When B6 extracts were added, CT binding approached that of B6 cells. These results further support the implication that ganglioside expression is a key factor in establishing the ETS phenotype.

GENETIC STUDIES OF THE ETS PHENOTYPE

Yasuhiro Hashimoto and his colleagues at the Tokyo Metropolitan Institute of Medical Science first reported in 1983 that the locus controlling ganglioside expression in mice is on chromosome 17 linked to the *H-2* complex (22a). In a subsequent series of papers, they described the phenotypes associated with the locus and mapped the gene to a region 1 centimorgan centromeric to the *K* end of the *H-2* domain. They named the gene *Ggm-1,* for ganglioside expression. Further work by other groups has indicated that multiple genes are involved (14, 35, 36). Our own work, described above, has led us to believe that the *ets* gene or genes are part of the *Ggm-1* complex.

FUTURE DIRECTIONS

Southern blots of mouse chromosomal DNA restriction enzyme digests have been analyzed with a battery of gene probes known to be linked to the *K* end of the *H-2* region to identify segments containing putative *ets* genes. Some of the probes are directed toward genes coding for enzymes known to be involved in ganglioside and glycolipid synthesis (14, 31, 36). Potential *ets*-containing DNA fragments will be electroeluted from gels subcloned and transfected into appropriate host cells.

Using this strategy, we hope to elucidate the mechanisms controlling the ETS⁺ phenotype at the gene level. Because of the strong homology between mouse and human DNAs, this model should eventually enable us to develop probes to screen for the comparable ETS⁺ phenotype in humans.

CURRENT STATUS OF HOST SUSCEPTIBILITY TO CHOLERA

The plea by Barua and Paguio for more research into factors important in host resistance to cholera has generated a lot of smoke but not much fire. To review, there is clearly a definite correlation between blood group O and increased risk of cholera gravis. The effect on severity seems to be restricted to El Tor disease. In contrast, group AB persons seem to be at less risk. In terms of susceptibility and infection, the O blood group does not seem to play a role, but group O persons respond less well to whole-cell and CTB vaccines with respect to protective efficacy. There are small but interesting structural differences between CTs from classical and El Tor *V. cholerae* that would bear investigation in an in vivo model. There are large differences in mechanisms and components employed for colonization by classical and El Tor vibrios that need to be investigated both in vitro and in vivo. One or more of these differences may hold keys to some of the blood group O phenomenology. Finally, there is evidence suggesting that distinct micro- and macrodifferences in glycolipid expression occur within a single individual's intestinal tract. Some of these differences may be reflected in the ability of *H. pylori* to colonize and cause chronic hypochlorhydria in certain individuals. When more information is available in all these areas, we may be able to clearly identify that subpopulation at risk for cholera gravis and let them be the first to benefit from the new generation of vaccines now in testing or in development.

Acknowledgments. I am grateful to Teresa Hire for help with manuscript preparation.

Development of the SAM model and progress in elucidating the ETS phenotype were supported by NIH grants AI 20205 and DK 38783 as part of the U.S.-Japan Cooperative Medical Sciences Program (R22).

REFERENCES

1. **Barra, J. L., C. G. Monferran, L. E. Balanzino, and F. A. Cumar.** 1992. *Escherichia coli* heat-labile enterotoxin preferentially interacts with blood group A-active glycolipids from pig intestinal mucosa and A- and B-active glycolipids from human red cells compared to H-active glycolipids. *Mol. Cell. Biochem.* **115**:63–70.

2. **Barua, D., and A. S. Paguio.** 1977. ABO blood groups and cholera. *Ann. Hum. Biol.* **4**:489–492.

3. **Bennun, F. R., G. A. Roth, C. G. Monferran, and F. A. Cumar.** 1989. Binding of cholera toxin to pig intestinal mucosa glycosphingolipids: relationship with the ABO blood group system. *Infect. Immun.* **57**:969–974.

4. **Cater, R. E., II.** 1992. *Helicobacter* (aka Campylobacter) *pylori* as the major causal factor in chronic hypochlorhydria. *Med. Hypoth.* **39**:367–374.

5. **Chaudhuri, A., and S. De.** 1977. Cholera and blood-groups. *Lancet* ii:404.

6. **Clemens, J. D., D. A. Sack, J. R. Harris, J. Chakraborty, M. R. Khan, S. Huda, F. Ahmed, J. Gomes, M. R. Rao, A.-M. Svennerholm, and J. Holmgren.** 1989. ABO blood groups and cholera: new observations on specificity of risk and modification of vaccine efficacy. *J. Infect. Dis.* **159**:770–773.

7. **Clemens, J. D., D. A. Sack, J. R. Harris, M. R. Khan, J. Chakraborty, S. Chowdhury, M. R. Rao, F. P. van Loon, B. F. Stanton, M. Yunus, M. Ali, M. Ansaruzzaman, A.-M. Svennerholm, and J. Holmgren.** 1990. Breast feeding and the risk of severe cholera in rural Bangladeshi children. *Am. J. Epidemiol.* **131**:400–411.

8. **Cover, T. L., S. G. Vaughn, and P. Cao.** 1992. Potentiation of *Helicobacter pylori* vacuolating toxin activity by nicotine and other weak bases. *J. Infec. Dis.* **166**:1073–1078.

9. **Donta, S. T., T. Tomicle, and R. K. Holmes.** 1992. Binding of class II *Escherichia coli* enterotoxins to mouse Y1 and intestinal cells. *Infect. Immun.* **60**:2870–2873.

10. **Dubey, R. S., M. Lindblad, and J. Holmgren.** 1990. Purification of El Tor cholera enterotoxins and comparisons with classical toxin. *J. Gen. Microbiol.* **136**:1839–1847.

11. **Elson, C. O.** 1992. Cholera toxin as a mucosal adjuvant: effects of *H-2* major histocompatibility complex and *lps* genes. *Infect. Immun.* **60**:2874–2879.

12. **Elson, C. O., and W. Ealding.** 1985. Genetic control of the murine immune response to cholera toxin. *J. Immunol.* **135**:930–932.

13. **Elson, C. O., and W. Ealding.** 1987. Ir gene control of the murine secretory IgA response to cholera toxin. *Eur. J. Immunol.* **17**:425–428.

14. **Ernst, L. K., V. P. Rajan, R. D. Larsen, M. M. Ruff, and J. B. Lowe.** 1989. Stable expression of blood group H determinants and GDP-L-fucose: beta-D-galactoside 2-alpha-L-fucosyltransferase in mouse cells after transfection with human DNA. *J. Biol. Chem.* **264**:3436–3447.

15. **Finkelstein, R. A., M. Boesman-Finkelstein, Y. Chang, and C. C. Hase.** 1992. *Vibrio cholerae* hemagglutinin/protease, colonial variation, virulence, and detachment. *Infect. Immun.* **60**:472–478.

16. **Galen, J. J., A. Ketley, A. Fasano, S. Richardson, S. Wasserman, and J. Kaper.** 1992. Role of *Vibrio cholerae* neuraminidase in the function of cholera toxin. *Infect. Immun.* **60**:406–415.

17. **Glass, R. I., J. Holmgren, C. E. Haley, M. R. Khan, A.-M. Svennerholm, B. J. Stoll, K. M. B.**

Hossain, R. E. Black, M. Yunus, and D. Barua. 1985. Predisposition for cholera of individuals with O blood group. *Am. J. Epidemiol.* **121**:791–796.

18. Glass, R. I., J. Holmgren, M. R. Khan, K. M. B. Hossain, M. I. Huq, and W. B. Greenough. 1984. A randomized, controlled trial of the toxin-blocking effects of B Subunit in family members of patients with cholera. *J. Infect. Dis.* **149**:495–500.

19. Glass, R. I., A.-M. Svennerholm, M. R. Khan, S. Huda, M. I. Huq, and J. Holmgren. 1985. Seroepidemiological studies of El Tor cholera in Bangladesh: association of serum antibody levels with protection. *J. Infect. Dis.* **151**:236–242.

20. Glass, R. I., A. M. Svennerholm, B. J. Stoll, M. R. Khan, S. Huda, and M. I. Huq. 1989. Effects of undernutrition on infection with *Vibrio cholerae* O1 and on response to oral cholera vaccine. *Pediatr. Infect. Dis. J.* **8**:105–109.

21. Hakomori, S. 1991. Immunochemical and molecular genetic basis of the histo-blood group ABO(H) and related antigen system. *Baillieres Clin. Haematol.* **4**:957–974.

22. Hall, R. H., G. Losonsky, A. P. D. Silveira, R. K. Taylor, J. J. Mekalanos, N. D. Witham, and M. M. Levine. 1991. Immunogenicity of *Vibrio cholerae* O1 toxin-coregulated pili in experimental and clinical cholera. *Infect. Immun.* **59**:2508–2512.

22a. Hashimoto, Y., A. Suzuki, T. Yamakawa, N. Miwashita, and K. Moriwaki. 1983. Expression of GM1 and GD1a in mouse liver is linked to the H-2 complex on chromosome 17. *J. Biochem.* **94**:2043–2048.

23. Holgersson, J., N. Strömberg, and M. E. Breimer. 1988. Glycolipids of human large intestine: difference in glycolipid expression related to anatomical localization, epithelial/non-epithelia tissue and the *ABO, Le* and *Se* phenotypes of the donors. *Biochimie* **70**:1565–1574.

24. Jonson, G., J. Holmgren, and A.-M. Svennerholm. 1991. Epitope differences in toxin-coregulated pili produced by classical and El Tor *Vibrio cholerae* O1. *Microb. Pathog.* **11**:179–188.

25. Johnson, G., J. Holmgren, and A.-M. Svennerholm. 1991. Identification of a mannose-binding pilus on *Vibrio cholerae* El Tor. *Microb. Pathog.* **11**:433–441.

26. Jonson, G., J. Holmgren, and A.-M. Svennerholm. 1992. Analysis of expression of toxin-coregulated pili in classical and El Tor *Vibrio cholerae* O1 in vitro and in vivo. *Infect. Immun.* **60**:4278–4284.

27. Kaper, J. Personal communication.

28. Katelaris, P. H., G. H. K. Tippett, P. Norbu, D. G. Lowe, R. Brennan, and M. J. G. Farthing. 1992. Dyspepsia, *Helicobacter pylori,* and peptic ulcer in a randomly selected population in India. *Gut* **33**:1462–1466.

29. Lange, S., and I. Lönnroth. 1986. Bile and milk from cholera toxin treated rats contain a hormone-like factor which inhibits diarrhea induced by the toxin. *Int. Arch. Allergy Appl. Immunol.* **79**:270–275.

30. Levine, M. M., D. R. Nalin, M. B. Rennels, R. B. Hornick, S. Sotman, G. Van Blerk, T. P. Hughes, S. O'Donnell, and D. Barua. 1979. Genetic susceptibility to cholera. **6**:369–374.

31. Mandel, U., T. White, J. Karkov, S. Hakomori, H. Clausen, and E. Dabelsteen. 1990. Expression of the histo-blood group ABO gene defined glycosyltransferases in epithelial tissues. *J. Oral Pathol. Med.* **19**:251–256.

32. Moyenuddin, M., K. Wachsmuth, S. H. Richardson, and W. L. Cook. 1992. Enteropathogenicity of non-toxigenic *Vibrio cholerae* O1 for adult mice. *Microb. Pathog.* **12**:451–458.

33. Nalin, D. R., M. M. Levine, E. Bergquist, J. Libonati, R. J. Levine, D. Hoover, J. McLaughlin, J. Alam, and R. B. Hornick. 1978. Cholera, non-vibrio cholera, and stomach acid. *Lancet* **ii**:856–859.

34. Nalin, D. R., J. Rhead, M. Rennels, S. O'Donnell, M. M. Levine, E. Bergquist, T. Hughes, and R. B. Hornick. 1978. Cannabis, hypochlorhydria, and cholera. *Lancet* **ii**:859–862.

35. Odelgah, P. G. 1990. Influences of blood group and secretor genes on susceptibility to duodenal ulcer. *East Afr. Med. J.* **67**:487–500.

36. Oriol, R., R. Mollicone, P. Coulin, A. M. Dalix, and J. J. Candelier. 1992. Genetic regulation of the expression of ABH and Lewis antigens in tissues. *APMIS Suppl.* **27**:28–38.

37. Palmer, D. L., and S. N. Zaman. 1979. Depression of cell-mediated immunity in cholera. *Infect. Immun.* **23**:27–30.

38. Ramirez-Ramos, A., R. Gilman, W. Spira, S. Recavarren, J. Watanabe, R. Leon-Barua, C. Rodriguez, C. Guevara, G. Gago, J. Bonilla, J. Cok, H. Huidobro Ruiz, E. Gomez, E. Olivares, R. Garcia, G. Garrido Klinge, G. Vargas, M. Astete, M. Valdivia, and R. Berendson. 1992. Ecology of *Helicobacter pylori* in Peru: infection rates in coastal, high altitude, and jungle communities. *Gut* **33**:604–605.

39. Richardson, S. H., J. C. Giles, and K. S. Kruger. 1984. Sealed adult mice: a new model for enterotoxin evaluation. *Infect. Immun.* **43**:482–486.

40. Richardson, S. H., and J. V. Kaspar. 1990. Genetic studies on *V. fluvialis* and its enterotoxins, p. 159–171. *In* A. E. Pohland et al. (ed.), *Microbial, Toxins in Foods and Feeds.* Plenum Press, New York.

41. Richardson, S. H., M. Kothary, and T. Sexton. 1990. H-2 regulated responses to enterotoxins in isolated cells, cell lysates and membranes, p. 235–

246. *In* Y. Zinnaka and R. B. Sack (ed.), *Advances in Research on Cholera and Related Diarrheas*, vol. 7. KTK Scientific Publishers, Tokyo.

42. **Richardson, S. H., R. Kuhn, J. Estrada, M. J. Black, and D. Pardi.** 1988. In vivo and in vitro studies on the genetic and cellular control of the response to cholera toxin in inbred mice, p. 297–306. *In* S. Kuwahara and N. Pierce (ed.), *Advances in Research on Cholera and Related Diarrheas*, vol. 5. KTK Scientific Publishers, Tokyo.

43. **Richardson, S. H., and R. E. Kuhn.** 1986. H-2 modulation of cholera enterotoxin effects in SAM: (sealed adult mice), p. 305–316. *In* S. Kuwahara and N. F. Pierce (ed.), *Advances in Research on Cholera and Related Diarrheas*, vol. 3. KTK Scientific Publishers, Tokyo.

44. **Richardson, S. H., and R. E. Kuhn.** 1986. Studies on the genetic and cellular control of sensitivity to enterotoxins in the sealed adult mouse model. *Infect. Immun.* **54:**522–528.

45. **Richardson, S. H., T. Sexton, J. Rader, and C. Kajs.** 1993. The biochemical and genetic basis for the ETS (enterotoxin sensitivity) phenotype in mice. *Clin. Infect. Dis.* **16**(Suppl. 2)**:**S83–S91.

46. **Richardson, S. H., T. Y. Sexton, C. Kajs, and J. Rader.** The enterotoxin sensitivity phenotype (ETS) is controlled by host cell ganglioside composition. *In* Y. Takeda and R. B. Sack (ed.), *Advances in Research on Cholera and Related Diarrheas*, vol. 9, in press. KTK Scientific Publishers, Tokyo.

47. **Rodman, D. M., and S. Zamudio.** 1991. The cystic fibrosis heterozygote—advantage in surviving cholera? *Med. Hypoth.* **36:**253–258.

48. **Sheinfeld, J., A. J. Schaffer, C. Cordon-Cardo, A. Ragatko, and W. R. Fair.** 1989. Association of the Lewis blood-group phenotype with recurrent urinary tract infections in women. *N. Engl. J. Med.* **320:**773–777.

49. **Sircar, B. K., P. Dutta, S. P. De, S. N. Sikdar, B. C. Deb, and S. C. Pal.** 1981. ABO blood group distributions in diarrhoea cases including cholera in Calcutta. *Ann. Hum. Biol.* **8:**289–291.

50. **Sugarman, B., and S. T. Donta.** 1979. Specificity of attachment of certain enterobacteriaccae to mammalian cells. *J. Gen. Microbiol.* **115:**509–512.

51. **van Loon, F. P., J. D. Clemens, M. Shahrier, D. A. Sack, C. B. Stephen, M. R. Khan, G. H. Rabbani, M. R. Rao, and A. K. Banik.** 1990. Low gastric acid as a risk factor for cholera transmission: application of a new non-invasive gastric acid field test. *J. Clin. Epidemiol.* **43:**1361–1367.

52. **Walser, B. L., R. D. Newman, A. A. M. Lima, and R. L. Guerrant.** 1992. Pathogen and host differences in bacterial adherence to human buccal epithelial cells in a northeast Brazilian community. *Infect. Immun.* **60:**4793–4800.

53. **Yamamoto, F., H. Clausen, T. White, J. Marken, and S. Hakomori.** 1900. Molecular genetic basis of the histo-blood group ABO system. *Nature* (London) **345:**229–233.

IV. EPIDEMIOLOGY AND SURVEILLANCE

Vibrio cholerae and Cholera: Molecular to Global Perspectives
Edited by I. Kaye Wachsmuth, Paul A. Blake, and Ørjan Olsvik
© 1994 American Society for Microbiology, Washington, DC 20005

Chapter 18

Historical Perspectives on Pandemic Cholera

Paul A. Blake

The history of cholera has been reviewed extensively by Pollitzer (8) and Barua (1). Rather than repeating their work, I will give an overview of the history of cholera to set the scene for understanding the epidemiology and surveillance of cholera since 1970.

The modern history of cholera began in 1817, when cholera spread out of India in what the literature describes as the first of seven pandemics (epidemics affecting many countries). Medical historians differ on whether cholera existed outside Asia before 1817 (1, 8). Even the number of pandemics since 1817 and when each pandemic began and ended are debated; in fact, the divisions between the pandemics are vague and may be somewhat artificial. The most commonly used dates of the first six pandemics (the dates used here) are those proposed by Pollitzer (8): pandemic 1, 1817 to 1823; pandemic 2, 1829 to 1851; pandemic 3, 1852 to 1859; pandemic 4, 1863 to 1879; pandemic 5, 1881 to 1896; and pandemic 6, 1899 to 1923.

During these six pandemics, epidemic cholera spread from the Indian subcontinent to much of the rest of the world, including the

Paul A. Blake • Foodborne and Diarrheal Diseases Branch, Division of Bacterial and Mycotic Diseases, National Center for Infectious Diseases, Centers for Disease Control and Prevention, 1600 Clifton Road, Atlanta, Georgia 30333.

Americas. Major incursions into the United States occurred in 1832, 1848, and 1866 (9). After 1869, cholera began to abate in the Americas and Europe. By 1900 it had disappeared from the Americas and most of Africa and Europe, and by 1950 it was found only in Asia.

The increasing speed and ease of travel and the increasing number of travelers probably contributed to the pandemic spread of cholera. European colonization of the Indian subcontinent increased contact between India and other parts of the world. The invention of the steamship and the 1869 opening of the Suez Canal greatly shortened travel time between India and Europe. Accounts of the spread of pandemic cholera in the 19th century are largely summaries of apparent transfer from place to place by steamship. The 20th century, with more rapid surface transportation and the advent of air travel, has brought the potential for dissemination of epidemic strains around the world in days or months rather than the years it has taken in the past.

Accounts of cholera-like disease go back to the times of Hippocrates and Lord Buddha (1) and perhaps even earlier. However, before 1883, when Robert Koch showed that the disease is caused by a bacterium, the diagnoses depended on clinical observations and are open to question. We know that the fifth and sixth pandemics were caused by *V. cholerae*

O1 of the classical biotype, but the nature of the strains causing the first four pandemics is unknown (1). If the pre-1817 cholera-like illnesses in India, Greece, England, and elsewhere were indeed cholera, a variety of strains could have caused the illnesses. Choleragenic strains are now known to include not only *V. cholerae* O1 of the classical biotype but also *V. cholerae* O1 of the El Tor biotype and *V. cholerae* O139 (10).

Before 1961, the El Tor biotype of *V. cholerae* O1 was known to cause cases and outbreaks of cholera-like disease, but it was not generally considered to have epidemic potential. However, in 1961, the El Tor biotype apparently spread out of Indonesia to other countries in Asia, beginning the seventh cholera pandemic, which still continues in 1993. For a decade, this pandemic was largely confined to Asia and the Middle East, but in the 1970s it spread to Africa, southern Europe, and several Pacific islands, and in 1991 it appeared in South America (11).

A third organism that might have caused cholera before 1883 was identified in 1993. In late 1992 and early 1993, epidemic disease clinically indistinguishable from cholera appeared in India and Bangladesh (4, 10). It proved to be caused by a cholera toxin-producing *V. cholerae* that had never been recognized as a cause of epidemic cholera. The organism was given a new serotype designation: O139. The disease spread rapidly and attacked all age groups in populations where *V. cholerae* O1 was endemic. It seems possible that the organism will continue to spread internationally and give rise to an eighth cholera pandemic (10). The discovery that the O1 antigen is not essential to producing epidemic cholera opens up the possibility that there are *V. cholerae* of additional O groups that can cause epidemic cholera and that may in fact already have caused epidemic cholera in the days before the strains could be identified.

It may be reasonable to consider the first six cholera pandemics as a single continuing pandemic, even though we have no bacteriologic proof that *V. cholerae* O1 of the classical biotype caused the first four pandemics as well as the fifth and sixth. First, the divisions between the six pandemics are indistinct, and it is not clear that they were separated by cholera-free intervals. Second, there is recent evidence that *V. cholerae* O1 can persist indefinitely as a free-living organism in some natural environments (2, 3, 5–7); pandemics 2 through 6 may have stemmed from organisms persisting in the environment in many countries as well as from small foci of human disease. Third, the seventh pandemic, caused by a different biotype, has persisted for 33 years, longer than any of the previous pandemics; it had explosive expansions in 1970 and 1991 that might have been considered new pandemics according to methods used to describe the first six pandemics. Finally, there may now be an eighth pandemic caused by *V. cholerae* of a different O group. In the future, it may appear in retrospect that we have had three pandemics caused by global spread of distinctive *V. cholerae* strains: one that began in 1817, caused by the classical biotype of *V. cholerae* O1; a second that began in 1961, caused by the El Tor biotype of *V. cholerae* O1; and a third that began in 1992, caused by *V. cholerae* O139.

REFERENCES

1. **Barua, D.** History of cholera, p. 1–36. *In* D. Barua and W. B. Greenough III (ed.), *Cholera*. Plenum, New York.
2. **Blake, P. A., D. T. Allegra, J. D. Snyder, T. J. Barrett, L. McFarland, C. T. Caraway, J. C. Feeley, J. P. Craig, J. V. Lee, N. D. Puhr, and R. A. Feldman.** 1980. Cholera—a possible endemic focus in the United States. *N. Engl. J. Med.* **302:**305–309.
3. **Bourke, A. T. C., Y. N. Cossins, B. R. W. Gray, T. J. Lunney, N. A. Rostron, R. V. Holmes, E. R. Griggs, D. J. Larsen, and V. R. Kelk.** 1986. Investigation of cholera acquired from the riverine environment in Queensland. *Med. J. Aust.* **144:**229–234.
4. **Cholera Working Group.** 1993. Large epidemic of cholera-like disease in Bangladesh caused by *Vibrio cholerae* O139 synonym Bengal. *Lancet* **342:**387–390.
5. **Colwell, R. R., and W. M. Spira.** 1992. The ecol-

ogy of *Vibrio cholerae,* p. 107–127. *In* D. Barua and W. B. Greenough III (ed.), *Cholera.* Plenum, New York.

6. **Lowry, P. W., A. T. Pavia, L. M. McFarland, B. H. Peltier, T. J. Barrett, H. B. Bradford, J. M. Quan, J. Lynch, J. B. Mathison, R. A. Gunn, and P. A. Blake.** 1989. Cholera in Louisiana: widening spectrum of seafood vehicles. *Arch. Intern. Med.* **149:**2079–2084.

7. **Miller, C. J., R. G. Feachem, and B. S. Drasar.** 1985. Cholera epidemiology in developed and developing countries: new thoughts on transmission, seasonality and control. *Lancet* **i:**261–263.

8. **Pollitzer, R.** 1959. *Cholera,* p. 11–50. World Health Organization monograph no. 43. World Health Organization, Geneva.

9. **Rosenberg, C. E.** 1962. *The Cholera Years: the United States in 1832, 1849, and 1866.* University of Chicago Press, Chicago.

10. **Swerdlow, D. L., and A. A. Ries.** 1993. *Vibrio cholerae* non-O1—the eighth pandemic? *Lancet* **342:**382–383.

11. **Tauxe, R. V., and P. A. Blake.** 1992. Epidemic cholera in Latin America. *JAMA* **267:**1388–1390.

Vibrio cholerae and Cholera: Molecular to Global Perspectives
Edited by I. Kaye Wachsmuth, Paul A. Blake, and Ørjan Olsvik
© 1994 American Society for Microbiology, Washington, DC 20005

Chapter 19

The Epidemiology of Cholera in Africa

David L. Swerdlow and Margaretha Isaäcson

In 1970, the seventh cholera pandemic reached the continent of Africa on two fronts, striking West Africa and North Africa with devastating consequences. Epidemic cholera quickly spread throughout the continent, and the disease is now endemic in many areas. Transmission occurred through contaminated food and water. Inadequate resources and limited access to health care led to high death rates, and massive population movements contributed to both the rapid spread and the high mortality. This chapter traces the epidemic in Africa, describes epidemiologic investigations that determined prevalent modes of transmission, and discusses efforts to control the epidemic and prevent deaths.

PANDEMIC CHOLERA: 19TH CENTURY

During the 19th century, all six pandemics of cholera reached Africa. Extension was thought to have been facilitated by caravans and ships and to have occurred along trans-portation routes. All six pandemics affected East Africa (31), the first in 1821 when a slave ship carrying people infected with cholera arrived from Zanzibar via Muscat and Oman. The second reached East Africa in 1836 or 1837, when pilgrims to Mecca returned with the illness. Small trading boats (dhows) spread the infection along the East African coast, and caravans carried the pandemic inland to Central Africa (31). Mortality was high, as demonstrated in Zanzibar in 1869, when 70,000 persons died from cholera (26, 31).

Five of the 19th century pandemics affected North Africa. Cholera was introduced into Egypt by pilgrims returning from Mecca, and the infection subsequently spread up the Nile into Upper Egypt and Sudan. Mediterranean trading ships also brought the infection to North African ports (26, 31). Outbreaks of cholera in West Africa in 1868 and in 1893 to 1894 were thought to have followed introduction of the disease by caravans.

THE SEVENTH PANDEMIC

During the 20th century, cases of cholera were mostly limited to areas in Asia and the Indian subcontinent where cholera is endemic. An outbreak with over 32,000 cases and 20,000 deaths occurred in Egypt in 1947, but the outbreak diminished quickly (31). No

David L. Swerdlow • Foodborne and Diarrheal Diseases Branch, Centers for Disease Control and Prevention, Mailstop C-09, Atlanta, Georgia 30333. *Margaretha Isaäcson* • School of Pathology, Department of Tropical Diseases, The South African Institute for Medical Research, University of the Witwatersrand, P. O. Box 1038, Johannesburg 2000, South Africa.

cases were reported from West Africa through the mid-20th century. The seventh cholera pandemic, caused by the El Tor biotype, began in Indonesia in 1961 and spread through Asia and into the Middle East. In August 1970, the West African cholera epidemic began in Guinea. It was caused by *Vibrio cholerae* O1, biotype El Tor, serotype Ogawa. It is not known how it was introduced, although some have blamed a symptom-free traveler returning from Asia or the Middle East. By September 4, 1970, over 2,000 cases and more than 60 deaths were reported (14, 37). Initially, cases concentrated along coastal regions, and spread was thought to occur as fishermen and boatmen traveled to contiguous areas (37). Epidemics occurred in Sierra Leone, Liberia, Cote d'Ivoire, Ghana, Togo, Benin (Dahomey), Nigeria, and southern Cameroon successively between September 1970 and February 1971 (37). The epidemic also spread inland along rivers and routes taken by traders and travelers. Modern-day rapid transport of ill or asymptomatic infected individuals facilitated a jump of 1,000 km in November 1970, when the epidemic suddenly appeared in Mopti, Mali. After striking Mali, the epidemic radiated outward, following the movements of people and the occurrence of large gatherings such as markets and funerals along the River Niger, its branches, and other trade routes (2, 14). Epidemics soon affected Burkina Faso (Upper Volta), Nigeria, Cote d'Ivoire, Ghana, and Mauritania (14). Mortality rates of 20 to 50% were noted in some areas. In Chad and northern Cameroon, the epidemic affected desert-like regions, proving that cholera could spread in areas without plentiful water, rivers, or lagoons (14). The consequences of the epidemic were enormous; some have estimated that 150,000 cases and 20,000 deaths occurred in West Africa alone during the first year (14). At the same time in North Africa (Libya, Tunisia, Algeria, and Morocco) and East Africa (Djibouti [Afars and Issas], Ethiopia, and Somalia), there were epidemics of cholera

thought to have originated in the Middle East (Fig. 1). By the end of 1971, 25 countries in Africa had reported cases to the World Health Organization (37). The overall case fatality rate for 1971 was 16%.

Between 1972 and 1991, cholera spread throughout much of the remainder of Africa (Fig. 2). The number of countries reporting cases of cholera ranged from 11 to 21 each year. During some years such as those between 1972 and 1977, only a few thousand cases per year were reported, but in other years, such as 1982, over 40,000 cases were reported (37) (Fig. 3). During this period the case fatality rate ranged between 4 and 12% (37) (Fig. 4). Forty-three countries reported cases on several occasions; cholera had become endemic in Africa (37).

In 1991 there was a recrudescence of cholera in many parts of Africa (Fig. 3). At least 20 countries were affected, with 153,367 cases and 13,998 deaths reported, the highest totals ever recorded and, given the incompleteness of reported cases, indicative of a massive epidemic in the continent (39). Epidemics appeared in parts of the population that had previously been spared, particularly in periurban areas (39). The causes of the large epidemics of 1991 are not fully understood but may be due to the poor living conditions of populations affected by drought and civil disturbance. Urban centers have become increasingly overcrowded, with progressive stressing of water and sanitation facilities (39). There were two main epidemic foci, in the southern and eastern part of the continent (around Zambia, Mozambique, Malawi, and Angola) and in West Africa (39). Some countries (Benin, Burkina Faso, Chad, and Togo) that had not reported cases for several years reported cases in 1991 (37). In Chad, 13,915 cases and 1,344 deaths were reported from 12 of the 14 prefectures. In 1990, Nigeria reported no cases of cholera, but in 1991, 59,478 cases and 7,654 deaths were reported from 19 of 21 states in that country (37, 39). The case fatality rate in Nigeria of 13% was the highest reported in 1991 (37). The high

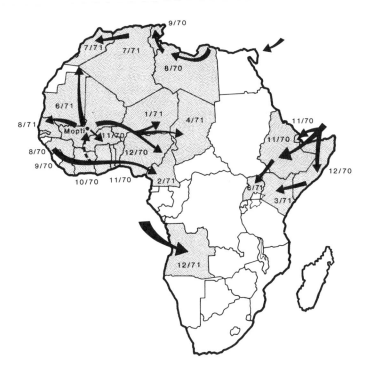

Figure 1. Emergence of cholera in Africa. Routes of transmission, as well as the date of emergence (month/year), are shown. Reprinted from reference 37 with permission.

case fatality rates in Africa were in marked contrast to the fatality rates of <1% reported from Latin America during the same time period (33) and are thought to be due to the lack of access to adequate oral and intravenous rehydration facilities in many parts of the continent (34).

In 1992 the number of reported cases of cholera decreased to 91,081 with 5,291 deaths (40); however, the epidemics of 1991 demonstrated that large outbreaks could occur against a background of endemic cholera and underscored the need to avoid complacency even after cholera had been present intermittently for decades.

MODES OF TRANSMISSION

Both foodborne and waterborne transmission of cholera have been documented in Africa. It had recently been recognized that cholera could be transmitted through contaminated food. Incriminated foods included leftover rice in Micronesia (15) and Singapore (13); undercooked shellfish in Portugal (5), Italy (1), and the United States (4); and unsalted dry fish in the Pacific (20). During the 1984 cholera epidemic in Mali, eating leftover millet gruel was associated with illness in a drought-affected village. Millet gruel was prepared once a day and eaten communally without being reheated, often after being kept at ambient temperature for many hours. Traditionally curdled goat's milk, an acid condiment, was added to the millet. However, at the time of the outbreak, goat's milk was in short supply because of the drought. In laboratory studies, the epidemic strain of *V. cholerae* O1 survived more than 24 h in millet gruel prepared without goat's milk but less than 6 h in gruel to which curdled goat's milk had been added. Apparently, the curdled goat's milk lowered the pH of the

Figure 2. Countries in Africa reporting cholera from 1961 to 1991 (World Health Organization, May 1993).

gruel and prevented survival of the vibrio (34).

In 1986, during an investigation of cholera in Guinea, persons with cholera were more likely than controls to have eaten leftover peanut sauces used as condiments on rice and were less likely than controls to have eaten tomato sauces. In laboratory experiments, the peanut sauces (pH 6 to 7) supported growth of *V. cholerae* O1, but the tomato sauce (pH 4.5 to 5.0) did not (30). Therefore, investigations in Mali and Guinea demonstrated that subtle changes in food preparation practices can alter food ecology enough to allow growth of *V. cholerae* O1 in previously safe foods.

Seafood, implicated as a vehicle of cholera transmission in other parts of the world, was associated with illness in Guinea-Bissau in 1987. A matched case-control study deter-

mined that eating small crabs, particularly leftover crabs, was associated with being hospitalized with cholera (27).

Large gatherings and funerals also played a role in the spread of cholera in Africa. In Guinea, an outbreak of cholera occurred after a funeral. A village survey determined that cases of cholera-like illness occurred in 8 of 16 households in which persons ate a rice meal served at the funeral but in only 1 of 24 households in which no household member ate the rice meal. Three women who had cleaned the bedsheets and helped administer enemas to the deceased had helped prepare the rice meal for the funeral shortly thereafter (30). Similarly, in Guinea-Bissau, an outbreak occurred after the body of a 25-year-old male who died of cholera was transported to his home village. One hundred eleven hospitalized cases and 11 deaths were reported in

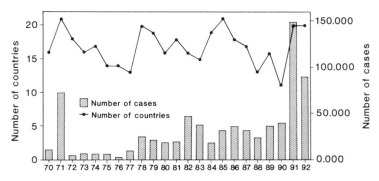

Figure 3. Cholera in Africa. The number of cases and number of countries reporting cases from 1970 to 1992 are shown. Reprinted from reference 40 with permission.

the village after a 5-day funeral attended by several hundred villagers and visiting guests from surrounding areas. Illness was associated with eating a rice-based meal made by persons who had previously prepared the body for burial (27).

Waterborne transmission of cholera has also been documented in Africa. In a case-control study in Mali, drinking unboiled water from one well (of 50 in the village) was associated with diarrheal illness (34). Cases of cholera were associated with using water from impure sources during an outbreak in South Africa in 1981 to 1982 (29). In 1981, during an outbreak in Lebowa, South Africa, persons could obtain water supplies from a river, a canal, a borehole, or rain water. In a matched case-control study, persons who drank river water were more likely to become

ill with cholera than persons who drank water from "safer" sources. Treatment of water with bleach or by boiling protected against illness (28). Similarly, during an outbreak in Nigeria in 1982, piped water was in short supply, and families resorted to drinking water from poorly constructed wells built without casings. These wells may have become contaminated from nearby pit latrines. All nine wells from the compounds of persons with cholera were contaminated with more than 540 coliform bacteria per 100 ml of water, and all had fecal coliform bacteria detected on culture. In contrast, only one of four piped water samples had coliform bacteria, with a count of 6 coliform bacteria per 100 ml of water (36).

During an outbreak in a Transvaal gold mine, transmission was thought to be due to

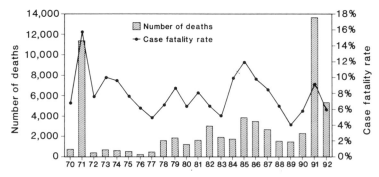

Figure 4. Cholera in Africa. The number of deaths and case fatality rate, by year, from 1970 to 1992 are shown (World Health Organization, May 1993).

contamination of drinking water kept in open containers on the floor of acclimatization centers. It was hypothesized that *V. cholerae* O1 might have been carried by perspiration from the perianal region of asymptomatically infected persons to the floor (17).

Water may also have become contaminated while in storage containers. During an outbreak of cholera in a refugee camp in Malawi in 1990, persons who placed their hands into the water of the household water storage container were more likely to develop cholera than were matched controls who did not place their hands into their drinking water. *V. cholerae* O1 was also isolated from a water sample from household water storage containers during this investigation (32). Water contaminated while in storage has also been associated with spread of cholera in India and Peru (9, 10, 33).

The survival of *V. cholerae* O1, biotype El Tor, in the types of water storage containers used in Africa has been studied. Vibrios survived longest in corroded iron drums (up to 22 days) and shortest in clay pots (up to 5 days) (25), and the presence of particulate iron increased vibrio survival in dechlorinated tap water (25a). However, use of clay pots may not be safer than use of iron drums, because the concentration of vibrios after inoculation was higher in the clay pots than in noncorroded iron drums or plastic containers. Clay pots also allowed outward seepage of vibrio-containing water, a possible mechanism for contamination of food and the hands of food handlers (25).

Use of proper hygienic practices may decrease the risk of becoming infected with *V. cholerae* O1. In Guinea-Bissau, the absence of hand-washing soap in the home was associated with illness (27), and in Guinea, cholera occurred less frequently in families whose members washed their hands with soap before meals. (30).

Several authors have reported what they believed to be person-to-person spread of cholera during cholera outbreaks in Africa. Three outbreaks occurred among hospitalized patients in

Dar es Salaam, Tanzania, between 1977 and 1983. A total of 216 persons (mostly children) developed cholera 1 to 5 days after being admitted to the hospital with other conditions. Two of the outbreaks involved multiply-antibiotic-resistant *V. cholerae* O1 strains that were not reported from other parts of Tanzania during the same period (22). Conditions were extremely overcrowded, and two to three patients occupied a single cot. A single sink served the entire ward, and it was used both to wash soiled bed clothes and to provide water for drinking and cooking. However, other than culturing food and water served to patients, no epidemiologic investigation into the route of transmission was conducted. Since children ate with their fingers from communal plates and powdered milk was reconstituted on the ward, transmission may have been by foodborne routes rather than by person-to-person direct contact (22).

An outbreak of cholera also occurred in a children's ward for severe diarrhea in Maputo Central Hospital, Mozambique, in 1980 to 1981. Upon admission to the ward, children with diarrhea and the adults accompanying them had rectal swabs obtained for culture for *V. cholerae* O1. Fifty-seven persons acquired infection during hospitalization, as documented by conversion of a culture-negative swab to a culture-positive swab. Most cases were among infants less than 4 years of age, but accompanying adults also became ill and developed positive stool cultures. The isolate was an antibiotic-resistant strain. Conditions in the ward were extremely overcrowded, and there were shortages of wash basins and soap. Different families shared food from the same plate and ate with their hands. Cultures of tap water, food, and formula were culture negative for *V. cholerae* O1. However, since culturing was limited to samples obtained at the time of serving and not of consumption, contamination after serving was not evaluated (7).

Similarly, in a South African hospital, eight hospitalized children had stool cultures that yielded *V. cholerae* O1 1 to 16 days after negative stool cultures had been collected on admission. As in Tanzania and Mozambique, patients were hospitalized with two persons in each cot, and the exact mechanism of infection was not determined (6).

It therefore remains controversial whether person-to-person spread of cholera by direct contact occurs. Some authorities consider person-to-person transmission of cholera to be rare, because the rate of infection among hospital workers taking care of ill cholera patients is low (3, 18).

However, person-to-person spread by direct contact may occur in situations of sanitary breakdown, although such spread has not been demonstrated conclusively. Some of the controversy may be due to differences in the definition of person-to-person spread. Some authors consider spread through contaminated family food or water to be person-to-person spread (22), but others would consider this to be foodborne or waterborne transmission (15).

Studies to determine modes of cholera transmission in Africa have been crucial for local governments and international agencies preparing for outbreaks of cholera in Africa and throughout the world (12). Preparations to ensure clean water, treatment of sewage, and safe food-handling practices remain critical to preventing the spread of cholera.

CHOLERA IN REFUGEE CAMPS

Since the 1970s, there have been large population movements in Africa, and millions of persons have become refugees. Outbreaks of cholera have occurred among refugees or displaced persons in Somalia, Ethiopia, Sudan, Malawi, and Mozambique (23, 35). The plight of refugees is complicated because their new host countries are poor and already have high mortality rates. Refugees and displaced persons are at particular risk for cholera because of overcrowding, inadequate sanitary facilities and water supplies, and intercurrent malnutrition. Case

fatality rates during cholera epidemics in refugee camps have usually been about 2 to 3% but have ranged from less than 1% of cases to 25% in one unplanned Somali refugee camp (23, 24, 35).

Low death rates can be achieved with careful preparation. In 1985, nearly half a million Ethiopian refugees were located in Sudan during a period of extreme famine. An outbreak of cholera with 1,175 cases (attack rate, 4.5%) occurred in two adjacent refugee camps in eastern Sudan. The death rate among patients who came to medical attention was low, about 1% (13 deaths per 1,175 cases). An additional 38 persons died before receiving medical attention. Patients were brought into cholera referral centers staffed 24 h a day by trained persons (24). Patients received intravenous fluids (an average of 8 liters per patient), and oral rehydration solutions were administered rapidly. Transportation was available at all hours to facilitate rapid transfer of patients from referral centers to the main cholera treatment facilities.

Management of cholera case treatment was evaluated during an outbreak among Mozambican refugees in Malawi in 1990. Inadequate staffing, frequent shortages of oral rehydration salts, late presentations of cases due to limitations in transport facilities, and extreme heat led to overreliance on intravenous rather than oral rehydration. Many persons who died during this outbreak had febrile illnesses after being in the cholera tents for several days and probably had secondary infections caused by prolonged use of intravenous catheters. Pediatric patients had the highest death rates during this outbreak, possibly because of the difficulty of adequately assessing fluid status among younger patients; some died from dehydration, and others died from overhydration (32). Special care should be taken with treatment of patients, especially children, in similar settings in the future.

There appear to be many transmission routes for cholera in refugee camps. During the outbreak in two camps in 1985 among Ethiopian refugees in Sudan, water for

Declaration

At a meeting held on 7 May 1992 in Geneva during the Forty-fifth World Health Assembly, the Ministers of Health of the WHO Member States in Africa and their representatives:

Considering

- that the cholera situation in Africa is extremely serious and is worsening year by year, and that the situation is likely to be exacerbated by the drought conditions currently affecting parts of the continent,

- that the high and rising incidence of epidemic and endemic cholera is an indicator of the poor and deteriorating environmental and sanitary conditions in which too many people in Africa are living and which underlie many of the health problems suffered by the people,

- that cholera is a disease whose control requires actions that address problems underlying major health and development issues, and

- that such action will only be possible with major national and community efforts and extensive and sustained financial input,

In Pursuance of

resolutions of the Forty-fourth World Health Assembly and the Forty-first WHO Regional Committee for Africa; recommendations of the Ministers of Health and Ministers of Interior at subregional meetings in Benin and Zambia; and the initative of the African Ministers of Health meeting in September 1991, called for a general mobilization for health in recognition of the fact that Africa was in the grips of a health crisis of unprecedented magnitude, brought about by economic, social and political constraints, declared their decision to propose to their Governments the establishment of an integrated, multisectoral strategy for all African countries for the control and prevention of cholera.

This strategy will include:

- *The establishment or updating, with the participation of all concerned national sectors, of cholera control and prevention plans, following the guidelines accepted in the region, as part of national programmes for the control of diarrhoeal disease and improvement of community water supplies and sanitation, within the context of district health systems in which the community can participate fully;*

- *the formulation, with the collaboration of WHO, UNICEF, and multi- and bilateral partners, of coordinated plans to meet the needs of individual countries in areas of case management, surveillance, outbreak investigation and control (including laboratory support), public education, epidemic preparedness, and long-term prevention through the improvement of water supplies, sanitation and food safety, focusing on populations most at risk;*

- *the establishment with support from WHO, of plans of technical cooperation between countries in the planning and implementation of control activities for diarrhoeal diseases and communicable diseases including cholera;*

- *the strengthening of exchange between countries with the assistance of WHO of information on cholera and other communicable dieseases to assist control and prevention measures; and*

- *on the basis of ongoing and planned activities, seeking support from the international community for the agreed coordinated strategy, so as to ensure maximum regional impact from its input.*

The Ministers look to the WHO Regional Directors to provide technical support and coordination for this strategy and to act to mobilize the necessary resources.

Figure 5. A strategy for the control and prevention of cholera in Africa. Reprinted from reference 38 with permission.

drinking was taken from a potentially contaminated lagoon and supplied by tanker truck to the camp. Chlorination was inadequate, and < 10 liters of water was available per person per day. This was an insufficient quantity of water for drinking and washing. Trench latrines were underutilized, and most persons defecated in open spaces. During rains, in areas previously used for defecation, trenches filled with water. Because of long lines for water rations, some persons used the contaminated water from the trenches (24).

In 1988, 951 cases of cholera were reported at a cholera treatment center in a Mozambican refugee camp in Malawi. The attack rate was 2.6% for the camp, and the case fatality rate was 3.3%. A matched-pair case-control study determined that cases were more likely than controls to use shallow or surface wells rather than deep boreholes. Fecal coliform bacteria were present in the shallow wells but not in 10 deep borehole water sources sampled (23).

In a retrospective cohort study conducted during another cholera outbreak among Mozambican refugees in Malawi, characteristics of 38 cholera households were compared with those of control households selected during a nutritional survey. Risk factors for illness included family size greater than five, having no water containers or cooking pots, or receiving dry fish during ration distribution (14a). However, lack of cooking pots or water containers may have been a marker of lower socioeconomic status or of recent arrival in the camp (also new arrivals may lack the immunity to cholera that established residents had) and may not have been in itself associated with illness. In a subsequent study during a cholera outbreak in the same refugee camp in 1990, three factors were associated with developing diarrheal illness: drinking river water, placing hands into the water of storage containers holding drinking water, and, for those with inadequate wood to reheat food, eating cooked peas that had been kept overnight (32).

Since refugees and displaced persons are at particular risk for becoming ill with cholera, control measures to provide sufficient quantities of clean chlorinated water should be undertaken, and an adequate number of latrines should be readied before placement of refugees. All refugees should have access to medical facilities stocked with intravenous and oral rehydration solutions and manned by personnel trained to treat cholera and prevent deaths in an epidemic setting.

ANTIMICROBIAL RESISTANCE

Isolates of *V. cholerae* O1 resistant to multiple antimicrobial agents were reported in Africa during the 1970s (8, 16, 19, 21). The rapidity with which resistant isolates could become prevalent was demonstrated in 1977 in Tanzania. Strains of *V. cholerae* O1 isolated during the first month of an epidemic were sensitive to tetracycline. However, after 5 months of extensive tetracycline use (1,788 kg were used by the Ministry of Health for treatment and prophylaxis), 76% of isolates were resistant to the drug. During the same period, resistance to chloramphenicol increased from 0 to 52% of isolates (21).

Similarly, in Kenya, mass tetracycline prophylaxis campaigns were carried out from 1981 to 1988. *V. cholerae* O1 isolates resistant to tetracycline, ampicillin, and trimethoprim-sulfamethoxazole emerged in 1982 (16). Resistance was mediated by a single plasmid, which differed from a plasmid found in resistant strains from Tanzania, Nigeria, and Bangladesh (11). Antibiotic resistance has not been a problem in South Africa, where mass chemoprophylaxis has not been practiced (16a).

CONTROL

Investigators throughout Africa have thus determined that cholera is spread through improperly prepared food, leftover food contaminated after cooking, contaminated shallow wells and river water, and water contaminated while in storage. Preventing cholera transmission will require improvements in sanitary conditions and water quality. Education and access to health facilities with well-trained personnel and adequate rehydration supplies is crucial to preventing loss of life during an epidemic. In Africa, as elsewhere, mass chemoprophylaxis and vaccination programs and quarantine measures have been ineffective and divert vital resources from life-saving activities such as providing treatment and improving the safety of water and food supplies. A strategy for the control and prevention of cholera in Africa was set forth in a declaration presented at the Forty-fifth World Health Assembly (Fig. 5). Since cholera has become endemic throughout much of Africa, local and foreign governments and relief organizations must coordinate their efforts to prevent unnecessary loss of life throughout that continent.

REFERENCES

1. **Baine, W. B., A. Zampieri, M. Mazzotti, G. Angioni, D. Greco, M. DiGioia, E. Izzo, E. J. Gangarosa, and F. Pocchiari.** 1974. Epidemiology of cholera in Italy in 1973. *Lancet* ii:1370–1374.

2. **Barua, D.** 1992. History of cholera, p. 1-36. *In* D. Barua and W. B. Greenough III (ed.), *Cholera.* Plenum Medical Book Co. New York.

3. **Benenson, A. S., S. Z. Ahmad, and R. O. Oseasohn.** 1965. Person-to-person transmission of cholera, p. 332-336. *In* O. A. Bushnell and C. S. Brookhyer (ed.), *Proceedings of the Cholera Research Symposium, Honolulu, Hawaii, 1965.* Public Health Service publication no. 1328. U.S. Government Printing Office, Washington, D. C.

4. **Blake, P. A., D. T. Allegra, J. D. Snyder, T. J. Barrett, L. McFarland, C. T. Caraway, J. C. Feely, J. P. Craig, J. V. Lee, N. D. Puhr, and R. A. Feldman.** 1980. Cholera: a possible endemic focus in the United States. *N. Engl. J. Med.* 302:305-309.

5. **Blake, P. A., M. L. Rosenberg, J. B. Costa, P. S. Ferreira, C. L. Guimaraes, and E. J. Gangarosa.** 1977. Cholera in Portugal, 1974. I. Modes of transmission. *Am. J. Epidemiol.* 105:337-343.

6. **Chapman, J. A., and L. P. Collocott.** 1985. Cholera in children at Eshowe Hospital. *S. Afr. Med. J.* 68:249-253.

7. **Cliff, J. L., P. Zinkin, and A. Martelli.** 1986. A hospital outbreak of cholera in Maputo, Mozambique. *Trans. R. Soc. Trop. Med. Hyg.* 80:473-476.

8. **Colaert, J., E. van Dyck, J. P. Ursi, and P. Piot.** 1979. Antimicrobial susceptibility of *Vibrio cholerae* from Zaire and Rwanda. *Lancet* ii:849.

9. **Deb, B. C., B. K. Sircar, P. G. Sengupta, S. P. De, S. K. Mondal, D. N. Gupta, N. C. Saha, S. Ghosh, U. Mitra, and S. C. Pal.** 1986. Studies on interventions to prevent eltor cholera transmission in urban slums. *Bull W.H.O.* 64:127-131.

10. **Deb, B. C., B. K. Sircar, P. G. Sengupta, S. P. De, D. Sen, M. R. Saha, and S. C. Pal.** 1982. Intra-familial transmission of *Vibrio cholerae* biotype El Tor in Calcutta slums. *Indian J. Med. Res.* 76:814-819.

11. **Finch, M. J., J. G. Morris, Jr., J. Kaviti, W. Kagwanja, and M. M. Levine.** 1988. Epidemiology of antimicrobial resistant cholera in Kenya and East Africa. *Am. J. Trop. Med. Hyg.* 5:484-490.

12. **Glass, R. I., M. Claeson, P. A. Blake, R. J. Waldman, and N. F. Pierce.** 1991. Cholera in Africa: lessons on transmission and control for Latin America. *Lancet* 338:791-795.

13. **Goh, K. T., S. Lam, S. Kumarapathy, and J. L. Tan.** 1984. A common source foodborne outbreak of cholera in Singapore. *Intl. J. Epidemiol.* 13:210-215.

14. **Goodgame, R. W., and W. B. Greenough III.** 1975. Cholera in Africa: a message for the West. *Ann. Intern. Med.* 82:101-106.

14a. **Hatch, D.** Unpublished data.

15. **Holmberg, S. D., J. R. Harris, D. E. Kay, N. T. Hargrett, R. D. R. Parker, N. Kansou, N. U. Rao, and P. A. Blake.** 1984. Foodborne tramsmission of cholera in micronesian households. *Lancet* i:325-328.

16. **Ichinose, Y., M. Ehara, S. Watanbe, S. Shimodori, P. G. Waiyaki, A. M. Kibue, F. C. Sang, J. Ngugi, and J. N. Kaviti.** 1986. The characterization of *Vibrio cholerae* isolated in Kenya in 1983. *J. Trop. Med. Hyg.* 89:269-276.

16a. **Isaäcson, M.** Unpublished data.

17. **Isaäcson, M., K. R. Clarke, G. H. Ellacombe, W. A. Smit, P. Smit, H. J. Koornhof, L. S. Smith, and L. J. Kriel.** 1974. The recent cholera outbreak in the South African gold mining industry: a preliminary report. *S. Afr. Med. J.* 48:2557-2560.

18. **MacKay, D. M.** 1980. Cholera: the present pandemic. *Public Health* 94:283-287.

19. **Maimone, F., A. Coppo, C. Pazzani, S. O. Ismail, R. Guerra, P. Procacci, G. Rotigliano, and K. H. Omar.** 1986. Clonal spread of multiply resistant strains of *Vibrio cholerae* O1 in Somalia. *J. Infect. Dis.* 153:802-803.

20. **McIntyre, R. C., T. Tira, T. Flood, and P. A. Blake.** 1979. Modes of transmission of cholera in a

newly infected population on an atoll: implications for control measures. *Lancet* **i:**311–314.

21. **Mhalu, F. S., P. W. Mmari, and J. Ijumba.** 1979. Rapid emergence of El Tor *Vibrio cholerae* resistant to antimicrobial agents during the first six months of fourth cholera epidemic in Tanzania. *Lancet* **i:**345–347.

22. **Mhalu, F. S., F. D. E. Mtango, and A. E. Msengi.** 1984. Hospital outbreaks of cholera transmitted through close person-to-person contact. *Lancet* **ii:**82–84.

23. **Moren, A., S. Stefanaggi, D. Antona, D. Bitar, M. G. Etchegorry, M. Tchatchioka, and G. Lungu.** 1991. Practical field epidemiology to investigate a cholera outbreak in a Mozambican refugee camp in Malawi, 1988. *J. Trop. Med. Hyg.* **94:**1–7.

24. **Mulholland, K.** 1985. Cholera in Sudan: an account of an epidemic in a refugee camp in eastern Sudan, May–June 1985. *Disasters* **9:**247–258.

25. **Patel, M., and M. Isaäcson.** 1989. Survival of *Vibrio cholerae* in African domestic water storage containers. *S. Afr. Med. J.* **76:**365–367.

25a. **Patel, M., and M. Isaäcson.** Unpublished data.

26. **Pollitzer, R.** 1959. *Cholera.* Monograph no. 43. World Health Organization, Geneva.

27. **Shaffer, N., P. Mendes, C. M. Costa, E. Larson, P. A. Anderson, N. Bean, R. V. Tauxe, and P. A. Blake.** 1988. Epidemic cholera in Guinea-Bissau: importance of foodborne transmission, abstr. 1459. *Program Abstr. 28th Intersci. Conf. Antimicrob. Agents Chemother.*

28. **Sinclair, G. S., M. Mphahlele, H. Duvenhage, R. Nichol, A. Whitehorn, and H. G. Küstner.** 1982. Determination of the mode of transmission of cholera in Lebowa. *S. Afr. Med. J.* **62:**753–755.

29. **Sitas, F.** 1986. Some observations on a cholera outbreak at the Umvoti Mission Reserve, Natal. *S. Afr. Med. J.* **70:**215–218.

30. **St. Louis, M. E., J. D. Porter, A. Helal, K. Drame, N. Hargrett-Bean, J. G. Wells, and R. V. Tauxe.** 1990. Epidemic cholera in West Africa: the role of food handling and high-risk foods.

Am. J. Epidemiol. **131:**719–727.

31. **Stock, R. F.** 1976. Cholera in Africa: diffusion of the disease 1970–1975 with particular emphasis on West Africa. Special report 3. International African Institute, London.

32. **Swerdlow, D. L., G. Malanga, G. Begokyian, A. Jonkman, M. Toole, R. V. Tauxe, R. Waldman, N. Puhr, and P. A. Blake.** 1991. Epidemic of antimicrobial resistant *Vibrio cholerae* O1 infections in a refugee camp—Malawi, abstract 529. *Program abstr. 31st Intersci. Conf. Antimicrob. Agents Chemother.*

33. **Swerdlow, D. L., E. D. Mintz, M. Rodriguez, E. Tejada, C. Ocampo, L. Espejo, K. D. Greene, W. Saldana, L. Seminario, R. V. Tauxe, J. G. Wells, N. H. Bean, A. A. Ries, M. Pollack, B. Vertiz, and P. A. Blake.** 1992. Waterborne transmission of epidemic cholera in Trujillo, Peru: lessons for a continent at risk. *Lancet* **340:**28–32.

34. **Tauxe, R. V., S. D. Holmberg, A. Dodin, J. G. Wells, and P. A. Blake.** 1988. Epidemic cholera in Mali: high mortality and mulitple routes of transmission in a famine area. *Epidemiol. Infect.* **100:**279–289.

35. **Toole, M. J., and R. J. Waldman.** 1990. Prevention of excess mortality in refugee and displaced populations in developing countries. *JAMA* **263:**3296–3302.

36. **Umoh, J. U., A. A. Adesiyun, and J. O. Adekeye.** 1983. Epidemiological feature of an outbreak of gastroenteritis/cholera in Katsina, Northern Nigeria. *J. Hyg.* (Cambridge) **91:**101–111.

37. **World Health Organization.** 1991. Cholera in Africa. *Weekly Epidemiol. Rec.* **66:**305–311.

38. **World Health Organization.** 1992. A strategy for the control and prevention of cholera in Africa—declaration. *Weekly Epidemiol. Rec.* **67:**241.

39. **World Health Organization.** 1992. Cholera in 1991. *Weekly Epidemiol. Rec.* **67:**253–260.

40. **World Health Organization.** 1993. Cholera in 1992. *Weekly Epidemiol. Rec.* **68:**149–155.

Vibrio cholerae and Cholera: Molecular to Global Perspectives
Edited by I. Kaye Wachsmuth, Paul A. Blake, and Ørjan Olsvik
© 1994 American Society for Microbiology, Washington, DC 20005

Chapter 20

Endemic Cholera in Australia and the United States

Paul A. Blake

Until recently, humans were believed to be the only reservoir for toxigenic *Vibrio cholerae* O1. Although the organism could be isolated from ponds and streams in Asian countries with endemic cholera, these bodies of water were subject to pollution by human feces, and the organism could well have come from human intestines. However, during the last 2 decades, evidence has accumulated from Australia and the United States that shows that both of these countries have environmental reservoirs for toxigenic *V. cholerae* O1.

INDIGENOUS CHOLERA IN AUSTRALIA

Discovery and Investigation of the First Two Cases

Australia recorded cholera for the first time in 1832, just 15 years after the first pandemic began in India (14). However, the disease persisted only briefly, and in the 20th century Australia had no cases of cholera detected except for a few imported cases (50). Thus, Australian health officials were surprised in

Paul A. Blake • Foodborne and Diarrheal Diseases Branch, Division of Bacterial and Mycotic Diseases, National Center for Infectious Diseases, Centers for Disease Control and Prevention, 1600 Clifton Road, Atlanta, Georgia 30333.

February 1977 when toxigenic *V. cholerae* O1, biotype El Tor, serotype Inaba, was isolated from the blood and subsequently the stool of a 56-year-old woman who was in hypovolemic shock caused by severe gastroenteritis (46). She was achlorhydric, which made her more susceptible to infection, and she had a history of ethanol abuse, which could indicate underlying liver disease and thus greater likelihood of bacteremia with normally enteric organisms. However, the source of infection was an enigma, as she had not left Australia for 13 years (46).

Investigation showed that the patient lived in a trailer park in Beenleigh, Queensland, approximately 35 km southeast of Brisbane, near Australia's eastern coast (Fig. 1). A man who lived in the same trailer with the patient developed mild diarrhea 10 days after the onset of her illness, and *V. cholerae* O1, serotype Inaba, was isolated from his feces (47). The trailer park's water came from Beenleigh's piped municipal water system. Cultures revealed that piped water in the trailer park and the town contained small numbers of *V. cholerae* O1 (47).

Because of a drought, Beenleigh's usual water supply had been supplemented with water from an auxiliary intake on the Albert River, 12 km to the south (20). *V. cholerae* O1 was recovered from the Albert River and the Logan River, one of nine other rivers in the area that were screened. The organism was detected in the rivers for a period of weeks and then intermittently over a 22-month period (47). The Albert and Logan rivers drain sparsely populated farming country, and significant continuing pollution by human feces was felt to be unlikely. Astutely, Bourke in 1978 (19), Rogers et al. in 1980 (47), and Sutton in 1980 (3) concluded that these findings suggested that *V. cholerae* O1 could grow in river water and that given the right conditions, water could be a long-term reservoir for cholera and could maintain endemic infection.

It should be noted that before 1983, the Queensland publications refer simply to *V. cholerae,* not *V. cholerae* O1, but it is clear from a publication by the Queensland State Health Laboratory (25) that the laboratory refers to non-O1 *V. cholerae* as noncholera vibrios. With the exception of the isolate from the index case's blood, which was toxigenic (46), testing for production of cholera toxin was not noted until later.

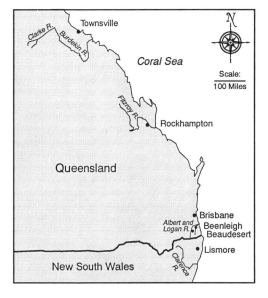

Figure 1. Rivers in eastern Australia associated with cholera cases in nine persons, 1977 to 1987.

Subsequent Cases in Queensland and New South Wales

The third case of domestically acquired cholera in Australia occurred in February 1980 in a child from Beaudesert, Queensland, approximately 60 km south of Brisbane (Fig. 1). His mother was infected but asymptomatic (3). The source of infection was believed to be the Albert River (20).

The fourth and fifth cases of cholera occurred in March 1981 in Lismore, New South Wales, 150 km south of Brisbane (Fig. 1). The suspect vehicle was cooked prawns from the Clarence River; they had been harvested from the river, boiled for 2 to 3 min, and then cooled in river water (4, 20). The more severely ill patient had a history of a gastric shunt for a gastric ulcer.

The sixth case occurred in January 1983 in a 32-year-old diabetic resident of Rockhampton, Queensland, 500 km northwest of Brisbane. His infection was attributed to eating lettuce wet with water from a creek that drained into the Fitzroy River (20).

The seventh case occurred in March 1983 in a child who lived near Beaudesert and had gone swimming in the Logan River (7, 20), ducking his head into the water.

The eighth case occurred in March 1984 in a 38-year-old man who went camping in the upper reaches of the Burdekin and Clarke River systems near Townsville, Queensland, 1,000 km northwest of Brisbane (20). He drank river water (8).

A ninth case of cholera occurred in February 1987 in a 24-year-old man in a rural area inland from Townsville, Queensland. He had drunk water from a creek in the Burdekin River watershed (9). The organism isolated from his stool, like those from the eight previous cases, was biotype El Tor and serotype Inaba.

The Riverine Reservoir

All of these cases were directly or indirectly associated with river water, and all occurred during January, February, or March, which are summer months in the southern hemisphere. The association of cholera cases with river water led to extensive studies of rivers in Queensland. Eleven different rivers over an expanse of more than 1,000 km from north of Townsville to south of Brisbane yielded toxigenic strains of *V. cholerae* O1 El Tor Inaba. Some Ogawa strains were recovered, but all were nontoxigenic. A 3-year study of the Albert, Logan, and Brisbane rivers showed that the occurrence of *V. cholerae* O1 in river water was highly seasonal, with 79% of the strains being isolated during the warm months of January, February, and March (20).

Studies have also been done in rivers in New South Wales (5, 26), but as of 1983, all of the *V. cholerae* O1 strains isolated from the Clarence and Georges rivers were serotype Ogawa and nontoxigenic (6). The cholera cases associated with the Clarence River indicate that it probably does contain toxigenic Inaba strains, but that river is very close to Queensland.

Relationship of Australian Strains to Others Worldwide

New methods that have become more and more capable of reproducibly subtyping *V. cholerae* O1 strains have been developed. The results of studies using these techniques suggest that while there are minor differences between Australian strains, the strains as a group are different from other El Tor strains worldwide and may have existed in the Australian environment for eons.

Desmarchelier and Senn compared the restriction endonuclease digest patterns of the chromosomal DNAs of 13 toxigenic *V. cholerae* O1 Australian strains from humans ($n = 5$) and the environment ($n = 8$) (27). They found differences between strains from the Beenleigh (1977), Lismore (1981), Rockhampton (1983), and Burdekin River (1984) cases. Although some environmental strains were identical to the Beenleigh and Burdekin

River human strains, in general there appeared to be more-marked differences among the environmental strains (27).

Olsvik et al. determined the DNA sequences of cholera toxin subunit B structural genes from 45 *V. cholerae* O1 strains isolated in 29 countries over a period of 70 years. They found only three genotypes; the two strains from Australia were unique, constituting one genotype (41).

Popovic et al. analyzed the restriction fragment length polymorphisms (RFLPs) of rRNA genes to characterize toxigenic *V. cholerae* O1 strains isolated worldwide over 60 years. Three human strains from Australia fell into two ribotypes, which were different from the ribotypes of all other strains (44).

Wachsmuth et al. used multilocus enzyme electrophoresis to characterize 197 *V. cholerae* O1 isolates from around the world. The isolates fell into six electrophoretic types. The five Australian isolates tested were all of the same electrophoretic type, and no other isolates were of that type (52).

So, according to multilocus enzyme electrophoresis, DNA sequencing of cholera toxin B subunit genes, and RFLP of rRNA genes, the Australian toxigenic *V. cholerae* O1 strains are unique and may be clonal. Additional well-defined strains from Australia need to be examined by a variety of techniques to further define possible diversity within Australia and gain a better understanding of the relatedness of the strains and their possible evolutionary implications.

INDIGENOUS CHOLERA IN THE UNITED STATES

Cholera before 1978

Cholera was first recognized in the United States on June 23, 1832, when it appeared in New York, having crossed the Atlantic soon after its extension from Asia into Europe in 1829 (14). The disease spread rapidly and caused catastrophic epidemics. It appeared intermittently between 1832 and 1875 and was particularly persistent in New Orleans (14). However, by 1900 the disease had disappeared. Thereafter, the literature noted no cases that could not be explained by importation (2) or laboratory accident until 1941, when cholera was diagnosed in a New Orleans longshoreman (32). His disease was typical of severe cholera, and a specimen from autopsy yielded an "El Tor vibrio." The source of infection was unknown. We do not know whether the organism was *V. cholerae* O1 or whether it was toxigenic.

The next detected case of unexplained and possibly indigenous cholera in the United States occurred in a 51-year-old Port Lavaca, Texas, shrimp fisherman in August 1973. Toxigenic *V. cholerae* O1, biotype El Tor, serotype Inaba, was isolated from his stool. There was no history of travel outside the United States, and no source of infection could be found (55).

Outbreak in Louisiana, 1978

The first hint that cholera might be endemic in the United States came in 1978 with the discovery of 11 cases of toxigenic *V. cholerae* O1 infection in Louisiana. The first of these cases was in a 44-year-old man in Abbeville with no history of foreign travel who developed dehydrating diarrhea on August 10 and was hospitalized (15). Cholera was not suspected, but a stool culture yielded an organism that was initially identified as *Serratia liquefaciens*. The laboratorians were uncomfortable with that identification, because the organism was not multiply resistant, and they sent it on to the state laboratory. Ultimately, the organism was identified as toxigenic *V. cholerae* O1, biotype El Tor, serotype Inaba (28).

Discovery of this case led to an investigation that expanded after September 12, when Abbeville sewage yielded *V. cholerae* O1. A second case was discovered in a 52-year-old Abbeville woman who was hospitalized in shock and developed acute renal tubular necrosis following an acute diarrheal illness that began on September 15. Ultimately, the investigation detected a total of 11 toxigenic *V. cholerae* O1 infections with onset between August 10 and September 27, a period of 49 days (15). Three of the infections were in asymptomatic family members of symptomatic persons. The infections occurred in five clusters in four communities up to 80 km apart in southwestern Louisiana. Identical organisms were isolated from the municipal sewage systems of three other towns, showing that these towns had had cholera cases that had not been diagnosed.

Discovery of an Environmental Reservoir

The investigation initially searched for possible routes of importation of cholera, but the vehicle of transmission ultimately implicated by a case-control study proved to be crabs harvested from multiple points in the Louisiana coastal marshes (15). Patients in the five clusters of cases had all eaten crabs caught at five different sites across a 130-km expanse of the coastal marshes and over a period of 47 days. This was evidence that toxigenic *V. cholerae* O1 was present in the marsh over a wide area and for a relatively long period.

Crabs were not initially suspect food items because they were eaten boiled or steamed, and it was thought that vibrios would be unlikely to survive cooking. However, crabs artificially inoculated with a toxigenic Louisiana *V. cholerae* O1 strain were used in laboratory cooking experiments that showed that the strain could be isolated from crabs boiled for up to 8 min or steamed for up to 25 min. Crabs cooked for these periods of time were red, and their meat was firm (15).

Moore swabs (13) of marsh water and samples of seafoods from the marshes were cultured for toxigenic *V. cholerae* O1, but the strain proved to be elusive. The organism was found only three times: in water and shrimp from one remote canal and in water from another canal (15).

All of the toxigenic *V. cholerae* O1 strains from patients, sewage, and the environment were biotype El Tor, serotype Inaba, and strongly hemolytic, as was the 1973 Port Lavaca, Texas, strain. By 1978, most toxigenic *V. cholerae* O1 strains found worldwide were nonhemolytic (12). To further investigate the relationship between the United States strains and strains collected worldwide, they were phage typed by a system used by John Lee of the Maidstone, England, Public Health Laboratory. He found that the 1973 Texas and 1978 Louisiana strains were identical by his phage typing system but different from all other *V. cholerae* O1 strains in his extensive collection (15).

Dr. Lee's results led to speculation that the Texas and Louisiana cases might not have been the result of a recent importation or the existence of a chronic carrier (10) who was seeding the coastal environment. Rather, there might be an environmental reservoir for the Texas-Louisiana (Gulf Coast) strain in the coastal marshes that had existed for decades or even for thousands of years.

The Gulf Coast strain might be indigenous, persisting as an autochthonous free-living organism. It could have caused the 1941 case in a longshoreman (32) as well as other sporadic cases and small outbreaks that were not detected because nobody was looking for cholera in that area. Its relationship to the pandemics of the mid-1800s is unknown, as no strains from those pandemics exist. The Gulf Coast strain might be a pandemic strain that has persisted since then, or it might even have existed in the marshes before 1832 but without the epidemic potential of pandemic strains.

Confirmation that Cholera Is Indigenous in the United States

Speculation that the Gulf Coast strain was persisting in an environmental reservoir was supported by the reappearance in May and June 1981 of cholera caused by the same strain. Two cases occurred in southeastern Texas just 6 weeks and 70 km apart. This time, both patients were alcoholics and the histories of exposures were unreliable, so no specific vehicles of transmission were incriminated (49).

In September 1981, a 23-year-old oil rig worker consulted his physician in Louisiana after 8 days of an illness that included profuse watery diarrhea and orthostatic syncopal episodes. The physician suspected a *Campylobacter jejuni* infection and ordered a stool culture. Fortunately, the laboratory routinely used thiosulfate-citrate-bile salts-sucrose (TCBS) agar for stool cultures and isolated a vibrio from the patient's stool that proved to be the Gulf Coast strain of *V. cholerae* O1. Interview of the patient revealed that many of his oil rig coworkers had also been ill with diarrhea. The subsequent investigation of the Texas oil rig barge and its workers showed that there were 16 *V. cholerae* O1 infections in all.

Feces from the 23-year-old patient, who was the first patient, had contaminated the nonpotable saltwater system, which in turn had contaminated the drinking-water system through a cross-connection. The drinking water was used to moisten boiled rice being held warm for 8 h before being served to oil rig workers (34). A case-control study showed that the rice was the vehicle of transmission for most of the cases.

This was the largest cholera outbreak and the only known instance of secondary transmission of cholera in the United States in this century (34). The outbreak illustrates the ease with which cholera can escape diagnosis: although three patients sought medical attention and two were hospitalized, cholera was not suspected, and the outbreak was discovered only because one laboratory happened to use TCBS routinely.

Cholera, 1982 to 1992

Indigenous cases of cholera have continued to be detected both as sporadic cases and in small clusters at irregular intervals since

1981, with cases occurring in 1984, 1986, 1987, 1988, 1990, and 1992 (Fig. 2). All were caused by the Gulf Coast strain, and the dominant vehicle of transmission was crabs (22, 29, 35, 37, 38, 42, 53).

In 1983, the Gulf Coast strain was isolated from a tourist returning from Cancun, Mexico (17, 21). The date of onset showed that she was infected while in Cancun. She had eaten ceviche (a dish prepared with raw seafood) in Cancun. Since seafood eaten in Cancun commonly came from the Mexican coast of the Gulf of Mexico (Tampico and Campeche), the environmental reservoir of the Gulf Coast strain may extend south around the Gulf of Mexico. Giono et al. have reported that their laboratory investigations of cholera in Mexico have detected 31 hemolytic strains; most of these strains came from areas along the Mexican Gulf Coast (30). The strains have not yet been tested to determine whether they are identical to the U.S. Gulf Coast strain.

In 1984, a case of cholera in Maryland was traced to crabs eaten in a restaurant; the crabs had been shipped in from Texas (37). Thus far, this is the only instance in which commercially prepared crabs have caused cholera in the United States; all of the other crab-associated cases were caused by crabs cooked privately.

The peak year for indigenous cholera in the United States was 1986, when Louisiana had 18 infections in 12 family clusters traced to eating cooked crabs and raw or cooked shrimp (38). In the same year, both Georgia (42) and Florida (35) had single cases in persons who had eaten raw Gulf Coast oysters.

In 1987, five cases of cholera in the United States were caused by the Gulf Coast strain. All five occurred in Louisiana, and four of the five patients had eaten boiled crabs.

In 1988, for unknown reasons, just one case was associated with eating crabs, while raw oysters were the dominant vehicle of transmission, associated with six cases in six states over a period of 55 days (22). Three lots of oysters were traced to their state of origin; all three had come from Louisiana.

Between 1989 and 1992, four indigenous cases of cholera were detected, with three in

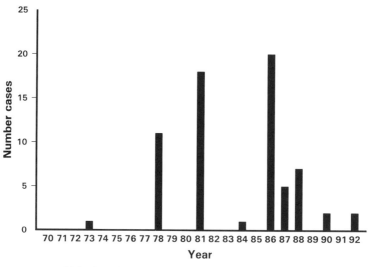

Figure 2. Indigenous cases of infection with the Gulf Coast strain of toxigenic *V. cholerae* O1, United States, 1970 to 1992.

Louisiana and one in Texas. Three of the four patients recalled eating crabs.

Crabs versus Oysters as Vehicles of Transmission

It is interesting that crabs have been the dominant vehicle for transmission of cholera in the United States even though they are eaten cooked, while oysters, which are often eaten raw, have been of lesser importance. Worldwide, bivalve shellfish like oysters have been exceedingly important vehicles for transmission of cholera (11, 16, 39, 48, 54). Furthermore, foodborne *V. vulnificus* infections in the United States are almost all caused by eating raw oysters (18, 51), with crabs playing little if any role.

Several factors may contribute to the relative unimportance of oysters in causing cholera in the United States. First, toxigenic *V. cholerae* O1 must be present in food in relatively large numbers to cause infection. Thus far, when large outbreaks of cholera in other countries have been caused by bivalve shellfish, there was apparently heavy contamination by sewage-laden water (Portugal [16], Kiribati [39], and Italy [11, 48]). Extensive programs exist in the United States to protect commercially harvested oysters from fecal contamination; in contrast, there is no control of the quality of the water from which crabs are harvested, because crabs are cooked before being eaten. Thus, crabs should be more likely than oysters to come from water that is heavily contaminated with toxigenic *V. cholerae* O1.

The *V. cholerae* O1 organisms available to oysters may be more likely to be free-living strains, and these appear to be present in the environment in culturable forms in very small numbers; in fact, the three isolates from water and shrimp in 1978 (15) were the only toxigenic *V. cholerae* O1 strains isolated from the natural environment in the United States between 1973 and 1990. Strains of the classical biotype detected in Louisiana coastal waters (33) probably represented contamination

of equipment, since they were identical to the investigators' laboratory reference strain.

Second, although crabs are cooked, a study in Louisiana showed that many cholera patients handle crabs in ways that could allow multiplication of vibrios (38). The persons who had cholera and had eaten crabs were significantly more likely than crab-eating controls to have undercooked their crabs and mishandled them after cooking (38). Cooked seafood is a good growth medium for *V. cholerae* O1, allowing rapid multiplication (36).

Third, Lowry et al. suggested that additional factors contributing to the relative importance of crabs as vehicles for cholera could include the facts that *V. cholerae* O1 adheres to crab exoskeleton, which may be ingested while eating crab, and that crabs have a wider range of feeding depths than oysters, possibly increasing their likelihood of encountering an ecologic niche for *V. cholerae* O1 (38).

Seasonal and Year-to-Year Variation of Cholera

Fifty percent of indigenous cholera cases in the United States occur during August and September (53), which are analogous to Australia's February and March, when eight of the nine Australian indigenous cases occurred. U.S. Gulf Coast waters are warmest during August and September, and the marked seasonality of cholera in the United States may correspond to optimal conditions for multiplication of autochthonous toxigenic *V. cholerae* O1 strains in the coastal waters from which seafoods are harvested (40). Likewise, the year-to-year variation in the incidence of indigenous cholera in the United States may reflect subtle changes in the coastal environment affecting the numbers of free-living toxigenic *V. cholerae* O1 (24, 40).

Relationship of U.S. Strains to Others Worldwide

The new subtyping methods used on the Australian strains have also been applied to strains from the United States. All toxigenic *V. cholerae* O1 strains from the Gulf Coast isolated between 1973 and 1990 have been identical (1, 23, 45) except for one strain in which the cholera toxin probe hydridized with one rather than two DNA restriction fragments (35).

Wachsmuth et al. used multilocus enzyme electrophoresis to characterize *V. cholerae* O1 isolates collected worldwide and found that all of the Gulf Coast strains tested fell into one of six electrophoretic types. The only other strains in that electrophoretic type were nontoxigenic strains from Latin America (52). Chen et al., working only with toxigenic and nontoxigenic strains isolated in the western hemisphere before the cholera invasion of Latin America in 1991, found that some of the nontoxigenic United States strains were of the same electrophoretic type as the toxigenic Gulf Coast strains (23).

Popovic et al. analyzed the RFLPs of rRNA genes to characterize toxigenic *V. cholerae* O1 strains isolated worldwide (44). They showed that all of the Gulf Coast strains tested were of a single ribotype, and no other strains were of that ribotype.

When Olsvik et al. examined the DNA sequences of cholera toxin subunit B structural genes from *V. cholerae* O1 strains isolated worldwide, they found only three genotypes (41). One genotype included only Australian strains, another included seventh pandemic and recent Latin American El Tor biotype strains, and the third included all strains of the classical biotype and the U.S. Gulf Coast strains. The finding of identical cholera toxin subunit B structural genes in classical biotype and U.S. Gulf Coast strains was totally unexpected. The authors suggested a possible explanation. They pointed out that nontoxigenic *V. cholerae* O1 strains of the same multilocus enzyme electrophoretic type as the Gulf Coast strain have been detected in the United States (23) and suggested that such environmental strains acquired toxin genes from classical strains during cholera epidemics in the 1800s and became the Gulf Coast strain (41).

Thus, the Gulf Coast strain is unique by RFLP of rRNA genes (44), shares a single multilocus enzyme electrophoretic type only with a few nontoxigenic western hemisphere strains (23, 52), and has cholera toxin subunit B structural genes that are unlike those of other El Tor strains but identical to those of strains of the classical biotype (41). Clearly, the Gulf Coast strain, like the Australian strains, is not recently imported but rather is an indigenous strain that has been persisting in the Gulf Coast environment for a long time.

CONCLUSION

Endemic cholera in Australia and the United States is characterized in both countries by reservoirs in water, hemolytic strains that are unique to each country, occurrence during the summer and fall, and very few detected cases. The small number of cases probably reflects the generally good state of sanitation in both countries, but it is also possible that these strains are highly adapted to environmental persistence and have less epidemic potential in humans than the strains responsible for epidemic cholera.

There are also important differences in the characteristics of cholera in Australia and the United States. First, the environmental reservoir in Australia is in freshwater rivers, while in the United States the reservoir is in brackish coastal swamps, bayous, and lakes. This difference could be explained by as yet unknown differences between the endemic organisms or between the conditions in rivers and estuaries in the two countries.

Second, the Australians have been highly successful in detecting toxigenic *V. cholerae* O1 in their rivers (20), while Americans have detected the organism in only three specimens taken from coastal marshes in 1978 (15). To some extent this difference may reflect a more determined effort by the Australian scientists (20), but it may also mean that the organisms are present in greater numbers in Australian rivers than in U.S. Gulf Coast marshes.

Third, most of the Australian cases are attributed to consumption of river water, while most of the U.S. cases are attributed to consumption of boiled crabs or raw oysters. Americans probably try to avoid drinking water from U.S. coastal marshes because it is brackish, while Australians may not normally eat raw or undercooked animals from the upper reaches of their rivers. The apparently greater numbers of toxigenic *V. cholerae* O1 in the Australian rivers may also play a role.

Although Australia and the United States have the strongest evidence for environmental reservoirs of toxigenic *V. cholerae* O1, similar reservoirs may exist in many countries. For example, in Matlab District, Bangladesh, Glass et al. (31) observed that cholera was highly seasonal, with few cases detected during some months of the year. However, when cholera reemerged each year, it appeared simultaneously throughout the study area, and the early cases were of different phage types. This phenomenon could be interpreted as resulting from some change in the environment favoring the growth of a variety of toxigenic *V. cholerae* O1 strains that were persisting in an environmental reservoir.

Paradoxically, it is more difficult to demonstrate an environmental reservoir in areas like Bangladesh, which have a great deal of cholera, than in areas like Australia and the United States, which have very few cases. Areas where cholera is common have relatively poor sanitation and heavy contamination of the environment with human feces; thus, it is exceedingly difficult to determine whether toxigenic *V. cholerae* O1 strains isolated from the environment are of human or environmental origin. The advent of powerful new molecular tools (23, 41, 44, 52) for subtyping *V. cholerae* O1 may well lead to an explosion of knowledge about the interrelationship of *V. cholerae* O1 reservoirs in the environment and cholera in humans in areas like Bangladesh, India, and Peru that have hyperendemic cholera.

REFERENCES

1. **Almeida, R. J., D. N. Cameron, W. L. Cook, and I. K. Wachsmuth.** 1992. Vibriophage VcA-3 as an epidemic strain marker for the U.S. Gulf Coast *Vibrio cholerae* O1 clone. *J. Clin. Microbiol.* **30:**300–304.
2. **Anonymous.** 1911. The cholera situation—New York. *Public Health Rep.* **26:**1133.
3. **Anonymous.** 1980. Cholera in Queensland. *Communicable Dis. Intell. Bull.* **80/**3:1.
4. **Anonymous.** 1981. Cholera—New South Wales. *Communicable Dis. Intell. Bull.* **81/**6:7.
5. **Anonymous.** 1981. Detection of *Vibrio cholerae* O1 in the Georges River estuary—New South Wales. *Communicable Dis. Intell. Bull.* **81/**18:2.
6. **Anonymous.** 1983. Indigenous cholera—Queensland. *Communicable Dis. Intell. Bull.* **83/**2:2.
7. **Anonymous.** 1983. Cholera surveillance—Queensland. *Communicable Dis. Intell. Bull.* **83/**5:6.
8. **Anonymous.** 1984. Indigenous cholera—Queensland. *Communicable Dis. Intell. Bull.* **84/**8:1.
9. **Anonymous.** 1987. A cholera case—Queensland. *Communicable Dis. Intell. Bull.* **87/**12:7.
10. **Azurin, J. C., K. Kobari, D. Barua, M. Alvero, C. Z. Gomez, J. J. Dizon, E.-I. Nakano, R. Suplido, and L. Ledesma.** 1967. A long-term carrier of cholera: Cholera Dolores. *Bull. W.H.O.* **37:**745–749.
11. **Baine, W. B., A. Zampieri, M. Mazzotti, G. Angioni, D. Greco, M. Di Gioia, E. Izzo, E. J. Gangarosa, and F. Pocchiari.** 1974. Epidemiology of cholera in Italy in 1973. *Lancet* **ii:**1370–1381.
12. **Barrett, T. J., and P. A. Blake.** 1981. Epidemiologic usefulness of changes in hemolytic activity of *Vibrio cholerae* El Tor during the seventh cholera pandemic. *J. Clin. Microbiol.* **13:**126–129.
13. **Barrett, T. J., P. A. Blake, G. K. Morris, N. D. Puhr, H. B. Bradford, and J. G. Wells.** 1980. Use of Moore swabs for isolating *Vibrio cholerae* from sewage. *J. Clin. Microbiol.* **11:**385–388.
14. **Barua, D.** 1992. History of cholera, p. 1–36. *In* D. Barua and W. B. Greenough III (ed.), *Cholera.* Plenum, New York.
15. **Blake, P. A., D. T. Allegra, J. D. Snyder, T. J. Barrett, L. McFarland, C. T. Caraway, J. C. Feeley, J. P. Craig, J. V. Lee, N. D. Puhr, and R. A. Feldman.** 1980. Cholera—a possible endemic focus in the United States. *N. Engl. J. Med.* **302:**305–309.
16. **Blake, P. A., M. L. Rosenberg, J. B. Costa, P. S. Ferreira, C. L. Guimaraes, and E. J. Gangarosa.** 1977. Cholera in Portugal, 1974. I. Modes of transmission. *Am. J. Epidemiol.* **105:**337–343.
17. **Blake, P. A., K. Wachsmuth, B. R. Davis, C. A. Bopp, B. P. Chaiken, and J. V. Lee.** 1983. Toxigenic *V. cholerae* O1 strain from Mexico identical to United States isolates. *Lancet* **ii:**912.

18. Blake, P. A., R. E. Weaver, and D. G. Hollis. 1980. Diseases of humans (other than cholera) caused by vibrios. *Annu. Rev. Microbiol.* **34**:341–367.

19. Bourke, A. T. C. 1978. The Queensland cholera incident—1977. Part II. A review. *Aust. Health Surv.* **April/May**:131–133.

20. Bourke, A. T. C., Y. N. Cossins, B. R. W. Gray, T. J. Lunney, N. A. Rostron, R. V. Holmes, E. R. Griggs, D. J. Larsen, and V. R. Kelk. 1986. Investigation of cholera acquired from the riverine environment in Queensland. *Med. J. Aust.* **144**:229–234.

21. Centers for Disease Control. 1983. Cholera in a tourist returning from Cancún, Mexico—New Jersey. *Morbid. Mortal. Weekly Rep.* **32**:357.

22. Centers for Disease Control. 1989. Toxigenic *Vibrio cholerae* O1 infection acquired in Colorado. *Morbid. Mortal. Weekly Rep.* **38**:19.

23. Chen, F., G. M. Evins, W. L. Cook, R. Almeida, N. Hargrett-Bean, and K. Wachsmuth. 1991. Genetic diversity among toxigenic and nontoxigenic *V. cholerae* O1 isolated from the Western Hemisphere. *Epidemiol. Infect.* **107**:225–233.

24. Colwell R. R., and W. M. Spira. 1992. The ecology of *Vibrio cholerae*, p. 107–127. *In* D. Barua and W. B. Greenough III (ed.), *Cholera.* Plenum, New York.

25. Cossins, Y. M., D. M. Murphy, M. G. Leverington, and B. R. W. Gray. 1978. The Queensland cholera incident—1977. Part III. Laboratory aspects. *Aust. Health Surv.* **April/May**:133–136.

26. Davey, G. R., J. K. Prendergast, and M. J. Eyles. 1982. Detection of *Vibrio cholerae* in oysters, water and sediment from the Georges River. *Food Technol. Aust.* **34**:334–336.

27. Desmarchelier, P. M., and C. R. Senn. 1989. A molecular epidemiological study of *Vibrio cholerae* in Australia. *Med. J. Aust.* **150**:631–634.

28. Donner, V., and B. Domingue. 1978. Cholera. *Bayou Tech.* **23**:10.

29. Gergatz, S. J., and L. M. McFarland. 1989. Cholera on the Louisiana Gulf Coast: historical notes and case report. *J. La. State Med. Soc.* **141**:29–34.

30. Giono, S., D. Gaytan, I. Cortez, and L. Gutierrez. 1993. Haemolytic *Vibrio cholerae* O1 strains from Gulf Coast of Mexico, abstr. C-271, p. 494. *Abstr. 93rd Gen. Meet. Am. Soc. Microbiol.*

31. Glass, R. I., S. Becker, M. I. Huq, B. J. Stoll, M. U. Khan, M. H. Merson, J. V. Lee, and R. E. Black. 1982. Endemic cholera in rural Bangladesh, 1966–1980. *Am. J. Epidemiol.* **116**:959–970.

32. Gordon, A. M. 1970. Cholera in the New World. *J. Ky. Med. Assoc.* **68**:657.

33. Hranitzky, K. W., A. D. Larson, D. W. Ragsdale, and R. J. Siebeling. 1980. Isolation of O1 serovars

of *Vibrio cholerae* from water by serologically specific method. *Science* **210**:1025–1026.

34. Johnston, J. M., D. L. Martin, J. Perdue, L. M. McFarland, C. T. Caraway, E. C. Lippy, and P. A. Blake. 1983. Cholera on a Gulf Coast oil rig. *N. Engl. J. Med.* **309**:523–526.

35. Klontz, K. C., R. V. Tauxe, W. L. Cook, W. H. Riley, and I. K. Wachsmuth. 1987. Cholera after the consumption of raw oysters. *Ann. Intern. Med.* **107**:846–848.

36. Kolvin, J. L., and D. Roberts. 1982. Studies on the growth of *Vibrio cholerae* biotype eltor and biotype classical in foods. *J. Hyg.* (Cambridge) **89**:243–252.

37. Lin, F. Y. C., J. G. Morris, Jr., J. B. Kaper, T. Gross, J. Michalski, C. Morrison, J. P. Libonati, and E. Israel. 1986. Persistence of cholera in the United States: isolation of *Vibrio cholerae* O1 from a patient with diarrhea in Maryland. *J. Clin. Microbiol.* **23**:624–626.

38. Lowry, P. W., A. T. Pavia, L. M. McFarland, B. H. Peltier, T. J. Barrett, H. B. Bradford, J. M. Quan, J. Lynch, J. B. Mathison, R. A. Gunn, and P. A. Blake. 1989. Cholera in Louisiana: widening spectrum of seafood vehicles. *Arch. Intern. Med.* **149**:2079–2084.

39. McIntyre, R. C., T. Tira, T. Flood, and P. A. Blake. 1979. Modes of transmission of cholera in a newly infected population on an atoll: implications for control measures. *Lancet* **i**:311–314.

40. Miller, C. J., R. G. Feachem, and B. S. Drasar. 1985. Cholera epidemiology in developed and developing countries: new thoughts on transmission, seasonality and control. *Lancet* **i**:261–263.

41. Olsvik, Ø., J. Wahlberg, B. Petterson, M. Uhlen, T. Popovic, I. K. Wachsmuth, and P. I. Fields. 1992. Use of automated sequencing of polymerase chain reaction-generated amplicons to identify three types of cholera toxin subunit B in *Vibrio cholerae* O1 strains. *J. Clin. Microbiol.* **31**:22–25.

42. Pavia, A. T., J. F. Campbell, P. A. Blake. J. D. Smith, T. W. McKinley, and D. L. Martin. 1987. Cholera from raw oysters shipped interstate. *JAMA* **258**:2374.

43. Pollitzer, R. 1959. *Cholera*, p. 11. World Health Organization monograph no. 43. World Health Organization, Geneva.

44. Popovic, T., C. A. Bopp, Ø. Olsvik, and I. K. Wachsmuth. 1993. Epidemiologic application of a standardized ribotype scheme for *Vibrio cholerae* O1. *J. Clin. Microbiol.* **31**:2474–2482.

45. Popovic, T., Ø. Olsvik, P. A. Blake, and I. K. Wachsmuth. Cholera in the Americas: foodborne aspects. *J. Food. Prot.*, in press.

46. Rao, A., and B. A. Stockwell. 1980. The Queensland cholera incident of 1977. 1. The index case. *Bull. W.H.O.* **58**:663–664.

47. Rogers, R. C., R. G. C. J. Cuffe, Y. M. Cossins, and D. M. Murphy. 1980. The Queensland chol-

era incident of 1977. 2. The epidemiological investigation. *Bull. W.H.O.* **58:**665–669.

48. **Salamaso, S., D. Greco, B. Bonfiglio, M. Castellani-Pastoris, G. De Felip, A. Bracciotti, G. Sitzia, A. Congiu, G. Piu, G. Angioni, L. Barra, A. Zampieri, and W. B. Baine.** 1980. Recurrence of pelecypod-associated cholera in Sardinia. *Lancet* **ii:**1124–1128.

49. **Shandera, W. X., B. Hafkin, D. L. Martin, J. P. Taylor, D. L. Maserang, J. G. Wells, M. Kelly, K. Ghandi, J. B. Kaper, J. V. Lee, and P. A. Blake.** 1983. Persistence of cholera in the United States. *Am. J. Trop. Med. Hyg.* **32:**812–817.

50. **Sutton, R. G. A.** 1974. An outbreak of cholera in Australia due to food served in flight on an international aircraft. *J. Hyg.* (London) **72:**441–451.

51. **Tacket, C. O., F. Brenner, and P. A. Blake.** 1984. Clinical features and an epidemiological study of *Vibrio vulnificus* infections. *J. Infect. Dis.* **149:**558–561.

52. **Wachsmuth, I. K., G. M. Evins, P. I. Fields, Ø. Olsvik, T. Popovic, C. A. Bopp, J. G. Wells, C. Carrillo, and P. A. Blake.** 1993. The molecular epidemiology of cholera in Latin America. *J. Infect. Dis.* **167:**621–626.

53. **Weber, J. T., W. Levine, D. Hopkins, and R. V. Tauxe.** Cholera in the United States, 1965–1991: risks at home and abroad. *Arch. Intern. Med.*, in press.

54. **Weber, J. T., E. Mintz, R. Canizares, A. Semiglia, I. Gomez, R. Sempertegui, A. Davila, K. Greene, N. Puhr, D. Cameron, T. Barrett, N. Bean, C. Ivey, R. Tauxe, and P. Blake.** Epidemic cholera in Ecuador: multidrug resistance and transmission by water and seafood. *Epidemiol. Infect.*, in press.

55. **Weissman, J. B., W. E. DeWitt, J. Thompson, C. N. Muchnick, B. L. Portnoy, J. C. Feeley, and E. J. Gangarosa.** 1975. A case of cholera in Texas, 1973. *Am. J. Epidemiol.* **100:**487–498.

Vibrio cholerae and Cholera: Molecular to Global Perspectives
Edited by I. Kaye Wachsmuth, Paul A. Blake, and Ørjan Olsvik
© 1994 American Society for Microbiology, Washington, DC 20005

Chapter 21

The Latin American Epidemic

Robert Tauxe, Luis Seminario, Roberto Tapia, and Marlo Libel

The cholera epidemic in Latin America entered its third year in early 1993. It spread swiftly from its beginning point in Peru; as of August 1993, cholera cases have been reported from all countries in South and Central America except Uruguay. The epidemic has both strained and strengthened public health resources in the hemisphere and has renewed emphasis on preventing cholera and other diseases by providing safer water and food and improved sanitation. After the initial epidemic spread, it seems likely that cholera will persist in some areas for years at a lower-incidence endemic state. All countries affected in 1991 continued to report cases in 1992 and 1993. The challenge for Latin America in preventing recrudescent epidemics is reminiscent of the challenge that epidemics of cholera posed to Europe and North America 150 years ago.

Epidemic cholera spread widely through Central and South America in the 19th century but did not persist for more than a few years following each introduction. Epidemics were reported in Mexico and Cuba as early as 1833, and cholera appeared throughout much of South America in the 1860s (33). The last outbreaks were reported in Brazil, Argentina, and Uruguay in 1895, after which transmission appears to have ceased altogether.

Concern that cholera would be reintroduced into Latin America awakened after the pandemic wave of cholera caused by the El Tor biotype extended through Africa in the 1970s. This pandemic, which began in 1961 in Southeast Asia, reached sub-Saharan Africa in 1970 and then spread rapidly from there throughout the continent, with high mortality (23). It was feared that cholera might be introduced into Latin America as a consequence of movement of refugees from the former Portuguese colonies of Mozambique and Angola to Brazil. This seemed imminent after 1974, when cholera appeared in Portugal itself near a military base housing troops returning from Africa (3). Microbiologic surveillance was heightened, and educational materials on the diagnosis, treatment, and prevention of cholera were prepared and distributed in Latin America. Environmental monitoring detected strains of nontoxigenic *Vibrio cholerae* O1 and non-O1 *V. cholerae* in Brazil and Peru (1, 5, 27); however, no toxigenic O1 strains were isolated. For reasons that remain unclear, cholera did not spread to Latin America at that time.

Robert Tauxe • Foodborne and Diarrheal Diseases Branch, Division of Bacterial and Mycotic Diseases, Centers for Disease Control and Prevention, 1600 Clifton Road, Atlanta, Georgia 30333. *Luis Seminario* • General Office of Epidemiology, Ministry of Health, Lima, Peru. *Roberto Tapia* • Dirección General de Epidemiologia, Col. Lomas de Plateros, Mexico. *Marlo Libel* • Pan American Health Organization, Washington, D.C. 20037.

In the interim, two major international initiatives to improve public health emerged, driven in part by global concerns about the spread of epidemic cholera. The first was an international effort to improve water and sanitation, pursued throughout the 1980s. This effort, known as the International Drinking Water and Sanitation Decade, focused development efforts on extending and upgrading municipal and village water and sanitation systems in Latin America and elsewhere. By 1988, 88% of the urban population and 55% of the rural population of Latin America were served by public water sources, and 49% of urban populations were served by sanitary sewerage systems (32). However, the promise of the Water Decade was not fulfilled. Although many communities were provided with water systems, far fewer actually received water that was safe to drink. In 1988, 17 of 25 countries in Latin America and the Caribbean reported that water service was intermittent. Intermittent water delivery leads to episodic suction in water lines that draws external contaminants into the pipes. Water that became contaminated in the piping was likely to stay that way, as informal surveys in 1984 indicated that 75% or more of water systems either did not disinfect water at all or were unable to provide continuous disinfection (32). Despite efforts to build sewerage systems, less than 10% of all sewage is treated correctly, and most is discharged directly into rivers that are the source of drinking water for downstream communities. Most efforts of the Water Decade went into building new systems or extending existing systems, and far less attention was paid to maintaining and repairing existing systems.

A second international effort, led by the World Health Organization, focused on improving clinical management of diarrheal illness by promoting treatment by rehydration, including that with intravenous and oral rehydration solutions (ORS). Promoting rehydration therapy was more successful than building water systems. Before 1980, only three Latin American countries had national programs for promoting rehydration treatment of diarrheal illness. By 1990, all 20 countries in Latin America had established such programs (45). In the nations of Latin America and the Caribbean, the estimated proportion of childhood diarrheal episodes treated with ORS or other appropriate home-based fluids increased from 12% in 1984 to 54% in 1991 (45). Fortuitously, by 1991, local manufacture of oral rehydration salts had begun in Peru, and an extensive nationwide campaign to promote their use had been launched. Thus, in the years between the first alarm and the actual arrival of cholera, the potential for providing rehydration treatment had greatly improved, though sanitation lagged behind.

APPEARANCE OF CHOLERA IN PERU

On January 29, 1991, an outbreak of severe watery diarrhea was reported to the Peruvian Office of Epidemiology (7). The outbreak was in the agricultural village of Candelaria, near Chancay, 80 km north of Lima. Investigation by a Peruvian field epidemiology team identified over 100 patients in the village, several of whom had died of acute dehydration. All ages and both sexes appeared to be affected more or less equally; 81% of cases were in persons over the age of 5 years. Stool specimens collected in the field were cultured at the National Public Health Laboratory. These yielded *V. cholerae* O1 strains, which were rapidly confirmed at the Centers for Disease Control (CDC) as toxigenic El Tor, serotype Inaba. The source of this outbreak was not clearly defined. Many affected villagers lived near a stream that was their usual source of drinking water, and some purchased fish and other fresh foods in Lima markets.

The same week, hundreds of cases of a similar illness affecting adults and children alike were reported from the cities of Callao, Chimbote, Trujillo, and Piura, along a 900-km coastline (Fig. 1). Review of clinical records suggested that the earliest cases oc-

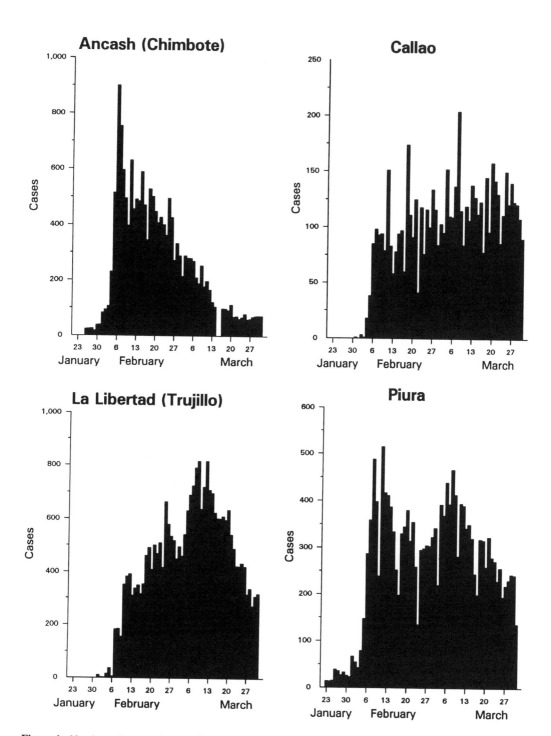

Figure 1. Numbers of reported cases of cholera by day of presentation in four departments of Peru, January 23 to March 31, 1991. Data are taken from OGE, Lima, Peru.

curred in the harbor city of Chimbote; none of these patients had traveled outside of Peru. The earliest probable case occurred in a patient who fell ill on January 19. Similar illnesses appeared in the other cities within a week. In Piura, the northernmost town to be affected initially, the first recognized cases occurred in simultaneous outbreaks which began on January 25, 1991, at two government institutions. Although the sources of these two outbreaks were not determined, the two institutions had divided a single truckload of fish shortly before the outbreaks occurred. By March 8, 6 weeks after recognition of the epidemic in Chancay, 65,198 cases of suspected cholera, 16,754 hospitalizations, and 363 associated deaths were reported to the Pan American Health Organization by the Ministry of Health in Lima.

In response to the epidemic, the Peruvian Ministry of Health distributed instructions and supplies for treatment to all affected areas. In a departure from usual billing practices, the government provided cholera treatment to all without charge. The entire medical community, including medical and nursing students, was trained in methods of rehydration. Waiting rooms, clinics, and hallways in hospitals were converted to emergency cholera treatment wards. Using mass media, the population was advised to boil their drinking water, seek treatment rapidly, and avoid eating raw seafood. A rapid surveillance network was created: the number of persons seeking care for cholera in hospitals was telephoned daily from local health units to regional and national offices. This surveillance was used to direct shipments of treatment supplies, and no prolonged shortages occurred, despite the remarkably high attack rates. The appearance and spread of the epidemic was reported swiftly to neighboring countries and the Pan American Health Organization. In contrast to what occurred during earlier epidemics in Asia and Africa, little public health effort was wasted on ineffective control measures such as vaccine or chemoprophylaxis cam-paigns, border closing, or concealment of the epidemic.

In the weeks following its introduction, the epidemic spread rapidly in Peru (Fig. 2a). By the fifth week of the epidemic, cases had been reported from 24 (83%) of the 29 Peruvian departments (provinces), including Cajamarca, in the Northern Andean mountains, and Iquitos, east of the Andes on the Amazon River. As in the first cities affected, cholera affected all age groups and both sexes. The epidemic in Peru continued at the highest incidence of more than 15,000 cases per week (a weekly incidence of over 70 cases per 100,000 population) through March and April and then began to diminish (see Fig. 4). After a relative lull corresponding to the cooler season, cholera incidence increased again sharply in the coastal region in January 1992, reaching an incidence similar to that of the first year even in areas heavily affected in 1991. A total of 322,562 cases and 2,909 deaths were reported in 1991, and 206,565 cases and 709 deaths were reported in 1992. Surveillance data from early 1993 indicate that cholera is increasing in many areas of the country though at substantially lower levels than in 1992. The distribution by month exhibits a sharp seasonality on the coast, even though Peru is near the equator.

The initial case fatality ratio (CFR) reached 1.5% on the coast and 5% in the mountains and jungle (Fig. 2b). Subsequently, the CFRs were below 1% except in the mountainous areas, where logistics were complicated by difficult terrain and terrorist activities. The overall CFRs for Peru were 0.9% in 1991 and 0.34% in 1992 (Table 1).

The cumulative number of cases reported in 1991 and 1992 represents 2.5% of the population of Peru. These cases represent only those who sought medical care for suspected cholera and received rehydration treatment and do not include those with milder diarrheal illness or asymptomatic infection. There is little reason to think that cases were overreported. In Trujillo in March of 1991, 79% of persons presenting for treatment with

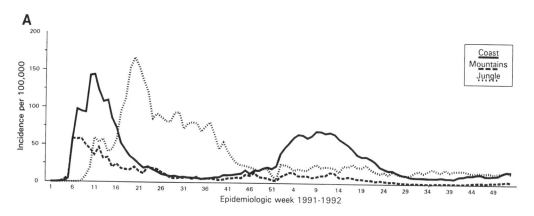

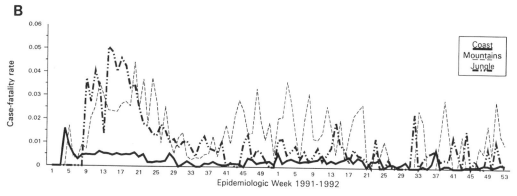

Figure 2. (A) Reported weekly incidence of cholera by physiographic region in Peru from January 1, 1991, to December 31, 1992. Peru has three regions: the dry coastal plains, the Andean mountain chain, and the Amazon jungle. (B) Reported weekly CFRs by region in Peru from January 1, 1991, to December 31, 1992. Data are taken from OGE, Lima, Peru.

diarrhea who had not already taken an antimicrobial agent had *V. cholerae* O1 isolated from their stools, and similar percentages were found in 1992 and 1993 (12a, 37). Results were similar in Piura, Peru, in a study also conducted in March 1991 (35). *V. cholerae* O1 was isolated from 79% of adults hospitalized in Piura for diarrheal illness, 79% of adults with mild diarrhea treated as outpatients, and 86% of children less than 5 years of age hospitalized for diarrheal illness (35). Asymptomatic infections appear to have been common. One to two months after the beginning of the epidemic, 29 and 39% of healthy control subjects in Trujillo and Piura, Peru, respectively, exhibited vibriocidal antibody

titers of >1:40, suggestive of recent infection (35, 37).

The infection rates in some areas were extremely high. A serosurvey conducted in one neighborhood in Trujillo at the end of March 1991 found that 53% of the inhabitants had serologic evidence of recent infection, 37% had had some diarrheal illness, 21% had been treated with rehydration for a diarrheal illness, and 4% had been hospitalized for diarrheal illness since the beginning of the epidemic 2 months earlier (38). Although these high attack rates might not be surprising in a population drinking heavily contaminated water, the relatively large proportion of infections that were associated with severe illness

Table 1. Cholera cases and deaths by country as reported January 1, 1991, through June 10, 1993, to the Pan American Health Organization, with date of first report

Country	Date of first report	No. of reported cases			No. of reported deaths		
		1991	1992	1993	1991	1992	1993
Argentina	Feb. 5, 1992	0	553	1,515	0	15	25
Belize	Jan. 9, 1992	0	159	14	0	4	0
Bolivia	Aug. 26, 1991	206	22,260	7,652	12	383	192
Brazil	Apr. 8, 1991	2,101	30,054	5,103	26	359	81
Chile	Apr. 12, 1991	41	73	28	2	1	0
Colombia	Mar. 10, 1991	11,979	15,129	182	207	158	4
Costa Rica	Jan. 3, 1992	0	12	4	0	0	0
Ecuador	Mar. 1, 1992	46,320	31,870	3,291	697	208	29
El Salvador	Aug. 19, 1992	947	8,106	3,191	34	45	8
French Guiana	Dec. 14, 1991	1	16	2	0	0	0
Guatemala	July 24, 1991	3,674	15,395	1,237	50	207	19
Guyana	Nov. 5, 1992	0	556	58	0	8	2
Honduras	Oct. 13, 1991	11	384	22	0	17	2
Mexico	June 13, 1991	2,690	8,162	1,415	34	99	25
Nicaragua	Nov. 12, 1991	1	3,067	701	0	46	34
Panama	Sept. 10, 1991	1,178	2,416	42	29	49	4
Paraguay	Jan. 25, 1993	0	0	3	0	0	0
Peru	Jan. 23, 1991	322,562	212,642	58,565	2,909	727	295
Suriname	Mar. 6, 1992	0	12	1	0	1	0
United States	Apr. 9, 1991	26	102	9	0	1	0
Venezuela	Nov. 29, 1991	13	2,842	63	2	68	3
Total		391,750	353,810	75,175	4,002	2,396	544

is also related to a specific genetic predisposition. Blood group O, the predominant blood group among Native Americans and the Latin American population in general, was strongly associated with more-severe disease; among those with diarrheal illness and serologic evidence of infection, 39% of those with blood group O and only 5% of those with other blood groups were treated with intravenous rehydration (38).

National age- and sex-specific data are not available for the epidemic in Peru. In general, males and females appear to have been equally affected. In the first 3 weeks of the epidemic in Piura, the cumulative incidence was 4.1% among children aged 0 to 4 years, 1.3% among those aged 5 to 14 years, and 1.8% for those over 15 years of age (35). The overall death-to-case ratio was 0.22%, being 0.04, 0.33, and 0.22% in the same age groups, respectively.

Given the extraordinarily high incidence rates observed in this epidemic and assuming that many infections were asymptomatic, it seems likely that in some parts of Peru, most of the population was infected by the end of 2 years.

THE UNSOLVED RIDDLE

The origin of the epidemic remains a mystery. The epidemic strain theoretically could have been introduced into Peru via any of several possible routes or could represent the change of a previously nonpathogenic local strain to one capable of causing an epidemic. The nearly simultaneous appearance of epidemic cholera in several port cities suggested a relationship to international shipping. Review of harbormasters' logs in Chimbote, the port with the earliest recognized cases, showed that in January 1991, only two ships had arrived in Chimbote from ports outside of Latin America. Both arrived from Hong Kong, one on January 2 and one on January 18. Neither had reported cholera-like illness, though the captain of the second ship requested cholera vaccine before the epidemic was recognized. After that ship left Peru, serum samples were collected from crew members and tested; all had high vibriocidal titers and no antitoxic antibodies. These results were consistent with recent vaccination in the absence of natural infection, so this investigation was inconclusive. It remains possible that the organism was introduced from another harbor in ballast water from some ship, as subsequently was demonstrated for freighters arriving in Mobile Bay (10).

Although the initial focus in coastal towns suggested a maritime source, an infected person arriving by air also could have introduced *V. cholerae* O1 from another part of the world. As an illustration of this possibility, in 1992 a man was hospitalized with cholera in Hawaii after falling ill on a flight from his home in the Philippines en route to Panama. Had his illness begun 24 h later, he would have fallen ill while back at work in a Panamanian harbor (12). The possibility also cannot be excluded that *V. cholerae* was introduced weeks or even months in advance of the epidemic and that it persisted silently in the environment for some period of latency and increased suddenly after a shift in season.

The mysterious origin of the Latin American epidemic has been only partially clarified by molecular characterization of Latin American strains of *V. cholerae* O1 (42). Regardless of the mechanism of introduction, one expects an introduced strain to resemble strains from some other part of the world. However, Latin American strains differ from all other strains tested to date. The comparisons are incomplete, as strains are not available from all countries likely to have cholera. For example, cholera is not reported by China to the World Health Organization, and

no *V. cholerae* O1 strains directly from China have been available for comparison.

Though the Latin American strains do not match strains from elsewhere in all characteristics, they resemble other strains of the current global pandemic more than they do strains of nontoxigenic *V. cholerae* O1 isolated in Latin America before the epidemic began. This makes the introduction from elsewhere more likely than the alternative, i.e., that a nontoxigenic organism already present in Latin America acquired a cholera toxin gene and emerged in epidemic form (42). Thus, the laboratory evidence suggests that the Latin American epidemic is an evolutionary offshoot of the global pandemic.

Explanations for the nearly simultaneous appearance of epidemics in several cities also remain speculative. The Peruvian coastline has a well-developed and heavily traveled net of highways. If a perishable product such as fish or ice was initially contaminated, it could have been rapidly distributed through normal commerce. It is also possible that after introduction of the organism into one harbor, coastal freighter traffic rapidly transferred it to other ports via ballast water.

EPIDEMIC SPREAD IN LATIN AMERICA

Following its explosive appearance in Peru, the epidemic spread rapidly into other countries at the rate of nearly one country per month (Table 1; Fig. 3). In early March, migrant workers returning to shrimp farms in southern Ecuador from homes in northern Peru brought cholera with them, further spread to Colombia occurred in that same month, and travelers on the Amazon brought cholera to Brazil by April. Spread was not always contiguous. The first case imported into the United States was documented on April 11, 1991, in a traveler returning from Lima; there was no subsequent spread (8). A traveler also apparently brought cholera to

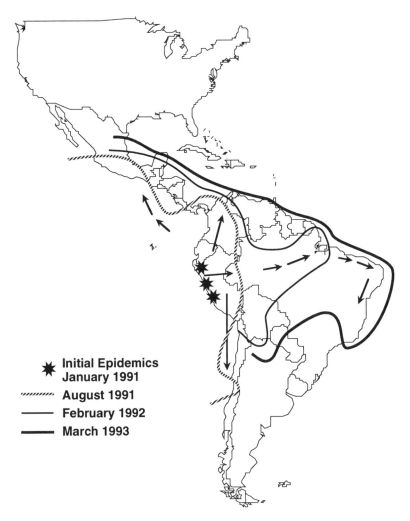

Figure 3. Geographic extent of the Latin American epidemic over time. Lines represent the advancing front of the epidemic at different dates. By March 1993, all Latin American countries except Uruguay had reported cholera, and no cases had been reported from the Caribbean.

Santiago, Chile, a distance of over 1,500 km from Peru, leading to a subsequent cluster of cases there in April. Mexican public health authorities prepared to identify cholera in harbor towns or at border crossing points were surprised when it first appeared on June 17, 1991, as an outbreak of 27 cases in the small village of San Miguel Totolmayloya, in the rural and mountainous municipio of Sultepec, just south of Mexico City itself (25). The

outbreak strain may have been introduced via nearby clandestine airstrips presumably used by coca smugglers arriving from South America.

The epidemic spread throughout Central America, northward from Colombia and southward from Mexico. In South America, it reached the mouth of the Amazon by December 1991 and spread southward along the Atlantic coast of Brazil, reaching Rio de Janeiro in February 1993 (Fig. 3). Outside Peru, where cities were the initial focus, rural areas have tended to be more heavily affected than urban areas. Indigenous populations pursuing more or less traditional lifestyles, including Amazonian Native American tribes, Aymara-speaking groups in the Andes, and traditional native cultures of the east coast of Central America, have been at particularly high risk.

The appearance of the epidemic in countries north of the equator in mid-1991 coincided with summer there, so that large numbers of cases continued to occur in the warm season there even while a cool-season decrease affected Peru and other countries south of the equator. Similarly, by January 1992, when the number of cases was decreasing in countries of the northern hemisphere, it was rapidly increasing again south of the equator (Fig. 4). The capacity to maintain this oscillating seasonality within the same continent means that reintroduction from the other hemisphere can perpetuate the epidemic efficiently from year to year.

CHOLERA IN MEXICO

After cholera was reported by Peru in 1991, the Mexican Ministry of Health prepared to detect entry of the disease into Mexico and to minimize subsequent spread by local intervention. Surveillance of diarrheal disease was increased, and environmental monitoring was begun in busy ports and border points. As noted above, the first cases were detected in a central mountain village in June 1991. Cholera subsequently spread widely; by the end of 1991, 2,690 laboratory-confirmed cases and 34 associated deaths had been identified in 17 states, principally in southern Mexico (Fig. 5). Of these cases, 57% were in men; over 90% were in persons over the age of 5 years (Table 2). Most cases occurred in late summer and fall, with the peak in October (Fig. 4).

In 1992, most cases again occurred between July and October, and the prepon-derance of cases were in adults. The highest rates were again reported from southern states and the states on the Gulf of Mexico (Fig. 5). Of 8,162 patients with laboratory-confirmed cases reported in 1992, 99 (1.22%) died, 49.4% were hospitalized, and 26.4% were asymptomatic carriers detected by microbiologic surveillance. As of week 19 of 1993, 1,415 cases had been reported; 79.6% of these patients required hospitalization, and 1.8% died.

Surveillance data were reported daily from the 31 states and the Federal District to the national cholera center. Once a case was detected through surveillance, a task force within the Ministry of Health tried to detect and limit local spread. This intensive effort required substantial resources and is generally possible only where relatively few cases are reported. Its efficacy and cost-effectiveness remain to be documented. This effort included visiting suspect cholera patients' homes, culturing the patient and household contacts, and giving them prophylactic antibiotics. The household was provided with a packet with education materials, ORS, a bar of soap, and chloramine water treatment tablets.

When a case was confirmed, other cases of diarrhea were sought within a five-block area for urban cases or within a 5-km radius around the house of the confirmed case in rural areas. Samples from each patient with an identified case of diarrhea were cultured, and the patients were given antibiotic treatment. More health packets were distributed, the use of ORS was reinforced, health education was conducted, sources of potable water were checked for chlorine, and chlorination was augmented where necessary. In this effort, between June 1991 and April 1993, public health workers visited 8,717 communities, about 2 million households, and 7 million people, or approximately 8% of the Mexican population. Public health officials also increased attention given to marketplaces, town festivals, and pilgrimages, providing disinfected water and temporary latrines at these

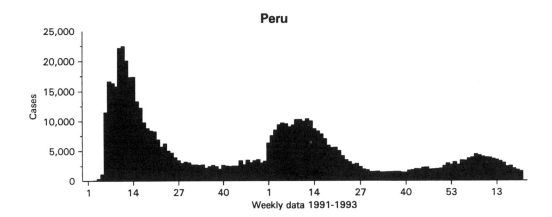

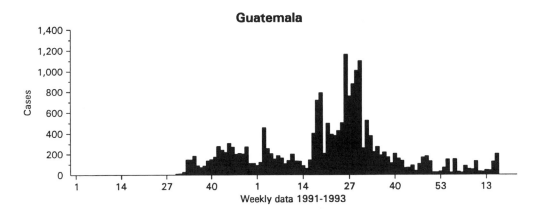

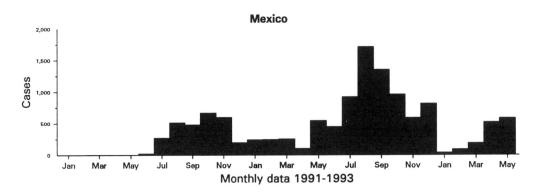

Figure 4. Numbers of reported cases of cholera by week (Peru, Guatemala) or by month (Mexico) from January 1991 through early 1993. In each country, incidence peaks in the summer and fall. Since Peru is south of the equator, the warm seasonal peaks occur in January through March. Data were obtained from the Pan American Health Organization.

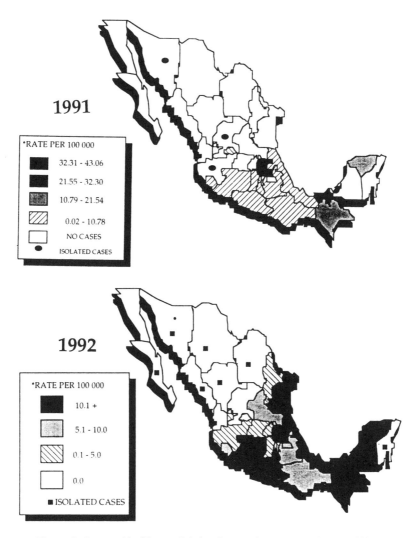

Figure 5. Reported incidence of cholera by state in Mexico, 1991 and 1992.

mass gatherings when possible. Mass media education targeted the general public.

Other measures may reduce cholera transmission. In 1991, testing for residual chlorine in drinking water was essentially nil. By 1993, an estimated 8,000 communities were testing chlorine levels routinely. The area on which raw sewage was used for irrigation decreased from 27,000 ha in 1991 to 1,159 ha in early 1993. A government program has begun to expand provision of piped treated water and complete sewerage systems.

EPIDEMIOLOGIC ASSESSMENT

Comparison of reported cholera incidence among countries or epidemics is complicated by variation in case definitions and completeness of reporting. In Mexico, for example, culture-confirmed asymptomatic infections are reported as cases. Nonetheless, the CFRs reported from Latin American countries during the first 2 years of this epidemic are extraordinarily low. This ratio was 1.02% for the entire hemisphere in 1991 and 0.68% in

Table 2. Reported cases and incidence of cholera by age and sex, Mexico, 1991

Age group (yr)	Males		Females		Total	
	No. of cases	Incidence[a]	No. of cases	Incidence	No. of cases	Incidence
0–4	131	2.4	93	1.8	224	2.1
5–14	200	1.6	149	1.3	350	1.5
15–24	259	2.8	198	2.2	457	2.5
25–44	492	4.8	368	3.6	860	4.2
45–64	292	6.8	227	4.9	519	5.8
>65	138	10	104	5.1	242	7.1
Total	1,533	3.5	1,157	2.6	2,690	3.1

[a]Per 100,000 population.

1992. By comparison, the comparable cholera CFR for Africa in 1991 was 9.1%, and that for Asia was 2.6% (44). The low mortality figures may reflect relatively thorough case counting in countries like Peru, Ecuador, and Colombia, but they also indicate the growing experience and skill of the Latin American medical community in treating cholera.

High CFRs, indicating the greatest challenge to providing effective treatment, occurred in affected parts of Amazonia. In an investigation of cholera-like illness reported from 12 villages in the Peruvian Amazon, the combined incidence was 6% and the CFR was 13.5% (34). Patient travel time to the nearest treatment facility was at least 4 h by canoe or commercial riverboat. Comparison of those who died with others in the same village who survived a bout of severe watery diarrhea occurring at the same time showed that survival was associated with reaching a health facility. CFRs were 5.6% in the villages with a local health care provider and 24% in the villages without a provider. Even though 62% of decedents had received ORS at home, the amount received (a median of three packets) was insufficient to treat cholera. The best option for decreasing the mortality rate in such remote areas appeared to be promotion of village-based treatment of diarrhea, including providing ORS and training residents in each village in how to use it.

The high CFRs observed where treatment options were limited are similar to rates observed in remote parts of Africa (40). This shows that without access to adequate treatment, the *V. cholerae* O1 strains causing the Latin American epidemic are as lethal as those that spread through Africa.

The moment of introduction of cholera into previously unaffected areas of Latin America was not often documented. However, in the United States, 109 cases related to South American travel were investigated and reported in 1991 and 1992. Each case represents a potential introduction and provides insight into how the organism could have been introduced into other countries. These cases also illustrate the futility of attempts to halt introductions through general border quarantines.

In 1991 and 1992, twenty-three travelers returned home with symptomatic or incubating infection acquired during travel in Latin America. Most were persons of Latin American origin and had been visiting relatives or friends in Latin America. None were following standard tourist itineraries, and few were likely to have used travel health clinics.

Travelers also became infected during the return trip itself. Seventy-five cases occurred among passengers from an airliner that flew from Buenos Aires to Los Angeles by way of Lima, Peru. The passengers fell ill with cholera after the flight landed in California (11, 12a). Illness was associated with eating a cold seafood salad loaded onto the plane in Lima. It led CDC to recommend that airliners add ORS to their in-flight emergency kits and that they not serve cold cooked foods prepared in cities experiencing cholera epidemics. No other episodes of airline food contamination have been reported in connection with the Latin American epidemic, but it is possible that they would not be recognized if the flight were shorter, if passengers were more dispersed before they became ill, and if the destination were a country that already had an ongoing cholera epidemic.

Travelers also unwittingly brought *V. cholerae* O1 home with them in their baggage. In 1991, two small cholera outbreaks occurred when travelers brought back cooked crabs in their suitcases as gifts for family and friends in the United States (9, 20). Although not illegal, this practice is not wise, and an education campaign to discourage bringing perishable souvenir seafood in luggage was mounted. No further such incidents in the United States occurred in 1992.

Finally, *V. cholerae* O1 was transported from harbor to harbor in the ballast water of freighters. Chance isolation of the Latin American strain of *V. cholerae* O1 in oyster beds of Mobile Bay, Alabama, led to discovery of this surprising means of introduction (10, 19, 30). After investigations failed to identify a land-based human source for the strains isolated repeatedly from the oysters, toxigenic *V. cholerae* O1 was isolated from water that was being discharged into Mobile Bay from the ballast tanks of freighters arriving from Latin American ports. Freighters routinely take on and discharge ballast water in port to maintain stability as they load and unload cargos of various densities. This practice transfers huge volumes of harbor water from port to port, along with anything that might be in the water. In Peru, concentrations of toxigenic *V. cholerae* O1 of 10,000/dl were measured repeatedly in samples of polluted seawater in 1991 (39). A single ship taking on 100,000 gal (ca. 378.5 kl) of such water would transport 10 billion organisms.

Ships have brought other harmful species into new areas this way, including the zebra mussel and stinging jellyfish (36). Because of growing concern about the introduction of harmful species, the International Maritime Organization and the U.S. Coast Guard now instruct freighters to flush their ballast tanks at sea before entering port (16).

As of mid-1993, no spread has occurred following the introduction of the Latin American strain by travelers into the United States. However, cholera might spread from contiguous areas of Mexico to populations living in substandard housing ("colonias") on the U.S. side of that border (6). In 1991 and 1992, northern Mexico reported only sporadic cases, with no apparent sustained transmission. The threat of spread in the colonias has stimulated long-overdue construction of water and sewerage systems, but large populations still live without these basic necessities.

MICROBIAL CHANGES IN THE COURSE OF THE EPIDEMIC

The strains isolated early in the Peruvian epidemic were uniform: they were susceptible to antimicrobial agents, were serotype Inaba, and shared other microbiologic characteristics consistent with spread from a single introduction (35, 37, 42). Later, changes were observed; the first strains isolated from Bolivia were serotype Ogawa, and by the second year of the epidemic in Trujillo, Peru, serotype Ogawa predominated (12a). By July 1991 in Guayaquil, Ecuador, 36% of strains were multiply resistant, having acquired resistance to tetracycline and trimethoprim-sulfamethoxazole among other agents (43). Additional changes can be expected as the organism moves into new ecologic niches throughout Latin America and as evolutionary pressures of population immunity, antimicrobial use, and chlorination of water supplies increase. In 1993, a strain of toxigenic *V. cholerae* O1, serotype Ogawa, with laboratory characteristics resembling those of

Asian seventh-pandemic strains rather than Latin American strains was identified in travelers returning to the United States from Mexico (12a). This suggests that there has been another introduction with subsequent spread in addition to the epidemic that started in Peru. In the future, other newly introduced strains may be identified, including the *V. cholerae* O139 strain spreading through Asia in 1993 as well as strains that have a previously undetected endemic focus in the Americas, like the U.S. Gulf Coast strain.

FIELD INVESTIGATIONS OF ROUTES OF TRANSMISSION

Successful efforts to halt ongoing transmission or to prevent it from happening depend on adequate understanding of the routes of transmission where it is occurring. This requires epidemiologic investigations that are specifically focused on this issue. Cholera can be acquired from a variety of food and water sources, even within a single epidemic or outbreak. Unsuspected vehicles have often been discovered in the past when cholera appeared in newly affected areas. Routinely collected surveillance data can indicate the magnitude of the epidemic, and simple case interviews and microbiologic testing of potential sources can suggest possible sources of infection. However, determining the actual routes of transmission usually requires an analytic field investigation such as a case-control study. In these investigations, systematic differences between ill and well persons in the frequency of a particular exposure, such as drinking water from a certain well, provide strong evidence that this was the source of infection. By contrast, simply demonstrating *V. cholerae* in a suspected water source does not by itself prove that the water was the source of infection for people; it may be a coincidental finding, occurring because persons infected from another source also happened to contaminate the water. Demonstrating a strong statistical association between illness and drinking water from that source before becoming ill is more convincing.

In Latin America, analytic field investigations have determined the importance of various modes of transmission, have indicated specific control measures that can be taken immediately, and are guiding long-term interventions.

Waterborne Transmission

In each of six analytic field investigations in which CDC has collaborated, waterborne transmission played an important role (Table 3). In three of the six, all conducted during large urban epidemics, drinking unboiled municipal water was associated with cholera. In two of these three, the municipal water in the system itself was probably contaminated, as the systems were in extremely poor repair; water delivery was intermittent, with frequent drops in pressure permitting entry of sewage; and the water systems were unchlorinated. These conditions are typical of many water systems in Latin America.

In one of these investigations, in the city of Trujillo, Peru, a case-control study was conducted 1 month after the beginning of the epidemic in that city (37). This study showed that cholera was associated with drinking unboiled municipal water and was independently associated with drinking water stored in a container into which hands had been introduced. Early in the epidemic, toxigenic *V. cholerae* O1 was isolated from the municipal water distribution system. The municipal water system was supplied by wells, which provided insufficient water to make continuous delivery possible. Though water outages, system leaks, and clandestine connections were common, the water was not chlorinated. Studies of municipal water quality documented some fecal coliforms in specimens collected at wellheads, suggesting possible contamination of the aquifer. However, both total coliform and fecal coliform counts were higher in specimens collected from peripheral water taps than in those from well-head

Table 3. Risk factors for cholera identified in six collaborative case-control investigations in Latin America, 1991 to 1992

Risk factor	Presence in given location, setting, date (mo/yr), reference					
	Peru			Guayaquil, Ecuador; urban, 7/91, 12a	El Salvador, rural, 3/91, 12a	Bolivia, rural, 2/92, 32
	Trujillo, urban, 3/91, 10	Piura, urban, 3/91, 11	Iquitos, urban, 7/91, 33			
Drinking unboiled or untreated water from:						
Municipal system	+	+	+	+		
River					+	+
Putting hands in water storage vessel	+	+				
Consuming:						
Street-vended food		+				
Street-vended beverage		+				
Uncooked seafood				+		
Cooked crabs				+	+	
Leftover rice				+		
Unwashed fruits or vegetables		+				
Protective factors						
Boiling water	+	+		+	+	+
Adding acidic juice to water			+			
Having soap in home				+		

samples, indicating progressive contamination during distribution. Because of the intermittent delivery of water, nearly all households stored water in a variety of containers. Fecal coliform counts from water in household storage containers were 10-fold higher than levels in the taps from which they had been drawn, indicating that water was being contaminated further during storage in the households.

A parallel study was conducted in the city of Piura, in northern Peru (35). In this study as well, drinking unboiled municipal water was associated with cholera, and drinking stored water into which persons had put their hands was again a risk factor. This municipal water supply was also drawn from underground aquifers and delivered intermittently through a system of pipes that had been incompletely repaired after flood damage. There were many indications of water system contamination. Fecal coliforms were detected in samples collected at well heads. Toxigenic

V. cholerae O1 was isolated from a cistern storing municipal water and from a sample of tap water. Although chlorination equipment was present, only 12% of water samples collected from the municipal system had residual chlorine. Attempts to chlorinate water were limited by lack of reliable chlorine supplies, lack of continuous electrical power, and damaged equipment.

Cholera in Piura was also significantly associated with consumption of beverages purchased from street vendors. Further investigation showed that the risk of illness did not depend on the particular type of beverage but rather on whether or not there was ice in it. The ice was produced locally from untreated municipal water under unsanitary conditions and was primarily used for the transport of fish. Before this investigation, the possibility that ice could serve as the source of contamination had not been seriously considered.

In July 1991, a case-control study conducted in Guayaquil, Ecuador, also implicated unboiled water as a major source (43). In this case, the source of drinking water was centrally chlorinated municipal water delivered by tank trucks to a neighborhood built on stilts over a tidal marsh. Although adequate levels of chlorine were detected in water samples from truck loading points, there were many points at which recontamination could occur as the water was being carried to the home and stored there. The fact that drinking boiled water was still protective indicates that the water was probably contaminated after delivery or perhaps that water tankers also brought water from alternative unsafe sources. An independent assessment conducted in December 1992 also found that municipal water supplies were well chlorinated in general, that disinfection was not always reliable because of the intermittent nature of the water delivery, and that opportunities for recontamination after collection were extremely common (13).

In a fourth investigation, conducted in the Amazonian city of Iquitos, Peru, water from a variety of sources, including the municipal system and the river, was consumed and in general was not boiled. In this setting, the risk of acquiring cholera was not linked to a particular water source. However, healthy controls were more likely to have added the juice of an acidic citrus fruit ("toronja," similar to a grapefruit) to their drinking water (31). The pH of toronja juice was 4.1, sufficient to inactivate *V. cholerae*. Thus, this traditional local practice appears to have protected against cholera, though the intent was simply to impart an agreeable taste to the water.

Two rural investigations implicated drinking untreated river water as a source of cholera. In a rural lowland community in Bolivia, cholera was associated with drinking untreated river water, while drinking boiled water or rainwater was protective (22). Many of those drinking river water were farmers whose fields were far from potable-water sources. In another case-control field investigation in rural El Salvador, cholera was associated with drinking water outside the home, which in that setting meant from a river (12a). These observations suggest that provision of safe domestic water alone may be insufficient if safe drinking water is not also available to workers in fields or other workplaces.

These investigations identified a number of interventions to prevent waterborne transmission in the short, medium, and long terms. They documented the efficacy of emergency instructions to boil water and supported efforts to disinfect municipal water supplies. Chlorination equipment and chlorine supplies were obtained from international donors and added to some systems. For the medium range, the data also indicate that central chlorination of water supplies is by itself likely to be insufficient in the setting of intermittent water supplies, defective piping, and universal storage of water in households. The documented increase in contamination from water taps to household water containers in Trujillo, Peru, is likely to have public health significance, as cholera in both Trujillo and Piura was significantly associated with drinking water stored in a container into which someone had placed their hands. In Guayaquil Ecuador, soap was more frequently present in the homes of healthy controls than in the homes of cholera patients, suggesting that handwashing was more likely among the controls (43). The entry of unwashed hands into the water may be an important mechanism of contamination, or it may simply indicate that the water is in a wide-mouthed water container that is generally unprotected against contamination from hands, kitchen debris, or other sources of vibrios. The observation in Ecuador that boiling water was still protective even when the central supply was adequately chlorinated also indicates that central chlorination is not sufficiently protective, either because intermittent contamination nonetheless occurred or because the water became recontaminated in the home.

Making sustainable improvements in domestic water treatment and storage is likely to be an important midrange control measure for cholera. The problems are linked: safe water storage prevents recontamination from occurring, and effective treatment means that water from a variety of sources can be used. Water stored in the home can be treated by several methods, including boiling or disinfection with bleach or other halide solutions. Of these, boiling is probably the most reliable method and the least sustainable. Boiling takes considerable time and fuelwood or kerosene, which are in short supply in many urban areas. Heated water may also be perceived as less refreshing or even less usable in some cultures. Because boiling drives off any residual chlorine that may be present in the water at the point of collection, boiled water is easily recontaminated unless it is stored safely. Methods of chemical treatment applicable at the domestic level need further exploration but could provide a sustained residual protective effect, need not degrade the environment, and may be economically sustainable if the treatment reagent can be locally and inexpensively produced. Chlorination is the most obvious route to disinfection, but the possibility of using acidic fruits to treat water, as traditionally practiced in Iquitos, Peru, needs further exploration as well. In many parts of Latin America, citrus fruits are readily available and inexpensive year-round. Like chlorine and unlike boiling, acidification would also provide sustained residual protection against recontamination. Solar inactivation of *V. cholerae* either by solar heating or UV inactivation may be a feasible alternative at high altitudes (28).

A critical link in the strategy to improve the safety of stored household water is the storage container itself. Use of a traditional narrow-necked water vessel was protective against cholera in Calcutta (17). However, explorations of alternate designs and methods for testing them in the field are just beginning. An appropriate container should be easy to fill with water, easy to carry and pour from, sturdy, inexpensive, and locally manufactured and should have a standard volume

of about 20 liters. CDC began field tests of improved water containers combined with locally produced chlorine solutions in 1992.

For the long run, the threat of waterborne cholera is likely to persist in Latin America until safe, continuous water supplies and sewage treatment are available for all. The Pan American Health Organization has proposed extending these public utilities to the entire population of Latin America at an estimated cost of $200 billion over 12 years, an expenditure similar to the combined defense budgets of the affected nations (18). The effort extends beyond cholera to preventing other diseases transmitted by impure water and sewage-contaminated foods, such as typhoid fever, hepatitis, and many pediatric diarrheas. Just as repeated cholera epidemics forced Europe and North America to invest in basic public health infrastructure in the 19th century, Latin America can benefit greatly from a similar "sanitary revolution" in coming years.

Foodborne Transmission

Investigations in Latin America have also demonstrated transmission through foods, which means that control measures focused on water alone will be insufficient and that other intervention points need further research.

Seafood

Seafood was implicated as a source of cholera in some parts of Latin America. In Guayaquil, Ecuador, a case-control study implicated eating raw seafood and cooked crab as important sources; raw seafood was eaten by 26% of patients compared to only 7% of controls (43). One specimen of pooled raw concha purchased at a local market yielded *V. cholerae* O1. In El Salvador, a case-control investigation implicated eating raw or cold seafood as a source of cholera (12a). In the two large case-control studies conducted early in the epidemic in Peru, the emergency warning to avoid raw seafood was almost universally heeded. No conclusions regarding the safety of seafood could be drawn in those studies, because neither patients nor healthy controls reported eating seafood (35, 37). A case-control study conducted among children hospitalized in Lima associated cholera with drinking unboiled water and eating raw fish and uncooked vegetables (21).

Other investigations also illustrate the potential hazard of seafood in Latin America. As noted above, two outbreaks of cholera occurred in the United States in 1991 after travelers brought crabs in their suitcases back from Guayaquil, Ecuador (9, 20). In both instances, the crabs were cooked in Ecuador and transported to the United States, where they were used in cold salads. Even seafoods prepared in more carefully supervised conditions can be risky; as mentioned above, in 1992, cholera affecting passengers from an international airliner was associated with eating a cold seafood salad prepared in Lima (11, 12a). Toxigenic *V. cholerae* O1 has been isolated from fish and shellfish in Peru and Ecuador, and toxin gene sequences were detected in leftover seafood that caused an outbreak in New York (9, 39, 43). Before the epidemic began in Latin America, a case of cholera acquired in Mexico caused by the Gulf Coast strain of toxigenic *V. cholerae* was related to eating ceviche, a marinated seafood made with raw fish and shellfish marinated in citrus juice and/or vinegar (4).

Seafood may pose a continuing hazard for cholera in some parts of Latin America for several reasons. The general ecologic association between *V. cholerae* and shellfish demonstrated elsewhere in the world suggests that *V. cholerae* may persist for years in some Latin American estuarine environments, many of which are also important fisheries. The export seafood industry uses modern production techniques, importing countries have not detected toxigenic *V. cholerae* O1 in the seafood, and no cases of cholera have been traced to such products. However, seafood for domestic consumption is often pro-

duced in unhygienic conditions: shellfish are harvested near harbor towns where waters are often contaminated with sewage; small fishing boats ply polluted waters and use harbor water to rinse, unload, and process the fish (24, 41). In Peru, coastal seawater had high densities of *V. cholerae* O1 in 1991 as a result of the discharge of untreated urban sewage at the height of the epidemic (39). Cholera was in one instance associated with consuming ice made to ship fish, showing that even seafood that was safely harvested and processed could still become contaminated during distribution (35). In the long run, domestically consumed seafood can be made safer in Latin America through efforts to monitor contamination of shellfish beds and to prevent contamination in fish processing and distribution.

Variation in seafood cuisine is likely to contribute to the risk. Cooked seafood may still be contaminated if it is not cooked sufficiently to kill *V. cholerae* present on the raw food or if it is recontaminated after cooking. Cooked shellfish are a good growth medium for *V. cholerae* O1, and the organism can grow to a billion organisms per gram of tissue in less than 12 h (26). Variation in cooking times and holding times after cooking may be critical to the safety of cooked cold seafood.

Varieties of marinated raw-seafood dishes known collectively as ceviche are popular in much of Latin America. As vibrios are rapidly inactivated by low pH, acidic marinades should theoretically lower the risk of transmission greatly. In studies using artificially contaminated fish (incubated in vibrio broth for 90 min), 30 min of marination in lemon juice was sufficient to inactivate 8 logs of *V. cholerae* O1; Mata concluded that correctly prepared ceviche should pose little risk for cholera transmission (29). However, further investigations are needed to define the limits of safety for variant ceviches. The locations of vibrios in naturally contaminated seafood are likely to make a difference. In shellfish, vibrios are likely to be present in the intestinal tract and other deep tissues, and thus longer marination times would be required than would be needed to inactivate vibrios present only on the surface of the food. There are substantial regional variations in marination times of ceviche: marination may occur for minutes rather than hours in some Latin American kitchens. A variety of marinating agents may be used, some substantially less vibriocidal than tart lemon juice. Determining the limits of time, pH, agent, and type of seafood within which marination is safe for naturally contaminated seafoods is important for preventing future seafood-borne cholera in Latin America.

Street-vended foods

The street vendor is a central feature of large poor urban communities throughout the developing world. In Piura, Peru, cholera was associated with consumption of foods purchased from street vendors independently of its association with street-vended beverages contaminated with ice (35). In Ecuador, cholera was associated with street-vended beverages and specifically with a street-vended orange juice (43). Simple measures may improve the safety of street-vended foods. In some countries, educating street vendors in food safety and licensing them to enforce education may be effective measures. Even in less-developed areas, vendors compete with each other to meet market expectations. If the public demands vendors with clean hands, vendors' carts with visible soap and washing basin, foods that are cooked and hot, beverages that are carbonated or acidic and have no ice in them, and fruits that are unpeeled, vendors will provide them.

Other foods

In the investigation in Piura, Peru, cholera was independently associated with eating cold leftover rice. Rice is a good growth medium for *V. cholerae*, and leftover rice that is contaminated after cooking and eaten without reheating is a plausible vehicle (26). Encouraging better hygiene in the kitchen and re-

heating leftovers just before eating them are potential prevention measures, limited by availability of soap and fuel.

In an investigation in Iquitos, Peru, cholera was associated with consumption of unwashed fruits and vegetables brought to market in small open boats and likely to have been splashed with contaminated river water during transport (31). In Trujillo, Peru, cholera was weakly associated with eating cabbage, which had been irrigated with fresh sewage (37). In Chile, a link between typhoid fever and eating unwashed vegetables grown in fresh sewage had previously been suspected (2). The appearance of cholera in Santiago among persons who had eaten such foods led to a ban on irrigation of fields with raw sewage and accelerated the construction of sewage treatment plants.

Although vegetables and nonacidic fruits are easily contaminated with sewage during harvest or distribution, education efforts can encourage cooking vegetables thoroughly, washing fruits before peeling them, and eating them immediately after peeling or slicing so that vibrios do not have the opportunity to multiply on the cut surfaces.

CHOLERA VACCINES—AN OPTION IN THE FUTURE?

The existing parenteral cholera vaccine confers limited protection for 3 to 6 months after two doses and does not reduce asymptomatic carriage. Development of an effective and inexpensive vaccine would be an important advance, and several experimental vaccines are being tested in Latin American populations. Although much remains to be learned, several observations caution against predicting early and dramatic success there. Vaccine efficacy depends on blood group. In Bangladesh, the experimental whole-cell–B-subunit vaccine (WC-BS) gave much less protection to recipients with blood group O than to recipients with other blood groups (14). Blood group O is much more common in Latin America than in other parts of the world. Vaccine efficacy also varies with the strain of *V. cholerae* that causes disease. WC-BS showed less protection against El Tor than against classical biotype, and among those infected with El Tor, the vaccine was least effective against serotype Inaba, the serotype initially dominant in much of Latin America (15). Efficacy appears to be less effective in populations without prior immunologic experience with cholera. In young Bangladeshi children, who were less likely than their elders to have been previously exposed to cholera, WC-BS vaccine showed little or no lasting protection (15). No vaccine has yet been shown to reduce asymptomatic carriage, which suggests that effects on transmission could be limited.

LESSONS OF THE LATIN AMERICAN CHOLERA EPIDEMIC

Epidemic cholera in Latin America provides several lessons for the world community. The beginning of the epidemic was identified and reported with remarkable speed and candor. Rapid epidemiologic field investigations determined the cause of the epidemic within days and clarified the modes of transmission within weeks. The level of hemispheric cooperation has remained high, and case reporting to the Pan American Health Organization has been rapid and relatively open despite economic consequences. With minor exceptions, the nations of the Americas have kept borders open and have avoided futile epidemic control strategies such as quarantine, broad import restrictions, mass vaccination, or mass chemoprophylaxis. Open surveillance allowed the hemisphere to deliver the life-saving activities of treatment, surveillance, and education swiftly. Other countries that conceal their cholera are not likely to have the same success in treatment and are missing the international support received by Latin America for prevention.

The remarkably low mortality rate reported throughout Latin America reflects effective use of rehydration therapy. There is little evidence that this infection was itself less severe than those in other parts of the world. In fact, the genetic predominance of blood group O in Latin America means that cholera is more likely to be severe than in Asia or Africa, with a greater need for hospitalization and intravenous treatment. The general success of treatment for cholera means that the Latin American medical system is now more skilled in treating any dehydrating diarrhea and that the populace is more confident about the benefit of such treatment. One future impact of cholera is likely to be reduced mortality for a variety of diarrheal illnesses. Emergency cholera treatment was often implemented in makeshift wards by persons initially unfamiliar with cholera. Central hospital locations tended to be used more than neighborhood clinics. The success of Latin American interventions may suggest models of treatment delivery that would be applicable in other parts of the world where cholera mortality remains high.

Cholera spread rapidly through the intercontinental transportation network of Latin America, appearing unexpectedly in persons and in the environment far from previously known cases. The likelihood of repeated introductions means that rapid and effective surveillance for cholera is needed everywhere. The persistence of cholera anywhere in the hemisphere means that reintroduction anywhere else may be an airplane flight or a freighter voyage away. It also means that for years to come, physicians throughout the hemisphere will need to consider cholera in the differential diagnosis of traveler's diarrhea and will need to know how to treat cholera appropriately. Microbiologists will need to be prepared to isolate and identify the organism, and public health officials will need to be familiar with methods of investigating and preventing infections with *V. cholerae* O1. The precautions suggested for travelers to Latin America in their selection of food and drink are similar to the advice given to many local populations.

The risk of repeated introductions from countries where cholera persists also means that investment in sustained prevention efforts to eliminate transmission of the organism in one country is likely to benefit all countries. The risk of continued epidemic cholera is particularly high in periurban slums, where growing urban populations far outstrip municipal infrastructure, and inadequate water supplies and unsafe street-vended food are a way of life. In this setting, a defective municipal water system can become a particularly efficient source of disease. It is a peculiar irony of modern life in the developing world that large and poorly maintained water systems may be more hazardous than traditional water sources if the community believes that water is safe just because it comes from a pipe. Although municipal water may be intermittently highly contaminated, the population may not see the need to treat it further, while the same people might think water from a river is inherently suspect.

In remote mountainous and jungle locations, providing adequate education, treatment, and prevention remains challenging because of cultural and linguistic barriers as well as geographic ones.

The challenge of developing sustainable prevention measures is a fertile arena for research. In many affected areas, preventing cholera begins with emergency public education to boil drinking water, cook foods well, and select only safe foods from street vendors. Such short-term preparations are likely to limit morbidity initially, but they are difficult to sustain and are insufficient by themselves where epidemic cholera is likely to persist or recur. Longer-term prevention measures that are focused on the specific sources of infection are needed. The impetus of cholera can lead to the adoption of relatively simple interventions to disinfect and protect drinking water and to make foods safer.

Epidemiologic investigations are critical to guiding the development of control measures in a given epidemic. Epidemic cholera is often transmitted by more than one route simultaneously, so efforts focused on a single route may fail to stop an epidemic. Careful investigation may identify unsuspected vehicles that require specific new control measures. Such investigations need not be expensive nor time-consuming; the case-control technique has proved useful in many locations.

Successful efforts to monitor, treat, and prevent cholera depend on a functioning public health infrastructure. In Latin America, the advent of cholera has strengthened public health mechanisms in many countries. Laboratory capacity to identify the organism has improved. More efficient and rapid surveillance systems have been implemented to identify the beginning and end of epidemics, to direct the distribution of treatment materials, and to monitor the success of control measures. Resources of central governments and of foreign donors are needed to continue these developments. The collaborative links that are developing between epidemiologists, microbiologists, and other public health officials throughout the hemisphere are likely to lead to better surveillance and control of other diseases as well as cholera.

The epidemic provides opportunities to advance global knowledge about cholera and its prevention. Investigations of the ecology of persistence of *V. cholerae* O1 in the environments of Latin America may help determine the natural reservoir for this organism. Field trials of experimental vaccines in Latin American populations will accelerate our knowledge of the potential usefulness of such vaccines. Microbiologic characterization of emerging subtypes of the epidemic strain of *V. cholerae* O1 offers a unique opportunity to observe the evolutionary changes in a single strain of bacteria as it enters a variety of new environments. Clinical studies may clarify the unexplained association of severe illness with blood group O. Other potential cofactors such as the effect of hypochlorhydria caused by infection with *Helicobacter pylori* may be identified. Applied research to develop new prevention efforts may lead to sustainable community-based prevention measures such as domestic water chlorination and a safer water storage vessel. This in turn may prevent a variety of illnesses in addition to cholera and thereby benefit many populations of the developing world.

Cholera seems likely to remain in Latin America for decades, as it has in Africa and Asia, establishing new sites of environmental persistence and causing occasional epidemics. Cholera will remain a risk where people live below a sanitary threshold, without safe water and food, sewage disposal, or elementary hygiene. The epidemic is a metaphor for underdevelopment, a marker for many diseases that are transmitted by contaminated food and bad water. The scope of the epidemic extends beyond health and clinical medicine to include the fisheries, agriculture, shipping, and tourist industries of many nations. Just as cholera spurred the sanitary reform movement in 19th century Europe and North America, epidemic cholera is likely to be less and less tolerable in the growing democracies of Latin America.

REFERENCES

1. **Batchelor, R. A., and S. F. Wignall.** 1988. Nontoxigenic 01 *Vibrio cholerae* in Peru: a report of two cases associated with diarrhea. *Diagn. Microbiol. Infect. Dis.* **10**:135–138.

2. **Black, R. E., L. Cisneros, M. M. Levine, A. Banfi, H. Lobos, and H. Rodriguez.** 1985. Case-control study to identify risk factors for paediatric endemic typhoid fever in Santiago, Chile. *Bull. W.H.O. Health Org.* **63**:899–904.

3. **Blake, P. A., M. L. Rosenberg, J. B. Costa, P. S. Ferreira, C. L. Guimaraes, and E. J. Gangarosa.** 1977. Cholera in Portugal, 1974. I. Modes of transmission. *Am. J. Epidemiol.* **105**:337–343.

4. **Blake, P. A., I. K. Wachsmuth, B. R. Davis, C. A. Bopp, B. P. Chaiken, and J. V. Lee.** 1983. Toxigenic *Vibrio cholerae* O1 strain from Mexico identical to United States isolates. *Lancet* **ii**:912.

5. **Blake, P. A., R. E. Weaver, and D. G. Hollis.** 1980. Diseases of humans (other than cholera) caused by vibrios. *Annu. Rev. Microbiol.* **34**:341–367.

6. **Busby, J.** 1991. Cholera: is Texas at risk? *Tex. Med.* **887**:64–65.

7. **Centers for Disease Control.** 1991. Cholera—Peru, 1991. *Morbid. Mortal. Weekly Rep.* **40**:108–109.

8. **Centers for Disease Control.** 1991. Importation of cholera from Peru. *Morbid. Mortal. Weekly Rep.* **40**:258–259.

9. **Centers for Disease Control.** 1991. Cholera—New York, 1991. *Morbid. Mortal. Weekly Rep.* **40**:516–518.

10. **Centers for Disease Control.** 1992. Isolation of *Vibrio cholerae* 01 from oysters—Mobile Bay, 1991–1992. *Morbid. Mortal. Weekly Rep.* **42**:91–93.

11. **Centers for Disease Control.** 1992. Cholera associated with an international airline flight, 1992. *Morbid. Mortal. Weekly Rep.* **41**:134–135.

12. **Centers for Disease Control.** 1992. Cholera associated with international travel, 1992. *Morbid. Mortal. Weekly Rep.* **41**:664–667.

12a. **Centers for Disease Control and Prevention.** Unpublished data.

13. **Chudy, J. P., E. Arneiella, and E. Gil.** 1993. Water quality assessment in Ecuador. WASH field report no. 390. Water and Sanitation for Health Project, Arlington, Va.

14. **Clemens, J. D., D. A. Sack, J. R. Harris, J. Chakraborty, M. R. Khan, S. Huda, F. Ahmed, J. Gomes, M. R. Rao, A.-M. Svennerholm, and J. Holmgren.** 1989. ABO blood groups and cholera: new observations on specificity of risk and modification of vaccine efficacy. *J. Infect. Dis.* **159**:770–773.

15. **Clemens, J. D., D. A. Sack, J. R. Harris, F. van L, J. Chakraborty, F. Ahmed, M. R. Rao, M. R. Klan, M. Yunus, N. Huda, B. F. Stanton, B. F. Kay, S. Walter, R. Eeckels, A.-M. Svennerholm, and J. Holmgren.** 1990. Field trial of oral cholera vaccines in Bangladesh: results from three-year follow-up. *Lancet* **335**:270–273.

16. **Coast Guard, U.S. Department of Transportation.** 1991. International Maritime Organization ballast water control guidelines. *Fed. Regist.* **56**:64831–64836.

17. **Deb, B. C., B. K. Sircar, P. G. Sengupta, S. P. De, S. K. Momdal, D. N. Gupta, N. C. Saha, S. Ghosh, U. Mitra, and S. C. Pal.** 1986. Studies on interventions to prevent eltor cholera transmission in urban slums. *Bull. W.H.O.* **64**:127–131.

18. **de Macedo, C. G.** 1991. Presentation of the PAHO regional plan, p. 39–44. *In Proceedings of the Conference: Confronting Cholera, the Development of a Hemispheric Response to the Epidemic.* North-South Center, University of Miami, Miami, Fla.

19. **DePaola, A., G. M. Capers, M. L. Motes, Ø. Olsvik, P. I. Fields, J. Wells, I. K. Wachsmuth,**

T. A. Cebula, W. H. Koch, F. Khambaty, M. H. Kothary, W. L. Payne, and B. A. Wentz. 1992. Isolation of Latin American epidemic strain of *Vibrio cholerae* O1 from US Gulf Coast. *Lancet* **339:**624.

20. **Finelli, L., D. Swerdlow, K. Mertz, H. Ragazzoni, and K. Spitalny.** 1992. Outbreak of cholera associated with crab brought from an area with epidemic disease. *J. Infect. Dis.* **166:**1433–1435.

21. **Fukuda, J., L. Chapparo, A. Yi, E. Chea, M. Campos, R. Martorrel, and E. Sanchez.** 1992. Low lethality with cholera in pediatric patients, abstr. 942, p. 267. *Program Abstr. 32nd Intersci. Conf. Antimicrob. Agents Chemother.*

22. **Gonzales, O., A. Aguilar, D. Antunez, and W. Levine.** 1992. An outbreak of cholera in rural Bolivia. Rapid identification of a major vehicle of transmission, abstr. 937, p. 266. *Program Abstr. 32nd Intersci. Conf. Antimicrob. Agents Chemother.*

23. **Goodgame, R. W., and W. B. Greenough III.** 1975. Cholera in Africa: a message for the West. *Ann. Intern. Med.* **82:**101–106.

24. **Granda, L. P., and M. C. Diaz de Juarez.** 1988. Procesamiento y calidad sanitario de la guitarra (Rhinobatus planiceps Garman, 1880) Deshidrata en la Bahia de Sechura-Piura, abstr. 336. *Program Abstr. (Libro de Resumen) IX Congr. Nac. Biol. "Peru: El reto hacia el ecodesarrollo."* Colegio de Biólogos del Peru, Piura, Peru.

25. **Instituto Nacional de Diagnóstico y Referencia Epidemiológicos.** 1991. Situación de cólera en México. *Bol. Quincenal Colera/Diarreas Infecc.* **1**(5):1.

26. **Kolvin, J. L., and D. Roberts.** 1982. Studies on the growth of *Vibrio cholerae* biotype eltor and biotype classical in foods. *J. Hyg.* (Cambridge) **89:**243–252.

27. **Levine, M. M., R. E. Black, M. L. Clements, L. Cisneros, A. Saah, D. R. Nalin, D. M. Gill, J. P. Craig, C. R. Young, and P. Ristaino.** 1982. The pathogenicity of nonenterotoxigenic *Vibrio cholerae* serogroup O1 biotype eltor isolated from sewage water in Brazil. *J. Infect. Dis.* **145:**296–299.

28. **MacKenzie, T. D., R. T. Ellison III, and S. R. Mostow.** 1992. Sunlight and cholera. *Lancet* **340:**367.

29. **Mata, L.** 1992. Efecto del jugo y de la pulpa de frutas ácidas sobre el *Vibrio cholerae*, p. 275. *In El Cólera: Historia, Prevención y Control.* Editorial Universidad Estatal a Distancia-Editorial de la Universidad de Costa Rica, San Jose, Costa Rica.

30. **McCarthy, S. A., R. M. McPhearson, and A. M. Guarino.** 1992. Toxigenic *Vibrio cholerae* O1 and cargo ships entering Gulf of Mexico. *Lancet* **339:**624.

31. **Mujica, O., R. Quick, A. M. Palacios, L. Biengolea, R. Vargas, D. Moreno, L. Seminario, N. Bean, and R. Tauxe.** 1992. Epidemic cholera in the Amazon: transmission and prevention by food, abstr. 936, p. 266. *Program Abstr. 32nd Intersci. Conf. Antimicrob. Agents Chemother.*

32. **Pan American Health Organization.** 1991. Environmental health conditions and cholera vulnerability in Latin America and the Caribbean. *Epidemiol. Bull.* **12:** 5–10.

33. **Pollitzer, R.** 1959. *Cholera,* p. 38–41. World Health Organization monograph no. 43. World Health Organization, Geneva.

34. **Quick, R. E., R. Vargas, D. Moreno, O. Mujica, L. Beingolea, A. M. Palacios, L. Seminario, and R. V. Tauxe.** 1993. Epidemic cholera in the Amazon: the challenge of preventing death. *Am. J. Trop. Med. Hyg.* **93:**597–602.

35. **Ries, A. A., D. J. Vugia, L. Beingolea, A. M. Palacios, E. Vasquez, J. G. Wells, N. G. Baca, D. L. Swerdlow, M. Pollack, N. H. Bean, L. Seminario, and R. V. Tauxe.** 1992. Cholera in Piura, Peru: a modern urban epidemic. *J. Infect. Dis.* **166:**1429–1433.

36. **Roberts, L.** 1990. Zebra mussel invasion threatens U.S. waters. *Science* **249:**1370–1372.

37. **Swerdlow, D. L., E. D. Mintz, M. Rodriguez, E. Tejada, C. Ocampo, L. Espejo, K. D. Greene, W. Saltana, L. Seminario, R. V. Tauxe, J. G. Wells, N. H. Bean, A. A. Ries, M. Pollack, B. Vertiz, and P. A. Blake.** 1992. Waterborne transmission of epidemic cholera in Trujillo, Peru: lessons for a continent at risk. *Lancet* **340:**28–32.

38. **Swerdlow, D. L., E. D. Mintz, M. Rodriguez, E. Tejada, C. Ocampo, L. Espejo, N. Bean, L. Seminario, T. Barrett, J. Pelzelt, R. Tauxe, and P. Blake.** 1992. Severe life-threatening cholera in Peru: predisposition for persons with blood group O, abstr. 941, p. 267. *Program Abstr. 32nd Intersci. Conf. Antimicrob. Agents Chemother.*

39. **Tamplin, M. L., and C. Carrillo Parodi.** 1991. Environmental spread of *Vibrio cholerae* in Peru. *Lancet* **338:**1216.

40. **Tauxe, R. V., S. D. Holmberg, A. Dodin, J. V. Wells, and P. A. Blake.** 1988. Epidemic cholera in Mali: high mortality and multiple routes of transmission in a famine area. *Epidemiol. Infect.* **100:**279–289.

41. **Tresierra, A. A., M. Z. Culquichicon, and A. T. Alvarado.** 1988. Situación social y económica de la pesqueria artesanal en la caleta constante (Piura, Peru), abstr. 338. *Program Abstr. (Libro de Resumen) IX Congr. Nac. Biol. "Peru: El recto hacia el ecodesarrollo."* Colegio de Biólogos del Peru, Piura, Peru.

42. **Wachsmuth, I. K., G. M. Evins, P. I. Fields, O. Olsvik, T. Popovic, C. A. Bopp, J. G. Wells, C.**

Carrillo, and P. A. Blake. 1993. The molecular epidemiology of cholera in Latin America. *J. Infect. Dis.* **167**:621–626.

43. **Weber, J. T., E. Mintz, R. Canizares, A. Semiglia, I. Gomez, R. Sempertegui, A. Davila, K. Greene, N. Puhr, D. Cameron, T. Barrett, and P. A. Blake.** 1992. Multiply-resistant epidemic cholera in Ecuador: transmission by water and seafood, abstr. 944, p. 268. *Program Abstr. 32nd Intersci. Antimicrob. Agents Chemother.*

44. **World Health Organization.** 1992. Cholera in 1991. *Weekly Epidemiol. Rec.* **67**:253–260.

45. **World Health Organization, Programme for Control of Diarrhoeal Diseases.** 1992. *Eighth Programme Report; 1990–1991.* World Health Organization, Geneva.

Vibrio cholerae and Cholera: Molecular to Global Perspectives
Edited by I. Kaye Wachsmuth, Paul A. Blake, and Ørjan Olsvik
© 1994 American Society for Microbiology, Washington, DC 20005

Chapter 22

Transmission of *Vibrio cholerae* O1

Eric D. Mintz, Tanja Popovic, and Paul A. Blake

I think the ways in which cholera can spread in a place are extremely diverse, and that, as almost every place has its own peculiar conditions, which must be thoroughly searched out, the measures which are of use for protecting the particular place from the pestilence must correspond to these conditions.

—Robert Koch, 1884 (38)

Infection with *Vibrio cholerae* O1 is thought to occur only in humans and only after ingestion of the bacterium (3). The tremendous number of bacteria excreted in the feces of infected persons—as many as 10^{13} *V. cholerae* O1 organisms per day (24)—can cause massive environmental pollution and provides the primary source of infection in cholera epidemics. In addition, there is strong evidence for long-term persistence of *V. cholerae* O1 in environmental reservoirs not directly contaminated by human feces (see chapter 20). Consumption of water or food from such reservoirs may put a person at risk for acquiring cholera.

BACKGROUND

Case-Control Studies

Most of our knowledge about the vehicles of cholera transmission stems from case-con-

trol investigations. Patients with cholera are interviewed about their food and drink exposures during the incubation period (1 to 5 days) that preceded the onset of their symptoms. Persons who did not become ill (controls), often matched to cases by sex, age group, and neighborhood, are interviewed with the same questionnaire about exposures during the same period. Often, laboratory evidence such as fecal cultures or vibriocidal antibody titers is used to better define cases and controls. Reported frequencies of the various suspect exposures among cases and controls are compared. Statistical tests can identify exposures that are reported significantly more frequently by patients than by controls, and such exposures are considered risk factors for illness. Conversely, exposures that are significantly more frequent among controls than among patients (e.g., drinking treated water) may be protective factors. Confirmation of epidemiologically implicated vehicles may be sought through culture for *V. cholerae* O1. However, it is difficult to evaluate the importance of sources from which *V. cholerae* O1 has been recovered without reasonable epidemiologic evidence regarding the role of these sources in transmission.

Molecular Epidemiologic Markers

Past investigations of cholera outbreaks were hindered by limited laboratory typing

Eric D. Mintz, Tanja Popovic, and Paul A. Blake • Foodborne and Diarrheal Diseases Branch, Division of Bacterial and Mycotic Diseases, Centers for Disease Control and Prevention, Atlanta, Georgia 30333.

systems that gave little strain discriminatory information and therefore may have been unable to differentiate an epidemic strain from other simultaneously present strains. Furthermore, a single biotype or serotype often persists for a prolonged period (30), hindering detailed epidemiologic dissection of smaller outbreaks within the pandemic or epidemic. However, molecular techniques such as ribotyping, multilocus enzyme electrophoresis, and pulsed-field gel electrophoresis that directly examine DNA or RNA can distinguish strains within the bio- and serotypes. Application of these molecular techniques to the analysis of strains causing the present Latin American epidemic has enhanced epidemiologic investigations and understanding of the sources and vehicles of transmission (2, 26, 76, 77).

RESERVOIRS

Cholera is exclusively a human disease, and no animal species has been found to be consistently infected (3). The primary source of infection in cholera epidemics is feces from persons acutely infected with *V. cholerae* O1 (Fig. 1). Chronic carriers are rare and are not known to play any role in cholera persistence. Free-living toxigenic strains of *V. cholerae* O1 have been recovered from remote rivers in northeast Australia and salty marshes along the U.S. Gulf Coast (4, 7, 20, 21, 47). These natural reservoirs are described in more detail in chapter 20. Similar environmental reservoirs are likely to exist in many other areas, such as estuaries in Bangladesh (22, 23, 29).

TRANSMISSION

Water

Early observations

In 1855, John Snow provided convincing epidemiologic evidence for the role of piped

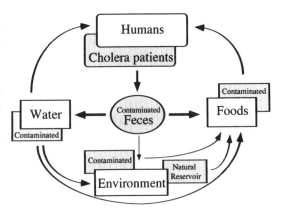

Figure 1. Major transmission routes of cholera. Reprinted with permission (Popovic et al., J. Food Prot. **56:**811–821, 1993).

Thames River water in the transmission of cholera during the London epidemic (61, 66). Nearly 30 years later, Robert Koch reported the isolation of the cholera bacillus from pond water during a cholera outbreak, lending bacteriologic plausibility to waterborne cholera transmission (38). Although improvements in municipal water systems were partially credited with the disappearance of epidemic cholera from Europe and the Americas at the turn of the century, water from various sources continues to be an important vehicle of cholera transmission (Table 1). Figure 2 illustrates the locations where waterborne transmission of cholera has been documented. These incidents are briefly discussed in the following paragraphs.

Seventh pandemic

More than a century after John Snow's seminal work, untreated river water remains a source of cholera transmission in Africa (64, 71) and Latin America (32, 50). Studies in these locations have demonstrated both the risk associated with drinking untreated river water and the protection afforded by various methods of water disinfection, including boiling (32, 50, 64), chlorination (64), and acidification with local fruit juice (50). In areas

Table 1. Water implicated by epidemiologic studies as vehicles of transmission of *V. cholerae* O1

Isolation site	Yr	Implicated water	Reference(s)
Seventh pandemic			
Portugal	1974	Commercially bottled uncarbonated spring water	6
Portugal	1974	Well water	5
Malawi	1990	Water stored in home, river water	71
Mali	1984	Well water	73
South Africa	1981	River water	64
Thailand	1980	Ice	48
Latin American epidemic			
Bolivia	1991	River water	32
Ecuador	1991	Street vendor drinks	78
Peru	1991	Street vendor drinks	58
Peru	1991	Municipal water, water stored in home	58, 72
Peru	1991	Ice	58

such as rural Bangladesh, where surface water is used for bathing, cooking, washing utensils, and other activities that may result in ingestion but where drinking water may be obtained from other sources, the role of surface water in cholera transmission has been difficult to evaluate (30).

Water from wells and springs can be contaminated with, support the survival of, and ultimately transmit *V. cholerae* O1. In the outbreak setting, *V. cholerae* O1 has been isolated from well water in Mali and spring water in Portugal; both of these sources were implicated by case-control studies (5, 73). In Portugal, cholera was transmitted by contaminated fresh and commercially bottled spring water. The investigation showed no association between cholera and drinking carbonated water of the same brand and from the same source as the uncarbonated spring water;

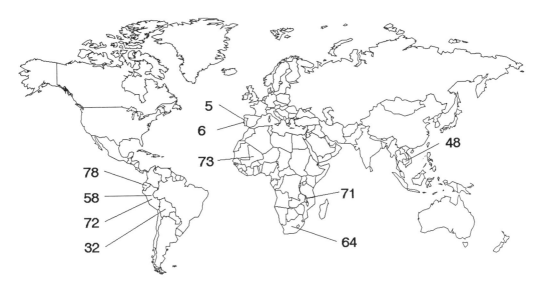

Figure 2. Water as a vehicle of transmission of *V. cholerae* O1 by country. Numbers correspond to reference numbers in Table 1.

these results and the laboratory observation that *V. cholerae* survives less than a day in carbonated water (25) suggest that carbonation of water is protective (6).

Water may become contaminated with *V. cholerae* O1 during household storage. This is particularly likely to occur when household drinking water is stored in wide-mouthed, open containers that allow entry of hands and other objects. Evidence supporting this conclusion was first demonstrated in Calcutta, where substitution of narrow-necked water storage vessels ("sorais") for wide-mouthed vessels was associated with a decrease in cholera infections among household members of cholera patients (18). Survival of *V. cholerae* O1 for up to 27 days in African domestic water storage containers was reported by Patel and Isaäcson in 1989 (54). In a refugee camp in Malawi, persons with cholera were more likely than controls to have used the household water storage container for hand washing (71). *V. cholerae* O1 is not destroyed by freezing (52), and ice was implicated as a vehicle of cholera transmission in Thailand (48).

Latin American epidemic

Municipal water systems that lack well-maintained facilities for water disinfection and distribution provide a ready means of disseminating *V. cholerae* O1 to a much larger population than is usually served by a single contaminated source. This situation contributed greatly to transmission of epidemic cholera in Peru (58, 72). In Trujillo, the second largest city in Peru, persons with cholera were much more likely than controls to have drunk unboiled municipal water. The municipal water system was not routinely chlorinated, and contamination could have been introduced at many points: fecal coliform bacteria were detected in water samples from wells that supplied the system, antiquated distribution lines were riddled with breaks and clandestine connections, and back-siphonage of contaminants was made possible

through low and intermittent water pressure. Fecal coliform counts were higher in water samples from the distribution system than from source wells, and *V. cholerae* O1 was isolated from water samples from the distribution system (72).

In Piura, Peru, drinking unboiled water from the municipal water system was also associated with cholera. As in Trujillo, the wells in Piura were contaminated, chlorination was not routinely practiced, and the distribution system suffered from breaks and inadequate water pressure (58).

In Peru and in much of the developing world, a variety of beverages are prepared from water and sold by street vendors. Beverages that are boiled during preparation, such as coffee and tea, are unlikely to transmit cholera. Juices and other beverages whose preparation does not require boiling are potential vehicles of cholera transmission if they are prepared from unsafe water. For example, in Guayaquil, Ecuador, persons with cholera were more likely than healthy controls to have purchased beverages from a street vendor. Orange juice, which may have contained water and ice, was the one beverage prepared by street vendors that was independently associated with illness (78). In Piura, Peru, beverages sold by street vendors were also implicated. Street beverages containing ice were specifically incriminated: street vendors in Piura made beverages from grains boiled in water to which sugar and ice were added after cooling. The ice was made from contaminated municipal water under poor hygienic conditions (58).

In Trujillo, where running water was available for only 1 or 2 h a day in many neighborhoods, families participating in the case-control study typically stored their drinking water in wide-mouthed containers. Water samples from household containers had much higher fecal coliform counts than water samples from the distribution system or from source wells. In both Trujillo and Piura, persons with cholera were more likely than controls to have drunk water from a household container into which hands had reportedly been introduced (58, 72). These findings suggest that a simple, inexpensive intervention—use of household water storage vessels like the "sorai" used in Calcutta, into which hands and other objects cannot be introduced—could reduce waterborne transmission of cholera in many areas.

Food

Survival and growth of *V. cholerae* O1 in food are enhanced by low temperatures, high organic content, near-neutral pH, high moisture, and absence of competing flora (19, 52, 53, 65). *V. cholerae* O1 survives for between 2 and 14 days on most foods, and survival is increased when foods are cooked before contamination (39, 53). Cooking eliminates competing organisms, destroys some heat-labile growth inhibitors, and produces denatured proteins that *V. cholerae* O1 uses for growth. The El Tor biotype appears better adapted to foodborne transmission than the classical biotype; El Tor *V. cholerae* O1 multiplies more rapidly and survives longer than classical strains on many foods (39). This environmental advantage may explain in part why the El Tor biotype has largely displaced the classical biotype and has enjoyed such widespread microbial success during the current pandemic. Documented instances of foodborne transmission of cholera are considered in the following paragraphs and are listed by location in Table 2 and Fig. 3.

Seafood

Fish and shellfish have long been recognized as frequent vehicles of cholera transmission (57). They may become contaminated by environmentally persistent *V. cholerae* O1 or after contamination of the aquatic environment by untreated human feces. In many parts of the world, fish and shellfish are consumed raw, a situation that favors transmission of *V. cholerae* O1 and other pathogens.

V. cholerae O1 produces chitinase, an enzyme that dissolves chitin, a component of the exterior skeletons of copepods (zooplankton). It has been suggested that the gastrointestinal tracts of mollusks, crustaceans, and fish may be colonized by *V. cholerae* O1 as a result of copepod ingestion by these animals (17). In any case, *V. cholerae* O1 persists for many weeks in crabs and mollusks, whether free-living or refrigerated (19, 43), and multiplies rapidly in shellfish that are left at ambient temperature (39, 43). In vitro, *V. cholerae* O1 adheres to chitin, and survival of chitin-adsorbed *V. cholerae* in a dilute hydrochloric acid solution of approximately the same pH as human gastric acid is prolonged (51). Chitin is also a component of crustacean shells and may protect adsorbed *V. cholerae* O1 from destruction by heat as well as by gastric acid. Crabs boiled for less than 10 min or steamed for less than 30 min may still harbor viable *V. cholerae* O1 organisms (4). The locations where documented transmission of cholera by seafood has occurred are illustrated in Fig. 3, and these incidents are briefly discussed in the following paragraphs.

In the Pacific Islands and the Asian subcontinent, where seafood is a major component of the diet, outbreaks of cholera have been attributed to consumption of raw shrimp (36), raw shellfish (44), raw fresh fish (70), salted raw fish (45), partially dried salt fish (44), and cooked squid (31).

In Europe, mussels (1), cockles (5), and other shellfish (60) have been implicated as vehicles. In Guinea-Bissau, West Africa, cooked crabs were the apparent source of a large outbreak (62). Oysters, shrimp, and crabs harvested along the Gulf Coast of the United States have engendered numerous episodes of *V. cholerae* O1 transmission (9, 28, 37, 40, 41, 55, 63, 79). (These are discussed in detail in chapter 20.)

Shortly after cholera was detected in Peru, the Minister of Health of that nation advised the public to avoid eating raw seafood, including the highly popular dish ceviche, a

Table 2. Foods implicated by epidemiologic studies as vehicles of transmission of *V. cholerae* O1

Isolation site (country of origin[a])	Yr	Implicated food	Reference
Seventh pandemic			
Australia	1972	Hors d'oeuvres	68
Italy	1973	Raw shellfish (mussels)	1
Portugal	1974	Raw and undercooked shellfish (cockles)	5
Guam	1974	Home preserved padas (salted raw fish)	45
Gilbert Islands	1977	Raw and salt fish and clams	44
Singapore	1982	Sambal sotong (cooked squid)	31
Truk	1982	Food prepared by ill food handlers	34
Mali	1984	Millet gruel	73
Guinea	1986	Leftover cooked rice	67
Thailand	1987	Laembo (raw pork)	69
Guinea-Bissau	1987	Cooked crabs	62
Malawi	1990	Cooked pigeon peas	71
Chuuk (formerly Truk)	1990	Raw fish	70
Maryland (Thailand)	1991	Frozen coconut milk	74
Latin American epidemic			
Ecuador	1991	Seafood	78
Peru	1991	Cooked rice	58
Peru	1991	Raw vegetables and fruit	50
Peru	1991	Street vendor food	58
Associated with Latin American epidemic			
New Jersey (Ecuador)	1991	Cooked crabs in cold salad	11
New York (Ecuador)	1991	Cooked crabs in cold salad	12
California (Peru)	1992	Shrimp and fish in cold salad	13
Associated with U.S. Gulf Coast			
Louisiana	1978	Cooked crabs	4
Texas	1981	Cooked rice	35
Louisiana	1986	Cooked crab, cooked or raw shrimp	41
Colorado and five other states	1988	Raw oysters	9

[a]If different from site of isolation.

cocktail of raw fish or shellfish marinated in lemon juice. However, although *V. cholerae* O1 was later isolated from fish and shellfish in Peru, very few persons enrolled in the case-control studies conducted there reported eating raw seafood, so its role in cholera transmission could not be defined (58, 72). Ceviche is popular in other Latin American countries, and questions regarding its safety have led to considerable debate. Although laboratory investigations suggest that ceviche prepared with adequately acid juice and marinated for a sufficient period may be safe to eat (42), contamination of fish or shellfish parts that are not exposed to acid, marination for too little time, or marination in juice that is not sufficiently acidic may allow *V. cholerae* O1 to survive. In Ecuador, where warnings about ceviche were less strident, persons with cholera were more likely than uninfected controls to report consumption of raw fish, raw seafood, and cooked crabs (78). Crabs transported from Ecuador have been associated with outbreaks of cholera in the United States on two occasions (11, 12, 27). In both instances, the crab was purchased and boiled in Guayaquil, brought to the United States in personal baggage, and served cold at small parties more than 24 h later. *V. cholerae* O1 may have survived the initial boiling or may have been subsequently introduced during handling.

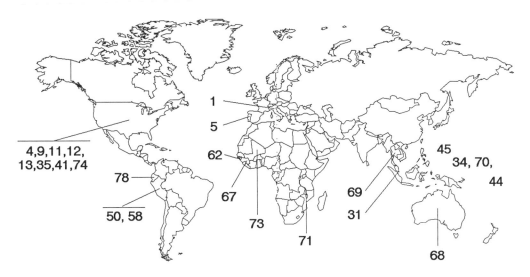

Figure 3. Food as a vehicle of transmission of *V. cholerae* O1 by country. Numbers correspond to reference numbers in Table 2.

A history of seafood consumption is often obtained from travelers who acquire cholera during visits to areas where cholera is endemic (10, 11, 14). In February 1992, cold seafood salad prepared in Lima, Peru, and served to passengers on board a flight from Buenos Aires, Argentina, through Lima to Los Angeles, California, was incriminated in an outbreak of cholera affecting over 75 airline passengers (13).

Other foods

Cooked grains and legumes support the growth of *V. cholerae* O1 quite well (39) and have been repeatedly implicated in cholera transmission. In Africa, cooked grains and legumes are a dietary staple (8), and several studies suggest that they contributed to the spread of epidemic cholera. Cooked pigeon peas kept overnight and eaten the following morning without being reheated were implicated in a case-control study in a refugee camp in Malawi (71). In a village in Mali, consumption of leftover steamed millet gruel was reported more frequently by persons with cholera than by well persons (73). Further,

adding curdled goat's milk (an acid food) to gruel gave protection against cholera. Tauxe et al. (73) demonstrated survival of *V. cholerae* O1 for 24 h in millet gruel that had been experimentally prepared and innoculated. *V. cholerae* O1 survival was decreased in the presence of curdled goat's milk.

An investigation in Guinea further defined the role of sauces served with cooked grains (67). In Guinea, boiled rice is typically prepared in the morning, held unrefrigerated, and eaten with sauce at the noonday meal and in the evening. Often the rice is not reheated before consumption. In a case-control study in Conakry, tomato sauces with a pH of 4.5 to 5.0 were protective against illness; less acidic sauces (pH 6.0 to 7.0) prepared from ground peanuts were associated with illness. Both types of sauce were subsequently prepared in the laboratory, inoculated with 10^4 *V. cholerae* O1 organisms per g, and held warm for 10 h. In the more alkaline ground-peanut sauce, *V. cholerae* O1 had multiplied to over 10^7 g, but fewer than 10^4 *V. cholerae* O1 per g were recovered from the tomato sauce.

The investigation in Guinea also clarified the association between funerals and cholera transmission in Africa (67). Previously, cholera out-breaks following funerals were attributed to person-to-person transmission (75). In Guinea, cooked rice was implicated in a case-control study of a village-wide cholera outbreak following a funeral. The rice had been prepared by women after they had cleaned the bedsheets and body of the index cholera victim. Later, the rice was served to persons who attended the funeral. A case-control study in Guinea-Bissau also implicated cooked rice as a vehicle of cholera transmission at a funeral (62).

Cooked rice was responsible for the largest outbreak and the only documented instance of secondary transmission of cholera in the United States in this century (35). This occurred on a Gulf Coast oil rig in 1981, when cooked rice was moistened with water contaminated by human feces and then held warm for 8 h after cooking (35). Sixteen *V. cholerae* O1 infections (one asymptomatic) were detected.

Rice that had been cooked and held for three or more hours before being eaten was implicated as a vehicle of *V. cholerae* O1 transmission in a case-control study in Piura, Peru (58). It is likely that many other grains, legumes, and staple foods consumed in Latin America can transmit cholera under appropriate conditions.

Meat supports the growth of *V. cholerae* O1 (39), and two instances of cholera transmission by meat have been documented. Both occurred in Thailand and involved meat that was held for hours at ambient temperature before being served raw (69, 69a).

The first commercially imported product known to have caused a cholera outbreak, frozen coconut milk, was imported from Thailand (74). Four persons in Maryland became infected after eating a rice pudding dessert topped with contaminated frozen coconut milk. Toxigenic *V. cholerae* O1 was isolated from an unopened package of the same brand of frozen coconut milk, and the product was recalled.

V. cholerae O1 survives for a number of days on certain fruits and vegetables (25, 53), but its survival is greatly decreased in a dry environment or in an environment with a pH of less than 4.5, as in citrus fruits. In many areas, water used for irrigation of fruits and vegetables is contaminated with human feces. Produce that grows close to the ground and is eaten raw, such as lettuce and melons, could easily be contaminated by irrigation water and might subsequently transmit *V. cholerae* O1. Contaminated water may also be used for moistening or cleaning fruits and vegetables or may be injected into larger fruits to increase their weight or turgor (22).

An outbreak in Israel in 1970 was attributed to raw vegetables irrigated with inadequately treated wastewater. *V. cholerae* O1 was isolated from irrigation water, but a case-control study was not conducted, so the epidemiologic evidence implicating raw vegetables is open to question (16). In Chile, vegetables watered with sewage-laden river water were thought to have caused an outbreak of cholera, but epidemiologic data have not been published (80).

Two case-control studies suggest that fruits and vegetables may have played a role in the spread of epidemic cholera in Peru (50, 72). In Iquitos, on the headwaters of the Amazon River, consumption of unwashed fruits and vegetables was reported more often by persons with cholera than by well persons. River water near Iquitos was contaminated with *V. cholerae* O1; produce brought in and out of the city by canoe could easily have become contaminated (50). A borderline association between eating cabbage and illness was found in Trujillo, where raw sewage is used for irrigation and cabbage is cooked only briefly (72).

Even epidemiologic investigations that fail to identify specific food items as vehicles of cholera transmission may bring to light certain high-risk situations in which food is served or prepared. In Chuuk (formerly Truk), a detailed study of intrafamilial transmission of cholera found higher attack rates in homes where the index case was a food handler (34). This suggests that foodborne transmission played an important role in the secondary "domestic" spread of cholera and that interventions designed to improve hygienic conditions and food preparation practices in homes may reduce transmission. In Piura, Peru, foods prepared by street vendors were associated with illness (58). Again, specific interventions, such as education and regulation of street vendors who prepare and serve foods, may reduce cholera transmission.

Other

Direct person-to-person transmission of cholera is not expected, because the infectious dose is high. Although numerous published articles refer to "person-to-person" transmission of cholera, spread of cholera from one person to another by contact has never been demonstrated by rigorous scientific studies; the evidence for such transmission consists of anecdotes for which alternative explanations are possible (15, 46, 49, 56). Some of the confusion in the literature on person-to-person transmission can be attributed to differing definitions of the term, with some publications using it only for transmission by contact, some including human fecal contamination of food or water, and some not defining the term at all.

A careful attempt to show a statistical association between close personal contact and cholera transmission occurred in Chuuk, Micronesia (34). There, Holmberg et al. showed that in a setting in which intrafamilial cholera transmission was occurring, persons who cared for cholera patients in the household and thus had the most contact with cholera patients did not have a significantly increased risk of infection with *V. cholerae* O1. Rather, there was evidence that cholera transmission was occurring through food contaminated in the home by the food handler; the attack rates were higher when the index patient was a food handler. Similarly, as was described above in the section on foodborne cholera transmission, the alleged person-to-person transmission at funerals in Africa has been shown to be foodborne in the two instances

that have been examined epidemiologically (62, 67).

The evidence for transmission of cholera by fomites (e.g., on clothes or bedsheets) and by flies has been reviewed recently by Glass and Black (30). Although *V. cholerae* O1 has been isolated from flies (33) and theoretically these flies could contaminate food in which the vibrios could multiply and reach an infective dose, flies have never been shown epidemiologically to play an important role in transmission of *V. cholerae* O1. Regarding fomites, one intriguing study in Bangladesh showed that cholera was more likely to occur in families with asymptomatically infected breast-feeding children in the household. The authors of that study speculated that a cloth pad used to absorb such children's feces might play a role in transmission (59). Small children were carried about naked with a cloth pad held against their buttocks to absorb feces, and the soiled cloth was sometimes folded and used several times before being washed, permitting contamination of the hands of the caregiver.

CONCLUSION

Transmission of cholera can occur through a wide variety of food and water vehicles, and sometimes the vehicles have been items that were thought to be highly unlikely at the time, such as cooked crabs in Louisiana or bottled water in Portugal. Furthermore, the important vehicles of transmission and thus the possible preventive measures can vary markedly from one place to another, being affected by local customs and practices as well as local conditions; in Mali, adding fermented goat's milk to millet gruel can be an important preventive measure, while in the headwaters of the Amazon, adding toronja juice to drinking water decreases the risk of cholera. Although what we have already learned about cholera enables us to make general recommendations about cholera prevention and control, it is not humanly possi-

ble to follow all of the recommendations indefinitely with complete rigor. Rather, it will
be important to continue to examine the
modes and vehicles of transmission of cholera and the possible protective factors in order to help officials focus their prevention
and control efforts on measures that are most
important locally.

REFERENCES

1. **Baine, W. B., M. Mazzotti, D. Greco, E. Izzo, A. Zampieri, G. Angioni, M. D. Gioia, E. J. Gangarosa, and F. Pocchiari.** 1974. Epidemiology of cholera in Italy in 1973. *Lancet* **ii:**1370–1381.

2. **Barrett, T. J., D. N. Cameron, and I. K. Wachsmuth.** 1992. Molecular typing of *V. cholerae* O1 by pulsed-field gel electrophoresis, p. 453. *Abstr. 92nd Gen. Meet. Am. Soc. Microbiol. 1992.*

3. **Benenson, A.** 1991. Cholera, p. 207–225. *In* A. S. Evans and P. S. Brachman (ed.), *Bacterial Infections of Humans.* Plenum Medical Book Co., New York.

4. **Blake, P. A., D. T. Allegra, J. D. Snyder, T. J. Barrett, L. McFarland, C. T. Caraway, J. C. Feeley, J. P. Craig, J. V. Lee, N. D. Puhr, and R. A. Feldman.** 1980. Cholera—a possible endemic focus in the United States. *N. Eng. J. Med.* **302:**305–309.

5. **Blake, P. A., M. L. Rosenberg, J. B. Costa, P. S. Ferreira, C. L. Guimaraes, and E. J. Gangarosa.** 1977. Cholera in Portugal, 1974. I. Modes of transmission. *Am. J. Epidemiol.* **105:**337–343.

6. **Blake, P. A., M. L. Rosenberg, J. Florenicia, J. B. Costa, L. P. Quintino, and E. J. Gangarosa.** 1977. Cholera in Portugal, 1974. II. Transmission by bottled mineral water. *Am. J. Epidemiol.* **105:**344–348.

7. **Bourke, A. T. C., Y. N. Cossins, B. R. W. Gray, T. J. Lunney, N. A. Rostron, R. V. Holmes, E. R. Griggs, D. J. Larsen, and V. R. Kelk.** 1986. Investigation of cholera acquired from the riverine environment in Queensland. *Med. J. Aust.* **144:**229–234.

8. **Cameron, M., and Y. Hofvander.** 1976. *Manual on Feeding Infants and Young Children,* 2nd ed. United Nations, New York.

9. **Centers for Disease Control.** 1989. Toxigenic *Vibrio cholerae* O1 infection acquired in Colorado. *Morbid. Mortal. Weekly Rep.* **38:**19–20.

10. **Centers for Disease Control.** 1991. Importation of cholera from Peru. *Morbid. Mortal. Weekly Rep.* **40:**258–259.

11. **Centers for Disease Control.** 1991. Cholera—New Jersey and Florida. *Morbid. Mortal. Weekly Rep.* **40:**287–289.

12. **Centers for Disease Control.** 1991. Cholera—New York. *Morbid. Mortal. Weekly Rep.* **40:**516–518.

13. **Centers for Disease Control.** 1992. Cholera associated with an international airline flight, 1992. *Morbid. Mortal. Weekly Rep.* **41:**134–135.

14. **Centers for Disease Control.** 1992. Cholera associated with international travel, 1992. *Morbid. Mortal. Weekly Rep.* **41:**664–667.

15. **Cliff, J. L., P. Zinkin, and A. Martelli.** A hospital outbreak of cholera in Maputo, Mozambique. 1986. *Trans. R. Soc. Trop. Med. Hyg.* **80:**473–476.

16. **Cohen, J., T. Schwartz, R. Klasmer, D. Pridan, H. Galayini, and A. M. Davies.** 1971. Epidemiological aspects of cholera El Tor outbreak in a nonendemic area. *Lancet* **ii:**86–89.

17. **Colwell, R. R., and W. M. Spira.** 1992. The ecology of *Vibrio cholerae,* p. 107–127. *In* D. Barua and W. B. Greenough III (ed.), *Cholera.* Plenum Medical Book Co., New York.

18. **Deb, B. C., B. C. Sircar, P. G. Sengupta, S. P. De, S. K. Mondal, D. N. Gupta, N. C. Saha, S. Ghosh, U. Mitra, and S. C. Pal.** 1986. Studies on interventions to prevent El Tor cholera transmission in urban slums. *Bull. W.H.O.* **64:**127–131.

19. **DePaola, A.** 1981. *Vibrio cholerae* in marine foods and environmental waters: literature review. *J. Food Sci.* **46:**66–70.

20. **Desmarchelier, P. M., and C. R. Senn.** 1989. A molecular epidemiological study of *V. cholerae* in Australia. *Med. J. Aust.* **150:**631–634.

21. **Epstein, P. R.** 1992. Cholera and the environment. *Lancet* **339:**1167–1168.

22. **Feachem, R. G.** 1981. Environmental aspects of cholera epidemiology. I. A review of selected reports of endemic and epidemic situations during 1961–1980. *Trop. Dis. Bull.* **78:**675–698.

23. **Feachem, R. G.** 1981. Environmental aspects of cholera epidemiology. II. Occurrence and survival of *Vibrio cholerae* in the environment. *Trop. Dis. Bull.* **78:**865–880.

24. **Feachem, R. G.** 1982. Environmental aspects of cholera epidemiology. III. Transmission and control. *Trop. Dis. Bull.* **79:**1–47.

25. **Felsenfeld, O.** 1967. *The Cholera Problem.* Warren H. Green Inc., St. Louis.

26. **Fields, P. I., T. Popovic, K. Wachsmuth, and Ø. Olsvik.** 1992. Use of polymerase chain reaction for detection of toxigenic *Vibrio cholerae* O1 strains from the Latin American cholera epidemic. *J. Clin. Microbiol.* **30:**2118–2121.

27. **Finelli, L., D. Swerdlow, K. Mertz, H. Ragazzoni, and K. Spitalney.** 1992. Outbreak of cholera associated with crab brought from an area with epidemic disease. *J. Infect. Dis.* **166:**1433–1435.

28. **Gergatz, S. J., and L. M. McPharland.** 1989. Cholera on the Louisiana Gulf Coast. Historical notes and case report. *J. La. State Med. Soc.* **141:**29–34.

29. **Glass, R. I., S. Becker, M. I. Huq, B. J. Stoll, M. U. Khan, M. H. Merson, J. V. Lee, and R. E.**

Black. 1982. Endemic cholera in rural Bangladesh. *Am. J. Epidemiol.* **116:**959–970.

30. Glass, R. I., and R. E. Black. 1992. Epidemiology of cholera, p. 129–154. *In* D. Barua and W. B. Greenough III (ed.), *Cholera.* Plenum Medical Book Co., New York.

31. Goh, K. T., S. Lam, S. Kumarapathy, and J. L. Tan. 1984. A common source foodborne outbreak of cholera in Singapore. *Int. J. Epidemiol.* **13:**210–215.

32. Gonzales, O., A. Anquilar, D. Antunez, and W. Levine. 1992. An outbreak of cholera in rural Bolivia. Rapid identification of a major vehicle of transmission, abstr. 937, p. 266. *Program Abstr. 32nd Intersci. Conf. Antimicrob. Agents Chemother.*

33. Greenberg, B. 1973. Cholera, p. 167–173. *In Flies and Disease,* vol. II. *Biology and Disease Transmission.* Princeton University Press, Princeton, N.J.

34. Holmberg, S. D., J. R. Harris, D. E. Kay, N. T. Hagrett, R. D. R. Parker, N. Kansou, N. U. Rao, and P. A. Blake. 1984. Foodborne transmission of cholera in Micronesian households. *Lancet* **i:**325–328.

35. Johnston, J. M., D. L. Martin, J. Perdue, L. M. McFarland, C. T. Caraway, E. C. Lippy, and P. A. Blake. 1983. Cholera on a Gulf Coast oil rig. *N. Engl. J. Med.* **309:**523–526.

36. Joseph, P. R., F. Tamayo, W. H. Mosley, M. G. Alvero, J. J. Dizon, and D. A. Henderson. 1965. Studies of cholera El Tor in the Philippines. 2. A retrospective investigation of an explosive outbreak in Bacolod City and Talisay, November 1961. *Bull. W.H.O.* **33:**637–643.

37. Klonz, K. C., R. V. Tauxe, W. L. Cook, W. H. Riley, and K. Wachsmuth. 1987. Cholera after consumption of raw oysters. *Ann. Intern Med.* **107:**846–848.

38. Koch, R. 1884. An address on cholera and its bacillus. *Br. Med. J.* **2:**403–407, 453–459.

39. Kolvin, J. L., and D. Roberts. 1982. Studies on the growth of *Vibrio cholerae* biotype El Tor and biotype classical in foods. *J. Hyg.* (Cambridge) **89:**243–252.

40. Lin, F. C., G. Morris, Jr., J. B. Kaper, T. Gross, J. Michalski, C. Morrison, J. P. Libonati, and E. Israel. 1986. Persistence of cholera in the United States: isolation of *V. cholerae* O1 from a patient with diarrhea in Maryland. *J. Clin. Microbiol.* **23:**624–626.

41. Lowry, P. W., A. T. Pavia, L. M. McFarland, B. M. Peltier, T. J. Barrett, H. B. Bradford, J. M. Quan, J. Lynch, J. B. Mathison, R. A. Gunn, and P. A. Blake. 1989. Cholera in Louisiana—widening spectrum of seafood vehicles. *Arch. Intern. Med.* **149:**2079–2084.

42. Mata, L. 1992. Efecto del jugo y de la pulpa de frutas ácidas sobre el *Vibrio cholerae*, p. 275. *In El Cólera: historia, prevención, y control.* Editorial Universidad Estal a Distancia—Editorial de la Universidad de Costa Rica, San José, Costa Rica.

43. McCormack, W. M., M. S. Islam, M. Fahimuddin, and W. H. Mosley. 1969. Endemic cholera in rural East Pakistan. *Am. J. Epidemiol.* **89:**393–404.

44. McIntyre, R. C., T. Tira, T. Flood, and P. A. Blake. 1979. Modes of transmission of cholera in a newly infected population on an atoll: implications for control measures. *Lancet* **i:**311–314.

45. Merson, M. H., W. T. Martin, J. P. Craig, G. K. Morris, P. A. Blake, G. F. Craun, J. C. Feeley, J. C. Camacho, and E. J. Gangarosa. 1977. Cholera on Guam 1974. Epidemiologic findings and isolation of non-toxigenic strains. *Am. J. Epidemiol.* **105:**349–361.

46. Mhalu, F. S., F. D. E. Mtango, and A. E. Msengi. 1984. Hospital outbreaks of cholera transmitted through close person-to-person contact. *Lancet* **ii:**82–84.

47. Miller, C. J., R. G. Feachem, and B. S. Drasar. 1985. Cholera epidemiology in developed and developing countries: new thoughts on transmission, seasonality and control. *Lancet* **i:**261–263.

48. Morris, J. G., G. R. West, S. E. Holck, P. A. Blake, P. D. Echeverria, and M. Karaulnik. 1982. Cholera among refugees in Rangsit, Thailand. *J. Infect. Dis.* **145:**131–133.

49. Mosley, W. H., M. G. Alvero, P. R. Joseph, J. F. Tamayo, C. Z. Gomez, T. Montague, J. J. Dizon, and D. A. Henderson. 1965. Studies of cholera El Tor in the Philippines. 4. Transmission of infection among neighbourhood and community contacts of cholera patients. *Bull. W.H.O.* **33:**651–660.

50. Mujica, O., R. Quick, A. Palacios, L. Beingolea, R. Vargas, D. Moreno, L. Seminario, N. Bean, and R. Tauxe. 1992. Epidemic cholera in the Amazon: transmission and prevention by food, abstr. 936, p. 266. *Program Abstr. 32nd Intersci. Conf. Antimicrob. Agents Chemother.*

51. Nalin, D., V. Daya, A. Reid, M. M. Levine, and L. Cisneros. 1979. Adsorption and growth of *Vibrio cholerae* on chitin. *Infect. Immun.* **25:**768–770.

52. Pan American Health Organization. 1991. Cholera situation in the Americas. *Bull. Pan Am. Health Organ.* **12:**10–24.

53. Pan American Health Organization. 1991. Risk of cholera transmission by foods. *Bull. Pan Am. Health Organ.* **25:**274–276.

54. Patel, M., and M. Isaäcson. 1989. Survival of *Vibrio cholerae* in African domestic water storage containers. *South Afr. Med. J.* **76S:**365–367.

55. Pavia, A. T., J. F. Cambell, P. A. Blake, J. D. Smith, T. W. McKinley, and D. L. Martin. 1987. Cholera from raw oysters shipped interstate. *JAMA* **258:**2374.

56. Philippines Cholera Committee. 1970. Study on the transmission of El Tor cholera during an outbreak in Can-Itom community in the Philippines. *Bull. W.H.O.* **43:**413–419.

57. Pollitzer, R. 1959. *Cholera.* Monograph no. 43. World Health Organization, Geneva.

58. Ries, A. A., D. J. Vugia, L. Beingolea, A. M. Palacios, E. Vasquez, J. G. Wells, N. G. Baca, D. L. Swerdlow, M. Pollack, N. H. Bean, L. Seminario, and R. V. Tauxe. 1992. Cholera in Piura, Peru: a modern urban epidemic. *J. Infect. Dis.* **166:**1429–1433.

59. Riley, L. W., S. H. Waterman, A. S. G. Faruque, and M. I. Huq. 1987. Breast-feeding children in the household as a risk factor for cholera in rural Bangladesh: an hypothesis. *Trop. Geogr. Med.* **39:**9–14.

60. Salamaso, S., D. Greco, B. Bonfiglio, M. Castellani-Pastoris, G. De Felip, A. Bracciotti, G. Sitzia, A. Congiu, G. Piu, G. Angioni, L. Barra, A. Zampieri, and W. B. Baine. 1980. Recurrence of pelecypod-associated cholera in Sardinia. *Lancet* **ii:**1124–1128.

61. Schoenberg, B. E. 1974. Snow on the water of London. *Mayo Clin. Proc.* **49:**680–684.

62. Shaffer, N., P. Mendes, C. M. Costa, E. Larson, P. A. Anderson, N. Bean, R. V. Tauxe, and P. A. Blake. 1988. Epidemic cholera in Guinea-Bissau: importance of foodborne transmission, abstr. 1459, p. 360. *Program Abstr. Intersci. Conf. Antimicrob. Agents Chemother.*

63. Shandera, W. X., B. Hafkin, D. L. Martin, J. P. Taylor, D. L. Maserang, J. G. Wells, M. Kelly, K. Ghandi, J. B. Kaper, J. V. Lee, and P. A. Blake. 1983. Persistence of cholera in the United States. *Am. J. Trop. Med. Hyg.* **32:**812–817.

64. Sinclair, G. S., M. Mphahlele, H. Duvenhage, R. Nichol, A. Whitehorn, and H. G. V. Kustner. 1982. Determination of the mode of transmission of cholera in Lebowa. *S. Afr. Med. J.* **62:**753–755.

65. Singleton, F. L., R. Attwell, S. Jangi, and R. R. Colwell. 1982. Effects of temperature and salinity on *V. cholerae* growth. *Appl. Environ. Microbiol.* **44:**1047–1058.

66. Snow, J. 1936. *Snow on Cholera.* A reprint of two papers with biographical memoirs by B. W. Richardson and introduction by Wade Hampton Frost. Oxford University Press, London.

67. St. Louis, M. E., J. D. Porter, A. Helal, K. Drame, N. Hargrett-Bean, J. G. Wells, and R. V. Tauxe. 1990. Epidemic cholera in West Africa: the role of food handling and high-risk foods. *Am. J. Epidemiol.* **131:**719–727.

68. Sutton, R. G. 1974. An outbreak of cholera in Australia due to food served in flight on an international aircraft. *J. Hyg.* (Cambridge) **72:**441–451.

69. Swaddiwudhipong, W., P. Akarasewi, T. Chayaniyayodhin, P. Kunasol, and H. M. Foy. 1990. A cholera outbreak associated with eating uncooked pork in Thailand. *J. Diarrhoeal Dis. Res.* **8:**94–96.

69a. Swaddiwudhipong, W., R. Jirakanuisun, and A. Rodklai. 1992. A common source foodborne outbreak of El Tor cholera following the consumption of uncooked beef. *J. Med. Assoc. Thail.* **75:**413–416.

70. Swerdlow, D. L., P. Effler, K. Aniol, S. J. Manea, K. Iohp, K. D. Greene, C. A. Bopp, E. K. Pretrick, and P. A. Blake. 1991. Reemergence of cholera in Chuuk: a new endemic focus?, abstr. 529, p. 187. *Program Abstr. 31st Intersci. Conf. Antimicrob. Agents Chemother.*

71. Swerdlow, D. L., G. Malanga, G. Begokyan, A. Jonkman, M. Toole, R. V. Tauxe, R. Waldman, N. Puhr, and P. A. Blake. 1991. Epidemic of antimicrobial resistant *Vibrio cholerae* O1 infections in a refugee camp, Malawi, abstr. 529, p. 187. *Program Abstr. 31st Intersci. Conf. Antimicrob. Agents Chemother.*

72. Swerdlow, D. L., E. D. Mintz, M. Rodriguez, E. Tejada, C. Ocampo, L. Espejo, K. D. Greene, W. Saldana, L. Seminario, R. V. Tauxe, J. G. Wells, N. H. Bean, A. A. Ries, M. Pollack, B. Vertiz, and P. A. Blake. 1992. Waterborne tranmission of epidemic cholera in Trujillo, Peru: lessons for a continent at risk. *Lancet* **340:**28–32.

73. Tauxe, R. V., S. D. Holmberg, A. Dodin, J. G. Wells, and P. A. Blake. 1988. Epidemic cholera in Mali: high mortality and multiple routes of transmission in a famine area. *Epidemiol. Infect.* **100:**279–289.

74. Taylor, J. L., J. Tuttle, T. Pramukul, K. O'Brien, T. J. Barrett, B. Jolbitado, Y. L. Lim, D. Vugia, J. G. Morris, R. V. Tauxe, and D. M. Dwyer. 1993. An outbreak of cholera in Maryland associated with imported commercial frozen fresh coconut milk. *J. Infect. Dis.* **167:**1330–1335.

75. Viguelloux, J., and G. Gausse. 1971. Reflexions sur l'epidemiologie du cholera en Afrique occidental. *Med. Trop.* **31:**678–684.

76. Wachsmuth, I. K., C. A. Bopp, P. I. Fields, and C. Carillo. 1991. Difference between toxigenic *V. cholerae* O1 from South America and US Gulf Coast. *Lancet* **337:**1097–1098.

77. Wachsmuth, I. K., G. M. Evans, P. I. Fields, Ø. Olsvik, T. Popovic, J. G. Wells, C. Carillo, and P. A. Blake. 1993. The molecular epidemiology of cholera in Latin America. *J. Infect. Dis.* **167:**621–626.

78. Weber, J. T., E. D. Mintz, R. Canizares, A. Semiglia, I. Gomez, R. Sempertegui, A. Davila, K. D. Greene, N. D. Puhr, D. N. Cameron, T. Barrett, N. H. Bean, C. Ivey, R. V. Tauxe, and P. A. Blake. Epidemic cholera in Ecuador: multidrug-resistance and transmission by water and seafood. *Epidemiol. Infect.*, in press.

79. Weissman, J. B., W. E. DeWitt, J. Thompson, C. N. Muchnick, B. L. Portnoy, J. C. Feeley, and E. J. Gangarosa. 1975. A case of cholera in Texas, 1973. *Am. J. Epidemiol.* **100:**487–498.

80. World Health Organization. 1992. Cholera in the Americas. *Weekly Epidemiol. Rec.* **67:**33–38.

Vibrio cholerae and Cholera: Molecular to Global Perspectives
Edited by I. Kaye Wachsmuth, Paul A. Blake, and Ørjan Olsvik
© 1994 American Society for Microbiology, Washington, DC 20005

Chapter 23

Molecular Epidemiology of Cholera

Kaye Wachsmuth, Ørjan Olsvik, Gracia M. Evins, and Tanja Popovic

Epidemiologic surveillance of cholera has been limited in the past by rather insensitive laboratory typing systems. Traditionally, toxigenic *Vibrio cholerae* serogroup O1 strains, which are the major cause of epidemic cholera, were differentiated by bacteriologic methods into two serotypes, Inaba and Ogawa, and two biotypes, classical and El Tor. The use of only four subtypes provided little discrimination, particularly because these phenotypic traits have not been stable over time. The antigenic shift of a strain between the Inaba and Ogawa serotypes and the general shift of El Tor strains from hemolytic to nonhemolytic have been well documented (3, 37, 41, 43, 47). In this chapter, we describe the newer molecular subtyping approaches as they have evolved and been applied to investigations of cholera since the 1970s. We present specific applications of subtyping in outbreak and epidemic settings and in retrospective studies of well-defined culture collections of *V. cholerae* O1, including some preliminary information on the serogroup O139. We also present a larger global distribution and perspective of *V. cholerae* based on data from these investigations.

Our increasing knowledge of the molecular epidemiology of cholera has been stimulated by advances in technology and by unexpected occurrences of the disease from 1978 to 1993. The first outbreak of cholera in the United States in this century occurred in Louisiana in 1978; small outbreaks persist along the Gulf Coast to this day (6, 27, 28, 35, 49). The use of toxin gene probes together with the Southern blot hybridization technique indicated that the Gulf Coast isolates of toxigenic *V. cholerae* O1 were clonal and that they were different from other seventh-pandemic isolates (23). In 1982, the classical biotype of *V. cholerae* O1 appeared and replaced the El Tor biotype in outbreaks of cholera in Bangladesh (39). Again, molecular approaches to subtyping isolates indicated that the newly isolated classical strains were closely related to older isolates from the sixth pandemic; it seems that these strains were persisting undetected over decades in the Indian subcontinent (16). In 1991, when cholera invaded Latin America, molecular approaches were used to define the epidemic strain and show that it was related to seventh-pandemic isolates from other parts of the world (47). In 1992 and 1993, cholera epidemics caused by *V. cholerae* O139 occurred and are still occurring in India and Bangladesh, and molecular techniques are being used to determine the relationship of O139 and O1 isolates. Newer, more discriminating,

Kaye Wachsmuth, Ørjan Olsvik, Gracia M. Evins, and Tanja Popovic • Division of Bacterial and Mycotic Diseases, National Center for Infectious Diseases, Centers for Disease Control and Prevention, Atlanta, Georgia 30333.

and more informative molecular methods have been applied to answer epidemiologic questions; our general knowledge of the organism and the disease have expanded greatly as a consequence.

Although plasmid profile analysis was one of the first subtyping tools and has been an effective epidemic marker system for many diarrheal pathogens (48), it has been of limited value for *V. cholerae* O1; most seventh-pandemic isolates, which are generally fully susceptible to antimicrobial agents, appear to be plasmidless. One of the first effective subtyping methods applied to cholera was a Southern blot analysis of whole-cell DNA that used a gene probe derived from the A subunit of the heat-labile enterotoxin gene of *Escherichia coli* (23, 30), which hybridized at moderate stringency with the gene for cholera toxin (30). The genes for A subunit (*ctxA*) and B subunit (*ctxB*) have since then been sequenced, and a variety of gene probes targeting *ctxA* or *ctxB* have been used in molecular subtyping studies (6, 7, 16, 19, 25, 46; see also chapter 3). Although the bacterial chromosome contains only one or two copies of the cholera toxin genes, this technique detected epidemiologically useful differences in both the numbers and genomic locations of those genes.

Additional Southern blot approaches have used the genomic DNA of mutator vibriophages VcA1 and VcA3 and highly conserved rRNA genes as probes of whole-cell DNA restriction digests (2, 16, 22). Although the vibriophages can insert and delete without classical recombination, their restriction fragment length polymorphism (RFLP) patterns indicated that both VcA1 and VcA3 are relatively conserved in sequence and insertion sites. However, multiple copies of the internally conserved rRNA genes showed great diversity in their chromosomal locations, and a standardized ribotyping schema has been proposed for *V. cholerae* O1 (26, 33). More recently, pulsed-field gel electrophoresis (PFGE) has proven to be a very discriminating way to measure RFLPs, and it does not require the additional time, labor, and reagents necessary for hybridization (unpublished data).

Multilocus enzyme electrophoresis (MEE) is not only a useful epidemic marker system but also a way to measure strain relatedness and diversity that has been used extensively by population geneticists to study prokaryote and eukaryote species. Differences in the electrophoretic mobilities of essential cellular enzymes, which correlate with the occurrence of different alleles at the enzyme structural gene locus, have been used to show that several closely related genetic clones exist within the *V. cholerae* O1 serogroup.

A recent breakthrough in the evolution of molecular techniques is the advent of automated and rapid DNA sequencing that can be used to sequence and compare 600-bp gene sequences from 10 to 100 isolates in a few days. This approach has been used to study the *ctx* genes (31) and more recently the genes that determine the O1 antigen of *V. cholerae* (unpublished data). Until it can be applied to the entire bacterial chromosome, this approach is focused and can address most gene characteristics but not organism characteristics. Knowing the genetic basis for differences observed in cholera toxin and O1 antigenic shifts will probably be more important to questions of pathogenicity and protection than to epidemiologic studies.

In this chapter, we look at epidemiologically and phenotypically defined groups of *V. cholerae* isolates and discuss the use of their molecular genetic characteristics and relationships to address specific questions regarding the occurrence and spread of cholera, i.e., the molecular epidemiology of cholera.

CLASSICAL BIOTYPE

Although it is thought that the first six pandemics of cholera, which occurred between 1817 and 1925, were caused by the classical biotype of *V. cholerae* O1, the organism was not isolated until 1883, during the fifth pandemic. The sixth pandemic (beginning in 1899) was caused by the classical biotype (32),which began to disappear in the 1960s; it was eventually replaced by the El Tor biotype (21), which is causing the current (seventh) pandemic. One exception to this phenomenon occurred in Bangladesh, where the classical biotype was occasionally isolated from patients over the next 2 decades and emerged in epidemic form to replace the El Tor biotype in 1982 (42). This resurgence of the classical biotype raised scientific interest in its relationship to sixth-pandemic strains (39). The El Tor biotype was initially thought to survive better in the environment; however, if the 1982 isolates of the classical biotype were similar or identical to sixth-pandemic isolates, this theory would be challenged (32).

Cook et al. used plasmid profiles and Southern blot analysis with the *ctxA* probe and the VcA1 probe to show that 42 classical biotype strains isolated as early as 1916 and 1921 (sixth pandemic) were usually indistinguishable from the 1982 isolates (16). Unlike toxigenic El Tor isolates, which had no plasmid DNA, most classical isolates had 4.8- and 33.6-kb plasmids. The *ctxA* RFLP patterns, based on a *Hind*III restriction digest, showed that all classical biotype isolates had two copies of *ctx* and that the restriction fragments varied from 2 to 3 kb in length among strains. The 1982 isolates were indistinguishable from the 1933 isolates; a typical banding pattern is seen in lane 2 of Fig. 1. Robot-assisted DNA sequence analysis of a 460-bp polymerase chain reaction amplification product from *ctxB* indicated that classical biotype strains isolated from 1921 through 1970 had identical DNA sequences (31). The *ctxB* sequences of 45 *V. cholerae* O1 strains showed only three nucleotide differences among all strains, and there were three *ctxB* genotypes; the four classical biotype isolates belong to genotype 1 (31, 47; see also chapter 3).

Cook et al. also showed that all 42 classical biotype strains hybridized with the VcA1 probe and gave an array of 8 to 10 fragments, with an 8-fragment pattern common to all strains (16). That study concluded that there were only minor differences among classical isolates, which might be explained by the age, multiple subculture, and general treatment of these cultures over long periods. The classical biotype of *V. cholerae* O1, which caused the sixth pandemic, appeared to re-emerge in 1982, for reasons that have never been fully understood. Perhaps there is something unique about Bangladesh that offers this biotype an environmental advantage (42).

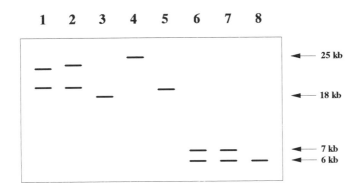

Figure 1. Schematic representation of numbers and sizes of chromosomal DNA *Hind*III fragments from human *V. cholerae* O1 strains containing *ctx* genes. Lanes: 1, classical, Japan, 1916; 2, classical, Bangladesh, 1982; 3, El Tor, Peru, 1991; 4, El Tor, Thailand, 1990; 5, El Tor, Rwanda, 1988; 6, El Tor, Louisiana, 1978; 7, El Tor, New Jersey (Cancun), 1983; 8, El Tor, Florida, 1986.

In agreement with the results of Cook et al. are the findings of Koblavi et al. (26), who developed an rRNA RFLP, or ribotyping, assay based on *BglI* cleavage of whole-cell DNA for studying a collection of 89 *V. cholerae* O1 isolates, 17 of which were of the classical biotype (26). They found that classical isolates fell into four ribotypes that were similar to one another but distinct from El Tor ribotypes. Six classical isolates from Bangladesh in the 1980s and one isolate from India in 1949 belonged to the same ribotype, B1. These authors concluded, as Cook et al. had previously, that "classical cholera in 1982 in Bangladesh is the reemergence of a strain involved in the sixth pandemic" (26). Eight isolates belonging to ribotype B2 again showed persistence over time (1942 to 1962). The remaining two isolates were ribotypes B3 and B4. Additionally, in a recent study by Faruque et al., 27 classical strains isolated in Bangladesh over a period of 30 years were of the same distinct ribotype (20).

In a similar ribotype study based on digestion with *BglI* endonuclease, Popovic et al. proposed a standardized ribotype scheme and extended the work of Koblavi et al. to include 16 classical biotype isolates that appeared clonal by plasmid profile, *ctxA* RFLP, and VcA1 RFLP (16, 33). They found that the 16 isolates belonged to a closely related ribotype group; i.e., they had an overall similar pattern. The strains did, however, belong to seven subgroups that have been labeled ribotypes 1a through 1g; the subgroup hybridization patterns differed by one or two bands of eight or nine general bands, as seen in Fig. 2. Six of the 16 isolates in that study belonged to ribotype 1b; they were isolated between 1921 and 1970 and were from Japan, Iran, Bangladesh, Hong Kong, India, and Pakistan. Again, this suggests that the classical biotype clone(s) was widely disseminated and persisted over decades.

In more recent studies using MEE (14, 19a), investigators have again shown that there is some diversity among the classical biotype strains but that the majority, includ-

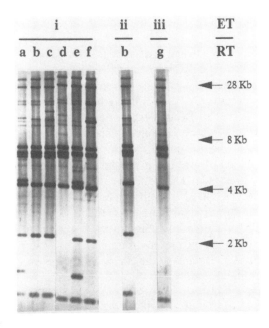

Figure 2. *BglI*-digested chromosomal DNA hybridized with digoxigenin-labeled 16S and 23S rRNAs from *E. coli*, showing relationships of ET and ribotype (RT) in classical human *V. cholerae* O1 strains. ET i, RT 1a, India, 1949; ET i, RT 1b, Pakistan, 1959; ET i, RT 1c, India, 1941; ET i, RT 1d, Bangladesh, 1972; ET i, RT 1e, India, 1960; ET i, RT 1f, India, 1940; ET ii, RT 1b, India, 1941; ET iii, RT 1g, India, 1960.

ing isolates from 1939 and 1982, belong to a single electrophoretic type (ET) (15). Data presented here for the first time show that 15 classical biotype isolates fall into three ETs, i through iii; this same group of isolates belongs to seven related ribotypes. Table 1 lists the 16 enzymes assayed and gives the relative migration differences that correspond to allelic differences among isolates. Figure 2 presents the relationship of ribotype to ET and generally emphasizes the sameness rather than the diversity of most classical biotype isolates. Ribotyping and MEE both detect diversity within conserved genes of the whole organism but measure entirely different cell parameters. Presumably, neutral selection mutations within highly conserved genes for the "housekeeping enzymes" must be significant enough to cause a structural change in

Table 1. Allele profiles of *V. cholerae* O1[a]

ET[b]	No. of isolates	THD	IDH	ADK	MDH	ME	G6P	GOT	LAP	PGI	ALD	LDH	GPT	DA1	DA2	DA3	NSP
i	13	1	1	1	1	1	1	2	2	2	1	2	2	1	1	1	1
ii	1	1	1	1	1	1	1	2	2	2	1	1	2	1	1	1	1
iii	1	1	1	1	1	1	1	2	2	2	1	0	1	1	1	1	1
1	5	1	1	1	1	1	1	1	1	1	1	2	2	2	1	1	1
2	29	1	1	1	1	1	1	1	2	1	1	2	2	2	1	1	1
3	63	1	1	1	1	1	1	1	1	1	1	2	2	3	1	1	1
4	102	1	1	1	1	1	1	1	3	1	1	2	2	3	1	1	1

[a]Abbreviations: THD, threonine dehydrogenase; IDH, isocitrate dehydrogenase; ADK, adenylate kinase; MDH, malate dehydrogenase; ME, malic enzyme; G6P, glucose 6-phosphate dehydrogenase; GOT, glutamic-oxalacetic transaminase; LAP, leucine aminopeptidase; PGI, phosphoglucose isomerase; ALD, alanine dehydrogenase; LDH, L-lactate dehydrogenase; GPT, glutamic-pyruvic transaminase; DA1, NADPH diaphorase 1; DA2, NADPH diaphorase 2; DA3, NADH diaphorase 3; NSP, nucleoside phosphorylase.
[b]i to iii, classical biotype; 1 to 4, El Tor biotype.

the enzyme that affects the enzyme's electrophoretic mobility in the MEE assay, while random and silent point mutations affecting the restriction enzyme target sites in DNA flanking the highly conserved genes for rRNA might occur more frequently.

In conclusion, the classical biotype isolates available for study from 1917 through 1983 belong to several very closely related groups, and isolates that are indistinguishable from one another might have persisted over decades in the Indian subcontinent, the Middle East, and the Far East. These subclones might once have derived from one pandemic strain that spread globally, or they might have persisted in separate niches throughout the world. Some of the diversity observed in these studies may simply reflect the ages of isolates and their treatment over time and subculture. No isolates from the North and South American epidemics before 1900 were available for study. It remains speculation that the sixth pandemic in the western hemisphere and previous pandemics were caused by this same clonal group of classical biotype strains.

EL TOR BIOTYPE

Several studies (15, 18, 38, 47) have analyzed the toxigenic El Tor biotype strains of *V. cholerae* O1 by MEE, which groups these organisms into four major ETs or clonal groups: (i) the seventh pandemic, (ii) the U.S. Gulf Coast, (iii) Australia, and (iv) Latin America. The four clones seem to reflect broad geographic and epidemiologic associations. The dendrogram in Fig. 3 shows that all of the El Tor and the classical biotype isolates are related at a genetic distance of less than 0.3. These data also indicate that each biotype reflects a closely related group of organisms at a genetic distance of less than 0.15. Within the El Tor biotype, the U.S. Gulf Coast and Australian clones are closely related, and the seventh-pandemic and Latin American strains are closely related. As with the classical biotype, MEE is less discriminating than many of the other subtyping methods for El Tor isolates, but it seems to be a reproducible approach to addressing the evolutionary significance of isolate relationships (40). The following discussion of this biotype will be generally organized on the basis of MEE clonal groups.

Seventh Pandemic

As shown in Table 2, ET3, or the seventh-pandemic clone, is composed of temporally and geographically diverse isolates that were obtained as early as 1961 and as late as 1993. As illustrated in Fig. 1, *ctx* RFLP indicates that most seventh-pandemic isolates from

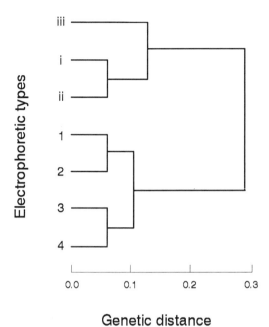

Figure 3. Genetic relationships of seven ETs of 214 isolates of *V. cholerae* O1. ETs and strain designations are given in Tables 1 and 2.

Asia, Africa, and the Pacific Islands carry one copy of the *ctx* gene, and the *Hin*dIII fragment carrying the gene ranges in size from 18 to 25 kb (23, 24, 46, 47; see also chapter 3). Historically, this assay has been more helpful in differentiating strains of the seventh-pandemic ET from other *V. cholerae* O1 isolates than in subtyping isolates with the clonal group. However, *ctx* RFLP was used successfully by investigators in Hong Kong to distinguish between indigenous and imported cases and to identify a cholera outbreak due to infection with a unique strain in a temporary Vietnamese refugee camp (50, 51).

Olsvik et al. showed that 25 seventh-pandemic isolates (as defined by MEE) all have identical *ctxB* sequences and belong to genotype 3 (31). In genotype 3, the DNA sequence of *ctxB* varies from the classical biotype *ctxB* sequence (genotype 1) by two nucleotides that code for a difference in two amino acids. The automation of this sequencing approach to strain identification has made it possible to look at large numbers of isolates, but the apparent genetic homology among toxin genes with *V. cholerae* O1, particularly within seventh-pandemic isolates, limits the use of the toxin gene itself in differentiation or subtyping.

Although they did not use MEE to define the seventh pandemic, Koblavi et al. (26) subtyped 72 El Tor isolates from Asia, Af-

Table 2. Characteristics of ETs of *V. cholerae* O1

ET[a]	No. of isolates	Yr isolated	Geographic source(s)
i	13	1921–1980	Bangladesh, Burma, Hong Kong, India, Iran, Japan, Pakistan
ii	1	1941	India
iii	1	1965	India
1	5	1977–1988	Australia
2	29	1973–1993	Gulf Coast,[b] Peru,[c] El Salvador[c]
3	63	1961–1993	Algeria, Bangladesh, Cameroon, China, Ethiopia, Gilbert Islands, Guam, Guinea/Bissau, Hong Kong, India, Indonesia, Iran, Iraq, Kuwait, Lebanon, Libya, Macao, Malawi, Malaya, Mali, Nauru, New Zealand, Philippines, Romania, Rwanda, Saudi Arabia, Spain, Thailand, Truk, Tunisia, Zaire
4	102	1991–1993	Bolivia, Chile, Colombia, Ecuador, El Salvador, Guatemala, Mexico, Peru, United States[d]

[a]i to iii, classical biotype; 1 to 4, El Tor biotype.
[b]Includes isolates from persons who ate seafood from the Gulf Coast region but live elsewhere and includes the 1983 imported cholera case from Mexico.
[c]Nontoxigenic isolates.
[d]Imported from Latin America.

rica, and the Pacific Islands into 13 ribotypes and showed that the predominant ribotype, B5, contained isolates from many parts of Africa and Asia. They also found that some ribotypes were unique to a region; e.g., ribotype B6 was unique to Ivory Coast, although Ivory Coast was also the source of B5 isolates. This study did not include isolates from Australia, the U.S. Gulf Coast, or Latin America, so it did not address the relationship of ribotypes in those groups. The authors concluded that "epidemic waves of biotype El Tor could be due to the emergence of new clones" (26).

Popovic et al. studied a group of 173 El Tor isolates, 39 of which were from Africa, Asia, and the Pacific Islands. Those 39 isolates were all in ET3, as shown in Table 1 and Fig. 4 (33). There were seven ribotypes within ET3, and one of these, ribotype 5, was also present in ET4, the Latin American clonal group. There was no other overlap among the groups defined by MEE. Like Koblavi et al., these investigators found that the majority of isolates throughout Africa belonged to a single and unique ribotype (ribotype 8), whereas Asian isolates belonged to ribotypes 3, 5, and 6.

Figure 4 illustrates typical ribotype patterns and shows how ET and ribotype data can be used to complement each other. Both approaches were used to analyze isolates from a human and from food associated with a small outbreak of cholera in Maryland (11). The meal included steamed crabs from U.S. waters. However, epidemiologic data incriminated a dish prepared with frozen coconut milk, and subsequent cultures by the Food and Drug Administration yielded *V. cholerae* O1 and a variety of other bacterial species from an unopened package of coconut milk (different production lot, however). The MEE analysis of isolates indicated that both patient and food isolates belonged to the typical seventh-pandemic ET. Although ribotyping data indicated that the human and the food isolates were different, both ribotypes might be expected to occur in the Asian coun-

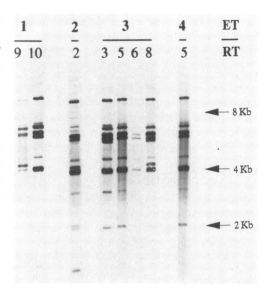

Figure 4. *Bgl*I-digested chromosomal DNA hybridized with digoxigenin-labeled 16S and 23S rRNAs from *E. coli*, showing relationships of ET and ribotype (RT) in El Tor human *V. cholerae* O1 strains. ET1, RT 9, Australia, 1984; ET1, RT 10, Australia, 1977; ET2, RT 2, Louisiana, 1978; ET3, RT 3, Philippines, 1963; ET3, RT 5, Uganda, 1992; ET3, RT 6, India, 1990; ET3, RT 8, Rwanda, 1988; ET 4, RT 5, Peru, 1991.

try where the frozen coconut milk was processed. This outbreak represents the first cholera outbreak occurring in the United States that was due to a contaminated commercially imported food (11).

Although data are preliminary, PFGE seems to be more discriminating than ribotyping for *V. cholerae* O1 (4). The large number of subtypes may be a problem in establishing a standardized PFGE subtyping system; however, within a defined epidemic setting, PFGE is probably the assay of choice, because it can distinguish within both ET and ribotype.

Gulf Coast

Almost a century after the last epidemic, cholera appeared in the United States as a single case occurring on the Gulf Coast of

Texas in 1973. The epidemic investigation could not find a likely explanation or evidence for the source of infection, and this was reported as the first naturally acquired case of cholera in the United States since 1911 (49). The organism was described as an El Tor vibrio "characteristic of the biotype isolated in the seventh pandemic" (49); it was hemolytic, as were the early-1960 isolates worldwide (49), while other 1970s isolates around the world were nonhemolytic (3).

When an outbreak of cholera occurred on the Louisiana Gulf Coast in 1978, new molecular subtyping approaches were used to study not only the outbreak isolates but also the 1973 Texas isolate. Using a *ctx* RFLP assay, Kaper et al. showed that U.S. isolates had two copies of the *ctx* genes and that these genes were located on 6- and 7-kb-long *Hin*dIII fragments (23, 24). As shown in Fig. 1, lanes 6 and 7, this pattern was easily recognized, and it differentiated U.S. Gulf Coast isolates from both seventh-pandemic and classical biotype isolates. The *ctx* RFLP assay has been valuable in identifying the persistence of the Gulf Coast clone from 1973 to the present. It has also been used to investigate sporadic cases of cholera occurring throughout the United States and has successfully differentiated the indigenous U.S. Gulf Coast strains from the background of imported seventh-pandemic cases of cholera.

An equally discriminating assay is the VcA3 RFLP, which looks at the presence of the vibriophage genome in the chromosome of *V. cholerae* O1 and analyzes its *Hin*dIII hybridization pattern (22). All Gulf Coast strains contain the VcA3 phage in its characteristic chromosomal location and with a characteristic pattern or group of related patterns (2); it has been a helpful epidemiologic marker, but the assay for it is relatively labor intensive. Neither phage production nor a colony blot assay for VcA3 were specific enough to identify the Gulf Coast clone; only the VcA3 RFLP assay was statistically sensitive and specific (2).

In June 1983, toxigenic *V. cholerae* O1 was isolated from a cholera patient in New Jersey who returned ill from Cancun, Mexico; *ctx* RFLP clearly showed that the isolate belonged to the Gulf Coast clone (7). It seems that this organism has a niche that extends along the U.S. and Mexican Gulf Coasts. Another unexpected observation was documented in 1986, when the toxigenic *V. cholerae* O1 isolate from a Florida cholera patient had only the 6-kb *Hin*dIII *ctx* fragment (Fig. 1, lane 8) (25). The isolate was otherwise typical of other Gulf Coast-associated isolates and was later shown to belong to the same MEE clonal group (15). This isolate might represent an intermediate form of the clone, either a precursor or a derivative of the strain carrying two copies of *ctx*. This finding and the fact that some nontoxigenic *V. cholerae* O1 isolates also belong to ET2 lend support to the notion that a nontoxigenic El Tor strain indigenous to the Gulf Coast region may have acquired *ctx* and emerged as a pathogen.

A third unexpected finding was that the DNA sequences of the *ctxB* genes from U.S. Gulf Coast strains were identical to the sequences of *ctxB* in classical biotype strains; i.e., all Gulf Coast-associated isolates belonged to genotype 1 (31). It seems unusual that these isolates have both the copy number and the gene sequence of *ctx* in common yet differ by MEE, ribotype, PFGE, and so many phenotypic characteristics. These data certainly complicate the already complex issues of strain relationships and evolution. Perhaps nontoxigenic *V. cholerae* O1 indigenous to the U.S. Gulf Coast acquired *ctx* from the sixth-pandemic classical biotype strains. Perhaps the Gulf Coast clone did cause early epidemics in the United States; isolation and biotype data from that time were not available for this study. By MEE, ribotype, and PFGE, all of the toxigenic U.S. Gulf Coast isolates belong to one clone, which is unique. PFGE does discriminate within ET2 and ribotype 2, but PFGE results seem to correspond to differences in the presence or copy number of the *ctx* genes.

Australia

Desmarchelier et al. have published two separate, large studies of *V. cholerae* O1 and non-O1 strains isolated from patients, food, and the environment in Australia since 1977. In the first study, which used *Hin*dIII *ctx* RFLP to characterize the strains, they reported that the 14 toxigenic *V. cholerae* O1 isolates from Australia had one copy of the gene on a 23-kb restriction fragment (19). They also found that the non-O1 *V. cholerae* isolates from Australia were a more diverse group of isolates. The second study used MEE to analyze a similar collection of isolates and again observed diversity among nontoxigenic *V. cholerae* O1 and non-O1 isolates, while the 22 toxigenic isolates belonged to one zymovar, designated Z14 (18). A recent study, which included five toxigenic O1 Australian strains isolated from 1977 through 1988, confirmed that these isolates belonged to ET1 (equivalent to a zymovar), as seen in Table 1 and Fig. 3, and that ET1 was separate from all other ETs of *V. cholerae* O1 (47). The riverine source and the persistence of an indigenous clone of *V. cholerae* O1 that caused sporadic and occasional outbreaks of disease have led many cholera epidemiologists to liken the Australian and U.S. Gulf Coast circumstances.

Popovic et al. included three toxigenic Australian *V. cholerae* O1 isolates from cholera patients in their ribotype studies (33); these isolates had been described in studies by Desmarchelier and Senn (19) and Bourke et al. (8). Popovic et al. found that two isolates from 1977 and 1984 belonged to ribotype 9, while the other from 1977 belonged to ribotype 10. The two ribotype patterns differ by five bands. They seem to represent distinct subclones and to indicate some diversity within the Australian clone. This is in contrast to the persisting clonality of the U.S. Gulf Coast isolates.

The *ctxB* genes of the 1984 isolate from ribotype 9 and the 1977 isolate from ribotype 10 were sequenced and found to be identical to each other (genotype 2) and different from those of all other *V. cholerae* O1 isolates (31). Genotype 2 differs from the genotype 1 group (classical biotype and the U.S. Gulf Coast strains) by one nucleotide and from the genotype 3 group (the seventh-pandemic isolate) by three nucleotides. These data support MEE and *ctx* RFLP results and epidemiologic data that suggest the clonality of endemic Australian *V. cholerae* O1 strains.

Latin American Epidemic

When cholera reappeared in Latin America in 1991 after a century, it spread rapidly through Peru and then throughout all of Latin America (44). Only Uruguay has not reported cholera in 1993, according to the Pan American Health Organization (13). Toxigenic *V. cholerae* O1 strains isolated in 1991 from eight Latin American countries were essentially clonal according to MEE, rRNA RFLP, *ctx* RFLP, and *ctxB* genotype (47). Data in Table 2 suggest that this trend extends into 1993, and unpublished PFGE data are in agreement with that (4).

The origin of the Latin American strain remains a mystery surrounded by speculation. Perhaps the organism was imported by travelers from areas where cholera is endemic, or perhaps it had been indigenous to Peru's coastal waters, where it was free-living until some environmental factor caused it to "bloom" to an infective number or state. Maybe it had been causing undetected disease in Latin America for years. There is no apparent epidemiologic or microbiologic evidence to support importation or blooms. However, prior to 1991, there were several good studies of diarrheal diseases, particularly in Peru, that were clearly negative by culture and/or serology for *V. cholerae* O1 (5, 45, 47).

We have attempted to address the importation theory on the basis of subtyping data.

Data in Table 2 and published elsewhere indicate that the Latin American clone is indistinguishable from some seventh-pandemic isolates by ribotyping, *ctx* RFLP, and *ctxB* genotype but is unlike any seventh-pandemic isolates by MEE and PFGE (4, 46, 47). The difference in MEE grouping is based on diversity at 1 of 16 enzyme loci, and it seems reasonable that this mutation(s), which might result from one mutation affecting enzyme atomic charge and migration, could have occurred over time during the 30-year spread of the seventh pandemic. It may not be possible to collect or assay enough 1990 or earlier isolates to find a member of the Latin American clone in Africa or Asia. However, it should be possible to learn more about divergence and the basis for subtyping data by continuing to monitor the Latin American clone as it spreads over time and by sequencing those genes that are unique to the clone, e.g., the gene(s) for leucine aminopeptidase.

In 1991, scientists with the Food and Drug Administration isolated strains belonging to the Latin American clone from shellfish off the U.S. Gulf Coast (17). Further investigations into the origin of those isolates indicated that bilge and ballast waters on ships coming from Latin America were contaminated with the organism (29). It is certainly possible that cholera is moved about the world in this way; however, the Latin American clone did not persist at a detectable level in Gulf Coast waters and caused no detectable human illness along the U.S. Gulf Coast.

Subtyping data can also be used to address the idea that indigenous *V. cholerae* O1 strains acquire toxin genes or some other property to become pathogenic. Chen et al. used MEE to look at 23 toxigenic and 23 nontoxigenic *V. cholerae* O1 isolates from the western hemisphere collected before the 1991 epidemic (15). At the time of that study, all of the toxigenic isolates, including one from Mexico in 1983, belonged to the Gulf Coast clone. Among the 23 nontoxigenic isolates, seven (including two from Peru) belonged to the Gulf Coast clone; two from Brazil and one

from Mexico belonged to two unique ETs, which are separate from the Latin American ET (15, 47). As mentioned above for evidence of importation, the sample size and sources are suboptimal for accurately addressing the issue. However, none of the available isolates from Latin America collected before 1991 belong to the current Latin American clone.

The Latin American clone has caused illness detected in the United States on several occasions. Travelers returning from areas where cholera is endemic have acquired their illness in Latin America and/or have "imported" contaminated seafood that was then consumed (9, 10, 12). A combination of epidemiology, which included travel and eating histories, and subtyping clearly showed that the Latin American clone was the cause of at least 10 episodes of illness in Georgia, Florida, New Jersey, New York, Connecticut, and California (35). One episode involved the acquisition of cholera by a number of people while in transit from Argentina and Peru to Los Angeles, California; isolates from many of the 76 known cases were subtyped and found to be the Latin American clone (12). No secondary spread within the United States has been noted.

To date, the Latin American isolates have been clonal by all of the genotypic assays mentioned above, but there has been an apparent phenotypic shift from the Inaba serotype to the Ogawa serotype (47). This phenomenon was reported earlier in studies with gnotobiotic mice, but its molecular basis was not identified (37). In addition, an in vivo Inaba-to-Ogawa shift was observed during the course of a laboratory-acquired cholera infection (41). More-recent studies indicate that serotype conversion in *V. cholerae* O1 can be caused by a point mutation, a single-base-pair deletion or insertion in the *rfbT* sequence, that generated a stop codon and resulted in a truncated *rfbT* gene product (43). Such single-base-pair mutations in the genes coding for the O1 antigen would be unlikely to result in detectable ribotype, *ctx* RFLP, *ctxB* sequence, or ET differences.

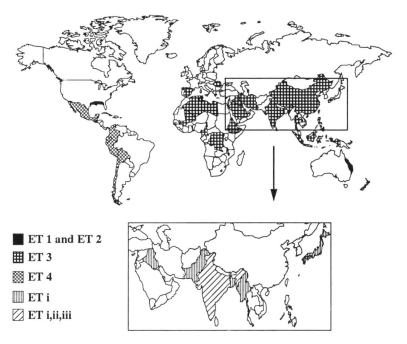

ET 1 and ET 2
ET 3
ET 4
ET i
ET i,ii,iii

Figure 5. Geographic distribution of *V. cholerae* O1 strains of different ETs.

V. cholerae O139

In what may be the eighth pandemic of cholera, a new *V. cholerae* strain has emerged from the Indian subcontinent to cause large outbreaks of disease (1, 36); this is the first time that a non-O1 *V. cholerae* has shown such epidemic potential. In preliminary studies, subtyping data show that *V. cholerae* serogroup O139 isolates may be most closely related to (and in some cases indistinguishable from) seventh-pandemic isolates (34). *V. cholerae* O139 isolates belong to the seventh-pandemic ET3 and *ctx* genotype 3 and are similar to other members of the group according to other methods (34). It seems reasonable to assume that the same environmental pressures (host immune factors, etc.) that may have influenced the Inaba-to-Ogawa shift could also influence an O1-to-O139 shift. It will be important to determine the genetic and/or biochemical similarities and differences between these two antigens.

CONCLUSIONS

Molecular subtyping of *V. cholerae* can identify specific outbreak strains and can often determine the geographic origins of strains. This has been a valuable epidemiologic approach in the United States, where each year residents are infected while traveling in Africa, Asia, and Latin America; by eating contaminated seafood from the U.S. Gulf Coast that is distributed throughout the country; or by eating contaminated seafood from Latin America brought into the United States by travelers. Additionally, both the U.S. Gulf Coast and the Latin American strains of *V. cholerae* O1 have been found in U.S. coastal waters and city sewer systems. Through a molecular definition for each *V. cholerae* O1 isolate, U.S. public health officials have been able to quickly identify sources and vehicles of transmission for cholera and control its spread.

Latin American countries, particularly those along the Gulf Coast, will probably encounter the same importation and environmental sources of cholera experienced by the United States. It is well documented that cholera occurring in countries of the Indian subcontinent is caused by at least three different clones of *V. cholerae*. The use of molecular subtyping in these circumstances should allow health officials to more accurately monitor the spread of disease and to assess prevention programs such as chlorination of water supplies and mass vaccination.

Molecular subtyping combined with epidemiologic studies and surveys has also given us a global perspective in cholera. This perspective is illustrated in Fig. 5, which is based on studies presented here and clonal groups as defined by MEE. As technology advances, developments in the field of molecular epidemiology will result in even better understanding of the evolution of *V. cholerae* strains and their geographic spread.

REFERENCES

1. **Albert, M. J., A. K. Siddique, M. S. Islam, A. S. G. Faruque, M. Ansaruzzaman, S. M. Faruque, and R. B. Sack.** 1993. Large outbreak of clinical cholera due to *Vibrio cholerae* non-O1 in Bangladesh. *Lancet* **341**:704.
2. **Almeida, R. J., D. N. Cameron, W. L. Cook, and I. K. Wachsmuth.** 1992. Vibriophage VcA-3 as an epidemic strain marker for the U.S. Gulf Coast *V. cholerae* O1 clone. *J. Clin. Microbiol.* **30**:300–304.
3. **Barrett, T. J., and P. A. Blake.** 1981. Epidemiological usefulness of changes in hemolytic activity of *Vibrio cholerae* biotype El Tor during the seventh pandemic. *J. Clin. Microbiol.* **13**:126–129.
4. **Barrett, T. J., D. N. Cameron, and I. K. Wachsmuth.** 1992. Molecular typing of *V. cholerae* O1 by pulsed-field gel electrophoresis, p. 453. *Abstr. 92nd Gen. Meet. Am. Soc. Microbiol. 1992.*
5. **Black, R. E., G. L. DeRomana, K. H. Brown, N. Bravo, O. G. Bazalar, and H. C. Kanashiro.** 1989. Incidence and etiology of infantile diarrhea and major routes of transmission in Huascar. *Am. J. Epidemiol.* **129**:785–798.
6. **Blake, P. A., D. T. Allegra, J. D. Snyder, T. J. Barrett, L. McFarland, C. T. Caraway, J. C. Feeley, J. P. Craig, J. V. Lee, N. D. Puhr, and R. A. Feldman.** 1980. Cholera—a possible endemic focus in the United States. *N. Engl. J. Med.* **302**:305–309.
7. **Blake, P. A., K. Wachsmuth, B. R. Davis, C. A. Bopp, B. P. Chaiken, and J. V. Lee.** 1983. Toxigenic *V. cholerae* O1 strains from Mexico identical to United States isolates. *Lancet* **ii**:912.
8. **Bourke, A. T. C., Y.N. Cossins, B. R. W. Gray, T. J. Lunney, N. A. Rostron, R. V. Holmes, E. R. Groggs, D. J. Larsen, and V. R. Kelk.** 1986. Investigation of cholera acquired from the riverine environment in Queensland. *Med. J. Aust.* **144**:229–234.
9. **Centers for Disease Control.** 1991. Importation of cholera from Peru. *Morbid. Mortal. Weekly Rep.* **40**:258–259.
10. **Centers for Disease Control.** 1991. Cholera—New York. *Morbid. Mortal. Weekly Rep.* **40**:516–518.
11. **Centers for Disease Control.** 1991. Cholera associated with imported frozen coconut milk. *Morbid. Mortal. Weekly Rep.* **40**:844–845.
12. **Centers for Disease Control.** 1992. Cholera associated with an international airline flight, 1992. *Morbid. Mortal. Weekly Rep.* **41**:134–135.
13. **Centers for Disease Control.** 1993. Update: cholera—Western Hemisphere, 1992. *Morbid. Mortal. Weekly Rep.* **42**:89–91.
14. **Chen, F.** 1992. Genetic structure and diversity among *Vibrio cholerae* populations. Doctoral dissertation. Georgia State University, Atlanta.
15. **Chen, F., G. M. Evins, W. L. Cook, R. Almeida, N. Hargrett-Bean, and K. Wachsmuth.** 1991. Genetic diversity among toxigenic and nontoxigenic *Vibrio cholerae* O1 isolated from the Western Hemisphere. *Epidemiol. Infect.* **107**:225–233.
16. **Cook, W. L., K. Wachsmuth, S. R. Johnson, K. A. Birkness, and A. R. Samadi.** 1984. Persistence of plasmids, cholera toxin genes and prophage DNA in classical *V. cholerae* O1. *Infect. Immun.* **45**:222–226.
17. **DePaola, A., G. M. Capers, M. L. Motes, Ø. Olsvik, P. I. Fields, J. G. Wells, I. K. Wachsmuth, T. A. Cebula, W. H. Koch, F. Khambaty, M. H. Kothary, W. L. Payne, and B. A. Wentz.** 1992. Isolation of Latin American epidemic strain of *Vibrio cholerae* O1 from the U.S. Gulf Coast. *Lancet* **339**:624.
18. **Desmarchelier, P. M., H. Momen, and C. A. Salles.** 1988. A zymovar analysis of *Vibrio cholerae* isolated in Australia. *Trans. R. Soc. Trop. Med. Hyg.* **82**:914–917.
19. **Desmarchelier, P. M., and C. R. Senn.** 1989. A molecular epidemiological study of *V. cholerae* in Australia. *Med. J. Aust.* **150**:631–634.
19a. **Evins, G. M.** Unpublished data.
20. **Faruque, S. M., A. R. M. A. Alim, M. M. Rahman, A. K. Siddique, R. B. Sack, and M. J. Albert.** 1993. Clonal relationships among classical

Vibrio cholerae O1 strains isolated between 1961 and 1992 in Bangladesh. *J. Clin. Microbiol.* 31:2513–2516.

21. **Glass, R. I., S. Becker, M. I. Huq, B. J. Stoll, M. U. Khan, M. H. Merson, J. V. Lee, and R. E. Black.** 1982. Endemic cholera in rural Bangladesh, 1966–1980. *Am. J. Epidemiol.* 116:959–970.

22. **Goldberg, S., and J. R. Murphy.** 1983. Molecular epidemiological studies of United States Gulf Coast *Vibrio cholerae* strains: integration site of mutator vibriophage Vc-A3. *Infect. Immun.* 42:224–230.

23. **Kaper, J. B., H. B. Bradford, N. C. Roberts, and S. Falkow.** 1982. Molecular epidemiology of *Vibrio cholerae* in the U.S. Gulf Coast. *J. Clin. Microbiol.* 16:129–134.

24. **Kaper, J. B., S. L. Moseley, and S. Falkow.** 1981. Molecular characterization of environmental and nontoxigenic strains of *Vibrio cholerae. Infect. Immun.* 32:661–667.

25. **Klontz, K. C., R. V. Tauxe, W. L. Cook, W. H. Riley, and I. K. Wachsmuth.** 1987. Cholera after consumption of raw oysters. *Ann. Intern. Med.* 107:846–848.

26. **Koblavi, S., F. Grimont, and P. A. D. Grimont.** 1990. Clonal diversity of *Vibrio cholerae* O1 evidenced by rRNA restriction patterns. *Res. Microbiol.* 141:645–657.

27. **Lin, F. C., G. Morris, Jr., J. B. Kaper, T. Gross, J. Michalski, C. Morrison, J. P. Libonati, and E. Israel.** 1986. Persistence of cholera in the United States: isolation of *V. cholerae* O1 from a patient with diarrhea in Maryland. *J. Clin. Microbiol.* 23:624–626.

28. **Lowry, P. W., A. T. Pavia, L. M. McFarland, B. H. Peltier, T. J. Barrett, H. B. Bradford, J. M. Quan, J. Lynch, J. B. Mathison, R. A. Gunn, and P. A. Blake.** 1989. Cholera in Louisiana—widening spectrum of seafood vehicles. *Arch. Intern. Med.* 149:2079–2084.

29. **McCarthy, S. A., R. M. McPhearson, A. M. Guarino, and J. L. Gaines.** 1992. Toxigenic *Vibrio cholerae* O1 and cargo ships entering Gulf of Mexico. *Lancet* 339:624–625.

30. **Moseley, S. L., and S. Falkow.** 1980. Nucleotide sequence homology between the heat-labile enterotoxin of *Escherichia coli* and *Vibrio cholerae* deoxyribonucleic acid. *J. Bacteriol.* 144:444–446.

31. **Olsvik, Ø., J. Wahlberg, B. Petterson, M. Uhlen, T. Popovic, I. K. Wachsmuth, and P. I. Fields.** 1993. Use of automated sequencing of PCR-generated amplicons to identify three types of cholera toxin subunit B in *Vibrio cholerae* O1 strains. *J. Clin. Microbiol.* 31:22–25.

32. **Pollitzer, R.** 1959. *Cholera.* Monograph no. 43. World Health Organization, Geneva.

33. **Popovic, T., C. A. Bopp, Ø. Olsvik, and K. Wachsmuth.** 1993. Epidemiologic application of a standardized ribotype scheme for *Vibrio cholerae* O1. *J. Clin. Microbiol.* 31:2474–2482.

34. **Popovic, T., P. I. Fields, Ø. Olsvik, J. G. Wells, G. M. Evins, D. N. Cameron, K. Wachsmuth, and J. C. Feeley.** Molecular characterization of the *Vibrio cholerae* O139 strains associated with epidemic cholera in India and Bangladesh. *In Proceedings of the 29th U.S.-Japan Joint Conference on Cholera and Related Diarrheal Diseases,* submitted for publication.

35. **Popovic, T., Ø. Olsvik, P. A. Blake, and K. Wachsmuth.** 1993. Cholera in the Americas: foodborne aspects. *J. Food Prot.,* 56:811–821.

36. **Ramamurthy, T., S. Garg, R. Sharma, S. K. Bhattacharya, G. B. Nair, T. Shimada, T. Takeda, T. Karasawa, H. Kurazano, A. Pal, and Y. Takeda.** 1993. Emergence of novel strain of *Vibrio cholerae* with epidemic potential in southern and eastern India. *Lancet* 341:703–704.

37. **Sack, R. B., and C. E. Miller.** 1969. Progressive changes of *Vibrio* serotypes in germ-free mice infected with *Vibrio cholerae. J. Bacteriol.* 99:688–695.

38. **Salles, C. A., and H. Momen.** 1991. Identification of *Vibrio cholerae* by enzyme electrophoresis. *Trans. R. Soc. Trop. Med. Hyg.* 85:544–547.

39. **Samadi, A. R., N. Shahid, A. Eusof, M. Yunus, M. I. Huq, M. U. Khan, A. S. M. M. Rahman, and A. S. G. Faruque.** 1983. Classical *Vibrio cholerae* biotype displaces El Tor in Bangladesh. *Lancet* i:805–807.

40. **Selander, R. K., T. K. Korhonen, V. Vaisanen-Rhen, P. H. Williams, P. E. Pattison, and D. A. Caugant.** 1986. Genetic relationships and clonal structure of strains of *Escherichia coli* causing neonatal septicemia and meningitis. *Infect. Immun.* 52:213–222.

41. **Sheehy, T. W., H. Sprintz, W. S. Augerson, and S. B. Formal.** 1966. Laboratory *Vibrio cholerae* infection in the United States. *JAMA* 197:99–104.

42. **Siddique, A. K., A. H. Baqui, A. E. K. Haider, M. A. Hossain, I. Bashir, and K. Zaman.** 1991. Survival of classic cholera in Bangladesh. *Lancet* 337:1125–1127.

43. **Stroeher, U. H., L. E. Karageorgos, R. Morona, and P. A. Manning.** 1992. Serotype conversion in *V. cholerae* O1. *Proc. Natl. Acad. Sci. USA* 89:2566–2570.

44. **Tauxe, R. V., and P. A. Blake.** 1992. Epidemic cholera in Latin America. *JAMA* 267:1388–1390.

45. **Varela, G., J. Olarte, A. Peres-Miravete, and L. Filloy.** 1971. Failure to find cholera and noncholera vibrios in diarrheal disease in Mexico City. *Am. J. Trop. Med. Hyg.* 20:925–926.

46. **Wachsmuth, I. K., C. A. Bopp, and P. I. Fields.** 1991. Difference between toxigenic *Vibrio cholerae* O1 from South America and the US Gulf Coast. *Lancet* i:1097–1098.

47. Wachsmuth, I. K., G. M. Evins, P. I. Fields, Ø. Olsvik, T. Popovic, C. A. Bopp, J. G. Wells, C. Carrillo, and P. A. Blake. 1993. The molecular epidemiology of cholera in Latin America. *J. Infect. Dis.* **167**:621–626.

48. Wachsmuth, I. K., J. A. Kiehlbauch, C. A. Bopp, D. N. Cameron, N. A. Strockbine, J. G. Wells, and P. A. Blake. 1991. The use of plasmid profiles and nucleic acid probes in epidemiologic investigations of foodborne diarrheal diseases. *Int. J. Food Microbiol.* **12**:77–90.

49. Weissman, J. B., W. E. DeWitt, J. Thompson, C. N. Muchnick, B. L. Portnoy, J. C. Feeley, and E. J. Gangarosa. 1975. A case of cholera in Texas, 1973. *Am. J. Epidemiol.* **100**:487–498.

50. Yam, W. C., M. L. Lung, K. Y. Ng, and M. H. Ng. 1989. Molecular epidemiology of *Vibrio cholerae* in Hong Kong. *J. Clin. Microbiol.* **27**:1900–1902.

51. Yam, W. C., M. L. Lung, K. Y. Ng, and M. H. Ng. 1991. Restriction fragment length polymorphism analysis of *Vibrio cholerae* strains associated with a cholera outbreak in Hong Kong. *J. Clin. Microbiol.* **29**:1058–1059.

Vibrio cholerae and Cholera: Molecular to Global Perspectives
Edited by I. Kaye Wachsmuth, Paul A. Blake, and Ørjan Olsvik
© 1994 American Society for Microbiology, Washington, DC 20005

Chapter 24

Cholera Surveillance

Duc J. Vugia

In 1851, immediately after the second cholera pandemic, the first International Sanitary Conference, the forerunner of the World Health Organization (WHO), convened in Paris to focus on cholera. The 12 attending nations struggled to find an agreement on maritime quarantine requirements in hope of controlling the spread of the disease (14). The requirement for international notification of cholera was not clearly set forth until 1951, when the International Sanitary Regulations (later International Health Regulations) were first approved by the then-nascent WHO after four additional cholera pandemics had passed over the globe (14). According to these regulations, every member country is required to notify WHO of the presence of cholera as soon as a case is identified (15). In practice, notification of WHO about cholera cases by many countries has been delayed, incomplete, or stopped altogether because of the fear of restrictions placed on trade and traffic by other countries (2, 3). We now know, however, that because many infected persons are asymptomatic, restrictions such as quarantine and cordon sanitaire are ineffective measures of control. In addition, we now have the means to make cholera prevent-

able, treatable, and no longer necessarily a deadly scourge. The global resurgence and persistence of cholera underscore the importance of effective cholera control in the modern world. This control is dependent in part on effective national and international surveillance. Presently, cholera surveillance varies from country to country and is not always smooth during an epidemic (13). Cholera control can be enhanced if the benefit of rapid and accurate reporting of cases is recognized and if attention is given to optimizing national and international surveillance.

PURPOSES AND OBJECTIVES OF CHOLERA SURVEILLANCE

Surveillance is a key component in an ideal plan for cholera control. Other integral components include health education, environmental sanitation, clinical management, laboratory diagnosis, and epidemiologic investigation. Whether a country is threatened by epidemic cholera or has endemic cholera, surveillance is important for the early detection of a cholera outbreak. The case fatality rate is often high in the initial period of an outbreak, when awareness of cholera's presence is low and access to proper treatment is lacking. During an outbreak or epidemic, surveillance is essential to estimate the incidence and fatality rate, to assess the movement of the epidemic,

Duc J. Vugia • Foodborne and Diarrheal Diseases Branch, Division of Bacterial and Mycotic Diseases, National Center for Infectious Diseases, Centers for Disease Control and Prevention, Atlanta, Georgia 30333.

epidemic, to plan the distribution of supplies for treatment and prevention, to plan timely epidemiologic investigations, and to determine the effectiveness of control measures.

Little has been written specifically on cholera surveillance. In its 1992 updated guidelines for cholera control (17), WHO emphasized that cholera surveillance should be part of diarrheal disease control programs, such as WHO's Control of Diarrhoeal Diseases Programme already established in many countries where cholera is endemic, epidemic, or threatening. Monitoring cholera should appropriately be part of any country's diarrheal diseases surveillance and should be linked to local Control of Diarrhoeal Diseases programs if they are available.

To be effective, a cholera surveillance system should be simple, widely accepted, and well communicated. It should have (i) simple case definitions, (ii) adequately staffed regional and central laboratories that can isolate and confirm *Vibrio cholerae* O1, (iii) rapid reporting of basic patient information to regional and central offices, and (iv) timely communication of surveillance results nationally and internationally to WHO. These components of a cholera surveillance system are discussed here, and an optimal system is proposed.

COMPONENTS OF A CHOLERA SURVEILLANCE SYSTEM

Case Definitions

A case definition is a set of objective criteria (symptoms, signs, and laboratory data) that lead to a reliable, reproducible report of the disease. For surveillance of cholera, a case definition should be brief and simple to facilitate uniform and rapid reporting of cases. To simplify case reporting, cholera case definitions should be limited to two categories, the confirmed case and the clinical case. A confirmed cholera case should be laboratory-confirmed *V. cholerae* O1 infection of any person who has diarrhea. A clinical case of cholera should be defined as acute watery diarrhea affecting a person 5 years of age or older.

Establishing a case definition is a common problem in cholera surveillance (13). In defining cases of cholera, one weakness arises because many countries use multiple case definitions within each category of clinical and confirmed. One country, for example, uses three definitions of a clinical case of cholera: (i) profuse diarrhea with severe dehydration affecting a person ≥ 5 years of age, (ii) acute diarrhea in an area with confirmed cholera, and (iii) acute diarrhea affecting a person who had traveled through an infected area within 5 days before onset of illness. Multiple definitions such as these increase the difficulty of reporting and are likely to confuse the analysis of surveillance data.

Another area of difficulty is in finding agreement on the lower age limit for a case definition. Many countries use 5 or 10 years as the cutoff age for cholera surveillance. Five years is a useful lower age limit, since it corresponds with the transition from preschool to school age; school children may be more likely to be exposed to communicable diseases, including cholera, than are preschool children, who usually stay home. For most countries, much of the morbidity from all causes of diarrhea occurs in the <5-year age group (6) and is commonly reported in diarrheal disease surveillance. However, because of other diarrheal diseases also occurring in this age group, surveillance of diarrheal cases in children under 5 years of age would not be helpful in detecting the beginning of a cholera epidemic and monitoring its movement. Using 5 and not 10 years as the lower age limit in a cholera case definition will improve both the sensitivity in identifying cases and the specificity in reducing the number of diarrhea cases that are not cholera.

Further problems stem from the practice of defining asymptomatic persons who have laboratory-confirmed *V. cholerae* as cases. In some countries, routine culturing of specimens from family members and close contacts of previously confirmed patients identified some persons with asymptomatic infections. Reporting persons without diarrhea as confirmed cholera patients can distort surveillance data.

Overall, the major difficulty with simple case definitions for cholera lies in the broad spectrum of illness associated with this infection. Over 70% of infected persons are asymptomatic, and an additional 15 to 23% of infected persons have mild or moderate nonbloody diarrhea similar to diarrhea from other causes. Some persons who meet the definition of a clinical case may not have cholera, although they are likely to represent only a small proportion of the reported cases in an outbreak or epidemic setting. For example, in the initial months of the epidemic in Peru in 1991, surveys of persons with diarrhea seeking medical care confirmed *V. cholerae* O1 in the stool samples of at least 79% (9, 10). While a case definition based solely on an adult cholera patient having dehydrating diarrhea (approximately 2 to 5% of those infected) will be more specific than a case definition based on patients with any diarrhea, it will also miss many infected persons. In the context of public health action, an accurate report of the number of symptomatic infections is a more useful measure. No single case definition is perfect; a balance is needed between sensitivity and specificity to provide a representative picture of the epidemic in any given area.

Laboratory Confirmation

The laboratory is a central component of cholera surveillance. It is essential for confirming the arrival of *V. cholerae* O1, monitoring its continued presence or disappearance, determining its antimicrobial susceptibilities, and identifying its presence in the environment. To confirm a case, the most commonly used laboratory method is stool culture, with confirmation that the isolate is *V. cholerae* O1. Where infection is extremely rare, it can be helpful to demonstrate that the isolate produces cholera toxin,

because some nontoxigenic O1 strains of *V. cholerae* have been documented (7). Serologic diagnosis based on measurement of acute- and convalescent-phase titers of vibriocidal or antitoxin antibodies is available although not as commonly used as stool culture. Ideally, there should be adequately staffed regional and central laboratories for cholera surveillance. In regional laboratories, trained personnel are needed to confirm *V. cholerae* O1 infection by using appropriate culture media such as thiosulfate-citrate-bile salts-sucrose (TCBS) agar and polyvalent antisera (5). At a central or national laboratory, trained personnel are also needed to confirm field isolates.

Initially, for an area threatened with cholera, all persons with clinical cases should have specimens taken for culture. However, if laboratories were required to confirm all cases throughout an epidemic or a large outbreak, many laboratories would be overwhelmed, and reporting would slow because of this increased amount of work. Instead, in a given area, after a sufficient number (e.g., 20 to 30) of clinical cases have been confirmed to establish the occurrence of a cholera outbreak, local laboratories may elect to reduce the frequency of cholera stool culture (e.g., to 10 specimens per month) to document the continuing outbreak and to monitor antibiotic susceptibility of the bacteria. Every laboratory that identifies *V. cholerae* O1 should report weekly the total number of patient isolates to designated regional and central offices.

When *V. cholerae* O1 is identified in a country for the first time, the isolate should be referred to an international reference laboratory for confirmation and further characterization (e.g., using molecular biological techniques), which may be helpful in determining its origin (see chapter 1).

Stages of Surveillance

In a cholera-threatened area, available diarrhea surveillance data can be reviewed to detect trends suggesting early cholera outbreaks. Any report of acute dehydrating diarrhea in a person 5 years of age or older should immediately alert local public health workers to investigate for possible cholera.

When epidemic cholera occurs in a region, two stages of surveillance are observed: an early stage, when cultures are obtained from most patients with diarrhea in order to diagnose cholera, and a later stage, when the epidemic is firmly established in the region, larger numbers of people are ill, and cultures are no longer routinely obtained from most patients.

Early in a cholera epidemic, the number of persons with confirmed cases may be small and may represent a minor proportion of the persons tested for clinical cases. A region may report only culture-confirmed cases at this time. However, when the number of confirmed cases becomes sufficiently high, surveillance should shift to using the definition of a clinical case, because it allows simpler, more timely, more accurate reporting and does not overwhelm laboratory resources. For a region in this later stage of cholera surveillance, it may be appropriate to stop or greatly limit culturing and simply report clinical cases. Unfortunately, many countries continue to report only culture-confirmed cases in this later stage of cholera surveillance, fearing panic and adverse economic consequences if the larger numbers of clinical cases were reported.

There is no simple answer as to when the number of confirmed cases becomes sufficiently high that officials should switch from reporting only confirmed cases to reporting all cases. Since most cholera outbreaks are large and often well established when confirmed, an early shift to reporting clinical cases may be most appropriate. These decisions should be made promptly on an individual basis after all implications, such as availability of laboratory resources, have been assessed.

When an epidemic wanes and the number of infections decreases to a low level like that of the early stage of the epidemic, it is reasonable to switch back to reporting only confirmed cases. In many regions of the world, however, cholera appears seasonally, and each region should be ready to shift back to using the clinical case definition when the number of confirmed cases again increases. Adjusting cholera surveillance this way by using mostly one or the other case definition depending on the seasonal prevalence of cholera in a region will help keep cholera reporting representative and timely without exhausting resources.

Environmental Surveillance

Surveillance of community sewage by using Moore swabs is useful for determining the presence of *V. cholerae* O1 in an area where no cases have been detected from human surveillance (1). In areas where cholera has not been confirmed, especially those bordering areas with cholera, Moore swabs can be placed in the sewage effluents of a limited number of sentinel towns and cities every 1 to 3 weeks. The sensitivity of the Moore swab technique in detecting *V. cholerae* O1 in water sources such as rivers has not been established. Since confirmation of cholera in a symptomatic person is crucial in the early stage of cholera surveillance, increasing the vigilance of local public health workers should take precedence over implementing environmental surveillance. The latter, if activated, should focus primarily on placing Moore swabs in common sewer lines or main sewage effluents. Once toxigenic *V. cholerae* O1 is isolated from sewage or from a person in a given area, the organism's presence has been established, and surveillance with Moore swabs can be discontinued. Thereafter, laboratory-based surveillance in that area should focus on patients with diarrhea.

Communication, Analysis, and Timely Reporting of Surveillance Data

Surveillance and laboratory information is of little value unless it is communicated effectively, that is, clearly and promptly. Timely reporting of laboratory results to a regional epidemiology office will allow early identification of cholera-affected areas and permit immediate investigations; timely reporting to the central laboratory will allow early confirmation of the isolates; and, to complete the information loop, timely feedback to the original source of the isolates will help validate diagnoses and improve patient care. Similarly, prompt reporting of the total number of cases and basic analysis results from a central epidemiology office back to the laboratories and regional epidemiology offices is essential to characterizing the epidemic and ensuring continued cooperation at all levels. In some locations, communication between these major components in the surveillance process has sometimes been unclear or delayed. Early in the cholera epidemic in Latin America, for example, some countries instituted emergency daily reporting of the number of cases to a central office yet did not communicate the total and cumulative number of cases by region back to their constituents until months later. The suitable form of information flow among local and central levels must be worked out country by country.

A proposed communication system includes both reporting to a central office and feedback to regional offices (Fig. 1).

Initially, treatment centers and regional offices may wish to report to the central office daily by a rapid method (e.g., radio, telephone, telegram) the number of clinical cases, number of confirmed cases, number of patients who were hospitalized, and number who died. Shortly thereafter, a switch to weekly reporting can greatly lower the burden of work created by daily reporting without compromising the main goals of surveillance. At the local level, including all treatment centers, information gathered about patients who meet the cholera surveillance case definition should be basic, including age, sex, date treated, and home address. The information transmitted to regional and cen-

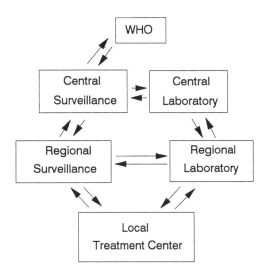

Figure 1. Proposed communication system for cholera surveillance.

tral levels may include age, sex, and location, but it should focus primarily on the number of cases, hospitalized patients, and deaths. The administrative level to which each treatment center's report should be sent must be clearly identified.

Surveillance data should be analyzed promptly and frequently. Basic tabulation and comparison of data should be performed at the local and regional levels if trained personnel are available. At the central level, epidemiologists should analyze reports of clinical and confirmed cholera cases by region and by week to track the spread of the epidemic, to determine whether unexpectedly large numbers of cases are occurring in any region, and to evaluate the impact of interventions. The results should be disseminated to all levels and used to estimate resources needed at local levels and to decide whether epidemiologic investigations are needed. When surveillance data suggest an increase in the number of cases in an area, an epidemiologic investigation should be conducted to determine the modes of transmission and to identify further prevention measures. The results of these investigations should also be reported to all levels of participants in the surveillance loop.

Some countries have administered lengthy, detailed questionnaires to every patient with cholera. Although detailed information may be helpful in describing a sample of cholera cases clinically, it is of epidemiologic interest only in the earliest phase of an epidemic. Thereafter, as the number of cholera cases increases, lengthy forms may be incomplete or ignored, and much of the data will be unmanageable and unanalyzed. Exposure information collected from patients alone will not determine the modes of transmission, because cholera may be transmitted by common foods or beverages or by multiple vehicles that vary from place to place. Well-designed and well-administered case-control investigations in affected areas are a more effective approach to identifying vehicles of transmission. In sum, exposure histories are best reserved for investigations, and lengthy surveillance questionnaires waste scarce human resources and impede handling of surveillance data.

International Surveillance

Cholera is a global public health problem that requires international cooperation for effective control. Presently, cholera is one of three internationally notifiable diseases (the other two being plague and yellow fever), and countries are requested to report cases to WHO promptly. The concern of many countries with cholera remains that reporting a large number of cases could have a detrimental impact on economic factors such as tourism and food exportation. However, travelers to countries with cholera can be instructed on simple methods of prevention and recognition of this highly treatable disease; food import businesses can be reminded that except for a recent outbreak involving three symptomatic persons and contaminated frozen fresh coconut milk (12), no other outbreak of cholera due to imported commercial foods across international borders is known (16). There are good reasons why prompt and complete reporting of cholera cases is more beneficial to an affected country. Principally, delayed and incomplete reporting actually hampers national and international cholera control efforts. In today's world of instant communication and same-day air travel, the presence of cholera in any one country cannot remain hidden for long; an asymptomatic traveler may also carry the infection to another part of the world, where it may spread. Delaying or suppressing case reports may allow an outbreak to spread to epidemic proportions and become a major problem. Prompt and accurate reporting to WHO brings clear and timely attention to this global problem and helps improve national and international efforts to allocate material and technical resources to control cholera morbidity and mortality.

SUMMARY

In its seventh pandemic, cholera still causes morbidity and mortality throughout the world. Modern transportation increases the speed and range for potential spread, making cholera a truly global problem. Given international cooperation and a public health commitment from every country, relative control of epidemic cholera could be attained with current medical supplies and technology. Surveillance, both national and international, is an essential part of cholera control. The detection of a cholera outbreak, the movement of the cholera epidemic, the need for supplies, and the effectiveness of control measures are better assessed with current and representative surveillance data. Such data can be obtained if a simple, well-communicated surveillance system is in place during an epidemic.

Ideally, cholera surveillance should be in place in every country, developed and developing alike, and should be part of diarrheal disease surveillance in countries with such a system. Until sanitary reforms improve conditions in developing countries and an effective cholera vaccine is implemented, cholera will continue to plague the globe and require constant surveillance for control. An optimal surveillance system that could facilitate prevention and treatment efforts is summarized below.

Case definitions
- Use only two case definitions, "clinical" and "confirmed."
- Define a clinical case as acute watery diarrhea affecting a person 5 years of age or older.
- Define a confirmed case as laboratory-confirmed *V. cholerae* O1 infection of any person who has diarrhea.

Laboratory confirmation and environmental surveillance
- Use trained personnel in regional and central laboratories to isolate and confirm *V. cholerae* O1 isolates.
- Confirm the diagnosis bacteriologically in clinical cases in newly threatened areas.
- During an epidemic, use stool cultures only to confirm the continuing presence of *V. cholerae* O1 and to monitor its antibiotic susceptibility.
- Consider using Moore swabs to identify *V. cholerae* O1 in sewage in cholera-threatened areas where cholera has not been confirmed.

Stages of surveillance
- In a cholera-threatened area, investigate cases of acute dehydrating diarrhea affecting persons 5 years of age or older.
- At the beginning of an outbreak or epidemic, when few persons have culture-confirmed cholera, report mainly confirmed cases.
- When an epidemic is established and the number of confirmed cases rises, report primarily clinical cases.
- When the number of cases wanes to a low level comparable to that in the early stage of the epidemic, report primarily confirmed cases.

Communication, analysis, and timely reporting of surveillance data
- Collect basic information on patients at the local level. Analyze data locally if trained personnel are available.
- Transmit summary data (primarily the numbers of cases, hospitalizations, and deaths) to the central level by a rapid method.
- Analyze national surveillance data, and disseminate surveillance reports to all levels and to WHO in a timely manner.
- Conduct epidemiologic investigations in areas with increasing numbers of cases.

ADDENDUM

Since December 1992, epidemic cholera due to *V. cholerae* O139 has been reported in Bangladesh and India and may herald the coming of the eighth pandemic (4, 8, 11). This *V. cholerae* strain and any future epidemic strain should be added to the laboratory confirmation component of cholera surveillance.

REFERENCES

1. **Barrett, T. J., P. A. Blake, G. K. Morris, N. D. Puhr, H. B. Bradford, and J. G. Wells.** 1980. Use of Moore swabs for isolating *Vibrio cholerae* from sewage. *J. Clin. Microbiol.* **11:**385–388.
2. **Barua, D.** 1992. History of cholera, p. 1–36. *In* D. Barua and W. B. Greenough III (ed.), *Cholera.* Plenum Publishing Corp., New York.
3. **Barua, D., and M. H. Merson.** 1992. Prevention and control of cholera, p. 329–349. *In* D. Barua and W. B. Greenough III (ed.), *Cholera.* Plenum Publishing Corp., New York.
4. **Cholera Working Group, International Centre for Diarrhoeal Diseases Research, Bangladesh.** 1993. Large epidemic of cholera-like disease in Bangladesh caused by *Vibrio cholerae* O139 synonym Bengal. *Lancet* **342:**387–390.
5. **Gangarosa, E. J., W. E. Dewitt, M. I. Huq, and A. Zarifi.** 1968. Laboratory methods in cholera: isolation of *Vibrio cholerae* (El Tor and classical) on TCBS medium in minimally equipped laboratories. *Trans. R. Soc. Trop. Med. Hyg.* **62:**693–699.
6. **Guerrant, R. L., J. M. Hughes, N. L. Lima, and J. Crane.** 1990. Diarrhea in developed and devel-

oping countries: magnitude, special settings, and etiologies. *Rev. Infect. Dis.* **12:**S41–S50.

7. **Morris, J. G., J. L. Picardi, S. Lieb, J. V. Lee, A. Roberts, M. Hood, R. A. Gunn, and P. A. Blake.** 1984. Isolation of nontoxigenic *Vibrio cholerae* O group 1 from a patient with severe gastrointestinal disease. *J. Clin. Microbiol.* **19:**296–297.

8. **Ramamurthy, T., S. Garg, R. Sharma, S. K. Bhattacharya, G. B. Nair, T. Shimada, T. Takeda, T. Karasawa, H. Kurazano, A. Pal, and Y. Takeda.** 1993. Emergence of novel strain of *Vibrio cholerae* with epidemic potential in southern and eastern India. *Lancet* **341:**703–704. (letter.)

9. **Ries, A. A., D. J. Vugia, L. Beingolea, A. M. Palacios, E. Vasquez, J. G. Wells, N. G. Baca, D. L. Swerdlow, M. Pollack, N. H. Bean, L. Seminario, and R. V. Tauxe.** 1992. Cholera in Piura, Peru: a modern urban epidemic. *J. Infect. Dis.* **166:**1429–1433.

10. **Swerdlow, D. L., E. D. Mintz, M. Rodriguez, E. Tejada, C. Ocampo, L. Espejo, K. D. Greene, W. Saldana, L. Seminario, R. V. Tauxe, J. G. Wells, N. H. Bean, A. A. Ries, M. Pollack, B. Vertiz, and P. A. Blake.** 1992. Waterborne transmission of epidemic cholera in Trujillo, Peru: lessons for a continent at risk. *Lancet* **340:**28–33.

11. **Swerdlow, D. L., and A. A. Ries.** 1993. *Vibrio cholera* non-O1—the eighth pandemic? *Lancet* **342:**382–383. (Commentary.)

12. **Taylor, J. L., J. Tuttle, T. Pramukul, K. O'Brien, T. J. Barrett, B. Jolbitado, Y. L. Lim, D. Vugia, J. G. Morris, Jr., R. V. Tauxe, and D. M. Dwyer.** 1993. An outbreak of cholera in Maryland associated with imported commercial frozen fresh coconut milk. *J. Infect. Dis.* **167:**1330–1335.

13. **Vugia, D. J., J. E. Koehler, and A. A. Ries.** 1992. Surveillance for epidemic cholera in the Americas: an assessment. *Morbid. Mortal. Weekly Rep.* **41**(SS-1):27–34.

14. **World Health Organization.** 1958. The first ten years of the World Health Organization. World Health Organization, Geneva.

15. **World Health Organization.** 1966. *International Sanitary Regulations,* annotated ed. World Health Organization, Geneva.

16. **World Health Organization.** 1991. Small risk of cholera transmission by food imports. *Weekly Epidemiol. Rec.* **66:**55–56.

17. **World Health Organization.** 1992. *Guidelines for Cholera Control,* revised ed. Report WHO/CDD/ SER/80.4 REV.4 (1992). World Health Organization, Geneva.

V. VACCINES

Vibrio cholerae and Cholera: Molecular to Global Perspectives
Edited by I. Kaye Wachsmuth, Paul A. Blake, and Ørjan Olsvik
© 1994 American Society for Microbiology, Washington, DC 20005

Chapter 25

Induction of Serum Vibriocidal Antibodies by O-Specific Polysaccharide–Protein Conjugate Vaccines for Prevention of Cholera

Shousun C. Szu, Rajesh Gupta, and John B. Robbins

Despite a century of study, there is no consensus about what host mechanism(s) confers protective immunity to cholera (3, 6–8, 13, 19, 20, 26, 28, 29, 32, 33, 35, 36, 41, 42, 52, 54, 55, 65–74, 80–82, 93, 99, 104, 106). There is a wealth of evidence that *Vibrio cholerae* O1, the causative organism, is localized in the small intestine and that the symptoms of the disease are mediated largely by pharmacologic actions of cholera toxin (32, 33, 41). Only strains of *V. cholerae* O1 that secrete this toxin cause cholera, and the disease can be mimicked by feeding the purified protein to volunteers (33). Bacteremia and the systemic complications of a gram-negative bacterial infection, such as fever and shock, are rarely encountered in cases of cholera (112). Another important observation is that there are no inflammatory or other morphologic changes in the intestine of patients with cholera. It comes as a surprise, then, that neither serum nor intestinal secretory antitoxin confers protection to cholera in humans (8, 13, 17, 20, 33, 36, 41, 42, 66, 75, 76, 85). The experimental evidence to date indicates that the level of serum vibrioci-dal antibodies is correlated with the resistance of humans to cholera (29, 33, 36, 66, 67). Serum antibodies to the lipopolysaccharide (LPS) of *V. cholerae* are the main component that exerts this vibriocidal activity (1, 6, 20, 28, 31, 36, 61, 73). Most workers, however, feel that the titer of serum vibriocidal antibodies represents a "marker" only of the immune state of humans and that only secretory immunity exerts protection (20, 33, 41, 55, 65, 104, 111). We propose that a critical level of serum vibriocidal antibodies alone can prevent cholera and have described a mechanism by which this may occur (37, 58). On the basis of this hypothesis, we have designed vaccines to elicit antibodies, especially of the immunoglobulin G (IgG) class, to the LPS of *V. cholerae*. Our progress has been limited by deficiencies in our knowledge about the immunoglobulin classes and specificities of the serum vibriocidal activity that has been correlated with resistance to cholera and about the structures of the two LPSs of *V. cholerae* O1 that account for their serologic differences.

CELLULAR AND LPS VACCINES

Shousun C. Szu and John B. Robbins • Laboratory of Developmental and Molecular Immunity, National Institute of Child Health and Human Development, Bethesda, Maryland 20891-0001. *Rajesh Gupta* • Massachusetts Public Health Biologic Laboratories, Jamaica Plain, Massachusetts 02130.

The strongest evidence that serum vibriocidal antibodies may confer protective immunity comes from clinical studies of cholera vaccines composed of inactivated *V. cholerae* O1 (cellular vaccines) (3, 6–8, 13, 14, 20,

28, 32, 66–71, 81). Parenterally administered cellular vaccines containing both serotypes as well as partially purified LPS induced statistically significant protection (~60%) against cholera in adults for ~6 months (3, 6, 7, 13, 32, 33, 44, 66, 68–71, 81, 105, 106). These types of vaccines elicited mostly serum LPS antibodies with vibriocidal activity but did not elicit serum antitoxin (66) or, by analogy with similar products, secretory antibodies (104). Cellular vaccines elicited lower levels of vibriocidal antibodies than crude LPS preparations and were less protective in children than in adults (6, 7, 66, 67). Further, cellular vaccines were ineffective for control of outbreaks of cholera (98). Similar limitations of age-related immunogenicity and effectiveness were obtained with orally administered vaccines composed of inactivated *V. cholerae* O1 of both serotypes and classical and El Tor biotypes (8, 19, 69, 93); addition of the B subunit of cholera toxin (CT) to this vaccine did not recruit additional protection (19). A longer-lasting and higher (approximately twofold) level of protection was achieved with addition of an oil-water adjuvant to the parenterally injected whole-cell vaccine, but this formulation caused severe local reactions (3). Alum-adsorbed vaccines induced higher levels and longer duration of vibriocidal antibodies in animals, but this formulation was never evaluated for its clinical efficacy (43). The age-related immunogenicity and protection of cellular vaccines are serious limitations, especially in areas where cholera is endemic, since the majority of cases and the highest attack rates are in children. Mosley stated (66),

From the 1966–67 field trial, using a commercial cholera vaccine in 1- and 2-dose schedules, it is evident that this vaccine produces an inadequate serological response, particularly in young children, both in terms of antibody titer and in duration of antibody response. Thus, future vaccine development should be directed towards a product that will produce a high and sustained antibody response in children. . . .

Both parenterally administered cellular and LPS vaccines elicited high rates of local and systemic adverse reactions. Although experience with methods of standardization for both cellular and LPS vaccines is limited, the mouse protection test developed by Feeley et al. (29, 69) gave encouraging results.

Since an LPS vaccine elicited a serologic response similar to that elicited by the cellular vaccine, the protective moiety elicited by these products is probably serum LPS antibody with vibriocidal activity.

It is not commonly appreciated that the serum antibody responses elicited by cellular bacterial vaccines injected intramuscularly into humans do not elicit a booster effect and are not as immunogenic as purified saccharides (23, 50, 97).

ASSAY AND CHARACTERIZATION OF SERUM VIBRIOCIDAL ANTIBODIES

The titers of serum vibriocidal antibodies can be reliably measured by several assays (1, 6, 31, 36, 61, 65, 69, 73), and results among several laboratories have been consistent (51). However, neither the antigenic specificity nor the immunoglobulin classes of these vibriocidal antibodies have been analyzed and correlated with protective immunity to cholera.

It is difficult for two reasons to derive an estimate of the "protective level" of serum vibriocidal antibodies, as has been done for systemic bacterial infections (90). First, the variability of the inoculum size of *V. cholerae* may vary. When the pathogen is spread by contaminated food or prepared drinks, the inoculum may be very high compared to that in drinking water. Second, in areas where malnutrition is common, the hosts may have defective gastric acid secretion, and the inoculum that reaches the duodenum may be higher than in otherwise healthy individuals (52, 72). Nevertheless, in the seroepidemiologic studies conducted with vaccinees and controls in East Pakistan in 1966 to 1967, almost all cases occurred in individuals with a serum vibriocidal titer of < 160 (1, 66, 67, 69). In these studies there was an approximately 44% reduction in the case rate of cholera with each doubling of the vibriocidal titer, and 96% of the cases occurred in individuals with titers of ≤ 160.

Serum vibriocidal antibodies also occur in nonvaccinated adults living in areas with no known contact with *V. cholerae* O1, although they occur at a lower level than in adults in areas of endemicity such as Bangladesh (1, 31, 32, 34, 74). The specificity of these "natural" vibriocidal antibodies has not been studied, but judging from information about other pathogens with surface polysaccharides, it is likely that the stimuli were induced cross-reacting bacteria of the indigenous flora (91, 92). Natural vibriocidal antibodies in residents of areas where cholera is not endemic were of the IgM class only (1, 64, 66). Both the incidence and the vibriocidal titers in individuals living in areas where cholera is endemic are higher than those in individuals not in contact with the pathogen. The immunoglobulin class composition of serum vibriocidal activity in adults vaccinated with cellular vaccines or following cholera was IgM and IgG (1, 64, 66). The failure to identify IgG antibodies in nonexposed adults may be quantitative, since the specific vibriocidal activity of IgM antibodies is likely to be higher than that of IgG (1, 89). Vibriocidal activity, as expected for a complement-dependent reaction, has not been demonstrated to be induced by IgA (24, 110). On the basis of absorption experiments, many sera seem to have vibriocidal antibodies directed to non-LPS antigens that may compose up to 15% of the observed activity (31, 73, 74, 79, 84).

Surprisingly, cholera is not always an effective stimulus for eliciting protective immunity. In a study of convalescent patients, the rates of primary infection (0.23%) and reinfection (0.22%) were similar: almost all of the reinfections were of the heterologous serotype, implying that disease-associated protective immunity was LPS specific (111).

Most of the vibriocidal activity in convalescent-phase sera from adult cholera patients from Mexico was IgM (103). Absorption of these sera with LPS or deacylated LPS removed the vibriocidal activity. This result is consistent with published data and provides an explanation for the comparatively short-lived protective immunity that follows cholera (111): the disease does not consistently elicit maturation of the B cells required for IgG LPS antibody synthesis.

A correlation was shown between the levels of vibriocidal antibodies and the ages of patients with cholera (66, 67). Of particular importance to our program was the inverse relation observed between the high incidence of cholera in 2- to 3-year-old children (31%) and the low level of vibriocidal antibodies in this age group. It is not commonly appreciated that the attack rate of cholera is highest in young children under endemic as well as epidemic conditions (66). During the latter, the total number of adult cases may be higher because of the greater number of individuals in this age group. Interestingly, infants < 1 year old rarely contract cholera (3, 66). In areas where cholera is endemic, cord serum contains vibriocidal antibodies of the IgG class (1). These data provide an explanation for results obtained recently in Bangladesh during a vaccine trial of orally administered cellular vaccines (8, 19). Infants of immunized mothers were protected against cholera for ~ 6 months. The vaccine-related resistance of the infants was not related to the titers of antitoxin or to LPS antibodies in milk samples from their mothers (17). Unfortunately, the titers of serum vibriocidal antibodies in these mothers were not reported. We suggest that the relatively low incidence of cholera in infants in these studies and in areas where cholera is endemic is due to placentally transmitted serum IgG vibriocidal antibodies (1).

We predict that the level of IgG rather than the total level of vibriocidal antibodies may be correlated more accurately with protection to cholera. There are two reasons for this: (i) the synthesis of IgG antibodies is predictive of long-lived immunity and probably reflects the successful induction of T-helper cells to the critical antigen-specific B-cell population, and (ii) IgG antibodies penetrate into the extracellular spaces and the lumen of the small intestine more effectively than IgM (4, 47, 82, 104, 108).

LPS OF *V. CHOLERAE* O1

The two serologically distinct LPSs of *V. cholerae* O1 distinguish serotypes Inaba and Ogawa (32, 33, 39). The structural basis for the serologic difference between these two serotypes remains unknown.

The LPS of serotype Inaba has been studied most extensively. It has the three domains found on the LPSs of other gram-negative members of the family *Enterobacteriaceae*: (i) lipid A, (ii) core oligosaccharide, and (iii) an O-specific polysaccharide (O-SP) (11, 39, 49, 58, 62, 86–88, 105, 106). The composition but not the structure of the core regions

of the two serotypes has been reported. The carboxyl group of the ketodeoxyoctanoic acid at the lipid A-core oligosaccharide juncture is phosphorylated (11). The core oligosaccharide is composed of 4-amino-4-deoxy-L-arabinose, quinovosamine, D-glucose, D-fructose, and heptose (39). It has been suggested that D-fructose is present as a branch and provides an antigenic determinant (45, 88). The O-SP of *V. cholerae* is a linear homopolymer composed of ~ 12 residues of $\alpha 1 \rightarrow 2$-linked perosamine, whose amino groups are acylated by 3-deoxy-L-*glycero*-tetronic acid (49, 87). We confirm previous reports (56) that the ^{13}C and ^{31}P nuclear magnetic resonance (NMR) spectra of the O-SPs of the two serotypes are indistinguishable. The ^{13}C NMR spectra of the LPSs of both serotype Inaba and serotype Ogawa show identical major peaks and some differences in the minor peaks, probably from the core region (Fig. 1). The ^{31}P spectra of the two serotypes show a similar pattern (Fig. 2). Circular dichroism spectra (unpublished data) of acetic acid-hydrolyzed LPS of both serotypes

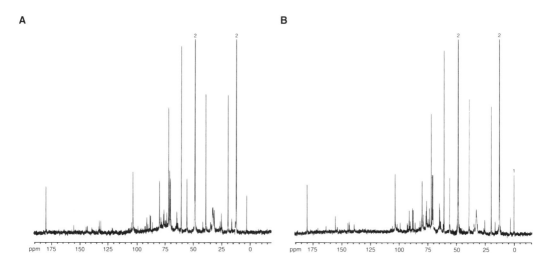

Figure 1. ^{13}C NMR spectra (125 MHz) of *V. cholerae* O1 LPS, serotypes Inaba (A) and Ogawa (B), recorded on a JEOL GSX-500 spectrometer. Approximately 10 mg of each LPS was dissolved in 0.5 ml of D$_2$O and 1 μl of trimethylamine. The spectra were recorded at ambient probe temperature (ca. 20°C), and 40,000 free induction decay signals were averaged. Chemical shifts are relative to external sodium 3-(trimethylsilyl) propionate-2,2,3,3-d4.

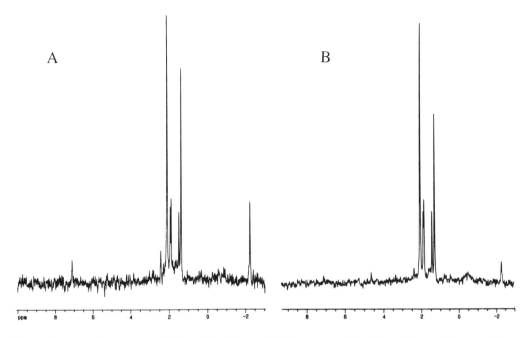

Figure 2. ^{31}P NMR spectra (200 MHz) of *V. cholerae* O1 LPS serotypes Inaba (A) and Ogawa (B), recorded on a JEOL GSX-500 spectrometer. Approximately 10 mg of each LPS was dissolved in 0.5 ml of D_2O and 1 μl of trimethylamine and recorded at ambient probe temperature (20°C).

were indistinguishable, providing evidence that the serologic difference was not due to isomerism of the O-SP. Several studies show that cellular and LPS vaccine elicit serotype-specific LPS and vibriocidal antibodies; in some studies, the vaccine- and disease-induced protective immunity was also serotype specific (3, 6, 7, 13, 28, 66, 71, 111). Since isolates from patients in areas where cholera is both endemic and epidemic are usually of both serotypes, conjugate vaccines will probably require the detoxified LPSs of both antigenic variants.

In summary, the structural basis for the immunologic differences between the two serotypes remains unknown. This deficiency in our knowledge is important, because there is still no explanation to the serotype-specific protective immunity elicited by Inaba and Ogawa strains in animal and human vaccine studies (3, 6, 7, 33, 69).

CT AS A CARRIER PROTEIN

We have reconsidered the data that suggest that serum-neutralizing antibodies to CT (antitoxin) do not confer protection against cholera (59, 100). *V. cholerae* O1 is a facultatively anaerobic gram-negative rod adapted to grow in simple media. Most strains will grow in a seawater basal medium at 20 to 45°C. Survival of *V. cholerae* O1 in the human intestinal tract does not require the secretion of CT: strains that have been engineered so that they do not secrete the holotoxin survive in the intestinal tract (46, 55, 65). The selective advantage conferred by CT on *V. cholerae* O1 may be the profuse diarrhea, which increases the chance that the organism will be ingested by another human (33). This differs from the selective advantage conferred by diphtheria toxin on *Corynebacterium diphtheriae* (78) and by pertussis toxin on

Bordetella pertussis (63, 95). In these cases, the exotoxin inactivates local resistance mechanisms such as phagocytic cells by virtue of its metabolic actions and permits survival of the pathogen in the host.

Serum antitoxin confers protection against cholera on laboratory animals (80, 83, 99). In these experiments, the immunogen (inactivated CT or toxoid) was purified from the strain of *V. cholerae* (569B) used to challenge the animals. The notion that serum antibodies elicited by one CT exert the same neutralizing properties against all CTs was based on data obtained from laboratory animals injected with toxoid in complete Freund's adjuvant (16, 32). These hyperimmune sera had neutralizing activities similar to those of CTs from other strains of *V. cholerae*. Later, it was shown that there are structural and antigenic variants of CT (5, 12, 33, 48). Antisera prepared by subcutaneous injections of these antigenic variants without adjuvants elicited neutralizing activity that was largely specific to the toxin used as the immunogen. The latter mode of immunization is more akin to that used as toxoids in humans than to antisera produced with adjuvants. Clinical trials showed that the toxoids were ineffective in preventing cholera (3, 75, 76, 85). In the trials, however, sera from vaccinees were assayed for antitoxin from the vaccine strain (569B) but not for toxins purified from the strains isolated from the patients. Last, the evaluation of cholera in the clinical trials of cholera toxoids did not distinguish between degrees of severity of the diarrhea. *V. cholerae* produces at least two other toxins that elicit diarrhea in experimental animals (5, 33). The massive watery diarrhea, the most serious symptom of cholera, is due largely to the action of CT. If expression of CT is eliminated by genetic manipulation or reduced significantly by serum antitoxin, diarrhea can still occur, as has been observed in normal volunteers fed mutant strains of *V. cholerae* (46, 55). The diarrhea elicited by these CT-negative strains was clinically unimportant. As Finkelstein has commented (30), the se-

verity as well as the occurrence per se of the diarrhea in the recipients of cholera toxoids may have been considerably less than in the controls.

We used CT for our conjugate vaccines because this protein has served as a good carrier in rhesus monkeys (96, 100) and because it has potential for inhibiting the symptoms caused by CT. Further, induction of an immune response to the cross-reactive heat-labile toxin, as has been shown for the orally administered B subunit of CT, may prevent diarrhea caused by enterotoxigenic *Escherichia coli* (16, 18, 22, 59, 85, 100). The use of the heat-labile toxin of enterotoxigenic *E. coli* as a carrier protein for one of the LPS subtypes has been considered for a bivalent cholera vaccine containing the O-SP of serotypes Inaba and Ogawa.

SYNTHESIS AND IMMUNOLOGIC PROPERTIES OF CONJUGATE VACCINES

When we started our study, the predominant organism causing cholera in South America was *V. cholerae* O1, serotype Inaba, biotype El Tor, which produces CT-2 (36, 48, 53). LPS was purified from serotype Inaba and Ogawa strains by the method of Westphal and Jann (109). We detoxified the LPS by two methods. (i) Acid hydrolysis was done in 1% acetic acid at 100°C for 90 min, and lipid A was removed (15). This method removes lipid A at the juncture of its glucosamine dimer with the core oligosaccharide. This product was denoted O-SP, had a molecular mass of ~ 5,900 Da, and failed to precipitate with *V. cholerae* hyperimmune antiserum (37). (ii) The LPS was treated with the organic base hydrazine at 37°C for 2 h, which resulted in hydrolysis of ester bonds and left a product, denoted DeA-LPS, with amide-linked fatty acid and two phosphorylated saccharides with molecular masses of 16,000 and 6,000 Da (37, 57). DeA-LPS precipitated with hyperimmune *V. cholerae* antiserum. Both types of products had only trace amounts of protein, nucleic acids, and intact LPS (102), and both passed Food and Drug Administration-prescribed tests for pyrogenicity and safety. Since the acid-treated LPS had a low molecular mass and was not antigenic, we chose to use the DeA-LPS to conjugate with proteins.

Two schemes were used to synthesize the conjugates. The first used the heterobifunctional linker *N*-succinimidyl 3-(2-pyridyldithio) propionate (SPDP) (101) to thiolate both the protein and the terminal amino group on the core of the DeA-LPS (27, 100, 101). The saccharide SPDP derivative was reduced with dithiothreitol to generate an active SH group and then mixed with an SPDP derivative of the protein (Fig. 3A). This resulted in the formation of a disulfide with a single point attachment between the saccharide and the protein in a carbohydrate/protein ratio of 0.7 (wt/wt) (Fig. 3A).

The second synthetic scheme produced multipoint attachments between the saccharide and the protein. The saccharide was activated with cyanogen bromide and derivatized with adipic acid dihydrazide (Fig. 3B) as described elsewhere for the polysaccharides of *Haemophilus influenzae* type b and *Shigella dysenteriae* type 1 (15, 91). This derivative was conjugated to the protein by carbodiimide-mediated activation to form amide-like bonds between the hydrazide moiety of the detoxified LPS (DeA) and the carboxyls of the protein (Fig. 3B).

The saccharide/protein ratio in this type of conjugate was ~ 1. The LPS contents of the conjugates were ~ 2 endotoxin units (EU)/μg of conjugate by the *Limulus* lysate assay (40), and the CT activity was reduced ~ 10^3- to 10^{10}-fold in the CHO cell assay (25). All conjugates passed the general safety tests in mice and guinea pigs (21).

Saline solutions of the conjugates at 1/10 the proposed human dose were injected subcutaneously into general-purpose mice of the National Institutes of Health (NIH) colony and into BALB/c mice (15). The DeA-LPS-CT conjugates prepared by method 2 elicited

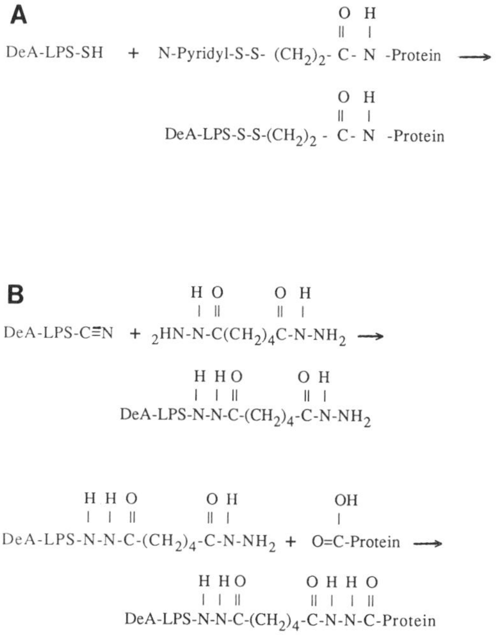

Figure 3. (A) DeA-LPS was thiolated with SPDP and mixed with *N*-pyridyl dithio-derivatized protein. (B) Cyanogen-derivatized DeA-LPS was linked with adipic acid dihydrazide. The hydrazide-derivatized polysaccharide was then linked to proteins in the presence of carbodiimide.

the highest level of LPS antibodies. Table 1 shows that the DeA-LPS-CT elicited LPS antibodies after the first injection and that the level of antibodies increased with subsequent injections. The conjugates elicited higher vibriocidal activity against the homologous serotype (Inaba) than against the heterologous serotype (Ogawa) (Table 2). The cellular vaccine composed of both serotypes elicited levels of vibriocidal antibodies to Inaba similar to those elicited by the conjugates, but levels of antibodies to Ogawa were higher. The DeA-LPS alone did not elicit LPS or vibriocidal antibodies in the mice (not shown). All vibriocidal activity was removed from either the conjugate-induced antibodies or cellular vaccine induced by absorption with LPS, DeA-LPS, or O-SP extracted from

Table 1. Serum LPS antibodies elicited in mice by three injections of DeA-LPS-CT, LPS, or cellular cholera vaccines[a]

Vaccine[b]	Amt injected	Mouse strain	Geometric mean titer	
			IgG	IgM
DeA-LPS-CT	2.5 μg	GPM[c]	80	150
DeA-LPS-CT	2.5 μg	BALB/c	46	130
			32[d]	50
Whole cell	0.1 ml	BALB/c	1,742	742
			90	101
LPS	2.5 μg	GPM	320	640

[a]The table is based on data from Gupta et al. (37).
[b]Conjugate (amount based on saccharide content), LPS from strain Inaba or cholera vaccine (courtesy of Wyeth Corp., Marietta, Pa.) was injected subcutaneously at 2-week intervals into groups of 10 mice. Serum samples were taken 7 days or 5 months after the second and third injections. O-SP and DeA-LPS alone did not elicit serum LPS antibodies (data not shown).
[c]GPM, general-purpose mice from the NIH colony.
[d]A 10-μg dose was used for these mice.

serotype Inaba strains. Absorption of either the conjugate-induced or the cellular vaccine-induced antiserum with the LPS from serotype Ogawa removed ~90% of the vibriocidal activity against strain Inaba. As expected, absorption of the conjugate-induced or cellular vaccine-induced serum with CT did not remove vibriocidal activity.

The conjugates elicited neutralizing antibodies to CT after the first injection, and rises

in titer were statistically significant (booster responses) after each of the two subsequent injections.

This immunization scheme, using 1/10 of the proposed human dose in saline injected subcutaneously into young mice, was predictive of the immunogenicity of *H. influenzae* type b and pneumococcus type 6 conjugates in infants and children (15, 91, 94). The low levels of both in vitro and in vivo "endotoxin" have also been predictive of the safety of these conjugates in infants and children (91). Recently, Watson et al., using a similar immunization and dosage scheme with O-SP–protein conjugates, showed protection of mice against *Salmonella typhimurium* (107).

ADVANTAGE OF CONJUGATE VACCINES

Simply stated, conjugates represent a safer, easily standardized, potentially more immunogenic product than cellular vaccines (91). Conjugate vaccines have been shown to elicit protective levels of polysaccharide antibodies in infants and children (2, 91), the age groups with the highest attack rate of cholera. Further, conjugate vaccines for cholera could be

Table 2. Vibriocidal antibody titers of pooled sera from mice immunized with cholera conjugates, LPS, or cellular vaccines[a]

Immunogen	Amt injected	Mouse strain	Challenge serotype	Geometric mean titer after injection:		
				1	2	3
DeA-LPS-CT[b]	2.5 μg	GPM	Inaba	100	25,000	50,000
DeA-LPS-CT	2.5 μg	GPM	Ogawa	<10	1,000	25,000
DeA-LPS-CT	2.5 μg	BALB/c	Inaba	250	5,000	100,000
Dea-LPS-CT	2.5 μg	BALB/c	Ogawa	100	500	50,000
Whole cell[c]	0.1 ml	BALB/c	Inaba	2,500	50,000	100,000
Whole cell	0.1 ml	BALB/c	Ogawa	25,000	500,000	1,000,000
LPS[d]	10.0 μg	GPM	Inaba	ND[e]	50,000	500,000
LPS	10.0 μg	GPM	Ogawa	ND	500	50,000

[a]Data are from Gupta et al. (37).
[b]Conjugate (amount based on saccharide content) containing Inaba O-SP only was injected subcutaneously at 2-week intervals into groups of 10 mice (general purpose mice from the NIH colony or BALB/c). Serum samples were taken 2 weeks after the first injection and 7 days after the second and third injections. O-SP and DeA-LPS alone did not elicit serum vibriocidal activity (data not shown).
[c]United States-licensed cholera cellular vaccine courtesy of Wyeth.
[d]LPS was purified from serotype Inaba strains.
[e]ND, not done.

injected into infants concurrently with diphtheria-tetanus-pertussis vaccine, as *H. influenzae* type b has been (91). The action of our detoxified LPS-conjugate vaccines is to elicit serum vibriocidal activity. If shown to be effective against cholera in adults, then the protective level of LPS antibodies could be estimated, as was done for whole-cell vaccines (1, 66, 67). Estimates of protective levels of the LPS and vibriocidal titers of conjugate vaccines could permit the use of these products in infants and young children without additional trials of efficacy: the serologic responses engendered by these products could provide sufficient evidences for their incorporation into routine immunization programs. The experience gained with *H. influenzae* type b polysaccharide conjugate vaccines could be applied to considerations of cholera conjugate vaccines for all age groups (90).

THEORETIC BASIS OF PROTECTIVE ACTION

We theorized about how serum vibriocidal antibodies could prevent cholera (37). Our theory is concerned only with the early events of cholera and does not involve symptoms of the disease. It describes a series of mechanisms and characteristics acting in concert. First, serum antibodies, especially those of the IgG class, penetrate into the lumen of the small intestine (4, 47, 78, 84, 91, 108); serum complement proteins are likely also present. Second, peristalsis ensures that the walls of the intestine are fully in contact with each other. Third, the inoculum of *V. cholerae* that survives the gastric acidity is low ($\leq 10^3$) (52, 72). Fourth, *V. cholerae* is highly susceptible to the cidal actions of antibodies and complement (31, 73). This susceptibility is probably due to the comparatively low molecular mass of their O-SP. Thus, serum vibriocidal antibodies could prevent cholera by lysis of the ingested pathogen in the small intestine. This proposed mechanism could explain the disappearance of poliovirus in Sweden and Holland (9, 10). In these two countries, infants and young children are injected with Formalin-inactivated poliovirus vaccine (Salk type); compliance with poliovirus vaccination is almost 100%. This type of immunogen and the route of immunization do not elicit secretory antibodies, yet they have eliminated poliovirus (and paralytic poliomyelitis) in these two countries. The best explanation for this effect is that vaccine-induced serum antibodies eliminate poliovirus in the intestine.

It should be noted that the level of serum vibriocidal antibodies is unrelated to the severity of cholera in patients with preexisting antibodies at the protective level (60). In a small fraction of individuals, the inoculum size of *V. cholerae* O1 that reaches the small intestine may be large, as in patients with low gastric acidity, and overwhelm the preexisting immunity (52, 72). As has been observed for many viral and bacterial diseases, serum antibodies are effective preventive but not therapeutic agents. This principle is probably also true for cholera.

CURRENT CLINICAL STUDIES

At the time this chapter was written, only preliminary data on the safety of our detoxified LPS-CT vaccines in 50 adults were available. None of the recipients experienced fever or serious local reactions.

SUMMARY

Clinical, epidemiologic, and experimental data support the hypothesis that serum vibriocidal antibodies confer protective immunity to cholera. On the basis of epidemiologic and clinical data, a protective titer of serum vibriocidal antibodies of $\geq 1/160$ was estimated. We developed conjugate vaccines composed of the detoxified LPS of *V. cholerae* O1 serotype Inaba bound to CT that were designed to elicit protective levels of vibriocidal antibodies, especially in infants as part of their routine immunization. The conjugates had only trace levels of endotoxin or CT, which has been predictive of their safety. Our yields were ~70%, making commercial production possible. The composition of conjugate vaccines can be standardized by unambiguous physiochemical assays. In young mice, our cholera conjugates injected at a dosage and in a fashion that are clinically acceptable elicited vibriocidal antibodies at similar levels and of longer duration than cellular vaccines did. Cholera conjugate vaccines have the potential to elicit protection in infants when injected along with diphtheria-typhoid-pertussis vaccine and the *H. influenzae* type b conjugate as part of routine immunization. If successful, their potency may be reliably predicted by their physiochemical and serologic properties. The concentration, specificity, and immunoglobulin class(es) of the vibriocidal activity should be standardized.

Acknowledgment. We are grateful to William Egan, Center for Biologics Evaluation and Research, Food and Drug Administration, for NMR analyses of *V. cholerae* O1 LPS and its derivatives.

REFERENCES

1. **Ahmed, A., A. K. Bhattacharjee, and W. H. Mosley.** 1970. Characteristics of the serum vibriocidal and agglutinating antibodies in cholera cases and in normal residents of the endemic and nonendemic cholera areas. *J. Immunol.* **105:**431–441.

2. **Anderson, P. W., M. E. Pichichero, R. A. Insel, R. Betts, R. Eby, and D. H. Smith.** 1986. Vaccines consisting of periodate-cleaved oligosaccharide from the capsule of *Haemophilus influenzae* type b coupled to a protein carrier: structural and temporal requirements for priming in the human infant. *J. Immunol.* **137:**1181–1186.

3. **Azurin, J. C., A. Cruz, T. P. Persigan, M. Alvero, T. Camena, R. Suplido, L. Ledesma, and C. A. Gomez.** 1967. A controlled field trial of the effectiveness of cholera and cholera El Tor vaccines in the Philippines. *Bull. W.H.O.* **37:**703–727.

4. **Batty, I., and J. J. Bullen.** 1961. The permeability of the sheep and rabbit intestinal wall to antitoxin present in the circulation. *J. Pathol. Bacteriol.* **81:**447–458.

5. **Baudry, B., A. Fasano, J. Ketley, and J. B. Kaper.** 1992. Cloning of a gene (*zot*) encoding a new toxin produced by *Vibrio cholerae*. *Infect. Immun.* **60:**428–434.

6. **Benenson, A. S.** 1976. Review of experience with whole-cell and somatic antigen vaccines, p. 228–242. *In* H. Fukimi and Y. Zinnaka (ed.), *Proceedings of the 12th Joint Conference of U.S.-Japan Cooperative Medical Science Program, Cholera Panel.* National Institute of Health, Tokyo.

7. **Benenson, A. S., P. R. Joseph, and R. O. Oseasohn.** 1968. Cholera vaccine field trials in East Pakistan. *Bull. W.H.O.* **38:**347–357.

8. **Black, R. E., M. M. Levine, M. L. Clements, C. R. Young, A.-M. Svennerholm, and J. Holmgren.** 1987. Protective efficacy in humans of killed whole-*Vibrio* oral cholera vaccine with and without B subunit of cholera toxin. *Infect. Immun.* **55:**1116–1120.

9. **Böttiger, M.** 1984. Long-term immunity following vaccination with killed poliovirus vaccine in Sweden, a country with no circulating virus. *Rev. Infect. Dis.* **6**(Suppl. 2):S548–S551.

10. **Böttiger, M., and E. Herrström.** 1992. Isolation of polioviruses from sewage and their characteristics: experience over two decades in Sweden. *Scand. J. Infect. Dis.* **24:**151–155.

11. **Brade, H.** 1985. Occurrence of 2-keto-deoxyoctonic acid 5-phosphate in lipopolysaccharides of *Vibrio cholerae* Ogawa and Inaba. *J. Bacteriol.* **161:**795–798.

12. **Burnette, W. N., V. L. Mar, B. W. Platler, J. D. Schotterbeck, M. D. McGinley, K. S. Stoney, M. F. Rohde, and H. R. Kaslow.** 1991. Site-specific mutagenesis of the catalytic subunit of cholera toxin: substituting lysine for arginine 7 causes loss of activity. *Infect. Immun.* **59:**4266–4270.

13. **Cash, R. A., S. I. Music, J. P. Libonati, J. P. Craig, N. F. Pierce, and R. B. Hornick.** 1974. Response of man to infection with *Vibrio cholerae*. II. Protection from illness afforded by previous disease and vaccine. *J. Infect. Dis.* **130:**325–333.

14. **Centers for Disease Control.** 1988. Cholera vaccine. *Morbid. Mortal. Weekly Rep.* **37:**617–624.

15. **Chu, C. Y., B. Liu, D. Watson, S. C. Szu, D. Bryla, J. Shiloach, R. Schneerson, and J. B. Robbins.** 1991. Preparation, characterization, and immunogenicity of conjugates composed of the O-specific polysaccharide of *Shigella dysenteriae* type 1 (Shiga's bacillus) bound to tetanus toxoid. *Infect. Immun.* **59:**4450–4458.

16. **Clemens, J. D., and R. A. Finkelstein.** 1978. Demonstration of shared and unique immunological determinants in enterotoxins from *Vibrio cholerae* and *Escherichia coli*. *Infect. Immun.* **22:**709–713.

17. **Clemens, J. D., D. A. Sack, J. Chakraborty, M. R. Rao, F. Ahmed, J. R. Harris, F. van**

Loon, M. R. Khan, Md. Yunis, S. Huda, B. A. Kay, A.-M. Svennerholm, and J. Holmgren. 1990. Field trial of oral cholera vaccines in Bangladesh: evaluation of anti-bacterial and anti-toxic breast-milk immunity in response to ingestion of the vaccines. *Vaccine* **8:**469–472.

18. **Clemens, J. D., D. A. Sack, J. R. Harris, J. Chakraborty, P. K. Neogy, B. Stanton, N. Huda, M. U. Khan, B. A. Kay, M. R. Khan, M. Anaruzzaman, M. Yunis, M. Rao, A.-M. Svennerholm, and J. Holmgren.** 1988. Cross-protection by B subunit-whole cell cholera vaccine against diarrhea associated with heat-labile toxin-producing enterogenic *Escherichia coli:* results of a large-scale field trial. *J. Infect. Dis.* **158:**372–377.

19. **Clemens, J. D., D. A. Sack, J. R. Harris, F. van Loon, J. Chakraborty, F. Ahmed, M. R. Rao, M. R. Khan, M. Yunis, N. Huda, B. F. Stanton, B. Kay, S. Walter, R. Eckels, A.-M. Svennerholm, and J. Holmgren.** 1990. Field trial of oral cholera vaccines in Bangladesh: results from three year-follow-up. *Lancet* **335:**270–273.

20. **Clemens, J. D., F. van Loon, D. A. Sack, J. Chakraborty, M. R. Rao, F. Ahmed, J. R. Harris, M. R. Khan, M. Yunis, S. Huda, B. A. Kay, A.-M. Svennerholm, and J. Holmgren.** 1991. Field trial of oral cholera vaccines in Bangladesh: serum vibriocidal and antitoxic antibodies as markers of the risk of cholera. *J. Infect. Dis.* **163:**1235–1242.

21. **Code of Federal Regulations.** 1990. C, 610.13(b).

22. **Dafni, Z., and J. B. Robbins.** 1976. Purification of heat-labile enterotoxin from *Escherichia coli* O78:H11 by affinity chromatography with antiserum to *Vibrio cholerae* toxin, *J. Infect. Dis.* **133:**S138–S141.

23. **DeMaria, A., M. A. Johns, H. Berbeerich, and W. R. McCabe.** 1988. Immunization with rough mutants of *Salmonella minnesota:* initial studies in human subjects. *J. Infect. Dis.* **158:**301–311.

24. **Eddie, D. S., M. L. Schulkind, and J. B. Robbins.** 1971. The isolation and biologic activities of purified secretory IgA and IgG anti-*Salmonella typhimurium* O antibodies from rabbit intestinal fluid and colostrum. *J. Immunol.* **106:**181–190.

25. **Elson, C. O., and W. Ealding.** 1984. Generalized systemic and mucosal immunity in mice after mucosal stimulation with cholera toxin. *J. Immunol.* **132:**2736–2741.

26. **Enteric Diseases Branch.** 1991. Update: cholera—Western Hemisphere 1991. *Morbid. Mortal. Weekly Rep.* **40:**860.

27. **Fattom, A., C. Lue, S. C. Szu, J. Mestecky, G. Schiffman, D. Bryla, W. F. Vann, D. Watson, L. M. Kimzey, J. B. Robbins, and R. Schneer-**son. 1990. Serum antibody response in adult volunteers elicited by injection of *Streptococcus pneumoniae* type 12F polysaccharide alone or conjugated to diphtheria toxoid. *Infect. Immun.* **58:**2309–2312.

28. **Feeley, J. C., and E. J. Gangarosa.** 1980. Field trials of cholera vaccine, p. 204–210. *In* O. Ouchterlony and J. Holmgren (ed.), *Cholera and Related Diarrheas.* 43rd Nobel Symposium. Karger, Basel.

29. **Feeley, J. C., and M Pittman.** 1962. A mouse protection test for assay of cholera vaccine, p. 92–94. *In SEATO Conference on Cholera.* Publishing Co., Bangkok, Thailand.

30. **Finkelstein, R. A.** Personal communication.

31. **Finkelstein, R. A.** 1962. Vibriocidal antibody inhibition (VAI) analysis: a technique for the identification of the predominant vibriocidal antibodies in serum and for the detection and identification of *V. cholerae* antigens. *J. Immunol.* **89:**264–271.

32. **Finkelstein, R. A.** 1973. Cholera. *Crit. Rev. Microbiol.* **2:**553–623.

33. **Finkelstein, R. A.** 1984. Cholera, p. 107–136. *In* R. Germainer (ed.), *Bacterial Vaccines.* Academic Press, Inc., New York.

34. **Finkelstein, R. A., C. J. Powell, Jr., J. C. Woodrow, and J. R. Krevans.** 1965. Serological responses in man to a single small dose of cholera vaccine with special reference to the lack of influence of ABO blood groups on natural antibody or immunological responsiveness. *Bull. Johns Hopkins Hosp.* **116:**152–160.

35. **Glass, R. I., I. Huq, A. Arma, and M. Yunis.** 1980. Emergence of multiple antibiotic resistant *Vibrio cholerae* in Bangladesh. *J. Infect. Dis.* **142:**929–942.

36. **Glass, R. I., A.-M. Svennerholm, M. R. Khan, S. Huda, M. I. Huq, and J. Holmgren.** 1985. Seroepidemiological studies of El Tor cholera in Bangladesh: association of serum antibody levels with protection. *J. Infect. Dis.* **151:**236–242.

37. **Gupta, R. K., S. C. Szu, R. A. Finkelstein, and J. B. Robbins.** 1992. Synthesis, characterization, and some immunological properties of conjugates composed of the detoxified lipopolysaccharide of *Vibrio cholerae* O1 serotype Inaba bound to cholera toxin. *Infect. Immun.* **60:**3201–3208.

38. **Hardy, A. V., T. DeCapito, and S. P. Halbert.** 1948. Studies of the acute diarrheal diseases. XIX. Immunization in Shigellosis. *Public Health. Rep.* **63:**685–688.

39. **Hisatsune, K., Y. Haishima, T. Iguchi, and S. Kondo.** 1990. Lipopolysaccharides of non-cholera vibrios possessing common antigen factor to O1 *Vibrio cholera. Adv. Exp. Med. Biol.* **256:**189–197.

40. **Hochstein, H. D.** 1990. Role of the FDA in regulating the Limulus amoebocyte lysate test, p. 38–

49. In R. B. Prior (ed.), *Clinical Applications of the Limulus Amoebocyte Lysate Test*. CRC Press, Boca Raton, Fla.

41. **Holmgren, J., and A.-M. Svennerholm.** 1977. Mechanism of disease and immunity in cholera: a review. *J. Infect. Dis.* **136:**S105–S108.

42. **Jertborn, M., A.-M. Svennerholm, and J. Holmgren.** 1986. Saliva, breast milk, and serum antibody responses as indirect measures of intestinal immunity after oral cholera vaccination or natural disease. *Infect. Immun.* **24:**203–209.

43. **Joo, I., Z. Csizer, and V. P. Juhasz.** 1972. Immune response to plain and adsorbed cholera vaccines. *Prog. Immunobiol. Stand.* **5:**350–354.

44. **Kabir, S.** 1987. Preparation and immunogenicity of a bivalent cell-surface protein-polysaccharide conjugate of *Vibrio cholerae*. *J. Med. Microbiol.* **23:**9–18.

45. **Kaca, W., L. Brade, E. T. Rietschel, and H. Brade.** 1986. The effect of removal of D-fructose on the antigenicity of the lipopolysaccharide from a rough mutant of *Vibrio cholerae* Ogawa. *Carbohydr. Res.* **149:**293–298.

46. **Kaper, J. B., and M. M. Levine.** 1990. Recombinant attenuated *Vibrio cholerae* strains used as live oral vaccines. *Res. Microbiol.* **141:**901–906.

47. **Kaur, J., W. Burrows, and M. A. Furlong.** 1971. Immunity to cholera: antibody response in the lower ileum of the rabbit. *J. Infect. Dis.* **124:**359–366.

48. **Kazemi, M., and R. A. Finkelstein.** 1990. Study of epitopes of cholera enterotoxin-related enterotoxins by checkerboard immunoblotting. *Infect. Immun.* **58:**2352–2360.

49. **Kenne, L., B. Lindberg, P. Unger, B. Gustafsson, and T. Holme.** 1982. Structural studies of the Vibrio cholerae O-antigen. *Carbohydr. Res.* **100:**341–349.

50. **Landy, M., S. Gaines, J. R. Seal, and J. E. Whiteside.** 1954. Antibody responses of man to three types of antityphoid immunizing agents: heat-phenol fluid vaccine, acetone-dehydrated vaccine and isolated Vi antigens. *Am. J. Public Health* **44:**1572–1579.

51. **Levine, M. M.** Personal communication.

52. **Levine, M. M.** 1980. Immunity to cholera as evaluated in volunteers, p. 195–203. *In* O. Ouchterlony and J. Holmgren (ed.), *Cholera and Related Diarrheas*. 43rd Nobel Symposium. Karger, Basel.

53. **Levine, M. M.** 1991. South America: the return of cholera. *Lancet* **i:**45–46.

54. **Levine, M. M., R. E. Black, M. L. Clements, L. Cisneros, D. R. Nalin, and C. R. Young.** 1981. The quality and duration of infection-derived immunity to cholera. *J. Infect. Dis.* **143:**818–820.

55. **Levine, M. M., R. E. Black, M. L. Clements, C. Lanata, S. Sears, T. Honda, C. R. Young, and R. A. Finkelstein.** 1984. Evaluation in humans of attenuated *Vibrio cholerae* El Tor Ogawa strain Texas Star-SR as a live oral vaccine. *Infect. Immun.* **43:**515–522.

56. **Lindberg, B.** 1993. *Vibrio cholerae* polysaccharide studies, p. 64–65. *In* P. J. Garegg and A. A. Lindberg (ed), *Carbohydrate Antigens*. American Chemical Society, Washington, D.C.

57. **Lugowski, C., and E. Romanowska.** 1974. Chemical studies on *Shigella sonnei* lipid A. *Eur. J. Biochem.* **48:**319–324.

58. **Manning, P. A., M. W. Heuzenroeder, J. Yeadon, D. I. Leavesley, P. R. Reeves, and D. Rowley.** 1986. Molecular cloning and expression in *Escherichia coli* K-12 of the O antigens of the Inaba and Ogawa serotypes of the *Vibrio cholerae* O1 lipopolysaccharides and their potential for vaccine development. *Infect. Immun.* **53:**272–277.

59. **Marchlewicz, B. A., and, R. A. Finkelstein.** 1983. Immunological differences among the cholera/coli family of enterotoxins. *Diagn. Microbiol. Infec.* **1:**129–138.

60. **McCormack, W. M., A. S. M. Rahman, A. K. M. Chowdhury, W. H. Mosley, and R. A. Philips.** 1969. 1965–67 cholera vaccine trial in rural East Pakistan. 3. The lack of effect of prior vaccination or circulating antibody on the severity of clinical cholera. *Bull. W.H.O.* **60:**199–204.

61. **McCormack, W. M. K., J. Chakraborty, A. S. M. M. Rahman, and W. H. Mosley.** 1969. Vibriocidal antibody in clinical cholera. *J. Infect. Dis.* **120:**192–201.

62. **McGroarty, E. J., and M. Rivera.** 1990. Growth dependent alterations in production of serotype specific and common antigen lipopolysaccharides in *Pseudomonas aeruginosa* PAO1. *Infect. Immun.* **58:**1030–1037.

63. **Meade, B. D., P. D. Kind, J. B. Ewell, P. McGrath, and C. R. Manclark.** 1984. In-vitro inhibition of murine macrophage migration by *Bordetella pertussis* lymphocytosis-promoting factor. *Infect. Immun.* **45:**718–725.

64. **Merritt, C. B., and R. B. Sack.** 1970. Sensitivity of agglutinating and vibriocidal antibodies to 2-mercaptoethanol in human cholera. *J. Infect. Dis.* **121:**S25–S30.

65. **Migasena, S., P. Pitisuttitham, B. Prayurahong, P. Suntharasamal, W. Supanaranond, V. Desakorn, U. Vongsthongsri, B. Tall, J. Ketley, G. Losonsky, S. Cryz, Jr., J. Kaper, and M. M. Levine.** 1989. Preliminary assessment of the safety and immunogenicity of live oral cholera vaccine strain CVD 103-HgR in healthy Thai adults. *Infect. Immun.* **57:**3261–3264.

66. **Mosley, W. H.** 1969. The role of immunity in cholera. A review of epidemiological and serological studies. *Tex. Rep. Biol. Med.* **27:**S227–S241.

67. **Mosley, W. H., S. Ahmed, A. S. Benenson, and A. Ahmed.** 1968. The relationship of vibriocidal antibody titre to susceptibility of cholera in family contacts of cholera patients. *Bull. W.H.O.* **38:**777–785.

68. **Mosley, W. H., K. J. Bart, and A. Sommer.** 1972. An epidemiological assessment of cholera. 1. Control programs in rural East Pakistan. *Int. J. Epidemiol.* **1:**5–11.

69. **Mosley, W. H., J. C. Feeley, and M. Pittman.** 1968. The interrelationships of serological responses in humans, and the active mouse protection test to cholera vaccine effectiveness. *Ser. Immunobiol. Stand.* **15:**185–196.

70. **Mosley, W. H., W. M. McCormack, A. Ahmed, A. K. M. Alauddan Chowdhury, and R. K. Barui.** 1969. Report of the 1966–1967 cholera vaccine field trial in rural East Pakistan. *Bull. W.H.O.* **40:**187–197.

71. **Mosley, W. H., W. E. Woodward, K. M. A. Aziz, A. S. M. Mizanur Rahman, A. K. M. Chowdhury, A. Ahmed, and J. C. Feeley.** 1970. The 1968–1969 cholera-vaccine field trial in rural East Pakistan. Effectiveness of monovalent Ogawa and Inaba vaccines and a purified Inaba antigen, with comparative results of serological and animal protection tests. *J. Infect. Dis.* **121:**S1–S9.

72. **Nalin, D. R., R. J. Levine, M. M. Levine, D. Hoover, E. Bergquist, J. McLaughlin, J. Libonati, J. Alam, and R. B. Hornick.** 1978. Cholera, non-vibrio cholera, and stomach acid. *Lancet* **ii:**856–859.

73. **Neoh, S. H., and D. Rowley.** 1970. The antigens of *Vibrio cholerae* involved in the vibriocidal action of antibody and complement. *J. Infect. Dis.* **121:**505–513.

74. **Neoh, S. H., and D. Rowley.** 1972. Protection of infant mice against cholera by antibodies to three antigens of *Vibrio cholera. J. Infect. Dis.* **126:**41–47.

75. **Noriki, H.** 1977. Evaluation of toxoid field trial in the Philippines, p. 302–310. *In* H. Fukimi and Y. Zinnaka (ed.), *Proceedings of the 12th Joint Conference U.S.-Japan Cooperative Medical Science Program.* National Institute of Health, Tokyo.

76. **Ohtomo, N.** 1977. Safety and potency tests of cholera toxoid Lot II in animals and volunteers, p. 286–296. H. Fukimi and Y. Zinnaka (ed.), *Proceedings of the 12th Joint Conference U.S.-Japan Cooperative Medical Sciences Program.* National Institute of Health, Tokyo.

77. **Ørskov, F., and I. Ørskov.** 1978. Serotyping of enterobacteriacae, with special emphasis on K antigen determination. *Methods Microbiol.* **11:**1–77.

78. **Pappenheimer, A. M., Jr.** 1983. Studies on the molecular epidemiology of diphtheria. *Lancet* **ii:**293–295.

79. **Parsot, C., E. Taxman, and J. J. Mekalanos.** 1991. ToxR regulated the production of lipoproteins and the expression of serum resistance in *Vibrio cholerae. Proc. Natl. Acad. Sci. USA* **88:**1641–1645.

80. **Peterson, J. W.** 1979. Synergistic protection against experimental cholera by immunization with cholera toxoid and vaccine. *Infect. Immun.* **26:**528–533.

81. **Philippines Cholera Committee.** 1973. A controlled field trial on the effectiveness of the intradermal and subcutaneous administration of cholera vaccine in the Philippines. *Bull. W.H.O.* **49:**389–394.

82. **Pierce, N. F., W. C. Cray, Jr., J. B. Kaper, and J. J. Mekalanos.** 1988. Determinants of immunogenicity and mechanisms of protection by virulent and mutant *Vibrio cholerae* O1 in rabbits. *Infect. Immun.* **56:**142–148.

83. **Pierce, N. F., E. A. Kaniecki, and R. S. Northrup.** 1972. Protection against experimental cholera by antitoxin. *J. Infect. Dis.* **126:**606–616.

84. **Pierce, N. F., and H. Y. Reynolds.** 1974. Immunity to experimental cholera. I. Protective effect of humoral IgG antitoxin demonstrated by passive immunization. *J. Immunol.* **113:**1017–1023.

85. **Rappaport, R. S., and G. Bonde.** 1981. Development of a vaccine against experimental cholera and *Escherichia coli* and diarrheal disease. *Infect. Immun.* **32:**534–542.

86. **Raziuddin, S.** 1978. Toxic and immunologic properties of the lipopolysaccharides (O-antigens) from Vibrio EL-TOR. *Immunochemistry* **15:**611–614.

87. **Redmond, J. W.** 1979. The structure of the O-antigenic side chain of the lipopolysaccharide 22 of *Vibrio cholerae* 569B (Inaba). *Biochim. Biophys. Acta* **584:**346–352.

88. **Redmond, J. W., and M. J. Korsch.** 1973. Immunochemical studies of the O-antigens of *Vibrio cholerae:* partial characterization of an acid-labile antigenic determinant. *AJEBAK* **51:**229–235.

89. **Robbins, J. B., K. Kenny, and E. Suter.** 1965. The isolation and biological activities of rabbit gamma-M and gamma-G anti-*Salmonella typhimurium* antibodies. *J. Exp. Med.* **122:**385–402.

90. **Robbins, J. B., J. C. Parke, R. Schneerson, and J. K. Whisnant.** 1973. Quantitative measurement of "natural" and immunization-induced *Haemophilus influenzae* type b capsular polysaccharide antibodies. *Pediatr. Res.* **7:**103–110.

91. **Robbins, J. B., and R. Schneerson.** 1990. Polysaccharide-protein conjugates: a new generation of vaccines. *J. Infect. Dis.* **161:**821–832.

92. Robbins, J. B., R. Schneerson, M. P. Glode, W. F. Vann, M. S. Schiffer, T.-Y. Liu, J.C. Parke, Jr., and C. Huntley. 1975. Cross-reactive antigens and immunity to diseases caused by encapsulated bacteria. *J. Allergy Clin. Immunol.* **56**:141–151.

93. Sack, D. A., J. D. Clemens, S. Huda, J. R. Harris, M. R. Khan, J. Chakraborty, M. Yunis, J. Gomes, O. Siddique, F. Ahmed, B. A. Kay, F. P. L. van Loon, M. R. Rao, A.-M. Svennerholm, and J. Holmgren. 1991. Antibody responses after immunization with killed oral cholera vaccines during the 1985 vaccine field trial in Bangladesh. *J. Infect. Dis.* **164**:407–411.

94. Santosham, M., R. Reid, D. M. Ambrosino, M. C. Wolf, J. Almeido-Hill, C. Priehs, K. M. Aspery, S. Garrett, L. Croll, S. Foster, G. Burge, P. Page, B. Zacher, R. Moxon, and G. Siber. 1987. Prevention of *Haemophilus influenzae* type b infections in high-risk infants treated with bacterial polysaccharide immune globulin. *N. Engl. J. Med.* **317**:923–929.

95. Sato, H., Y. Sato, and I. Ohishi. 1987. Effect of monoclonal antibody to pertussis toxin on toxin activity. *Infect. Immun.* **55**:909–915.

96. Schneerson, R., J. B. Robbins, C.-Y. Chu, A. Sutton, W. Vann, J. C. Vickers, W. T. London, B. Curfman, M. C. Hardegree, J. Shiloach, and S. C. Rastogi. 1984. Serum antibody responses of juvenile and infant rhesus monkeys injected with *Haemophilus influenzae* type b and pneumococcus type 6A polysaccharide-protein conjugates. *Infect. Immun.* **45**:582–591.

97. Schwartzer, T. A., D. V. Alcid, V. Numsuwan, and D. J. Gocke. 1988. Characterization of the human antibody response to an *Escherichia coli* O111:B4 and J5 vaccine. *J. Infect. Dis.* **158**:1135–1136.

98. Sommer, A., and W. H. Mosley. 1973. Ineffectiveness of cholera vaccination as an epidemic control measure. *Lancet* **i**:1232–1235.

99. Svennerholm, A.-M., and J. Holmgren. 1976. Synergistic protective effect in rabbits of immunization with *Vibrio cholerae* lipopolysaccharide and toxin/toxoid. *Infect. Immun.* **13**:735–740.

100. Szu, S. C., X. Li, R. Schneerson, J. H. Vickers, D. Bryla, and J. B. Robbins. 1989. Comparative immunogenicities of Vi polysaccharide-protein conjugates composed of cholera toxin or its B subunit as a carrier bound to high- or lower-molecular-weight Vi. *Infect. Immun.* **57**:3823–3827.

101. Szu, S. C., R. Schneerson, and J. B. Robbins. 1986. Rabbit antibodies to the cell wall polysac-charide of *Streptococcus pneumoniae* fail to protect mice from lethal infection with encapsulated pneumococci. *Infect. Immun.* **54**:448–455.

102. Tsai, C.-M. 1986. The analysis of lipopolysaccharide (endotoxin) in meningococcal polysaccharide vaccines by silver staining following SDS-polyacrylamide gel electrophoresis. *J. Biol. Stand.* **14**:25–33.

103. Valdespino Gómez, J. L. Unpublished data.

104. Waldman, R. H., Z. Benzic, B. C. Deb, R. Sakazaki, K. Tamura, S. Mukerjee, and R. Gguly. 1972. Cholera immunology. II. Serum and intestinal antibody response after naturally occurring cholera. *J. Infect. Dis.* **126**:401–407.

105. Watanabe, Y., and W. F. Verwey. 1965. The preparation and properties of a purified mouse-protective lipopolysaccharide form the Ogawa subtype of the El Tor variety of *Vibrio cholerae,* p. 253–259. O. A. Bushell and C. S. Brookhyer (ed.), *Proceedings of the Cholera Research Symposium.* Public Health Service publication no. 1328. U.S. Government Printing Office, Washington, D.C.

106. Watanabe, Y., and W. F. Verwey. 1965. Protective antigens from El Tor vibrios. 1. The preparation and properties of a purified protective antigen from an El Tor Vibrio (Ogawa subtype). *Bull. W.H.O.* **32**:809–821.

107. Watson, D. C., J. B. Robbins, and S. C. Szu. 1992. Protection of mice against *Salmonella typhimurium* with an O-specific polysaccharide-protein conjugate vaccine. *Infect. Immun.* **60**:4679–4686.

108. Wernet, P., H. Breu, J. Knop, and D. Rowley. 1971. Antibacterial action of specific IgA and transport of IgM, IgA and IgG form serum into the small intestine. *J. Infect. Dis.* **124**:223–226.

109. Westphal O., and K. Jann. 1965. Bacterial lipopolysaccharide extraction with phenol:water and further application of the procedure. *Methods Carbohydr. Chem.* **5**:83–91.

110. Winner, L., III, J. Mack, R. Weltzin, J. J. Mekalanos, J.-P. Kraehenbuhl, and M. R. Neutra. 1991. New model of analysis of mucosal immunity: intestinal secretion of specific monoclonal immunoglobulin A from hybridoma tumors protects against *Vibrio cholerae* infection. *Infect. Immun.* **59**:977–982.

111. Woodward, W. E. 1971. Cholera reinfection in man. *J. Infect. Dis.* **123**:61–66.

112. Woodward, W. E., W. H. Mosley, and W. W. McCormack. 1970. The spectrum of cholera in rural East Pakistan. *J. Infect. Dis. Suppl.* **121**:S10–S16.

Vibrio cholerae and Cholera: Molecular to Global Perspectives
Edited by I. Kaye Wachsmuth, Paul A. Blake, and Ørjan Olsvik
© 1994 American Society for Microbiology, Washington, DC 20005

Chapter 26

Recombinant Live Cholera Vaccines

Myron M. Levine and Carol O. Tacket

An ideal cholera vaccine would provide a high level of long-term protection following administration of a single dose, and the onset of protection would commence within days of vaccination. Such a vaccine would be given orally for practicality and for optimal stimulation of the mucosal immune system of the intestine. An ideal cholera vaccine would satisfactorily protect groups such as young children and individuals of blood group O that have not been well protected by earlier cholera vaccines such as the parenteral or oral inactivated vaccines (5, 7, 39). Lastly, an ideal cholera vaccine would be packaged in a practical formulation that would facilitate mass vaccination, including vaccination of young children, and would be inexpensive. During the past decade, considerable progress has been made in constructing recombinant bacteria and adapting them for use as live oral cholera vaccines. In extensive clinical studies carried out so far, one such recombinant strain, CVD 103-HgR, exhibits many of the features of an ideal cholera vaccine (34). This chapter reviews developments and progress in the use of recombinant bacteria as live oral cholera vaccines.

RATIONALE FOR THE USE OF LIVE CHOLERA VACCINES

Several facts constitute the basis for the use of recombinant bacteria, particularly attenuated *Vibrio cholerae* O1, as live oral cholera vaccines. These include the following.

1. A single clinical infection due to wild-type *V. cholerae* O1 confers significant protection against cholera upon subsequent exposure to wild-type *V. cholerae* O1 (9, 16, 31, 33–35, 38, 39).
2. While many virulence properties contribute to the pathogenesis of cholera, the in vivo expression of cholera enterotoxin is a prerequisite for the profuse purging of voluminous rice water stools that is characteristic of cholera gravis (39).
3. The critical protective immunity to cholera is antibacterial rather than antitoxic in nature (38, 39), although in the short term, antitoxic immunity may synergistically enhance antibacterial immunity (38, 48).
4. Curiously, the degree of stimulation of serum vibriocidal antibody following ingestion of a live oral cholera vaccine or following infection with wild-type *V. cholerae* O1 constitutes the best correlate of the elicitation of antibacterial immunity in the intestine (8, 17, 34, 39, 47, 48).

Myron M. Levine and Carol O. Tacket • Center for Vaccine Development, University of Maryland School of Medicine, 10 South Pine Street, Baltimore, Maryland 21201-1116.

5. Although many antigens on the surface of *V. cholerae* O1 have been identified, the antigens or combination of antigens that constitutes the protective repertoire is still the subject of study and debate (4, 20, 22, 38, 49, 52, 54, 57, 59–61, 66, 67).

Infection-Derived Immunity

The extent to which a prior clinical infection with wild-type enterotoxigenic *V. cholerae* O1 stimulates protection against cholera in the face of subsequent challenge with wild-type *V. cholerae* O1 comes from two sources, volunteer studies and epidemiologic studies in endemic areas. As summarized in Table 1, rechallenge studies with volunteers clearly demonstrate that an initial clinical infection with either classical biotype or El Tor biotype *V. cholerae* O1 confers 90% (El Tor) to 100% (classical) protection against clinical illness when the volunteers are rechallenged with *V. cholerae* O1 of the same biotype (33–35, 38). Protection is equal against rechallenge with either the homologous or the heterologous serotype (i.e., either Inaba or Ogawa) within the same biotype. In studies with classical biotype organisms, protection persisted for as long as 3 years, the longest interval tested (31). Regrettably, no data are available from studies of cross-biotype rechallenge of volunteers with wild-type vibrios to ascertain the degree of cross-biotype immunity evident in

this model. One notable difference observed in relation to biotype was the extent of excretion of *V. cholerae* O1 (Table 1). The level of excretion of *V. cholerae* O1 was so low in volunteers rechallenged with classical biotype vibrios that direct cultures of stool onto thiosulfate-citrate-bile salts-sucrose (TCBS) medium were all negative. In contrast, *V. cholerae* O1 biotype El Tor could be recovered on direct cocultures from approximately 30% of individuals who were rechallenged with that biotype. These data from volunteer studies suggest that classical biotype vibrios stimulate a more potent immunity than El Tor vibrios.

Three reports address infection-derived immunity in persons living in Matlab Bazaar, Bangladesh, an area where cholera is highly endemic. Woodward (75) described repeat infections in 14 subjects, of whom 6 had experienced clinical cholera infections both times. In five of these instances, the initial clinical infection occurred in a child 5 years of age or younger. One or the other infection was asymptomatic in the other eight cases. Woodward concluded that cholera is not a highly immunizing disease, but a lack of precise denominators prevented him from adequately comparing the risk of the general population with that of persons who had experienced an earlier episode of cholera.

One decade later, Glass et al. (16) reviewed data from Matlab Bazaar and concluded that an initial clinical infection dimin-

Table 1. Protective efficacy in volunteers of prior clinical infection with pathogenic *V. cholerae* O1 of classical or El Tor biotype against biotype-homologous challenge

Immunizing *V. cholerae* biotype	Attack rate for diarrhea[a]		% Protective efficacy	Isolation of *V. cholerae* from direct cocultures[b]	
	Controls	Veterans[c]		Controls	Vaccinees
Classical	24/27 (89%)	0/16 (0%)	100	26/27 (96%)	0/16 (0%)[d]
El Tor	32/37 (86%)	2/22 (9%)	90	34/37 (92%)	8/22 (36%)[d]

[a]Challenge with 10^6 pathogenic *V. cholerae* O1 organisms given with $NaHCO_3$. Numbers are number with diarrhea/number challenged.
[b]Number showing *V. cholerae*/number challenged.
[c]Includes both serotype-homologous and serotype-heterologous challenges. Volunteers who developed diarrhea following ingestion of *V. cholerae* O1 on initial challenge were rechallenged 4 to 6 weeks later with *V. cholerae* O1 of either the same or the heterologous serotype within the identical biotype.
[d]$P = 0.012$.

ishes by approximately 90% the risk of experiencing a subsequent episode of clinical cholera (Table 2). The data of Glass et al. (16) concerned mainly classical biotype infections. On the basis of life table analysis, 29 clinical cholera infections were expected, but only 3 were observed in persons who had an initial clinical cholera infection. Two represented El Tor-El Tor recurrences, and one was a classical-classical recurrence. All three individuals who experienced these recurrent episodes of cholera had their first illness when they were 5 years of age or younger.

Finally, Clemens et al. (9) reviewed the frequency of repeat clinical cholera infections in Matlab. As summarized in Table 2, they observed that an initial clinical infection due to *V. cholerae* O1 of classical biotype conferred 100% protection against subsequent cholera due to either biotype. In contrast, an initial infection with El Tor biotype was nonprotective against subsequent cholera due to the heterologous classical biotype and poorly protective (29% efficacy) even against subsequent cholera due to the homologous El Tor biotype (Table 2).

The volunteer model and the field studies concur in demonstrating the superior protective immunity that follows an initial clinical infection with classical biotype *V. cholerae* O1. The discrepancy over the degree of immunity following clinical infection due to biotype El Tor in volunteers versus field studies may be explained by the duration of follow-up. The high level of protection attributed to El Tor cholera infection in volunteers was tested after only 2 months. In the epidemiologic studies in Bangladesh, the El Tor infection-derived immunity was examined over years rather than several months. It is conceivable that the protective immunity conferred by El Tor vibrios wanes over time. Under any circumstances, one might conclude after reviewing the above epidemiologic data that if the intention is to utilize recombinant *V. cholerae* O1 strains as live oral vaccines, it would be wise to administer strains of the classical biotype.

Cholera Enterotoxin Is Required for Cholera Gravis

The molecular pathogenesis of cholera infection is one of the best-studied bacterial infections and is characterized by cascades of coordinately regulated virulence properties (13). Many virulence properties play a role as *V. cholerae* O1 converts itself from an environmental niche in brackish water to survival in and colonization of the mucosa of the human intestine. Of the many virulence properties that *V. cholerae* O1 possesses (38), one that must be expressed to produce the syndrome of cholera gravis is the elaboration in vivo of cholera enterotoxin. Indeed, administration of minute quantities of purified cholera enterotoxin to healthy adult North American volunteers induced severe purging and a syndrome that resembles cholera gravis (38). Ingestion of as little as 5.0 μg of purified cholera enterotoxin resulted in diarrhea in four of five volunteers, one of whom passed a total diarrheal stool volume of >5.0 liters. (Note: A total purge of ≥5.0 liters is the

Table 2. Infection-derived immunity conferred by clinical infection with different biotypes of *V. cholerae* O1[a]

| Study (reference) | Biotype of: | | % Protective efficacy |
	Initial infection	Subsequent infection	
Glass et al. (16)	Mostly classical	Mostly classical	90
Clemens et al. (9)	Classical	Classical	100
	Classical	El Tor	100
	El Tor	El Tor	29
	El Tor	Classical	0

[a]Observations are from field studies in an area of Bangladesh where cholera is endemic.

criterion that defines severe cholera in the volunteer model.) Two volunteers who ingested a mere 25.0 μg of purified cholera enterotoxin each purged more than 20 liters (38).

These data emphasize the fundamental importance of eliminating cholera toxin expression from candidate live vaccine strains. Whatever residual diarrhea may be elicited by such strains would not be sufficient to cause cholera gravis.

Antibacterial Immunity

A wealth of evidence supports the contention that the main mechanism of protection against *V. cholerae* O1 involves antibacterial rather than antitoxic immunity.

1. Parenteral whole-cell inactivated vaccines that elicit serum vibriocidal antibody but not antitoxin confer significant protection, albeit for only short periods (several months) (39).
2. Parenteral toxoids that stimulate high levels of serum antitoxin do not confer credible protection, even short term (12, 48, 50).
3. Over 3 years of surveillance in a controlled field trial in Bangladesh, three spaced doses (6 weeks apart) of an oral vaccine consisting of only inactivated *V. cholerae* O1 conferred protection (52% efficacy) virtually identical to that conferred by three doses of the same inactivated *V. cholerae* O1 vaccine given in combination with B subunit (50% efficacy) (6). Although the large doses of B subunit stimulated strong antitoxic serologic responses in the intestine and in serum (6), these were not accompanied by any longterm enhancement of the protection conferred by the inactivated bacteria alone (7).
4. A single oral dose of a recombinant A$^-$B$^-$ *V. cholerae* O1 El Tor candidate vaccine strain (JBK 70) conferred on volunteers 89% protection against ex-

perimental challenge with wild-type *V. cholerae* El Tor (38). Whatever mechanisms were operative in conferring the high level of protection, antitoxin was not among them, since this strain elaborated neither the A nor the B subunit of cholera toxin and did not stimulate an antitoxic immune response.

Serum Vibriocidal Antibody as a Correlate of Protection

Vibriocidal antibody is measured by bacterial lysis when serial dilutions of serum are incubated with a large standardized inoculum of *V. cholerae* O1 in the presence of guinea pig complement (2). Following natural or experimental infection of humans, one observes manyfold rises in the titer of serum vibriocidal antibody. In areas where cholera is endemic, the repeated ingestion of *V. cholerae* O1 gives rise to long-lived elevated titers of immunoglobulin G (IgG) vibriocidal antibody; indeed, in such areas, the prevalence and geometric means titer (GMT) of vibriocidal antibody increase with age of the host (8, 17, 47, 48). Such elevated titers are correlated with protection (8, 17, 47, 48). Vibriocidal responses following experimental challenge of adult volunteers tend to be mainly IgM and are short-lived, typically falling to a level twofold over baseline within 6 to 12 months. Nevertheless, the initial short-lived serum vibriocidal response is a marker for the elicitation of long-lived intestinal immunity, since such volunteers were protected as long as 3 years after initial challenge despite the fact that they no longer had elevated vibriocidal titers (31). Thus, the potent serum vibriocidal antibody response that appears following the ingestion of live oral antigens, be they wild-type vibrios or attenuated strains, serves as a marker for the stimulation of a potent intestinal immunity that endures long after the serum vibriocidal antibody titers have returned to near-baseline levels (31, 34, 40).

For these reasons, serum vibriocidal antibody is monitored as the proxy for elicitation of a protective intestinal immune response following the administration of candidate live oral vaccines (34). In general, the more potent the serum vibriocidal immune response following oral vaccination, the greater the protective immunity induced.

Protective Surface Antigens of *V. cholerae* O1

Many surface antigens have been proposed to play a role as protective antigens (4, 20, 38, 49, 54, 57, 59–61, 66). These include lipopolysaccharide (LPS) O antigen, pili (fimbriae), outer membrane proteins, and certain hemagglutinins. However, the relative importance of the various recognized antigens is disputed. To further complicate the issue, there exist antigens that are expressed in vivo but are not readily observed when *V. cholerae* O1 is cultured in vitro, despite attempts to mimic in vivo conditions (23, 62).

Approximately 85 to 90% of the vibriocidal antibodies stimulated by wild-type infection are directed against the LPS O antigen (4, 38, 60). However, important protective antibodies are believed to reside among the vibriocidal antibodies that are directed against antigens other than LPS (4, 38, 60). Toxin-coregulated pili must be expressed for classical biotype *V. cholerae* O1 to elicit strong vibriocidal responses (21), yet the immune response to the pili themselves is minimal (20).

TWO APPROACHES TO DEVELOPMENT OF RECOMBINANT BACTERIA LIVE VACCINES AGAINST CHOLERA

The main approach followed in developing live recombinant vaccines against cholera has been to disarm known pathogenic strains of certain specific virulence properties, thereby presumably rendering them incapable of causing cholera yet leaving intact the various surface antigens (known and unknown) in-

volved in protection. This approach has been pioneered by research groups at the Center for Vaccine Development of the University of Maryland School of Medicine (24–26, 34) and at Harvard Medical School (43, 55).

A radically different approach has been taken by investigators in Australia, who have utilized an attenuated *Salmonella typhi* vaccine strain to express a putative protective antigen of *V. cholerae* O1 and to deliver that vibrio antigen to the human intestinal immune system (15). Each of these approaches is discussed below.

First- and Second-Generation Attenuated *V. cholerae* O1 Recombinant Vaccine Candidates

A series of early vaccine candidates was constructed from wild-type *V. cholerae* strains known to be pathogenic for volunteers by introducing deletions in the chromosomal genes encoding the A (ADP-ribosylating "toxic") subunit (25, 43) or both the A and B (nontoxic binding) subunits (26) of cholera enterotoxin. It was observed that these first-generation vaccine strains, such as JBK 70, CVD 101, and 395N1, were markedly attenuated compared to their wild-type parents (32, 37) (Table 3). These strains were clearly incapable of causing severe diarrhea (37). Nevertheless, residual reactogenicity remained (37). Approximately one-half of the volunteers suffered adverse reactions such as combinations of malaise, nausea, vomiting, abdominal cramps, low-grade fever, headache, and mild diarrhea. Although these first-generation strains were unacceptably reactogenic, a single oral dose of these vaccine strains containing as few as 10^3 CFU and given with sodium bicarbonate buffer elicited potent vibriocidal responses. The objective of subsequent research was to diminish reactogenicity while retaining immunogenicity with a single oral dose.

Among the second generation of attenuated candidates were strains with deletions in

Table 3. Summary of salient characteristics of various live oral cholera vaccines evaluated in phase 1 and 2 clinical trials for safety, immunogenicity, and efficacy

Vaccine	Wild-type parent	Characteristics of strain	Adverse reactions[a]	Immunogenicity	Protective[b]	Reference(s)
JBK 70	ETI N16961	$\Delta ctxAB$	Yes	4+	Yes	37
CVD 101	CO 395	$\Delta ctxA$	Yes	4+	NT	37
CVD 102	CO 395	$\Delta ctxA\ thyA$	No	2+	NT	37
CVD 104	ETI N16961	$\Delta ctxAB\ \Delta hlyA$	Yes	3+	NT	37
CVD 105	CO 395	$\Delta ctxA\ \Delta hlyA$	Yes	3+	NT	37
395N1	CO 395	$\Delta ctxA$	Yes	4+	Yes	21
JJM43	CO 395	$\Delta ctxA\ \Delta toxR$	No	1+	NT	21
TCP2	CO 395	$\Delta ctxA\ \Delta tcpA$	No	0+	NT	21
CVD 103	CI 569B	$\Delta ctxA$	No	4+	Yes	36, 37
CVD 103-RM	CI 5698	$\Delta ctxA\ \Delta recA$	No	2+	NT	27
CVD 103-HgR	CI 569B	$\Delta ctxA\ \Delta hlyA{::}mer$	No	4+	Yes	34–37
CVD 110	ETO E7946	Deletion of ctx virulence cassette[c] (zot ace), $\Delta hlyA{::}mer\ ctxB$	Yes	4+	NT	70
Bahrain-3	ETO E7946	Deletion of ctx virulence cassette,[c] $\Delta recA{::}$ P_{hap}-ctxB	Yes	4+	NT	72
Peru-3	ETI C6709	Deletion of ctx virulence cassette,[c] $\Delta recA{::}$ P_{hap}-ctxB	Yes[d]	4+	Yes	72
Peru-5	ETI C6709	Deletion of ctx virulence cassette,[c] $lacZ{::}$ P_{ctx}-ctxB	Yes	4+	Yes	72

[a]Any combination of diarrhea, abdominal cramps, nausea, vomiting, or fever.
[b]NT, not tested.
[c]This results in deletion of the region of the chromosome that contains ctxAB, zot, ace, cep, and orfU.
[d]Relatively mild symptoms.

hlyA, the gene that encodes the El Tor hemolysin. In animal models, this El Tor hemolysin exhibits enterotoxic activity (46). The resultant strains, CVD 104 and CVD 105, representing further derivatives of JBK 70 and CVD 101, respectively, appeared to be somewhat further attenuated but were still unacceptably reactogenic (37) (Table 3).

Another second-generation vaccine candidate, CVD 102, was a thymine-dependent auxotroph derived from CVD 101 (37). Although a live vaccine, CVD 102 was unable to proliferate in the human intestine. CVD 102 did not cause adverse reactions when fed to volunteers at a dose of 10^7 CFU. However, the vibriocidal responses were so diminished compared to those elicited by CVD 101 that this strain was not considered capable of immunizing with a single-dose vaccine, and so further clinical studies with CVD 102 were abandoned.

Two additional second-generation vaccine strains provided invaluable information in directing further live cholera vaccine development (21). JJM43 is a further derivative of strain 395N1 (*ctxA* mutant of Ogawa 395) in which the regulatory gene *toxR* is inactivated, making the vaccine candidate unable to respond to environmental signals and unable to coordinate the regulation of virulence gene expression (46). TCP2 is another derivative of 395N1 in which *tcpA,* the gene encoding the structural subunit of cholera toxin-coregulated pilus, has been inactivated (73). Each of these strains exhibited a greatly diminished ability to colonize the intestines of volunteers and to elicit vibriocidal antibody responses (21). Results of these studies emphasize the importance of retaining intact the fimbrial colonization factors and the master regulatory gene (*toxR*) of *V. cholerae* O1 in attenuated *V. cholerae* candidate live vaccines.

Attenuated *V. cholerae* O1 Vaccine Strains CVD 103 and CVD 103-RM

It was not clear whether the residual reactogenicity of the first generation of recombinant *V. cholerae* O1 vaccine candidate strains was due to the expression of as-yet-uncharacterized enterotoxins or whether the very act of adherence of vibrios to enterocytes somehow results in secretion or diminished absorption by the enterocytes (37). In the mid-1980s, O'Brien et al. (51) reported that most wild-type strains of *V. cholerae* O1 elaborate minute quantities of a Shiga-like toxin that is cytotoxic for HeLa cells in tissue culture and that is neutralized, at least partially, by Shiga antitoxin. One wild-type strain known to be pathogenic for volunteers that was negative for this putative Shiga-like toxin was classical Inaba strain 569B (36). It was not known whether this Shiga-like toxin played a role in the residual reactogenicity encountered in volunteers fed vaccine candidates CVD 101, CVD 104, CVD 105, JBK 70, and 395N1, all of which were derived from wild-type parent strains that elaborated the Shiga-like toxin. Accordingly, there was interest in testing vaccine candidate CVD 103, an A^-B^+ derivative of classical 569B. Whatever the underlying explanation, the clinical studies of CVD 103 constituted a breakthrough; this vaccine strain was well tolerated yet highly immunogenic and protective (36, 37). A single dose of CVD 103 vaccine elicited seroconversions of vibriocidal antibody in 90% of vaccinees and of antitoxin in >80% and conferred upon volunteers significant protection against challenge with pathogenic *V. cholerae* O1 of either serotype or biotype (36, 37).

Ketley et al. (27) prepared a derivative of CVD 103 in which *recA* was inactivated. The purpose of this introduced mutation was to interfere with the ability of the vaccine strain to recombine foreign DNA into its chromosome. As a consequence, the theoretical possibility that the vaccine strain would acquire wild-type *ctxA* via a recombinational event involving a wild-type donor would be even more remote. However, the *recA* mutation, which was introduced into CVD 103 for purely theoretical reasons, notably diminished the immunogenicity of the vaccine strain (27). As a result, further work with CVD 103-RM was abandoned.

Construction of Definitive Attenuated *V. cholerae* O1 Vaccine Strain CVD 103-HgR

An Ad Hoc Advisory Committee that included representation from the National Institute of Allergy and Infectious Diseases, the Food and Drug Administration, the U.S. Agency for International Development, and the World Health Organization recommended that a unique marker be introduced into CVD 103 before clinical studies outside physical containment in less-developed countries were undertaken. The purpose of such a marker would be to allow the vaccine strain to be readily differentiated from wild-type strains. Accordingly, Ketley and others (24, 28, 36) introduced a gene encoding resistance to Hg^{2+} (*mer*) or encoding the production of urease into the *hlyA* locus of the chromosome of CVD 103. The derivative encoding resistance to Hg^{2+} was designated CVD 103-HgR (24, 28, 36). CVD 103-HgR is phenotypically resistant to Hg^{2+}, a property not found in wild-type *V. cholerae* O1 strains recovered in nature.

CLINICAL AND FIELD STUDIES WITH CVD 103-HgR

Overview

To date, approximately 4,000 subjects ranging in age from 24 months to 50 years have participated in clinical trials with CVD 103-HgR. These trials have been carried out in industrialized countries (United States, Switzerland, Italy) (10, 11, 29, 34, 36, 71) and in developing countries with endemic cholera (Indonesia, Thailand) (45, 63–65), epidemic cholera (Peru, Colombia) (19), or little or no cholera (Chile, Costa Rica) (30). For these studies, a practical formulation of the vaccine was used. It consisted of two aluminum foil sachets, one containing lyophi-

lized vaccine (and aspartame as sweetener) and the other containing buffer (to protect the vaccine strain from gastric acid) (Fig. 1). The two sachets are mixed in a cup containing 100 ml of water, and the resultant suspension is ingested by the subject (Fig. 2). In 1993, a large-scale field trial was initiated with approximately 66,000 subjects in North Jakarta, Indonesia, to assess the efficacy of a single dose of CVD 103-HgR in preventing cholera under natural conditions of challenge.

Safety of CVD 103-HgR

Table 4 summarizes the results of clinical studies that have assessed the safety of CVD 103-HgR. Initial studies in adult U.S. volunteers did not include placebo control groups. However, subsequently, multiple studies in adults, school-age children, and preschool children were of randomized, double-blind, and placebo-controlled design (10, 11, 29, 45, 64, 65, 71). Since CVD 103-HgR was derived from a pathogenic parent strain (classical Inaba 569B) that is known to be capable of causing cholera gravis in volunteers, con-

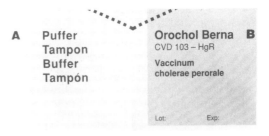

Figure 1. Practical formulation of live oral cholera vaccine CVD 103-HgR consisting of two sachets. One sachet contains lyophilized vaccine, and the other holds a buffer powder. The contents of the buffer powder sachet and then those of the lyophilized vaccine sachet are put into a cup containing approximately 100 ml of water and stirred (either chlorinated or nonchlorinated water may be used; the ascorbic acid in the buffer powder inactivates the antibacterial effect of residual chlorine). The resultant suspension, consisting of vaccine bacteria in buffer (the "oral vaccine cocktail"), is then ingested by the subject to be vaccinated.

Figure 2. Two young children in Area Norte, Santiago, Chile, ingesting a "vaccine cocktail" containing a dose of CVD 103-HgR live oral cholera vaccine.

siderable attention has been directed to ascertaining whether CVD 103-HgR causes diarrhea. As summarized in Table 4, a large clinical experience now attests to the safety of this vaccine strain in all age groups. Diarrheal adverse reactions have not been observed more often in vaccines than in placebo recipients in controlled trials. Placebo-controlled studies will begin shortly in infants and toddlers 7 to 23 months of age.

Immunogenicity of CVD 103-HgR

Since vibriocidal antibody is currently considered the best correlate of protection and the best measure of the successful stimulation of antibacterial immunity, the serum vibriocidal antibody response has been used to assess the immunogenicity of CVD 103-HgR (10, 11, 29, 30, 34, 36, 40, 45, 64, 65, 71); a fourfold or greater rise is considered significant (i.e., seroconversion).

Immunogenicity studies in populations in industrialized countries

In clinical studies with North American, Swiss, and Italian volunteers, a single dose of CVD 103-HgR (5×10^8 CFU) has elicited fourfold or greater rises in vibriocidal antibody in approximately 90% of vaccinees. Table 5 summarizes serum vibriocidal antibody responses in the first 156 Maryland adult volunteers who received CVD 103-HgR; these data comprise results from the vaccination of numerous groups who received different lots of vaccine. Ninety-two percent of individuals who ingested a single dose containing 5×10^8 CFU of CVD 103-HgR seroconverted, with 50% attaining reciprocal titers of $\geq 2,560$ (arbitrarily considered a high titer based on earlier studies with other cholera vaccines). Table 5 also shows the serologic responses of a cohort of 94 young adults who were immunized with a single dose of CVD 103-HgR (5×10^8 CFU) from a single representative lot. Once again, a high rate of seroconversion (97%) and a high GMT are observed.

Also summarized in Table 5 are Inaba vibriocidal responses in Marylanders who ingested three spaced doses (2 weeks apart) of the inactivated oral cholera vaccines (WCV and WCV/BS) from a collaborative study (3) with Holmgren and Svennerholm of Goteborg University, who developed the oral inactivated vaccines. A single dose of the live vaccine elicited vibriocidal antibody GMTs three- to fivefold higher than those elicited by three doses of the oral inactivated vaccines. Few subjects who received the inactivated vaccines attained high vibriocidal titers.

It is appropriate to hypothesize what factors account for the markedly greater antibacterial immunogenicity of the live oral vaccine. We propose two. First, the initiation of the intestinal immune response begins with uptake of antigen by the dome-like epithelial cells (so-called M cells) that cover the Peyer's patches and other organized lymphoid tissue of the gut (42). Owen et al. (53) have shown

Table 4. Rate of diarrheal adverse reactions in recipients of CVD 103-HgR live oral cholera vaccine versus placebo in randomized, double-blind trials with adults and children

Age group and site	Vaccine dose (CFU)	Rate of diarrhea (%)[a]		Reference
		Vaccinees	Controls	
Adults				
United States	5×10^8	1/200 (0.5)		34
United States	5×10^8	1/94 (1.1)	0/94 (0)	29
Switzerland	5×10^8	1/25 (4)	2/25 (8)	10
Thailand	5×10^8	0/12 (0)	0/12 (0)	45
Thailand	5×10^8	11/102 (11)	13/104 (13)	64
Thailand	5×10^8	0/103 (0)		64
Thailand	5×10^9	5/119 (4)	2/89 (2)	64
Peru	5×10^8	2/41 (4.9)	3/40 (7.5)	19
	5×10^9	5/41 (12.2)	3/40 (7.5)	19
Chile	5×10^9	0/40 (0)	1/41 (2.4)	30
5–9 yr old				
Indonesia	5×10^6–5×10^8	10/209 (4.8)	4/65 (6.2)	65
	5×10^9–5×10^{10}	4/124 (3.2)	4/32 (12.5)	
Chile	5×10^9	2/178 (1.1)	1/171 (0.6)	
Costa Rica	5×10^9	2/196 (1.0)	4/193 (2.1)	
Peru	5×10^8	1/40 (2.5)	1/40 (2.5)	
	5×10^9	1/40 (2.5)	1/40 (2.5)	
24–59 mo old				
Indonesia	5×10^9	18/155 (11.6)	12/148 (8.1)	63
Costa Rica	5×10^9	2/118 (1.7)	2/118 (1.7)	

[a]Number with diarrhea/number vaccinated. P = not significant in all cases.

that in rabbit ileal loops containing a Peyer's patch, M cells take up live *V. cholerae* O1 more readily than inactivated bacteria. Second, and perhaps more important, *V. cholerae* O1 infection is a model of coordinate regulation of numerous tightly controlled virulence properties that are activated or depressed by regulatory genes that respond to environmental signals (13, 46). For example, ToxR, the product of the *toxR* regulatory gene, directly or indirectly (through *toxT*) regulates 18 other genes, including those for cholera toxin and the toxin-coregulated pilus colonization factor (13, 46). With inactivated vaccine, the surface antigens present during growth in the fermentor are the surface antigens that will be

Table 5. Inaba vibriocidal responses in adult Maryland volunteers who received inactivated or live oral vaccines

Vaccine	No. of doses	n	% with:		Peak GMT
			4-fold rise in titer	Titer of > 2,560	
CVD 103-HgR[a]	1	155	92	50	1,699
CVD 103-HgR[b]	1	94	97	60	2,656
BS-WCV[c]	3	19	89	16	565
WCV[d]	3	14	56	6	306

[a]Results from the first 155 recipients of a single dose of vaccine (5×10^8 CFU). Multiple lots were used.

[b]Results of a cohort study in which college students received a single dose of vaccine (5×10^8 CFU) from a single vaccine lot.

[c]Volunteers received three doses of vaccine (2 weeks apart), each containing 2×10^{11} inactivated *V. cholerae* O1 bacteria (of both biotypes and serotypes) and 5 mg of purified B subunit.

[d]Volunteers received three doses of vaccine (2 weeks apart), each containing 2×10^{11} inactivated *V. cholerae* O1 bacteria (of both biotypes and serotypes).

exposed to the intestinal immune system; the inactivated vibrios cannot change in response to environmental stimuli in the intestine. In contrast, in CVD 103-HgR, *toxR* is intact, and this live vaccine organism can modify and express the new proteins that are required to survive in and colonize the intestine. Some of these in vivo-activated gene products are immunogenic and may play a role in protection. A clear-cut demonstration of the effect of the *toxR* regulation of virulence properties on the degree of stimulation of vibriocidal antibodies was reported by Herrington et al. (21) and is summarized in Table 6. Herrington et al. (21) fed volunteers a series of isogenic vaccine strains constructed by Mekalanos and coworkers (73) that included 395N1, JJM43, and TCP2. Strain 395N1 is an A^-B^+ derivative of wild-type classical Ogawa 395, JJM43 is a further mutant of 395N1 in which *toxR* is inactivated, and TCP2 is a derivative of 395N1 in which *tcpA* (the gene encoding the structural subunit of toxin-coregulated pilus) is inactivated. When adult Maryland volunteers were immunized orally with these three strains, the vibriocidal responses were markedly diminished in recipients of the derivatives harboring mutations in *toxR* or *tcpA* (Table 6).

Immunogenicity studies in populations in less-developed countries

When community studies of adults and children began in less-developed countries where cholera was endemic or epidemic, a dosage level of 5×10^8 CFU, which is highly immunogenic in subjects in industrialized countries (>90% seroconversion), elicited seroconversions of vibriocidal antibody in only ca. 25% of Thai soldiers (64) and in only 16% of Indonesian children (65). It was subsequently found that many individuals in endemic areas already have elevated vibriocidal titers and are presumably at least partially immune (64, 65). In several studies it has been noted that the baseline vibriocidal GMT in subjects who do not seroconvert

Table 6. Effect of inactivation of *toxR* or *tcpA* on degree of stimulation of serum vibriocidal antibodies by adult vaccinees[a]

Vaccine strain	No. seroconverting/ no. vaccinated	Peak postvaccination GMT of vibriocidal antibody
395N1[b]	7/8	2,153
JJM43[c]	2/7	230
TCP2[d]	0/8	52

[a]*toxR* coordinates the regulation of numerous virulence properties, and *tcpA* encodes the structural subunit of toxin-coregulated pilus.
[b]A^-B^+ derivative of wild-type strain classical Ogawa 395 in which a deletion has been made in *ctxA*.
[c]Further derivative of 395N1 in which *toxR* has been inactivated.
[d]Further derivative of 395N1 in which *tcpA* has been inactivated.

(i.e., at least a fourfold rise in titer) is significantly higher than the baseline GMT of individuals who do seroconvert (64, 65). In such persons with elevated baseline titers, a further rise in serum vibriocidal antibody is not a good measure of vaccine "take" or boosting. In these populations, administering a single dose of vaccine containing a 1-log-higher number of vaccine organisms (i.e., 5×10^9 CFU) markedly increases the vibriocidal seroconversion. With a single 5×10^9-CFU dose of vaccine, high rates of seroconversion of vibriocidal antibody (usually 75 to 85%) have been achieved in the less-developed-country populations studied so far (64, 65). Table 7 shows vibriocidal antibody responses in adult populations in Thailand, Peru, and Chile and in pediatric populations in Indonesia and Chile.

Efficacy

Experimental challenge studies to assess vaccine efficacy

A series of challenge studies have been carried out in the volunteer model of experimental cholera. Eight such challenges have been performed, three with volunteers who received CVD 103 and five with volunteers who received CVD 103-HgR (34, 36). In all but one study, only a single dose of vaccine was administered; in one study (El Tor

Table 7. Serum vibriocidal antibody response of adult and pediatric populations in different geographic areas who were immunized with a single oral dose (5×10^9 CFU)

Site of study and population	No. seroconverting/ no. vaccinated (%)	GMT		Reference
		Prevaccination	Peak postvaccination	
Thailand				
Adults	39/80 (48)	82	368	64
Chile				
Adults	34/40 (85)	15	222	30
5–9 yr old	127/171 (74)	20	328	
Peru				
Adults (high SEL[a])	31/40 (78)	14	374	19
Adults (low SEL)	28/39 (72)	16	202	19
Indonesia				
5–9 yr old (lot C)	27/31 (87)	43	800	65
5–9 yr old (lot F)	21/28 (75)	51	453	65
2–4 yr old	94/125 (75)	24	243	63

[a]SEL, socioeconomic level.

Ogawa challenge), the volunteers ingested two doses of vaccine 1 week apart. Subsequent immunogenicity studies have detected no difference in the vibriocidal antibody responses of subjects who receive a single dose of CVD 103-HgR and those who get two doses 1 week apart (64).

In each of these eight challenge studies (Table 8), CVD 103-HgR (or its parent, CVD 103) conferred significant protection against challenge with fully enterotoxigenic wild-type *V. cholerae* O1 (strains that caused cholera diarrhea in 78 to 100% of unimmunized control volunteers) (34). Overall protection against challenge was quite high with classical biotype (82 to 100% irrespective of serotype) and moderate with El Tor biotype (62 to 67% irrespective of serotype). However, a more relevant way to consider the challenge data is from the perspective of prevention of severe and moderate diarrhea, since it is the syndrome of cholera gravis that establishes cholera as a public health problem. Table 9 summarizes the eight challenge studies with respect to protection against diarrhea of different severity determined by total diarrheal stool volume. CVD 103 and CVD 103-HgR provide complete protection against severe (≥ 5.0-liter total purge) and moderate (≥ 3.0-liter total purge) diarrhea. In fact, of the 11

vaccinees who manifested diarrhea following challenge with wild-type *V. cholerae* O1, only two subjects purged a total diarrheal stool volume of 2.0 liters or more. Thus, CVD 103-HgR is highly protective against severe and moderate cholera of the type that would lead to dehydration.

A single dose of CVD 103-HgR provided 100% protection in volunteers who were challenged with wild-type *V. cholerae* O1

Table 8. Efficacy of CVD 103 and 103-HgR live oral cholera vaccines evaluated in experimental challenge studies in volunteers[a]

Vaccine	Interval between vaccination and challenge (wk)	% Protective efficacy
CVD 103		
Classical Inaba	4	87
Classical Ogawa	4	82
El Tor Inaba	4	67
CVD 103-HgR		
Classical Inaba	4	100
Classical Inaba	24	100
Classical Inaba	1	100
El Tor Inaba	4	62
El Tor Ogawa	4	64

[a]In all but one study, subjects received a single dose of vaccine; before challenge with El Tor Ogawa, vaccinees received two doses 1 week apart. This summary includes data from Levine et al. (34, 36) and Tacket et al. (71) and previously unpublished data.

Table 9. Efficacy of CVD 103 and CVD 103-HgR live oral cholera vaccines against experimental challenge with wild-type *V. cholerae* O1

Stool vol (liters)[a]	Controls	Vaccinees	% Efficacy	P
≥5.0	7/80[b]	0/82[b]	100	<0.01
≥3.0	16/80	0/82	100	<0.001
>2.0	24/80	2/82	92	<0.001
≥1.0	36/80	4/82	89	<0.001
Any	63/80	11/82	83	<0.001

[a]During episode of experimental cholera.
[b]Number with diarrhea of indicated stool volume/number challenged.

(classical Inaba) 6 months after immunization (the longest interval so far tested) (71). Moreover, when a group of volunteers were challenged with classical Inaba a mere 8 days after ingesting a single dose of CVD 103-HgR, they were also 100% protected against cholera (71). A challenge organism of the homologous biotype and serotype was used in these studies to maximize the chance of finding a beneficial effect, because these are the shortest and longest intervals between vaccine and challenge that have ever been tested for any candidate vaccine in the volunteer model.

Field trials of efficacy in areas where cholera is endemic

In 1993, a large-scale, randomized, placebo-controlled field trial was initiated in about 68,000 subjects 2 to 42 years of age in North Jakarta to assess the efficacy of a single oral dose of CVD 103-HgR in protecting against cholera in a population exposed to natural challenge in an area where cholera is endemic. The aims of this study include measuring vaccine efficacy in persons younger than 6 years of age (an age group not well protected by earlier vaccines) and comparing efficacy in persons of blood group O versus those of other blood groups. Surveillance will be maintained for at least 3 years before data are analyzed.

Newer Attenuated *V. cholerae* O1 Vaccine Strains

Investigators at the Center for Vaccine Development of the University of Maryland have identified two new toxins in addition to cholera toxin that are elaborated by virulent *V. cholerae* O1. The genes that encode these two toxins, zonula occludens toxin (Zot) (1, 14) and accessory cholera toxin (Ace) (74), are located on the "cholera toxin element" of the cholera chromosome contiguous to *ctxAB*, which encodes cholera toxin. This region in El Tor strains is found between two insertion-like sequences, RS, that make this transposon-like region of the chromosome unstable and subject to duplication or deletion (18). Accordingly, it has been possible to construct new vaccine candidate strains from wild-type El Tor parent strains such that they lack the entire cholera toxin virulence cassette that falls between the two RS elements (34, 44, 56). Thus, these new candidates lack Zot and Ace as well as cholera toxin. The genes encoding B subunit were then introduced into another chromosomal locus such as *hlyA* or *recA*. The candidate constructed by Michalski et al. (44) also lacks El Tor hemolysin, a protein that exhibits enterotoxic activity (41). Clinical studies have been carried out with several of these El Tor candidate strains, including strain CVD 110 (70). Surprisingly, the candidate vaccine strains of this nature tested so far, including CVD 110, have caused mild (and occasionally moderate) diarrhea accompanied by headache, malaise, abdominal cramps, or low-grade fever in a proportion of volunteers (70, 72). These are important findings because they demonstrate that *V. cholerae* O1 candidate vaccine strains that are derived from known virulent strains by deleting the capacity to express cholera toxin, Zot, Ace, and El Tor hemolysin still retain the propensity to cause mild (but definite) diarrheal illness. The pathophysiologic explanation for the residual diarrheagenic potential of these strains has yet to be elucidated. One possibility is that additional, so-far-unidentified enterotoxins are elaborated in vivo. An alternative plausible explanation is that the act of adherence of vaccine strains to enterocytes in the proximal small intestine per se results in net secretion.

VACCINE STRAIN TRANSMISSIBILITY

In some quarters, concern has been expressed over the excretion of recombinant vaccine strains and their possible persistence in the environment, where they might acquire wild-type genes that could theoretically restore their virulence. Objectively, there is no inherent reason to be more concerned about mutations prepared by recombinant techniques than about those prepared by other modalities, such as chemical mutagenesis. Nevertheless, steps have been taken to address such concerns. These have included selecting attenuated strains that are minimally excreted, attempting to introduce mutations that diminish the likelihood of recombinational events, and looking for evidence of persistence of the vaccine strain in the environment around the households of vaccinated subjects in less-developed countries where clinical trials have been carried out.

A striking contrast between CVD 103-HgR and its parent, CVD 103, is that while the latter was recovered from >80% of vaccinees (36), the former was isolated from stool cultures of only 5 to 25% of vaccinees (29, 30, 36, 45, 65, 71). By culturing all stools passed by volunteers on a research ward and using enrichment procedures, CVD 103-HgR was recovered from 25% of vaccinees in the United States (36) and 12% of adult volunteers in Thailand (45). In phase 2 field studies, where only once-daily stool cultures have been obtained for only a few days, the vaccine strain has been recovered from 0 to 12% of vaccinees (29, 34, 65).

Because of the minimal excretion of CVD 103-HgR, it is not surprising that the vaccine strain has been recovered from only one contact control among several hundred residing within the households of vaccinees whose specimens were cultured (63).

During the vaccination of several hundred children in North Jakarta, Indonesia, environmental bacteriologic sampling in ditches and open sewers around the houses of vaccinees failed to yield the vaccine strain. In contrast, wild *V. cholerae* (non-O1 serogroups) were recovered from 96% of the environmental samples.

Attempts have been made to alter attenuated cholera strains so that they have a diminished likelihood of acquiring wild-type *ctxA* by recombinational events. This was the objective of the construction by Ketley et al. (27) of a variant of CVD 103 with a deletion in *recA*. More recently, Pearson et al. (56) have prepared attenuated cholera vaccine strains that are deleted of the 18-bp sequence (so-called attRS1) that is the "hot spot" for recombination with RS1 elements. The objective of these constructs is to reduce the possibility that a cholera toxin virulence gene would enter the chromosome of the vaccine strain at the attRS1 site in a transposon-like manner.

ATTENUATED *S. TYPHI* EXPRESSING *V. CHOLERAE* O1 ANTIGENS

Strain Construction

Investigators in Adelaide, Australia, cloned the genes required for expression of the Inaba serotype O antigen of *V. cholerae* O1 onto a plasmid that also carried *thyA*, which encodes thymine independence. They introduced this plasmid into a rifampin-resistant variant of attenuated *S. typhi* Ty21a (a licensed live oral typhoid vaccine) carrying a deletion mutation in *thyA;* strain Ty21a was further altered in that the *rfa* chromosomal region involved with LPS expression was replaced with *rfa* from *Escherichia coli* K-12. The plasmid corrected the *thyA* defect in *trans,* thereby stabilizing the plasmid in the *S. typhi* background. The resultant transconjugant, *S. typhi* EX645, expressed the O antigen of *V. cholerae* O1 Inaba (15).

Reactogenicity and Immunogenicity

In phase 1 studies with EX645 carried out in Adelaide (15) and in Baltimore, Maryland (68), the vaccine was well tolerated when subjects were administered three large doses (every-other-day schedule), each containing approximately 5×10^{10} CFU.

Immunogenicity

In preliminary phase 1 studies with adult volunteers in Adelaide, EX645 was only modestly immunogenic in eliciting cholera antibodies. Following administration of three spaced doses (every-other-day interval) containing ca. 10^{10} CFU per dose, most vaccinees exhibited detectable but minimal immune responses to Inaba LPS either as serum antibody measured by enzyme-linked immunosorbent assay or as gut-derived circulating peripheral blood mononuclear cells that secrete IgA antibody when exposed to Inaba LPS. In contrast, only 5 of 10 vaccinees developed fourfold or greater rises in serum Inaba vibriocidal antibody (15).

Among 14 adult volunteers immunized with three doses (10^{10} CFU per dose) of EX645 in Maryland, only 2 (14%) of 14 manifested significant rises in serum IgG or IgA antibody to Inaba LPS, and 5 had significant rises in serum Inaba vibrocidal antibody. In contrast, all 14 vaccinees manifested significant rises in serum IgG antibody to *S. typhi* LPS. Five (36%) of 14 vaccinees developed significant rises in serum vibriocidal antibody (68).

The intestinal secretory IgA (sIgA) antibody responses of the 14 Maryland vaccinees to *S. typhi* LPS were strong, but those to Inaba LPS were weak (68). Twelve (86%) of 14 subjects exhibited fourfold or greater rises in jejunal fluid sIgA antibody to *S. typhi* LPS, whereas only 1 (7%) of 14 manifested sIgA rises to Inaba LPS ($P < 0.0001$). Similar results were observed when specific gut-derived trafficking IgA antibody-secreting cells were looked for among peripheral blood mononuclear cells that secreted antibody when incubated in the presence of *V. cholerae* Inaba and *S. typhi* O antigens.

Efficacy in Preventing Experimental Cholera

One month following vaccination, eight subjects who received three doses of EX645 and 13 unvaccinated controls participated in an experimental challenge study to assess vaccine efficacy (68). The volunteers were challenged with 10^6 CFU of wild-type strain N16961 (El Tor Inaba). The attack rate for diarrhea overall in the vaccinees (6 of 8; 75%) was not significantly lower than that observed in the controls (13 of 13; 100%) (overall, 25% vaccine efficacy). However, the severity of the diarrhea was significantly less in the vaccinees (mean, 867 ml per ill volunteer) than in the controls (mean, 2,603 ml per ill volunteer) ($P < 0.05$). The excretion of wild-type *V. cholerae* O1 following challenge was also significantly lower in vaccinees than in controls ($P < 0.05$).

Although the EX645 vaccine was notably less protective than a number of other live and inactivated oral vaccines evaluated in the experimental challenge model as used at the Center for Vaccine Development, it nevertheless clearly demonstrated a significant biologic effect by ameliorating the severity of illness and by curtailing excretion of vibrios. The fact that this protective effect was accomplished by a live vector vaccine that expressed only a single *V. cholerae* O1 antigen and was only modestly immunogenic should be considered highly encouraging with respect to this general approach. Presumably, if multiple additional *V. cholerae* O1 bacterial and toxoid antigens could be expressed with greater efficiency in a new-generation attenuated *S. typhi* live vector that is inherently more immunogenic (69), this strategy to elicit protective immunity might yet prove fruitful.

EFFICACY OF *V. CHOLERAE* O1 VACCINES AGAINST DISEASE CAUSED BY SEROGROUP O139 AND PROSPECTS FOR O139 VACCINES

In late 1992 and early 1993, there appeared in the Indian subcontinent an epidemic cholera of notable clinical severity that was caused by a *V. cholerae* strain of a serogroup other than O1. In its general epidemiologic and clinical behavior, this infection had the characteristics of typical cholera. It caused explosive outbreaks of a profuse diarrheal illness that rapidly led to severe dehydration. The epidemic was extensive, involving numerous areas of India and Bangladesh. One unusual epidemiologic feature of the epidemic in these areas where cholera is endemic was the age-specific incidence pattern, whereby high attack rates were observed in adults as well as in children, suggesting the immunologic susceptibility of the population. Heretofore, dogma and orthodoxy have taught that only *V. cholerae* of serogroup O1 is capable of causing large-scale explosive epidemics and pandemic spread. Nevertheless, the preliminary epidemiologic behavior of the new epidemic strain, which is of a previously unrecognized serogroup that has been tentatively assigned the designation O139, is typical of epidemic O1 strains.

Ordinarily, in areas of Bangladesh where cholera is endemic, the incidence of cholera is highest in toddlers and young children and markedly diminishes with increasing age; this distribution is evidence that immunity is increasingly acquired from repeated contact with cholera vibrios. Viewed from this perspective, the occurrence of high incidence rates of O139 cholera in adults in Bangladesh suggests that prior immunity elicited by *V. cholerae* O1 offers little protection against the new strain. As a corollary, there is little reason to believe that the new generation of oral cholera vaccines will protect against O139 disease.

Rapid vaccine development programs have been initiated in several centers worldwide to prepare candidate vaccines, including live oral vaccines, against the O139 strain. The strategies being followed are those that have been used to develop vaccines against *V. cholerae* O1, as described earlier in this chapter. Aside from its different O antigen and the presence of a polysaccharide capsule, the new epidemic strain closely resembles an El Tor strain in its virulence properties. Clinical trials with candidate live oral vaccines against *V. cholerae* O139 are expected to commence in 1994.

REFERENCES

1. **Baudry, B., A. Fasano, J. Ketley, and J. B. Kaper.** 1992. Cloning of a gene (*zot*) encoding a new toxin produced by *Vibrio cholerae*. *Infect. Immun.* **60**:428–434.
2. **Benenson, A. S., A. Saad, and W. H. Mosley.** 1968. Serological studies in cholera. 2. The vibriocidal antibody response of cholera patients determined by a microtechnique. *Bull W.H.O.* **38**:277–285.
3. **Black, R. E., M. M. Levine, M. L. Clements, C. R. Young, A.-M. Svennerholm, and J. Holmgren.** 1987. Protective efficacy in humans of killed whole-vibrio oral cholera vaccine with and without the B subunit of cholera Toxin. *Infect. Immun.* **55**:1116–1120.
4. **Chongsa-Nguan, M., W. Chaicumpa, T. Kalambaheti, H. Soejedi, P. Luxananil, S. Swasdikosa, P. Thanasiri, and S. Mayurasakorn.** 1986. Vibriocidal antibody and antibodies to *Vibrio cholerae* lipopolysaccharide, cell-bound haemagglutinin and toxin in Thai population. *Southeast Asian J. Trop. Med. Public Health* **17**:L558–566.
5. **Clemens, J. D., D. A. Sack, J. Harris, J. Chakraborty, M. Khan, S. Huda, F. Ahmed, J. Gomes, M. Rao, A.-M. Svennerholm, and J. Holmgren.** 1989. ABO blood groups and cholera: new observations on specificity of risk and modifications of vaccine efficacy. *J. Infect. Dis.* **159**:770–773.
6. **Clemens, J. D., D. A. Sack, J. R. Harris, J. Chakraborty, M. R. Khan, B. F. Stanton, B. A. Kay, M. U. Khan, M. Yunus, and W. Atkinson.** 1986. Field trial of oral cholera vaccines in Bangladesh. *Lancet* **ii**:124–127.
7. **Clemens, J. D., D. A. Sack, J. Harris, F. Van Loon, J. Chakraborty, F. Ahmed, M. R. Rao, M. R. Khan, M. Yunus, and N. Huda.** 1990. Field trial of oral cholera vaccines in Bangladesh: results from three year follow-up. *Lancet* **335**:270–273.

8. Clemens, J. D., F. Van Loon, D. A. Sack, J. Chakraborty, M. R. Rao, F. Ahmed, J. R. Harris, M. R. Khan, M. Yunus, and S. Huda. 1991. Field trial of oral cholera vaccines in Bangladesh: serum vibriocidal and antitoxic antibodies as markers of the risk of cholera. *J. Infect. Dis.* **163:**1235–1242.

9. Clemens, J. D., F. Van Loon, D. A. Sack, M. R. Rao, F. Ahmed, J. Chakraborty, B. A. Kay, M. R. Khan, M. Yunus, J. R. Harris, A.-M. Svennerholm, and J. Holmgren. 1991. Biotypes as determinant of natural immunising effect of cholera. *Lancet* **337:**883–884.

10. Cryz, S. J., Jr., M. M. Levine, J. B. Kaper, E. Furer, and B. Althaus. 1990. Randomized double-blind placebo-controlled trial to evaluate the safety and immunogenicity of the live oral cholera vaccine strain CVD 103-HgR in Swiss adults. *Vaccine* **8:**577–580.

11. Cryz, S. J., Jr., M. M. Levine, G. Losonsky, J. B. Kaper, and B. Althaus. 1992. Safety and immunogenicity of a booster dose of *Vibrio cholerae* CVD 103-HgR live oral cholera vaccine in Swiss adults. *Infect. Immun.* **60:**3916–3917.

12. Curlin, G., R. Levine, K. M. S. Aziz, A. S. M. Mizanur Rahman, and W. F. Verwey. 1975. Field trial of cholera toxoid, 314–329. *In Proceedings of the 11th Joint Conference on Cholera*. National Institutes of Health, Bethesda, Md.

13. DiRita, V. J., C. Parsot, G. Jander, and J. J. Mekalanos. 1991. Regulatory cascade controls virulence in *Vibrio cholerae. Proc. Natl. Acad. Sci. USA* **88:**5403–5407.

14. Fasano, A., B. Baudry, D. W. Pumplin, S. S. Wasserman, B. D. Tall, J. M. Ketley, and J. B. Kaper. 1991. Vibrio cholerae produces a second enterotoxin, which affects intestinal tight junctions. *Proc. Natl. Acad. Sci. USA* **88:**5242–5246.

15. Forrest, B. D., J. LaBrooy, S. R. Attridge, G. Boehm, L. Beyer, R. Morona, D. J. Shearman, and D. Rowley. 1989. A candidate live oral typhoid/cholera hybrid vaccine is immunogenic in humans. *J. Infect. Dis.* **159:**145–146.

16. Glass, R. I., S. Becker, I. Huq, B. J. Stoll, M. U. Khan, M. H. Merson, J. V. Lee, and R. E. Black. 1982. Endemic cholera in rural Bangladesh, 1966–1980. *Am. J. Epidemiol.* **116:**959–970.

17. Glass, R. I., A.-M. Svennerholm, M. R. Khan, S. Huda, M. I. Huq, and J. Holmgren. 1985. Seroepidemiological studies of El Tor cholera in Bangladesh: association of serum antibody levels with protection. *J. Infect. Dis.* **151:**236–242.

18. Goldberg, I., and J. J. Mekalanos. 1986. Effect of *recA* mutation on cholera toxin gene amplification and deletion events. *J. Bacteriol.* **165:**723–731.

19. Gotuzzo, E., B. Butron, C. Seas, M. Penny, R. Ruiz, G. Losonsky, C. F. Lanata, S. S. Wasserman, E. Salazar, J. B. Kaper, S. Cryz, and M. M. Levine. 1993. Safety, immunogenicity and excretion pattern of single-dose live oral cholera vaccine CVD 103-HgR in Peruvian adults of high and low socioeconomic level. *Infect. Immun.* **61:**3994–3997.

20. Hall, R. H., G. Losonsky, A. D. P. Silveira, R. K. Taylor, J. J. Mekalanos, N. D. Witham, and M. M. Levine. 1991. Immunogenicity of *Vibrio cholerae* O1 toxin coregulated pili in experimental and clinical cholera. *Infect. Immun.* **59:**2508–2512.

21. Herrington, D. A., R. H. Hall, G. A. Losonsky, J. J. Mekalanos, R. K. Taylor, and M. M. Levine. 1988. To, toxin-coregulated pili, and the *tox*R regulon are essential for *Vibrio cholerae* pathogenesis in humans. *J. Exp. Med.* **168:**1487–1492.

22. Jonson, G., J. Holmgren, and A.-M. Svennerholm. 1991. Epitope differences in toxin-coregulated pili produced by classical and El Tor *Vibrio cholerae* O1. *Microb. Pathog.* **11:**179–188.

23. Jonson, G., A.-M. Svennerholm, and J. Holmgren. 1989. *Vibrio cholerae* expresses cell surface antigens during intestinal infection which are not expressed during in vitro culture. *Infect. Immun.* **57:**1809–1815.

24. Kaper, J. B., and M. M. Levine. 1990. Recombinant attenuated *Vibrio cholerae* strains used as live oral vaccines. *Res. Microbiol.* **141:**901–906.

25. Kaper, J. B., H. Lockman, M. M. Baldini, and M. M. Levine. 1984. Recombinant live oral cholera vaccine. *Bio/Technology* **2:**345–349.

26. Kaper, J. B., H. Lockman, M. M. Baldini, and M. M. Levine. 1984. Recombinant nontoxigenic *Vibrio cholerae* strains as attenuated cholera vaccine candidates. *Nature* (London) **308:**655–658.

27. Ketley, J., J. B. Kaper, D. A. Herrington, G. Losonsky, and M. M. Levine. 1990. Diminished immunogenicity of a recombination-deficient derivative of *Vibrio cholerae* vaccine strain CVD 103. *Infect. Immun.* **58:**1481–1484.

28. Ketley, J. M., J. Michalski, J. Galen, M. M. Levine, and J. B. Kaper. 1993. Construction of genetically-marked *Vibrio cholerae* O1 vaccine strains. *FEMS Microbiol. Lett.,* **111:**15–21.

29. Kotloff, K., S. S. Wasserman, S. A. O'Donnell, G. A. Losonsky, S. J. Cryz, and M. M. Levine. 1992. Safety and immunogenicity in North Americans of a single dose of live oral cholera vaccine CVD 103-HgR: results of a randomized, placebo-controlled, double-blind crossover trial. *Infect. Immun.* **60:**4430–4432.

30. Lagos, R., A. Avendaño, I. Horwitz, V. Prado, C. Ferreccio, G. Losonsky, S. S. Wasserman, S. Cryz, J. B. Kaper, and M. M. Levine. 1993. Tolerancia e immunogenicidad de ina dosis oral de la cepa de *Vibrio cholerae* O1, viva-atenuada, CVD 103-HgR: estudio de doble ciego en adultos Chilenos. *Rev. Med. Chile* **121:**857–863.

31. **Levine, M. M., R. E. Black, M. L. Clements, L. Cisnero, D. R. Nalin, and C. R. Young.** 1981. The quality and duration of infection-derived immunity to cholera. *J. Infect. Dis.* **143**:818–820.

32. **Levine, M. M., R. E. Black, M. L. Clements, and J. B. Kaper.** 1984. Present status of cholera vaccines. *Biochem. Soc. Trans.* **12**:200–202.

33. **Levine, M. M., R. E. Black, M. L. Clements, D. R. Nalin, L. Cisnero, and R. Finkelstein.** 1981. Volunteer studies in development of vaccines against cholera and enterotoxigenic *Escherichia coli*: a review, p. 443–459. *In* T. Holme and J. Holmgren (ed.), *Acute Enteric Infections in Children: New Prospects for Treatment and Prevention.* Elsevier, Amsterdam.

34. **Levine, M. M., and J. B. Kaper.** 1993. Live vaccines against cholera: an up-date. *Vaccine* **11**:207–212.

35. **Levine, M. M., J. B. Kaper, R. E. Black, and M. L. Clements.** 1983. New knowledge on pathogenesis of bacterial enteric infections as applied to vaccine development. *Microbiol. Rev.* **47**:510–550.

36. **Levine, M. M., J. B. Kaper, D. Herrington, J. Ketley, G. A. Losonsky, C. O. Tacket, B. D. Tall, and S. J. Cryz.** 1988. Safety, immunogenicity and efficacy of recombinant live oral cholera vaccine CVD 103 and CVD 103-HgR. *Lancet* **ii**:467–470.

37. **Levine, M. M., J. B. Kaper, D. A. Herrington, G. A. Losonsky, J. G. Morris, M. L. Clements, R. E. Black, B. D. Tall, and R. Hall.** 1988. Volunteer studies of deletion mutants of *Vibrio cholerae* O1 prepared by recombinant techniques. *Infect. Immun.* **56**:161–167.

38. **Levine, M. M., D. Nalin, J. P. Craig, D. Hoover, E. J. Bergquist, D. Waterman, H. P. Holley, R. B. Hornick, N. F. Pierce, and J. P. Libonati.** 1979. Immunity to cholera in man: relative role of antibacterial versus antitoxic immunity. *Trans. R. Soc. Trop. Med. Hyg.* **73**:3–9.

39. **Levine, M. M., and N. F. Pierce.** 1992. Immunity and vaccine development, p. 285–327. *In* W. B. Greenough III and D. Barua (ed.), *Cholera.* Plenum Press, New York.

40. **Losonsky, G. A., C. O. Tacket, S. S. Wasserman, J. B. Kaper, and M. M. Levine.** 1993. Secondary *Vibrio cholerae* specific cellular antibody responses following wild-type homologous challenge in people vaccinated with CVD 103-HgR live oral cholera vaccine: changes with time and lack of correlation and protection. *Infect. Immun.* **61**:729–733.

41. **Madden, J. M., B. A. McCardell, and D. B. Shah.** 1984. Cytotoxin production by members of genus *Vibrio. Lancet* **i**:1217–1218.

42. **McGhee, J. R., J. Mestecky, M. T. Dertzbaugh, J. H. Eldridge, M. Hirasawa, and H. Kiyono.** 1992. The mucosal immune system: from fundamental concepts to vaccine development. *Vaccine* **10**:75–88.

43. **Mekalanos, J. J., S. J. Swartz, G. D. N. Pearson, N. Harford, F. Groyne, and M. de Wilde.** 1983. Cholera toxin genes: nucleotide sequence, deletion analysis and vaccine development. *Nature* (London) **306**:551–557.

44. **Michalski, J., J. Galen, A. Fasano, and J. B. Kaper.** 1993. An attenuated *Vibrio cholerae* O1 El Tor live oral vaccine strain. *Infect. Immun.* **61**:4462–4468.

45. **Migasena, S., P. Pitisuttitham, B. Prayurahong, P. Suntharasami, W. Supanaranond, V. Desakorn, U. Vongsthongsri, B. Tall, J. Ketley, G. Losonsky, S. Cryz, J. B. Kaper, and M. M. Levine.** 1989. Preliminary assessment of the safety and immunogenicity of live oral cholera vaccine strain CVD 103-HgR in healthy Thai adults. *Infect. Immun.* **57**:3261–3264.

46. **Miller, J. M., J. J. Mekalanos, and S. Falkow.** 1989. Coordinate regulation and sensory transduction in the control of bacterial virulence. *Science* **243**:916–922.

47. **Mosley, W. H., S. Ahmad, A. S. Benenson, and A. Ahmed.** 1968. The relationship of vibriocidal antibody titre to susceptibility to cholera in family contacts of cholera patients. *Bull. W.H.O.* **38**:777–785.

48. **Mosley, W. H., A. S. Benenson, and R. Barui.** 1968. A serological survey for cholera antibodies in rural east Pakistan. I. The distribution of antibody in the control population of a cholera vaccine field-trial area and the relation of antibody titer to the pattern of endemic cholera. *Bull. W.H.O.* **38**:327–334.

49. **Neoh, S. H., and D. Rowley.** 1970. The antigens of Vibrio cholerae involved in the vibriocidal action of antibody and complement. *J. Infect. Dis.* **121**:505–513.

50. **Noriki, H.** 1976. Evaluation of toxic field trial on the Philippines, p. 302–310. *In* H. Fukumi and Y. Zinnaka (ed.), *Proceedings of the 12th Joint Conference on Cholera.* U.S.-Japan Cooperative Medical Science Program. Fuji, Tokyo.

51. **O'Brien, A. D., M. E. Chen, R. K. Holmes, J. B. Kaper, and M. M. Levine.** 1984. Environmental and human isolates of *Vibrio cholerae* and *Vibrio parahaemolyticus* produce a *Shigella dysenteriae* 1 (Shiga)-like cytotoxin. *Lancet* **i**:77.

52. **Osek, J., A.-M. Svennerholm, and J. Holmgren.** 1992. Protection against *Vibrio cholerae* El Tor infection by specific antibodies against mannose-binding hemagglutinin pili. *Infect. Immun.* **60**:4961–4964.

53. **Owen, R. L., N. F. Pierce, R. T. Apple, and W. D. Cray, Jr.** 1986. M Cell transport of Vibrio cholerae from the intestinal lumen into Peyer's patches: a mechanism for antigen sampling and for micro-

bial transepithelial migration. *J. Infect. Dis.* **153:**1108–1118.

54. **Parsot, C., E. Taxman, and J. J. Mekalanos.** 1991. ToxR regulates the production of lipoproteins and the expression of serum resistance in *Vibrio cholerae. Proc. Natl. Acad. Sci. USA* **88:**1641–1645.

55. **Pearson, G. D., V. J. DiRita, M. B. Goldberg, S. A. Boyko, S. B. Calderwood, and J. J. Mekalanos.** 1990. New attenuated derivatives of *Vibrio cholerae. Res. Microbiol.* **141:**893–899.

56. **Pearson, G. D. N., A. Woods, Chiang, and J. J. Mekalanos.** 1993. CTX genetic element encodes a site-specific recombination system and an intestinal colonization factor. *Proc. Natl. Acad. Sci USA* **90:**3750–3754.

57. **Richardson, K., J. B. Kaper, and M. M. Levine.** 1989. Human immune response to *Vibrio cholerae* O1 whole cells and isolated outer membrane protein antigens. *Infect. Immun.* **57:**495.

58. **Roberts, A., G. Pearson, and J. Mekalanos.** 1992. Cholera strains derived from a 1991 Peruvian isolate of *Vibrio cholerae* and other El Tor strains. *Abstr. 28th Joint Conf. Cholera Related Diarrheal Dis. Panel, U.S.-Japan Cooperative Medical Science Program.*

59. **Sciortino, C. V.** 1989. Protection against infection with V. cholerae by passive transfer of mononclonal antibodies to outer membrane antigens. *J. Infect. Dis.* **160:**248–252.

60. **Sharma, D. P., S. Attridge, J. Hackett, and D. Rowley D.** 1987. Nonlipopolysaccharide protective antigens shared by classical and El Tor biotypes of *Vibrio cholerae. J. Infect. Dis.* **155:**716–723.

61. **Sharma, D. P., C. Thomas, R. H. Hall, M. M. Levine, and S. R. Attridge.** 1989. Significance of toxin-coregulated pili as protective antigens of *Vibrio cholerae* in the infant mouse model. *Vaccine* **7:**451–456.

62. **Sigel, S. P., and S. M. Payne.** 1982. Effect of iron regulation on growth, siderophore production, and expression of outer membrane proteins of *Vibrio cholerae. J. Bacteriol.* **150:**148–155.

63. **Simanjuntak, C. H., P. O'Hanley, N. H. Punjabi, F. Noriega, G. Pazzaglia, P. Dykstra, B. Kay, Suharyono, A. Budiarson, A. R. Rifai, S. S. Wasserman, G. Losonsky, J. Kaper, S. Cryz, and M. M. Levine.** 1993. Studies of the safety, immunogenicity and transmissibility of single-dose live oral cholera vaccine CVD 103-HgR in 24 to 59 month old Indonesian children. *J. Infect. Dis.* **168:**1169–1176.

64. **Su-Arehawatana, P., P. Singharaj, D. Taylor, C. Hoge, A. Trofa, K. Kuvanont, S. Migasena, P. Pitisuttitham, Y.-L. Lim, G. Losonsky, J. B. Kaper, S. S. Wasserman, S. Cryz, P. Echeverria, and M. M. Levine.** 1992. Safety and immunogenicity of different immunization regimens of CVD 103-HgR live oral cholera vaccine in soldiers and civilians in Thailand. *J. Infect. Dis.* **165:**1042–1048.

65. **Suharyono, C. Simanjuntak, N. Witham, N. Punjabi, D. G. Heppner, G. A. Losonsky, H. Totosudirjo, A. R. Rifai, J. Clemens, Y.-L. Lim, D. Burr, S. S. Wasserman, J. B. Kaper, K. Sorenson, S. Cryz, and M. M. Levine.** 1992. Safety and immunogenicity of single-dose live oral cholera vaccine CVD 103-HgR in 5–9-year-old Indonesian children. *Lancet* **340:**689–694.

66. **Sun, D., D. M. Tillman, T. N. Marion, and R. K. Taylor.** 1990. Production and characterization of monoclonal antibodies to the toxin coregulated pilus (TCP) of *Vibrio cholerae* that protect against experimental cholera in infant mice. *Serodiagn. Immunother. Infect. Dis.* **4:**73–81.

67. **Svennerholm, A.-M., M. M. Levine, and J. Holmgren.** 1984. Weak serum and intestinal antibody responses to *Vibrio cholerae* soluble hemagglutinin in cholera patients. *Infect. Immun.* **45:**792–794.

68. **Tacket, C. O., B. Forrest, R. Morona, S. R. Attridge, J. LaBrooy, B. D. Tall, M. Reymann, D. Rowley, and M. M. Levine.** 1990. Safety, immunogenicity, and efficacy against cholera challenge in man of a typhoid-cholera hybrid vaccine derived from *Salmonella typhi* Ty21a. *Infect. Immun.* **58:**1620–1627.

69. **Tacket, C. O., D. M. Hone, G. Losonsky, L. Guers, R. Edelman, and M. M. Levine.** 1992. Clinical acceptability and immunogenicity of CVD 908 *Salmonella typhi* vaccine strain. *Vaccine* **10:**443–446.

70. **Tacket, C. O., G. Losonsky, J. P. Nataro, S. J. Cryz, R. Edelman, A. Fasano, J. Michalsi, J. B. Kaper, and M. M. Levine.** Safety, immunogenicity and transmissibility of live oral cholera vaccine candidate CVD 110, a ΔctxA Δzot Δace derivative of El Tor Ogawa *Vibrio cholerae. J. Infect. Dis.,* in press.

71. **Tacket, C. O., G. Losonsky, J. Nataro, S. J. Cryz, R. Edelman, J. B. Kaper, and M. M. Levine.** 1992. Rapid onset and duration of protective immunity in challenged volunteers after vaccination with live oral cholera vaccine CVD 103-HgR. *J. Infect. Dis.* **166:**837–841.

72. **Taylor, D., and J. J. Mekalanos.** Personal communication.

73. **Taylor, R. K., C. Shaw, K. Peterson, P. Spears, and J. J. Mekalanos.** 1988. Safe, live *Vibrio cholerae* vaccines. *Vaccine* **6:**151–154.

74. **Trucksis, M., J. E. Galen, J. Michalski, A. Fasano, and J. B. Kaper.** 1993. Accessory cholera enterotoxin (Ace), the third member of a Vibrio cholerae virulence cassette. *Proc. Natl. Acad. Sci. USA* **90:**5267–5271.

75. **Woodward, W.** 1971. Cholera reinfection in man. *J. Infect. Dis.* **123:**61–66.

Vibrio cholerae and Cholera: Molecular to Global Perspectives
Edited by I. Kaye Wachsmuth, Paul A. Blake, and Ørjan Olsvik
© 1994 American Society for Microbiology, Washington, DC 20005

Chapter 27

Protective Oral Cholera Vaccine Based on a Combination of Cholera Toxin B Subunit and Inactivated Cholera Vibrios

J. Holmgren, J. Osek, and A.-M. Svennerholm

There is an increasing need for an effective cholera vaccine. Cholera continues to be an important cause of illness and death in large parts of Asia, Africa, and the Middle East, and since 1991, cholera is also rapidly spreading in South and Central America as well. It has been estimated that at least 150,000 to 200,000 people, both children and adults, die from cholera each year. Although cholera can be treated successfully by oral and intravenous rehydration and electrolyte replacement therapy (2), diarrhea treatment centers are still scarce in many areas where cholera is now endemic, and use of an effective vaccine may in many places be the most promising possibility for controlling the disease or at least be a valuable complement to other control strategies (26). Parenteral whole-cell vaccines against cholera have existed for almost a century, but these vaccines are no longer considered useful from a public health standpoint, mainly because of the short duration of protection they provide. Several field trials since 1960 have shown that these vaccines usually protect older people for only a few months and fail to protect many young children at all (8).

The identification of the critical role of a toxin in the pathogenesis of cholera (10) led to great expectations that the use of a toxoid would result in a more effective cholera vaccine. However, a large field trial with a parenterally administered glutaraldehyde toxoid vaccine in Bangladesh in the mid-1970s failed to give any better protection than that achieved by previous parenteral whole-cell vaccines (8). This did not come as too much of a surprise, however, since during the time it took to organize, execute, and evaluate the trial, it had become apparent from animal studies that protective cholera immunity does not depend on the type of serum antibodies that are mainly stimulated by the parenteral vaccination but rather on mucosal secretory immunoglobulin A (IgA) antibodies produced locally in the gut and that these latter antibodies are only inefficiently stimulated by parenteral antigen administration (13). Therefore, since the late 1970s, attention has turned to development of oral vaccines that could stimulate intestinal immunity more efficiently.

The intense vaccine development and research efforts during the last 10 years have now provided several oral vaccine candidates based on either a combination of nonliving bacteria and purified B-subunit antigen, as described below, or on live attenuated mutants of *Vibrio cholerae* O1 that produce the B subunit (11).

J. Holmgren, J. Osek, and A.-M. Svennerholm • Department of Medical Microbiology and Immunology, University of Göteborg, S-413 46 Göteborg, Sweden.

RATIONALE FOR THE ORAL B-SUBUNIT–WHOLE-CELL VACCINE

The different "modern" approaches to development of cholera vaccines have taken their departure from the new insights into the mechanisms of disease and immunity in cholera achieved during the 1970s and early 1980s (see, e.g., references 10 and 13).

Cholera Toxin and Antitoxic Immunity

The clarification of the subunit structure of cholera toxin and the roles of different subunits in toxin action (Fig. 1) immediately suggested a way to prepare a safe and highly immunogenic "toxoid" consisting of the purified cholera B subunits. Studies showed that the purified B-subunit portion of cholera toxin was entirely devoid of toxicity, yet it contained selectively the protective toxin epitopes against which neutralizing antitoxin antibodies were directed (14). Indeed, the B subunit is particularly well suited to oral immunization not only because it is stable in the intestinal milieu but also because it retains the ability to bind to the intestinal epithelium (including the M cells of the Peyer's patches), both of which properties have been shown in animals to be important for stimulating

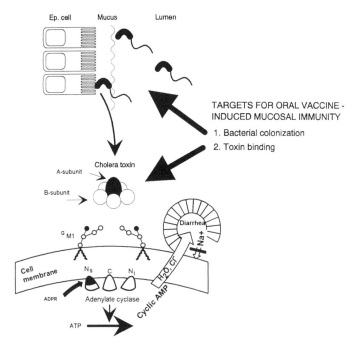

Figure 1. Pathogenesis of cholera and main targets of vaccine-induced immune protection. As illustrated, *V. cholerae* O1 bacteria have special (and apparently in part biotype-specific) means of adhering to, colonizing, and multiplying to large numbers in the small intestine. During these processes, the bacteria produce and release significant amounts of cholera enterotoxin. After binding (by means of the five B subunits) to GM_1 ganglioside receptors and ADP-ribosylating (ADPR) one of the regulatory subunits ($N_s[G_s]$) of adenylate cyclase (by means of the A subunit), cholera toxin stimulates excessive formation in affected enterocytes of cyclic AMP, which in turn through the associated secretion of electrolytes and water into the gut lumen may cause severe diarrhea and dehydration. Protective immunity is mainly mediated by locally produced secretory IgA antibodies interfering with bacterial colonization and/or toxin binding, and the goal of vaccination should be to effectively stimulate these types of antibodies (together with a long-lasting immunologic memory for anamnestic local antibody production).

mucosal immunity, including local immunologic memory (23).

Mucosal Immunity and Oral Vaccination

Another important line of vaccine research has focused on the gut mucosal immune system and how it can be stimulated by immunogens. The intestine is the largest immunological organ in the body, comprising 70 to 80% of all immunoglobulin-producing cells and producing a greater amount of secretory IgA (50 to 100 mg/kg of body weight per day) than the whole body produces of IgG (ca. 30 mg/kg/day). An important basis for local immunity is the migration of specific, antigen-activated B and T cells from the Peyer's patches to the intestinal lamina propria and epithelium. Antigen administered orally is taken up in the Peyer's patches by modified epithelial cells, known as M cells, and then transported to the lymphoid tissue of the Peyer's patches, where the initial mucosal immune response occurs. After undergoing antigen-induced proliferation and partial differentiation in the Peyer's patches, both B and T cells enter the regional mesenteric lymph nodes and then, after further differentiation, are transported through the thoracic duct into the circulation. By means of their selective affinity for endothelial cells in mucosal and glandular tissues, mucosa-derived B and T cells eventually return to and extravasate into these tissues in which the final differentiation takes place (in the case of B cells, into plasma cells predominantly producing antibody of the IgA isotype). Most of the B and T cells activated in the intestine "home" back to the lamina propria of the intestine, and another population of T cells returns to the intestinal epithelium, but a proportion, perhaps 10 to 20%, ends up in mucosal tissues outside the intestine instead (for reviews, see reference 17).

Studies of the immune mechanisms operating in cholera and the immune response to various cholera antigens, especially cholera toxin and its B-subunit moiety, have been central themes in much of the recent mucosal immunity research (12). These studies have defined the importance of locally produced IgA antibodies and IgA immunologic memory for protection against cholera, and they have shown that the oral route of vaccine administration is usually superior to the parenteral route in both priming and boosting the mucosal immune system (12, 13). It is now almost axiomatic that in order to be efficacious, vaccines against at least noninvasive enteric infections such as cholera must be able to stimulate the local gut mucosal immune system and that this goal is usually much better achieved by administering the vaccines orally rather than parenterally. Furthermore, oral vaccines are in general easier to produce and to quality control than parenteral vaccines, and they are also easier (and safer) to administer to large groups, since they do not require any medically trained personnel or sterile supplies. Indeed, based on the concept of a common mucosal immune system through which activated lymphocytes from the gut can disseminate immunity to other mucosal and glandular tissues, there is currently also much interest in the possibility of developing oral vaccines against other mucosal infections, e.g., those in the respiratory and urogenital tracts (12, 19).

Antitoxic and Antibacterial Synergy

A third important observation guiding the design of the new cholera vaccines has concerned the synergistic cooperation between antitoxic and antibacterial immune mechanisms in cholera. Two main protective antibodies have been identified, one directed against the *V. cholerae* O1 cell wall lipopolysaccharide and the other directed against the cholera toxin B subunits. Either of these two types of antibody can confer protection against disease by inhibiting bacterial colonization and toxin binding, respectively, and when present together in the gut, they can have a strongly synergistic protective effect (14, 28).

Composition of Vaccine

The oral B-subunit–whole-cell vaccine contains, per dose, 1 mg of purified cholera B-subunit together with 10^{11} heat- or Formalin-killed *V. cholerae* bacteria of three different strains (Table 1). The B-subunit–whole-cell vaccine is given orally together with an alkaline buffer (provided by an effervescent tablet) to protect the vaccine during passage through the stomach. The vaccine was designed to safely provide the key antigens for evoking protective antitoxic and antibacterial mucosal immunity against *V. cholerae* O1 of different serotypes (Inaba and Ogawa) and biotypes (classical and El Tor). As mentioned above, the B-subunit component is completely nontoxic and has proved to be an exceptionally potent oral immunogen because of its stability and its ability to bind to the intestinal mucosa. The whole-cell component in its turn provides protective Inaba and Ogawa lipopolysaccharide antigens through its heat-killed organisms and various heat-labile bacterial antigens through its Formalin-killed cells that may add further to the antibacterial immunogenic effect.

CLINICAL TRIALS INCLUDING A LARGE FIELD TRIAL

The B-subunit–whole-cell vaccine has been extensively tested in several phase 1 and phase 2 clinical trials in Swedish, Bangladeshi, and American volunteers. In these studies, the vaccine has proved to be completely safe with no adverse reactions and, after either two or three doses, to stimulate a gut mucosal IgA antitoxic and antibacterial immune response (including memory) comparable to that induced by cholera disease itself (24, 29). In American volunteers who received three oral immunizations with either the B-subunit–whole-cell vaccine or the whole-cell vaccine component alone, the vaccination protected against challenge with a dose of live cholera vibrios that caused disease in 100% of concurrently tested unvaccinated controls (1). Protection against cholera of any severity was significant (protective efficacy, 64%; $P < 0.01$), and it was even more impressive against "clinically significant" cholera illness (defined as diarrheal fluid loss exceeding 1 liter), which was completely prevented by vaccination (1).

The B-subunit–whole-cell vaccine and the whole-cell component without any B subunit (both vaccines were produced by Institut Mérieux, France, in collaboration with the National Bacteriological Laboratory of Sweden [SBL]) have also been evaluated in a large, randomized, placebo-controlled field trial involving almost 90,000 vaccinated adults and children in Bangladesh. The trial was performed by the International Centre for Diarrhoeal Disease Research, Bangladesh (ICDDR, B) (together with the Ministry of Health, Bangladesh, and the World Health Organization) in the years 1985 to 1988.

Table 1. Composition of B-subunit–whole-cell cholera vaccine

Component	Description	Amt
Cholera toxin B subunit		1 mg
Whole cell	Heat-killed *V. cholerae* Inaba, classical biotype (strain Cairo 48)	2.5×10^{10} cells
	Heat-killed *V. cholerae* Ogawa, classical biotype (strain Cairo 50)	2.5×10^{10} cells
	Formalin-killed *V. cholerae* Inaba, El Tor biotype (strain Phil 6973)	2.5×10^{10} cells
	Formalin-killed *V. cholerae* Ogawa, classical biotype (strain Cairo 50)	2.5×10^{10} cells
Phosphate-buffered saline		3 ml

Sixty-three thousand persons (children aged 2 to 15 years and adult women) received, at 6-week intervals, three oral doses of the B-subunit–whole-cell vaccine, whole-cell vaccine alone, or a placebo, and another 26,000 participants took one or two doses of the three agents. Each dose of cholera vaccine contained 10^{11} killed whole vibrios representing three different strains belonging to the Inaba and Ogawa serotypes and the classical and El Tor biotypes; the B-subunit–whole-cell vaccine also contained 1 mg of purified B subunit from cholera toxin. The placebo preparation consisted of 10^{11} heat-killed *Escherichia coli* K-12 bacteria.

The main results during 3 years of follow-up are summarized in Table 2. Both vaccines protect against cholera for at least 3 years. The B-subunit–whole-cell vaccine had the advantage over the whole-cell vaccine alone of a significantly higher efficacy level (85% versus 58%) for the initial 4- to 6-month period (4). However, after the first 8 to 12 months, the efficacies of the two vaccines were similar (6). For the 3 years of follow-up, the B-subunit–whole-cell vaccine conferred 50% overall protection, and the whole-cell vaccine conferred 52% protection against culture-proven cholera; protection after two doses was similar to that after three doses (Table 2). For both vaccines, protective efficacy was of shorter duration in children vaccinated at the ages of 2 to 5 years than in older groups. After a high initial level of protection (100%) for B-subunit–whole-cell vaccine during the first 6 months of follow-up, protection declined gradually, and no protection was observed in the 2- to 5-year-old age group in the third year of follow-up. In contrast, the level of protection in persons aged >5 years was about 65% for both vaccines and was sustained during the entire follow-up period. The very high protective efficacy observed in all age groups during the first 6 months after vaccination comprised El Tor and classical cholera to the same degree. In contrast, the longer-term protection was slightly higher against classical than El Tor

Table 2. Results from randomized placebo-controlled field trial of oral B-subunit–whole-cell cholera vaccine and whole-cell vaccine alone given in three (or two) doses[a]

Follow-up time and group	Vaccine efficacy (%)[b]	
	B-WC	WC
After 6 mo		
All ages	85	58
>5 yr old	77	62
<5 yr old	100	53
After 3 yr		
All ages	50	52
>5 yr old	63	68
<5 yr old	26	23
Recipients of only two doses, all ages	64	39
Overall protection after two or three doses, all ages[c]	62	60

[a]Results are from studies by the ICDDR, B in Matlab, Bangladesh.
[b]B-WC, B-subunit–whole-cell vaccine; WC, whole-cell vaccine.
[c]Including estimate for adult men.

cholera but did not differ with serotype (Inaba or Ogawa) (6). Assuming that protection in adult males would have equaled that seen in adult females, the overall protective efficacy of two or three doses over 3 years of follow-up can be estimated at 62% for the B-subunit–whole-cell vaccine and 60% for the whole-cell vaccine.

The B-subunit–whole-cell vaccine, through its B-subunit component, also provided substantial protection (ca. 67%) for a few months against diarrhea caused by enterotoxigenic *E. coli* (ETEC) producing either heat-labile toxin (which cross-reacts immunologically with cholera toxin) alone or together with heat-stable toxin (5). Protection was 86% against severe disease and 54% against milder illness (5). Interestingly, a similar protective effect (60 to 82% for different subcategories of ETEC-associated diarrhea) of the oral B-subunit–whole-cell cholera vaccine against ETEC diarrhea was recently reported by Peltola et al. (22) in Finnish travellers who had received two doses of the vaccine shortly before travel to Morocco.

Furthermore, in Bangladesh, during the large field trial, both the B-subunit–whole-cell and the whole-cell vaccines substantially and in a sustained manner reduced overall diarrhea morbidity among the vaccinees. Thus, over the 3-year follow-up period, there was a 25% reduction in admissions for "all diarrhea" and as much as a 50% reduction in admissions for life-threatening diarrhea in the vaccinated compared with the placebo group (2a, 3).

FURTHER DEVELOPMENTS

Work to further facilitate the production and use of the oral B-subunit–whole-cell cholera vaccine is in progress. By utilizing recombinant DNA technology, it may be possible to produce the B-subunit and whole-cell components in a single step. This could greatly simplify large-scale production of the B-subunit–whole-cell vaccine, reduce costs, and allow local production of this vaccine in developing countries. Other vaccine modifications to be considered for potentially further improved protective efficacy and stability could be to include or enhance the expression of relevant fimbrial antigens on the whole-cell component (especially the El Tor-associated mannose-sensitive hemagglutinin [MSHA] antigen) and to distribute the vaccine in a dry sachet or tablet formulation rather than, as now, as a liquid.

Recombinant B-Subunit Production

We have recently constructed overexpression systems in which the gene encoding the B subunit of cholera toxin has been placed under the control of tacP or other strong "foreign" promoters in a wide-host-range multicopy plasmid (16, 27). Recombinant nontoxigenic classical and El Tor *V. cholerae* strains of different serotypes harboring such plasmids have been made to excrete more than 500 mg of B subunit per liter (16) and may therefore be useful killed oral vaccine strains. Thus, a future B-subunit–whole-cell vaccine might be based on two strains, one of classi-

cal and the other of El Tor biotype and representing the Inaba and Ogawa serotypes, that through genetic manipulations lack the ability to produce cholera toxin but instead produce large amounts of cholera B subunit. Indeed, the SBL has already manufactured almost one million doses of B-subunit–whole-cell cholera vaccine based on such a recombinantly produced B subunit that it plans to use in various studies in Latin America. The oral B-subunit–whole-cell cholera vaccine, including the newer version using recombinantly produced B subunit, has been approved by Swedish regulatory authorities and has replaced the old parenteral cholera vaccine as a licensed regular vaccine product. Since the current cholera pandemic is mainly associated with El Tor vibrios, an important question is to what extent toxins produced by classical and El Tor vibrios are different or sufficiently similar to allow one type of toxin B subunit to induce good protection against both biotypes of cholera. Tamplin et al. (30) found five shared epitopes and one unshared epitope between classical and El Tor CTB. However, our previous findings (7) and those of others (18) demonstrated that the two toxins, despite their (minor) epitope differences, are similar from a practical immunologic point of view in that immune sera against either type of toxin have similar neutralization titers against the homologous and heterologous toxins. Likewise, rabbit antisera or human postvaccination sera against the (classical) B subunit of the oral B-subunit–whole-cell vaccine have neutralized El Tor and classical cholera toxins to a very similar extent (unpublished data). This suggests that the presence of unshared, biotype-specific epitopes would not have any major influence on the further improvement of oral cholera vaccines.

TCP and MSHA Fimbrial Antigens

It has also been considered whether the preparation of the whole-cell vaccine component should be modified such that strains and conditions of cultivation and inactivation would ensure the inclusion of two *V. cholerae* antigens that were missing in the original B-subunit–whole-cell vaccine and that might further add to protective immunogenicity: (i) TCP, which appears to be an important adhesive pilus involved in the colonization of at least classical cholera vibrios in the human intestine (9) (this antigen was not known at the time the original B-subunit–whole-cell vaccine was prepared); and (ii) the El Tor-associated cell-bound mannose-sensitive hemagglutinin (MSHA) that has also been implicated in cholera pathogenesis and immunity. Especially the latter antigen, which was recently shown to be of a fimbrial nature (15), is worth serious attention in this regard in consideration of the El Tor biotype nature of the current cholera pandemic. MSHA is being found on all strains of El Tor *V. cholerae,* and specific antibodies against MSHA protect animals against experimental cholera caused by *V. cholerae* of the El Tor but not the classical biotype (20). Conversely, toxin-coregulated pilus (TCP), which is present on almost all classical strains, is missing as a fimbrial or surface antigen on most El Tor strains, including those recently isolated in South America (even though El Tor strains can produce the TcpA subunit, they appear to be unable to assemble and translocate this protein to the cell surface), and, as shown in Table 3, antibodies to TCP protect baby mice against cholera caused by classical but not El Tor strains (21). Anti-MSHA and anti-TCP Fab fragments and monoclonal antibodies showed the same patterns of biotype-specific protection as their parent polyclonal antisera (20, 21).

Sachet Formulation

Yet another practically important further development of the oral B-subunit–whole-cell cholera vaccine could be to prepare a tablet or sachet formulation containing the vaccine in combination with alkaline buffer in a dry form. This would facilitate storage

Table 3. Biotype-specific protective effect of anti-MSHA and anti-TCP antibodies against El Tor and classical cholera, respectively, in the infant mouse cholera model[a]

Antibodies tested[b]	Protective efficacy (%)[c]	
	El Tor	Classical
MSHA	76–100[d]	0–33[e]
TCP	0–20[e]	78–100[d]

[a]Data are adapted from references 20 and 21.
[b]Preparations tested included (i) polyclonal antisera against the different fimbriae extensively absorbed with fimbria-negative whole bacterial cells to remove anti-LPS and other "contaminating" antibodies; (ii) Fab fragments of the same absorbed antisera; and (iii) monoclonal antibodies. Preimmune sera, Fab preparations, or monoclonal antibodies of irrelevant specificity were used as concomitantly tested controls.
[c]Challenge strains included different El Tor and classical organisms (two strains of each). Protective efficacy was estimated as follows: [1 − (rate of dead mice in the presence of tested specific antibodies versus in the presence of negative control immune preparation)] × 100. Ranges show protective efficacies in different experiments using the various specific antibody preparations and against different challenge organisms.
[d]$P < 0.01$.
[e]$P > 0.20$.

and distribution and probably also further increase the stability of these vaccines for use in areas where cholera is endemic. However, it should be mentioned that the B-subunit–whole-cell vaccine in its liquid form has already proved to be exceptionally stable, having shown intact immunogenicity in human volunteers after more than 8 years of storage (unpublished data).

CONCLUSIONS

The results with the oral B-subunit–whole-cell vaccine represent marked improvements over those achieved previously with parenteral cholera vaccines. In contrast to the parenteral vaccines, which cause frequent local side effects and occasional systemic adverse reactions (mainly fever), the oral B-subunit–whole-cell vaccine has not been associated with any side effects that would reduce the acceptability of the vaccine in areas where cholera is endemic. Of even greater importance, though, is the fact that the protection achieved by the B-subunit–whole-cell vaccine has been superior to that obtained with the parenteral vaccines with regard to both the level and the duration of protection. While parenteral cholera vaccination gives maximally 50% protection for no more than 3 to 6 months, the short-term protective efficacy of the oral B-subunit–whole-cell vaccine was 85% accompanied by long-term protection lasting for at least another 2 to 3 years, with 55 to 60% efficacy for the overall population and substantially higher efficacy in those older than 5 years. The main remaining deficiency in the oral B-subunit–whole cell cholera vaccine is obviously the relatively short duration of protection in many children under the age of 5. Since this age group experiences the highest rates of cholera in areas where cholera is endemic, either these young children will need booster doses or further improvements in the vaccines must be made, such as, for instance, including additional protective antigens as discussed above. These considerations, however, should not detract from the recognition that two-thirds of cholera cases and deaths in areas where cholera is endemic occur in those over age 5, who would benefit greatly from the already-existing B-subunit–whole-cell vaccine formulation. Indeed, the results of the field trial in Bangladesh indicate that in an area where cholera is endemic, vaccination could decrease the number of episodes of life-threatening cholera by more than 50% over 3 years. In an area with adequate intravenous as well as oral rehydration treatment facilities, rarely found in rural developing-country communities, vaccination could decrease the number of admissions to hospital, and in an area without adequate facilities, it could provide a proportionate decrease in cholera mortality. There is no question that cholera vaccination, if used in an area where cholera is endemic, has the potential to be a major life-saving intervention and most likely also a quite cost-effective one (25). Vaccination should be regarded as a useful complement to oral therapy and other components needed for effective public health control of cholera, and practical steps should be taken to develop in-

tegrated vaccination programs (26). The B-subunit–whole-cell vaccine has the potential to be both produced and delivered at a very low cost. This, together with its high degree of safety and storage stability, should facilitate its public health utility. Future production of this vaccine locally by developing countries in need of appropriate cholera vaccines is also fully realistic. This is exemplified by the current large-scale production of the whole-cell component of this vaccine in Vietnam for an estimated cost of $0.05 per dose (which is equivalent to the cost for 250 g of rice) (31, 32).

Acknowledgments. We thank our many coworkers at the University of Göteborg, ICDDR,B, SBL, Institut Mérieux, and ACO Ltd. (Stockholm) for valuable contributions in essential parts of the studies.

Our basic research was supported by the Swedish Medical Research Council (project 16X-3383), the Swedish Agency for Research Cooperation with Developing Countries, and the World Health Organization.

REFERENCES

1. **Black, R. E., M. M. Levine, M. L. Clements, C. R. Young, A.-M. Svennerholm, and J. Holmgren.** 1987. Protective efficacy in man of killed whole vibrio oral cholera vaccine with and without the B subunit of cholera toxin. *Infect. Immun.* **77:**1116–1129.
2. **Carpenter, C. C. J.** 1992. The treatment of cholera: clinical science on the bedside. *J. Infect. Dis.* **166:**2–14.
2a. **Clemens, J.** Personal communication.
3. **Clemens, J., D. A. Sack, J. R. Harris, J. Chakraborty, M. R. Khan, B. F. Stanton, M. Ali, F. Ahmed, M. D. Yunus, B. A. Kay, M. U. Khan, M. R. Rao, A.-M. Svennerholm, and J. Holmgren.** 1988. Impact of B subunit killed whole-cell and killed whole-cell-only oral vaccines against cholera upon treated diarrhoeal illness and mortality in an area endemic for cholera. *Lancet* i:1375–1379.
4. **Clemens, J., D. A. Sack, J. R. Harris, J. Chakraborty, M. R. Khan, B. F. Stanton, B. A. Kay, M. U. Khan, M. D. Yunus, A.-M. Svennerholm, and J. Holmgren.** 1986. Field trial of oral cholera vaccines in Bangladesh. *Lancet* i:124–127.
5. **Clemens, J., D. A. Sack, J. R. Harris, J. Chakraborty, P. K. Neogy, B. Stanton, N. Huda, M. U. Khan, B. A. Kay, M. R. Khan, M. Ansaruzzaman, M. Yunus, M. R. Rao, A.-M.**

Svennerholm, and J. Holmgren. 1988. Cross-protection by B subunit-whole cell cholera vaccine against diarrhea associated with heat-labile toxin-producing enterotoxigenic *Escherichia coli:* results of a large-scale field trial. *J. Infect. Dis.* **158:**372–377.

6. Clemens, J., D. A. Sack, J. R. Harris, F. van Loon, J. Chakraborty, F. Ahmed, M. R. Rao, M. R. Khan, M. D. Yunus, N. Huda, B. F. Stanton, B. A. Kay, S. Walter, R. Eeckels, A.-M. Svennerholm, and J. Holmgren. 1990. Field trial of oral cholera vaccines in Bangladesh: results from three-year follow-up. *Lancet* **i:**270–273.

7. Dubey, R. S., M. Lindblad, and J. Holmgren. 1990. Purification of El Tor cholera enterotoxins and comparisons with classical toxin. *J. Gen. Microbiol.* **136:**1839–1847.

8. Feeley, J. C., and E. J. Gangarosa. 1980. Field trials of cholera vaccine, p. 204–210. *In* Ouchterlony and J. Holmgren (ed.), *Cholera and Related Diarrheas.* 43rd Nobel Symposium, Stockholm. S. Karger, Basel.

9. Herrington, D. A., R. H. Hale, G. Losonsky, J. J. Mekalanos, R. K. Taylor, and M. M. Levine. 1988. Toxin, toxin-coregulated pili, and the *tox*R regulon are essential for *Vibrio cholerae* pathogenesis in humans. *J. Exp. Med.* **168:**1487–1492.

10. Holmgren, J. 1981. Actions of cholera toxin and the prevention and treatment of cholera. *Nature* (London) **292:**413–417.

11. Holmgren, J., J. Clemens, D. A. Sack, and A.-M. Svennerholm. 1989. New cholera vaccines. *Vaccine* **7:**94–96.

12. Holmgren, J., C. Czerkinsky, N. Lycke, and A.-M. Svennerholm. 1992. Mucosal immunity: implications for vaccine development. *Immunobiology* **184:**57–179.

13. Holmgren, J., and A.-M. Svennerholm. 1983. Cholera and the immune response. *Prog. Allergy* **33:**106–119.

14. Holmgren, J., A.-M. Svennerholm, I. Lönnroth, M. Fall-Persson, B. Markman, and H. Lundbäck. 1977. Development of improved cholera vaccine based on subunit toxoid. *Nature* (London) **269:**602–604.

15. Jonson, G., J. Holmgren, and A.-M. Svennerholm. 1991. Identification of a mannose-binding pilus on *Vibrio cholerae* El Tor. *Microb. Pathog.* **11:**433–441.

16. Lebens, M. R., J. Osek, S. Johansson, M. Lindblad, and J. Holmgren. Over-expression of cholera toxin B subunit for use in vaccines. Submitted for publication.

17. MacDermott, R. P., and C. O. Elson (ed.). 1991 and 1992. *Gastroenterology Clinics of North America.* The W. B. Saunders Co., Philadelphia.

18. Marchlewicz, B. A., and R. A. Finkelstein. 1983. Immunological differences among the chol-

era/coli family of enterotoxins. *Diagn. Microbiol. Infect. Dis.* **1:**129–138.

19. Mestecky, J., and J. R. McGhee. 1989. New strategies for oral immunization. *Curr. Top. Microbiol. Immunol.* **146:**185–237.

20. Osek, J., A.-M. Svennerholm, and J. Holmgren. 1992. Protection against *Vibrio cholerae* El Tor infection by specific antibodies against mannose-binding hemagglutinin pili. *Infect. Immun.* **60:**4961–4964.

21. Osek, J., A.-M. Svennerholm, and J. Holmgren. Role of antibodies against biotype-specific *Vibrio cholerae* pili in protection against experimental classical and El Tor cholera. Submitted for publication.

22. Peltola, H., A. Siitonen, H. Kyrönseppä, I. Simula, L. Mattila, P. Oksanen, M. J. Kataja, and M. Cadoz. 1991. Prevention of travellers' diarrhoea by oral B-subunit/whole cell cholera vaccine. *Lancet* **338:**1285–1289.

23. Pierce, N. F. 1978. The role of antigen form and function in the primary and secondary intestinal immune responses to cholera toxin and toxoid in rats. *J. Exp. Med.* **148:**195–206.

24. Quiding, M., I. Nordström, A. Kilander, G. Andersson, L.-Å. Hanson, J. Holmgren, and C. Czerkinsky. 1991. Intestinal immune responses in humans. Oral cholera vaccination induces strong intestinal antibody responses, gamma-interferon production, and evokes local immunological memory. *J Clin. Invest.* **88:**143–148.

25. Sack, D. A. 1990. Cost effectiveness model for cholera vaccination with oral vaccines, p. 82–89. *In* D. A. Sack and L. Freij (ed.), *Prospects for Public Health Benefits in Developing Countries from New Vaccines against Enteric Infections.* SAREC research symposium, SAREC documentation, conference report 1990:2.

26. Sack, D. A., and L. Freij (ed.). 1990. *Prospects for Public Health Benefits in Developing Countries from New Vaccines against Enteric Infections.* SAREC research symposium, SAREC documentation, conference report 1990:2.

27. Sanchez, J., and J. Holmgren. 1989. Recombinant system for overexpression of cholera toxin B subunit in *Vibrio cholerae* as a basis for vaccine development. *Proc. Natl. Acad. Sci. USA* **86:**481–485.

28. Svennerholm, A.-M., and J. Holmgren. 1976. Synergistic protective effect in rabbits of immunization with *Vibrio cholerae* lipopolysaccharide and toxin/toxoid. *Infet. Immun.* **13:**735–740.

29. Svennerholm, A.-M., M. Jertborn, L. Gothefors, A. M. Karim, D. A. Sack, and J. Holmgren. 1984. Mucosal antitoxic and antibacterial immunity after cholera disease and after immunization with a combined B subunit-whole cell vaccine. *J. Infect. Dis.* **149:**884–893.

30. **Tamplin, M. L., M. K. Ahmed, R. Jalali, and R. R. Colwell.** 1989. Variation in epitopes of the B subunit of El Tor and classical biotype *Vibrio cholerae* O1 cholera toxin. *J. Gen. Microbiol.* **135:**1195–1200.

31. **Trach, D. D.** 1990. The problem of cholera vaccination in Vietnam, p. 217–218. *In* D. A. Sack and L. Freij (ed.), *Prospects for Public Health Benefits in Developing Countries from New Vaccines against Enteric Infections.* SAREC research symposium, SAREC documentation, conference report 1990:2.

32. **Trach, D. D.** Personal communication.

Vibrio cholerae and Cholera: Molecular to Global Perspectives
Edited by I. Kaye Wachsmuth, Paul A. Blake, and Ørjan Olsvik
© 1994 American Society for Microbiology, Washington, DC 20005

Chapter 28

Public Health Considerations for the Use of Cholera Vaccines in Cholera Control Programs

John Clemens, Dale Spriggs, and David Sack

Interest in creating vaccines to control cholera dates virtually to the discovery of the cholera vibrio by Robert Koch in 1883 (42). However, modern experimental evaluations of parenteral vaccines against *Vibrio Cholerae* O1 did not take place until the early 1960s. Disappointingly, these evaluations found that parenteral vaccines usually conferred only modest protection that was short-lived (4). Conventional parenteral cholera vaccines have largely been abandoned as public health tools, and recent studies have elucidated the probable reason for their poor performance. As detailed in other chapters in this section, it is now recognized that immune protection against cholera resides primarily at the intestinal mucosal surface and that mucosal immunity is best elicited by oral, not parenteral, presentation of immunogens (32). This concept has refocused efforts on the development of orally administered cholera vaccines, and a recent field trial in Bangladesh has provided evidence that orally administered inactivated oral cholera vaccines can confer protection that is substantially

more prolonged than that observed for conventional parenteral vaccines (13).

Although newer oral cholera vaccines may well prove to confer greater protection and to be associated with fewer side effects than older parenteral cholera vaccines, a commitment to introducing new vaccines as public health tools requires a broad assessment of their utility. In this chapter, we review the reasons that older parenteral cholera vaccines have largely been abandoned as public health tools and also discuss several considerations that should be addressed when judging whether new cholera vaccines are sufficiently attractive to be used in public health programs. We restrict our focus to the use of cholera vaccines in populations residing in areas affected by cholera and do not specifically consider the problem of vaccinating travelers to such areas.

DEMISE OF CONVENTIONAL PARENTERAL CHOLERA VACCINES

As noted above, several properly controlled trials (4) demonstrated that acceptably nonreactogenic parenteral vaccines were not sufficiently protective to justify their use as tools in cholera control programs. These vaccines had specific limitations: (i) protection tended to be modest and short-lived; (ii) when protection was observed in populations

John Clemens • Division of Epidemiology, Statistics, and Prevention Research, National Institute of Child Health and Human Development, Bethesda, Maryland 20892. *Dale Spriggs* • Virus Research Institute, 61 Moulton Street, Cambridge, Massachusetts 02138. *David Sack* • Department of International Health, Johns Hopkins School of Public Health, Baltimore, Maryland 21205.

living in areas where cholera is endemic, it often was seen only in older children and adults, in whom vaccination presumably acted to boost preexisting natural immunity to cholera; (iii) when tested in outbreaks, vaccination had little effect in the prevention of infection or disease (47); and (iv) vaccination tended not to interrupt transmission of asymptomatic *V. cholerae* O1 infections, which account for the vast majority of infections in endemic areas (3). Although not a problem with vaccination per se, anecdotal evidence also suggested that attempts to vaccinate during outbreaks commonly deterred control efforts by consuming inordinate medical resources and by creating a sense of complacency about using safe water and practicing proper hygiene. Perhaps most compelling was evidence (to be discussed in greater detail later in the chapter) that it was more rational to allocate resources to case management of cholera with appropriate rehydration therapies than to immunization programs with parenteral vaccine (18, 39).

In 1973, the 26th World Health Assembly amended the International Health Regulations by abolishing requirements for a cholera vaccination certificate for international travel (54). In addition, the World Health Organization refocused its cholera control activities on national Diarrheal Disease Control Programs, whose primary emphasis was the proper case management of all diarrheal illnesses (without a specific focus on cholera) and proper practices related to water, food, and hygiene to prevent diarrheal illnesses. Cholera control efforts in such programs emphasize adequate surveillance for cholera through reporting in primary care settings and appropriate nonvaccine responses to cholera epidemics (55).

PUBLIC HEALTH CONSIDERATIONS FOR USING NEW CHOLERA VACCINES

No simple set of rules can provide a rational basis for the decision to incorporate a cholera vaccine into a country's public health armamentarium. Indeed, such a decision requires consideration of the complex interplay between the categories of factors listed in Table 1.

Epidemiologic Setting for Vaccination

In developing countries, cholera may occur in either epidemic or endemic form. The former is illustrated on a massive scale by the outbreak of cholera in Latin America, initially observed in 1991, after an apparent absence during the entire 20th century (52). Epidemics are also well documented in refugee camps as well as at fairs, feasts, and pilgrimages (26). Endemic cholera is well exemplified by cholera that appears in regular, predictable yearly cycles in populations living in the Ganges delta region of Bangladesh and India (24).

Table 2 summarizes several salient characteristics that distinguish epidemic and endemic cholera (25). First, in contrast to the occurrence of endemic cholera, the occurrence of epidemic cholera is not predictable in either time or place. Thus, whereas systematic and regular vaccination of the target population in an endemic setting would be a logical goal for a suitable vaccine, it would make little sense for a population that may or may not experience an epidemic. Second, whereas modes of transmission of cholera in an epidemic are relatively few (particularly early in an epidemic), those in endemic-disease settings tend to be multiple, entailing

Table 1. Issues to consider in making decisions to use cholera vaccine in public health programs

1. Epidemiologic setting for vaccination
2. Burden of cholera morbidity and mortality
3. Protective characteristics of the vaccine
4. Clinical effectiveness of the vaccine
5. Anticipated costs incurred or saved by vaccinating
6. Alternative or complementary nonvaccine control measures
7. Balance between costs and effects

Table 2. Epidemiologic setting for vaccination: epidemic versus endemic cholera

Feature	Setting	
	Epidemic	Endemic
Occurrence	Not predictable	Predictable
Modes of transmission	Few	Many
Nonhuman reservoirs	Uncommon	Common
Preexisting natural immunity	Uncommon	Common
Higher risk in children	No	Yes
Asymptomatic infections	Less common	More common

various combinations of waterborne, foodborne, and fecal-oral routes simultaneously. Accordingly, whereas interruption of an epidemic with a simply formulated intervention aimed at improving water quality and/or hygiene may be feasible, the complexity of transmission in an area where cholera is endemic may confound such simple approaches.

Third, in settings where the disease is endemic, cholera may be sustained in a nonhuman reservoir, such as phytoplankton or shellfish in brackish bodies of water (34), which can be perpetuated without repeated contamination by human feces. In contrast, no such reservoir is generally evident for areas with epidemic cholera. This implies that termination of an epidemic, either by vaccination or by nonvaccination approaches, may result in eradication of cholera from an affected population. In contrast, the perpetuation of cholera in nonhuman reservoirs in areas where it is endemic makes diminution of cholera morbidity and mortality, rather than cholera eradication, a more feasible goal.

Fourth, populations affected by epidemics generally will have had limited past exposures to *V. cholerae* O1, in contrast to the repeated past, exposures experienced by populations in areas where cholera is endemic. Populations experiencing epidemics thus tend to have sub-

stantially lower levels of background natural immunity to cholera, with the results that cholera in such populations affects both children and adults to a similar extent and that the clinical spectrum of disease is more severe than that seen in populations with background immunity. In contrast, background immunity to cholera in populations residing in areas where the disease is endemic yields a pattern in which the risk is greatest in young children, who have not yet had substantial previous exposures to cholera, and the clinical spectrum is more dominated by asymptomatic infections. These contrasting features suggest that attempts to control cholera with vaccination in epidemic-disease versus endemic-disease areas should contemplate these differences in age-related burdens of morbidity and mortality.

Burden of Cholera Morbidity and Mortality

A basic requirement for making a rational decision to launch a cholera vaccine program is that the burden of cholera morbidity and mortality be sufficient to warrant the diversion of scarce resources necessary for such a program. The current (seventh) pandemic of cholera, which began in Sulawesi, Indonesia, in 1961, spread through Asia, the Middle East, Africa, and most recently South and Central America (26). In its course, isolated epidemics were noted in both Europe and the United States. In total, approximately 100 countries have been affected, and in 1991, more than 570,000 cases and 17,000 cholera deaths were reported to the World Health Organization.

Although giving a general global picture of cholera, these data are not sufficient for decision making about the public health utility of cholera vaccines. First, the data are based on voluntary reporting of cholera and undoubtedly represent severe underestimates of the global problem. The completeness and accuracy of such reporting is often limited by the reluctance of countries to report cholera in view of the negative impact that such reports may have on trade and tourism. For example, Bangladesh, which is well documented to be a hyperendemic focus for cholera and likely experiences well in excess of 100,000 cases each year (24), currently does not report cholera to the World Health Organization. Even if a country is willing to report such statistics, the statistics themselves are limited by their dependence on being reported by health care facility workers, to whom only a fraction of the population may have access, and by the relatively uncommon availability of suitable microbiologic techniques for diagnosing cholera.

Second, a global summary does not convey the substantial heterogeneity of the burden of morbidity and mortality in different settings. Heterogeneity of the disease burden is clearly introduced by the regular annual occurrence of cholera in some areas where the disease is endemic, such as South Asia; the sporadic occurrence in other areas, such as many countries in Africa; and the unpredictable occurrence in areas newly affected by cholera epidemics where cholera is not endemic. Even within settings falling into these different epidemiologic categories, the burden of disease may differ notably.

Such differences were dramatically illustrated in the recent epidemic of cholera in South America, where rates of cholera varied enormously between countries and where case fatality rates also differed dramatically both within and between countries (52). However, such variations are also seen within countries experiencing stable endemic cholera. For example, in Bangladesh, cholera has been noted to cause death only rarely in the Matlab field studies area of the International Centre for Diarrhoeal Disease Research, Bangladesh, an area that is well served by a community-based oral rehydration program and by diarrheal treatment centers (24). In contrast, cholera outbreaks with high case fatality rates have been observed in parts of the country lacking accessible treatment facilities (46). Particularly dramatic examples of the burden that cholera can impose on a commu-

nity have been reported from outbreaks in refugee camps in Asia and Africa. In such settings, attack rates of clinical cholera have been noted to be as high as 15 cases per 1,000 people and to be associated with case fatality rates as high as 55% (37).

Third, these global statistics may not completely capture the total health impact of cholera, since the statistics imply that the importance of cholera is only as a disease that, if not immediately fatal, has no health implications beyond the treatment of the acute episode. Direct effects of cholera that may conceivably have long-term sequelae include the capacity of the disease to cause acute renal failure as well as hypoglycemia with neurological manifestations (5, 31). Later development of cataracts has also been reported as a long-term complication, although evidence for this complication is not entirely consistent (6). In addition, cholera may cause abortions and stillbirths in pregnant patients (30), a feature of potentially considerable importance in view of the observation that women of childbearing age are at particularly high risk in areas where cholera is endemic (24). Finally, as one of the few diarrheal agents that commonly affects adults, cholera includes in its spectrum of casualties young children who are rendered vulnerable by the cholera-caused death of a parent or guardian.

To be most useful, evaluations of the suitability of cholera vaccines as elements of cholera control programs should take into consideration the enormous diversity of the burden of disease in different settings. Vaccination is more appropriate in some countries than in others, and even within affected countries, targeted rather than nationwide vaccination may be the most rational strategy. Moreover, such evaluations should attempt to embrace the diversity of short- and long-term effects of cholera on the health of affected individuals as well as the indirect effects of cholera on the health of young children who are dependent on adults afflicted with cholera.

Protective Characteristics of the Vaccine

Several protective characteristics of a cholera vaccine must be considered when its potential utility in public health programs is evaluated (Table 3). These features include the magnitude and kinetics of protection, the clinical spectrum of protection and the heterogeneity of protection by characteristics of the host and the cholera pathogen. A recurrent theme of all of these features is that their relative importance depends to a great extent on the epidemiologic setting and the population targeted for receipt of cholera vaccine.

The magnitude of protection of a vaccine is conventionally expressed as protective efficacy, or the proportionate reduction of cholera incidence that vaccination can be expected to effect. Regardless of the level of protective efficacy of a vaccine, the public health utility of a vaccine will depend on the anticipated incidence of cholera in the target population, since this utility is best measured as the number of cholera cases prevented per vaccinated person (9). For example, if the anticipated incidence is 3 cases per 1,000 people, such as may be experienced during a typical year in some parts of Bangladesh, a 100% protective vaccine given to 1,000 persons will prevent 3 cases. On the other hand, in a refugee camp outbreak for which an incidence of 15 cases per 1,000 people might be anticipated, a vaccine with a protective efficacy of 20% administered to 1,000 persons will also prevent 3 cases.

The kinetics of protection, its onset and its duration, must also be considered, though the importance of these factors varies with the epidemiologic setting for vaccination. A vaccine that protects rapidly after immunization might be of great use in the control of an epidemic, though this property would be less crucial for a vaccine intended for the long-term control of cholera in a population with endemic disease. Conversely, a long duration of protection after completion of a vaccine regimen would be of great value in endemic-disease curcumstances, though this would be a less compelling requirement for use in an epidemic.

The clinical spectrum of vaccine protection is also relevant. One argument against the use of earlier parenteral vaccines was that they failed to confer significant protection against asymptomatic excretion of *V. cholerae* O1 (4). Because of uncertainties about the importance of asymptomatic excretion in the initiation and perpetuation of cholera epidemics, interruption of asymptomatic excretion may not be very relevant to the potential public health utility of a cholera vaccine. Conversely, protection against cholera of life-threatening severity (only a small minority of *V. cholerae* O1 infections [45]) seems to be of great importance in evaluating a cholera vaccine for public health use. In this regard, it is important to appreciate that a vaccine may confer different levels of protection against severe versus mild diarrhea associated with *V. cholerae* O1. The recent field trial of inactivated oral cholera vaccines in Bangladesh, which tested a vaccine consisting of cholera toxin B subunit and killed cholera whole cells (BS-WC) and a second vaccine made only of killed whole cells, demonstrated that the former but not the latter vaccine modified the clinical expression of cholera infection toward less severe disease (15).

Heterogeneity of vaccine protection, which depends on characteristics of the host and of the infecting vibrio, must also be contemplated, though again, the importance of such heterogeneity will vary with the epidemiologic setting for vaccination. The recent field trial of inactivated oral cholera vaccines in Bangladesh illustrated these considerations. This trial found that vaccine protection

Table 3. Protective characteristics relevant to use of cholera vaccines in public health programs

1. Magnitude of protection
2. Kinetics of protection
3. Clinical spectrum of protection
4. Heterogeneity of protection as affected by host and pathogen characteristics

varied by age at vaccination (being substantially lower in young children than in older persons), ABO blood group of the vaccinee (being lower in persons with blood group O), and biotype of the infecting vibrio (being lower against El Tor than against classical cholera) (10, 13).

These gradients of protection may have important public health implications. The gradient by biotype is important because the biotype for the current pandemic is El Tor. The age-specific gradient of protection might have several implications. If the poorer protection in children reflected poorer protection in persons who were immunologically naive with respect to past exposure to cholera, which at present is a matter of speculation, this gradient might suggest a limited applicability of these vaccines to populations without recent past exposure to cholera. The poorer protection in children might also suggest that the vaccines would have limited applicability to endemic-disease settings, where the risk of cholera is typically highest in young children. This line of reasoning, however, would not necessarily argue against their use in populations residing in areas where cholera is not endemic, where the risk is typically higher in older than in younger persons, nor does it consider the total burden of cholera in relation to age in populations with endemic cholera. An analysis of cholera in Bangladesh, for example, demonstrated that although the risk of cholera was clearly highest in the age group between 2 and 5 years, the total burden of cases of cholera in this population was primarily borne by older persons in the population (44).

Clinical Effectiveness of the Vaccine

Considerations outlined above address the intrinsic protective characteristics of a cholera vaccine, sometimes referred to as vaccine efficacy. Contemplation of a cholera vaccine for use in public health practice, however, requires additional consideration of factors that determine vaccine effectiveness, or the balance between beneficial and nonbeneficial outcomes when a vaccine is administered under the ordinary conditions of public health practice.

Clinical acceptability

Central to this balance is the clinical acceptability of a cholera vaccine. Clinical acceptability has several dimensions. A vaccine that is associated with frequent mild side effects or even rare severe side effects may not be acceptable from a public health perspective if the target disease for the vaccine is rare. Even if the level of side effects associated with a vaccine seems justifiable in relation to its anticipated benefit, these side effects may still be unacceptable to the persons for whom the vaccine is targeted, reducing compliance and compromising the overall impact of a vaccine program.

Cultural acceptability

Although a suitable level of clinical safety will be necessary for a vaccine to be culturally acceptable, cultural acceptability entails several additional considerations. The formulation of a vaccine may have an important influence on cultural acceptability. For example, in the field trial of inactivated oral cholera vaccines in Bangladesh (13), an effervescent buffer was designed in Sweden for coadministration with the vaccines. It was expected that the effervescence would enhance compliance among young children because it would make the buffer resemble a carbonated beverage. In fact, the buffer created noteworthy problems among young children, who had had little exposure to drinking carbonated beverages and who found the novelty of effervescence to be unpleasant.

Additional problems can be anticipated from the use of oral rather than injectable vaccines against cholera. While oral vaccines may be attractive because their administration does not entail the many risks associated with injections (particularly in less-developed settings, where sterility of needles and syringes may be a concern), societies that regard the act of injection as necessary for the potency of medications may greet oral vaccines with distrust and skepticism (28).

Also relevant to the cultural acceptability of cholera vaccines is the influence of demographic, socioeconomic, and behavioral characteristics of a population on acceptance of vaccination. Low socioeconomic status, poor educational attainment, social isolation, and migrant status are commonly cited as determinants of the risk of nonacceptance of conventional immunizations (1, 7, 21). Cholera tends to selectively affect these disadvantaged groups (24), in areas with both endemic and epidemic disease. Thus, populations that are easiest to reach with a new cholera vaccine, who can be expected to be more affluent and to have increased access to health care, will likely be those at lowest risk of cholera, and levels of vaccine coverage will serve as poor markers of the impact of vaccination. This phenomenon was illustrated by the field trial of inactivated oral cholera vaccines in Bangladesh, in which persons who refused to participate had a 20% higher risk of cholera than persons who participated and received a placebo, even after adjustment for sociodemographic factors (16).

Finally, potential problems with acceptability can be anticipated because of the fact that even in areas with endemic cholera, cholera accounts for only a small fraction of diarrheal episodes (20). As a result, even if the target population for immunization perceives cholera as important, the success of cholera vaccination may be derailed by the occurrence of severe diarrheal episodes due to agents other than *V. cholerae* O1 among vaccinees. These problems will only be compounded by the facts that it is not likely that any future vaccine will be completely protective and that genuine vaccine failures will be inevitable.

Logistical requirements

Logistics can profoundly affect effectiveness. The success of a vaccine program can be severely diminished if a vaccine is not available and accessible, if it is rendered inactive during storage, if it requires a complicated regimen entailing multiple visits in order to confer protection, or if it is difficult to administer properly. Availability of a vaccine in a less-developed setting may be enhanced if the vaccine can be produced locally rather than imported from an international source. However, even for vaccines that do not require sophisticated technology and may therefore be amenable to local production, development of suitable quality control capacities may constitute a formidable challenge.

Availability of vaccine also requires a well-organized health system that is capable of adequate planning for large-scale acquisition, storage, and timely distribution of vaccine and that offers service delivery in a manner that is acceptable and accessible to the target population. Dependence of vaccine potency upon maintenance of a cold chain greatly increases the likelihood that a fraction of vaccine will be thermally inactivated at some point during its movement from the site of production to the site of administration. Clearly, a vaccine that could be administered in a single dose, that would not require a cold chain during storage, and that could be given during an existing schedule of visits for the delivery of health services would be ideal. If, as seems likely, a single vaccine is not able to fulfill all of these characteristics, it still must be dependably deliverable without placing inordinate demands on a health care system.

Although appealing in many ways, oral administration of new cholera vaccines itself presents many challenges. It may be difficult to ensure complete ingestion of a dose, particularly in certain age groups. In the Bangladesh trial of inactivated oral cholera vaccines (13), for example, very young children and adolescent girls were the most difficult to vaccinate, the latter due to intentional and often surreptitious regurgitation of ingested doses. The importance of complete ingestion and proper administration and the challenges associated with oral administration will impose demanding requirements on the training and supervision of vaccinators.

"Real-life" modifiers of protection

The introduction of a cholera vaccine into public health practice per se may lead to levels of protection that are different from those expected on the basis of results from field trials of vaccine efficacy. Several factors may lead to lower levels of protection. When a vaccine is introduced in practice, the spectrum of recipients of vaccine inevitably widens from the rather restricted group typically selected for a vaccine efficacy trial. If this widened spectrum includes persons who may not respond well to vaccination, such as those with comorbid illnesses or malnutrition, protection will decline. Relevant in this regard is the observation that live oral vaccines are often less immunogenic in persons from less-developed settings than in persons residing in developed settings (41). This phenomenon, which is not well explained, was earlier noted for both live polio and certain rotavirus vaccines and more recently was described for CVD 103-HgR, an oral live attenuated cholera vaccine (51).

As noted above, the vagaries of improper vaccine storage and administration that inevitably occur in a routine vaccination program may act to diminish vaccine protection. Also of concern are "comaneuvers" that may diminish vaccine immunogenicity. For example, it is conceivable that the intake of antibiotics, whose use is widespread and often unregulated in less-developed settings, may diminish the immunogenicity of live oral vaccines. Another comaneuver that may interfere with vaccine immunogenicity is the intake of breast milk shortly before or after immunization, since breast milk is known to be highly protective against clinically severe cholera (27). For cholera vaccines introduced into routine childhood immunization programs, it is also possible that concurrently administered vaccines may interfere with immuno-

genicity. This sort of vaccine interference was observed earlier when parenteral cholera vaccine and yellow fever vaccine were administered within three weeks of one another (23).

On the other hand, at least two additional factors may tip the balance toward a higher-than-expected level of protection when a vaccine is introduced in public health practice. First, the utility of a vaccine may be greater if it also cross-protects against noncholera pathogens. For example, the BS-WC vaccine tested in Bangladesh conferred short-term cross-protection against diarrhea associated with heat-labile-toxin-producing enterotoxigenic *Escherichia coli*. This cross-protection was presumably due to the antigenic similarities of cholera toxin B subunit, included in the vaccine, and the B subunit of heat-labile enterotoxigenic *E. coli* toxin (11). Such cross-protection may be of great relevance in a public health program if the related pathogen is an important source of morbidity and mortality in the target population and if the age at vaccination allows the age group(s) most affected by this pathogen to be protected. Second, a cholera vaccine may confer protection against cholera in nonvaccinees by indirect mechanisms (40). In the Bangladesh trial, for example, vaccination of mothers was shown to protect their young, nonvaccinated children against life-threatening cholera (12). This protection was presumably due to interruption of transmission between mother and child.

Anticipated Costs Incurred or Saved by Vaccinating

Evaluation of the potential public health utility of cholera vaccines demands estimation both of the costs of vaccine production and administration on the one hand and of money saved from prevention of cholera on the other hand. The precise enumeration of these costs will vary depending on the perspective of the analysis. For example, costs included in such analyses might differ if the

pected for the widely variable settings in which cholera vaccination might be considered, such sensitivity analyses will be crucial for future economic analyses of cholera vaccines.

Third, all but one of the analyses (18) failed to contemplate the incremental effects of adding vaccination to different background combinations of treatment and nonvaccination preventive services. Such incremental analyses may yield different conclusions about the value of vaccination. For example, in a population with a particular incidence of cholera, health care costs spared by vaccination (in the numerator of the cost-effectiveness ratio) will be greater if the population is already well served by clinical treatment services for cholera, since prevention of cholera by vaccination will diminish the costs for these services. Conversely, for areas with fewer health services, the preexisting toll of cholera deaths is likely to be higher, and deaths prevented by vaccination (in the denominator of the cost-effectiveness ratio) will be greater. Since such incremental effects correspond to the pragmatic questions posed by decision makers, future analyses should be framed in a fashion that directly addresses these questions. Finally, it is worth emphasizing that regardless of how persuasive the economic rationale is for vaccinating, no vaccine has a hope of succeeding in a public health program if there is not sufficient political will to support the introduction of vaccination and ensure its sustainability.

CONCLUSION

Until recently, the demise of enthusiasm for use of parenteral cholera vaccines together with the demonstrable potential of oral and intravenous rehydration therapy to save the lives of patients with severe cholera seemed to relegate even new oral vaccines to the role of biological probes that were interesting primarily for their ability to unravel many mysteries about mucosal immunology.

This position was reinforced by economic analyses that found vaccination to be expensive in relation to its anticipated benefits and by the hopes for prevention offered by the programs of the International Drinking Water Supply and Sanitation Decade.

The emergence of cholera as a major health problem throughout Latin America in 1991, its subsequent ensconcement in numerous South and Central American countries, and the continuing problem of cholera outbreaks with unacceptably high mortality rates in Asia and in Africa have prompted many to reappraise whether expectations about the effectiveness of nonvaccine strategies to control cholera were realistic and whether cholera vaccines should again be considered as potential elements in the public health armamentarium.

This chapter has attempted to make the case that serious and rational decision making about the use of cholera vaccines must incorporate considerations of several types of scenarios for use of vaccine, several types of outcomes that might occur in these scenarios, and several types of costs both engendered and saved by the use of vaccine. Moreover, such decisions cannot be made in isolation; the incremental benefits of vaccinating in relation to a preexisting program of cholera prevention, case management, and outbreak control must also be weighed.

Both cost-benefit and cost-effectiveness analyses have been proposed as mechanisms for synthesizing relevant considerations and arriving at rational, quantitative decisions. Although providing a useful start, analyses published to date have not been adequate to inform decisions about the diverse array of circumstances and target groups for which vaccination might be considered, the full scope of outcomes that might be of relevance to such target groups, and the nonvaccine interventions that might be used in concert with or as alternatives to cholera vaccination. Further evaluations that model the specific applications of cholera vaccines, contemplate the uncertainties of estimates of costs and outcomes, and consider ranges of estimates of costs and outcomes that would make vaccination either acceptable or unacceptable constitute a high priority for future research.

REFERENCES

1. **Akesode, F.** 1982. Factors affecting the use of primary health care clinics for children. *J. Epidemiol. Commun. Health* **36:**310–314.
2. **Azurin, J., and M. Alverin.** 1974. Field evaluation of environmental sanitation measures against cholera. *Bull. W.H.O.* **51:**19–26.
3. **Bart, K., Z. Huq, M. Khan, and W. H. Mosley.** 1970. Seroepidemiological studies during a simultaneous epidemic of infection with El Tor Ogawa and classical Inaba *Vibrio cholerae. J. Infect. Dis.* **121:**S17–S24.
4. **Benenson, A.** 1976. Review of experience with whole-cell and somatic antigen vaccines, p. 228–252. *In Symposium on Cholera, U.S.-Japan Cooperative Medical Science Program.* Sapporo, Japan.
5. **Bennish, M., A. Azad, O. Rahman, and R. Phillips.** 1990. Hypoglycemia during diarrhea in childhood: prevalence, pathophysiology, and outcome. *N. Engl. J. Med.* **322:**1357–1363.
6. **Bhatnagar, R., K. West, S. Vitale, and A. Sommer.** 1991. Risk of cataract and history of severe diarrheal disease in southern India. *Arch. Ophthalmol.* **109:**696–699.
7. **Burt, R.** 1973. Differential impact of social integration on participation in the diffusion of innovations. *Soc. Sci. Res.* **2:**1–20.
8. **Carpenter, C. C. J.** 1976. Treatment of cholera—tradition and authority versus science, reason, and humanity. *Johns Hopkins Med. J.* **139:**153–162.
9. **Clemens, J.** 1990. Assessing the value of enteric vaccines: efficacy, effectiveness, and efficiency, p. 131–140. *In* D. Sack and L. Freij (ed.), *Prospects for Public Health Benefits in Developing Countries from New Vaccines against Enteric Infections. SAREC Documentation, Conference Report 2.* Gotab, Stockholm.
10. **Clemens, J., D. Sack, J. Harris, J. Chakraborty, M. R. Khan, S. Huda, F. Ahmed, J. Gomes, M. Rao, A.-M. Svennerholm, and J. Holmgren.** 1989. ABO blood groups and cholera: new observations on specificity of risk and modification of vaccine efficacy. *J. Infect. Dis.* **159:**770–773.
11. **Clemens, J., D. Sack, J. Harris, J. Chakraborty, P. Neogy, B. Stanton, N. Huda, M. U. Khan, B. A. Kay, M. R. Khan, M. Ansaruzzaman, M. Yunus, M. R. Rao, A.-M. Svennerholm, and J. Holmgren.** 1988. Cross-protection by B subunit-whole cell cholera vaccine against diarrhea associated with heat-labile toxin-producing enterotoxigenic *Escherichia coli*: results of a large-scale field trial. *J. Infect. Dis.* **158:**372–379.
12. **Clemens, J., D. Sack, J. Harris, M. R. Khan, J. Chakraborty, S. Chowdhury, M. R. Rao, F. P. L. van Loon, B. Stanton, M. Yunus, M. Ali, M. Ansaruzzaman, A.-M. Svennerholm, and J. Holmgren.** 1990. Breast feeding and the risk of severe cholera in rural Bangladeshi children. *Am. J. Epidemiol.* **131:**400–411.
13. **Clemens, J., D. A. Sack, J. Harris, F. van Loon, J. Chakraborty, F. Ahmed, M. R. Rao, M. R. Khan, M. Yunus, N. Huda, B. Stanton, B. Kay, S. Walter, R. Eeckels, A.-M. Svennerholm, and J. Holmgren.** 1990. Field trial of oral cholera vaccines in Bangladesh: results from long-term follow-up. *Lancet* **335:**270–273.
14. **Clemens, J. D., D. Sack, J. R. Harris, J. Chakraborty, M. R. Khan, B. Stanton, M. Ali, F. Ahmed, M. Yunus, B. Kay, M. U. Khan, M. R. Rao, A.-M. Svennerholm, and J. Holmgren.** 1988. Impact of B subunit killed whole-cell and killed whole-cell-only oral vaccines against cholera upon treated diarrhoeal illness and mortality in an area endemic for cholera. *Lancet* **2:**1375–1378.
15. **Clemens, J. D., D. A. Sack, M. R. Rao, J. Chakraborty, M. R. Khan, B. Kay, F. Ahmed, A. K. Banik, F. P. L. van Loon, M. Yunus, and J. R. Harris.** 1992. Evidence that inactivated oral cholera vaccines both prevent and mitigate *Vibrio cholerae* O1 infections in a cholera-endemic area. *J. Infect. Dis.* **166:**1029–1034.
16. **Clemens, J. D., F. P. L. van Loon, M. Rao, D. A. Sack, F. Ahmed, J. Chakraborty, M. R. Khan, M. Yunus, J. R. Harris, A.-M. Svennerholm, and J. Holmgren.** 1992. Nonparticipation as a determinant of adverse health outcomes in a field trial of oral cholera vaccines. *Am. J. Epidemiol.* **135:**865–874.
17. **Creese, A., N. Sriyabayya, G. Casabal, and G. Wiseso.** 1982. Cost-effectiveness appraisal of immunization programmes. *Bull. W.H.O.* **60:**621–632.
18. **Cvjetanovic, B., B. Grab, and K. Uemura.** 1978. Dynamics of acute bacterial diseases. *Bull. W.H.O.* **56**(Suppl 1):65–80.
19. **Deb, B., B. Sircar, P. Senoupta, S. De, S. Mondal, D. Gupta, N. Saha, S. Ghosh, U. Mitra, and S. Pal.** 1986. Studies on interventions to prevent El Tor cholera transmission in urban slums. *Bull. W.H.O.* **64:**127–131.
20. **de Zoysa, I., and R. Feachem.** 1985. Interventions for the control of diarrhoeal diseases among young children: rotavirus and cholera immunization. *Bull. W.H.O.* **63:**569–583.
21. **Dick, B.** 1985. Issues in immunization in developing countries—an overview. *EPC publication 7.* London School of Hygiene and Tropical Medicine, London.

22. **Esrey, S., J. Potash, L. Roberts, and C. Shiff.**
1990. Health benefits from improvements in water
supply and sanitation: survey and analysis of the
literature on selected diseases. *WASH technical report 66.* Water and Sanitation for Health Project.

23. **Felsenfield, O., R. H. Wold, and K. Gyr.** 1973.
Simultaneous vaccination against cholera and yellow fever. *Lancet* **1:**457–458.

24. **Glass, R., S. Becker, M. Huq, B. Stoll, M. U.
Khan, M. Merson, J. Lee, and R. Black.** 1982.
Endemic cholera in rural Bangladesh, 1966–1980.
Am. J. Epidemiol. **116:**959–970.

25. **Glass, R., and R. Black.** 1992. The epidemiology
of cholera, p. 129–154. *In* D. Barua and W. Greenough (ed.), *Cholera.* Plenum Press, New York.

26. **Glass, R., M. Claeson, P. Blake, R. Waldman,
and N. Pierce.** 1991. Cholera in Africa: lessons on
transmission and control for Latin America. *Lancet*
338:791–795.

27. **Glass, R., A.-M. Svennerholm, B. Stoll, M. R.
Khan, K. Hossain, M. Huq, and J. Holmgren.**
1983. Protection against cholera in breast-fed children by antibodies in breast milk. *N. Engl. J. Med.*
308:1389–1392.

28. **Heggenhougen, K.** 1990. The concept of acceptability: the role of communication and the social
sciences, p. 180–189. *In* D. Sack and L. Freij
(ed.), *Prospects for Public Health Benefits in Developing Countries from New Vaccines against Enteric Infections. SAREC Documentation, Conference Report 2.* Gotab, Stockholm.

29. **Hernborg, A.** 1985. Stevens-Johnson syndrome
after mass prophylaxis with sulfadoxine for cholera
in Mozambique. *Lancet* **2:**1072–1073.

30. **Hirschhorn, N., A. Chaudhury, and J. Lindenbaum.** 1969. Cholera in pregnant women. *Lancet*
1:1230–1232.

31. **Hirschhorn, N., J. Lindenbaum, W. Greenough, and S. Alam.** 1966. Hypoglycemia in children with acute diarrhea. *Lancet* **2:**128–132.

32. **Holmgren, J., and A.-M. Svennerholm.** 1983.
Cholera and the immune response. *Prog. Allergy*
33:106–119.

33. **Hughes, J., J. Boyce, R. Levine, M. Khan, K.
Aziz, M. Huq, and G. Curlin.** 1982. Epidemiology of El Tor cholera in rural Bangladesh: importance of surface water in transmission. *Bull.
W.H.O.* **60:**395–404.

34. **Huq, A., E. Small, P. West, M. Huq, and R.
Colwell.** 1983. Ecological relationships between
Vibrio cholerae and planktonic crustacean
copepods. *Appl. Environ. Microbiol.* **45:**275–283.

35. **Johnston, J., D. Martin, J. Perdue, L. McFarland, C. Caraway, E. Lippy, and P. Blake.** 1983.
Cholera on a Gulf Coast oil rig. *N. Engl. J. Med.*
309:523–526.

36. **Levine, R., M. R. Khan, S. D'Souza, and D.
Nalin.** 1976. Failure of sanitary wells to protect

against cholera and other diarrheas in Bangladesh.
Lancet **2:**86–89.

37. **Mandara, M., and F. Mhalu.** 1980. Cholera control in an inaccessible district in Tanzania: importance of temporary rural centers. *Med. J. Zambia*
15:10–13.

38. **McCormack, W., A. Chowdhury, N. Jahangir,
A. Ahmed, and W. H. Mosley.** 1968. Tetracycline
prophylaxis in families of cholera patients. *Bull.
W.H.O.* **38:**787–797.

39. **Mosley, W. H., K. M. A. Aziz, A. S. M. Rahman, A. K. Chowdhury, A. Ahmed, and M.
Fahimuddin.** 1972. Report of the 1966–67 cholera
vaccine trial in rural East Pakistan. *Bull. W.H.O.*
47:229–238.

40. **Nokes, D., and R. Anderson.** 1988. The use of
mathematical models in the epidemiological study
of infectious diseases and in the design of mass
immunization programmes. *Epidemiol. Infect.* **101:**
1–20.

41. **Patriarca, P., P. Wright, and T. J. John.** 1991.
Factors affecting the immunogenicity of oral poliovirus vaccine in developing countries: a review.
Rev. Infect. Dis. **13:**926–939.

42. **Pollitzer, R.** 1959. *Cholera.* World Health Organization, Geneva.

43. **Ries, A., D. J. Vugia, L. Beingolea, A. Palacios,
E. Vasquez, J. Wells, N. Garcia Baca, D.
Swerdlow, M. Pollack, and N. Bean.** 1992. Cholera in Piura, Peru: a modern urban epidemic. *J.
Infect. Dis.* **166:**1429–1433.

44. **Sack, D.** 1990. Cost effectiveness model for cholera vaccination with oral vaccines, p. 82–89. *In* D.
Sack and L. Freij (ed.), *Prospects for Public Health
Benefits in Developing Countries from New Vaccines against Enteric Infections. SAREC Documentation, Conference Report 2.* Gotab, Stockholm.

45. **Shahid, N., A. Samadi, M. Khan, and M. Huq.**
1984. Classical vs. El Tor cholera. A prospective
family study of a concurrent outbreak. *J. Diarrhoeal Dis. Res.* **2:**73.

46. **Siddique, A. K., K. Zaman, A. Baqui, K.
Akram, P. Mutsuddy, A. Eusof, K. Haider, S.
Islam, and R. Sack.** 1992. Cholera epidemics in
Bangladesh: 1985–1991. *J. Diarrhoeal Dis. Res.*
10:79–86.

47. **Sommer, A.** 1972. Ineffectiveness of cholera vaccination as an epidemic control measure. *Lancet*
1:1232–1235.

48. **Sommer, A., and W. Woodward.** 1972. The influence of protected water supplies on the spread of
classical/Inaba and El Tor Ogawa in rural East Bengal. *Lancet* **2:**985–987.

49. **Spira, W., M. Khan, Y. Saeed, and M. Sattar.**
1980. Microbiological surveillance of intra-neighborhood El Tor cholera transmission in rural Bangladesh. *Bull. W.H.O.* **58:**731–740.

50. **Stanton, B., M. Rowland, and J. Clemens.** 1987. Oral rehydration solution—too little or too much? *Lancet* **1**:33–34.

51. **Suharyono, C. Simanjuntak, N. Witham, N. Punjabi, D. Heppner, G. Losonsky, H. Totosudirjo, A. Rifai, J. Clemens, Y. Lim, D. Burr, S. Wasserman, J. Kaper, K. Sorenson, S. Cryz, and M. Levine.** 1992. Safety and immunogenicity of single-dose live oral cholera vaccine CVD 103-HgR in 5–9 year-old Indonesian children. *Lancet* **340**:689–694.

52. **Tauxe, R. V., and P. A. Blake.** 1992. Epidemic cholera in Latin America. *JAMA* **267**:1388–1390.

53. **Weinstein, M.** 1981. Economic assessments of medical practices and technologies. *Med. Decision Making* **1**:309–330.

54. **World Health Organization.** 1974. *International Health Regulations*, 2nd annotated ed. World Health Organization, Geneva.

55. **World Health Organization.** 1986. *Guidelines for Cholera Control.* WHO/CDD/SER 80.4 Rev. 1. World Health Organization, Geneva.

56. **World Health Organization.** 1992. *Guidelines for Cholera Control* (rev. 1992). WHO/CDD/SER 80.4 Rev. 3. World Health Organization, Geneva.

VI.   CHOLERA: THE FUTURE

Vibrio cholerae and Cholera: Molecular to Global Perspectives
Edited by I. Kaye Wachsmuth, Paul A. Blake, and Ørjan Olsvik
© 1994 American Society for Microbiology, Washington, DC 20005

Chapter 29

The Future of Cholera: Persistence, Change, and an Expanding Research Agenda

Robert Tauxe, Paul Blake, Ørjan Olsvik, and I. Kaye Wachsmuth

The emergence of epidemic cholera in Latin America in 1991 and *Vibrio cholerae* O139 epidemics in Asia in 1993 has focused new attention on many longstanding questions about cholera. We conclude this book with a few thoughts on the future of cholera and on issues ripe for research.

IS CHOLERA LIKELY TO BE ERADICATED?

The history of cholera provides a framework for assessing the potential for eradicating the disease. Recognizable cholera was described in India in the 16th century, but it may not have existed outside the Indian subcontinent before 1817. In 1817, an epidemic of cholera began in India; by 1823 it had spread to the borders of Europe. By 1832, the spreading epidemic had reached North America, and within a few years, it appeared in Central and South America. Toward the end of the 19th century, after recurrent worldwide waves, cholera began to recede, and as of 1990, cholera had disappeared from the Americas and most of Africa and

Robert Tauxe, Paul Blake, Ørjan Olsvik, and I. Kaye Wachsmuth • Foodborne and Diarrheal Diseases Branch, Division of Bacterial and Mycotic Diseases, Centers for Disease Control and Prevention, 1600 Clifton Road, Atlanta, Georgia 30333.

Europe. By 1950, cholera was again confined to areas of endemicity in the Indian subcontinent and the Asian countries west of India. Cholera had apparently been eliminated from the rest of the world, including areas with poor sanitation. After having had epidemic cholera, South and Central America, southern Africa, and Northern Europe became cholera free for a century, leaving us to wonder if cholera might disappear again.

It is appropriate to think of epidemic cholera as a modern phenomenon rather than an ancient scourge. Since the first cholera pandemic began, the disease has shown a special predilection for societies in the throes of industrial transformation. In the 19th century, European and American cities enlarged with poor populations drawn from farms were particularly afflicted, and fear of cholera epidemics propelled those cities to provide safer living conditions for all. Today, vast poor populations surround the cities of Latin America and Africa, as millions choose abysmal living conditions in the city rather than starvation or open warfare in the countryside. To some extent, the future of epidemic cholera will be defined by the course of modern periurban poverty.

The recent history of cholera suggests that both the organism and the epidemiologic patterns may have changed. The pandemics of cholera before 1950 are believed to have been caused by the classical biotype of *V.*

cholerae O1, though of course no confirmation was available before Koch discovered *V. cholerae* in 1883. In 1961, a new cholera pandemic began when the El Tor biotype spread out of Indonesia. This pandemic extended across Asia into the Middle East during the 1960s, reached Africa disastrously 10 years after it had first emerged, spread to southern Europe and the Pacific islands in the 1970s, and reached Latin America in 1991, 20 years after its first appearance. In Africa, cholera has persisted as a recurrent epidemic challenge to public health since its introduction in 1970, which suggests that the Latin American epidemic may not recede soon. Although cholera disappeared from Africa and South America a century ago, the current etiologic agent (the El Tor biotype) survives better in the environment than the classical biotype that presumably caused the earlier epidemics. The populations of Africa and Latin America are now much larger and more urbanized, the sheer quantity of fecal contamination is greater, and transport of the organism from place to place is more rapid.

The speed of spread is likely to increase as the transportation network of the developing world improves. In October 1992, a new toxigenic variant of *V. cholerae*, *V. cholerae* O139, emerged in south India and within 3 months spread more than 1,000 km along the coastline from Madras to Bangladesh, causing epidemic cholera in populations with extensive previous exposure to *V. cholerae* O1. In less than a year, it had extended to Thailand and Pakistan; further spread to Africa and Latin America is likely and may take far less time than the 20 years it took the seventh pandemic to move from Indonesia to Peru.

Factors Favoring the Eradication of Cholera

Beyond the historical considerations listed above, several characteristics of cholera favor the eventual eradication of this disease.

Fragility of the organism

V. cholerae O1 is easily killed by acid pH, heat, drying, and low concentrations of chlorine, and the organism does not survive well outside of the aquatic environment. In addition to large municipal water purification and sewage treatment efforts, simple intervention and control strategies that exploit the fragility of the organism may be practical.

High infectious dose and lack of person-to-person transmission

The infectious dose is high, usually 10^6 organisms or more, which means that transmission should be easier to interrupt than for organisms with lower infectious doses. Transmission by person-to-person contact, which is characteristic of organisms with a low infectious dose, has not been demonstrated convincingly for cholera.

Short duration of carriage by humans

Acutely ill cholera patients excrete 10^6 to 10^9 vibrios per ml of stool; a patient producing 10 liters of stool could theoretically excrete 10^{13} vibrios. However, vibrio excretion normally continues for only 5 to 8 days even in untreated patients. Treatment with antimicrobial agents can halt vibrio excretion more rapidly. Although antimicrobial resistance has developed in some areas, *V. cholerae* O1 is more often susceptible to antimicrobial agents. Chronic carriers are rare and are not known to play any role in cholera transmission or persistence. In epidemics, the main source of infection is human feces, which can, at least theoretically, be collected and treated. There are no other animal carriers on land.

Seasonal changes in prevalence

In most areas, cholera is seasonal, peaking during the summer and fall and dropping to very low levels in the winter and spring. The cool-season periods of low prevalence suggest environmental modulation of incidence occurs, so the vibrios may be more vulnerable to eradication efforts at some times of the year.

Likely impact of nonspecific control efforts

In the past, even modest improvements in food and water sanitation have been sufficient to render an area unreceptive to cholera. Improvements in food and water sanitation sufficient to stop epidemic cholera will also help control other diseases that spread through the fecal-oral route. Such improvements are also economically beneficial and esthetically pleasing and are likely to be made in many areas.

Motivation of the public

Eradication of cholera would be politically desirable to businesses, tourists, and the population in general. The public is highly motivated to do something to prevent cholera because the disease can occur in large epidemics, kill rapidly, and have a high death-to-case ratio. Even though few deaths occur if treatment is adequate, the disruption of social and economic life caused by cholera can still be enormous. In 1991, several Latin American countries temporarily lost their European markets for produce. The impact of the epidemic on tourism has not been estimated but is thought to have been substantial.

Factors Impeding Eradication of Cholera

However, a number of factors make eradication less likely. These include the following.

Persistence of *V. cholerae* O1 in the aquatic environment

V. cholerae O1 can survive in some aquatic environments for months, and at least some strains can persist indefinitely in environmental reservoirs. Studies in the United States and Australia have produced convincing evidence that toxigenic *V. cholerae* O1 El Tor can persist in the environment as a free-living organism for many years in the apparent absence of human sewage contamination. Environmental reservoirs exist in remote rivers in eastern Australia, in salty marshes along the coast of the Gulf of Mexico in the United States, and probably in many other locations as well. The fact that cholera disappeared from South America and southern Africa in the 19th century does not mean that it will disappear again, as the current El Tor biotype is better adapted to persistence in the environment.

Increasing density of human population

The density of population around the world has increased 5-fold since 1850, and that of Latin America has increased 14-fold, producing greater demands on food and water sources and greater fecal contamination of the environment. The large cities of the developing world provide new environments in which El Tor *V. cholerae* O1 is easily transmitted. In an epidemic, the majority of infections are likely to be asymptomatic, and this increases the potential for chance contamination of food or water by the silent transient carrier. Because vibrios can multiply rapidly in many foods at ambient temperatures, they are easily transmitted through the informal and unrefrigerated street vendor economy. Because they survive well in water, they can be transmitted efficiently by large and poorly maintained municipal water systems.

Lack of a good cholera vaccine and limited protection of natural immunity

No highly effective vaccine exists. The current cholera vaccine gives only limited protection for several months and does not prevent asymptomatic infections. Because even natural infection does not provide complete and long-lasting immunity against reinfection with the El Tor pandemic strain, an effective vaccine will have to be more immunogenic than natural infection. The development of an effective vaccine is also complicated by variation in target bacterial and human populations. The epidemic in Latin America gave vaccine development and evaluation new impetus. The unexplained link between O blood group, severe symptoms, and vaccine failure means that vaccines are likely to be less effective in the Latin American population, in which the prevalence of blood group O is high. In 1993, epidemic cholera caused by *V. cholerae* O139 appeared in populations in India and Bangladesh with high levels of natural immunity to cholera. This suggests that evolutionary changes in the bacterial O-group antigen may permit the organism to evade natural and vaccine-induced immunity, further increasing the challenge of vaccine development. Thus, vaccines may be unlikely to lead to eradication efforts in the near future, and vaccine campaigns may even be counterproductive if they decrease the motivation to improve sanitation.

Failure of some other traditional public health actions

Mass chemoprophylaxis, quarantine, and trade embargo have been ineffective in halting cholera epidemics, although they were tried in the Asian and African cholera epidemics of the 1970s. Mass chemoprophylaxis can produce antimicrobial resistance and may lead to the incorrect belief that an antimicrobial agent is a primary means of effective prevention and treatment. Quarantine stations, where travelers are detained or denied entry to a country because of suspected cholera, continue to be operated by some countries but are likely to have little effect except to encourage deception and delay in seeking treatment, because there are so many asymptomatic infections for each clinically recognized case. Broad embargoes of trade goods from countries reporting cholera have

wrought economic havoc without any perceptible public health benefit.

Although these traditional epidemic control measures have had no defined impact on actual transmission of disease, they do lead to concealment and denial of epidemic disease as the official policy of some countries. For example, Bangladesh does not report cholera to the World Health Organization, presumably out of concern for the impact this might have on its export economy, even though it has been the homeland of cholera for centuries and the site of major cholera vaccine trials. The same is true for several other large countries, so lack of officially reported cholera cases should not be mistaken for eradication.

Development of effective treatment

Ironically, the increasing availability of relatively inexpensive and highly effective therapy lessens the fear of cholera and public motivation to eradicate the disease. Once epidemic control is achieved, cholera may be just one more occasional case of severe endemic diarrhea.

Flexibility and variety in toxigenic *V. cholerae* strains

The flexibility of the organism is becoming better understood. Arrays of cellular proteins, including cholera toxin, and colonization factors, are produced by the vibrio in response to a variety of environmental stimuli such as salinity, pH, and oxygen content and are moderated by intricate internal control mechanisms. Some environmental stimuli may induce a "noncultivable state"; others may return the organisms to their usual physiologic state. Genetic flexibility includes the minor antigenic drift from Inaba to Ogawa, which appears to have occurred in many parts of Latin America within a year after the beginning of the epidemic there, and the major shift that the emergence of *V. cholerae* O139 appears to represent. The gene for cholera toxin itself is flanked by repeated DNA sequences indicative of transposable elements and is readily moved by bacteriophage in the laboratory. Variations in toxin gene sequences and gene copy numbers suggest that the toxin genes may have an evolutionary taxonomy separate from that of the organism. Strains that clearly had lost one or all of their cholera toxin genes have been found in epidemic settings. Extrachromosomal elements, including resistance plasmids and sex factors, that may facilitate the transfer of genetic information to and from other bacterial strains, even other bacterial species, have been documented in *V. cholerae* O1.

Epidemic cholera is caused by a group of related organisms. The combination of characteristics required for a strain to have epidemic potential is incompletely understood. Two strain groups, the *V. cholerae* O1 El Tor found in Australian rivers and that found along the U.S. Gulf Coast, produce cholera toxin, have persisted in their niches for many years, but have not been associated with epidemic spread. This could be because they lack the capacity to cause epidemics or because they have not yet been introduced into circumstances favorable for epidemic transmission. Four major strain groups have definite epidemic capacity: the El Tor biotype of *V. cholerae* O1 which appeared as a new global seventh-pandemic strain in the 1960s; the Latin American variant of the seventh-pandemic strain group; the classical biotype *V. cholerae* O1 strains associated with epidemic disease before 1950; and the new *V. cholerae* O139 strains causing cholera epidemics in the Indian subcontinent in 1993. The appearance of toxigenic *V. cholerae* O139 demonstrates that the complete O1 antigen itself is not required for epidemic potential. The diversity of organisms that have been documented as causing epidemics suggest that cholera epidemics that preceded the discoveries of Koch may have been caused by yet other strains.

Sensitive molecular subtyping techniques show substantial diversity among toxigenic *V. cholerae* O1 isolates within these broad groups. A pattern of local persistence and evolutionary differentiation punctuated by periodic widespread dissemination of epidemic strains could have produced the global patchwork of strains. These changes make it more likely that at least some strains would persist even in settings of limited transmission and high levels of natural immunity.

In sum, although large cholera epidemics may become less likely to occur as effective control measures are put into place, persistence of cholera as an epidemic disease in some regions of the world and its periodic epidemic emergence from natural reservoirs is likely for the foreseeable future. In the 19th century, recurrent cholera epidemics stimulated the development of modern public health, including development of permanent public health offices, a standard disease notification system, and integration of the microbiology laboratory into public health control efforts. Epidemic cholera will continue to bring change and improvement to the public health institutions of the world.

IMPACT OF CHOLERA ON PUBLIC HEALTH INSTITUTIONS

Development of Public Health Laboratory Capacities

The public health management of cholera requires a coordinated laboratory effort. At the most basic level, laboratory confirmation and determination of antimicrobial resistance are required to establish the existence of cholera in an area and to optimize treatment. Simple networks of surveillance points where clinical samples can be collected from suspect cases and either tested directly or carried in transport media to central laboratories can be effective. Once cholera is detected, periodic culturing of specimens from a few patients is useful for monitoring antimicrobial resistance and defining the end of the epidemic; however, there is no need to culture specimens from every patient. At a more central level, the laboratory services required during an epidemic include confirmation of toxin production, determination of specific strain characteristics such as biotype and serotype, identification of important variations in subtype, and support of targeted epidemiologic investigations with serodiagnosis and environmental sampling. In Latin America, epidemic cholera caused a hemispheric effort to train laboratorians in the basic skills critical for epidemic management and is stimulating development of an international network of public health laboratories linked to collaborating reference centers. Future goals include linking peripheral laboratories to national laboratories electronically and increasing communication among national laboratories. Rapid transfer of data and strains of interest among them would be a dramatic advance.

Developing an international public health laboratory network will require sustained programs of laboratory training and quality assurance coordinated through reference laboratories. There is no need to assume that one laboratory will be the reference center for all purposes, and it may well be more efficient to foster distributed "centers of excellence" with various specific skills among the various laboratories of the network. Standards for interpretation of serologic assays and antimicrobial resistance testing and for production of new antisera need to developed; standard procedures for subtyping and maintaining subtype collections by which laboratories can measure themselves need to be defined.

Intelligent laboratory algorithms for deciding which samples to collect and which isolates to characterize further will conserve laboratory resources and provide the most useful information in the least time. All too often, public health laboratories have been committed to characterizing all isolates or conducting massive environmental surveillance schemes. Once the presence of cholera has been confirmed, there is no need to reconfirm it in each new case; appropriate statistical selec-

tion of a sample of cases or isolates will answer the same questions for a fraction of the cost.

Development of Epidemiologic Capacities

In many countries, epidemic cholera has served as the focus for improving surveillance activities in general. Limited surveillance data can be communicated from peripheral surveillance points to central locations at least weekly in any part of the world. The critical breakthrough has been to link this centripetal surveillance data transmission with an outgoing response, using the incoming data to direct shipments of necessary supplies and target further investigations. Public health authorities in many countries have learned that surveillance is not an abstraction but a meaningful call to action. Increases in case fatality rates signal the need for specific attention to improving treatment. Increases in cholera incidence after preliminary control measures have been implemented tell public health authorities that further specific control measures are needed. The cholera epidemic has stimulated development of rapid surveillance networks to transmit all sorts of surveillance information electronically.

Effective surveillance generates more questions for epidemiologic investigation. In many parts of the world, cholera is stimulating public health authorities to acquire skills in analytic field epidemiology. Simple case investigation, case-control study techniques, cluster surveys, and community intervention studies are part of the epidemiologic repertoire of a growing number of ministries of health. The success of the Field Epidemiology Training Program in Peru in rapidly identifying the nature of the epidemic and defining the means of transmission and prevention, and the usefulness of a cadre of epidemiologists trained in a similar program in Mexico have encouraged other countries to create similar programs. Portable computers and epidemiologic software can put the analytic skills of the well-trained epidemiologist in the most remote location. Because cholera investigations require the collaborative efforts of epidemiologists and laboratorians, communication between these two groups is growing.

FUTURE RESEARCH QUESTIONS

Better understanding of the organism and of its natural and human environments is needed to improve control strategies, and fertile areas exist for future research to improve the treatment and prevention of cholera.

QUESTIONS ABOUT THE MICROBE

What Characteristics Are Necessary To Cause Disease?

Although the ability to produce cholera toxin is the principal virulence determinant of *V. cholerae* O1, other factors have been identified in laboratory strains developed as potential live vaccines that do not produce cholera toxin but still cause diarrheal illness. Cholera toxin itself is genetically heterogeneous, though the variants all seem to have the same pathogenic potential. Difficulties in developing reliably avirulent live vaccine strains suggest that more virulence factors remain to be identified.

What Characteristics Are Necessary for Epidemic Spread?

It may be possible to define the microbial characteristics of specific strains that are associated with epidemic behavior. The appearance on the Indian subcontinent of cholera epidemics caused by *V. cholerae* O139 may mean that immunologically important changes in the O antigen can occur without loss of epidemic potential. In contrast, Gulf Coast strains of El Tor *V. cholerae* O1 have persisted for at least 20 years on the United States coast of the Gulf of Mexico and are likely to have been present in Mexico as well, since a U.S. traveler to Can-

cun, Mexico, acquired infection with this strain in 1983. If so, then the Gulf Coast strains have been present in Mexico for many years without causing large epidemics in regions where the Latin American cholera epidemic strain now spreads rapidly. Perhaps the Gulf Coast strain lacks determinants that favor development of epidemics.

There may also be strain-specific determinants that favor environmental persistence. *V. cholerae* O1 El Tor strains seem to be more likely than classical strains to persist in the environment. The two groups of toxigenic El Tor *V. cholerae* O1 best documented to be capable of long-term environmental persistence (the U.S. Gulf Coast strains and the Australian river strains) are both consistently serotype Inaba, raising the possibility that this serotype is better suited to persistence than is the Ogawa serotype. Identification of determinants of persistence should lead to better understanding of environmental reservoirs and ultimately to modifications that make persistence less likely.

What Other Evolutionary Changes Should Be Expected?

Strategies of control that put intense selective pressure on the organism during an ongoing epidemic may lead to other evolutionary developments. For example, multiple antimicrobial resistance appeared rapidly in Ecuador in the setting of a policy of intensive chemoprophylaxis. Also, other Latin American *V. cholerae* O1 strains recently described by Glenn Morris of the University of Maryland autoagglutinate into large clumps of bacteria in vitro, enabling some bacteria in the interior of a clump to survive in vitro levels of acidity and chlorine that would otherwise kill them. While the public health significance of these strains remains to be determined, chlorine-resistant vibrios would be a potent challenge to control strategies. Similarly, the appearance of the O139 serogroup in the Indian subcontinent suggests that strains that are effectively resistant to vaccine-induced or natural immunity can emerge in areas with high levels of serogroup-specific immunity.

The introduction of *V. cholerae* O1 into new ecologic niches may itself provide opportunity for evolutionary change. The initial uniformity of the strains isolated in Peru was consistent with introduction of a single clone of serotype Inaba. As it spread across the continent, it developed variability in serotype patterns, switching to Ogawa in many areas for reasons that remain to be clarified.

QUESTIONS ABOUT THE CHOLERA PATIENT

Can Treatment Be Further Improved?

Current cholera treatment protocols represent a tremendous advance over the last 30 years, but further efforts may bring mortality still lower. In some areas of the developing world, the greatest difficulty still lies in the logistic challenge of getting sufficient treatment supplies to the affected areas swiftly; this requires better surveillance and rapid response to surveillance data. In other parts of the world, logistic support is not a constraint, but cholera treatment protocols may need to be adapted to local conditions. The current World Health Organization's cholera treatment protocols were optimized for use in poor developing countries, using end points of mortality, duration of hospitalization, and volumes of fluid required. Treatment often depends on the presence of a family member at bedside to give oral rehydration solution, and intravenous fluids are reserved for the most severe cases. Optimizing the cost-effectiveness of cholera treatment protocols in more-developed parts of the world needs to include consideration of economic end points such as the cost of staff needed to provide care when an unpaid family attendant is unavailable, of daily hospital bed fees, and of measures to prevent nosocomial infection; even the impact of insurance reimbursement policies may be a factor in some countries. Specific guidelines could speed the transition

from intravenous to oral treatment after the severely ill patient has been rehydrated. In Latin America, some cholera patients who present with advanced shock developed transient acute renal failure and receive temporary kidney dialysis along with vigorous rehydration. Characterization of how best to achieve rapid rehydration in this and other extreme clinical settings would be helpful.

What Are the Biological Determinants of Severe Illness and Immune Protection?

Severe cholera is more likely to occur when the infecting dose is high or when the gastric acid barrier to infection is absent, as in patients whose stomachs have been surgically removed. Other less well understood factors affect the host response to infection. Infection with *Helicobacter pylori* may lead to gastric hypochlorhydria and so may be a cofactor for more-severe illness. Patients with blood group O are more likely to have severe clinical manifestations of cholera than are patients with other blood groups. This phenomenon has now been documented in populations in Asia and Latin America and remains unexplained; it is probably related to the poor efficacy of experimental vaccines observed in persons with blood group O. Understanding this effect may lead to modulation of the clinical response to cholera and may be central to the development of vaccines that will be effective in Latin American populations, where blood group O is predominant.

The nature of the immunity induced by natural infection also remains unclear. Serologic markers of immunity have limited utility in predicting the degree of protection offered by experimental vaccines, suggesting that cell-mediated immunity or intestinal mucosal immunity may be more important than humoral response. The bacterial target of immunity is also unclear. The new epidemic of cholera caused by *V. cholerae* O139 in Asian populations with relatively high levels of preexisting immunity to *V. cholerae* O1 suggests that effective immunity is dependent on response to the O antigen. Although these populations also had antibodies to cholera toxin and the toxin produced by the new epidemic strain appears to be a typical cholera toxin, antitoxic antibodies seem not to protect against the new strain and perhaps have little protective effect in general.

QUESTIONS ABOUT CHOLERA AND THE NATURAL ENVIRONMENT

What Is the Ecology of Environmental Persistence?

Cycles of transmission linking human sewage, drinking water, and food supplies continue unbroken in many parts of the world. However, even with minimal contamination by human sewage, unique strains of toxigenic El Tor *V. cholerae* have persisted in estuarine bayous of the U.S. Gulf Coast since at least 1973 and in the rivers of northeastern Australia since at least 1977, strong evidence that some strains can persist in the environment indefinitely. Cases are often associated with eating shellfish, and there seems to be a specific link between crustacea and the persistence of *V. cholerae*. The great estuaries of Latin America, such as the mouths of the Amazon, Orinoco, and Guayaquil rivers, could also prove hospitable to long-term persistence of toxigenic *V. cholerae* O1.

Defining the ecologic determinants of persistence of toxigenic *V. cholerae* O1 in the environment will require clarification of the potential roles of copepods and other plankton in providing a niche for *V. cholerae* O1 and of the mechanism by which shellfish become contaminated in sewage-free environments. Control strategies may be more successful if the precise nature of the target is better defined. Vibrios living in colonies in the intestinal tracts of copepods that parasitize crabs may be harder to eliminate than single free-living bacteria. The role of dormant or noncultivable states of toxigenic *V. cholerae* O1 and the mechanisms by which the organism enters and leaves such states need better definition. Better understand-

ing of the ecology of the natural reservoir may suggest ways of improving prevention strategies by reducing or at least predicting the levels of contamination in foods related to those environments.

QUESTIONS ABOUT CHOLERA AND THE HUMAN ENVIRONMENT

What Are the Specific Routes and Mechanisms of Cholera Transmission in Endemic-Disease and Newly Affected Areas?

Defining specific routes of transmission is critical to developing successful control measures to interrupt them. Detailed epidemiologic investigations in the last 20 years have shown that cholera is transmitted through a variety of specific food and water vehicles. Because the dominant vehicle varies from place to place, successful cholera control may depend on understanding local transmission patterns. Mechanisms of transmission of cholera in the Americas, Oceania, and Africa are better documented than those in Asia, where few analytic investigations of transmission have been conducted. Surprisingly little is known about specific mechanisms of transmissions in the traditional homeland of cholera. It is possible that simple and modifiable behaviors could explain much of cholera transmission in Asia. For example, in Bangladesh, practices such as rinsing drinking water pots in the river before filling them with clean well water, holding cooked rice in water for many hours before eating it, and consuming partially cooked or partially dried seafood may be common. The role these practices play in cholera transmission could be investigated with analytic epidemiologic tools, which could lead to more specific prevention measures. In other unstudied parts of the world, identifying unsuspected vehicles of transmission may be important and may suggest direct control measures.

What Specific Aspects of Economic and Social Development Prevent Sustained Transmission of Epidemic Cholera?

The lack of widespread transmission following repeated introductions into the United States, Costa Rica, and Chile suggest that there is a threshold level of safe water supply and proper sewage disposal above which sustained cholera transmission becomes unlikely. In contrast, Portugal sustained a nationwide epidemic of cholera following probable introduction from Africa in 1974. Although chance is likely to play a role in determining the outcome of any introduction, mapping cholera incidence following repeated introductions against specific indices of development may provide better evidence for a threshold. As various control measures are implemented throughout Latin America in the coming decade, it may be possible to determine the impact of specific interventions on cholera transmission. Such information would be of immediate relevance to African and Asian countries contemplating selective cholera control measures.

What Is the Potential of New Experimental Vaccines in Various Populations?

Research on cholera vaccines had been accelerated by the emergence of cholera in Latin America. Development of an effective and inexpensive vaccine providing durable protection after a single dose would be of immediate use in reducing mortality and treatment costs throughout the world. If it also prevented asymptomatic infection, it could have additional value in disrupting cycles of transmission. It remains unclear whether even natural infection can provide such protection. Detecting cases of repeated cholera in Latin America as the epidemic continues in areas with good surveillance will be of particular interest.

Because vaccine efficacy may vary in different populations, the effectiveness of experimental vaccines should be measured directly in Latin American and other newly affected populations rather than extrapolated from experience in a single population.

The appearance of epidemic *V. cholera* O139 makes vaccine development more complex. It appears to be causing epidemics in populations with high levels of preexisting immunity due to *V. cholerae* O1 infections, suggesting that neither antitoxic antibodies nor O1-antigen-directed immunity is protective. If *V. cholerae* O139 persists and spreads, vaccines will need to be reengineered to include components of this organism.

How Can Safe Water and Sewage Treatment Systems Be Built Efficiently throughout the Developing World?

Cholera control will require substantially increased investment in repair and protection of existing water and sewage systems and extension of those systems to persons without services. The projected cost of extending full water treatment, sewage collection and treatment, and sanitary garbage disposal throughout Latin America alone is estimated to be $200 billion. The cost of these projects could be substantially less if the effort itself stimulates development of better technology and materials, such as lightweight durable plastic piping instead of concrete for sewer pipes, processed sewage sludge that can be sold as fertilizer to recover costs, and increased production efficiency of chlorine due to greater demand for chlorine and new local production facilities. The cost of waste disposal can be reduced by changes in population behavior, such as recycling garbage and putting soiled toilet paper into sewers instead of domestic garbage.

Complex urban utilities require constant maintenance and a community-wide effort to protect the integrity of the system. Investment in infrastructure declined during periods of sustained hyperinflation in many Latin American economies, and poorly maintained municipal water systems became dangerously efficient machines for transmitting cholera and other diseases to unsuspecting populations. Rebuilding such systems depends on the confidence and participation of people in their local governments.

While Efforts To Develop a Highly Effective Vaccine and To Build New Water and Sewage Systems Continue, How Can the Population Best Be Protected against Cholera in the Interim?

Cholera transmission can occur by a variety of waterborne and foodborne routes, and the specific route that predominates can change with time. This means that interventions focused on a single route of transmission, such as transmission through the municipal water supply, would not be expected to prevent all cholera cases, and a multipronged approach is usually needed. In the initial emergency response to a new epidemic, people can be instructed to boil their water, cook all food thoroughly, and limit their exposure to street-vended foods and beverages. However, these emergency behavior changes are often impossible to sustain long term. Once transmission routes are understood, interim prevention efforts that are locally appropriate, inexpensive, and easy to continue and that will have impact on other diseases in addition to cholera may be identified. In Latin America, simple strategies to improve the quality of domestic drinking water, such as the use of new water storage vessels, need to be evaluated along with domestic water disinfection schemes. This strategy is analogous to the efforts in Africa to eradicate Guinea worm by filtering drinking water in the household, and in some areas perhaps the two methods could be productively combined. Efforts to improve the safety of street-vended foods and beverages are needed and could begin with educating street vendors

and consumers to look for safer practices and products. This is an arena for an expanded global research effort. Improving the safety of domestic seafood, including monitoring of harvest waters, better sanitation during processing, and determination of how traditional seafood cuisines can be made safer, is also likely to have immediate benefits.

CONCLUSION

Eradication of choleras is unlikely, because it has environmental reservoirs that will probably continue to cause occasional cases of cholera indefinitely. *V. cholerae* strains that cause epidemic cholera have changed dramatically in the last 30 years, exhibit a broad range of genetic diversity now, and are likely to continue to evolve. However, epidemic cholera can be controlled. Some control measures, such as improvements in food and water sanitation, should be continued indefinitely even in cholera-free areas because of their other benefits. Other, more-specific measures, such as vaccination with a cheap and effective vaccine, could limit epidemics caused by specific *V. cholerae* strains.

Persons in wealthy countries with adequate food and water sanitation and members of the wealthier classes of poor countries are unlikely to experience cholera themselves and will be motivated to control the disease out of altruism or because everyone is affected by the economic impact of cholera. Cholera can be a net benefit to humanity by calling attention to deficiencies in basic systems and services that need to be corrected. Just as cholera spurred the sanitary reform movement and the development of the field of public health in 19th century Europe and North America, recurrent epidemic cholera will continue to drive constructive change in the developing world.

Index